D1229786

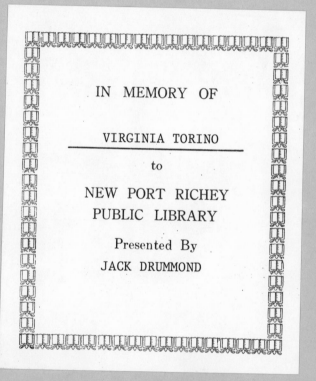

IN MEMORY OF

VIRGINIA TORINO

to

NEW PORT RICHEY
PUBLIC LIBRARY

Presented By

JACK DRUMMOND

A Flora of
Tropical Florida

REFERENCE

Robert W. Long and Olga Lakela

OCT 1 7 1995.

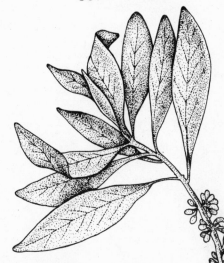

A Flora of
Tropical Florida

A Manual of the Seed Plants and Ferns
of Southern Peninsular Florida

Banyan Books
Miami, Florida

NEW PORT RICHEY PUBLIC LIBRARY

ROBERT W. LONG, Ph.D.
Professor of Botany, University of South Florida
and OLGA LAKELA, Ph.D.
Research Associate in Botany, University of South Florida (r

This new edition
Copyright © 1976 by
Robert W. Long and Olga Lakela
Library of Congress Card Number 76–402
ISBN 0–916224–06–6

Original edition
Copyright © 1971 by
University of Miami Press

Library of Congress Catalog Card Number 70–143460

ISBN 0–87024–201–6

All rights reserved, including rights of reproduction
and use in any form or by any means, including the making
of copies by any photo process, or by any electronic or
mechanical device, printed or written or oral, or recording
for sound or visual reproduction or for use in any knowledge
or retrieval system or device, unless permission in
writing is obtained from the copyright proprietors.

Illustrations by Linda Baumhardt

Designed by Bernard Lipsky and Mary Lipson

Manufactured in the United States of America

To John Kunkel Small (1869–1938),
botanical explorer and author
of Florida's first flora

Contents

Illustrations

Preface

The fascination that southern Florida has for many of its residents and visitors is in no small part due to its tropical plant life. This rich, varied, and often luxuriant flora is unique in the continental United States, and for decades it has attracted students and amateur botanists who come to see plants not found anywhere else in the country.

We have defined tropical Florida as the southernmost tip of the peninsula, which includes the political subdivisions of Collier, Dade, and Monroe counties. The tropical flora occupies roughly a U-shaped area south of Lake Okeechobee, and in this region are the Everglades National Park, the Florida Keys, the Big Cypress Swamp, the Ten Thousand Islands, and adjacent parts of southern Florida. It is an area that deserves a special manual of plants because of its importance not only to professional and amateur botanists and naturalists but also to the general public as well. Southern Florida is a pleasant place to live, as can be attested by its phenomenal growth in population in the past few decades. This increase in numbers of people has had a drastic effect on the vegetation, however, and the rate of change is steadily accelerating. Many of us hope we can persuade our fellow residents that conservation measures are desperately needed to preserve for future generations our marvelous tropical flora. This manual, we hope, will serve to acquaint a few more people with the plants around them and perhaps serve to stimulate an appreciation for our natural heritage.

Our objective has been to produce a book that will enable anyone with some knowledge of botany to identify the wild or naturalized ferns and seed plants of the area. Native plants make up the largest number of included species, but introduced plants that appear to be self-reproducing are also described. The manual consists of a general introduction to the flora followed by diagnostic keys for the identification of plant families. The descriptive text follows a systematic arrangement of families, genera, and species with diagnostic keys. The first author was responsible for families 18–21, 53–92, and 94–177; the second author was responsible for families 1–17, 22–52, and 93. Information is provided regarding the general characteristics of each plant, its habitat, ecology, geographical distribution, chromosome number (if known), and pertinent synonomy.

ROBERT W. LONG
OLGA LAKELA

Tampa, Florida
January, 1971

Acknowledgments

The research that has formed the basis for this book was supported continuously by the University of South Florida during the five-year duration of the project. We are also grateful for grants from the National Science Foundation and from the George Cooley Botanical Research Fund that provided additional funds for field and herbarium work and the employment of research assistants.

In the preparation of a flora there are many technical problems, and we have been most fortunate in obtaining encouragement and assistance from specialists who kindly agreed to review the preliminary taxonomic treatments given their groups. As a result of suggestions and corrections, numerous changes and improvements were made, and these are incorporated in the text. We wish to thank particularly Drs. Preston Adams (*Hypericum*), David Bates (MALVACEAE, BYTTNERIACEAE), C. Ritchie Bell (APIACAE), Lyman Benson (CACTACEAE), Derek Burch (EUPHORBIACEAE), Donovan Correll (ORCHIDACEAE), Wilbur Duncan (VITACEAE, SMILACACEAE), Murray Evans (FILICINEAE), Donald Farrar (HYMENOPHYLLACEAE), James Horton (*Polygonella*), Marshall Johnston (RHAMNACEAE), Rogers McVaugh (MYRTACEAE), Harold Moldenke (VERBENACEAE), C. V. Morton (FILICINEAE), Richard Pohl (POACEAE), Peter Raven (ONAGRACEAE), C. M. Rogers (*Linum*), Robert Read (*Thrinax*), Reed Rollins (BRASSICACEAE), Lloyd Shinners (CARYOPHYLLACEAE, CONVOLVULACEAE, LAMIACEAE), H. K. Svenson (CYPERACEAE), W. H. Wagner, Jr. (FILICINEAE), and Bernice Schubert (FABACEAE). These persons should not be held responsible, however, for any errors of judgment embodied in this book.

We also want to acknowledge a few of the many individuals who helped us with particular problems in the course of the research. Mr. George Avery and Dr. Frank C. Craighead, Sr., gave assistance and co-operation with field work in southern Florida and shared with us their keen observations on the local flora. Our appreciation is given to Drs. R. K. Godfrey, Tetsua Koyama, D. B. Ward, Robert Kral, H. E. Moore, Carroll Wood, John Reeder, Arthur Cronquist, Robert Thorne, Frank Gould, Calaway H. Dodson, John E. Fairey III, Taylor Alexander, F. J. Hermann, Charles Feddema, John Beckner, Richard Howard, Velva Rudd, and John Popenoe who have been helpful in answering questions and supplying us with special information. The description of the geology of southern Florida was checked for accuracy by University of South Florida geologists Drs. Wendel Ragan and W. H. Taft. Curators of many herbaria have been most helpful when records of collections or specimens were needed.

Members of the staff of the University of South Florida Herbarium have been called upon repeatedly during the preparation of this flora for aid in various bibliographic and other technical matters. We wish particularly to thank Mrs. Martha New, herbarium assistant and librarian, for her conscientious and valuable help. Some dissections for preparation of the figures were made under the direction of Dr. Derek Burch. The preliminary draft of the glossary, English equivalent of Latin species names, and the statistical analysis were prepared by Miss Rose Broome. Mrs. Martha Meagher served as reader and has aided us in a variety of ways. Other assistants helped especially in field collecting, and here we want to thank Mr. Frank Almeda, Mr. Larry Pardue, Mr. John Utley, Mr. Richard Vagner, Mr. Gary Mills, Mr. James De Boer, Miss Katherine Whitney, and Mrs. Rosalind Herbert.

Dr. George Cooley, long-time student of the Florida flora and honorary curator of the University of South Florida Herbarium, has supplied us with many of his collections and records as well as assisting us in our field work. Mr. Albert Greenberg has also given us valuable information on Florida plants.

Our typists have been Mrs. Thelma Fricks, who completed the first version of the manuscript, and Mrs. Carol Williams, who typed the final version.

We wish to acknowledge here the contribution of Professor Joseph Ewan, Tulane University, who wrote the History of Botanical Collecting in Southern Florida. We are most grateful for his willingness to prepare this especially for the flora.

Finally, we wish to thank the many undergraduate students of the

Systematic Botany class at the University of South Florida who over the years have used our preliminary keys for plant identification. Their suggestions as to how we might improve them are gratefully acknowledged.

*A Flora of
Tropical Florida*

Allan Hiram Curtiss (upper left) sold specimens of nearly 1,500 species of Florida plants. Abram Paschall Garber (upper right) was commemorated by Asa Gray in the name of the shrub *Garberia*. John Kunkel Small (center) was the foremost field botanist and author of Florida's flora. William Edwin Safford (lower left), all-around naturalist, explored and wrote along with his career in the Navy. Ellsworth Paine Killip (lower right) after his retirement made extensive collections on Big Pine Key.

History of Botanical Collecting
in Southern Florida,
by Joseph Ewan

Florida was scarcely inhabited by Europeans when the United States acquired the province from Spain in 1821 although Saint Augustine had been founded by the Spanish forty-two years before the English settled Jamestown. Documented botanical exploration did not begin for another decade. The Florida in 1542 of Alvar Núñez Cabeza de Vaca, who first reported the use of Black Vomit (*Ilex vomitoria*), of Jean Ribaut who reported the red and yellow varieties of maize in his *Discoverye of Terra Florida* in 1563, and of Jacques Le Moyne and René Laudonnière was to the north of our area. The Quaker Jonathan Dickinson who was shipwrecked near the mouth of the Loxahatchee River on September 23, 1696, told how he and his shipmates were refreshed by "seaside grapes," and mentioned "seaside coco-plums," "palm berries," and "pumpion vines." Francois Coreal's *Relation* of 1722 told of South Florida trees observed on a voyage to the West Indies, but Percy G. Adams identifies this as a compilation laced with fiction from the *Atlas Geographicus* published in London in 1717. Mark Catesby did not reach south of the Saint Johns River country, nor did John or William Bartram, or André Michaux. Thomas Drummond, who collected in Texas and Louisiana for the elder Hooker, had planned a thorough botanical exploration of Florida, but his death in Cuba in 1835 ended this plan.

The earliest serious botanical collecting in southern Florida was associated with the Seminole wars of the 1830s and the United States Army personnel garrisoned at Fort Brooke at the head of Hillsboro Bay and Fort King 130 miles northeast of Tampa. Engagements with the Indians restricted botanizing activities, and the specimens sent John Torrey must have been taken within range of the camps. Lieutenant Bradford

Ripley Alden (1811–1870), a West Pointer, arrived at Fort King in 1832 or 1833 and sent specimens to Torrey; these were later cited in Torrey and Gray's *Flora of North America.* Also cited in that *Flora* were collections made by Dr. Silas E. Burrows who evidently served at Fort King. On February 22, 1836, Dr. Melines Conkling Leavenworth (1796–1862), "one of the rising stars of the botanical firmament" upon his graduation from Yale in 1817, arrived at Fort King and followed the campaigns against the Seminoles for the next several months. His collections in Florida were made mostly in the northern counties. Another army surgeon who sent specimens to Torrey was Dr. Gilbert White Hulse (1808–1883) who served under Scott beginning in 1836 at Fort Brooke.

German-born Edward Frederick Leitner, naturalist and apprentice of Dr. J. E. Holbrook of Charleston, South Carolina, explored southern Florida for three or four years, supported by several subscribers during 1833 including Benjamin D. Greene of Boston and John Bachman. Leitner enlisted with Lieutenant L. M. Powell's regiment in 1836. The following year Leitner explored with Lieutenant Powell from Key West to Charlotte Harbor along swamps untracked by a naturalist until he was overtaken and scalped by Indians on January 15, 1838, near Jupiter Inlet. The yellow flowered waterlily, illustrated by Audubon with the "great white swan," was one of Leitner's discoveries. Its existence was disbelieved until it was rediscovered by Ferdinand Rugel in 1848 and later by Edward Palmer and was finally publicized by Mary Treat. A.W. Chapman commemorated Leitner when he described corkwood. Another casualty of the Seminole wars was Dr. Henry Perrine (1797–1840), a pioneer in plant introduction living on Indian Key who planned a program of testing subtropical crops from Mexico and the Bahamas for South Florida. *Agave sisalina,* now widely naturalized, was his introduction. The Reverend Alva Bennett sent plant collections to Torrey and Gray from Key West at least before October 1838.

By far the most important figure in South Florida's early botanical history was John Loomis Blodgett (1809–1853), physician to the Mississippi and Louisiana Colonization Society for two years in Liberia before he settled as druggist and physician at Key West in 1838. Blodgett's fortunate early correspondence with Nuttall about the West Indian species heretofore unknown for the Florida Keys made possible their inclusion in Nuttall's *Sylva.* Robert James Shuttleworth, English botanist and conchologist, resident of Berne, whose herbarium of 170,000 specimens was purchased by the British Museum in 1877, distributed Blodgett's plants with labels he had had printed. Nuttall's own herbarium incorporated in the British Museum (Natural History) contains specimens labeled "K. West," surely Blodgett's since Nuttall did not visit southern Florida. Torrey and Gray's *Flora* contains numerous Key West records attributed to Blodgett. The artist naturalist Titian Ramsey Peale (1800–1885) visited Key West on his Florida trip during the winter of 1824–1825. His collections were noticed by Nuttall and by Torrey and Gray.

One of Blodgett's visitors was Alvan Wentworth Chapman (1809–1899) of Apalachicola whose pioneer botanizing in west Florida led to the first identification manual for the South. It was published in 1860. Chapman also published a series of reports on Florida records, which included some from Blodgett's gatherings, in Coulter's newly founded *Botanical Gazette*.

Important in making known the plants peculiar to Florida was Ferdinand Rugel (1806–1874), whose specimens are well represented in the British Museum, Paris, Geneva, and Trinity College, Dublin. He had come to Florida in 1842 supported by Shuttleworth. His most extensive collections were made in the vicinity of Saint Marks. His localities have not been tabulated, but he was, for example, at Manatee in July 1845, at "Miami, Cape Florida," in September, and at Key West in February 1846.

William Henry Harvey (1811–1866), British botanist who had been a house guest of Asa Gray while he gave the Lowell Lectures in Boston, botanized down the coast in the summer of 1849 stopping at New York, Washington, Norfolk, Charleston, Savannah, and Key West. Professor Harvey spent the month of February 1850 at Key West, and wrote Hooker from there that he had made a "fine collection of algae," but very few other plants were flowering.

Samuel Ashmead of Philadelphia collected various *naturalia*, especially algae, at Key West during the winter of 1855–1856; some of these reached Professor Harvey, then at Dublin. Joseph Bassett Holder (1824–1888) took his M.D. degree at Harvard and, influenced by Agassiz and Baird, pursued natural history throughout his lifetime. He arrived as medical officer at Fort Jefferson, Dry Tortugas, in 1859 and stayed ten years. It was natural that he corresponded with Dr. Chapman. During the 1870s Louis Francois Pourtales (1824–1880) collected about Cape Sable with the United States Coast Survey and published a paper he titled "Hints on the Origin of the Flora and Fauna of the Florida Keys" in volume eleven of the *American Naturalist.* "Key West outranks even the famous Bairritz for number of species [of algae]" wrote Professor W. G. Farlow, and so James Cosmo Melvill (1845–1929), Manchester merchant who traveled from Florida to Canada, went to Key West in February 1872 to collect the algae Harvey had found there. Melvill also took some flowering plants now preserved in the Manchester Museum and proceeded to Cedar Keys in March.

After a season botanizing about Palatka, Gainesville, and Cedar Keys in 1876, Abram Paschall Garber (1838–1881) visited southern Florida the following year, hoping to check his tuberculosis. Dr. Garber wrote to George Vasey from Miami, May 16, 1877, that he believed he had some plants "new to our flora and possibly to science" but that he found drying specimens difficult. His paper on fern collecting in South Florida was published in volume three of *Botanical Gazette*. Asa Gray named the genus *Garberia* (Compositae) and several species in other families in recognition of his botanical acumen.

Captain John Donnell Smith (1829–1928), fastidious collector of Baltimore, botanized in 1878 with Coe Finch Austin (1831–1880), the New Jersey bryologist, along the Caloosahatchee River. His later studies on the Guatemalan flora are classic. Allan Hiram Curtiss (1845–1907), who succeeded in collecting and distributing as sets nearly 1,500 species in seven years, settled in Jacksonville and coasted about the Keys during 1880–1881. He reported on *Pseudophoenix sargentii* in the first volume of Sargent's *Garden and Forest*. His mother, Floretta Anna Allen Curtiss (1822–1899), had a long time interest in botany and collected Florida algae for the last twenty years of her life.

It was from talking with Alexander Agassiz that William Edwin Safford (1859–1926) determined to fit himself for scientific collecting as an adjunct to his career in the Navy. Safford relates his stay at Key West in April 1881 in a delightful essay in volume two of the *American Fern Journal*, and tells of the plants he observed there. Nearly forty years later his classic survey entitled "Natural History of Paradise Key and the nearby Everglades of Florida" appeared in the *Smithsonian Institution Annual Report* for 1917, published in 1919. In April of 1885 Charles Torrey Simpson (1846–1932) accompanied Pliny Reasoner from Tampa Bay to Cape Sable in a sharpie, Reasoner gathering nursery stock for his Oneco business, Simpson collecting especially shells. Simpson wrote in the *Plant World* (vol. 5, pp. 4–7, 1902) of royal palms about 125 feet high they saw at Evans Plantation on Rodgers River and of *Dendrophylax lindeni* epiphytic on the palms just as he had seen the orchid on the same tree in Honduras. Reasoner was to die at twenty-five of yellow fever three years later, but his younger brother Egbert Norman Reasoner (1869–1926) lived to make the Royal Palm Nursery famous. After twenty years as conchologist at the National Museum in Washington, D.C., Simpson's health failed, and he settled on fifteen acres of wild hammock about six miles north of the hamlet of Miami where he brought together on eight acres about 2,500 kinds of trees and plants. His *In Lower Florida Wilds* (1920) reads as a moving requiem. "Eaton waded in, and when about waist deep, stepped on a fourteen foot alligator. The gator got up and apologized and offered his seat. Eaton sat down, then Eaton arose. . . ." The fern enthusiast Alvah Augustus Eaton (1865–1908) had joined Simpson and John Soar on a foray into the Everglades. Simpson told the story in the *Fern Bulletin* (17:38–41, 1909). He wrote, "Before us lay Paradise Keys, the most lovely bit of tropical scenery I have ever beheld. It might have covered a hundred acres—a low rounded dome of giant trees, and rich tangled vegetation, punctuated here and there with magnificent royal palms, singly or in groups, rising from 60 to 120 feet in height, their beautiful plumy heads swaying low in the morning breeze."

Field botanists were aware of South Florida's vanishing flora before the turn of the century. Elizabeth Gertrude Britton (1858–1934), wife of the director of the New York Botanical Garden, a bryologist in her own right, botanized in 1904 and wrote that year in *Plant World* of the burning of the palms in the glades. The entomologist Willis Stanley Blatchley's field notes in *In Days Agone* (1932) identifies both the plants and insects for his old Cape Sable haunts. Herbert John Webber (1865–1946) addressed the 1898 meeting of the American Association for the Advancement of Science on the strand flora, Ernst Athearn Bessey (1877–1957) wrote on hammocks and the Everglades in the *Plant World* (14:268–276, 1911), and Mrs. Emilia Crane Anthony on the "jungles about Miami" and the glades in the *Fern Bulletin* (11:21–23, 1903).

Florida's most important name in botany is that of John Kunkel Small (1869–1938), field botanist, photographer, author, and, incidentally, an accomplished flutist. Beginning in 1901 and for the next thirty-five years Small published 55 papers in the *Bulletin of the Torrey Botanical Club,* and three identification manuals, the first in 1903, and the last edition (1933, of 1,554 pages) described 5,500 species from the Southeastern United States. The *Flora of Miami* and *Flora of the Florida Keys* both were published by the author in 1913. His travelogues, generally excellently illustrated, should someday be assembled in book form.

The Dutch botanist Hugo de Vries (1848–1935) visited Palm Beach, Key West, and the Everglades during his 1912 sojourn in the United States, and his botanical commentary, in Dutch, with photographs amounts to a foreigner's appraisal. Samuel Mills Tracy (1847–1920), botanist and agronomist who developed giant Bermuda and other forage grasses for the South, collected far and wide in the United States over the years, and about Miami in May 1905, and distributed specimens widely to foreign herbaria. *Florida Wild Flowers,* the best known popular book on the subject, was written by Mrs. Mary Francis Evans Baker, who settled in Florida in 1917. Her husband, Thomas R. Baker, taught the natural sciences at Rollins College, where her herbarium is preserved. Charles A. Mosier (d. 1936), "a born woodsman and accomplished naturalist," became warden at Royal Palm State Park about this time and collaborated with Small in his explorations.

"It was on the houseboat of Mr. Allison V. Armour, anchored off the Brickell wharf in the Miami River, that I agreed that if he would buy a boat and equip it for the collecting of living plants, I would go with him on an expedition." Thus the Fairchild Tropical Garden really set sail on the yacht *Utowana.* David Grandison Fairchild (1869–1954) relates the story in his delightful autobiography *The World was my Garden* (1943). The New York Botanical Garden's mycologist, William Alphonso Murrill (1869–1957), retired in Gainesville but collected and observed as far as Key West. His *Historic Foundations of Botany in Florida (and America)* (Gainesville, 1945) makes diverting reading. Hester Stansbury Ferguson (Mrs. James Alexander Henshall) painted about 400 native Florida wild flowers from the Tampa region before 1925. They are preserved in the Lloyd Library, Cincinnati. Among the many botanical visitors who wintered in South Florida was Ellsworth Paine Killip (1890–1968) of the Smithsonian Institution who made extensive collections on Big Pine Key and elsewhere beginning in 1935. Harold Norman Moldenke (1909–) reported on his Florida collections made jointly with his father in 1927, 1929, and 1930 in the *American Midland Naturalist* (32:529–590, 1944) with identifications provided by monographers. Thomas Alva Edison (1847–1931) studied goldenrods in connection with his search for rubber substitutes derived from native plant species. The corporation lawyer, bibliophile, and caricologist K. K. Mackenzie named *Solidago edisoniana,* and J. K. Small proposed the segregate asclepiadaceous genus *Edisonia* for the inventor. Walter Mardin Buswell (1866–1951) of Coral Gables came to Florida in 1918 and acquired a detailed knowledge of the local flora. Roland McMillan Harper (1878–1966) summarized the *Natural Resources of Southern Florida* (1928), providing a valuable bibliography, and later published on its endemic flora.

Of the parade of botanical explorers, then, there were army surgeons who were associated with the Indian and Civil wars: Hulse, Burrows, Leavenworth, Leitner, and Holder; of those who came for their health: Garber and Simpson; Blodgett came close to earning recognition as the first resident botanist; of the professional collectors who distributed their specimens in sets there were Rugel, Curtiss, and Tracy. South Florida has long attracted the tourist-botanist: importantly Harvey, Chapman, Melvill, Safford, J. D. Smith, A. A. Eaton, Killip, and many more. Of all the enthusiasts John Kunkel Small stands out as the botanist who tramped the most, described and wrote the most extensively, and left the classic literature on the flora of South Florida.

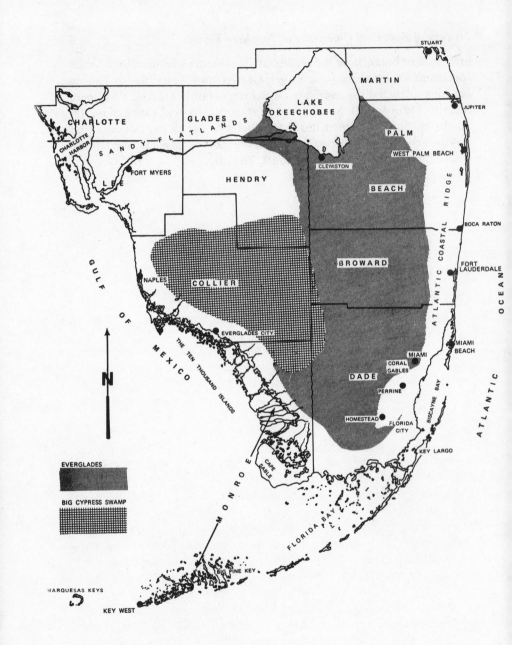

Map of southern peninsular Florida

Introduction

Geology of Southern Peninsular Florida

In a geologic sense, Florida is the youngest state in the Union. Early speculation was that southern peninsular Florida had been built up by successive growths of coral reefs in shallow seas, and that wind and waves had accumulated quartz sands and detritus on the reefs to a few feet above sea level. The discovery of thick shell beds of Pliocene age along the Caloosahatchee River conclusively disproved this theory.

Peninsular Florida is the emerged portion of a much broader projection from the continental land mass of North America. Geologists refer to this projection as the Floridan Plateau. Southern Florida is the southeastern corner of the Plateau and is underlain chiefly by marine limestones, clays, salt, and sandstones. The peninsula was separated from the continent by a seaway in the Oligocene, and peninsular Florida existed as an island in Miocene times before being connected with the mainland again in the Pliocene. During Pliocene times, about 10 million years ago, parts of southern Florida emerged from the sea concomitantly with crustal movements in eastern North America. Prior to Miocene times all of southern and eastern Florida had been submerged and the shoreline had extended south through Lake County to Sebring, then had circled west through Arcadia to Sarasota, and northwest across the Gulf to Tallahassee.

During the Pleistocene epoch, beginning about one million years ago, the sea level fell and rose repeatedly over the Floridan Plateau in response to major glaciation in North America. During interglacial times the sea stood upon the present land, and in southern Florida evidence of this submergence can easily be seen in the marine sedimentary deposits and shoreline features. Former shorelines may be identified by beach ridges, dunes, wave-cut benches, and by changes in slope. Apparently there were eight major shorelines ranging from 270 to 5 feet above the present sea level. During the submergences nearly all of Florida was covered by water, and the nearest land was a small group of islands in what is now Polk county. The last major coastal terraces still visible in southern Florida are the Penholoway at plus 70 feet above present sea level, the Talbot at plus 42 feet, and Pamlico at plus 25 feet. During the Sangamon interglacial period in the Pleistocene a distinctive limestone called the Miami oolite was deposited. This is commonly found south of the Tamiami Trail in the Everglades area and in the lower Florida Keys.

At the end of the last glaciation, the sea rose to its present level and dunes assumed their present form. Limestone deposits and associated formations, surface materials, slope, and vegetation caused the slow drainage to the south from the uneven surface of the interior of southern Florida. Saw grass grew where water was shallow enough, and their compacted remains made up much of the peat and muck of the Everglades. Large parts of southern Florida are underlain by limestone and other calcareous deposits, and surface water that is highly charged with organic acids plays an important role in the development of surface features. Surface water has corrosive effects that produce a threadlike network on the limestone, sometimes etching it to depths up to a foot and leaving uneven or jagged columns.

Solution holes of various sizes are common in southern Florida, and they often support a distinctive flora. Possibly they begin as vertical holes dissolved along taproots, along rock fractures, or as small pits that gradually deepen to become tubes that enlarge into holes. Trees blown over by hurricanes break up rock, and roots leave depressions where water can concentrate and new solution holes begin. Many solution holes or depressions, some up to several feet in diameter, can be seen in the pineland and wet prairies around Miami. Sinkhole lakes also occur in southern Florida, and one, Deep Lake in Collier County, has a depth of over 100 feet.

Southern Florida is flat and of low elevation, and drainage is very sluggish. The overflow from Lake Okeechobee formerly passed south more or less as a sheet into the Everglades before dikes and canals were built. Today this water drains chiefly to the east through the St. Lucie canal, westward through the Caloosahatchee canal and river, and during the rainy season southward through several canals that empty on the lower southwestern coast.

Geologists recognize five major natural physiographic areas in southern Florida: (1) the Sandy Flatlands; (2) the Everglades; (3) the Big Cypress Swamp; (4) the Atlantic Coastal Ridge; and (5) the Coastal Marshes and Mangrove Swamps. The areas are located on the map of southern Florida.

Sandy Flatlands. This area surrounds Lake Okeechobee on the west and east, passing without interruption on the west to beyond Naples. Inland from Naples it extends in an irregular pattern to the margin of Big Cypress Swamp. The area is poorly drained and often inundated during the rainy season. The main drainways are the Okaloachoochee Slough, the Devil's Garden, Fahkahatchee Slough, and the Allapattah Marsh. The vegetation is very heavy and tends to retard water movement. A shallow body of water, Lake Trafford, is found in this region west of Immokalee; the basin probably represents a depression in the Pleistocene sea bottom.

Everglades. This area extends south from Lake Okeechobee into the mangrove swamps and salt marshes of Florida Bay and the Gulf of Mexico. The basin of the Everglades was the Plio-Pleistocene sea bottom, and during interglacial times marine deposits were laid down. In recent times layers of freshwater marl and peat were deposited largely by the thick growth of the sedge *Cladium jamaicensis.* Organic soils up to eight feet in thickness were formed near the lake, but they tend to be thinner away from the basin. Lake Okeechobee itself probably represents an original hollow in the Plio-Pleistocene sea floor, but it is very shallow; its greatest depth is about at sea level.

In the Everglades are many elongated tree islands separated from each other by swales, sloughs, and runs. They are arranged generally in linear, parallel patterns in a northwest-southwest axis in the upper Everglades, and on a south or southwest axis in the lower region. Apparently this pattern is parallel to the drainage pattern of the gently sloping plain leading to Florida Bay and the southwestern Gulf coast.

Big Cypress Swamp. Alternating swampy and hammocky ground is found west of the Everglades that grades slowly to the south and southwest into coastal marshes. Generally, it is a low-lying area with poor drainage, and it is flooded in many parts during the rainy season. Rock occurs at the surface in some places, and small differences in elevations produce marked differences in plant life. Generally, the cypress forests cover the lower spots, and the trees are often stunted; pines or palms occupy the slightly elevated places.

Atlantic Coastal Ridge. The coastal ridge is found along the Atlantic side, rising slightly from the Everglades and lowering gradually to the ocean. It begins to disappear southwest of Florida City in a series of low "islands" referred to as the Everglade Keys. The ridge reappears in the lower Florida Keys, from Big Pine Key to Key West, and here Miami oolite is again bedrock.

Across the ridge, particularly south of Miami, occur transverse glades that support an assemblage of plants distinct from the surrounding areas, and in places these glades are conspicuous. Geologists theorize that the Coastal Ridge originated from a bar formed during Sangamon interglacial times. The bar lays between the deep sea at the edge of the Floridan Plateau and the broad basin of the Everglades. The transverse glades, then, may represent old drainage ways that probably were once tidal runways during the Pleistocene. The Atlantic Coastal Ridge may have served as a island refugium for plants during the Pleistocene similar to the islands in Polk County during the Pliocene and earlier.

Coastal Marshes and Mangrove Swamps. Along the southern end of the peninsula, from Fort Lauderdale to Naples, occurs the coastal vegetation in a narrow band behind the sandy beach ridge. The mangrove swamps fringe the coast, often forming impenetrable forests along the edge of Florida Bay, and they are also well developed on the drowned dunes of the Ten Thousand Islands area of the southwestern coast of the peninsula. Salt marshes are found also in the strip bordering the salt water and consist of characteristic salt-marsh plants that are found in similar habitats throughout the tropics of the world.

In summary, then, southern Florida has been repeatedly flooded by seawater during interglacial periods of the Pleistocene epoch. Marine deposits were made in the Everglades basin during these times, while freshwater marls, sands, and later organic deposits accumulated during glacial stages when sea levels were lower. The gently south-sloping floor of southern Florida, overlaid with these Pleistocene deposits, has been modified by vegetation, winds, rain, surface water and groundwater, and solution that have determined the modern topographic features. Natural processes have been disturbed, however, by the activities of man, who through drainage has reversed accumulation of peat and muck in the Everglades. The organic soils are rapidly being lowered by agricultural practices, fire, natural oxidation, and compaction. Unless these processes are altered, the Everglades basin as we know it today will be destroyed.

Origin and Composition of the Flora

The geological evidence suggests that southern Florida has been available for plant colonization only since Pleistocene times, and during this epoch only during periods of glaciation. Much of southeastern

United States and parts of northern and central Florida have had almost continuous vegetation since Pliocene times and earlier. Geography and climate have doubtless been more important than the historical factor in the determination of the vegetational composition of tropical Florida. The southern location, plus the proximity of the Gulf Stream, high summer rainfall and high humidity, and mild winters with a midwinter dry season give it an almost Antillean, insular climate.

Migration into newly emergent southern Florida in recent times apparently took place from three principal directions: (1) from the Caribbean region, the Yucatán peninsula, and other areas in tropical America; (2) from the Southeastern Coastal Plain and other temperate areas of northern and central Florida; (3) from Pleistocene island refugia such as central Florida and possibly from the Atlantic Coastal Ridge. The flora of southern Florida contains, therefore, a tropical American element, a temperate or coastal plain element, and an endemic element.

The flora is predominently tropical, and its most important relationship is with the flora of the Caribbean region of tropical America, namely the islands of Cuba, Hispaniola, Jamaica, Puerto Rico, and the Bahamas. Close floristic similarities are found with Cuba, and many fern, bromeliad, orchid, and seastrand species are common to both areas. Relatively few species of tropical affinities are found north of the latitude of Lake Okeechobee but rather are found in a broad U-shaped distribution pattern south of the lake. The frost line is an effective barrier to northward migration of the Caribbean species, just as the Everglades, the Lake District of central Florida, and the limestone outcrops are effective barriers for southward migration of coastal plain and temperate species at the present time.

Approximately 61% of the flora is tropical in relationship, and of this number 91% are species that occur in the Caribbean area. The temperate or nontropical element presumably has been derived through migration from continental United States. The flora is comprised of about 65% herbaceous species and about 35% woody species. Some 77% of the woody flora are tropical species, but only about 52% of the herbaceous species are of tropical origin.

The identification of endemic or localized species in Florida has been of interest to botanists for many years. Over 300 endemic flowering plant species are recorded for Florida, mostly found in the Lake District of central Florida, with a second area of concentration in southern Florida and the Florida Keys. About 9% of the flora of tropical Florida is endemic to Florida, and the majority of these are herbaceous dicotyledonous species. Presumably they are of recent origin. Some examples of endemics in our area are *Aristida patula, Cassia keyensis, Cereus gracilis* var. *aboriginum, Flaveria floridana, Linum carteri* var. *smallii,* and *Vernonia blodgettii.*

Plant Communities

The environmental combinations of climate, soil, drainage patterns, and recent geological history have produced a wide variety of ecological niches in southern Florida. The diversity of plant life here is one of the most conspicuous natural features. Plant species are found in various kinds of associations or formations with the composition determined by ecological factors and the inherited tolerances of the individual plant. The most conspicuous formations in tropical Florida are the *scrub forests, hammock and tree islands, freshwater swamps, dry pineland, wet or low pineland, the mangroves, salt marshes, wet prairies, dry prairies, coastal strands and dunes, pond and river margins, marine communities,* and *ruderal communities.*

Scrub Forests. These formations contain a number of low trees and shrubs with little herbaceous ground cover. Some examples of these associations can be found in the Sandy Flatlands, but they are more common in central Florida. Some examples of plants found here are *Aristida condensata, Galactia regularis, Heterotheca scabrella, Liatris tenuifolia, Palafoxia feayi, Pinus clausa, Quercus chapmanii, Quercus myrtifolia,* and *Quercus virginiana* var. *geminata.*

Hammocks and Tree Islands. Groups of broadleaf evergreen trees often associated with palms that form dense forests in relatively small areas are called hammocks. They are well developed locally in the lower Everglades although fire has destroyed many formerly large stands. Floristically they are rich and highly diversified associations. A few examples of plants found in these formations are *Bursera simaruba, Capparis cynophyllophora, Citharexylum fruticosum, Coccoloba diversifolia, Dipholis salicifolia, Eugenia,* ssp., *Lysiloma latisiliqua, Mastichodendron foetidissimum,* and *Simaruba glauca.*

Freshwater Swamps. These are forests which are flooded by shallow surface water most of the year, and they may be found in many parts of southern Florida, especially in the Big Cypress Swamp. Some examples of typical species are *Cornus foemina, Ilex cassine, Juncus polycephalus, Ludwigia palustris* var. *americana, Myriophyllum pinnatum, Proserpinaca palustris, Quercus nigra, Salix caroliniana, Saururus cernuus, Scirpus validus,* and *Taxodium distichum.*

Dry Pineland. Pine flatwoods are nearly level plains of open forests, grasses, and other herbs. They are well developed both in the Sandy

Flatlands and on the Atlantic Coastal Ridge. Some species found here are *Agalinis filifolia, Byrsonima cuneata, Diospyros virginiana, Elephantopus elatus, Eupatorium villosum, Galactia pinetorum, Guettarda scabra, Liatris spicata, Pinus elliottii* var. *densa, Serenoa repens, Stipulicida setacea,* and *Zamia integrifolia* Ait.

Wet Pineland. Wet areas that may be seasonally flooded occur in many low pinelands. These formations may be found on the margins of the Big Cypress Swamp and Atlantic Coastal Ridge if the surface drainage is poor. Some species found here are *Cordia sebestena, Cynoctonum mitreola, Erythrina herbacea, Forestiera segregata,* var. *pinetorum, Ipomoea tenuissima, Morus nigra, Pinus elliottii* var. *densa, Polygala nana, Spiranthes vernalis,* and *Xyris elliottii.*

Mangroves. Along the southwestern and southern coastlines of Florida are some of the largest and best developed mangrove swamps in the world. The seaward zone of the mangrove belt is usually occupied by the red mangrove, where the soils are nearly always flooded; the middle zone, where soils are exposed at low tides, is occupied by the black mangrove; the inner zone, with soils above the high tide, is occupied by white mangrove and buttonwood. Examples of characteristic species are *Avicennia germinans, Batis maritima, Caesalpinia crista, Conocarpus erecta, Dalbergia ecastophyllum, Hymenocallis latifolia, Laguncularia racemosa, Rhabdadenia biflora, Rhizophora mangle,* and *Salicornia perennis.*

Salt Marshes. These communities are more numerous northward from our area, but they cover large parts of southern Florida as well. They can be found around tidal estuaries, bays, and inlets, and may be in association with mangroves. Some plants found here are *Borrichia arborescens, Borrichia frutescens, Cyperus ligularis, Cyperus planifolius, Distichilis spicata, Hyptis pectinata, Limonium carolinianum, Spartina patens,* and *Spartina spartinae.*

Wet Prairies. The largest areas in southern Florida are covered by treeless wet prairies and are best developed in the Everglades. This vast formation is also the richest and most diverse floristically. The dominant species is *Cladium jamaicensis,* the saw grass. Other species found here are *Amphicarpum muhlenbergianum, Cyperus haspan, Erianthus giganteus, Eriocaulon decangulare, Hydrolea corymbosum, Ludwigia alata, Panicum nitidum, Pontederia lanceolata, Sabatia grandiflora, Sagittaria lancifolia, Thalia geniculata,* and *Typha latifolia.*

Dry Prairies. Formations referred to as saw palmetto prairies are quite variable in floristic composition. In southern Florida they appear to be ecotones and are marginal to pinelands or wet prairies. Some plants in this association are *Andropogon glomeratus, Aristida stricta, Cyperus*

*pollardii, Eriochloa michauxii, Lygodesmia aphylla, Paspalum gigan-
teum, Serenoa repens,* and *Sporobolus domingensis.*

Coastal Strand and Dunes. This is one of the most distinctive and
ancient plant formations in Florida, and the vegetation differs little
from that found in similar habitats in Cuba and other islands of the
Caribbean area. The sandy beaches and dunes are usually very narrow
because of lagoons, ponds, and lakes that often extend along the inland
side. A few examples of plants in this association are *Batis maritima,
Cakile fusiformis, Cenchrus tribuloides, Coccoloba uvifera, Cocos nucifera, Iva frutescens, Scaevola plumieri, Suriana maritima, Tournefortia
gnaphaloides, Uniola paniculata,* and *Yucca aloifolia.*

Pond and River Margins. Along the edges of freshwater lakes, canals,
and rivers are mixed formations that contain elements of several types
of plant associations. In some areas they may form dense vegetation, in
others they may occur as a narrow band of plants along a waterway. A
few of the plants found here are *Carya aquatica, Celtis laevigata, Cepha-
lanthus occidentalis, Fraxinus caroliniana, Magnolia virginiana, Myrica
cerifera, Persea borbonia, Schinus terebinthifolius,* and *Woodwardia
virginica.*

Marine Communities. Salt-water grasses occur in extensive formations
in shallow water of bays and near beaches where they occur in dense
submarine meadows. Storms and high tides often throw up masses of
these grasses on beaches. Characteristic species are *Halophila engelmannii, Ruppia maritima, Syringodium filiforme,* and *Thalassia testudinum.*

Ruderal Communities. Weed formations are highly variable in composition or in some areas may be composed of only a few species. They
are found in a great variety of kinds of habitats, but ecologically these
are disturbed sites, such as roadsides, burned-over areas, and excavations.
Soils such as these permit a number of plants to become established that
otherwise could not compete with the native plants. Some common
weeds in our area are *Amaranthus hybridus, Ambrosia artemisiifolia,
Argemone mexicana, Bidens pilosa* var. *radiata, Boerhavia erecta,
Cenchrus echinatus, Lepidium virginicum, Oxalis stricta,* and many
others.

Use of This Book

The chief objective of this book is to provide a means for identifying
the approximately 1,650 species of flowering plants, gymnosperms, ferns,
and fern allies known to grow without cultivation in southern Florida.

Diagnostic keys are provided to 179 families, to the genera within each family, and to the species of each genus if more than one occurs in our area. In some instances keys are provided for the infraspecific categories of variety or subspecies.

In using the keys, the plant in hand is compared to the descriptive statements in the first paragraph numbered "1," and if it does not correspond, then the plant is compared to the second paragraph numbered "1," and so on. The keys are indented, so paragraphs that are to be used when comparing the same characters of the plant will have the same margin along the left side of the page. Each paragraph leads to another that correctly describes the plant in hand, and is further indented, until the user is led eventually to the name of the taxon. The number of the family corresponds to its position in the text; the number for the genus and for the species corresponds to their position in the keys.

The General Key is to families of ferns and fern allies, gymnosperms, Liliatae (monocots), and Magnoliatae (dicots), and the indented paragraphs lead directly for the first three groups to the family name and page where this family is described. The general key to Magnoliatae has been grouped into 16 sections in order that one can run down a given plant more readily to its main category. The dichotomous keys are artificial, and a given family may appear in more than one section and more than once in a given section. We have attempted to construct keys based on obvious and easily recognizable characteristics and to provide a clearly understandable means of identifying the plant. We have also attempted to minimize the use of technical vocabulary in the keys. However, a Glossary is provided at the end of the book to provide immediate explanation of any term whose meaning is unclear.

The arrangement and sequence of families is that of the Engler and Prantl system. It corresponds to that found in *Gray's Manual of Botany*, 8th ed., and most other modern works on flora. This system is used by most botanists throughout the world and is adopted here. Family names are those accepted by the *International Code of Botanical Nomenclature*, 1966. Where two names for the same family are equally acceptable, both are given, but the alternate, usually older name is listed as "nomen alt." in the text. An example is the legume family which may correctly be called Fabaceae or Leguminosae. Within the family the genera and species are arranged artificially according to their sequence in the key, except in the larger families as the Poaceae (Gramineae), Fabaceae, and Asteraceae (Compositae) where they are arranged within tribes or subfamilies. No presumption of phylogenetic relationship can be assumed based merely on the juxtaposition of two or more taxa within a key.

Throughout this book each taxon has been given the botanical name which, in our judgment, is valid according to the *International Code of Botanical Nomenclature*, 1966. It will quickly be noted that the names used in this flora are often different from those found in Small's *Manual of the Southeastern Flora* and from certain other regional or local manuals. One reason for this is that changes have been made in the rules of nomenclature that have required modification of names. Small followed the American Code of Botanical Nomenclature, which has since been abandoned and is now obsolete. Another reason for name changes is that new information has been obtained that has influenced the classification of a plant, and this often results in a new name, or a shift of the plant to a different category. Detailed studies and recent publications on particular groups of plants, such as the *Generic Flora of the Southeastern United States,* have provided other criteria which have necessitated adoption of different names. In most instances the monographic work has been accepted wherever possible because particular investigators have certainly been able to devote more time to special problems than we have. It should be made clear, however, that perfect unanimity of opinion on all species names does not exist among botanists.

In many instances, in addition to the species name there is also provided a *subspecies* or *variety* name. The rules state that the infraspecific category that includes the type specimen for the species shall have the specific name repeated. For instance, *Ruellia caroliniensis* ssp. *caroliniensis* includes the type for the species. This taxon is spoken of as the typical subspecies. Similarly, if a variety has been described within a species, then the typical variety is the one containing the nomenclatural type on which the species was based, and becomes the variety with the repetition of the species name, *e.g., Quercus virginiana* var. *virginiana*. Subspecies or varieties differing from the typical ones are given different names, *e.g., Quercus virginiana* var. *geminata*. Questions may arise regarding the authors' position relative to the use of subspecies and variety because in one group we may use one category and in another the other category. There is no consistent policy followed in this book because various authors have often used either one, or the other, or both categories depending on their understanding of the terms. We have used the literature with as little change as possible and have accepted the author's own use of the categories. American botanists differ in their understanding and application of subspecies and variety, sometimes confusing the terms and considering them equivalent, or using the variety as an inferior category to the subspecies. It is, at present, impracticable to have a consistent policy as long as systematists differ in their use of the categories.

Most scientific names have a background, and doubtless many users of this book will want to know something about the meaning of Latin names or their derivation. A list of common English Equivalents of Latin Species Names is given in the appendix. For the meanings of many generic names found in this flora the reader is referred to *Gray's Manual of Botany* and other similar works.

In general, accentuation follows the classical Latin model. Throughout the book two accents are used, both indicating the syllable to be stressed and the sound to be given the vowel. The grave (`) indicates the long English sound of the vowel; the acute (´) indicates the short or modified sound. Only family and generic names are marked for pronunciation in the body of the text, but species names are also marked in the appendix list of English equivalents.

Following genera, species, and infraspecific names, the author's (or authors') name is given. These are the persons who validly named the taxon. We have standardized the abbreviations for authors as far as possible. The full name and date of birth (and death, if deceased) is given in the Flora Author List in the appendix.

Following the scientific name one or more common or colloquial names are often given. If more than one name is available, that one is given which is frequently used in Florida. The use of coined common names or translated Latin names has been avoided. Many common names are rooted in ancient tradition, medicinal tales, folklore, or aboriginal use, and their meaning now is seemingly nonsensical.

Formal, more or less standardized descriptions are given for each species, unless it is monotypic, and then the generic description includes that for the species. Generally, descriptions are brief and specific and comparable in detail for related taxa. In some instances, however, full descriptions are given where characters are critical and more detailed information is required. The descriptions always begin with reference to vegetative characters, followed by characteristics of the reproductive ones, perianth, androecium, gynoecium, and fruit. All measurements are given in the metric system. The apparent normal ranges of variation in morphological characters given are based on Florida material.

Chromosome numbers are given for those species in our area where numbers have been reported in the botanical literature, chiefly in Index to Plant Chromosome Numbers, M.S. Cave, editor. In most instances these counts are not based on Florida material. However, we believe that their inclusion with other characters of species may be useful and may serve to encourage botanists in Florida to increase our knowledge of chromosome numbers for indigenous plants. The 2n or diploid number is given throughout the manual where a diploid or a haploid count has been reported. If polyploidy is known to occur in genera or species and it is important in understanding the taxonomic problems in the group, this is also noted in the descriptions.

We have attempted to emphasize the plant community concept throughout the manual with the hope that it will be of value in understanding the flora. For each species ecological observations are included that place it in one or more plant communities. Often a given species can be shown to occur in quite distinctive environments. For instance, *Ernodea littoralis* is composed of two varieties, one found in the maritime strand, the other in the pineland. This kind of information may point out possibilities of different ecotypes within a species or subspecies in a manner that was not so apparent in earlier manuals, and may suggest to biosystematists, geneticists, and evolutionists possible groups for further investigation.

Two main groups of plants occur in our flora: *indigenous,* meaning those plants that are natural elements within the wild flora of the area; and *introduced* and *naturalized,* meaning those plants which were intentionally or unintentionally brought into our area from the outside. This latter category would include adventive species that have come in as weeds or vagrants. The introduced, naturalized species described in this flora are included because there is at least some evidence they are established as self-reproducing outside of cultivation. Admittedly, sometimes this evidence is doubtful. A few exotics, such as *Colocasia esculentum,* are included although they are probably not truly naturalized. This was done because they are common in southern Florida and may appear sometimes to be growing "wild."

The general range of distribution is given for species and infraspecific taxa as it exists outside of southern Florida. In general, the distribution is stated for Florida, then to the west, to the north, and to the midwest. Thus, "south Fla. to Tex., N.C." would mean that the species occurs from southern Florida around the Gulf coast to Texas, and from Florida north to North Carolina. Ranges of distribution have been obtained from published records, monographs, revisions, and regional floras.

The earliest and latest flowering date is based on southern Florida phenological records. Many species in our area, however, are recorded as flowering "all year," meaning that flowering material may be found at any time during the calendar year.

Synonyms are given in italics in the text following the description of the taxon to which they refer. We have attempted to include those names which likely will be encountered in other works dealing with this flora, particularly Small's *Manual of the Southeastern Flora.* The user can thereby relate the names accepted in this book with those that may be used in other works. By no means are all synonyms given. Modernization of the nomenclature of the flora has resulted in many necessary name changes. It may be discouraging to some who are already familiar with the plants of the area to discover that a "new" name has replaced an old familiar one.

Just as italics indicate synonomy, the other type faces in the text are significant. The accepted name of the plant being described is in boldface type, followed by the common name, if any, in capital and small capital letters. Within the descriptive matter, a reference to a genus is also in capitals and small capitals. Species names may appear in one of two ways. If the species being mentioned falls within another genus, the species name is in capitals and small capitals; if the species is within the same genus as the plant being described, the species name will be in boldface type.

Economic value and special features are mentioned where it seems appropriate under the description. Matters such as food value to man and animals, utility as timber, medicinal or poisonous qualities, use by Indians, etc., are included for many species.

The figures were drawn by Miss Linda Baumhardt, staff botanical artist, from living material or from specimens in the University of South Florida Herbarium. The plan was to include a figure for one species representative of each family and also characteristic of the flora of southern Florida. Dissections are given to aid in the identification not only for the particular species but also for the family. Although it would have been desirable to include more illustrations, this was not practicable within the scope of this work.

The writing of a manual of this kind cannot be completed without wishing we had additional time for study of each group of plants. Many species complexes need thorough study, and, in a work such as this, one cannot hope to state clearly the taxonomy of each group. Bio-systematic investigations have been completed for relatively few genera within our area, and workers do not always agree on classification even when detailed investigations have been completed. For these reasons it is to be expected that the treatments of various groups will be uneven, varying from one to the other. Many points still need checking. For instance, some species are included in this flora based on prediction they will be eventually found here. They are reported in areas so close to our range as to make their future discovery in southern Florida highly probable.

There has never been a systematic flora devoted entirely to southern Florida. The present book is offered in the hope that it will serve the purpose for which it was written until that time when new information will require a new flora.

Summary of the Flora of Tropical Florida

	Families	Genera	Species	Varieties or Subspecies
Pteridophyta	18	37	76	6
Gymnospermae	4	4	7	2
Angiospermae				
Liliatae	32	186	490	25
Magnoliatae	125	537	1074	157
Total	179	762	1647	190

Selected References

Bailey, L. H. *Standard Encyclopedia of Horticulture.* New York: Macmillan Co., 1942.

Bailey, L. H. *Manual of Cultivated Plants.* 7th printing. New York: Macmillan Co., 1963.

Britton, N. L. and Millspaugh, C. F. *Bahama Flora.* New York: published by authors, 1920. Reprint, Hafner Publishing Co.

Chapman, A. W. *Flora of Southern United States.* 3rd ed. New York: American Book Co., 1883.

Copeland, E. B. *Genera Filicum.* Waltham, Mass.: Chronica Botanica Co., 1947.

Correll, D. S. *Native Orchids of North America.* Waltham, Mass.: Chronica Botanica Co., 1950.

Fassett, Norman. *A Manual of Aquatic Plants.* New York: McGraw-Hill Book Co., 1940.

Fawcette, W., and Rendle, A. B. *Flora of Jamaica.* London: British Museum (Natural History), 1936.

Fernald, M. L. *Gray's Manual of Botany.* 8th ed. New York: American Book Co., 1950.

Gooding, E. G. B., Loveless, A. R., and Proctor, G. R. *Flora of Barbados.* London: Her Majesty's Stationery Office, 1965.

Grisebach, A. H. R. *Flora of British West Indian Islands.* Reprint. New York: Hafner Publishing Co., 1963.

Hitchcock, A. S. *Manual of the Grasses of the United States.* 2nd ed., revised by Agnes Chase. U.S.D.A. Misc. Publ. 200. Washington, D.C.: Government Printing Office, 1950.

Kingsbury, John M. *Poisonous Plants of the United States and Canada.* Englewood Cliffs, N.J.: Prentice-Hall, Inc., 1964.

Kurz, H., and Godfrey, R. K. *Trees of Northern Florida.* Gainesville: University of Florida Press, 1962.

Leon, H. *Flora de Cuba*. Vol. I. Havana: Cultural, S.A., 1946.

Leon, H., and Alain, H. *Flora de Cuba*. Vol. II. Havana: Cultural, S.A., 1951.

Leon, H., and Alain, H. *Flora de Cuba*. Vol. III. Havana: Cultural, S.A., 1953.

Leon, H., and Alain, H. *Flora de Cuba*. Vol. IV. Havana: Cultural, S.A., 1957.

Morton, Julia. *Wild Plants for Survival in South Florida*. Miami: Hurricane House, 1962.

Radford, A. E., Ahles, H. F., and Bell, C. R. *Manual of the Vascular Flora of the Carolinas*. Chapel Hill: University of North Carolina Press, 1968.

Small, John K. *Manual of the Southeastern Flora*. Reprint. Chapel Hill: University of North Carolina Press, 1933.

Small, John K. *Ferns of the Southeastern States*. Lancaster, Pa.: Science Press, 1938.

West, E., and Arnold, L. E. *Native Tress of Florida*. Gainesville: University of Florida Press, 1950.

Wherry, Edgar T. *Southern Fern Guide*. Garden City, N.Y.: Doubleday and Co., 1964.

Wood, Carroll E., Jr., ed. *Generic Flora of the Southeastern United States*. Cambridge, Mass.: Arnold Arboretum, Harvard University, 1958–. Issued in parts by *Journal of the Arnold Arboretum*.

General Keys to Families of Vascular Plants

PTERIDOPHYTA

Ferns and Fern Allies

1. Terrestrial, epiphytic, and aquatic plants.
    2. Roots and leaves lacking; stems with forking
        branches; sporangium three-chambered in
        axil of a cleft scale . . . .   1. *Psilotaceae*   p. 63
    2. Roots and leaves present; stem and sporangia
        variously disposed.
        3. Plants microphyllous; leaves with one vein;
            sporophylls aggregated into strobili or at
            stem tips when sporophylls similar to
            foliage leaves.
            4. Leaves without ligule; strobili cylindri-
                cal, spores uniform, all of one kind
                (homosporous) . . . .   2. *Lycopodiaceae*   p. 63
            4. Leaves with ligule; strobili usually four-
                sided, spores of two types, large and
                small in the same strobilus (hetero-
                sporous).
                5. Terrestrial plants; leaves in four axial
                    rows; stems conspicuous . . . .   3. *Selaginellaceae*   p. 67
                5. Aquatic plants; leaves imbricated cone
                    fashion on an abbreviated stem .   4. *Isoetaceae*   p. 69
        3. Plants macrophyllous; leaves with complex
            venation; spores uniform (homosporous).

6. Sporangia without annulus; leaves erect or
   inclined in bud; sporangia two-seriate;
   epiphytes with simple leaves three- to
   six-lobed . . . .   5. *Ophioglossaceae*   p. 71
6. Sporangia with annulus; leaves circinate
   in bud.
  7. Annulus a complete ring of thickened
     cells, transverse, oblique, or terminal,
     or a mere patch of cells.
    8. Sporangia on ultimate segments;
      leaves small, delicate, exstipulate.
      9. Leaves pellucid; stipe herbaceous;
        sporangia indusiate . . . .   6. *Hymenophyllaceae*   p. 71
      9. Leaves opaque; stipe wiry; sporan-
        gia nonindusiate . . . . .   7. *Schizaeaceae*   p. 73
    8. Sporangia nonsoral in panicles; leaves
      large, stipulate; annulus a patch of
      thick-walled cells . . . . . .   8. *Osmundaceae*   p. 77
  7. Annulus an incomplete vertical ring of
     thickened cells interrupted by sto-
     mium.
    10. Leaves subsessile, linear; sori in linear
      series with intramarginal groove on
      each side of the midvein . . . .   9. *Vittariaceae*   p. 77
    10. Leaves differentiated into blade and
      stipe; sori variously disposed.
      11. Stipe articulate to the rhizome (ex-
        cepting *Nephrolepis*).
        12. Sori exindusiate, circular; blades
          simple or pinnatifid; veins free
          or anastomosing . . . .   10. *Polypodiaceae*   p. 79
        12. Sori indusiate, attached by base
          or sides, terminal on veinlets;
          blades pinnate . . . . .   11. *Davalliaceae*   p. 84
      11. Stipe inarticulate.
        13. Erect or creeping terrestrials; if
          scandent not acrostichoid.
          14. Pinnae jointed to the rachis,
            auriculate on upper margin   11. *Davalliaceae*   p. 84
          14. Pinnae not jointed to the rachis.
            15. Sporangia protected by re-
              flexed margins, continu-
              ous, or lobed, with or with-
              out indusia or acrostichoid   12. *Pteridaceae*   p. 85
            15. Sporangia variously disposed
              on lamina, indusiate.
             16. Sporangia in elongate sori,
                indusium not peltate.
              17. Sori parallel to a vein
                on each side of the

SPERMATOPHYTA

1. Ovules naked, not included within an ovary of a flower; woody plants with needlelike leaves, or fernlike with pinnate leaves, or scalelike leaves; producing seeds in cones or pulpy disk, not producing true flowers and fruits . . . . GYMNOSPERMAE I. p. 29
1. Ovule or ovules enclosed by the ovary of a true flower, the ovary becoming the fruit at maturity; woody or herbaceous plants with leaves or true leaves absent . . . . . ANGIOSPERMAE II. p. 30

I GYMNOSPERMAE

Trees or shrubs, monoecious or dioecious, usually with needlelike, scalelike, or slender leaves. Ovules not within an enclosing ovary but rather borne naked on a scale within a cone or on a disk.

1. Leaves large, pinnately compound, in a terminal whorl or crown . . . . 19. *Cycadaceae* p. 108
1. Leaves linear, needlelike, or scalelike.
  2. Leaves evergreen, needlelike, in fascicles; seeds produced in elongate cones over 3 cm wide . . . . 20. *Pinaceae* p. 110
  2. Leaves evergreen or deciduous, needlelike or linear, not in fascicles, or scalelike; seeds produced in globular cones less than 3 cm wide or "cone" fleshy and berrylike.
    3. Leaves needlelike or linear; cone globular, dry . . . . 21. *Taxodiaceae* p. 111
    3. Leaves scalelike; cone fleshy, berrylike . 22. *Cupressaceae* p. 113

II ANGIOSPERMAE

1. LILIATAE (MONOCOTYLEDONEAE)

8. Culms solid throughout, trigonous, often roundish; leaves three-ranked; sheaths tubular; floret subtended by a simple or saccate scale; anthers basifixed . . . . *33. Cyperaceae*   p. 202
7. Leaves without ligules.
  9. Inflorescence a fleshy spadix with or without subtending spathe . . . *35. Araceae*   p. 247
  9. Inflorescence not a spadix.
    10. Perianth segments scarious or of bristles.
      11. Terrestrials, marsh plants, stem scapose, flowers unisexual.
        12. Inflorescence buttonlike, leaves radical in dense rosettes . . *38. Eriocaulaceae*   p. 259
        12. Inflorescence a terete spike, leaves straplike, basal . . *23. Typhaceae*   p. 114
      11. Submersed aquatics or sometimes floating-leaf aquatic with dilated blades.
        13. Ovary superior.
          14. Flowers unisexual in axillary spathes.
            15. Achenes falcate, beaked . *25. Zannichelliaceae*   p. 118
            15. Achenes not falcate.
              16. Perianth present in staminate flowers, pistillate with sterile stigmoid appendages; leaves cauline; freshwater plants . . . *27. Najadaceae*   p. 119
              16. Perianth lacking, pistillate flowers without sterile rudiments; leaves basal; marine plants . . . . . *28. Cymodoceaceae*   p. 121
          14. Flowers bisexual.
            17. Stamens four with sepaloid anther connectives; carpels four, sessile . . . *26. Potamogetonaceae*   p. 118
            17. Stamens two, anther connectives not sepaloid; carpels long stipitate . *24. Ruppiaceae*   p. 116
        13. Ovary inferior; plants dioecious   *31. Hydrocharitaceae*   p. 129
    10. Perianth segments not scarious or bristlelike.
      18. Ovary superior.
        19. Sepals and petals similar, petals sometimes lacking; plants rushlike.

20. Fruit three- to six-carpellate, one- to
two-seeded follicle, severally deciduous from the stigmas; sepals and
petals greenish . . . .   29. *Juncaginaceae*   p. 123
20. Fruit three-carpellate loculicidal cap-
sule, persistent, many seeded; sepals
and petals greenish brown or pur-
plish . . . .   42. *Juncaceae*   p. 275
19. Sepals and petals differentiated in form
and color or sometimes alike.
21. Flowers with radial symmetry.
22. Pistils simple, free; flowers bisexual;
androgynous if unisexual . . .   30. *Alismataceae*   p. 123
22. Pistil united into a compound ovary.
23. Flowers in conelike head with
closely imbricated bracts covering
the flower; leaves basal, scape
elongate . . . .   37. *Xyridaceae*   p. 254
23. Flowers in cymes, compound pan-
icles, or solitary.
24. Leaves soft, sheathing; flowers
small, ephemeral, filaments
often fimbriate . . . . .   40. *Commelinaceae*   p. 269
24. Leaves firm, not sheathing; flowers
not ephemeral, filaments gla-
brous . . . .   43. *Liliaceae*   p. 280
21. Flowers with bilateral symmetry.
25. Sepals and petals united into a basal
tube, emersed aquatic or marsh
plant . . . .   41. *Pontederiaceae*   p. 274
25. Sepals and petals free, unlike in form
and color; plant terrestrial . .   40. *Commelinaceae*   p. 269
18. Ovary inferior (partially superior in No. 43).
26. Flowers with radial symmetry.
27. Plant a twining vine with net-veined
leaves; flowers small, unisexual in axil-
lary spikes . . . .   48. *Dioscoreaceae*   p. 295
27. Plant not a vine.
28. Leaves fibrous; plants with terminal
compound panicles, pungent leaf
tips, prickly leaf margins . . . .   45. *Agavaceae*   p. 287
28. Leaves not fibrous.
29. Perianth densely pubescent outside;
juice red; flowers in terminal
corymbs . . . .   46. *Haemodoraceae*   p. 290
29. Perianth not densely pubescent out-
side; juice colorless; flowers variously disposed, not in terminal
corymbs.
30. Stamens six . . . .   47. *Amaryllidaceae*   p. 291

30. Stamens three.
    31. Small plants less than 20 cm
        tall; leaves minute, scalelike
        with or without chlorophyll   53. *Burmanniaceae*   p. 305
    31. Large plants over 20 cm tall;
        leaves sword shaped, equi-
        tant with chlorophyll . .   49. *Iridaceae*   p. 296
26. Flowers with bilateral symmetry.
    32. Pollen-bearing stamens five to six; cau-
        lescent treelike herbs . . . . .   50. *Musaceae*   p. 300
    32. Pollen-bearing stamens one or two;
        caulescent herbs, never treelike.
        33. Petallike staminodes five, in addition
            to fertile stamen.
            34. Ovary and fruit with warty excres-
                cences; capsule three-locular, each
                with many seeds . . . . .   51. *Cannaceae*   p. 301
            34. Ovary and fruit smooth; capsule
                baccate, with two aborted locules,
                the fertile locule one-seeded . .   52. *Marantaceae*   p. 303
        33. Petallike staminodes not present;
            anther, stigma, and style united to
            form a column . . . . . .   54. *Orchidaceae*   p. 308

2. MAGNOLIATAE (DICOTYLEDONEAE)

Key to Major Sections of the Dicotyledons

1. Plants parasitic, growing attached to stems of
   host plant and without connection with soil   Section 1.   p. 35
1. Plants rooting in soil, growing in water, or
   epiphytic, but not parasitic.
   2. Herbaceous plants, the above-ground parts
      not continuing their growth in successive
      seasons.
      3. Flowers bisexual, with stamens and pistils
         in the same flower.
         4. Flowers with a perianth (sepals or petals
            or both).
            5. Flowers with both sepals and petals
               present, the sepals sometimes repre-
               sented by hairs, bristles, or scales.
               6. Gynoecium composed of two or
                  more free carpels . . . . .   Section 2.   p. 35
               6. Gynoecium composed of one carpel,
                  or a compound ovary of two or
                  more fused carpels.
                  7. Ovary superior.

15. Flowers not in catkins; in most (not all) species the flowers bisexual or individually large and conspicuous, with calyx and corolla both present. *Section 16.*   p. 57

Section 1

Parasitic Plants Growing Attached to Stems of Host Plant

1. Stems slender and twining over host plant; leaves none, plants appearing orange or yellow in color, never green.
   2. Flowers cymose; fruit a capsule . . . . 160. *Convolvulaceae*   p. 711
   2. Flowers spicate; fruit a drupe . . . . . 85. *Lauraceae*   p. 422
1. Stems not twining; leaves well developed, green   67. *Loranthaceae*   p. 366

Section 2

Herbaceous Plants With Bisexual Flowers, Perianth of Both Sepals and Petals, These Sometimes Fused, and Gynoecium Composed of Two or More Free or Apparently Free Carpels

1. Style one per flower, style often divided above into two or more branches.
   2. Pistils two.
      3. Styles separate; anthers fused to stigmas; pollen in waxy masses (pollinia) . . . 159. *Asclepiadaceae*   p. 703
      3. Styles fused; stamens free, distinct; pollen granular . . . . 158. *Apocynaceae*   p. 697
   2. Pistils four, five, or more.
      4. Pistils four; petals fused; stamens free, never monadelphous.
         5. Leaves alternate; stamens five . . . 163. *Boraginaceae*   p. 726
         5. Leaves opposite; stamens two or four . 165. *Lamiaceae*   p. 743
      4. Pistils five; petals not fused or partly so at the base; stamens numerous, fused (monadelphous) . . . . . . 121. *Malvaceae*   p. 589
1. Styles as many as the pistils, or the styles poorly developed or absent.
   6. Sepals and petals each three.
      7. Aquatics with submersed dissected leaves and floating, entire leaves . . . . 79. *Nymphaeaceae*   p. 407
      7. Terrestrials with opposite, entire leaves . 90. *Crassulaceae*   p. 436
   6. Sepals and petals each four or more.
      8. Petals partly fused.
         9. Leaves opposite . . . . . . 158. *Apocynaceae*   p. 697
         9. Leaves alternate.
            10. Stems trailing or creeping; flowers solitary in leaf axils . . . . . 160. *Convolvulaceae*   p. 711
            10. Stems erect; flowers terminal . . . 158. *Apocynaceae*   p. 697
      8. Petals free, not fused.

11. Leaves peltate or cordate; flowers soli-
    tary, up to 12–25 cm wide; aquatics .   79. *Nymphaeaceae*   p. 407
11. Leaves not peltate; flowers much less
    than 12 cm wide; terrestrial plants . .   90. *Crassulaceae*   p. 436

Section 3

Herbaceous Plants with Bisexual Flowers, Perianth of Both Sepals and Petals,
Superior Ovary, Stamens More Numerous Than the Petals or Lobes of the
Corolla, and Flowers with Radial Symmetry

1. Leaves reduced to scales or bracts, leaves and
   stems without green color . . . . . .   123. *Hypericaceae*   p. 605
1. Leaves not scalelike or bractlike, stems and
   leaves green.
   2. Calyx of two sepals.
      3. Leaf blade serrate, relatively thin . . .   86. *Papaveraceae*   p. 425
      3. Leaf blade entire, thick and succulent . .   76. *Portulacaceae*   p. 397
   2. Calyx of three or more sepals.
      4. Stamens twice as many as petals or fewer.
         5. Stamens twice the number of petals.
            6. Sepals six or more; petals six or more.
               7. Styles as many as the petals; leaves
                  thick and fleshy; hypanthium
                  poorly developed . . . . .   90. *Crassulaceae*   p. 436
               7. Styles one; leaves thin; hypanthium
                  tubular . . . .   136. *Lythraceae*   p. 632
            6. Sepals four or five; petals four or five.
               8. Leaves trifoliolate, pinnately com-
                  pound, or divided nearly to the
                  base.
                  9. Corolla yellow or greenish yellow.
                     10. Leaves trifoliolate . . . .   97. *Oxalidaceae*   p. 502
                     10. Leaves pinnately compound .   100. *Zygophyllaceae*   p. 507
                  9. Corolla some other color.
                     11. Leaves opposite . . . . .   96. *Geraniaceae*   p. 502
                     11. Leaves alternate.
                        12. Style one.
                           13. Leaves pinnately or twice-
                               pinnately compound .   95. *Fabaceae*   p. 445
                           13. Leaves palmately com-
                               pound . . . . .   88. *Capparaceae*   p. 433
                        12. Styles five . . . . . .   97. *Oxalidaceae*   p. 502
               8. Leaves simple, never deeply divided.
                  14. Style one.
                     15. Hypanthium present, cup
                         shaped or tubular.
                        16. Hypanthium cup shaped;
                            anthers opening by termi-
                            nal pores . . . . .   142. *Melastomataceae* p. 649

16. Hypanthium tubular; anthers opening longitudinally . . . . . .   136. *Lythraceae*   p. 632
    15. Hypanthium absent . . .   125. *Cistaceae*   p. 610
  14. Styles two or more.
    17. Ovary definitely lobed and each lobe bearing a style .   90. *Crassulaceae*   p. 436
    17. Ovary smooth, not lobed, styles arising together at the top of the ovary.
      18. Corolla yellow . . . .   123. *Hypericaceae*   p. 605
      18. Corolla another color . .   78. *Caryophyllaceae*   p. 403
5. Stamens more than the number of petals but fewer than twice as many.
  19. Styles one.
    20. Sepals four; petals four.
      21. Flowers subtended by conspicuous bracts . . . . . .   88. *Capparaceae*   p. 433
      21. Flowers not subtended by bracts   87. *Brassicaceae*   p. 428
    20. Sepals or calyx lobes five or six; petals three to six.
      22. Sepals six; petals six.
      22. Sepals or calyx lobes five; petals three or five.
        23. Leaves simple . . . . .   125. *Cistaceae*   p. 610
        23. Leaves compound . . . .   95. *Fabaceae*   p. 445
  19. Styles two to five.
    24. Stamens in three groups or fascicles, three stamens in each group . .   123. *Hypericaceae*   p. 605
    24. Stamens not in groups or fascicles .   78. *Caryophyllaceae*   p. 403
4. Stamens more than twice as many as the petals.
  25. Aquatics; leaves large, entire, cleft to the petiole . . . .   79. *Nymphaeaceae*   p. 407
  25. Terrestrials; leaves not large and cleft to the petiole.
    26. Leaves simple.
      27. Ovary one-locular.
        28. Sepals three, ephemeral; leaves lobed . . . .   86. *Papaveraceae*   p. 425
        28. Sepals five, persistent; leaves not lobed . . . .   125. *Cistaceae*   p. 610
      27. Ovary two- to many-locular.
        29. Leaves opposite . . . . .   123. *Hypericaceae*   p. 605
        29. Leaves alternate . . . . .   121. *Malvaceae*   p. 589
    26. Leaves compound.
      30. Leaves trifoliolate, pinnately or twice-pinnately compound . .   95. *Fabaceae*   p. 445
      30. Leaves palmately compound . .   88. *Capparaceae*   p. 433

Section 4

Herbaceous Plants with Bisexual Flowers, Perianth of Both Sepals and Petals,
Superior Ovary, Stamens More Numerous Than the Petals or Lobes of the
Corolla, and Flowers with Bilateral Symmetry

1. Sepals petallike in size and color.
  2. None of the sepals prolonged into a spur;
    leaves not peltate . . . .   107. *Polygalaceae*   p. 525
  2. One sepal prolonged into a spur or sac;
    leaves peltate . . . .   98. *Tropaeolaceae*   p. 504
1. Sepals not petallike, mostly green.
  3. Lower two petals fused along their lower
    margins, appressed to each other and en-
    closing ten or five stamens . . . . .   95. *Fabaceae*   p. 445
  3. Lower two petals free and not enclosing sta-
    mens, or petal one.
    4. Flowers in dense terminal spikes or head-
      like clusters . . . .   95. *Fabaceae*   p. 445
    4. Flowers variously arranged but never in
      spikes or heads.
      5. Leaves simple, entire . . . . . .   136. *Lythraceae*   p. 632
      5. Leaves compound.
        6. Petals and sepals each four; stems
          climbing with tendrils . . . .   116. *Sapindaceae*   p. 572
        6. Petals and sepals each five . . . .   95. *Fabaceae*   p. 445

Section 5

Herbaceous Plants with Bisexual Flowers, Perianth of Both Sepals and Petals,
Petals Free, not Fused, Superior Ovary, and Stamens as Many as or Fewer
Than the Petals

1. Leaves simple, entire to deeply lobed but never
  divided or dissected.
  2. Leaves opposite.
    3. Sepals two; petals three or five . . . .   76. *Portulacaceae*   p. 397
    3. Sepals four to twelve; petals four to twelve.
      4. Leaves deeply palmately lobed . . .   96. *Geraniaceae*   p. 502
      4. Leaves entire or toothed.
        5. Style one.
          6. Hypanthium well developed, cup
            shaped to tubular . . . . .   136. *Lythraceae*   p. 632
          6. Hypanthium absent.
            7. Stamens opposite the petals . .   151. *Primulaceae*   p. 674
            7. Stamens alternate with petals .   157. *Gentianaceae*   p. 692
        5. Styles two to five.
          8. Ovary and fruit four- to five-locular   99. *Linaceae*   p. 504
          8. Ovary and fruit one-locular.

9. Corolla yellow . . . . . .   *123. Hypericaceae*   p. 605
9. Corolla another color.
  10. Styles two . . . . . .   *157. Gentianaceae*   p. 692
  10. Styles three to five . . . .   *78. Caryophyllaceae*   p. 403
2. Leaves alternate or all clustered at the base of the plant.
  11. Leaves shallowly to deeply palmately lobed.
    12. Climbing or trailing vines with tendrils; flowers radially symmetric . . . .   *131. Passifloraceae*   p. 618
    12. Plants not vinelike or with tendrils; flowers bilaterally symmetric . . .   *128. Violaceae*   p. 614
  11. Leaves entire, serrate or crenate, or pinnately lobed.
    13. Styles two or more.
      14. Leaves densely covered over with glandular hairs, the blades often reddish colored . . . . . .   *89. Droseraceae*   p. 434
      14. Leaves glabrous or nearly so, not covered with glandular hairs.
        15. Leaves chiefly basal . . . . .   *152. Plumbaginaceae*   p. 676
        15. Leaves cauline, not chiefly basal.
          16. Erect herbs, often with woody base; styles five.
            17. Corolla light blue or yellow; leaves entire . . . . .   *99. Linaceae*   p. 504
            17. Corolla reddish, violet, or purple; leaves toothed . . .   *122. Byttneriaceae*   p. 602
          16. Succulent vines; styles three . .   *77. Basellaceae*   p. 400
    13. Styles one or none.
      18. Hypanthium well developed, tubular   *136. Lythraceae*   p. 632
      18. Hypanthium none or poorly developed, or very short and not tubular.
        19. Flowers radially symmetrical.
          20. Sepals four; petals four . . .   *87. Brassicaceae*   p. 428
          20. Sepals five; petals five.
            21. Stamens united; flowers in axillary clusters; leaves serrate . . . .   *122. Byttneriaceae*   p. 602
            21. Stamens distinct; flowers in terminal racemes; leaves entire . . . .   *151. Primulaceae*   p. 674
        19. Flowers bilaterally symmetrical .   *128. Violaceae*   p. 614
1. Leaves compound, dissected, or deeply divided.
  22. Flowers solitary; plants scapose . . .   *128. Violaceae*   p. 614
  22. Flowers in inflorescences on a leafy stem, not scapose.
    23. Leaves deeply divided . . . . .   *87. Brassicaceae*   p. 428
    23. Leaves compound . . . .   *95. Fabaceae*   p. 445

Section 6

Herbaceous Plants with Bisexual Flowers, Perianth of Both Sepals and Petals,
Petals Fused, Superior Ovary, Stamens as Many as the Lobes of the Corolla,
and the Corolla Radially Symmetrical

1. Leaves basal and covered with glandular hairs,
   the blades often reddish in color . . . .   89. *Droseraceae*   p. 434
1. Leaves cauline, or, if basal, then not covered
   with glandular hairs.
   2. Stamens opposite the lobes of the corolla.
      3. Style one . . . .   151. *Primulaceae*   p. 674
      3. Styles five (rarely only three) . . . .   152. *Plumbaginaceae*   p. 676
   2. Stamens alternate with the lobes of the
      corolla.
      4. Ovary deeply lobed, appearing almost like
         tv ɔ or four separate ovaries.
         5. Style one; ovary four-lobed.
            6. Leaves alternate . . . . . .   163. *Boraginaceae*   p. 726
            6. Leaves opposite.
               7. Style cleft at the apex, arising be-
                  tween the lobes of the ovary . .   166. *Lamiaceae*   p. 743
               7. Style not cleft at the apex, arising
                  from top of the ovary . . . .   165. *Verbenaceae*   p. 732
         5. Styles two; ovary two-lobed . . . .   160. *Convolvulaceae*   p. 711
      4. Ovary not prominently lobed.
         8. Ovary one-locular or three-locular; stems
            not twining.
            9. Ovary one-locular.
               10. Styles three . . . . . .   130. *Turneraceae*   p. 616
               10. Styles one or two.
                  11. Style one or none . . . . .   157. *Gentianaceae*   p. 692
                  11. Styles two . . . .   162. *Hydrophyllaceae*   p. 725
            9. Ovary three-locular.
               12. Stems twining . . . . . .   160. *Convolvulaceae*   p. 711
               12. Stems erect, not twining . . .   161. *Polemoniaceae*   p. 724
         8. Ovary two-locular or four-locular.
            13. Leaves basal in rosettes; inflorescence
                a scapose spike; corolla scarious,
                four-lobed . . . . . .   172. *Plantaginaceae*   p. 790
            13. Leaves cauline, not all basal.
               14. Leaves opposite.
                  15. Leaves with stipules or stipules
                      connected at base by a stipular
                      line; flowers in terminal clusters   156. *Loganiaceae*   p. 689
                  15. Leaves without stipules; flowers
                      in axillary heads or spikes . .   165. *Verbenaceae*   p. 732
               14. Leaves alternate, or at least the up-
                   per leaves alternate.

16. Stamens four, the upper pair differing in appearance from the lower pair; or, stamens one or two . . . . 168. *Scrophulariaceae* p. 760
16. Stamens five, more or less all alike.
   17. Corolla obviously lobed, to or below the middle.
      18. Style and stigma one.
         19. Flowers in scorpioid spikes or racemes; fruit separating into two or four nutlets . . . . . 163. *Boraginaceae* p. 726
         19. Flowers not in scorpioid spikes or racemes; fruit a berry . . . . . . 167. *Solanaceae* p. 752
      18. Styles and stigmas two or four; fruit a capsule . . 162. *Hydrophyllaceae* p. 725
   17. Corolla unlobed or shallowly lobed.
      20. Styles one or two, branched or lobed; ovules two per locule; fruit a capsule . . 160. *Convolvulaceae* p. 711
      20. Style one, stigma one; ovules numerous in each locule; fruit a berry . . . . 167. *Solanaceae* p. 752

Section 7

Herbaceous Plants with Bisexual Flowers, Perianth of Both Sepals and Petals, Petals Fused, Ovary Superior, Stamens as Many as or Fewer Than the Lobes of the Corolla, and Corolla Bilaterally Symmetrical

1. Ovary distinctly four-lobed.
   2. Leaves opposite; flowers not in one-sided cymes.
      3. Style cleft at the apex, arising between the lobes of the ovary . . . . . . 166. *Lamiaceae* p. 743
      3. Style not cleft, arising from the top of the ovary . . . . 165. *Verbenaceae* p. 732
   2. Leaves alternate; flowers in one-sided cymes that unroll from the end . . . . . 163. *Boraginaceae* p. 726
1. Ovary not distinctly four-lobed.
   4. Corolla spurred or saclike at base.
      5. Calyx deeply five-parted.
         6. Leaves all basal or absent, scapes one-flowered . . . . 170. *Lentibulariaceae* p. 777
         6. Leaves cauline . . . . 168. *Scrophulariaceae* p. 760
      5. Calyx deeply two-parted, apparently of two sepals; mostly aquatics . . . . 170. *Lentibulariaceae* p. 777

4. Corolla base not spurred or saclike.
  7. Stamens two.
    8. Leaves in basal rosette, flowers in sca-
      pose spikes . . . .   172. *Plantaginaceae*   p. 790
    8. Leaves cauline, not all clustered at base.
      9. Leaves with cystoliths, seeds less than
        twenty per capsule . . . . .   171. *Acanthaceae*   p. 781
      9. Leaves without cystoliths; seeds more
        than twenty per capsule . . . .   168. *Scrophulariaceae*   p. 760
  7. Stamens four or five.
    10. Some or all the leaves opposite.
      11. Ovules one per locule; fruit splitting
        at maturity into one-seeded nutlets   165. *Verbenaceae*   p. 732
      11. Ovules more than one per locule;
        fruit a capsule.
        12. Leaves with cystoliths; seeds less
          than twenty per capsule . . .   171. *Acanthaceae*   p. 781
        12. Leaves without cystoliths; seeds
          more than twenty per capsule .   168. *Scrophulariaceae*   p. 760
    10. Leaves alternate . . . .   168. *Scrophulariaceae*   p. 760

Section 8

Herbaceous Plants with Bisexual Flowers, Inferior Ovary, Perianth of Both
Sepals and Petals, Sepals Sometimes Represented by Hairs, Bristles, or Scales

1. Stamens more numerous than petals.
  2. Stamens more than twice as many as sepals
    or petals, usually twenty or more.
    3. Stems becoming thick, fleshy, mostly
      spiny; leaves absent or reduced to
      scales . . . .   135. *Cactaceae*   p. 623
    3. Stems not thick and fleshy; leaves present,
      alternate . . . .   133. *Loasaceae*   p. 622
  2. Stamens twice as many as the petals or sepals
    but not more than twelve.
    4. Style one.
      5. Leaves simple; terrestrial plants . . .   143. *Onagraceae*   p. 652
      5. Leaves pinnately dissected or toothed;
        aquatics or plants of marshy places .   144. *Haloragaceae*   p. 657
    4. Styles two or more.
      6. Leaves simple, entire; terrestrial plants .   76. *Portulacaceae*   p. 397
      6. Leaves pinnately dissected into capillary
        divisions or pinnately toothed; aquat-
        ics or plants of marshy places . . .   144. *Haloragaceae*   p. 657
1. Stamens of the same number or fewer than the
  petals or corolla lobes.
  7. Petals free, not fused.
    8. Petals four.

9. Submersed leaves pinnately dissected
    into capillary divisions . . . . .   144. *Haloragaceae*   p. 657
9. Submersed leaves none, or, if present,
    their blades entire, not pinnately
    dissected . . . .   143. *Onagraceae*   p. 652
8. Petals five . . . .   146. *Apiaceae*   p. 659
7. Petals fused.
  10. Flowers small, tightly arranged in involu-
     crate heads . . . .   179. *Asteraceae*   p. 820
  10. Flowers not in involucrate heads.
    11. Cauline leaves alternate.
      12. Corolla radially symmetrical.
        13. Corolla white or pink . . . .   151. *Primulaceae*   p. 674
        13. Corolla blue . . . .   177. *Campanulaceae*   p. 818
      12. Corolla bilaterally symmetrical.
        14. Corolla glabrous within, more or
          less two-lipped . . . . . .   177. *Campanulaceae*   p. 818
        14. Corolla woolly within, open to base
          on one side . . . .   178. *Goodeniaceae*   p. 818
    11. Cauline leaves opposite, whorled, or in
      basal clusters.
      15. Stipules present; corolla radially sym-
        metrical or nearly so . . . . .   173. *Rubiaceae*   p. 792
      15. Stipules absent; corolla radially or bi-
        laterally symmetrical.
        16. Corolla bilateral, tube open to the
          base on one side . . . . .   178. *Goodeniaceae*   p. 820
        16. Corolla radial, funnelform . . .   175. *Valerianaceae*   p. 871

Section 9

Herbaceous Plants with Bisexual Flowers, Perianth with either Sepals or Petals but not with Both Present, Ovary Superior

1. Ovaries more than one per flower, each ovary
   with style and stigma.
  2. Aquatics; leaves submersed and dissected or
    floating and entire, often peltate . . .   79. *Nymphaeaceae*   p. 407
  2. Terrestrials; leaves not submersed, floating
    or peltate.
    3. Carpels separate and distinct; herbaceous
      vines with compound or deeply cleft
      leaves . . . .   82. *Ranunculaceae*   p. 413
    3. Carpels united at the base; erect or reclin-
      ing herbs with entire leaves . . . .   74. *Phytolaccaceae*   p. 392
1. Ovaries one per flower (styles and stigmas may
   be more than one).
  4. Stamens more than twice as many as the
    parts or lobes of the perianth.

5. Perianth well developed and conspicuous.
  6. Leaves entire.
    7. Leaves large, hastate or deeply cor-
      date; aquatics . . . . . .   79. *Nymphaeaceae*   p. 407
    7. Leaves linear, succulent; terrestrials   76. *Portulacaceae*   p. 397
  6. Leaves compound, lobed, or spiny
    toothed.
    8. Perianth four- or eight-parted;
      leaves dissected, lobed or spiny
      toothed, not compound; sap milky
      or colored . . . .   86. *Papaveraceae*   p. 425
    8. Perianth five-parted; leaves twice-
      pinnately compound; sap not
      milky or colored . . . . .   95. *Fabaceae*   p. 445
5. Perianth small, less than 10 mm wide,
  inconspicuous, pale white, greenish,
  or purplish . . . .   75. *Aizoaceae*   p. 395
4. Stamens twice as many as the parts or lobes
of the perianth or fewer than twice as
many.
  9. Styles two or more.
    10. Leaves reduced to mere scales;
      succulent plants with jointed
      stems . . . .   71. *Chenopodiaceae*   p. 378
    10. Leaves not scalelike.
      11. Leaves tending to be clustered
        near the base or lower half
        of the stems . . . . .   76. *Portulacaceae*   p. 397
      11. Leaves all or mainly cauline.
        12. Leaves opposite or whorled.
          13. Ovary and fruit one-locular   78. *Caryophyllaceae*   p. 403
          13. Ovary and fruit three- to
            five-locular . . . .   75. *Aizoaceae*   p. 395
        12. Leaves alternate.
          14. Stamens ten; styles five or
            ten . . . . . .   74. *Phytolaccaceae*   p. 392
          14. Stamens fewer than ten;
            styles two to five.
            15. Stipules sheathing the
              stem above the base
              of each leaf; fruit an
              achene . . . . .   70. *Polygonaceae*   p. 372
            15. Stipules none . . .   71. *Chenopodiaceae*   p. 378
  9. Style one or none (two or more
    stigmas may be present).
    16. Stamens more numerous than the
      lobes or parts or the perianth.
      17. Perianth parts or lobes three or
        five.

18. Perianth parts five.
  19. Leaves simple.
    20. Perianth parts distinctly
      unequal, two smaller
      than other three . .   125. *Cistaceae*   p. 610
    20. Perianth parts equal or
      nearly so.
      21. Flowers axillary . .   75. *Aizoaceae*   p. 395
      21. Flowers in long pe-
        duncled, terminal
        clusters . . . .   76. *Portulacaceae*   p. 397
  19. Leaves compound . . .   95. *Fabaceae*   p. 445
18. Perianth parts three . . .   125. *Cistaceae*   p. 610
17. Perianth parts or lobes four.
  22. Leaves opposite . . . .   136. *Lythraceae*   p. 632
  22. Leaves alternate or basal .   87. *Brassicaceae*   p. 428
16. Stamens as many as the parts or
  lobes of the perianth or fewer.
  23. Leaves opposite or whorled.
    24. Flowers or flower clusters
      axillary, sessile or nearly so.
      25. Flowers several to many in
        heads or clusters; peri-
        anth thin membrana-
        ceous, subtended by con-
        spicuous bracts . . .   72. *Amaranthaceae*   p. 381
      25. Flowers solitary or in few-
        flowered axillary clusters;
        perianth herbaceous,
        green, not subtended
        by conspicuous bracts.
        26. Perianth four-lobed . .   136. *Lythraceae*   p. 632
        26. Perianth five-lobed . .   75. *Aizoaceae*   p. 395
    24. Flowers terminal or in ter-
      minal inflorescences.
      27. Flowers in capitate or
        umbellate clusters or in
        dense spikes.
        28. Perianth parts and sub-
          tending bracts mem-
          branaceous or scar-
          ious; flowers in dense
          spikes . . . . .   72. *Amaranthaceae*   p. 381
        28. Perianth parts not mem-
          branaceous or scar-
          ious, conspicuously
          five-lobed or with a
          five-parted involucre;
          flowers in capitate or
          umbellate clusters .   73. *Nyctaginaceae*   p. 388

27. Flowers in loose open
cymes or panicles.
29. Leaves broad, petiolate,
without stipules .   .   73. *Nyctaginaceae*   p. 388
29. Leaves linear to nar-
rowly elliptic, sessile,
often with stipules .   78. *Caryophyllaceae*   p. 403
23. Leaves alternate or in basal
clusters.
30. Stamens four or five, as many
as the sepals.
31. Flowers sessile in small, ax-
illary clusters.   65. *Urticaceae*   p. 361
31. Flowers with pedicels or
the inflorescence with
definite peduncle.
32. Flowers in dense spikes
or spikelike panicles .   72. *Amaranthaceae*   p. 381
32. Flowers in racemes .   .   74. *Phytolaccaceae*   p. 392
30. Stamens one to three, fewer
than the sepals, flowers sca-
pose, cleistogamous   .   .   128. *Violaceae*   p. 614

Section 10

Herbaceous Plants with Bisexual Flowers, Perianth with either Sepals or Petals
but not with Both Present, Sepals Sometimes Represented by Hairs, Bristles,
Scales, or None, and Ovary Inferior

1. Stamens more numerous than the lobes or divisions of the perianth.
2. Leaves opposite; herbs often succulent .   .   75. *Aizoaceae*   p. 395
2. Leaves alternate, blades entire or pinnatifid.
3. Erect aquatic herbs; ovary three- to four-
locular .   .   .   .   144. *Haloragaceae*   p. 657
3. Twining terrestrial herbs; ovary six-locular   69. *Aristolochiaceae* p. 370
1. Stamens as many as or fewer than the lobes or
divisions of the perianth.
4. Anthers fused into a tube surrounding the
style; flowers small, in involucrate heads   179. *Asteraceae*   p. 820
4. Anthers free and distinct, not fused; flow-
ers not in involucrate head.
5. Leaves opposite or whorled.
6. Leaves whorled .   .   .   .   173. *Rubiaceae*   p. 792
6. Leaves mostly opposite.
7. Stems conspicuously succulent .   .   75. *Aizoaceae*   p. 395
7. Stems not succulent .   .   .   .   .   143. *Onagraceae*   p. 652
5. Leaves alternate or basal.
8. Stamens four or fewer; perianth three-
or four-parted.

9. Perianth divisions three; stamens
   three . . . . 144. *Haloragaceae* p. 657
9. Perianth divisions four; stamens
   four . . . . 143. *Onagraceae* p. 652
8. Stamens five; perianth five-parted; in
   umbels or capitate clusters . . . 146. *Apiaceae* p. 659

Section 11

Herbaceous Plants with Bisexual Flowers but without a Perianth

1. Ovary superior; flowers numerous on long
   spikes; epiphytic, terrestrial, marsh, or shal-
   low-water plants with stems erect, arising
   above the surface.
   2. Inflorescence creamy white; stamens four to
      six . . . . 56. *Saururaceae* p. 339
   2. Inflorescence green; stamens two . . . . 57. *Piperaceae* p. 341
1. Ovary inferior; flowers axillary or in spikes;
   aquatics, either submersed or in wet soils . 144. *Haloragaceae* p. 657

Section 12

Herbaceous Plants with Unisexual Flowers

(This section includes plants which also may have bisexual flowers. If any
bisexual flowers are present, they should be traced through a different section.)

1. Flowers in an involucrate head, the bracts of
   the involucre resembling the sepals of a typical calyx; usually only some of the flowers of
   the head are unisexual . . . . 179. *Asteraceae* p. 820
1. Flowers not in an involucrate head.
   2. Leaves highly modified; either lacking, scale-
      like or reduced to small scales.
      3. Leaves pinnately or palmately dissected
         into narrow or filiform segments; aquat-
         ics or of marshy places.
         4. Leaves pinnately dissected . . . . 144. *Haloragaceae* p. 657
         4. Leaves palmately dissected . . . . 80. *Ceratophyllaceae* p. 411
      3. Leaves lacking or reduced to mere scales.
         5. Stems with fleshy joints; ovary superior;
            usually saline soils . . . . . . 71. *Chenopodiaceae* p. 378
         5. Stems without fleshy or jointed stems;
            ovary inferior; aquatics of fresh water
            or moist places . . . . 144. *Haloragaceae* p. 657
   2. Leaves present, not dissected or reduced to
      scales.
      6. Leaves simple.
         7. Leaves all basal.
            8. Flowers in spikes . . . . . . 172. *Plantaginaceae* p. 790

8. Flowers in panicles . . . ⌊ . . . .   70. *Polygonaceae*   p. 372
7. Leaves chiefly cauline, not basal.
  9. Leaves opposite or whorled.
    10. Leaves densely stellate pubescent or
        scaly . . . .   108. *Euphorbiaceae*   p. 532
    10. Leaves glabrous or pubescent, never
        stellate, pubescent, or scaly.
      11. Flowers solitary . . . . .   109. *Callitrichaceae*   p. 558
      11. Flowers in axillary or terminal
        clusters.
        12. Leaf blade unequilateral at
          base; involucre with four
          glands . . . .   108. *Euphorbiaceae*   p. 532
        12. Leaf blade equilateral at base;
          inflorescence glandless.
          13. Inflorescence axillary.
            14. Leaves entire or nearly so .   72. *Amaranthaceae*   p. 381
            14. Leaves serrate . . . .   65. *Urticaceae*   p. 361
          13. Infloresence terminal; petals
            absent . . . . . .   72. *Amaranthaceae*   p. 381
  9. Leaves alternate.
    15. Leaves densely stellate pubescent or
        scaly . . . .   108. *Euphorbiaceae*   p. 532
    15. Leaves glabrous, or, if pubescent,
        then not stellate pubescent or
        scaly.
      16. Calyx and corolla both present,
        the latter mostly colored white.
        17. Stems twining or climbing with
          tendrils present; flowers ra-
          dially symmetrical.
          18. Stems twining; stamens six;
            ovary superior.   83. *Menispermaceae*   p. 415
          18. Stems climbing by tendrils or
            trailing and with tendrils;
            stamens three; ovary infe-
            rior . . . . . .   176. *Cucurbitaceae*   p. 812
        17. Erect plants; flowers with petals
          unequal . . . . . .   134. *Begoniaceae*   p. 622
      16. Calyx present or absent; corolla
        absent.
        19. Flowers in small axillary clus-
          ters.
          20. Flowers pistillate.
            21. Style one, unbranched .   65. *Urticaceae*   p. 361
            21. Styles two or three.
              22. Styles three, each of these
                branched or two-lobed   108. *Euphorbiaceae*   p. 532
              22. Styles two or three, un-
                branched.

23. Sepals or bracts herbaceous (sepals
    often lacking) . . . . . .   71. *Chenopodiaceae*   p. 378
23. Sepals or bracts membranaceous or
    scarious, acute . . . . . .   72. *Amaranthaceae*   p. 381
20. Flowers staminate.
24. Flowers or flower clusters subtended and
    often exceeded by the bracts.
25. Bracts serrate, lobed, or cleft . . .   108. *Euphorbiaceae*   p. 532
25. Bracts entire.
26. Sepals and bracts membranaceous
    or scarious, acute . . . . .   72. *Amaranthaceae*   p. 381
26. Sepals and bracts herbaceous . .   65. *Urticaceae*   p. 361
24. Flowers or flower clusters not bracted.
27. Leaves with stipules . . . . .   108. *Euphorbiaceae*   p. 532
27. Leaves without stipules . . . . .   71. *Chenopodiaceae*   p. 378
19. Flowers in spikes, racemes, or panicles, usu-
    ally terminal, but in some forms opposite
    the leaves or basal.
28. Sepals petal-like in size and color.
29. Leaves lobed; stem prickly . . . .   108. *Euphorbiaceae*   p. 532
29. Leaves entire, not lobed; stem glabrous.   74. *Phytolaccaceae*   p. 392
28. Sepals minute, green or pale, not petal-like,
    or absent.
30. Perianth parts six, or two series . . .   70. *Polygonaceae*   p. 372
30. Perianth parts five, or fewer, or absent.
31. Sepals acute, membranaceous, mingled
    with similar acute scarious bracts .   72. *Amaranthaceae*   p. 381
31. Sepals not acute, or not membrana-
    ceous, or not subtended by scarious
    bracts.
32. Flowers pistillate or plant in fruit.
33. Ovary three-locular; fruit a three-
    locular capsule with three or
    six seeds . . . .   108. *Euphorbiaceae*   p. 532
33. Ovary, one-locular; fruit a one-
    seeded utricle . . . . .   71. *Chenopodiaceae*   p. 378
32. Flowers staminate.
34. Stamens numerous; leaves large,
    palmately veined and lobed .   108. *Euphorbiaceae*   p. 532
34. Stamens numerous; leaves large,
    palmately veined and lobed .
35. Sepals free . . . . . .   71. *Chenopodiaceae*   p. 378
35. Sepals fused into a two- to five-
    lobed calyx . . . . .   108. *Euphorbiaceae*   p. 532
6. Leaves compound.
36. Leaves trifoliolate or deeply cleft; perianth
    conspicuous, stamens numerous . . .   82. *Ranunculaceae*   p. 413
36. Leaves pinnately or twice-pinnately com-
    pound or ternately compound.

37. Leaves even twice-pinnately compound .   95. *Fabaceae*   p. 445
37. Leaves odd compound, ternate, or biter-
    nate, or twice-pinnately compound.
  38. Leaves opposite; stems erect or creeping   82. *Ranunculaceae*   p. 413
  38. Leaves alternate; stem climbing .   .   .   116. *Sapindaceae*   p. 572

Section 13

Woody Plants with Opposite or Whorled Leaves and Leaf Scars

1. Flowers in an involucrate head .   .   .   .   .   179. *Asteraceae*   p. 820
1. Flowers not in an involucrate head.
  2. Leaf buds or blades not open or fully ex-
    panded at the time of flowering.
    3. Perianth of both calyx and corolla; ovary
      two-lobed; stamens five or more .   .   .   115. *Aceraceae*   p. 570
    3. Perianth of a single series of parts or ab-
      sent; ovary not lobed; stamens two to
      four .   .   .   .   155. *Oleaceae*   p. 685
  2. Leaf blades partly or fully expanded at time
    of flowering.
    4. Leaves mere scales 1–3 mm long, whorled,
      on slender jointed branches; flowers
      without a perianth .   .   .   .   .   .   55. *Casuarinaceae*   p. 338
    4. Leaves not scalelike, stems not jointed;
      flowers with a perianth.
      5. Leaves simple.
        6. Perianth of one series of parts, or if
          more than one series, then all much
          alike.
          7. Leaves palmately veined and lobed   115. *Aceraceae*   p. 570
          7. Leaves entire, or pinnately veined
            and serrate, unlobed.
            8. Flowers bisexual.
              9. Low shrubs; corolla yellow .   .   125. *Cistaceae*   p. 610
              9. Shrubs or trees; flowers another
                color.
                10. Ovary or fruit several-loc-
                  ular; leaves strongly aro-
                  matic, punctate .   .   .   140. *Myrtaceae*   p. 640
                10. Ovary or fruit one-locular;
                  leaves not aromatic, not
                  punctate .   .   .   .   .   139. *Combretaceae*   p. 636
            8. Flowers chiefly unisexual.
              11. Leaves succulent or flat.
                12. Stems prostrate or creeping;
                  leaves linear .   .   .   .   61. *Bataceae*   p. 350
                12. Stems erect; leaves broad .   139. *Combretaceae*   p. 636
              11. Leaves appearing needlelike,
                revolute, not succulent .   .   110. *Empetraceae*   p. 559
        6. Perianth of both calyx and corolla, the
          calyx inconspicuous in some species.

13. Stamens more numerous than the petals or lobes of the corolla.

  14. Fruits elongate, cylindrical, up to several dm long; maritime plants with branching aerial roots . . . .   138. *Rhizophoraceae*   p. 636

  14. Fruits not elongate-cylindrical.

    15. Stamens ten or fewer.

      16. Leaves lobed; fruit a samara . . .   115. *Aceraceae*   p. 570

      16. Leaves not lobed; fruit a drupe or berry.

        17. Petals clawed; anthers opening longitudinally . . . .   106. *Malpighiaceae*   p. 524

        17. Petals not clawed; anthers opening by pores . . . .   142. *Melastomataceae*   p. 649

    15. Stamens more than ten.

      18. Styles two to five, separate or somewhat united at the base . . . .   123. *Hypericaceae*   p. 605

      18. Style one.

        19. Low shrubs; flowers yellow . . .   125. *Cistaceae*   p. 610

        19. Shrubs or trees; flowers some other color.

          20. Ovary or fruit several-locular; leaves aromatic, punctate . .   140. *Myrtaceae*   p. 640

          20. Ovary or fruit one-locular; leaves not aromatic or punctate . .   139. *Combretaceae*   p. 636

13. Stamens as many as the petals or corolla lobes, or fewer.

  21. Petals free.

    22. Flowers in a terminal cyme . . . .   147. *Cornaceae*   p. 667

    22. Flowers axillary or in axillary clusters.

      23. Style short, unbranched; stamens alternate with the petals.

        24. Stamens four or five; shrubs or trees.   113. *Celastraceae*   p. 567

        24. Stamens three; trailing vines . .   114. *Hippocrateaceae*   p. 569

      23. Style forked; stamens opposite the petals . . . .   117. *Rhamnaceae*   p. 577

  21. Petals fused, sometimes only partially at the bases.

    25. Ovaries superior.

      26. Corolla radially symmetrical.

        27. Stems trailing or twining; ovaries two, separate.

          28. Filaments distinct . . . . .   158. *Apocynaceae*   p. 697

          28. Filaments fused into a tube . .   159. *Asclepiadaceae*   p. 703

        27. Stems erect or twining, ovary compound.

          29. Stamens two . . . . . .   155. *Oleaceae*   p. 685

          29. Stamens four or five.

            30. Stamens four.

              31. Fruit dry or drupaceous; shrubs . . . . . .   165. *Verbenaceae*   p. 732

31. Fruit a fleshy capsule; mari-
time trees or shrubs . .   164. *Avicenniaceae*   p. 732
30. Stamens five.
  32. Twining vines; staminodes
    absent . . . . . .   156. *Loganiaceae*   p. 689
  32. Shrubs or trees; staminodes
    present . . . . . .   149. *Theophrastaceae*   p. 670
26. Corolla bilaterally symmetrical.
  33. Ovary deeply four-lobed; shrubs .   166. *Lamiaceae*   p. 743
  33. Ovary not deeply four-lobed; trees,
  vines, or shrubs.
    34. Leaf blades thick; cymes pe-
    dunculate; corolla inconspicu-
    ous; maritime trees or shrubs;
    fruit one-seeded . . . . .   164. *Avicenniaceae*   p. 732
    34. Leaf blades thin; inflorescence
    various, not cymose; fruit more
    than one-seeded.
      35. Ovules one in each of two to
      four locules of ovary . . .   165. *Verbenaceae*   p. 732
      35. Ovules numerous in each locule
      of ovary . . . . . .   168. *Scrophulariaceae*   p. 760
25. Ovaries inferior.
  36. Leaves opposite and with stipules or
  whorled and without stipules . .   173. *Rubiaceae*   p. 792
  36. Leaves opposite and without stipules.
    37. Corolla radially symmetrical or
    nearly so . . . .   174. *Caprifoliaceae*   p. 809
    37. Corolla bilaterally symmetrical . .   178. *Goodeniaceae*   p. 820
5. Leaves compound.
  38. Stems climbing or trailing . . . .   169. *Bignoniaceae*   p. 772
  38. Stems erect, not climbing or trailing.
    39. Petals absent . . . .   155. *Oleaceae*   p. 685
    39. Petals well developed and conspicuous.
      40. Stamens four; leaves palmately com-
      pound . . . .   165. *Verbenaceae*   p. 732
      40. Stamens five; leaves pinnnately com-
      pound.
        41. Corolla white, petals fused . . .   174. *Caprifoliaceae*   p. 809
        41. Corolla blue, petals free . . . .   100. *Zygophyllaceae*   p. 507

Section 14

Woody Plants with Alternate Leaves and Leaf Scars, Dioecious

1. Flowers in an involucrate head . . . . .   179. *Asteraceae*   p. 820
1. Flowers not in an involucrate head.
  2. Stems climbing, twining, or trailing vines.
    3. Stems producing tendrils; leaves or lobes

dentate or leaves compound; petals four
to five, deciduous . . . .   118. *Vitaceae*   p. 582
3. Stems not producing tendrils; leaves pal-
mately veined . . . .   83. *Menispermaceae*   p. 415
2. Stems erect, ascending or nearly prostrate,
not climbing or vinelike.
  4. Flowers in catkins or catkinlike clusters,
clusters globose to ovoid or elongate;
calyx inconspicuous or absent, corolla
absent.
    5. Calyx present but minute; sap milky .   64. *Moraceae*   p. 359
    5. Calyx absent; sap not milky.
      6. Stems resin dotted; shrubs; some of the
leaves usually irregularly toothed;
ovary with one ovule . . . . .   59. *Myricaceae*   p. 345
      6. Stems not resin dotted; trees or shrubs
of moist places; leaves finely serrate;
ovary many-ovuled . . . . .   58. *Salicaceae*   p. 345
  4. Flowers not in catkins or catkinlike clusters; either calyx or corolla or both present; individual flowers may be large and
conspicuous.
    7. Leaves simple or not present during
flowering.
      8. Leaves needlelike, 8–12 mm long, revo-
lute; petals absent . . . . .   110. *Empetraceae*   p. 559
      8. Leaves much larger, not needlelike.
        9. Flowers pistillate.
        10. Perianth not differentiated into
calyx and corolla or perianth
absent.
          11. Perianth six-parted . . . .   85. *Lauraceae*   p. 422
          11. Perianth three- to five-lobed, or
three to five free segments, or
perianth absent.
          12. Style and stigma one.
            13. Style very short; stigma ses-
sile . . . . . .   112. *Aquifoliaceae*   p. 564
            13. Style elongate, curved or
coiled at apex.
              14. Style pubescent . . .   65. *Urticaceae*   p. 361
              14. Style glabrous . . .   141. *Nyssaceae*   p. 648
            12. Style divided above, stigmas
two to four.
              15. Leaf blades conspicuously
inequilateral at base .   63. *Ulmaceae*   p. 356
              15. Leaf blades essentially
equilateral.
                16. Leaves with conspicuous
sheathing stipules .   70. *Polygonaceae*   p. 372

16. Leaves without stipules .   117. *Rhamnaceae*   p. 577
10. Perianth differentiated into calyx
and corolla, the calyx often
small and inconspicuous.
  17. Flowers in terminal panicle.
    18. Petals free; ovary three- to
five-locular . . . . .   111. *Anacardiaceae*   p. 560
    18. Petals partly fused; ovary
four- to six-locular . . .   150. *Myrsinaceae*   p. 672
  17. Flowers solitary in leaf axils or
in axillary cymes or clusters.
    19. Style one; petals free or par-
tially fused.
      20. Petals partially fused; flow-
ers on spurs . . . .   150. *Myrsinaceae*   p. 672
      20. Petals free; flowers not on
spurs.   .
        21. Style divided above the middle;
stigmas two to four . . . . .   117. *Rhamnaceae*   p. 577
        21. Style simple, undivided; stigma one.
          22. Style very short; stigma sessile .   112. *Aquifoliaceae*   p. 564
          22. Style elongate, coiled or curved at
apex . . . .   141. *Nyssaceae*   p. 648
    19. Styles two to six; petals par-
tially fused . . .   154. *Ebenaceae*   p. 684
9. Flowers staminate.
  23. Stamens more numerous than the sepals
or petals, or perianth lacking.
    24. Perianth minute or absent . . . .   141. *Nyssaceae*   p. 648
    24. Perianth well developed and conspic-
uous.
      25. Stamens usually ten; perianth of
five similar divisions . . . .   132. *Caricaceae*   p. 621
      25. Stamens usually sixteen; perianth a
four-lobed calyx and a four- to
six-lobed corolla . . . . .   154. *Ebenaceae*   p. 684
  23. Stamens as many as calyx lobes and as
many as the petals if petals are present.
    26. Flowers in a terminal panicle.
      27. Stamens not fused to the corolla;
petals free . . . .   111. *Anacardiaceae*   p. 560
      27. Stamens fused to corolla; petals par-
tially fused . . . .   150. *Myrsinaceae*   p. 672
    26. Flowers solitary in leaf axils or in ax-
illary clusters or clustered on spurs.
      28. Stamens alternate with the sepals,
opposite petals (if petals are present).
        29. Petals free . . . . . .   117. *Rhamnaceae*   p. 577
        29. Petals partially fused . . . .   150. *Myrsinaceae*   p. 672

28. Stamens opposite the sepals, alternate with petals (if petals are present).
   30. Leaf blades very inequilateral at base . . . .   63. *Ulmaceae*   p. 356
   30. Leaf blades essentially equilateral   112. *Aquifoliaceae*   p. 564
7. Leaves compound, present during flowering.
  31. Leaves trifoliolate.
    32. Flowers in terminal cymes or compound cymes; fruit a berry . . . . . .   101. *Rutaceae*   p. 508
    32. Flowers in panicles from leaf axils or from axils of leaf scars . . . . .   111. *Anacardiaceae*   p. 560
  31. Leaflets more than three.
    33. Stems thorny; leaves once-pinnate . .   101. *Rutaceae*   p. 508
    33. Stems unarmed.
      34. Stamens eight to twelve; ovary three-locular; leaves once- or twice-pinnate.
        35. Leaflets five to seven; petals four to six, green or greenish brown, stamens eight to twelve; bark smooth, reddish . . . . .   104. *Burseraceae*   p. 520
        35. Leaflets four to twelve; petals four to five, green or white, stamens five to ten; bark not reddish . .   116. *Sapindaceae*   p. 572
      34. Stamens ten; ovary one-locular and becoming a legume or two- to five-lobed and becoming a fruit of one to five narrow samaras.
        36. Ovary deeply lobed; leaves once-pinnate . . . .   102. *Simaroubaceae*   p. 517
        36. Ovary not lobed; leaves twice-pinnate . . . .   95. *Fabaceae*   p. 445

Section 15

Woody Plants with Alternate Leaves and Leaf Scars, Monoecious or Bisexual, Flowers in Catkins, Racemes, or Heads and Unisexual, or Bisexual and with no Perianth or Perianth of a Single Series of Parts

1. Flowers or some of them, especially the staminate, in catkins, catkinlike clusters, or racemes, the flowers always unisexual and individually small.
  2. Pistillate flowers solitary or in small clusters.
    3. Leaves simple, but sometimes deeply lobed   62. *Fagaceae*   p. 352
    3. Leaves pinnately compound . . . .   60. *Juglandaceae*   p. 347
  2. Pistillate flowers in catkins, heads, or conelike structures.
    4. Leaves simple, broad, not scalelike, stems not jointed.

5. Sap milky; calyx present . . . . . 64. *Moraceae*   p. 359
5. Sap not milky; calyx absent.
  6. Leaves narrow to elliptic, not lobed . 59. *Myricaceae*   p. 345
  6. Leaves deeply five-lobed, star shaped   92. *Hamamelidaceae* p. 440
4. Leaves mere scales 1–3 mm long on slender
  stems, jointed . . . . 55. *Casuarinaceae*   p. 338
1. Flowers not in catkins, in many (not all) species
  bisexual or individually large; perianth none
  or of a single series of parts or not differentiated clearly into calyx and corolla.
  7. Leaves pinnately compound or pinnately
    parted.
    8. Leaves pinnately compound; ovary of two
      to five carpels; perianth white, yellow,
      or greenish . . . . 102. *Simaroubaceae*   p. 517
    8. Leaves pinnately parted; ovary of one car-
      pel; perianth orange or red . . . . 66. *Proteaceae*   p. 365
  7. Leaves simple or absent at time of flowering.
    9. Stamens more numerous than the lobes
      or divisions of perianth, or perianth
      nearly lacking.
      10. Climbing vines . . . . 70. *Polygonaceae*   p. 372
      10. Erect shrubs.
        11. Gynoecium of numerous free parts;
          stamens many; perianth parts
          nine to twelve, large . . . . 81. *Magnoliaceae*   p. 411
        11. Gynoecium of one part; stamens
          twelve or fewer; perianth lobes
          four to six.
          12. Perianth minute, inconspicuous . 141. *Nyssaceae*   p. 648
          12. Perianth well developed.
            13. Perianth of six parts . . .   85. *Lauraceae*   p. 422
            13. Perianth of four to five parts . 139. *Combretaceae* p. 636
    9. Stamens as many as the lobes or divisions
      of the perianth.
      14. Style one, simple or branched above.
        15. Vines or low, partially woody
          plants with elongate, vinelike
          stems.
          16. Vines with tendrils . . . . 118. *Vitaceae*   p. 582
          16. Vinelike, elongate, woody stems
            without tendrils . . . . .   74. *Phytolaccaceae* p. 392
        15. Shrubs or trees.
          17. Style very short and stigma one,
            almost sessile . . . . . 112. *Aquifoliaceae* p. 564
          17. Styles two- to four-lobed, stigmas
            two to four . . . . . . 117. *Rhamnaceae*   p. 577
      14. Styles two or three.
        18. Styles two . . . .   63. *Ulmaceae*   p. 356
        18. Styles three . . . . 108. *Euphorbiaceae* p. 532

Section 16

Woody Plants with Alternate Leaves and Leaf Scars, in Most (not all) Species
the Flowers Bisexual or Individually Large and Conspicuous, Perianth
Differentiated into Calyx and Corolla

1. Ovaries three to many, distinct or nearly so.
   2. Stamens ten or fewer; leaves simple or once-
      pinnately compound.
      3. Leaves simple, fleshy, clustered toward the
         branch tips . . . .   103. *Surianaceae*   p. 518
      3. Leaves once-pinnately compound . . .   102. *Simaroubaceae*   p. 517
   2. Stamens more than ten; leaves simple or
      compound.
      4. Sepals five; petals five; leaves simple or
         compound . . . .   93. *Rosaceae*   p. 442
      4. Sepals three or six; petals six or more;
         leaves simple.
         5. Stems ringed at each node by a stipular
            scar; flowers white; trees . . . .   81. *Magnoliaceae*   p. 411
         5. Stems not ringed at the node; shrubs or
            small trees . . . .   84. *Annonaceae*   p. 418
1. Ovary one, simple or compound (styles and
   stigmas may be more than one).
   6. Corolla radially symmetrical.
      7. Petals fused.
         8. Style or long style branches four or five.
            9. Corolla lobes four; stamens four to
               sixteen; leaves entire . . . .   154. *Ebenaceae*   p. 684
            9. Corolla lobes five or six; stamens
               numerous . . . .   121. *Malvaceae*   p. 589
         8. Style one; stigma one.
            10. Stamens more numerous than the
                corolla lobes.
               11. Ovary with two to seven locules;
                   anthers opening by terminal
                   pores . . . .   148. *Ericaceae*   p. 667
               11. Ovary with one locule; anthers
                   opening with longitudinal valves   139. *Combretaceae*   p. 636
            10. Stamens as many as the corolla lobes.
               12. Stamens nearly or quite free from
                   corolla.
                  13. Style very short, stigma nearly or
                      quite sessile . . . . .   112. *Aquifoliaceae*   p. 564
                  13. Style well developed . . . .   148. *Ericaceae*   p. 667
               12. Stamens definitely attached to the
                   corolla tube.
                  14. Stamens opposite the corolla
                      lobes.

29. Style one (stigmas may be more than one).
  30. Ovary densely pubescent . . .   120. *Tiliaceae*   p. 587
  30. Ovary glabrous or puberulent.
    31. Stamens united in a tube; corolla purple . . . . .   127. *Canellaceae*   p. 613
    31. Stamens not fused in a tube; corolla of another color.
      32. Stamens fused in bundles opposite the petals . . .   129. *Flacourtiaceae*   p. 616
      32. Stamens free, not fused.
        33. Stigma one; leaves equilateral.
          34. Corolla yellow; low shrubs or trees.
            35. Stems armed; corolla densely pubescent within . . . .   68. *Olacaceae*   p. 369
            35. Stems unarmed; corolla glabrous within; low shrubs . .   125. *Cistaceae*   p. 610
          34. Corolla white; trees or shrubs.
            36. Style terminating the ovary . . . .   93. *Rosaceae*   p. 442
            36. Style emerging from near the base of the ovary . . . .   94. *Chrysobalanaceae*   p. 444
        33. Stigmas two; leaves inequilateral at base . . .   119. *Elaeocarpaceae*   p. 586
    29. Styles or style branches five . . .   121. *Malvaceae*   p. 589
25. Stamens twice as many as the petals or fewer.
  37. Leaves compound.
    38. Leaves evenly pinnate or evenly twice-pinnate.
      39. Leaves twice-pinnate . . . .   95. *Fabaceae*   p. 445
      39. Leaves once-pinnate.
        40. Petals green or white . . . .   116. *Sapindaceae*   p. 572
        40. Petals blue . . . .   100. *Zygophyllaceae*   p. 507
    38. Leaves odd-pinnate, odd twice-pinnate, trifoliolate, or palmate.
      41. Flowers in terminal racemes, panicles, or cymes.
        42. Flowers in cymes or panicles, white or greenish.
          43. Flowers in open cymes; leaves punctate with translucent dots . . . . . .   101. *Rutaceae*   p. 508

43. Flowers in dense panicles;
leaves not punctate . . . 111. *Anacardiaceae* p. 560
42. Flowers in racemes, green or
greenish brown . . . . . 104. *Burseraceae* p. 520
41. Flowers in lateral or axillary clusters.
  44. Stamens ten to twelve; trees . . 105. *Meliaceae* p. 521
  44. Stamens four or five; shrubs or
vines.
    45. Stamens alternate with the
petals; if climbing, the terminal leaflet on a distinctly
longer petiolule than the
lateral ones . . . . . 111. *Anacardiaceae* p. 560
    45. Stamens opposite early decidu-
ous petals; rarely shrublike,
usually climbing by tendrils;
if shrubby, then leaves pin-
nate . . . . 118. *Vitaceae* p. 582
37. Leaves simple.
  46. Stamens more numerous than the
petals.
    47. Flowers unisexual . . . . . 108. *Euphorbiaceae* p. 532
    47. Flowers bisexual . . . . . . 148. *Ericaceae* p. 667
  46. Stamens as many as the petals.
    48. Styles more than one, separate to
the base.
      49. Styles three; corolla yellow . . 130. *Turneraceae* p. 616
      49. Styles five; corolla reddish, pur-
ple, or violet . . . . . . 122. *Byttneriaceae* p. 602
    48. Style one, in some genera very short,
in others three-lobed or cleft, in
others unbranched.
      50. Flowers in slender, elongate
racemes . . . . 91. *Saxifragaceae* p. 440
      50. Flowers solitary and axillary, or
in axillary clusters, or in terminal panicles.
        51. Stamens opposite the petals;
style three-cleft or three-
lobed . . . . . . 117. *Rhamnaceae* p. 577
        51. Stamens alternate with the
petals.
          52. Style very short; stigma al-
most sessile . . . . . 112. *Aquifoliaceae* p. 564
          52. Style elongate; stigma lobed 122. *Byttneriaceae* p. 602
6. Corolla bilaterally symmetrical.
  53. Stamens eight to ten; leaves compound.

54. Stamens eight; flowers in large panicles;
    leaves pinnately compound . . . 116. *Sapindaceae* p. 572
54. Stamens ten . . . . 95. *Fabaceae* p. 445
53. Stamens five; leaves simple . . . . . 178. *Goodeniaceae* p. 820

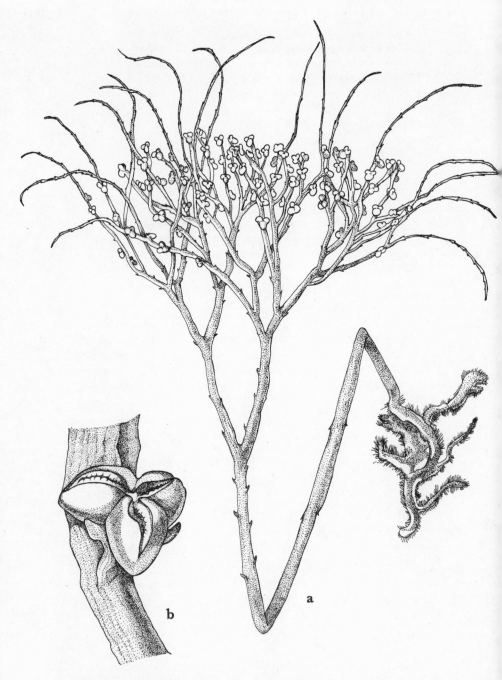

Family 1. **PSILOTACEAE.** *Psilotum nudum:* a, mature plant, × 1; b, sporangium, × 17½.

Descriptive Flora

1. PSILOTÀCEAE Psilotum Family

Plants mostly epiphytic. Stems forking. Rhizome rootless. Leaves small, subulate appendages in two to three rows on ridges of the stem and branches. Sporangia usually three-chambered, subtended by a cleft bractlet. Dehiscence by a radial slit from the apex of each chamber to the base; spores numerous. Only two genera and three species in tropical and subtropical regions of the Northern and Southern Hemispheres. Considered by many to be the most primitive existing group of vascular plants.

1. PSILÒTUM Sw. Whisk Fern

Characters of the family

1. P. nudum (L.) Beauv.—Plants 1–4 dm tall with three-angled stems, continuous with creeping rhizome, with delicate outgrowths simulating root hairs. Principal stem 2–3 mm in diameter, fleshy, ridged, decreasing toward the summit and repeatedly forking. True leaves lacking. Sporangia 2–3 mm wide, yellowish to pale buff when mature. Spores in tetrads, individually 60 μ long, bilateral, translucent, demarked by a line along the edge. $2n =$ c.420. Hammocks, commonly in moss mats on trunks of Sabal Palmetto, south Fla., Mexico, W.I., South America. Spores mature all year. *P. nudum* L., *P. triquetrum* Sw.

2. LYCOPODIÀCEAE Club Moss Family

Low homosporous plants with dichotomous branching of stems

and roots. Stems prostrate, creeping above or below the surface, forking into upright aerial stems. Leaves narrowly linear-lanceolate, in two to several series, numerous, spirally arranged, continuous with the stem. Sporangia adaxial at the base of sporophylls, these sometimes resembling the foliage leaves and disposed toward stem apex, or sporophylls specialized and imbricated into distinct strobili. Sporangia pouch shaped, dehiscing vertically across the summit. Spores numerous, yellowish. Two genera and about 450 species tropical, subtropical to arctic regions.

1. LYCOPÒDIUM L. Club Moss

Perennials, leaves one-veined, two- to sixteen-ranked. Sporangia coriaceous, 1–2.5 mm in diameter; spores sulphur color, inflammable. Nearly cosmopolitan, species tropics to arctic regions.

1. Terrestrials; sporophylls in strobili at stem
 tip.
  2. Strobili nodding, never stalked; foliage
   leaves uniform . . . . 1. *L. cernuum*
  2. Strobili erect, stalked; foliage leaves
   dimorphic. .
   3. Sporophylls deltoid, sporangia reniform 2. *L. carolinianum*
   3. Sporophylls lanceolate, sporangia glo-
    bose . . . . 3. *L. adpressum*
1. Epiphyte; sporophylls in zones at stem tips . 4. *L. dichotomum*

1. L. cernuum L.—Rhizome subterranean; aerial stems 8 dm long, short and thick when young, becoming reclined, trailing, wide-spreading with numerous branchlets. Leaves numerous, incurving or reflexed at tips; strobili 6–10 mm long, cylindric, slightly curved; sporophylls delicate, flabellate, 0.9 mm long, narrowed at base, the midvein excurrent, flanked on each side by vertical rows of cells ending in a fimbriolate margin. Sporangia 0.4 mm wide, spores very small. Canal banks, moist sites, south Fla., Miss., La., N.J., Mexico, South America.

2. L. carolinianum L.—Rhizome branching, prostrate, firmly rooted, dorsiventral, leafless on the lower side; in two rows in the upper side, blades sessile, widest at base tapering to apex, often terminating in curved tip. Erect branches 2–12 cm long, slender, pedunclelike with four rows of narrowly lanceolate pale green or stramineous leaves. Strobilus thicker than the peduncle, 1–5 cm long; sporophylls 2.2 mm long, deltoid-acuminate, firm, smooth, margin entire; sporangia reniform, 1 mm wide at base. Spores small, numerous. Canal banks, swampy ditches, south Fla., W.I., rare.

3. L. adpressum (Chapm.) Lloyd & Underw.—Rhizome elongate,

Family 2. **LYCOPODIACEAE.** *Lycopodium cernuum:* a, mature plant, × ¾;
b, strobilus, × 7½; c, sporophyll, abaxial surface, × 50.

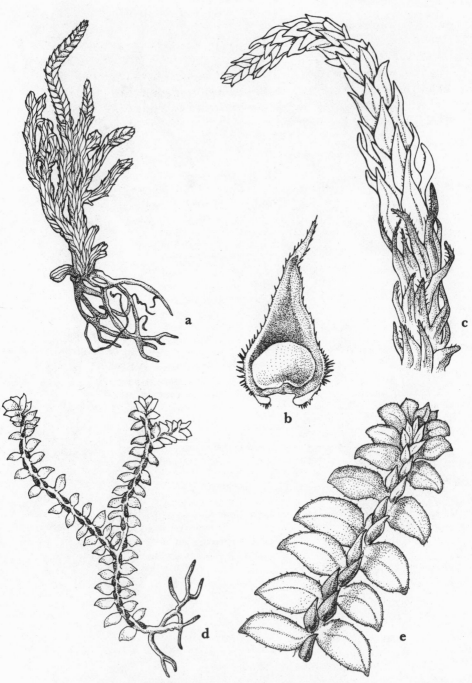

Family 3. **SELAGINELLACEAE.** *Selaginella arenicola:* a, mature plant, ×
1½; b. microsporophyll, adaxial surface, and microsporangium, × 50; c, strobilus, × 15; d, *Selaginella apoda*, mature plant, × 1½; e, leafy branch, × 7.

branching, leafy all-around, with sparingly denticulate blades; aerial stems to 3.5 dm tall; leaves numerous, narrowly lanceolate with incurving tips, entire; strobili 3–5 cm long; sporophyll linear-lanceolate, only slightly dilated at base. Sporangia globose about 1 mm wide, at bases of strongly appressed sporophylls; spores pale yellow. Spores all year. Sandy peat, about borders of hammocks, glades, Fla. to New England.

4. **L. dichotomum** Jacq.—Perennial, pendent epiphyte attached to trees by short, hard rhizome, roots villous at base. Principal stem 8–13 cm long, 2.5 mm thick above the crown, forking repeatedly into a stiffly spreading, fan-shaped plant. Leaves usually 1–2 cm long, linear, acicular, six- to eight-ranked, smooth, deciduous above the base. Sporophylls in zones in series on younger branches; sporangia reniform 1.6 mm wide; spores large, numerous. Big Cypress Swamp, southwestern Fla., W.I., rare. Florida plant reported but not seen.

3. SELAGINELLÀCEAE Spike Moss Family

Annual or perennial herbs with adventitious roots; stems dorsiventral or symmetrical branching. Leaves all alike or of two types, four- to twelve-ranked. Sporangia one-chambered, large and small, borne adaxially at the base of sporophylls aggregated in strobili. One genus, 700 species, tropical, subtropical, and temperate regions.

1. SELAGINÈLLA Beauv. Spike Moss

Leaves soft, green, the large lateral ones spreading from the stem in two rows; the small leaves dorsal in two rows, imbricate. Roots disposed at the ends of slender leafless shoots termed rhizophores or at bases of tufted stems. Strobili usually four-sided, of imbricated megasporophylls at the base and microsporophylls at apex. *Diplostachyum* Beauv.

1. Leaves of two types, disposed on lateral and dorsal sides of the stem, four-ranked.
  2. Plants mat-forming; branches simple; leaves pellucid.
    3. Sporophylls acute; internodes longer than the leaf width . . . . . 1. *S. apoda*
    3. Sporophylls caudate; internodes shorter than the leaf width . . . . . 2. *S. armata*
  2. Plant scandent; branches plumose; foliage lustrous . . . . 3. *S. willdenovii*
1. Leaves of one type, uniformly disposed around the stem, six- to fourteen-ranked 4. *S. arenicola*

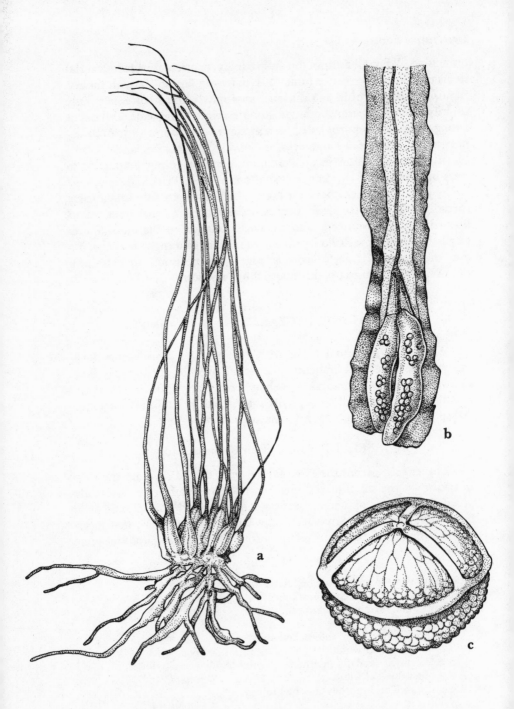

Family 4. **ISOETACEAE.** *Isoetes flaccida:* a, mature plant, × 1; b, megasporophyll, adaxial surface showing megaspores, × 7; c, megaspore, × 40.

1. **S. apoda** (L.) Fern.—Plants bright green; stems repent, prostrate, branching, forming mats to 1 dm wide or more. Lateral leaves at right angles from the stem, to 2 mm long, 1 mm wide, oblong-acute, serrulate, margined; dorsal leaves smaller, imbricate. Strobili with ascending sporophylls, 5–8 mm long. Megasporangia orange, 0.5 mm wide; microsporangia 0.3 mm wide. Hammocks, moist rocks, Fla., Tex., Midwest, Canada, South America. *Lycopodium apodum* L., *Diplostachyum apodum* (L.) Beauv.

2. **S. armata** Baker—Plants grayish green; stems 1–4 cm long, freely branching, matted. Lateral leaves 1–1.5 mm long, inequilateral, serrulate on margins; dorsal leaves smaller, acuminate. Strobili 2–5 mm long; megasporangia 0.4 mm wide, orange megaspores 0.22 mm wide, cleary visible with the naked eye. In moss mats, walls of lime sinks, and solution holes, Everglade Keys, south Fla., W.I. *Displostachyum eatonii* (Hieron.) Small

3. **S. willdenovii** (Desvaux) Baker—Stems shrubby, clambering up to crowns of high hammock trees; branches plumose with numerous leafy branchlets; leaves with contiguous margins, inequilateral, the smaller overlapping alternately, chain fashion. Strobili not seen. $2n = 36$. Introduced to south Fla. from Old World. *S. laevigata* Spring, *S. caesia arborea* Horton

4. **S. arenicola** Underw.—Forking from elongate, densely branched, spreading bases. Rhizophores frequent. Leaves narrowly linear, 1.6 mm long, ciliate with bristle-tipped apex; dorsal groove puberulent or glabrous; mature strobili thicker than the subtending branches, to 5 cm long or less; sporophylls 1.9 mm long, subcordate-lanceolate to rhombic-ovate, finely ciliate. Megaspore white, 0.3 mm long, smooth at least on upper half. Microspores small, orange. In thick mats among lichens, open pinelands, and scrub, Fla., Ga., Ala., Tex. *S. acanthonoda* Underw., *S, arenicola* Underw. var. *acanthonoda* (Underw.) Tryon

4. ISÒËTACEAE Quillwort Family

Aquatic or amphibious perennial herbs. Stem a depressed fleshy corm, two to three lobes, bearing dichotomous roots between the lobes. Apex crowded with quill-like leaves, narrowed to subterete tips, the bases enlarged with embedded sporangia. Megasporophylls surrounding the microsporophylls. Sporangia adaxial protected by velum, an outgrowth from the leaf surface. Megaspores triradiate, tetrahedral, chalky; microspores grayish, muricate. Represented by 2 genera, 75 species, cosmopolitan, north temperate regions.

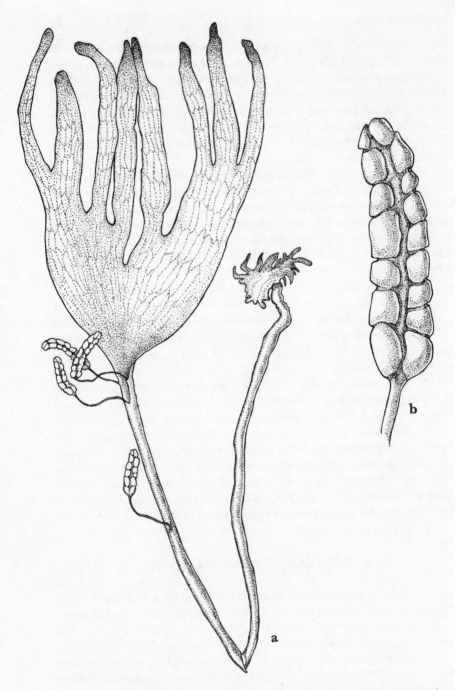

Family 5. **OPHIOGLOSSACEAE.** *Ophioglossum palmatum:* a, mature plant, × ¾; b, fertile stalk, × 7.

1. ISÒËTES L. Quillwort

Characters of the family

1. I. flaccida Shuttlew. ex A. Braun—Leaves 1–4 dm long, linear, dark green above, basal margins broadly hyaline. Megaspores 300 μ long, low tubercled; microspores 26 μ long. Sluggish waters, central and south Fla.

5. OPHIOGLOSSÀCEAE Adder's-Tongue Family

Perennial herbs from tuberous rootstocks. Leaves fleshy, stipe green, blade divided into a sterile abaxial and a fertile adaxial segment. Foliage segments usually one to two, entire; venation areolate. Fertile segment stalked, spikelike with biseriate sporangia, in marginal tissue. About 3 genera and 80 species, tropical and subtropical temperate regions.

1. OPHIOGLÓSSUM L.

Terrestrial or epiphytic. Foliage blade ovate or palmately lobed; fertile spikes, one or several rising ventrally from the base of the sterile leaf and apical portion of the stipe. Spores shed by a transverse slit, without annulus. About 45 species in temperate and tropical regions.

1. O. palmatum L. Hand Fern—Soft pendent epiphyte from tuberous rhizomes with exposed buds. Roots cordlike, fleshy. Foliage blade 20 cm long, palmately divided into three or six lobes; five to twenty fertile stalks. Stipe bases paleate. On Sabal palmetto, in hammocks, south Fla., tropical America, and Old World; rare. *Cheiroglossa palmata* (L.) Presl

6. HYMENOPHYLLÀCEAE Filmy Fern Family

Delicate, small, terrestrial, epiphytic, or epipetric ferns with velvety rhizomes. Leaves one cell thick between veinlets. Indusia marginal, funnelform, goblet shaped, or urceolate, bilabiate. Receptacle an exserted bristle bearing sessile sporangia. Sporangia with complete transverse annulus, opening vertically. Three genera and about 650 species, tropical and subtropical, chiefly Southern Hemisphere.

1. TRICHÓMANES L. Filmy Fern

Characters of the family, as to taxa of the range of this manual

Family 6. **HYMENOPHYLLACEAE.** *Trichomanes holopterum:* a, mature plant, × 2; b, fertile frond, × 2½; c, indusium with exserted bristle and sessile sporangia, × 30.

1. Venation pinnate; ferns epiphytic.
  2. Pinnules less than 2 mm wide, with spuri-
    ous veinlets and hairs . . . . .   1. *T. krausii*
  2. Pinnules more than 2 mm wide, without
    spurious veinlets and hairs . . . .   2. *T. holopterum*
1. Venation flabellate; ferns epipetric.
  3. Veins slender throughout to the margin .   3. *T. punctatum*
  3. Veins thickened toward the margin . .   4. *T. lineolatum*

1. **T. krausii** Hook. & Grev.—Leaves 2–4 cm long; pinnae 1–1.5 mm wide, including false, unconnected veinlets; margin with rounded sinuses, each with a stellate hair. Indusia tubular, two-lipped, the lips usually brown on edges. Bases of trees, south Fla., tropical America. *Didymoglossum krausii* Presl

2. **T. holopterum** Kunze—Rhizome erect; leaves tufted 3–6 (10) cm long, the pinnae 3–6 mm wide. Blade pinnatifid, pellucid, crisped, narrowed to a winged petiole; lobes obovate with three to four forking to the margin; indusium in the fork of the veins, commonly three per lobe, 1.5 mm long, vase shaped, flaring at the margins, not brown edged; receptacle protruding 2–3 mm beyond the indusial rim. Deep shady swamps, in colonies on decaying cypress stumps, south Fla., W.I.

3. **T. punctatum** Poir. ex Lam.—Rootstock slender, creeping; internodes manifest. Leaves to 2 cm long, narrowed to petiole; blades suborbicular, incised; the margins with scattered stellate hairs; veins not thickened toward the margin; indusium immersed, tubular, 1.5–2 mm long; lips narrow, the outer one or two rows of cells dark edged. Edges of lime sinks, on rocks, tree bases and roots, south Fla., tropical America. *Didymoglossum punctatum* Desvaux

4. **T. lineolatum** (Bosch) Hook.—Rootstock matted, slender, creeping, copiously bristly, with distinct nodes. Leaves to 2 cm long, narrowed to petioles; blades ovate, crenately incised; marginal hairs stellate, the thickened forking veins ending in the margin; the lip of the indusium broadly dark edged. In hammocks, edges of lime sinks, south Fla., tropical America. *Didymoglossum lineolatum* Bosch

7. SCHIZAEÀCEAE Ray Fern Family

Erect or scandent. Rootstocks bristly hairy. Leaves simple or pinnately compound. Sporangia biseriate, originating in margins of fertile segments. Fertile segments apical or lateral. Annulus a complete ring, splitting vertically across the thin-walled cells at summit. About 4 genera, 150 species, chiefly tropical.

1. Leaves simple; plant erect; spore-bearing
   segment terminal . . . . 1. *Schizaea*
1. Leaves pinnately compound.
   2. Climbing vines by means of elongate leaf
      rachis; pinnules distichous; ultimate
      segments bearing sporangia . . . . 2. *Lygodium*
   2. Stems upright, not climbing; the lowest
      pair of pinnae bearing sporangia . . 3. *Anemia*

1. SCHIZAÈA Sm. RAY FERN

Rhizome tuberous, bristly hairy. Leaves simple or repeatedly forking. Blade and stipe scarcely differentiated. Sporangia biseriate, or distal appendages of the leaf. Thirty species, tropical and temperate North America and the Southern Hemisphere.

1. **S. germanii** (Fee) Prantl—Leaves few, tufted to 12 cm tall, grass-like. Fertile leaves narrowly pinnate with segments crowded at summit; sporangia borne under reflexed margin, naked. South Fla., W.I. *Actinostachys germanii* Fee

2. LYGÒDIUM Sw. CLIMBING FERN

Rootstock wiry, creeping, branches dichotomous. Leaves monopodial with indefinite growth, adapted for twining by means of rachises. Pinnae fertile or sterile, variously compounded, stalked, spreading from the main rachis. Thirty-nine tropical and subtropical species.

1. **L. japonicum** (Thunb.) Sw.—Leaves to 5 m long, pinnae dimorphic. Foliage pinnae 10–18 cm long; 7–10 cm wide; with five to seven pinnatifid pinnules, incised, crenate margins with free, simple or forking veinlets. Fertile pinnules flabellate, with eight or more narrow lobes extending from the margin. Sporangia opening antrorsely on free veinlets. Spores tetrahedral. Established and naturalized in hammocks, creek banks, roadsides; of Asiatic origin.

3. ANÈMIA Sw. PINE FERN

Rootstock ascending, bristly hairy. Leaves dimorphic, stiff, erect, contiguous, pinnate or pinnatifid blades with long wiry stipes. The lowest pair of pinnae fertile. Ultimate fertile lobes narrow, crowded with biseriate sporangia on slender stalks. Spores tetrahedral. About 90 species, tropical and subtropical regions of southern Africa, India.

1. **A. adiantifolia** (L.) Sw.—Leaves 4–8 dm long; blades deltoid in outline, 12–28 cm long; stipes usually longer than the blades, retrorsely hirtellous at base; rachises fluted pubescent with incurved hairs; ultimate segments sparingly pubescent, glabrate. Ultimate foliage segments

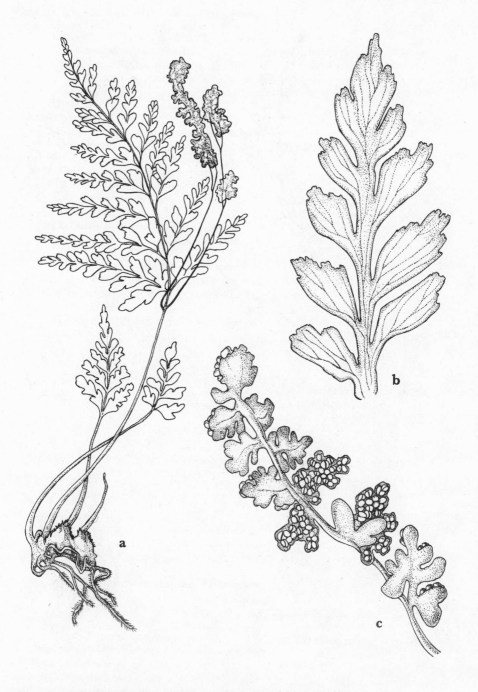

Family 7. **SCHIZAEACEAE.** *Anemia adiantifolia:* a, mature plant, × ½; b, foliage segments, × 2½; c, fertile pinnae, × 3.

Family 8. OSMUNDACEAE. *Osmunda regalis* var. *spectabilis:* a, leaf blade with terminal panicle of fertile pinnae, × ½; b, immature leaf croziers and rootstock, × ¼; c, paniculate clusters of mature sporangia, × 5.

cuneate, entire, or two- to three-lobed; the veins seemingly depressed into lamina, forking free to the margin. Stalks of fertile pinnae to 15 cm long. Sporangia naked, barely protected by the lamina margin. $2n = 76$. Crevices of oolitic rocks, south Fla., disjunct to Citrus County in central Fla.

8. OSMUNDÀCEAE ROYAL FERN FAMILY

Terrestrial ferns with erect massive rootstocks, barely subterranean. Leaves circinate in vernation, radially disposed around the crown. Blades once or twice pinnate-pinnatifid; stipes stout, winged at the base, stipulate, without scales. Leaves dimorphic. Sporangia-bearing leaves distinct from foliage leaves, or the apical and the middle pinnae of blades becoming fertile. Sporangia large in paniculate clusters. Annulus represented by a patch of thickened cells. Three genera, 21 species, tropical and subtropical regions.

1. OSMÚNDA L.

Spore-bearing pinnae contracted. Sporangia individually short stalked, obovoid, vertically dehiscent across the summit. Nineteen species, tropical, subtropical, and temperate regions.

1. **O. regalis** L. var. **spectabilis** (Willd.) Gray, ROYAL FERN—Leaves to 8 dm long or less in large tussocks, covered with old leaf bases. Stipes and rachis essentially glabrescent, reddish brown, polished; rachis dorsally sulcate; blades twice-pinnate, spreading, pinnules to 5 cm long, distant, oblong-lanceolate, oblique at base, serrulate. Spore-bearing pinnae in terminal panicles of foliage leaves; sporangia 1.5 mm long, spores brown. $2n = 44$. Wet shores, cypress swamps, Fla., North America, W.I., Mexico, Asia; very common. *O. spectabilis* Willd.

9. VITTARIÀCEAE SHOESTRING FERN FAMILY

Tufted epiphytes with congested rhizomes covered with clathrate paleae. Roots matted with villous branching hairs. Leaves contiguous, crowded, blades linear, sessile; sori in submarginal abaxial grooves, protected by paraphyses. About 8 genera, 120 species, tropical and subtropical regions.

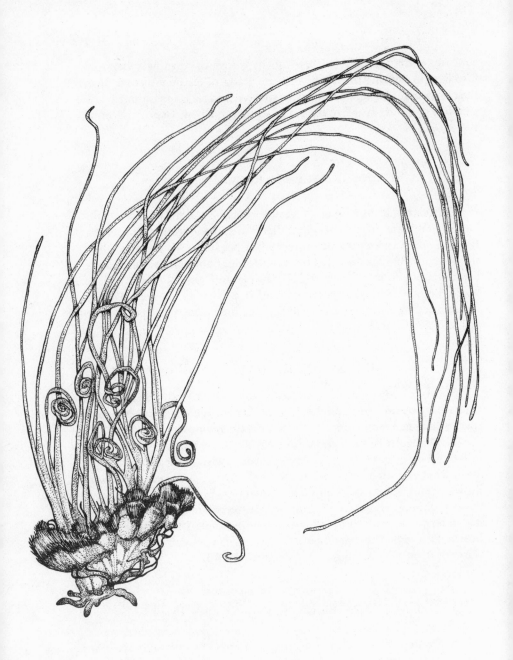

Family 9. **VITTARIACEAE.** *Vittaria lineata:* growth habit.

1. VITTÀRIA Sm.

Leaves to 1 m long, 2–3 mm wide, flexible. Blade margin entire, reflexed. Veins obscurely reticulate with a row of areolae between the midvein and the intramarginal fertile vein. About 80 tropical and subtropical species.

1. V. lineata (L.) Sm.—Roots often lodged in axils of SABAL leaf bases; rhizome abbreviated, context friable aerenchyma; paleae iridescent. Blades dark green, lustrous. Sori in a single series, the length of the blade on the submarginal vein on each side; annulus interrupted, sporangia dehiscent transversely. Ripe spores oval-elliptic, hyaline or colorless. Hammocks, south and central Fla., disjunct in Ga.

10. POLYPODIÀCEAE POLYPODY FERN FAMILY

Epiphytic, epipetric, or terrestrial ferns with creeping, branching rhizomes. Leaves simple or pinnatifid. Veins free or anastomosing. Sori exindusiate, orbicular, terminal on free veins, sometimes marginal or acrostichoid on the back of the blades. About 50 genera, nearly cosmopolitan distribution.

1. Veins mostly free, forking; sori terminal on
    free veinlets . . . .   1. *Polypodium*
1. Veins anastomosing; sori on veins in areolae
    or marginal.
    2. Leaves entire, areolate.
        3. Blades dimorphic; climbing vinelike
            epiphyte . . . .   2. *Microgramma*
        3. Blades uniform.
            4. Sori enclosed in areolae between pin-
                nate lateral veins of the blade . .   3. *Campyloneurum*
            4. Sori in marginal band around the
                apex of the blade . . . . .   4. *Paltonium*
    2. Leaves pinnatifid or nearly so.
        5. Rachis wing 5 mm or more wide . .   5. *Phlebodium*
        5. Rachis wing merely a ridge or absent .   6. *Goniophlebium*

1. POLYPÒDIUM L. POLYPODY FERN

Leaves in two rows on dorsal side, articulated to phyllopodial stubs of the scaly rhizome. Roots black, hairy, forking. Sori round, exindusiate in one or more rows on each side of the segment midvein or scattered in areoles over the lamina of simple blades. Veins free, two- to four-forked, ending below margin. Spores bilateral. About 75 species, temperate, tropical, and subtropical regions.

1. Stipe and rachis copiously peltate-scaly . . 1. *P. polypodioides*
1. Stipe and rachis without peltate scales.
  2. Pinnae bases long cuneate, with gradually
    decrescent segments; stipe, rachis, and
    veins dark maroon . . . . . . 2. *P. ptilodon*
                         var. *caespitosum*
  2. Pinnae bases not long cuneate; stipe,
    rachis, and veins blackish.
    3. Plants epiphytic, pendulous; basal seg-
      ments reduced to mere auricles, not
      deflexed; veins one-forked . . . 3. *P. plumula*
    3. Plants usually epipetric, erect; basal
      segments barely reduced, deflexed;
      veins two-forked . . . . . . 4. *P. dispersum*

    1. P. polypodioides (L.) Watt, RESURRECTION FERN—Evergreen creeping epiphytic fern with superficial scaly rhizome. Leaves 5–20 cm high, coriaceous; segments oblong-obtuse or rounded, with brown-centered scales on dorsal side. Sori on free tips of veins close to margins. Plants forming extensive colonies in open shade of trees, central and south Fla., tropical America. *Marginaria polypodioides* Tidestrom
    2. P. ptilodon Kunze var. **caespitosum** (Jenman) A. M. Evans—Fern terrestrial or on decaying logs, from thick abbreviated rootstocks; leaves to 9 dm long; 5–14 cm wide at midblade, stiff, loosely spreading; rachis inconspicuously paleate; segments commonly tapering to obtuse or rounded apex, ciliate becoming glabrous. Sori naked, arising from a patch of delicate colorless radially oriented hairs. Sporangia suborbicular. Spores 64, bilateral, ca. 56 μ long. $2n = 148$. Elegant ferns in hammocks, south and central Fla., W.I., Mexico, Honduras. *P. pectinatum* L.
    3. P. plumula Humb. & Bonpl. ex Willd.—Rhizome abbreviated, paleate with wide-spreading roots; leaves to 7 dm long, 4–6 cm wide at midblade; stipes, bases, rachis paleate with a mixture of multicellular hairs in young leaves, sparingly so in age. Blades oblong or linear-lanceolate in outline, segments chaffy on the veins, ciliate, linear or oblong, puberulent with acicular hairs on the surface. Sori round; sporangia suborbicular; spores 64, bilateral,. ca. 50 μ long. $2n = 148$. South and central Fla., tropical America.
    4. P. dispersum A. M. Evans—Epipetric or epiphytic fern with slender rhizome; roots wide-spreading, slender, villous. Leaves 2–6 dm long, slender; stipe base and rachis with conspicuous brown triangular

Family 10. **POLYPODIACEAE.** *Polypodium dispersum:* a, mature plant; b, fertile pinna with sori, × 3½; c, detail of sori, × 20.

scales; blade narrow-ovate, 10–35 cm long, 4.5–8 cm wide; segments 2–4 cm long, 2–3 mm wide, straight obtuse, entire, appressed pubescent on both surfaces or glabrate; sori round, sporangia with two (one to four) paraphyses; spores ca. 43 μ long, globose to ovoid with a variable scar, 32 per sporangium. $2n = 222$. South and central Fla., tropical America.

2. MICROGRÁMMA Presl

Slender epiphytic or epipetric fern creeping by rhizomes. Internodes covered with attenuate scales. Leaves articulate upon phyllopodial stubble. Sori 1.5 mm wide, in rows of areolae on each side of the midrib. About 20 species, tropical America, Africa. *Phymatodes* Presl; *Polypodium* L.

1. **M. heterophylla** (L.) Wherry—Rhizome minutely ridged, held close to bark by numerous delicate rootlets. Leaves simple, 3–12 cm long, with stipe 5–15 mm long; blade elliptic, narrowly oblanceolate, 4–10 mm wide, margins undulate or nearly entire; veins anastomosing, with areolate pattern, enclosing free terminally soriferous veinlets; annulus usually twelve to sixteen cells; spores reniform, reticulate, mucronate. Moist hammocks, south Fla., tropical America. *Phymatodes heterophyllum* (L.) Small, *Polypodium heterophyllum* L.

3. CAMPYLONEÙRUM Presl

Terrestrial or epiphytic ferns from stout or slender rootstocks. Blades entire, lanceolate, or linear, with prominently spaced lateral veins. Sori areolate, terminating free veinlets. About 25 species in tropical America.

1. Blades 1–2 cm wide, linear-attenuate; sori in
  a single row each side of the midrib . .   1. *C. angustifolium*
1. Blades 3–12 cm wide, usually long acuminate.
  2. Leaf tapering to winged stipe; leathery,
    veinlets manifest.
    3. Blade 4–12 cm broad, widest above the
      middle . . . .   2. *C. latum*
    3. Blade 3–4 cm broad; leaves essentially
      elliptic, coriaceous, semiopaque . .   3. *C. phyllitidis*
  2. Leaf rounded to a wingless stipe; abruptly
    acuminate at apex; blades fleshy; obscuring the veinlets . . . .   4. *C. costatum*

1. **C. angustifolium** (Sw.) Fee—Rootstock slender, paleate; roots wiry capillary branching in large mats; leaves glabrous, tufted, 3–4 dm

long, linear-elliptic, tapering to short compressed stipes, paleate at base; blades yellowish green, long attenuate at apex, margins inrolled. Sori numerous, 1 mm wide; sporangia with brown annulus of thickened cells. Hammocks of Everglade Keys, south Fla., W.I., Mexico, South America, rare.

2. C. latum Moore—Rootstock short, thickish with matted wiry roots. Leaves few, thick, to 1 m long, glabrous, linear-elliptic, lustrous; stipes rigid, sparingly paleate at base. Sori 2 mm wide, numerous, usually one per areole in the copiously anastomosing fertile surface. South Fla., W.I., Mexico, South America, rare.

3. C. phyllitidis (L.) Presl—Rhizome short, creeping, scales deciduous, roots villous, brown. Leaves to 9 dm long, broadly linear-elliptic, attenuate at apex, tapering to winged stipe, glabrous shiny, green; blade relatively thin, white midvein; margin repand; sori numerous, one to two in each areole, 1.5–2 mm wide, in lines between the lateral veins; annulus brown, of thickened cells; spores yellow. Hammocks, south Fla., W.I., Mexico, South America.

4. C. costatum (Kunze) Presl—Epiphytic from thick rootstock; leaves 3–4 dm long, few together, tapering to a rounded petiole; blade fleshy, repand, elliptic, abruptly narrowed at apex; veins inconspicuous; sori appearing scattered. Big Cypress Swamp, south Fla., W.I., rare.

4. PALTÒNIUM Presl RIBBON FERN

Epiphytic fern with short or elongate repent rootstocks with often densely matted roots. Rhizomes short-scaly; blades elliptic, entire; soral strips submarginal. One species, tropical America.

1. P. lanceolatum (L.) Presl—Leaves erect or usually pendent; blades 1–4 dm long, gradually narrowed to a short, winged stipe and an attenuate obtuse apex; midvein prominent, lateral veins obscure areolate, with inclusions of free veinlets, open to the margin. Sori contiguous, forming a distinct band on each side of the apical part of the blade. Known only in hammocks in the upper Keys, Fla.; W.I., Central America, rare. *Pteris lanceolata* L.

5. PHLEBÒDIUM (R. Brown) Sm. GOLDEN POLYPODY

Epiphytic, pendent ferns. Rhizome creeping; stipes articulate to phyllopodia on rhizome. Blades pinnatisect, minutely punctate, segments with noncontractile repand margin. Sori 1.5 mm wide, numerous, prominent, golden brown. Ten species in tropical America.

1. P. aureum (L.) Sm.—Rhizome to 15 mm in diameter, densely covered with attenuate golden-colored scales. Mature blades broadly ovate, 3–6 dm long, 1.5–4 dm wide, pendent or spreading, on shiny petioles; lobes 1.5–4 cm wide, papery, entire, undulate or shallowly toothed; veins anastomosing, areolae large, open submarginally and along the prominent midvein; medial veins, including one or usually two free veinlets. Sori compital, golden brown, borne at the united tips of free veinlets. Sporangia brown; annulus of twelve cells; spores oblong, colorless, tuberculate. $2n = 74$. On SABAL PALMETTO, south and central Fla., W.I., Mexico, South America. *Polypodium aureum* L.

6. GONIOPHLÈBIUM Presl

Epiphytic fern with elongate creeping rhizome, profusely covered with brown attenuate scales, leaves subpinnate; lateral segments nearly sessile; sori in areoles, two or more rows, parallel to midrib on each side. About 20 species, tropical Asia to Fiji Islands.

1. **G. triseriale** (Sw.) Wherry—Leaves 5–6 dm long, to 3 dm wide, blade barely pinnatisect; margin repand; lateral veins from midrib to margin obsolescent; sori 1.2 mm wide; spores smooth. South Fla., tropical America, rare. *Polypodium triseriale* Sw.; *G. brasiliense* Farw.

11. DAVALLIÀCEAE

Epiphytic and terrestrial ferns; creeping rhizomes. Leaves pinnate, stipes articulate or continuous with the rhizomes. Indusium   orbicular or reniform, basally affixed to the vein, opening toward the pinna apex. Seven genera and about 75 species, Asiatic regions.

1. NEPHRÓLEPIS Schott   BOSTON FERN

Stoloniferous ferns, sometimes with tubers. Stipe continuous with the rhizome. Pinnae articulate with the rachis. Sori terminal on the upper branch of a free veinlet. About 30 species, tropics, Japan, New Zealand.

1. Base of pinna cordate on the upper side of
the midvein; pinna 2 cm long .   .   .   .   1. *N. cordifolia*
1. Base of pinna truncate, or nearly so, or short
auricled.
  2. Pinna 3–5 cm long, rachis internode short   2. *N. exaltata*
  2. Pinna 6–12 cm long, rachis internodes
  distant .   .   .   .   3. *N. biserrata*

1. **N. cordifolia** (L.) Presl—Rootstock slender, creeping, tuberiferous. Leaves erect stiff, to 7 dm long, 4–5 cm wide at midblade; stipes

paleate-shaggy at base and on the rachis; blade linear in outline of close-set lanceolate-cordate pinnae with overlapping auricles, veins one-forked. Sori numerous between midvein and margin; indusium crescentic, 0.8 mm wide. $2n = 82$. Cosmopolitan in the tropics and subtropics, introduced as an ornamental in south Fla.

2. **N. exaltata (L.)** Schott—Leaves once-pinnate, jointed to rachis; blades relatively long and narrow, 1.5 m or less long; pinnae spreading and auriculate, at least on upper side of midrib; veins free; stipes channeled, smooth or scurfy. Sori reniform on tips of free veins. Foliage highly variable; rhizome stoloniferous. $2n = 82$. Escaped from cultivation and widely naturalized in Fla.

3. **N. biserrata** Schott—Leaves to 1.5 m tall, 10–30 cm wide, from stout rootstock. Blades arching, pinnae spreading, crenate-serrate, slightly oblique-auriculate upper side of midrib at base; stipe scurfy and hairy, becoming nearly glabrous. Indusia orbicular, with a narrow sinus, in mature sori with protruding sporangia. $2n = 82$. Handsome terrestrial fern in hammocks, south Fla., pantropical.

12. PTERIDÀCEAE Bracken Fern Family

Terrestrial ferns with creeping, ascending or erect, rootstock. Leaves pinnate, or pinnatifid, simple or decompound, continuous with the rachis. Sporangia in marginal, usually indusiate, sori; or nonsoral, acrostichoid on the network of veins over the entire abaxial surface, sometimes intermixed with waxy puberulence, or paraphysate. Annulus of thickened cells, sporangia transversely dehiscent; spores tetrahedral (except No. 3). Family includes some of the largest ferns. About 500 species, cosmopolitan.

1. Sporangia acrostichoid.
   2. Veins reticulate throughout, without free
      veinlets; sporangia intermixed with
      paraphyses . . . .   1. *Acrostichum*
   2. Veins free or a single row of costal areoles
      on each side; paraphyses not present.
      3. Pinnae uniform, simple or trifoliolate;
         sporangia intermixed with waxy in-
         dument . . . .   2. *Trismeria*
      3. Pinnae dimorphic; sporangia not pro-
         tected . . . .   3. *Stenochlaena*
1. Sporangia soral, contiguous, or distinct.
   4. Sorus marginal.
      5. Sori borne between inner indusial
         membrane and outer revolute seg-

ment margin . . . . 4. *Pteridium*
5. Sori borne on inner surface of indusioid
reflexed segment margin.
6. Soral bands continuous; leaves pinnate or pedate . . . . . . 5. *Pteris*
6. Soral bands discontinuous, sometimes
inconsistently so.
7. Sori indusiate, on enlarged tips of
pinnule veins . . . . . . 6. *Cheilanthes*
7. Sori truly exindusiate, protected by
reflexed scarious margins.
8. Pinnule lobes sinuous, pinnatifid; sori at the tips of sinus
vein . . . . 7. *Hypolepis*
8. Pinnules dimidiate; sori on the
veins of the inner surface of
a scarious reflexed lobe margin 8. *Adiantum*
4. Sorus terminal, surmounting the tip of a
clavate pinnule; pinnae linear forking 9. *Sphenomeris*

1. ACRÓSTICHUM L. Leather Fern

Rootstock erect, ligneous. Leaves simply pinnate; stipes stout, compressed at base. Pinnae stiff or coriaceous, without free veins. Sporangia-
bearing pinnae disposed toward the apex, above the sterile pinnae on
the same rachis. Three species in swamps and mangroves, pantropical.

1. Sterile pinnae oblong, rounded at apex . . 1. *A. aureum*
1. Sterile pinnae lanceolate, acute at apex . . 2. *A. danaeaefolium*

1. **A. aureum** L.—Leaves tufted to 2 m long; rhizome stout with
numerous long roots with thick cortex of aerenchyma. Stipe sulcate,
paleate 5–10 mm thick at base, glabrous above; pinnae 1–2 dm long,
oblong-oblique distant on rachis, midvein conspicuous. Fertile pinnae
smaller, dark brown on abaxial surface; sporangia intermixed with
capitate paraphyses. $2n = 60$. Coastal hammocks, shores of brackish
waters, Fla., W.I., Mexico, South America.
2. **A. danaeaefolium** Langsd. &. Fisch.—Leaves tufted to 3 m
long. Rhizome stout; roots fleshy, thick, and long. Stipe sulcate, paleate
at base 5–18 mm thick, glabrous above; pinnae lanceolate, 1–3 dm long,
stiff, cuneate at base, lamina with overlapping basal margins on distal
rachis. Fertile pinnae smaller; mature sporangia golden brown. $2n = 60$.
Fresh-water marshes, brackish swamps, Fla., W.I., Mexico, South
America.

2. TRISMÈRIA Fee

Rhizome congested, erect paleate. Leaves to 9 dm long, 10 cm wide, from ligneous rootstock; stipes 3–5 mm wide at base, castaneous, scaly below, glabrous and polished above; pinnae simple, linear, 7–8 mm long, 5–8 mm wide, serrulate at apex, three-cleft or trifoliolate below. Fertile pinnae narrower; bearing scattered, short-stalked, numerous sporangia on the veins of abaxial surface. One species, tropical America, subtropical Fla.

1. **T. trifoliata** (L.) Diels—Characters of the genus. Hammocks, south Fla., W.I.

3. STENOCHLAÉNA Sm. HOLLY FERN

Rhizome with radial symmetry, high climbing; vascular bundles of two or more, glabrous in age except the apex covered with roundish brown scales. Leaves dimorphous; the sterile, once-pinnate, the fertile, bipinnate. Pinnae, except the apical one, jointed to the rachis. Stipes well spaced with arching blades. Annulus twelve to twenty cells; spores bilateral, translucent papillose, without perispore. Five species, Africa to Pacific. Represented in the range of this manual by one species.

1. **S. tenuifolia** (Desv.) Moore—Rhizome climbing to 2 m or more on trees; sterile blades to 7 dm long, pinnae lustrous, spinulose-serrate, linear-lanceolate; veins simple or forked, free to the marginal teeth, arising from the single row of areoles flanking the midvein on each side. Introduced from southern Africa. Cultivated in gardens, escaped to hammocks, south Fla. *Lomaria tenuifolia* Desv., *Acrostichum tenuifolium* Baker

4. PTERÍDIUM Gled. ex Scop. BRACKEN FERN

Rhizome deep-seated in the ground, elongate, creeping and branching, invested with hairs. Roots velvety with fine, straight trichomes. Leaves upright, alternate, large, vascular bundles numerous; blades expansive, usually tripinnate, with numerous segments, veins free, soral band nearly continuous, protected by a reflexed indusium, wider than the revolute segment margin. Spores brown, tetrahedral, spinulose. Annulus of thirteen thickened cells. Single cosmopolitan species in temperate and tropical regions.

1. **P. aquilinum** (L.) Kuhn var. **caudatum** (L.) Sadebeck—Leaves usually 1–2.5 m long. The lower pinnae unrolling before the upper ones in vernation; indumentum of multicellular white, bright brown, or bi-colored hairs. Stipes ligneous, pilose-felted at base; puberulent, glabres-

cent above, dorsally grooved, decurrent lines prominent. Primary pinnae to 7 dm long, glabrous, spreading at right angles from the rachis;
rachises glabrous. Mature ultimate segments 1.4–2 mm wide, distant;
veins pinnate, forking, slightly thickened at tips. Indusial margin serrulate, the sori extending from the sinus up to three-fourths of the segment length. Terminal segment 2–5 cm long. $2n = 104$. Palm hammocks,
pinelands, white sand scrub, often very abundant, south Fla., W.I.,
South America. *P. caudatum* Maxon

5. PTÈRIS L.   BRAKE FERN

Terrestrial or epipetric ferns with erect or creeping rhizomes.
Leaves tufted, pinnate, decompound or pedately divided. Veins free on
the outer lamina, anastomosing along midveins. Indusium provided by
a single revolute margin; sori usually continuous. Annulus of thick-
walled cells. A cosmopolitan genus with about 250 species.

1. Blades compound, three to five parts, deltoid or mostly pentagonal in outline;
veins areolate along each side of the midveins . . . .   1. *P. tripartita*
1. Blades simple, pinnate- or pedate-pinnati-
fid; veins free.
  2. Leaves one kind, uniform, pinnate.
    3. Pinnae ascending, sessile, tapering to
caudate apex, stipe strongly paleate .   2. *P. vittata*
    3. Pinnae spreading, short stalked, oblong-
rounded, stipe inconspicuously paleate . . . .   3. *P. longifolia*
var. *bahamensis*
  2. Leaves two kinds, fertile and sterile, pe-
date-pinnatifid . . . .   4. *P. cretica*
var. *albolineata*

1. **P. tripartita** (Sw.) Presl—Rootstock erect, short, giving rise to
leaves 1–2 m long. Stipes erect, stout, blade three-parted, the divisions
branched into numerous pinnatifid pinnae with caudate tips. Ultimate
segments slightly falcate separated by rounded sinuses. Mature sori,
scarcely covered by the marginal indusium extending from the sinus to
about two-thirds toward the apex. Veins netted above the midvein, forking to the margin. Spores dark brown, tetrahedral. $2n = 116$. Swamps,
south Fla., Everglade Keys, naturalized from Old World tropics. *Litho-
brochia marginata* Presl, *P. marginata* Bory
2. **P. vittata** (L.) Small—Rhizomes paleate, suberect, giving rise to
leaves 3–9 dm long, usually in large erect tufts. Stipes ligneous at base,
4–6 mm in diameter at base, coarsely, retrorsely paleate below; rachis

sulcate decrescent, sparingly scaly throughout. Blades oblanceolate in outline, the lower pinnae gradually diminished, the middle pinnae to 20 cm long, apex abruptly terminating in a prolonged pinna. Margin of indusium usually hyaline in mature soral strips or sometimes cinnamon brown when young. Spores light brown. $2n = 116$. Canal banks, hammocks, disturbed sites, south Fla., Ala., La., Old World tropics.

3. **P. longifolia** L. var. **bahamensis** (Ag.) Hieron.—Rootstock short, knotty, branching; leaves 4–7 dm long with wiry stipes, sparingly paleate at base. Blades oblong in outline; the lower pinnae gradually diminished below the middle; the middle pinnae 3–6 cm long, stalk 1–2 mm long, base subcordate, slightly auriculate. Veins forking usually near the base, oblique to the margin; indusium serrulate or lacerate, brownish. Spores light brown. Crevices of oolitic rocks in pinelands, south Fla., W.I., tropical America. *Pycnodoria bahamensis* (Ag.) Small

4. **P. cretica** L. var. **albolineata** Hook.—Rootstock ascending, creeping, new growth sparingly scaly about the stipe bases; roots villous, stubble-forming. Leaves close-set, numerous, 2–5 dm long; stipes 2–5 dm long, slender, spreading, attenuated to a short rachis of three to seven pedately disposed pinnae; terminal pinnae 9–14 cm long, 2–2.5 cm wide, sharply serrate; lateral pinnae forking from the common apex of the petiolule; all pinnae distinguished by the usual silvery gray band along the midvein throughout on dorsal surface. Lime-rock grottoes. $2n = 174$. Introduced, south Fla., disjunct in Citrus County, rare.

6. CHEILÁNTHES Sw. Lip Fern

Terrestrial ferns; rhizome short, creeping. Leaves minutely glandular-pubescent, erect to spreading, bipinnate to tripinnate, ultimate segments entire or variously pinnatifid; stipes rigid, black, densely ferruginous, paleate, often longer than the blade. Sori borne on the enlarged tips of veinlets protected by the reflexed margin of pinnule lobes. Small ferns in dry situations. About 180 species in tropical and temperate regions of America.

1. **C. microphylla** Sw.—Leaves distichous, to 7 dm long, lanceolate in outline. Rachis glandular tomentose throughout, becoming glabrate in lines. Pinnules glabrous on the upper surface, pubescent below; soral band interrupted by lobe sinuses, indusium around the tips of pinnae. Annulus of fourteen to twenty-four cells. Spores minutely roughened. $2n = 116$. Shell mounds, south Fla., W.I., rare.

7. HYPÓLEPIS Bernh.

Rootstocks elongate, branching, wide-spreading, minutely paleate, pubescent. Leaves large, one- to four-pinnate, ultimate segments pinnatifid, veins simple or forking. Sori borne at the ends of the veinlets, protected by the revolute lobe of margins, 45 tropical and subtropical species.

1. **H. repens** (L.) Presl—Leaves to 1 m long or more; stipes reddish or stramineous, pubescent or sometimes spinescent. Blades deltoid in outline to 15 dm long and nearly as wide. Pinnae and pinnules distant, spreading. Ultimate segments oblong, crenate undulate, 4–9 mm long. Sorus in a sinus. Spores oblong, spinulose or glabrous. Annulus thirteen to fifteen thickened cells. Central and south Fla., W.I., South America.

8. ADIÁNTUM L. MAIDENHAIR FERN

Terrestrial, with creeping or ascending paleate rhizome. Stipes scaly at base, polished, black or brownish above, or rarely dull, pubescent. Blades pendulous, broad, pinnately decompound, or simply pinnate, pinnules dimidiate or flabellate on decrescent rachis. Veins free, forking to margin. Pinnule margins with separated reflexed outgrowths bearing and protecting sporangia on the inner surface. Annulus cells thickened; spores tetrahedral. A cosmopolitan genus of about 200 species, mostly in the tropics and subtropics.

1. Pinnules dimidiate, continuous with stalk 1
   mm long . . . . 1. *A. melanoleucum*
1. Pinnules flabellate, jointed with stalk 3–4
   mm long . . . . 2. *A. tenerum*

1. **A. melanoleucum** Willd.—Leaves to 4 dm long, 2–8 cm wide; stipe erect, finely hairy, glabrescent, dark reddish brown. Pinnules dusky green, lustrous, 10–15 mm long, rectangular in outline, unevenly serrate, oriented at right angles to the rachis. Veins flabellate. No indusium; sporangial folds, crescentic, usually two to four, on anterior margin. Hammocks, south Fla., tropical America.

2. **A. tenerum** Sw.—Leaves to 8 dm long; stipe erect, polished, paleate only at the base; rachis flexuous decrescent toward the apex. Blade bipinnate to tripinnate, pendulous, 3–4 dm wide; pinnules variously lobed; sporangial folds on margins crescentic to U-shaped, as many as the pinnule lobes. $2n = 60$. Hammocks, south Fla., tropical America.

9. SPHENÓMERIS (L.) Maxon PARSLEY FERN

Slender, glabrous rock plants with creeping rootstocks with profuse, multicellular, linear, hairlike scales. Leaves mostly erect, closely spaced on brownish rhizomes. Blades decompound, ultimate segments long

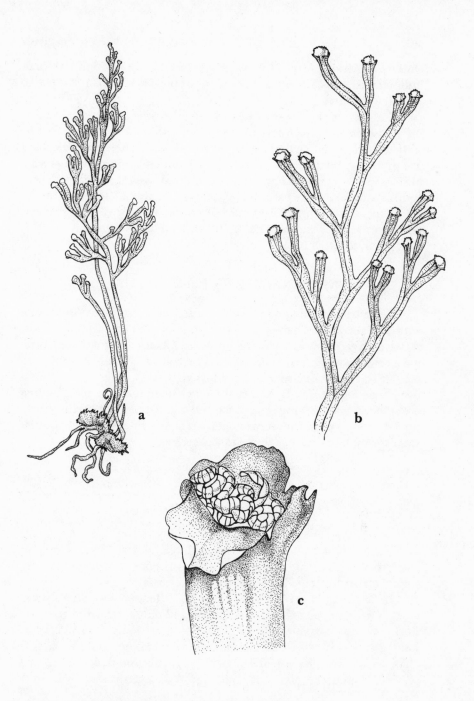

Family 12. **PTERIDACEAE.** *Sphenomeris clavata:* a, mature plant, × 1½;
b, fertile leaf segments with terminal sori, × 5; c, detail of indusium and sorus,
× 35.

cuneate; venation of free veins, sparingly forking distally. Sori terminal, indusiate. About 18 species, tropics, subtropics, Fla., Japan, Australia, and New Zealand.

1. **S. clavata** (L.) Maxon—Leaves to 4.5 dm long, from short branching rootstock and wiry roots. Stipes 1–2 mm in diameter, stramineous, scaly at base, often shorter than the rachis, with alternately disposed leaf divisions. Blade diffuse, three- to four-pinnate, forking into clavate ultimate segments 6–15 mm long, all directed toward the rachis apex. Indusia attached at the base and on sides; sori marginal, terminal on the veins; sporangia long stalked, annulus wide, of fourteen to eighteen cells, spores globose to tetrahedral. Pinelands, Fla., Everglade Keys, W.I. *Adiantum clavatum* L.

13. BLECHNÀCEAE BLECHNUM FERN FAMILY

Terrestrials from stout creeping rhizomes. Roots copiously hirsute with straight brown hairs. Leaves pinnate or pinnate-pinnatifid. Sori situated on each side of the midveins. Veins forking anastomosing, forming a row of areolae with secondary connective vein parallel to the midvein. Distal lamina veins simple or forking free to the margin. Protective linear indusium attached to the outer margin of the vein, folding over the sori and opening toward the midvein. Fertile and sterile pinnae similar or unlike in form. A cosmopolitan family chiefly of the Southern Hemisphere of about 5 genera and 220 species.

1. Pinnae simple . . . . 1. *Blechnum*
1. Pinnae pinnatifid . . . . 2. *Woodwardia*

1. BLÉCHNUM L.

Leaves from erect ligneous tips of creeping rhizomes. Pinnae fertile and sterile, essentially alike. Sporangia large; annulus twenty cells or more; spores bilateral, subglobose, with perispore. About 200 species, cosmopolitan but chiefly in the tropics and subtropics.

1. **B. serrulatum** Richard—Leaves 1.5 m long or less; stipes stout, dorsally grooved, densely paleate at base and on the crown of the rootstock; glabrous above the base. Pinnae 4–18 cm long, 5–15 mm wide, jointed to rachis at base, deciduous, the terminal one stalked; lamina oblong-acute, margins essentially parallel, cartilaginous. Spores mature all year. Hammocks, pine flatwoods, Fla., W.I., American tropics.

2. WOODWÁRDIA CHAIN FERN

Leaves pinnate-pinnatifid. Areolae in a single row on each side of the midveins. Indusium opening toward the midveins, protecting the

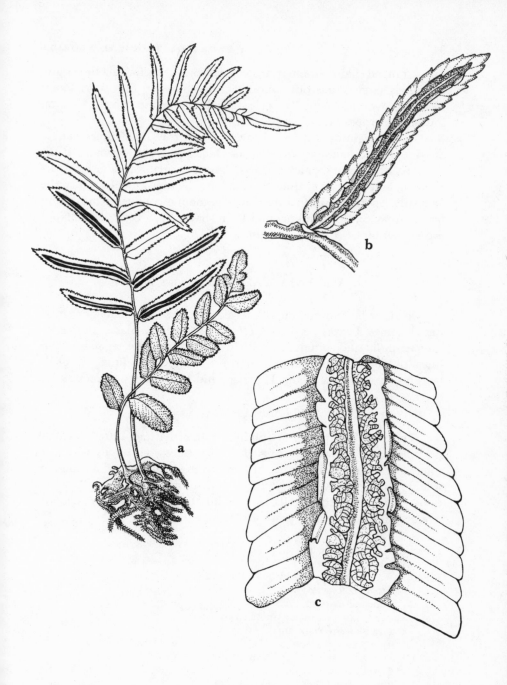

Family 13. **BLECHNACEAE.** *Blechnum serrulatum:* a, mature plant, × ¼;
b, fertile pinna with sori along each side of midvein, × 1; c, detail of sori and
indusia, × 6.

sori, arranged chain fashion in areolae. Lamina veins forking to margin. Sporangia large, numerous; annulus of up to twenty-four cells; spores bilateral, smooth. About 12 species, mostly temperate, North America, southern Europe, Japan.

1. W. virginica (L.) Sm.—Leaves in two rows, sometimes paired, from opposite sides of an elongate blackish stout rhizome; stipes purplish black, stout, paleate at base; greenish, glabrous above; blades thickish, lanceolate in outline, 3–7 dm long, 1–3 dm wide. Leaves sterile or fertile. Sporangia in mature blades becoming confluent throughout fertile areas. Swamps and hammocks in shade, Fla., Tex., Mich., Nova Scotia, Bermuda. *Anchistea virginica* Presl

14. ASPLENIÀCEAE　Spleenwort Family

Small or large, wood or rock ferns with erect, horizontal, or creeping rhizomes. Leaves persistent, erect or spreading. Blades simple, or pinnate-pinnatifid, often finely divided into ultimate segments. Sori linear-oblong or oval, indusiate, borne on veinlets, obliquely to blade margins, or midveins. About 11 genera and 1000 species, cosmopolitan.

1. ASPLÈNIUM L.

Terrestrial, epipetric, or epiphytic, often delicate, tufted, rosulate perennials. Stipe bases bearing clathrate scales continuous with rhizomes. Veins free; forking near the midveins, sometimes reforking, or joining ends submarginally without inclusion of free tips. Annulus twenty to twenty-eight cells, spores bilateral, spinulose or smooth. About 700 species, chiefly tropical.

1. Ferns with simple cuneate, short-stipitate
　　leaves to 7.5 dm long . . . . . .　1. *A. serratum*
1. Ferns with variously divided compound
　　leaves to 6 dm long or less.
　　2. Blade once-pinnate.
　　　3. Foliar axes maroon; pinnae sessile
　　　　auriculate, overlapping dorsal groove;
　　　　stipe bases maroon . . . . .　2. *A. platyneuron*
　　　3. Foliar axes green; stipe bases often
　　　　dark.
　　　　4. Pinnae 1 mm long or less, firm, uni-
　　　　　laterally crenate-dentate; stipe and
　　　　　rachis capillary . . . . . .　3. *A. dentatum*
　　　　4. Pinnae 3–7 cm long, membraneous,
　　　　　bimarginally cleft or dentate.

5. Bases of pinnae on upper margin
   entire, angles . . . . . .   4. *A. abscissum*
5. Bases of pinnae on upper margin
   incised, prominently auriculate   5. *A. auritum*
2. Blade bipinnate to tripinnate.
   6. Blades lanceolate in outline; ulti-
      mate segments with one simple and
      one forked veinlet, leaves in crowded
      tufts . . . .   6. *A. verecundum*
   6. Blades oblong in outline; ultimate seg-
      ments with one simple and two
      forked veinlets, leaves loosely tufted   7. *A.* × *biscayneanum*

1. A. serratum L.—Leaves from erect rhizome 4–7 dm long, 6–9 cm wide, oblanceolate, leathery, tapering to a short scaly stipe. Sori linear on lateral veinlets extending from the midrib to about one-fourth to the margin; indusium continuous less than 1 mm wide. Local in hammocks on logs or rocks, south Fla., W.I.

2. A. platyneuron (L.) Oakes—Ascending leaves tufted to 5 dm long, from rhizome with persistent stipe bases. Leaves functionally dimorphic; the sessile rosulate ones wholly vegetative; the elongate petioled ones reproductive; spore-bearing blades, cuneate, linear-elliptic or obovate, usually widest above the middle; stipe 5 cm long or less, with scattered scales and hairs, polished, reddish brown like the nearly terete rachis throughout to pinnatifid blade apex; pinnae 1–5 cm long, oblong, biserrate, with overlapping auricles; sori numerous, proximate to midveins. $2n = 72$. Hammocks, Fla., eastern North America. *A. ebeneum* Ait.

3. A. dentatum L.—Delicate glabrous crevice ferns with tufted capillary leaves to 15 cm long or less; stipe short or the leaf pinnae-bearing to the base. Pinnae short stalked, 8–10 mm long, obliquely rectangular or broadly elliptic, crenate-dentate at tip and upper margin or rarely incised on both margins; sori four to six or fewer. Hammocks, south Fla., W.I., rare and local.

4. A. abscissum Willd.—Tufted crevice ferns from short rhizome. Leaves 1–2 dm long, rachis nearly equaling the stipe in mature plants; pinnae biserrate, oblique at base, tapering to a narrow apex; terminal segment elongate, shallowly crenate on margins. Sori in the middle of veinlets. Spores mature all year. $2n = 72$. South Fla., local and rare. *A. firmum* Kunze

5. A. auritum Sw.—Tufted epiphytic and epipetric ferns from black crowns. Leaves 3–4 dm long, essentially glabrous; stipes wiry, polished, black at base; pinnae upward curving with crenately cleft margins, auricled at base, the sori oblong-elliptic, with indusium opening toward margin. South Fla., W.I., rare.

6. A. verecundum Chapm.—Delicate lacy crevice ferns with spreading pinnae 14 cm long or less. Leaves essentially glabrous or with clathrate scales on rhizomes; ultimate divisions bilobed, with oval sori as wide as the subtending segment. $2n = 144$. On walls of exposed lime-bearing rocks or on sides of solution holes, local, south Fla., W.I.

7. A. × biscayneanum (D. C. Eaton) A. A. Eaton (pro var.)—Tufted epipetric fern from short, erect, scaly rootstock. Leaves firm, 1–3.5 dm long, 1–3 cm wide, ascending; stipes castaneous, sparingly clathrate-paleate; blades once- to twice-pinnate, linear-oblanceolate; pinnae cuneate-oblong, incised into two to five pinnules, with two or more forking veinlets; rachis compressed, edged with green, with remote lower pinnae. Sori indusiate, linear to 3 mm long, one to two per pinnule. Growing in association with **A. verecundum** and **A. dentatum**, its supposed parents. Lime-rock formations in hammocks, endemic to south Fla. *A. rhizophyllum* var. *biscayneanum* D. C. Eaton

15. ASPIDIÁCEAE   ASPIDIUM FAMILY

Terrestrial ferns with erect, ascending or horizontal creeping rootstocks. Stipes rarely jointed. Leaves one- to three-pinnate or decompound; venation pinnate or anastomosing. Indumentum of scales or trichomes, unicellular, articulate, or branching; sori indusiate or exindusiate, orbicular, peltate, or reniform, on the veins or at tips of free veinlets. Annulus fourteen cells or more; spores bilateral, oblong, smooth or tuberculate. Represented in our area by three genera.

1. Veins free, rarely one to two pairs of segment veins united to the sinus.
   2. Leaves decompound; stipe base copiously covered with reddish scales; vascular bundles at base of stipe more than two   1. *Ctenitis*
   2. Leaves pinnate-pinnatifid; trichomes colorless or stramineous; vascular bundles, two, at the base of stipe; sori on veins ending free to the margin or on areole veins below the sinus . . . . .   2. *Thelypteris*
1. Veins anastomosing.
   3. Leaves one- to three-pinnate-pinnatifid; pinnae deltoid; sori indusiate on free veinlets within areolae or on areole veins . . . .   3. *Tectaria*
   3. Leaves once-pinnate; pinnae elliptic-lanceolate; sori exindusiate at the confluence of transverse veins . . . .   2. *Thelypteris*

Family 14. **ASPLENIACEAE.** *Asplenium* × *biscaynianum:* a, mature plant, × 1; b, mature pinnae, × 6; c, fertile pinna with mature sporangia, × 10.

1. CTENÍTIS C. Chr. & Ching

Rootstocks erect or ascending, paleate. Stipe bases massively covered with reddish-brown broad-based scales, gradually narrowed to flexuous silky filaments. Rachises pubescent above with ctenoid hairs, paleate below. Leaves bipinnate or decompound. Sori round-reniform, indusia with marginal glands; spores echinulate. About 150 species, pantropical.

1. Blades deltoid, ovate; ultimate segments
    2–3 mm wide . . . . 1. *C. sloanei*
1. Blades linear-oblong; ultimate segments 4–5
    mm wide . . . . 2. *C. submarginalis*

1. C. sloanei (Poepp.) Morton—Leaves 1.5 m long, spreading from the crown of upright elongate rootstock. Blades deltoid, pinnules attenuate, ultimate segments sparingly glandular-puberulent; sori about even distance from midvein to margins. $2n = 82$. Big Cypress Swamp, hammocks, south Fla., W.I., tropical America. *Dryopteris ampla* (Humb. & Bonpl.) Kuntze; *Ctenitis ampla* (HBW) Copel.

2. C. submarginalis (Langsd. & Fisch.) Copel.—Pubescence similar to the preceding species. Leaves to 1 m long from a ligneous crown of rootstock; rachis more or less fluted on four sides, appearing as "x" in transverse section; ultimate segments obscurely denticulate, ciliolate. Sori submarginal on veinlets; indusia fugacious. $2n = 82$. South Fla., Big Cypress Swamp, W.I., Mexico, South America. *Dryopteris submarginalis* Langsd. & Fisch.

2. THELÝPTERIS Schmidel

Rootstock slender, creeping or erect. Leaves of pinnate type, rarely dimorphic, herbaceous. Trichomes simple or branched, sometimes intermixed with scales. Venation pinnate or meniscioid. Sori indusiate or exindusiate, sometimes entirely covering the fertile surface. Stipe base with two or more vascular bundles. About 500 species, cosmopolitan, both temperate and tropical regions.

1. Venation pinnate; veins simple, free or forking.
  2. Rootstock horizontal, creeping.
    3. Leaves bipinnate-pinnatifid; stipe base
      copiously scaly . . . . . . 1. *T. torresiana*
    3. Leaves once-pinnate-pinnatifid.
      4. The lower pinnae much shorter
        than the middle ones . . . . 2. *T. resinifera*

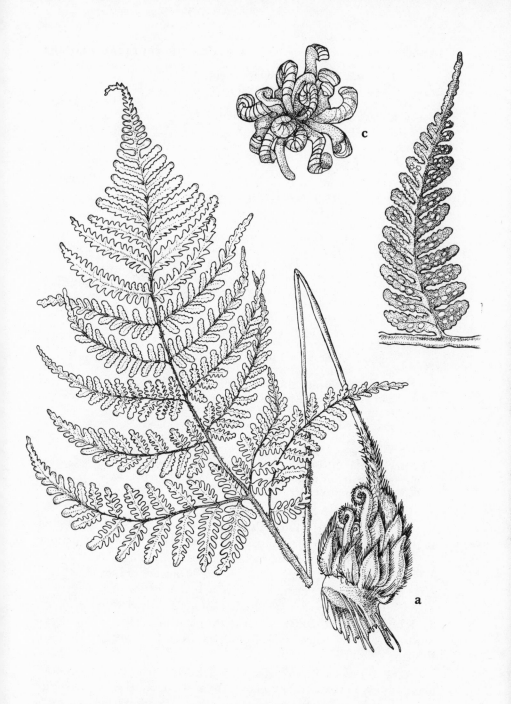

Family 15. **ASPIDIACEAE.** *Ctenitis sloanei:* a, mature plant, × ½; b, fertile pinna, × 1; c, sorus, × 32½.

4. The lower and middle pinnae about equal in length; segments eglandular.
   5. Plant glabrous; fertile leaves distinct from foliage leaves . . . 3. *T. palustris* var. *haleana*
   5. Plants pubescent; all leaves potentially fertile.
      6. Blades gradually narrowing to the terminal pinna; trichomes simple . . . . . . 4. *T. kunthii*
      6. Blades abruptly narrowing to the terminal segment; trichomes intermixed with paleae 5. *T. augescens*
2. Rootstock erect, crowded with close-set leaves; scales ovate-apiculate . . . 6. *T. patens*
1. Venation anastomosing, areolate, wholly or partly covering the lamina.
   7. Blades pinnate-pinnatifid; basal veins of adjacent lobes with a sinus vein.
      8. Plant nearly glabrous, with few scales on rachis; one to two pairs of veins ending in the sinus . . . . . 7. *T. totta*
      8. Plants puberulent; trichomes simple and branched; one pair of veins ending in the sinus.
         9. Plants creeping by proliferous, rooting rachis tips . . . . . 8. *T. reptans*
         9. Plants erect, tufted . . . . . 9. *T. sclerophylla*
   7. Blades simply pinnate, dorsally areolate.
      10. Pinnae margins undulate . . . . 10. *T. reticulata*
      10. Pinnae margin serrate . . . . . 11. *T. serrata*

1. **T. torresiana** (Gaud.) Alston—Leaves to 2 m long from a stout ascending or horizontal rootstock; scape copiously chaffy with ciliate scales at base, glabrate above; rachis and costae fluted, pubescent adaxially throughout. Blades decompound, divisions deltoid, caudate; ultimate segments crenate-dentate, pilose above and below with long white trichomes; veinlets pilose; sori medial, indusia early deciduous; spores light brown, smooth. Elegant ferns over oolitic rocks in hammocks, south Fla., W.I. Introduced from Old World. *Dryopteris setigera* Kuntze

2. **T. resinifera** Proctor—Leaves 5–8 dm long from stout rootstocks; blades thinly pubescent throughout; stipes short, scaly at the base. Pinnae 7–9 cm long, upward curving, resiniferous on dorsal surface, gradually reduced below the middle toward the base. Sori confluent, indusia pilose, glandular on margins, early deciduous; spores smooth. Hammocks, south Fla., W.I. *T. panamensis* (Presl) E. St. John

3. T. palustris Schott var. **haleana** Fern.—Colonial marsh fern with slender, elongate, black, creeping, nearly naked rhizomes; leaves to 9 dm long, broadly linear, minutely puberulent above, stipes black at bases with scattered broad thin scales. Rachis fluted on adaxial side; pinnae 5–8 cm long, linear, upward arching with recurving attenuate tips; fertile pinnae prominently contracted, margins revolute. Sori on the veins, confluent; indusia reniform; spores smooth. $2n = 70$. Moist glades, hammocks, Fla., W.I. *Dryopteris thelypteris* (L.) Gray

4. T. kunthii (Desvaux) Morton—Leaves 5–9 dm long in bilinear series on elongate rootstocks; rootstocks brown, scaly at growing apex; scales linear-lanceolate, appressed pubescent; stipes scaly at base, sparingly villous above. Blade broadly ovate, gradually abbreviated at apex; nonpaleate, moderately villous on rachis and costae; pinnae sessile, 6–14 cm long; often recurving at caudate tips. Sori nearly marginal with reniform, hispid indusia; spores dark brown, smooth. Hammocks, abundant throughout Fla. to S.C., Gulf states to Tex., W.I. *T. normalis* (C. Chr.) Small

5. T. augescens (Link) Munz & Johnston—Leaves to 1 m long in bilinear series on elongate, thick, moderately scaly rootstocks; scales setose ciliate at stipe bases; on rachis and costae mostly linear, few in intermixed dense hirsute indumentum. Blades broadly ovate-deltoid, hirtellous; pinnae distant, 7–14 cm long, usually abruptly abbreviated at apex. Sori submarginal; indusia reniform, copiously hispid. Spores black, smooth. Hammocks, lime rocks, south Fla.

T. serra (Sw.) St. John ex Small—No authentic specimens from Fla.

6. T. patens (Sw.) Small—Leaves to 1.5 m long, fasciculate on erect rootstock; stipes upright, close-set, covered at base with light-brown, ovate or oblong scales, each showing longitudinal rows of cells with clear lumen. Blades to 7 dm long, 3 dm wide, abruptly narrowed at apex; the basal pinnae slightly diminished and deflexed. Rachises grooved and puberulent on dorsal side; pinnae pinnatifid of entire falcate segments, widest at base, attenuate toward apex. Sori yellowish, contiguous, evenly spaced in rows, one on each side of the midvein between the curving margins of the segment. Indusia persistent, pubescent. Lime rocks, Ross Hammock, south Fla., W.I., Mexico, South America. Not recently collected in Fla.; its present occurrence is doubtful. *Polypodium patens* Sw., *Dryopteris stipularis* (Willd.) Maxon

7. T. totta (Thunb.) Schelpe—Colonial fern with creeping, branching, glabrous rootstock. Leaves 5–7 dm long, ascending to reclining; rachis and costae usually with small delicate scales on lower surface without hairlike tomentum; the fluted rachis on upperside, minutely pubescent; segment veins seven to nine pairs curving toward the rounded apex of the lobe, effecting half-cyclic pattern with geometric precision. Indusium glabrous; spores muricate. Hammocks, central and south Fla., W.I., Mexico, South America. *T. gongylodes* (Schk.) Small

8. T. reptans (J. F. Gmel.) Morton—Rhizome short. Leaves reclining, once-pinnate, becoming 4–5 dm long; proliferating rooted offsets successively, runner fashion; pinnae distant, oblong, crenate, ciliolate, commonly pubescent on both surfaces with branched and unicellular hairs. Sori medial on veinlets; spores smooth. $2n = 144$. Rock fern in lime sinks and limestone grottoes, central and south Fla., W.I., Mexico, Central and South America. *Goniopteris reptans* (J. F. Gmel.) Presl

9. T. sclerophylla Kunze—Rhizome thick, rigid, scaly, produced conspicuously above the ground; leaves in dense clusters, stellate-puberulent throughout; stipes to 15 cm long. Blades to 30 cm long, the lower five to six pairs of pinnae gradually reduced toward the base; pinnae distant, deeply cordate; petiolule about 1 mm long; sori medial. Dark green rock fern with pale veins. $2n = 72$. Oolitic rocks, south Fla., W.I. *Goniopteris sclerophylla* (Kunze) Wherry

10. T. reticulata (L.) Proctor—Leaves to 1 m tall or more from thick, fleshy rootstocks; stipes stout, scaly at base; finely pubescent or glabrate above; rachis pubescent, fluted above or glabrous. Pinnae often subopposite, short stalked, with oblong-lanceolate blades 10–15 cm long, 15–30 mm wide. Sori on outward-arching anastomosing veinlets, crescentic when young, becoming confluent; sporangia numerous. Hammocks, south Fla., W.I., Central and South America. *Meniscium reticulatum* (L.) Sw.

11. T. serrata (Cav.) Alston—Similar to preceding species. Leaves to 2 m long or less; rootstocks stout, scaly; stipes scaly at base; scattered scales throughout the blade and rachis; pubescence dense of coarse hairs. Fertile pinnae stalked, oblong-lanceolate, distant, subopposite, 13–20 cm long, 20–24 mm wide, prominently serrate; sori on outward-pointing angle of anastomosing veinlets, covering the entire surface when mature. $2n = 72$. Cypress sloughs, south Fla., W.I. *Meniscium serratum* Cav.

3. TECTÀRIA Cav. HALBERD FERNS

Terrestrial rock ferns with short, erect or creeping, rhizomes. Leaves variable in form and size, fasciculate on the crown; blades ovate-deltoid, hastate, pinnatifid, pinnate, or trifoliolate; hairs articulate. Sori on anastomosing veins or in areolae on free veins; indusia orbicular, reniform; spores low tuberculate. About 200 species, pantropical.

1. Blades hastate, 1.5–5 dm long, 1–4 dm wide;
    indusia orbicular, 2–2.4 mm in diameter .   1. *T. heracleifolia*
1. Blades ovate-lanceolate, 15 cm long or less,

6–10 cm wide; indusia reniform, sometimes with overlapping margins, 1.3 mm in diameter or less.

2. Rootstock slender, knotty; indusia 1.3 mm in diameter, orbicular or reniform . . 2. *T. lobata*

2. Rootstock thick; indusia 1 mm in diameter or less, reniform.

3. Blade with two pairs of pinnae below the terminal pinna; proliferous in axils; indusia caducous . . . . 3. *T. coriandrifolia*

3. Blade with one pair of subequal pinnae below the terminal pinna; not proliferous in axils; indusia persistent . 4. *T. amesiana*

1. **T. heracleifolia** (Willd.) Underw.—Caudex erect, strongly paleate at apex, studded with persistent stipe bases; stipes 5 dm long or less, rough and scaly below, glabrous toward summit. Roots velvety with delicate reddish brown trichomes. Blades firm, lustrous above, three to five pinnate-pinnatifid, or juvenile blades entire, acuminate; the basal pinnae deeply incised into falcate, long-acuminate segments, the lowermost simulating an auricle; the surface of pinnae glabrous above and below, only the midveins of segments glandular-puberulent on upper surface. Sori in regular rows or nearly so within paracostal areolae; indusia orbicular-peltate, centrally depressed with a transparent flaring rim. Limestone grottoes, hammocks, Everglade Keys, south Fla., W.I., Central and South America.

2. **T. lobata** (Poir.) Morton—Leaves in clusters on slender, creeping, copiously paleate rootstocks; stipes with linear-lanceolate scales below, glabrous or minutely puberulent above; juvenile blades simple, nearly glabrous, deeply pinnatifid, undulate on margins. Mature blades with puberulent rachis and costae, with a long pinnatifid terminal lobe, and ovate or acuminate unequal basal pinnae. Sori usually disposed regularly in paracostal areolae or irregularly in areoles toward blade margins; indusium orbicular, persistent. $2n = 80$. Rims of solution holes, Everglade Keys, south Fla., W.I., Cuba. *T. minima* Underw.

3. **T. coriandrifolia** (Sw.) Underw.—Leaves fasciculate on thick, short rootstocks; the growing apex densely clothed with linear-lanceolate filiform scales; stipes sparsely scaly below, puberulent above; rachis and costae puberulent. Blades pubescent above; terminal pinnae deeply pinnatifid with a pair of undulate lower segments; the lower pinnae, usually two or several pairs, distant, pinnatifid and proliferous, giving rise to offsets; paracostal areolae without free veinlets. Sori mostly on marginal forking veinlets; indusia early withering. Solution holes, wall crevices, Everglade Keys, south Fla., W.I., Cuba.

4. **T. amesiana** A. A. Eaton—Leaves fasciculate on short creeping

rootstocks stubbled with stipe bases and clothed at growing apex with linear-lanceolate scales; stipes sparsely scaly at castaneous base, puberulent above. Blades with terminal large pinnatifid pinnae and usually with one pair of basally distant subequal small pinnae; or blades simple pinnatifid, segments round, sinuses broad; lamina areolae without regular arrangement. Sori seldom disposed in regular rows in paracostal areolae; indusia reniform or semicircular. Everglade Keys, south Fla., Bahama Islands.

16. LOMARIOPSIDÀCEAE Lomariopsis Family

Creeping, climbing, or epiphytic ferns, with bilateral symmetry; leaves simple or pinnate, fertile pinnae acrostichoid; vascular strands plural, veins free or anastomosing, areoles not including free veins. Six genera, American tropics, South Pacific, Eurasia, Malaya.

1. LOMARIÓPSIS Fee

Tree- or rock-climbing ferns, by rhizome with dorsiventral symmetry; roots only on ventral side, leaves dorsal, with well-spaced stipes in two or more rows; stipes decurrent, scales dark brown, peltate to 1 cm long; blades 1–6 dm long, dark green, lustrous, rachis winged; lateral pinnae cuneate, abruptly acuminate, jointed to the rachis; terminal pinna continuous with the rachis. Venation pinnate of simple or forked veins, passing from the midvein to marginal teeth. Fertile pinnae narrow and small, covered with confluent sporangia. Annulus fourteen to twenty-two cells; spores bilateral, brownish with epispore. Pantropical genus of 40 species. Represented in our area by one species.

Characteristics of the genus
1. **L. kunzeana** (Presl ex Underw.) Holtt.—Hammocks, Everglade Keys, southern peninsular Fla. *Stenochlaena kunzeana* (Presl) ex Underw.

17. PARKERIÀCEAE Water Horn Fern Family

Aquatic or subaquatic annuals from erect, spongy, sparingly scaly, rooted rhizome. Leaves alternate, dimorphic, pinnately decompound, the sterile ones floating or emergent. Fertile blades erect, repeatedly pinnate-pinnatifid into linear ultimate segments; venation areolate, without included veinlets. Sporangia individually borne, usually in two

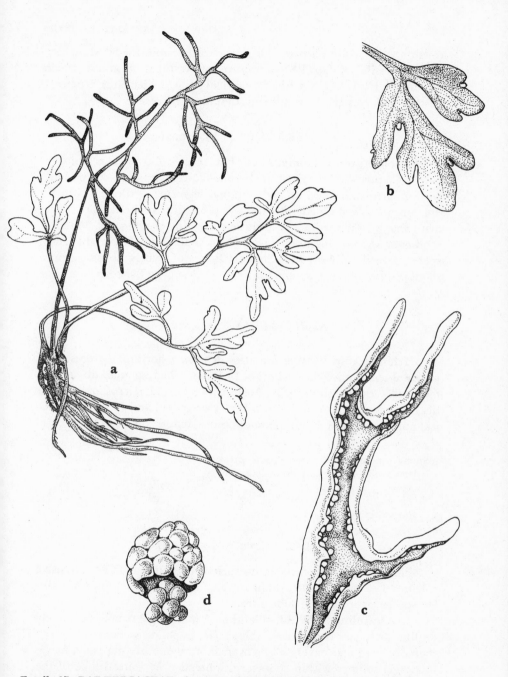

Family 17. **PARKERIACEAE.** *Ceratopteris pteridoides:* a, mature plant with narrow pinnatifid fertile segments and expanded foliage segments, × 1½; b, foliage segment, × 4; c, fertile segment with revolute margins bearing sporangia, × 7½; d, globular sorus showing individual sporangia, × 40.

rows on the veins of narrow, elongate areolae within the revolute margin, subsessile, thin-walled, complete or vestigial annulus; spores ephemeral, triplanate, the free face reticulate. A single genus, tropics of Eastern and Western Hemispheres. *Ceratopteridaceae*

1. CERATÓPTERIS Brongn.

Characters of the family. About four species of tropical regions.

1. C. pteridoides (Hook.) Hieron.—Foliage leaves 2–3 dm long, with inflated stalks and dilated terminal segment; fertile leaves to 3.5 dm long, freely divided into narrow pinnatifid segments. Sporangia maturing within revolute, translucent margins of segments. Walls of sporangia 0.8 mm long, fragile with obsolescent annulus, to eighty cells; spores sixteen to thirty-two. Shallow shores of swiftly flowing creeks, shore pools of moving waters, Fla., W.I. *C. deltoidea* Benedict

18. SALVINIÀCEAE SALVINIA FAMILY

Heterosporous floating aquatics. Rhizome horizontal, roots present or absent. Leaves in a whorl of three: two floating, one submerged; dissected, bearing sporocarps. Megasporangia and microsporangia in distinct sporocarps. Annulus wanting. Two genera and 16 species, tropical and subtropical regions extending into temperate zones.

1. Stems pinnately branched; roots present;
    leaves with upper and lower lobe; imbri-
    cate . . . . 1. *Azolla*
1. Stems mostly simple; roots lacking; leaves
    four-ranked . . . . 2. *Salvinia*

1. AZÓLLA Lam. MOSQUITO FERN

Plants small, free-floating at surface of still waters in extensive colonies. Leaves imbricate, pinnately arranged. Sporocarps in leaf axils. Six species, tropical and subtropical.

1. A. caroliniana Willd.—Plants 5–15 mm long, green or red, freely prolific, often covering all available surface of ponds, lagoons, lakes, and shores of streams. Leaves oblong-ovate, obtuse, minutely spongy; megasporangia solitary within sporocarp, subtended by ventral lobe of the leaves; microsporangia stalked, several in a sorus within sporocarp; apex of rootstock dorsally upward curving; roots with root hairs. Swamps, canals, south Fla., American tropics, naturalized in Old World.

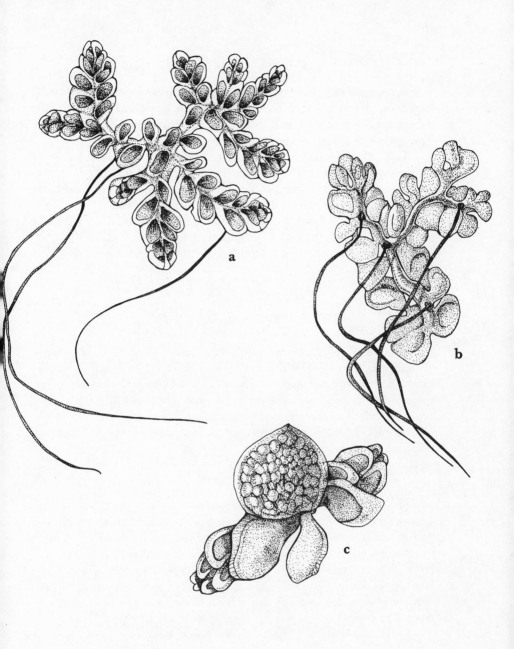

Family 18. **SALVINIACEAE.** *Azolla caroliniana:* a, mature plant, dorsal side, × 10; b, mature plant, ventral side, × 12½; c, sporocarp containing microsporangia, × 50.

2. SALVÍNIA Adans. WATER FERN

Leaves dimorphic, aerial and submerged from compressed floating rootstock, without roots. Sori indusiate; megasporangium with one megaspore, in contrast to microsporangium with numerous microspores. Ten species, tropical and subtropical regions.

1. **S. rotundifolia** Willd.—Floating leaves, two, suborbicular, 1–1.2 cm long or less, sessile, emarginate, cordate or truncate, spreading, usually with overlapping margins. Veinlets free from midvein, concealed by stiff, dense, branched hairs on upper surface, in contrast to looser dark pubescence below; sporocarps in fascicles borne on dissected hairy segments of submersed leaves; apex of rootstock horizontal. Still waters of shore pools and streams, Fla., tropical America, Africa. *S. auriculata* Aubl.

19. CYCADÀCEAE CYCAD FAMILY

Dioecious palmlike or fernlike shrubs or trees with compound leaves spirally arranged which form a terminal crown. Microsporophylls produced in conelike structures or on modified leaves, deciduous; megasporophylls in terminal cones, or ovules on modified leaves, persistent. Seeds enclosed within a cone or exposed on the margins of the megasporophylls. About 9 genera and about 90 species in tropics and subtropics of both hemispheres.

1. ZÀMIA L.

Shrubby plants with the stem wholly or largely subterranean. Leaves pinnately compound, coriaceous, pinnae entire or serrate, petioles smooth or armed. Cones short pedunculate, sporophylls densely compacted, oblong, cylindrical to subglobose; ovulate cone larger than the male cone, scales peltate, vertically superposed; scales of male cone bearing several pollen sacs, deciduous, scales of ovulate cone persistent. Seeds angulate, seed coat fleshy. About 20 species in tropics and subtropics of America.

1. **Z. integrifolia** Ait. FLORIDA ARROWROOT, COONTIE—Stem wholly or mostly underground. Leaves 5–9 dm long, in a terminal crown or whorl, pinnae 10–12 cm long, linear, stiff, and margins often revolute. Male cones mostly 10–12 cm long, narrowly cylindrical, often produced in clusters of four to five; ovulate cones 12–15 cm long, cylindric-ellipsoidal, solitary within the crown of leaves, persistent. $2n = 16$. Ham-

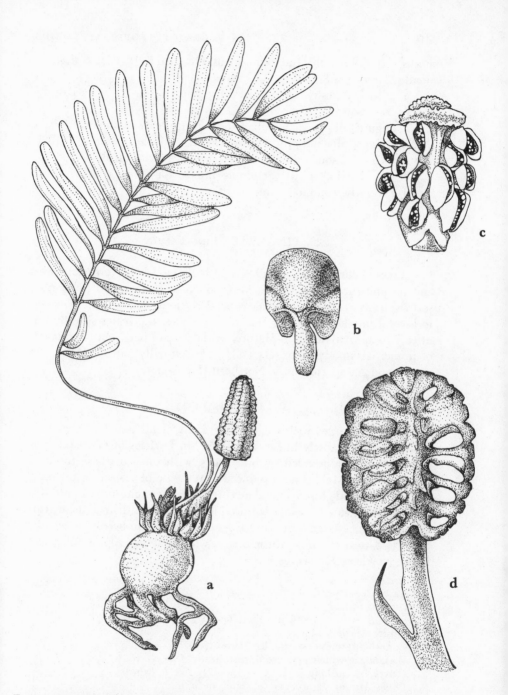

Fam19. CYCADACEAE. *Zamia integrifolia:* a, mature plant with tuberous base, pinnate leaf, and staminate strobilus, × ½; b, megasporophyll, adaxial surface, × 1½; c, microsporophyll, abaxial surface bearing many microsporangia (pollen sacs), × 5½; d, mature ovulate strobilus, × ½.

mocks, scrub, pineland, Indian middens, Fla., Fla. Keys, W.I. *Z. silvicola* Small; *Z. umbrosa* Small

The cooked fleshy rootstock was used by Florida Indians as an important food source.

Z. **angustifolia** Jacq., a Cuban species with leaves somewhat longer than *Z.* **integrifolia** and pinnae arranged rather distantly in the rachis, is also reported for south Fla. Its occurrence here is doubtful.

"Conti hateka" is the Seminole phrase for white root or white bread, hence the common name coontie.

20. PINÀCEAE   Pine Family

Trees or shrubs with needlelike, usually fascicled, evergreen leaves and resinous branches. Monoecious, pollen sacs subtended by a scale, scales spirally arranged in a male cone, deciduous; ovules solitary or two to several on the surface of the ovulate scales, the scales spirally imbricate in an ovulate cone. Mature, ovulate cones persistent, composed of numerous woody or papery scales, seeds usually winged. About 22 genera and 200 species in the Northern Hemisphere.

1. PÌNUS L.   Pine

Evergreen trees with excurrent growth, stems and branches resinous. Leaves narrowly linear, needlelike, in fascicles of two to five or rarely solitary, subtended by bud scales, with some of these fused to form a sheath. Male cones produced at the base of season's growth, in spirally arranged clusters. Ovulate cones solitary or clustered, composed of many imbricated scales, each producing ovules in their axils, maturing to become a large cone with scales elongating and becoming woody. Seeds two, near the base of each scale, winged above. About 85 species in the Northern Hemisphere.

1. Needles (leaves) only two per fascicle, 6–8
    cm long or less . . . .   1. *P. clausa*
1. Needles (leaves) two and three per fascicle,
    mostly over 15 cm long.
   2. Cones subterminal, sessile, less than twice
      as long as wide when closed; leaf sheath
      over 1.3 cm long . . . .   2. *P. palustris*
   2. Cones lateral, short pedunculate, over
      twice as long as wide when closed; leaf
      sheath less than 1.3 cm long . . . .   3. *P. elliottii*

1. **P. clausa** (Engelm.) Sarg. SAND PINE—Trees up to 25 m tall with bark rather smooth. Leaves mostly 4–8 cm long, two per fascicle, dark green, sheath 5–7 mm long. Cones 5–6 cm long, conic or becoming ovoid when open, each scale appendaged with a stout spine near the middle. Seed about 4 mm long, wing 1.5 cm long. Coastal and interior dunes, southwest Fla. to south Ala., unknown from the Everglades region and the Fla. Keys.

2. **P. palustris** Mill. SOUTHERN LONG-LEAF PINE—Tree up to 30 m tall with scaly bark. Leaves 2–40 cm long or more, two or three per fascicle, dark green, long, sheath long, needle clusters tending to crowd toward the tips of the branches. Male cones 6–8 cm long, ovulate cones 15–25 cm long, subterminal on the branches, sessile, less than twice as long as wide when closed, scales armed with a short recurved spine. $2n = 24$. Large trees have been reported as far south as Lee County and are possibly also in the range of this manual, sandy soil, Fla. to Tex., Va.

3. **P. elliottii** Engelm. SLASH PINE—Tree up to 30 m tall with scaly bark. 15–30 cm long or more, two or three per fascicle, dark green, sheath moderately long. Male cones 3–5 cm long, ovulate cones over twice as long as wide when closed, lateral, short pedunculate, scale appendage prominent. Seed about 6–8 mm long, wing 2–3 cm long. Two varieties occur in south Fla.

1. Open cones mostly with truncate base; seedlings with slender stems . . . . . 3a. var. *elliottii*
1. Open cones mostly with rounded base; seedlings with thick, shortened stems . . . 3b. var. *densa*

3a. P. elliottii var. **elliottii**—Flatwoods, Fla. to La., S.C. on the coastal plain. *P. palustris* sensu Small, not Mill. Not common in south Fla. where it is displaced by:

3b. P. elliottii var. **densa** Little & Dorman—Thin limestone soil, southern peninsular Fla. and Fla. Keys. *P. caribaea* sensu Small, not Morelet

This is the common pine in south Fla. with large stands in the Everglades National Park and on Big Pine Key in the Fla. Keys.

21. TAXODIÀCEAE BALD CYPRESS FAMILY

Evergreen or deciduous, monoecious trees with spirally arranged needlelike or scalelike leaves. Male cones with spirally arranged peltate sporophylls. Ovulate cones woody, ovule-bearing scales spirally ar-

Family 20. **PINACEAE.** *Pinus elliottii* var. *densa:* a, leafy branch, × ½; b, staminate cones, × 1½; c, mature ovulate cone, × 1; d, fascicles of leaves, × ½.

ranged. Seeds winged. About 8 genera and 15 species of wide distribution.

1. TAXÒDIUM Richard BALD CYPRESS

Large trees, deciduous, spirally arranged needlelike or linear leaves. Male cones many, in pendent, terminal panicles, microphylls usually six to eight; ovulate cones on previous year's growth, persistent, globose or subglobose, scales irregularly quadrangular, each with two ovules. Three species in eastern North America and Mexico.

1. Leaves linear, spreading; young branches
   wide-spreading . . . . 1. *T. distichum*
1. Leaves subulate, appressed; young branches
   tending to be ascending . . . . . 2. *T. ascendens*

1. T. distichum (L.) Richard, BALD CYPRESS—Trees up to 40 m tall with trunk often conspicuously buttressed below, bark flaky, younger branches tending to be spreading. Leaves about 1–1.5 cm long, linear, spreading, pale green, two-ranked. Male panicles about 10 cm long, appearing in the spring; ovulate cones pendulous, 2–3 cm wide, scales rugose. In fresh-water swamps flooded for at least part of the year, Fla. to Tex., Del., north in Miss. Valley to Ind.

One of the most common trees in southern peninsular Fla. but apparently not present in the Fla. Keys.

2. T. ascendens Brogn. POND CYPRESS—Trees up to 40 m tall, buttressed trunk or conical at the base, younger branches ascending or erect. Leaves about 5–10 mm long, subulate, appressed or somewhat incurved, two-ranked. Otherwise morphologically similar to **T. distichum**. Fla. to Ala., Va., on the coastal plain.

The specific distinction of this plant is questionable since it may represent a mere growth form or minor variant of **T. distichum**.

22. CUPRESSÀCEAE CYPRESS FAMILY

Evergreen trees or shrubs with opposite or whorled, sometimes dimorphic, leaves, three needlelike or more often scalelike, overlapping or separated by short internodes. Monoecious or dioecious, microsporophylls opposite, few; ovulate cones small, scales opposite or whorled, woody and distinct or more often fused to form a berrylike cone at maturity with one to many erect seeds at the base of the scales. About 15 genera and 125 species of wide distribution.

1. JUNÍPERUS L. Juniper

Dioecious or sometimes monoecious shrubs or trees, bark thin and tending to shred, with scalelike or needlelike leaves, opposite or whorled in threes. Male cones with many microsporophylls; ovulate cones berry-like or drupelike, usually bluish, the scales coalescent and rather fleshy or leathery at maturity. Seeds one to ten per cone, wingless. About 60 species, chiefly in the Northern Hemisphere. *Sabina* Hall.

1. **J. silicicola** (Small) Bailey, SOUTHERN RED CEDAR—Shrub or tree up to 30 m tall with thin, shredding bark. Leaves 1–4 mm long, acute, scalelike, opposite or three- to four-whorled. Male cones 4–5 mm long, ovulate cones ovoid or ellipsoid, 3–4 mm long. Occurring in drier sites, not common in south Fla., apparently absent in the Fla. Keys but planted throughout the range of this manual, Fla. to Tex. and S.C. on the coastal plain and W.I.

23. TYPHÀCEAE Cattail Family

Perennial rhizomatous herb with terete jointless flowering stem. Leaves sheathing, blades elongate-linear, about equaling the stem. Plant monoecious; flowers unisexual, disposed in two terete, contiguous, or separate congested spikes, each subtended by an early deciduous bract. The upper portion bearing staminate flowers; the lower, pistillate. Typical perianth lacking; stamens three to four, subtended by bristles, anthers linear, four-locular, longer than the connate filaments; fertile pistillate, flowers stipitate, one carpellate, with linear persistent styles. Sterile pistillate flowers with clavate hairs. Fruit an achene, airborne by basal tuft of hairs. One genus, ten species, swamps, shallow shores, temperate, tropical, and subtropical regions. Rhizomes, young shoots, and inflorescences may be utilized as sources of food.

1. TÝPHA L. Cattail

Characters of the family

1. Leaves plano–convex; pistillate and staminate spikes separated; pollen grains one-celled.
  2. Leaves strongly convex on the back; gynophore hairs acute . . . . 1. *T. angustifolia*
  2. Leaves slightly convex on the back; gynophore hairs clavate . . . . . 2. *T. domingensis*
1. Leaves flat; pistillate and staminate spikes not separated; pollen grains four-celled . 3. *T. latifolia*

Family 23. **TYPHACEAE.** *Typha domingensis:* a, mature inflorescence with staminate and pistillate spikes and mature leaves, × ½; b, stem base and rhizome, × ¼; c, staminate flower, × 20; d, pistillate flower, × 21.

1. T. angustifolia L. NARROW-LEAVED CATTAIL—Stems slender, rigid, 3 m tall or less. Leaves 4–9 mm wide, overtopping the spike. Spikes usually 2–3 cm apart; the pistillate, 10–20 cm long, 8–11 mm in diameter; pistillate flowers with bractlets, stigmas linear; hairs of sterile pistillate flower cuneate at tips. Fruiting spikes chestnut brown. $2n = 30$. Marshes, ponds, watery ditches, Fla., Fla. Keys, eastern U.S., W.I., Mexico, South America. Summer, fall.

2. T. domingensis Pers. SOUTHERN CATTAIL—Stems 2–3 m tall, bearing numerous leaves. Blades 0.8–1.5 cm wide, moderately rounded on the back, firm, exceeded by the spike. Pistillate spikes 1–2 dm long, 2 cm in diameter, flowers with bractlets, stigmas linear; hairs of sterile pistillate flowers spatulate at tips; mature spikes pale cinnamon brown. $2n = 30$. Brackish coastal marshes, ponds, south Fla. to Calif., coastal plain to Me., W.I., South America. Summer, fall.

3. T. latifolia L. COMMON CATTAIL—Plants robust to 3 m tall arising from strong, spreading rhizomes; leaves 8–22 mm wide, grayish green, overtopping the spikes. Mature pistillate spikes to 2 dm long, 3.5 cm in diameter, dark brown, without floral bractlets; stigmas spatulate, persistent; achene body, slender about 1 mm long, shed in hair tufts about 8 mm long. $2n = 30$. Swamps, marshy lands, shores of lakes and rivers throughout North America. Widely distributed in Eurasia. Summer, fall.

24. RUPPIÀCEAE WIDGEON GRASS FAMILY

Submerged perennial herbs with simple or branching stems from slender rhizome. Leaves filiform, alternate with enlarged stipular sheaths. Flowers two, bisexual, oppositely arranged on spadix. Stamens two, anthers sessile, pollen grains tubular, curved; pistils four, free, with hairlike stigmas, sessile, ovaries one-ovuled. Pollen grains off-white, shed at the surface during pollination; fruit an achene. Single genus, about three to four species, cosmopolitan distribution.

1. RÚPPIA L.

Characters of the family

1. R. maritima L. DITCH GRASS—Stems dark green, profusely branching, 6 dm long; leaves linear, 4–10 cm long, less than 1 mm wide, one-veined, flat; stipular sheaths free at tips, subtending the flowers within. Fruiting carpels 3 mm long, ovoid-oblique at base, shortbeaked, long stipitate, disposed in umbelliform clusters, lowered to the bottom

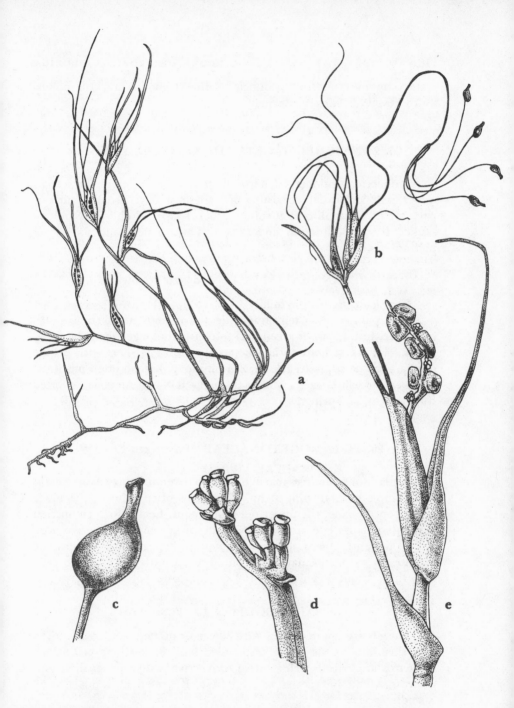

Family 24. **RUPPIACEAE.** *Ruppia maritima:* a, flowering plant, × 1; b, fruiting plant, × 2; c, mature fruit, × 16; d, pistillate flowers after pollination, × 8½; e, flowering spathe, × 5.

waters by the spiraling peduncle. Saline lagoons, brackish ponds, south Fla., W.I., Mexico. Spring.

25. ZANNICHELLIÀCEAE HORNED PONDWEED FAMILY

Monoecious submerged herbs with slender rhizomes; flowers in upper axils, unisexual, consisting of a single stamen, usually four carpels, all in a cupuliform involucre, maturing into obliquely oblong beaked fruits. Pollination underwater. Three genera and six cosmopolitan species.

1. ZANNICHÉLLIA L. HORNED PONDWEED

Characteristics of the family

1. **Z. palustris** L.—Stems fragile, leaves opposite, to 10 cm long, 0.5 mm wide, linear, bright green and soft, usually longer than the internodes. Flowers without perianth, carpels free, style short with peltate stigma. Achene stipitate, falcate, 2–3 mm long, keeled, often muricate. Fresh and brackish waters, almost throughout North America, reported from south Fla. Spring.

26. POTAMOGETONÀCEAE PONDWEED FAMILY

Aquatic or sometimes marsh or swamp perennial rhizomatous herbs with simple or branching stems. The stems often jointed, the lower nodes root bearing, the upper ones foliaceous. Leaves with submersed or floating blades, two-ranked. Flowers in spicate emersed inflorescences, typical perianth absent, flowers tetramerous, hypogynous; stamens one to four. Ovary one-locular, ovule one. Fruit a nutlet or drupelet, pericarp thin. About 4 genera and 100 species, cosmopolitan.

1. POTAMOGÈTON L. PONDWEED

Largely fresh-water herbs with linear or dilated leaf blades, stipulate, sheathing at the sessile or petioled base. Spikes long peduncled, stamens and pistils in cycles, anther connectives with four sepallike outgrowths. About 100 species, cosmopolitan. Represented in our area by a single species.

1. **P. illinoensis** Morong—Stem simple, floating leaf blades 4–12 cm long, 2–3 cm wide, elliptic-lanceolate, coriaceous, crisped on margins, mucronate, submersed blades narrowly elliptic, acute, 1.5–3.5 cm wide,

commonly seven- to fifteen-veined, lacunae two or three rows of cells along midvein; stipules prominently spreading. Fruiting spikes 2–4 cm long, sepaloid connective 3 mm wide. Fruit 2.5 mm wide, orbicular, compressed, strongly keeled with two lateral keels. Oolitic waters of ponds, canals, south Fla., throughout U.S., Canada, Mexico, Cuba. *P. angustifolius* B & P, *P. lucens* L.

27. NAJADÀCEAE NAIAD FAMILY

Small aquatic annual herbs, monoecious or rarely dioecious, with opposite sheathing leaves dilated at base. Flowers unisexual, staminate flower within a spathe. Perianth two-lipped, stamen single, pistillate flower naked, sessile. Ovary one; one-locular, one-ovulate; stigmas two to three, awl shaped. Pollen grains spherical or oblong; pollination occurring in water. Fruit indehiscent, drupaceous. One genus, 50 species, cosmopolitan.

1. NÀJAS L. NAIAD

Characters of the family

1. Monoecious; leaf margin entire, beset with
   minute spinules.
  2. Leaf apex acute, margins of dilated base
     rounded . . . . 1. *N. guadalupensis*
  2. Leaf apex pointed, margins of dilated
     base angled . . . . 2. *N. flexilis*
1. Dioecious; leaf margin conspicuously dentate . . . . 3. *N. marina*

    1. **N. guadalupensis** (Spreng.) Magnus—Stems leafy, branching to 5 dm long; leaves numerous, 1–3 cm long, linear, 0.5–1 mm wide, margins with about twenty to forty spinules. Fruit brownish purple in maturity, seed 2.5–3 mm long, distinguished with numerous rows of four-sided areolae. Fresh or brackish waters, Fla., Tex., to New England, to Midwest, Pacific states; W.I., Central and South America. Fall, spring.
    2. **N. flexilis** (Willd.) Rostk. & Schmidel—Stems capillary, branching, short or greatly elongate; leaves 1–22 mm long, acute, crowded and fasciculate in axils and stem tips, margins spinulose, often reddish throughout; fruit reddish; seed 1.8–2.2 mm long, compressed ellipsoid, areolae six-sided in numerous rows. $2n = 24,12$. Shallow waters, fresh or brackish, south Fla., N.Y. to Calif. Summer.

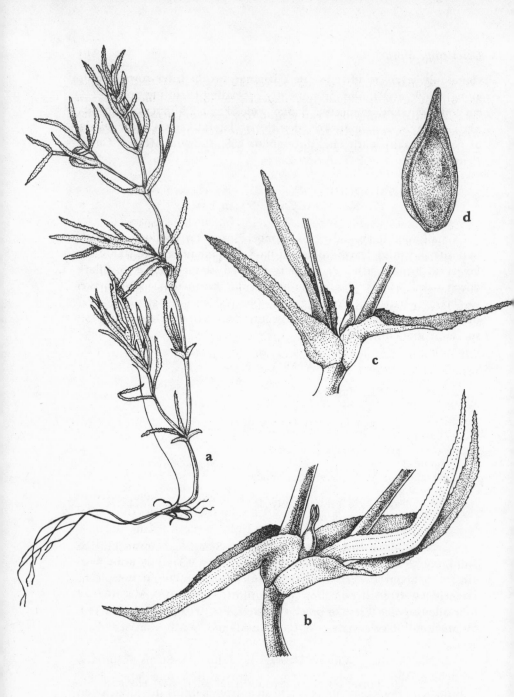

Family 27. **NAJADACEAE.** *Najas flexilis:* a, mature plant, × ¾; b, pistillate flower in axil of dilated leaf base, × 15; c, staminate flower in axil of dilated leaf base, × 12½; d, mature fruit, × 35.

3. N. marina L.—Stems and leaves visibly toothed; teeth triangular with subulate spinulose tips; blades linear to linear-oblong, 1–3 cm long, about 1–2 mm wide. Fruit ellipsoidal, 4–8 mm long. Seed 4–5 mm long, minutely reticulate. $2n = 12$. Shallow waters, south Fla., N.Y. to Calif., Mexico, Eurasia. Summer.

28. CYMODOCEÀCEAE Manatee Grass Family

Submerged marine perennials, monoecious or dioecious, rootstocks creeping, often rooting at the articulate nodes. Leaves linear, sheathing at the base. Flowers unisexual, without perianth, solitary or cymose, sometimes subtended by minute hyaline bracts. Staminate flowers of two long-pedicelled anthers equally or unequally attached. Anthers four-locular, longitudinally dehiscent. Pistillate flowers of two united carpels or a single carpel, sessile or stipitate. Style long or short, crowned by two, or one, threadlike stigmas. Mature fruit nutlike; seed one, flat, wholly filling the cavity, testa thick, turgid. Herbs of warm seas. Two genera, about 18 species.

1. Stigmas two, elongate; anthers evenly
   raised, leaves terete . . . .   1. *Cymodocea*
1. Stigma one, short; anthers unevenly raised,
   leaves flat . . . .   2. *Halodule*

1. CYMODOCEA Koenig Manatee Grass

Leaves terete, acute. Plants dioecious; flowers unisexual, borne in leaf sheaths, with two stigmas. Seed pendulous. One species of tropical and subtropical America, one species of India, western Pacific Ocean.

1. C. filiformis Kuetz—Acaulescent with creeping rootstocks; dioecious, leaves 35 cm long, terete, acute, the sheaths slightly auriculate. Flowers solitary or in dichotomous cymes. Pistillate flowers two-carpellate, subtended by hyaline bracts, style short, stigmas two. Mature fruit 3 mm long, beaked by the persistent style. Coastal waters of Fla., W.I. Summer, fall. *Cymodocea manatorum* Aschers.

2. HALODULE Endl.

Leaves flat, borne on erect sprouts from horizontal elongate rhizome. Flowers cymose, exserted from leaf sheaths. Staminate flower of two anthers on unequal pair of pedicels. Pistillate flower a solitary carpel, fruit small, globular, one-seeded. Seven species, mostly tropical. *Diplanthera* Thouars

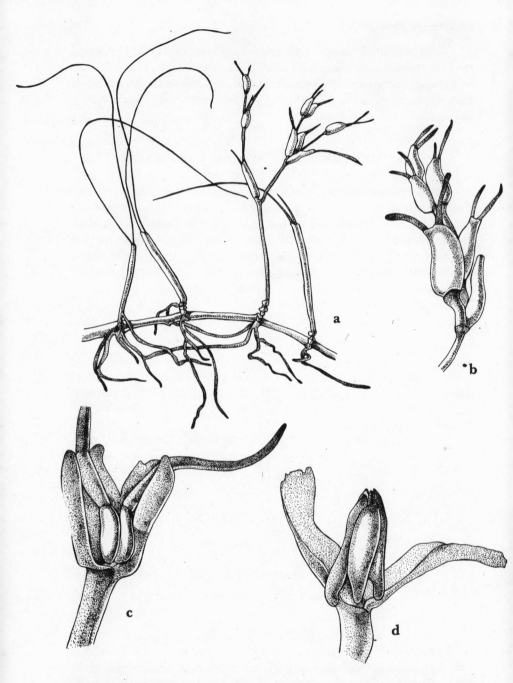

Family 28. **CYMODOCEACEAE.** *Cymodocea filiformis:* a, flowering plant,
× ½; b, pistillate flowers, × 2; c, pistillate flower, × 8; d, staminate flower,
× 6½.

1. **D. wrightii** (Aschers.) Aschers.—Rootstocks articulate profusely branching sprouts at the rooted nodes. Leaves to 40 cm long, 1–12 mm wide, with several fine veins; sheaths scarious, loose, one, fruit black. Coastal water of Fla., W.I., Cuba. Fall, winter. *Halodule wrightii* Aschers.

29. JUNCAGINÀCEAE Arrow Grass Family

Perennial scapose marsh herbs with deep-seated rhizomes. Flowers radially symmetrical, trimerous, bisexual or unisexual, in bracted or bractless racemes or spikes. Perianth of one whorl of three or two alternating whorls of similar segments. Stamens three or six, sessile or with short filaments. Carpels three or six, simple, one- or two-ovuled; stigma nearly sessile, plumose or papillose; fruit follicular. North and south temperate and subarctic regions; 5 genera, 25 species.

1. TRIGLÒCHIN L.

Leaves terete, fleshy, sheathing at base of a naked jointless scape. Flowers bisexual; perianth of three to six deciduous segments, each with a basally attached stamen. Anthers two-locular, sessile; ovaries three or six, closely connivent in anthesis, appearing three-locular, stigmas sessile. Style very short with minutely plumose stigma. Represented by a single species.

1. **T. striata** Ruiz & Pav.—Scape to 27 cm tall, one or several from stoloniferous rhizomes. Leaves to 22 cm long; blades green, narrowed-to-blunt tips; sheaths with broadly hyaline margins hooded at apex. Spike 6–13 cm long, to 5 mm wide; flowers in three alternating rows in the rachis; pedicels 1–1.2 mm long; sepals three, about 1 mm long, navicular, pale green, one-veined; petals none; anthers three. Mature fruit 2 mm long, three-angled, the carpels separating from the capsule summit, opening upwardly on the inner side; seed one cylindroid filling the cavity. Coastal strand, salt marshes, Fla. to Md., Calif., tropical America, Africa, Australia. Summer.

30. ALISMATÀCEAE Water Plantain Family

Rhizomatous fibrous-rooted marsh or aquatic monoecious herbs with scapelike stems; leaves long petioled with sheaths open at base; blades prominently veined, with transverse veinlets. Flowers unisexual, radially symmetrical, trimerous; perianth of distinct sepals and petals,

Family 29. **JUNCAGINACEAE.** *Triglochin striata:* a, mature plant, × ½; b, flower, × 20; c, fruiting raceme, × 1½; d, fruit, × 30.

free, imbricate in bud; stamens and pistils free, spirally arranged on dome-shaped torus; carpels one-ovuled, one-seeded, fruit beaked achene. Cosmopolitan, about 30 genera and 90 species.

1. SAGITTÀRIA L. ARROWHEAD, WAPATO

Plants acaulescent with corm-bearing stolons, milky latex. Leaves simple, with blades or straplike phyllodia, floating, submersed or emersed. Scape with terminal inflorescence; floral bracts spathaceous, embracing verticils of floral buds, splitting wholly or partly into three bracts, with free, often subulate tips. Staminate flowers with three green deciduous sepals, three white petals, six to indefinite stamens; anthers extrorse. Pistillate flowers with three green persistent sepals, three white petals, numerous simple pistils, torus often accrescent. Fruiting head an aggregate of compresssed achenes. About 20 species in America.

1. Filaments pubescent.
  2. Plant 10 dm or more tall; petioles stout;
    filaments linear, longer than anthers . 1. *S. lancifolia*
  2. Plant 0.8–6 dm tall; petioles slender; filaments dilated, about as long as anthers.
    3. Phyllodia flat to biconvex; achene beak
      spreading, subterminal . . . . 2. *S. graminea*
    3. Phyllodia semiterete; achene beak erect,
      terminal . . . . 3. *S. isoetiformis*
1. Filaments glabrous.
  4. Leaves typically sagittate, erect . . . 4. *S. latifolia*
  4. Leaves phyllodial or sometimes dilated, unlobed.
    5. Sepals of pistillate flowers appressed,
      spreading . . . . . . 5. *S. subulata*
    5. Sepals of pistillate flowers reflexed.
      6. Phyllodia prominently three-
        ribbed; leaves usually submersed 6. *S. kurziana*
      6. Phyllodia not ribbed; leaves float-
        ing at surface or emersed . . 7. *S. stagnorum*

1. **S. lancifolia** L.—Leaves to 1 m or more long, emersed, from stout rhizome; blades to 3 dm long, 2 dm wide, overtopped by scape. Raceme with five to twelve whorls of flowers, rarely branching from lower axils; floral bracts thickened, strongly striate, ribbed; sepals to 10 mm long, striate, hooded, petals twice as long, clawed; stamens twenty-eight to thirty, anthers extrorse. Fruiting torus to 2 cm in diameter; achenes 2–2.3 mm long, winged on both sides; facial wings low, often resinous;

Family 30. **ALISMATACEAE.** *Sagittaria graminea:* a, flowering plant, × ½;
b, staminate flower, × 3; c, mature fruit, × 20; d, single fruit, × 37½.

beak oblique, broad at base. Common in marshes, shores, sinkholes, coastal Atlantic and Gulf states to Fla., Fla. Keys, tropics. Summer. *S. angustifolia* L.

2. **S. graminea** Michx.—Perennial herbs 0.8–6 dm tall from well-developed cormose rhizomes, emersed or submersed; leaves phyllodial, strap shaped or tips dilated, bladelike, 2–15 cm long, 3–4 cm wide; scapes overtopping the leaves; racemes two to eight whorls of flowers; floral bracts shortly connate, pedicels slender, 2–4 cm long, ascending; stamens less than twenty. Fruiting heads 8–12 mm in diameter; achenes cuneate-oblong to 2 mm long, beak minute, subterminal; crests three, resin ducts two to three. Fla., Ala., Ga., to N.C., W.I. Represented in the area by two varieties.

1. Inflorescence racemose . . . . 2a. var. *graminea*
1. Inflorescence paniculate . . . . . . 2b. var. *chapmanii*

2a. **S. graminea** var. **graminea**—Inflorescence usually a simple raceme. Phyllodia 10 mm wide or less; strap shaped. Widely distributed in North America, Fla., Cuba, W.I. Summer, fall.

2b. **S. graminea** var. **chapmanii** Sm.—Inflorescence paniculate; phyllodia 3–25 mm wide, widest above the middle, attenuate at tips; emersed blades 2–4 cm wide, three- to five-veined. Panicle branches commonly from the lowest whorl; flowers in five to twelve whorls; bracts to 15 mm long, shortly connate, tips free. Fruiting heads to 10 mm in diameter; achene 2–2.3 mm long, winged, facial crests one to two, oil tubes two. Cypress swamps, Fla., Ala., Ga. Fall, winter. *S. cycloptera* (Sm.) Mohr

3. **S. isoetiformis** Sm.—Plants with slender rhizomes; phyllodia sheathing at base, semiterete, dilated at tips when emersed, flat linear, lax, 2–4 mm wide when submerged; scape 8–15 cm long with two or more whorls of flowers; floral envelope half-connate at base. Staminate flower to 18 mm wide, stamens nine to eighteen, pedicels filamentous; pedicels of pistillate flowers shorter and thicker; fruiting torus to 10 mm in diameter, globular. Achene to 1.8 mm long, dorsal wing undulate, facial crest one to two, oil duct one, beak minute or obscure in maturity. Shallow, marshy shores of ponds, ditches, persisting in dry lake beds, south Fla., Ga. Spring, summer.

4. **S. latifolia** Willd. Wapato—Plants 2–12 dm tall, partially or entirely emersed, rhizomes stout; leaves erect or ascendent; blades including lobes 7–25 cm long, 2–18 cm wide at sinus level; principal veins three to seven, with three veins branching into basal lobes, apex acute

or obtuse. Flowering scape commonly overtopping the leaves, angled, bearing two to thirteen whorls of flowers; bracts nearly free; petals in staminate flowers to 2 cm long, stamens twenty or more numerous; filaments longer than anthers. Fruiting torus to 2 cm in diameter; achenes winged, beak spreading at right angles from the summit. $2n = 22$. Swamps and shallow waters, Fla., Mexico, Central America, Nova Scotia to Pacific coast. Highly variable species. Summer. *S. pubescens* Muhl.; *S. viscosa* C. Mohr

    5. **S. subulata** (L.) Buch.—Submersed or emersed, annual, perennating by runners with corms; leaves phyllodial 2–12 cm long, 1–15 mm wide, or sometimes dilated into blades 1.5–5 cm long, 1–2 cm wide. Scapes 2–30 cm tall, erect or floating with one to three whorls of flowers. Staminate flowers with slender pedicels, stamens ten or fewer; pedicels of pistillate flowers stouter, recurved; corollas to 15 mm in diameter; fruiting torus 6–7 mm thick. Achenes few, to 2 mm long, crenulate winged, two to three facial crests, the beak lateral, porrect. Tidal waters, in ponds, east coast of North America, Fla. to N.Y. *S. lorata* (Chapm.) Small

    6. **S. kurziana** Glueck—Phyllodia 1–16 dm long, 5–15 mm wide, submersed, floating, tips rarely dilated, rising from vertical stoloniferous rhizome; roots fine, fibrous. Leaf margins thickened, lamina with three to five parallel, sharply elevated ribs on the lower side. Flowering scapes about as long as the phyllodia. Inflorescence 3–4 dm long; staminate pedicels 1–5 mm thick with flowers to 2 cm wide; carpels with prominent styles and lateral stigma. Fruiting heads to 12 mm wide, sepals reflexed; achenes 2–2.5 mm long with three irregularly scalloped crests; beak short, ascending. Swiftly flowing water, springs, central and south Fla. to Ala., New England. Fall. *S. subulata* (L.) Buch. var. *kurziana* Bogin

    7. **S. stagnorum** Small—Phyllodia 2–5 dm long, 3–10 mm wide, in tufts on slender crown, giving rise to white, creeping, cormose stolons; corms hard at flowering time with two to four internodes, scale leaves hyaline. Flowering scape about as long as the leaves or shorter; flowers borne in two to three distant whorls, open a little above the water level; floral bracts nearly free, subulate at tips; staminate flowers showy, white, stamens seven to fifteen, filaments subulate; petals of pistillate flowers 5–6 mm long, sepals and pedicels reflexed; fruiting torus to 8 mm in diameter; mature achene 2–2.3 mm long with two to three crenate crests, beak subterminal incurved, oil ducts one to two. $2n = 44$. Margins of shallow waters, pools and swamps, coastal plain, Fla., Ala. Spring. *S. natans* Michx., not Pall.

31. HYDROCHARITÁCEAE Frog's-Bit Family

Submerged aquatic herbs, stoloniferous or with rootstocks, monoecious or dioecious. Leaves simple radical, from short crowns, or cauline, opposite or whorled. Flowers unisexual, enclosed in sessile or long-penduncled spathes, the peduncle sometimes spirally twisted. Sepals three, separate or united; petals three, or sometimes lacking; stamens three to many, the filaments often connate. Ovary one, inferior, one-locular with parietal placentae or intruding toward the center; stigmas as many as placentae, usually trifid. Fruit indehiscent, baccate, or many-seeded berry, ripening underwater. About 16 genera and 80 species in waters of the warmer parts of the world.

1. Fresh-water herbs.
  2. Plant acaulescent; leaves radical; pistillate
     flower long peduncled . . . . . 1. *Vallisneria*
  2. Plant caulescent; leaves cauline; flowers
     sessile . . . . 2. *Hydrilla*
1. Marine herbs.
  3. Leaves cauline, blades dilated . . . . 3. *Halophila*
  3. Leaves basal, ribbonlike . . . . . 4. *Thalassia*

1. VALLISNÈRIA L. Tape Grass

Soft dioecious plants with rosulate leaves. Staminate spathe short peduncled, enclosing numerous small flowers in a spicate cluster individually with three sepals, three white petals, and one to three stamens, all separating at anthesis and floating to the surface. Carpellate spathe long peduncled with solitary flower with three sepals, three small petals, and three broad stigmas, one-locular ovary with three parietal placentae bearing numerous ovules; pollination effected by the floating staminate flowers contacting the stigmas; the spiraling peduncle lowering the spathe beneath the surface to mature the fruit. About ten temperate, tropical, and subtropical species.

  1. V. neotropicalis Marie-Victorin—Leaves 15–22 mm wide, linear, thin, apices obtuse, minutely denticulate, lamina veins more or less reticulate; vegetative reproduction by runners rooting at tips. Fresh-water lakes, central and south Fla., Cuba. Spring, summer.

2. HYDRÍLLA Rich.

Submerged, bottom-rooted aquatic herb of fresh or brackish waters. Stems profusely branching; leaves 6–15 mm long, commonly in whorls of four to eight, with red midvein, sessile, obovate-oblong, sharply serrulate. Staminate flower solitary in a sessile spathe; pistillate flower solitary, the spathe two-pointed, stalked. Plant prolific by tuberiferous rhizomes. Ovary one-locular, stigma fimbriate. Fruit subulate smooth or muricate; seeds two to three, minute, testa produced at each end. One species, Eurasia, Africa, Australia, North America.

1. **H. verticillata** Royle—Characters of the genus. $2n = 24$. Dominating the entire bottom growth of shallow water, naturalized, central and south Fla.

3. HALÓPHILA Aschers. SEA GRASS

Submerged marine aquatics with creeping rhizomes; dioecious or monoecious. Scales and leaves in pairs, sometimes appearing verticillate. Flowers unisexual; staminate peduncled; pistillate sessile; hypanthium flask shaped; sepals minute; petals none; stigmas filiform, sessile. Pollination by water. Fruit enveloped in a spathe of two free bracts; seeds numerous; cotyledons spirally twisted. About ten species, tropical waters, Indian, Pacific, and south Atlantic Oceans.

1. Petioles 5 mm long or longer, filiform;
   blades essentially entire . . . . . 1. *H. baillonis*
1. Petioles 4 mm long or less, winged; blades
   patently serrate-scabrous . . . . . 2. *H. engelmannii*

1. **H. baillonis** Aschers.—Plants delicate, minutely puberulent throughout. Stems and rootstocks filiform, the upright branches 3–4 cm long or shorter; blades obovate or spatulate, three-veined, the lateral veins uniting subterminally with the midvein. Flowers unisexual, subtended by a common spathe, or distinct in paired spathes. Capsule ovoid, 2–3.5 mm long, beaked. In lagoons, Fla. Keys, W.I.

2. **H. engelmannii** Aschers.—Erect shoots from slender rhizome to 10 cm long; leaves six, appearing whorled at summit; blades elliptic, cuneate, acute, fleshy, prominently veined; hypanthium flask shaped; capsule 2–3 mm long. Marly sand in bays and reefs, south Fla., W.I.

4. THALÁSSIA Banks TURTLE GRASS

Marine herbs with elongate articulate rootstock, dioecious. Leaves straplike, distichous tufted on short crowns of the rhizome with persistent, fibrous, scarious sheath remnants. Scape axillary; spathes tubular, stamens distinct, pistil solitary. One species, coast of Caribbean and one coast of Indian and Pacific Oceans.

1. **T. testudinum** Koenig & Sims—Submerged herb, colonial by

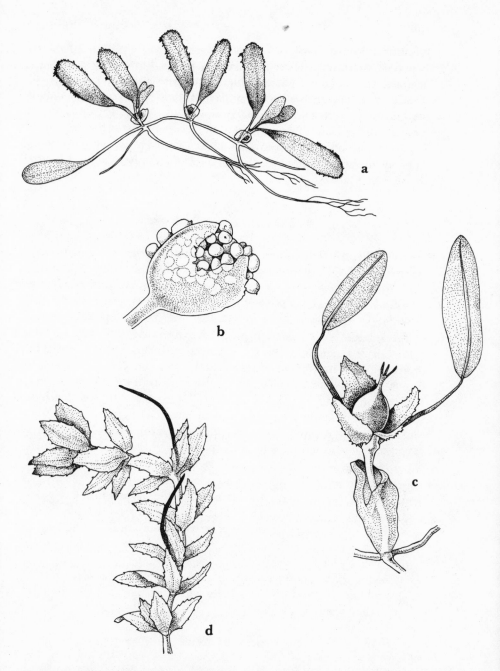

Family 31. **HYDROCHARITACEAE.** *Halophila baillonis:* a, leafy stem with adventitious roots, × 1½; b, mature fruit with seeds, × 25; c, pistillate flower in axils of mature leaves, × 12½; d, *Hydrilla verticillata:* leafy stem with immature fruits, × 1½.

creeping rhizome; leaves two to five per tuft, sheathing the scape at the base. Leaves linear, 5–30 cm long, 4–9 mm wide, finely veined, rounded at apex, not differentiated into blade and petiole. Flowers unisexual; staminate flower long pedicelled, sepals petaloid, stamens six, filaments very short. Pistillate flowers nearly sessile, ovary six- to nine-locular. Fruit echinate, beaked, seeds numerous. Occurs in vast submarine fields as shelter and food for marine animal life. Washed ashore in great quantities by the surf. Shallow waters of the coast of the Gulf of Mexico, south Fla.

32. POÀCEAE Grass Family

Annual or perennial herbs of varying habit, or treelike and dumose, woody plants. Culms with hollow internodes and solid nodes or sometimes solid throughout; roots fibrous, stolons often present. Leaves alternate, two-ranked, blades linear or lanceolate-confluent or rarely articulate with sheaths, enveloping the internodes with overlapping margins; ligule a collarlike appendage across the sheath summit within, consisting of a membrane, entire or fringed with hair, or a circle of hairs, or absent. The unit of inflorescence is the spikelet, variously arranged in panicles, racemes, or spikes. It usually consists of one or several two-ranked florets on a jointed axis, the rachilla. At the base of the rachilla subtending the floret are two alternate bracts, designated as the first and the second, or outer and inner glume; the third and fourth bracts called lemma and palea subtend the perianth of two or three lodicules, the three stamens and the pistil of two- to three-carpellate, one-locular, one-ovuled ovary with one style and one to three stigmas, or two styles and two plumose stigmas. Lemma enclosing the palea and the essential organs of the flower constitutes a floret; with the empty glumes below, it is called a spikelet. Fruit mainly a caryopsis, embraced by the lemma, or a naked achene, utricle, or berry. Seed free or adherent to pericarp; endosperm typically starchy, large.

Economically grasses constitute the most important plant family. About 500 genera and 10,000 species with cosmopolitan distribution. Nomen alt. Gramineae.

Key to Groups

1. Spikelets laterally compressed, mainly articulate above the glumes.
   2. Plants woody or herbaceous, spikelets many-flowered, glumes present; or one-

flowered, glumes reduced . . . .   I. BAMBUSOID GROUP
2. Plants herbaceous, spikelets several-
   flowered; if one-flowered, glumes present.
   3. Culms reedlike, tall, panicles large,
      plumose . . . .   II. ARUNDINOID-
                                       DANTHOID GROUP
   3. Culms not reedlike, panicles various.
      4. Lemmas five-or-more veined . . .   III. FESTUCOID GROUP
      4. Lemmas three (one)-veined . . .   IV. ERAGROSTOID-
                                              CHLORIOID GROUP
1. Spikelets dorsally compressed, articulate below the glumes . . . .   V. PANICOID GROUP

Key to Tribes

Group I

1. Arborescent or dumose bamboos; culm
   sheath present; lodicules three   . . .   Tribe 1. Arundinarieae
1. Herbaceous grasses; culms annual or peren-
   nial; culm sheath lacking; lodicules two .   Tribe 2. Oryzeae

Group II

Represented by single tribe . . . . . .   Tribe 3. Arundineae

Group III

1. Spikelets one-flowered; lemma veins ob-
   scure; floret with bearded callus . . .   Tribe 4. Stipeae
1. Spikelets two- to several-flowered; veins
   manifest; floret without callus.
   2. Lowest lemma, excluding awns if present,
      extended beyond the glumes . . .   Tribe 5. Festuceae
   2. Lowest lemmas as long as or shorter than
      the glumes.
      3. Lemmas awned from the back or bifid
         apex; rachilla not produced beyond
         the ultimate spikelet . . . . .   Tribe 6. Aveneae
      3. Lemma tapering to a long awn; rachilla
         produced to a short stipe beyond the
         ultimate floret . . . .   Tribe 7. Triticeae

Group IV

1. Spikelets in diffuse or contracted panicles.
   2. Floret one, lemmas persistent, usually
      shorter than paleae (except in Tribe
      9).

3. Lemmas one-veined; awnless; testa free
   from pericarp, seed falling free . .   Tribe 8. Sporoboleae
3. Lemmas three-veined; awn trifid; testa
   fused with pericarp, spikelet falling
   intact . . . .   Tribe 9. Aristideae
  2. Florets several, lemmas deciduous, about
     as long as paleae, rachis continuous, or
     articulate . . . .   Tribe 10. Eragrosteae
1. Spikelets in two rows on the same side of a
   continuous rachis . . . . . .   Tribe 11. Chlorideae

Group V

1. Spikelets with one sessile, bisexual terminal
   floret, and below it one staminate or neuter floret; lemmas indurated, margins inrolled over the paleae; never awned . .   Tribe 12. Paniceae
1. Spikelets in pairs, one sessile, bisexual floret,
   and one pedicellate, staminate, neuter or
   wholly reduced floret, on articulate rachis;
   lemmas awned . . . .   Tribe 13. Andropogoneae

I. BAMBUSOID GROUP

TRIBE 1. *Arundinarieae*

1. Branches several at each node . . . . 1. *Bambusa*
1. Branches solitary at each node . . . . 2. *Pseudosasa*

1. BAMBÙSA Retz. Bamboo

Plant cespitose in open or dense clumps, culms ascending, often arched or nodding at apex. Spikelets bisexual, rarely unisexual; usually many-flowered. Glumes shorter that the lemmas, the first usually small. Lemmas five- to many-nerved, awnless, similar to the glumes. Palea two-keeled and two-veined. Lodicules three; stamens six, rarely three, style one, stigmas three. Ovary one-locular, one-ovulate. Fruit a berry, nut, or utricle. Tropics and subtropics, about 70 species.

1. **B. vulgaris** Schrad. Bamboo—Culms ligneous from strong determinate rhizomes to 10 m tall or less, 8–10 cm in diameter at base; internodes hollow, solid at nodes. Branches leafy, numerous, fasciculate. Leaf blades pubescent on lower surface, narrowed to a petiole, articulate with the sheath. Blades to 15 cm long, 4 cm wide, ligule membranous. Panicle large, leafy; spikelets axillary in congested clusters. Flowering sporadic at intervals of years. On maturing the fruit, the culms die. Cultivated; persisting about abandoned sites, pantropical.

2. PSEUDOSÁSA Makino

Spikelets two- to nine-flowered, in loose panicles; florets with three to four stamens, stigmas three, style short. Three species of eastern Asia.

1. **P. japonica** Makino—Culms to 3 m tall, terete, fistulose, covered with waxy bloom; sheaths hispid; branches few; blades 3 dm long, 3–5 cm wide, sheaths persistent, large, lustrous. Doubtfully escaping from cultivation in south Fla., Japan. *Arundinaria japonica* Sieb. & Zucc.; *Bambusa metake* Sieb.

TRIBE 2. *Oryzeae*

1. Spikelets bisexual, bilaterally compressed;
   stamens one to six . . . .   3. *Leersia*
1. Spikelets unisexual; stamens six.
   2. Staminate and pistillate spikelets in sepa-
      rate inflorescences; floating-leaf aquatic   4. *Hydrochloa*
   2. Staminate and pistillate spikelets in the
      same inflorescence; rhizomatous terres-
      trial . . . .   5. *Zizaniopsis*

3. LEÉRSIA Sw.   RICE CUT-GRASS

Perennial rhizomatous grasses of wet habitats. Glumes lacking. Florets one, subtended by a five-nerved navicular, hispid-ciliate lemma; palea three-nerved, keeled, hispid-ciliate, the ribbed margins firmly conforming to those of lemma; stamens six. Caryopsis loosely invested within lemma and palea. About 15 species, tropical and subtropical. *Homalocenchrus* Mieg.

1. Plants stoloniferous; panicles contracted;
   spikelets 3–3.5 mm long . . . . .   1. *L. hexandra*
1. Plants tufted; panicle open; spikelets 1.5
   mm long . . . .   2. *L. monandra*

1. **L. hexandra** Sw.—Culms to 1 m long, geniculate, tufted on rhizomes and leafy creeping stolons. Blades to 15 cm long, 3–5 mm wide, linear-attenuate, folded, retrorsely scabrous throughout; ligule 3 mm long, hyaline membranous. Panicle branches distant, spikes floriferous nearly to the base. Fruiting lemmas falcate to 5 mm long, antrorsely scabrous on margins. $2n = 48$. Fla., Tex., pantropical. Summer. *H. hexandrus* Kuntze

2. **L. monandra** Sw.—Culms 1 m tall, densely tufted. Blades linear-

elongate, 3–5 mm wide, grayish green, firm; rhizomes lacking. Panicles to 15 cm long, branches remote, spreading, distally spikelet bearing; spikelets imbricate, broadly ovate to suborbicular, glabrous; stamens usually one. Pinelands, prairies, Fla. Keys, Tex., W.I. Summer. *H. monandrus* Kuntze

4. HYDRÓCHLOA Beauv. WATER GRASS

Bottom-rooted perennial aquatic with small terminal or axillary panicles. Spikelets monoecious, staminate lemmas seven-veined, opposed by two-veined paleae with six stamens. Pistillate lemmas seven-veined, with five-veined paleae, styles plumose, white, like anthers, long exserted in anthesis. One species, U.S.

1. **H. caroliniensis** Beauv.—Culms to 1 m long, slender, ascending in streamlined dense strands. Leaf blades flat, floating, 1–3 cm long, 2–5 mm wide. Fruiting spikelets 2–3 mm long, few. Often abundant in slow-moving waters of canals, south and central Fla., N.C. Summer.

5. ZIZANIÓPSIS Doell & Aschers. GIANT CUT-GRASS

Robust terrestrial, rhizomatous perennial. Spikelets in ample terminal panicles, pistillate above the staminate on common branches. Glumes lacking, florets deciduous from pedicels; pistillate lemmas short awned, seven-veined, paleae three-veined; staminate floret with six stamens; styles united, prominent, persistent in fruit. Two species southeastern U.S. and two species in tropical South America.

1. **Z. miliacea** (Michx.) Doell & Aschers.—Culms stout leafy to 3 m tall. Blades elongate, 1–3 cm wide, glabrous on surfaces, scabrous along margins; ligule 10–15 mm long, membranous. Panicle nodding, 3–5 dm long, racemes rebranching, numerous, fasciculate; florets 5–6 mm long, appressed-pubescent. Mature grain stramineous, 4–5 mm long with stipe and beak, the body often inflated by insect pests within pericarp. $2n = 24$. Marshy borders, woods, Fla., Tex., Okla. to Me. Summer.

II. ARUNDINOID-DANTHOID GROUP

TRIBE 3. *Arundineae*

1. Florets unisexual, plants dioecious; pistillate spikelets silky-pilose . . . . . 6. *Cortaderia*
1. Florets bisexual.
  2. Blades 1–2 cm wide; spikelets 6–8 mm
    long, stipitate . . . . 7. *Neyraudia*

2. Blades more than 2 cm wide; spikelets
   more than ten mm long.
3. Panicle flexuous; rachilla hairy . . . 8. *Phragmites*
3. Panicle erect; rachilla glabrous . . . 9. *Arundo*

6. CORTADÉRIA Stapf PAMPAS GRASS

Tall leafy dioecious perennial with spikelets two- to five-flowered; rachilla internodes jointed; florets stipitate; staminate and pistillate plants similar in aspect; the spreading basal leaves progressively shorter above. About 15 species, South America.

1. **C. selloana** (Schultes) Aschers. & Graebn.—Tussock-forming grass; culms to 3 m tall or taller; longer blades bunched to base; blades narrow, long attenuate, sharply serrate. Panicle 3–10 dm long, pinkish to lustrous white, feathery and fluffy. $2n = 72$. Ornamental grass in Fla., naturalized from South America.

7. NEYRÁUDIA Hook. f. SILK REED

Tall perennial with ample plumose panicles. Spikelets four- to several-flowered; rachilla jointed; florets stipitate; glumes unequal, one-veined; lemmas three-veined, slenderly awned between two teeth, pilose on margins. Two species, tropical Africa, China, Indomalaya.

1. **N. reynaudiana** (Kunth) Keng—Culms reedlike to 3 m tall; nodes blue; glabrous-striate; leaf blades deciduous, elongate, long attenuate, midrib whitish, thick; sheaths woolly at summit, ligule long hairy; spikelets delicately feathery with curving lemma awns; lowest one to two lemmas empty. $2n = 40$. Ornamental grass, south Fla., naturalized from southern Asia.

8. PHRAGMÌTES Trin.

Tall colonial reeds of swamps and shores. Culms stout from articulating spreading rhizomes. Spikelets several-flowered in lax panicles; glumes and lemmas semitransparent; rachilla disarticulate between florets and above the glumes. About three species, cosmopolitan.

1. **P. australis** (Cav.) Trin. ex Steud. REED—Culms terete, 4 m tall, 15 mm or more in diameter, leafy up to the panicle. Sheaths overlapping, pubescent or glabrate at the throat; blades to 5 cm wide, elongate, attenuate, slightly auriculate; ligule a short fringe of hair. Spikelets stipitate, tawny or purple; outer glume 4 mm long, one-veined, inner glume 7 mm long, three-veined; lemma 12–15 mm long, weakly awned, three-veined palea 5 mm long; rachilla hairs surpassing the spikelets in maturity. $2n = $ c.96. Distribution cosmopolitan, common throughout Fla. *P. communis* Trin.

9. ARÚNDO L. Giant Reed

Culms from knotty rhizomes with distichous leaves up to the plumose panicle. Spikelets few-flowered; florets reduced in size successively; rachilla articulating above the glumes and between the florets; glumes three-veined; lemmas three-veined, awn straight, flanked by short teeth. About 12 species, tropical and subtropical regions.

1. **A. donax** L.—Culms to 6 m tall, conspicuously clump-forming. Leaves numerous with overlapping sheaths; blades to 7 cm wide, auriculate, ligule short, membranous. Panicle to 7 dm long, spikelets often purplish, 12 mm long. Ornamental grass escaping from plantings, naturalized from warmer parts of the Old World, persistent on old home sites, south Fla.

III. FESTUCOID GROUP

TRIBE 4. *Stipeae*

10. STÌPA L. Needle Grass

Tufted grasses with convolute blades and open panicles. Spikelets one-flowered, articulating obliquely above the glumes; callus sharp, pointed, bearded. Glumes thin, membranous, long and narrow; lemma terete, indurated, convolute terminating in a long, twisted, bent awn. About 100 species in temperate regions.

1. **S. avenacioides** Nash—Culms glabrous, stiff to 1 m tall. Leaf blades rigid, involute, tapering to filiform tips; sheaths glabrous tight, ligule 2–3 mm long, membranous. Panicle to 18 cm long, branches filiform, rebranching, elongate, nodding, minutely scabrous, spikelet bearing at tips, glumes glabrous to 18 mm long; mature lemma blackish brown, 8–10 mm long, papillose roughened, ringed with stiff short hairs at summit; callus beard golden brown. Endemic to rocky pinelands, central and south Fla., Key Largo.

TRIBE 5. *Festuceae*

11. LÒLIUM L. Rye Grass

Spikelets solitary, several-flowered, disposed edgewise alternately in cavities of a continuous rachis. First glume wanting, the second glume three- to five-veined, commonly equaling or surpassing the second floret. Lemmas five- to seven-veined, convex; obtuse, acute, or awned. Spikes compressed. About 12 species, temperate Asia, northern Africa, Europe.

1. **L. perenne L.**—Short-lived perennial with culms decumbent or erect at base, 3.5 dm tall or taller. Foliage soft glossy green. Blades 2–5 mm wide, auriculate; sheaths shorter than the internodes. Spikelets to 15 mm long surpassing the glumes; lemmas awnless. $2n = 14$. Lawns, disturbed sites, roadsides, Fla., naturalized from Europe.

TRIBE 6. Aveneae

12. AVÈNA L.　OATS

Spikelets two- to three-flowered, rachilla bearded, jointed above the glumes and between florets. Glumes subequal, seven- to nine-veined, surpassing the florets; lemmas indurate, five- to nine-veined, awn dorsal, or awnless. Seventy species, temperate regions and mountains of tropics.

1. **A. sativa L.**—Annual with culms tufted, essentially glabrous up to 4 dm tall. Blades linear, ligule white, membranous; sheaths overlapping. Panicle exserted, flexuous, nodding; pedicels capillary, spikelets two-flowered, to 2 cm long, nodding. Fruit adherent to lemma and palea. $2n = 42$. Cover plant used along roads, not persisting, naturalized from Old World.

TRIBE 7. Triticeae

13. SECÀLE L.　RYE

Spikelets usually two-flowered sidewise on the spike rachis, rachilla prolonged beyond upper floret. Glumes narrow, stiff, one-veined; lemmas scabrous-ciliate on keels and margins, five-veined, tapering into long awns. About 20 species, Europe and western Asia.

1. **S. cereale L.**—Winter annual, in cultivation. Autumn growth sod-forming with soft linear leaves. Spring and summer growth with leafy flowering and fruiting culms. $2n = 14$. Appearing in roadside planting, naturalized from Old World.

IV. ERAGROSTOID-CHLORIOID GROUP

TRIBE 8. Sporoboleae

14. SPORÓBOLUS R. Brown　DROPSEED

Annual or perennial grasses, tufted or colonial by stolons, culms often wiry, leaves involute. Spikelets one-flowered; glumes and lemmas awnless, obtuse or acute, one-veined, seed readily dropping off. Panicle open or sometimes contracted. A polyploid complex involving species with various ploidy levels. About 96 species of warmer parts of America.

1. Plants tufted, usually in large bunches.
   2. Second glume as long as the lemma.
      3. Panicle open, branches whorled.
         4. Spikelet 3 mm long, red, fading brown; leaves strongly involute with needle-sharp tips . . . .   1. *S. junceus*
         4. Spikelet 1.5 mm long, green; leaves flat; not sharp at tips . . . .   2. *S. pyramidatus*
      3. Panicle more or less contracted; branches appressed; spikelets 2 mm long . . . .   3. *S. domingensis*
   2. Second glume shorter than the lemma.
      5. Spikelet 2 mm long; first glume oblong; blades involute . . . .   4. *S. poiretii*
      5. Spikelet 0.5–2 mm long; first glume orbicular; blades flat, attenuate . .   5. *S. indicus*
1. Plants creeping by rhizomes; sheaths overlapping; blades long attenuate; spikelets 2–2.5 mm long; glumes and lemma nearly equal . . . .   6. *S. virginicus*

1. **S. junceus** (Michx.) Kunth—Plants densely tufted in large bunches up to 6 dm tall from filiform fibrous roots, culms slender, wiry, usually overtopping the leaves. Blades narrow, folded tapering to sharp apex; sheaths short, loose, shreddy in age. Panicle maroon in full anthesis, pyramidal, spikelets 2.8–3 mm long, outer glume one-third as long as the inner glume and the acute lemma. Pinelands, Fla. to Va. and Tex. Summer. *S. gracilis* Trin.

2. **S. pyramidatus** (Lam.) Hitchc.—Culms 3–4 dm tall, tufted, decumbent, geniculate at spreading base. Blades 1.5–4 mm wide, 5–8 cm long; basal sheaths with hyaline margins, overlapping the internodes, pilose at throat; ligule a fringe of hairs. Panicle long exserted, 5–7 cm long, lower branches whorled, the upper gradually reduced, spreading, pyramid fashion; spikelets 1.4 mm long, outer glume minute, the inner as long as the lemma and palea. Grain 0.8 mm long, compressed. $2n = 24,54$. Alkaline soils, south Fla. to Tex., Kans., tropical America, naturalized from South America. Summer.

3. **S. domingensis** (Trin.) Kunth—Erect tufted perennial leafy at base; culms to 1 m tall, commonly much less. Blades 5–15 cm long, 3–8 mm wide; flat at base, involute at tips; sheaths mostly overlapping. Pan-

icle long exserted, dense with ascending branches or interrupted and lobulate in young plants; outer glume about one-half as long as the inner glume; grains viscid. Calcareous rocks and sandy beaches, south Fla., W.I. Summer.

4. **S. poiretii** (Roem. & Schultes) Hitchc.—Erect perennials, solitary or tufted. Culms strict to 1 m tall, glabrous, leafy to the panicle. Blades flat at base, attenuate at flexuous tips, becoming involute; ligule obsolescent. Panicle 2–3 dm long, interrupted, branches appressed, lobulate. Spikelets numerous, glumes subequal, one-half as long as the acute lemma. $2n = 36$. Fields and wastelands, Fla. to Va., Okla., South America. Summer.

5. **S. indicus** (L.) R. Brown—Habit similar to **S. poiretii** excepting a looser panicle with spreading branches of fewer flowers. Blades flattish at the base becoming long attenuate, involute, at flexuous tips. Inner glume obtuse-ovate, outer suborbicular. $2n = 24$. Roadsides, disturbed sites, pinelands, Fla.; ballast, Ala.; tropical America. Summer.

6. **S. virginicus** (L). Kunth—Plants widely creeping by leafy stolons, rooting at nodes and covering extensive stretches of coastal sands; culms erect to 4 dm tall, sheaths imbricate, blades prominently distichous 3–5 cm long, involute; ligule a short membrane, fringed with hairs. Inflorescence spicate, slender or thick, usually lobulate, spikelets 2–2.5 mm long, glumes and lemmas about equal, acute. $2n = 20,30$. Cover-forming grass in brackish sands of the Atlantic and the Gulf coasts, Fla., Va., Tex., Cuba, British Honduras. Summer.

TRIBE 9. *Aristideae*

15. ARÍSTIDA L.   THREE-AWN GRASSES

Annual or perennial grasses, tufted or rhizomatous with flat or involute blades. Contracted or open panicles, spikelets one-flowered, obliquely articulate above the glume; glumes commonly narrowly lanceolate, nearly equal, acute, or awn pointed, one-veined; lemmas narrow, convolute produced at summit into a column of varying length with three awns or one awn by reduction of the lateral ones; the base of lemmas indurated into a sharp hairy callus. Undesirable range grasses. Widely distributed by wind. About 330 species, temperate and subtropical regions.

1. Awns three, lemma beak straight.
  2. Rhizome spreading, scaly, palea reduced
    fimbriate; glumes 12–15 mm long; pani-
    cle open . . . . 1. *A. patula*

2. Rhizomes knotty in close tufts; panicles
   more or less contracted.
   3. Column 10–25 mm long, twisted,
      glumes awned . . . .   2. *A. spiciformis*
   3. Column 3–4 mm long, straight, glumes
      awnless.
      4. Blades flat at base, becoming involute
         at tips.
         5. Awns about equally divergent.
            6. Glumes 8–12 mm long; subequal,
               panicle many-flowered . .   3. *A. purpurascens*
            6. Glumes 6–7 mm long; equal,
               panicle few-flowered . . .   4. *A. simpliciflora*
         5. Awns spirally contorted.
            7. Panicle slender; lemma awns
               12–15 mm long . . . . .   5. *A. tenuispica*
            7. Panicle dense; lemma awns 10–
               15 mm long . . . . . .   6. *A. condensata*
      4. Blades involute; first glume 7–8 mm
         long; second glume 10–11 mm
         long; lemma awns spirally con-
         torted . . . .   7. *A. gyrans*
1. Awn one, lateral awns obsolescent; glumes
   subequal; lemma beak falcate . . . .   8. *A. floridana*

   1. A. patula Chapm. ex Nash—Erect perennial, rhizomatous grass
to 1 m tall. Blades flat at base, attenuate involute above; sheaths sparsely
short pilose at throat; minutely fringed ligular membrane less than 0.5
mm long. Panicle 3–5 dm long, open, nodding, branches naked below,
distally floriferous, pulvinate at base; spikelets relatively few, large;
glumes awn pointed; central lemma awn 2–2.8 cm long. Fruit 3.4–3.8
mm long; callus appressed, white hairy. Low pinelands, coastal strand,
endemic to peninsular Fla.

   2. A. spiciformis Ell.—Erect tufted perennials to 1 m tall. Blades
flat at base, becoming involute; sheaths hyaline on margins, abruptly
narrowed to blades; ligule less than 0.5 mm long, minutely ciliolate.
Panicle stiff, spicate, bristly, 12–15 cm long, dense, with numerous imbricate ascending branches; glumes unequal, the inner, 8–10 mm long,
surpassed by awns. Fruit with awns 5–6.5 cm long; callus slender, hairy,
1.3 mm long. Fla., Miss. to S.C., W.I.

   3. A. purpurascens Poir. ARROWFEATHER—Culms slender to 7 dm
long, somewhat geniculate, lax, becoming flexuous in age or tips attenuate, involute in smaller tufts; with short internodes at base. Blades flat,
sheaths auriculate, ligule very short; panicle nodding, overtopping the
leaves, often half as long as the plant. Panicle branches loose, ascending.

Mature fruit 20 mm long; callus white hairy, 0.5 mm long. Dry sandy soils, Fla. to Tex., Kans., Miss., Wis., Central America.

4. **A. simpliciflora** Chapm.—Culms erect to 6 dm tall, slender. Blades 1 mm wide, 5–15 mm long, flat. Panicle simple, slender, nodding; about one-third as long as the plant; florets few; lemma awns spreading, the central one with a semicircular bend, somewhat reflexed. Wet pineland, Fla., Miss.

5. **A. tenuispica** Hitchc.—Perennial, slender tufted plants to 7 dm tall. Leaves lax, blades becoming flexuous, 10 cm long or longer, 1–2 mm wide. Panicle slender, nearly half the length of the plant; mature fruit with awns 12–15 mm long; callus 1 mm long, slender, white hairy. Wet pineland, peninsular Fla., Central America.

6. **A. condensata** Chapm.—Robust tufted perennial to 1 m tall from knotty rhizomes; culms leafy to the inflorescence. Blades flat 1–2.5 dm long, 2–3 mm wide, becoming attenuate, involute, flexuous in age; sheaths pilose at throat. Panicle to 3 dm long, nodding, relatively dense with appressed-ascending branches. Fruit with awns 18 mm long; callus sharp, 0.8 mm long, white bearded. Fla. to N.C., Ala.

7. **A. gyrans** Chapm.—Densely tufted perennial to 6 dm tall or less with slender culms; blades 5–10 cm long, about 1 mm wide, setaceous. Branches ascending, many- or few-flowered, panicle lax, often longer than the culm. Mature fruit mottled with dark purple; glumes and lemmas bronzed purple; callus slender, 1.4 mm long, densely white bearded. Sandy pineland, dry soil, Fla. to Ga.

8. **A. floridana** (Chapm.) Vasey—Erect tufted perennial with simple, slender culms to 7 dm tall; sheath glabrous, pilose at orifice; blades elongate, flat 2–3 mm wide, involute tapering to filiform tips. Panicle narrow, to 30 cm long, rachis angled, scabrous; the filiform branches ascending, overlapping, floriferous nearly to the base; spikelets short pedicelled, glumes subequal, glabrous 7–10 mm long, the first acuminate, the second obtuse, mucronate, margin hyaline; lemma glabrous 7–8 mm long, prolonged into an angled, scabrous beak, with a filiform, terminal, sickle-shaped awn, lemma awn 15–20 mm long; lateral awns obsolete; callus white pilose, 1 mm long. Hammocks, Key West, Fla., the site of original collections, and Ramrod Key, one of the lower Fla. Keys. Summer. Endemic. *Streptachne floridana* Chapm.; *A. ternipes* of Small.

TRIBE 10. *Eragrosteae*

1. Florets unisexual, dioecious; plants colonial,
    rhizomes or stolons creeping.

2. Blades less than 7 mm long, fascicled; spikelets three- to five-flowered, obscured in leafy tips of branches . . .   16. *Monanthochloe*
2. Blades 6–10 cm long; spikelets eight- to fifteen-flowered in exserted panicles .   17. *Distichlis*
1. Florets bisexual.
  3. Spikelets one-flowered; lemmas awned, three-veined . . . .   18. *Muhlenbergia*
  3. Spikelets several- to many-flowered.
    4. Lemmas awnless, entire.
      5. Culms stout; glumes three- to seven-veined; spikelets 10 mm wide or less . . . .   19. *Uniola*
      5. Culms slender; glumes one-veined; spikelets 3.5 mm wide or less . .   20. *Eragrostis*
    4. Lemmas awned by excurrent veins from notched apex.
      6. Panicle ample, branches flexuous; lemma two-lobed with excurrent midvein; blades elongate, linear .   21. *Tridens*
      6. Panicle with few remotely spreading branches; lemma cleft, midvein an exserted awn, sometimes longer than the lobes; blades short, attenuate . . . .   22. *Triplasis*

16. MONANTHÓCHLOE Engelm. KEY GRASS

Creeping, wiry-stemmed perennial with rigid, linear, falcate leaves. Plants dioecious; spikelets three- to five-flowered, uppermost rudimentary; rachilla tardily disarticulating in pistillate spikelets. Glumes wanting, lemmas convolute, rounded, constricted above, three-veined, paleae two-veined, narrow. One species, North America, W.I.

1. **M. littoralis** Engelm.—Culms 2–3 dm tall, tufted, spreading from knotted rhizomes, with copious fibrous roots. Leaves 7–14 mm long prominently veined, gray green, conduplicate, with overlapping sheaths; ligules about 0.5 mm long, ciliolate, base of blades minutely puberulent. Spikelets terminating the branchlets; staminate floret with long-exserted filaments and white anthers 3–4 mm long; pistillate florets with two plumose style branches arising from the narrow summit of the ovary. Fruiting lemmas about 6 mm long, caryopsis 2–3 mm long, gibbous, beaked, reddish brown. Tidal flats and coastal marshes, thrives in dense colonies, Fla., Tex., southern Calif., Mexico, Cuba.

17. DISTICHLIS Raf. SALT GRASS

Aerial stems from horizontal underground rhizomes, forming extensive mats on salt flats and wet prairies. Dioecious; pistillate spikelets commonly five- to ten-flowered or fewer; staminate spikelets eight- to fourteen-flowered or fewer; the rachilla of fruiting spikelets disarticulating between the florets and above the glumes. Glumes three- to seven-veined; margins hyaline. Six species, mostly America.

1. **D. spicata** (L.) Greene—Plants essentially glabrous; culms from rhizomes to 8 dm tall or less, leafy up to the inflorescence. Leaves distichous with overlapping sheaths, the lowest bladeless; ligule very short membranous; blades 4–16 cm long, linear-attenuate, involute. Panicle 4–10 cm long, congested, conspicuously stalked; spikelets compressed, sessile, imbricate in clusters, lemmas 3–6 mm long; paleae winged on keels in staminate florets; anthers 2 mm long. $2n = 40$. Fla., Tex. to B.C., Calif., Mexico, W.I., Central America, South America.

18. MUHLENBÉRGIA Schreb. Muhly

Mostly perennial tufted or rhizomatous grasses. Culms simple or profusely branched. Inflorescence a diffuse or contracted panicle. Spikelets one-flowered, rarely two-flowered, rachilla disarticulating above the glumes, lemma awned, keeled or convex; three-veined, acute, with or without two filamentous teeth at apex. About 100 species, U.S., South America, Japan, India.

1. Glumes awnless, or sometimes awn pointed;
   lemma awns without filamentous teeth at
   apex . . . . 1. *M. capillaris*
1. Glumes awned; lemma awn flanked by two
   filamentous teeth at base . . . . . 2. *M. filipes*

1. **M. capillaris** (Lam.) Trin.—Culms 5–8 dm long, cespitose, erect, slender, puberulent or glabrescent, with one to two nodes; sheath summit with two membranous auricles 2–4 mm long, shreddy at base; blades 1–4 dm long, flat, tardily involute at tips; innovations involute, numerous. Panicle in anthesis about one-half the entire length of the culms, purple or rarely white. Spikelets long pedicelled, without awns 2–4 mm long, puberulent; lemma 5–12 mm long, awn straight, scaberulous; callus minutely hairy. Low pinelands, coastal strand in wet sands, prairie, Fla., Tex., Kans., Ind., W.I., Mexico. Fall. *M. trichodes* Steud.

2. **M. filipes** M. A. Curtis—Differing from **M. capillaris** in stouter glabrate, rigid, shiny, culms, strongly involute leaves, with firm membranous sheath bases; awned glumes and two delicate teeth at the apex of the lemma. Coastal strand, prairie, south Fla., Tex., Miss., Ga., S.C. Fall. *M. capillaris* var. *filipes* (M. A. Curtis) Chapm.

19. UNÍOLA L. SEA OATS

Tall colonial grasses with strong rhizomes, elongate nodding panicles. Spikelets three- to many-flowered, strongly compressed, glumes and lemmas keeled, one to four lemmas empty at base and at apex, the middle ten or more floret bearing, rachilla articulate between florets. Nine American species.

1. U. paniculata L.—Culms to 2 m tall, stout, glabrous, leafy throughout. Basal leaves numerous, elongate; blades narrowly lanceolate, involute, with long-attenuate flexuous tips; sheaths shorter than internodes; ligule a fringe of hair. Spikelets 2–3 cm long, 8–10 mm wide, lanceolate, ovate, stramineous or purplish; lemmas closely imbricate, polished; lateral veins three, keel scabrous; paleae sulcate with ciliolate margins. 2n = 40. Coastal plain, sea dunes, Fla., Tex., Va., W.I., tropical America.

20. ERAGRÓSTIS Host LOVE GRASSES

Annual or perennial grasses with flat leaves. Panicles usually open; rachilla articulate above the glumes and between the florets or continuous. Spikelets few- to many-flowered. Lemmas deciduous; paleae commonly persistent; glumes one-veined; lemmas prominently three-veined; paleae two-veined, glabrous or ciliate on the keels. About 100 species, widely distributed in temperate and tropical regions.

1. Rachilla disarticulating between the florets;
    palea keels ciliate, plants annual.
    2. Panicle contracted; blades 2–3 mm wide   1. *E. ciliaris*
    2. Panicle diffuse; blades 3–5 mm wide . .   2. *E. tenella*
1. Rachilla continuous; palea keels scabrous.
    3. Annuals.
        4. Stoloniferous; creeping by nodal roots;
            spikelets three- to ten-flowered . .   3. *E. hypnoides*
        4. Nonstoloniferous; erect or ascending;
            spikelets to twenty-flowered . . .   4. *E. simplex*
    3. Perennials.
        5. Panicles longer than wide; rachis and
            axis persistent.
            6. Panicles to 5.5 dm long; strict . .   5. *E. domingensis*
            6. Panicle to 15 cm long; usually more
               or less open.
               7. Spikelets closely imbricate, 2–2.2
                   mm wide, grayish . . . . .   6. *E. nutans*
               7. Spikelets relatively loosely imbricate, 1.2–1.8 mm long, greenish

suffused with purple . . . .   7. *E. bahiensis*
5. Panicles as long as wide or nearly so; rachis and axis fragile.
  8. Spikelets divergent; pedicels elongate.
    9. Sheaths hirsute; spikelets purple .   8. *E. spectabilis*
    9. Sheaths glabrous; spikelets pale green . . . .   9. *E. elliottii*
  8. Spikelets nearly sessile, appressed against the axis, dusky maroon. .   10. *E. refracta*

1. **E. ciliaris** (L.) R. Brown—Culms slender, rigid in aggregate tufts to 3 dm tall. Blades 3–5 dm long, involute tips; sheaths pilose at summit. Panicle purple, to 15 cm long, lobulate with congested spikelets to 4 mm long. $2n = 20$. Sandy soil, disturbed sites, south Fla., Gulf coast to Tex., S.C., W.I., Mexico, South America, Africa, Asia.

2. **E. tenella** (L.) Beauv. ex Roem. & Schultes—Similar in habit to **E. ciliaris.** Primary panicle branches and pedicels 5–10 mm long, floriferous at apices, panicle moderately diffuse, plumose, to 4.5 cm wide; mature spikelets reddish. Disturbed sites, dry soil, Fla. to Tex., tropical America, naturalized from the tropics. *E. amabilis* (L.) Wight et. Arn. ex Nees

3. **E. hypnoides** (Lam.) BSP, CREEPING LOVE GRASS—Culms creeping, rooting at the nodes; foliage and axes puberulent or glabrous. Blades 2–3 cm long, ligule a fringe of short hair. Peduncles solitary or naked, below the panicle, spikelets twenty-four- to forty-flowered in more- or less-capitate clusters on panicle branches, lemma 3 mm long, palea 1 mm long, tardily deciduous; caryopsis 0.5 mm long. Moist places, south Fla., W.I., South America.

4. **E. simplex** Scribn.—Plant to 3 dm tall, aggregated in tufts, culms decumbent, leafy below the middle. Blades 3–5 cm long, sheaths glabrous, hirtellous on margins, throat pilose, ligule of sparse hairs. Spikelets to 10 mm long, clustered on distant primary branches, florets twenty or more, lemma keels prominent. Disturbed areas, dry soil, south Fla., Ga., Ala.

5. **E. domingensis** (Pers.) Steud.—Perennial pale green, essentially glabrous. Culms to 1.5 m tall, purplish at nodes, pilose at throat; basal leaves few, cauline leaves three or more with overlapping sheaths, silvery on adaxial surface; blades to 4 dm long, 2–5 mm wide, flat at base, long attenuate, involute toward apex, stiff, veins flattened, minutely scaberulous on sides; ligule less than 0.5 mm long, panicle 3–5.5 dm long, strict with appressed racemes, the lowest one or two remote; axes wiry, minutely scabrous. Spikelets short pedicelled, appressed, nine- to twenty-flowered, glabrous; glumes 1 mm long, keeled, lemma 1.5–1.8

mm long, ovate-acute, deciduous with the grain; palea a little shorter, scaberulous on the veins, persistent. Fruit amber, 0.5–0.8 mm long, obliquely truncate at base, slightly contracted at apex, finely striate. Strand vegetation, Plantation Key, Fla., the only known site in North America; tropical America; Cuba; W.I.; Mexico.

6. **E. nutans** (Retz.) Nees ex Steud.—Plant to 6 dm tall, essentially glabrous; the lower nodes geniculate; internodes shorter than the sheaths; blades 7–12 cm long, tapering to attenuate tips; panicle long peduncled, nodding, racemes ascending; spikelets 5–10 mm long, eight-to twenty-flowered, slender pedicelled; lateral nerve of lemma terminating in margin. South and central Fla., naturalized from southeast Asia. *E. chariis* (Schultes) Hitchc.

7. **E. bahiensis** Schrad.—Culms to 6 dm tall, tufted in rhizomatous knotty clumps; sheaths overlapping or shorter than internodes, glabrous, hairy on the collar; blades pilose above; panicle branches condensed; spikelets to thirty-flowered, short pedicelled, glumes subequal, 2 mm long, acute; lemmas 2 mm long, narrow near the middle; paleae prominently persistent. South Fla., Ala., La.

8. **E. spectabilis** (Pursh) Steud. PURPLE LOVE GRASS—Culms leafy, in rhizomatous tufts to 6 dm tall, spreading and wiry; sheaths hirsute or glabrate, conspicuously pilose at throat; panicle purple, diffuse fragile, more than half the height of the plant; spikelets linear, six- to twelve-flowered, long pedicel; axils pilose; lemmas 1.8 mm long, scabrous, acute lateral nerves ending submarginally, paleae keels scabrous. Sandy soil, disturbed sites, Fla. to Ariz., Me., Minn., Mexico.

9. **E. elliottii** Wats.—Culms leafy, tufted, aggregated into large clumps, to 6 dm tall with panicles about two-thirds of the height; sheaths glabrous, overlapping; throat pilose, blades elongate; panicle diffuse, branches capillary fragile, floriferous at tips; spikelets to eighteen-flowered, to 12 mm long, pedicels elongate, spreading; lemmas inbricate, acute, margins scabrous; caryopsis 0.6 mm long. Fla. to Tex., N.C., W.I., Mexico.

10. **E. refracta** (Muhl.) Scribn.—Culms erect glabrous to 7 dm tall, tufted from knotty bases; sheaths glabrous, pilose at throat; blades pilose at base, distally glabrous; mature blades 8–20 cm long, 3–4 mm wide, attenuate, becoming flexuous in age, panicle diffuse, half as long as the culm or more; branches minutely scabrous, axils copiously pilose; spikelets disposed towards the ends of the branches; florets twenty-three or fewer, plumbaginaceous, or reddish; lemmas 2.5 mm long, acuminate, three-nerved, the lateral nerves usually not reaching the whitish margins; palea keels minutely scabrous; grain readily falling from the

spikelet with the lemma; seed amber, 0.8 mm long. Pine flatwoods in moist sand, Fla. to Tex., Del. on coastal plain.

21. TRÌDENS Roem. & Schultes

Perennial grasses with erect, tufted culms. Inflorescence a panicle, spikelets several-flowered, glumes unequal, the outer one-veined, the inner three- to five-veined; midvein scabrous, excurrent, or ending in margin; lemmas dorsally convex, apices erose-hyaline, two-lobed, with three excurrent veins; or obtuse, the veins ending in margin; paleae two-veined, glabrous or minutely scabrous-ciliate. About 30 species of temperate regions.

1. Lateral veins ending in margin; spikelets to
   5 mm long . . . . 1. *T. eragrostoides*
1. Lateral veins excurrent; spikelets to 8 mm
   long . . . . 2. *T. flavus*

1. **T. eragrostoides** (Vasey & Scribn.) Nash—Culms to 1 m tall, tufted. Blades 10–25 cm long, linear-attenuate, to elongate setaceous tips; sheaths longer than internodes, like the blades minutely scabrous-striate on ribs; ligule white hyaline, 2–3 mm long. Panicle long exserted; diffuse; with distant, spreading to reflexed branches; spikelets mostly six- to ten-flowered; lemmas 2 mm long, pubescent on marginal veins with excurrent midvein; caryopsis 1.2 mm long, compressed concave, with style remnants. South Fla., Key West to Tex., Ariz., Mexico, Cuba.

2. **T. flavus** (L.) Hitch. TALL REDTOP—Culms 1 m tall or taller, erect; basal sheaths keeled, compressed. Blades to 10 mm wide, linear-attenuate, glabrous. Panicle long exserted, spreading, open 15–30 cm long or less with pulvinate, pubescent axils; branchlets viscid; capillary flexuous with numerous appressed spikelets of eight or fewer florets; glumes acutish, mucronate, the first shorter than the second, semitranslucent, suffused with red; lemmas 4 mm long, pubescent on the callus, keel and margins below the middle with three excurrent veins; paleae shorter, patently gibbous in profile. Open hardwoods, pinelands, and old fields, south Fla. to Northeast, Nebr., Tex.

22. TRIPLÀSIS Beauv. SAND GRASSES

Slender tufted annuals fragile at the nodes. Panicle branches remote; spikelets Y-shaped, few-flowered, disarticulation above the glumes; rachis slender, glumes acute, the outer shorter; lemmas two-lobed, three-veined, the midvein excurrent as an awn between the lobes, margins white, silky; paleae hyaline, hirsute anthers and stigmas deep purple, cleistogamous floret within lower sheath. Three species in the U.S.

1. **T. purpurea** (Walt.) Chapm.—Culms slender to 8 dm tall from the crown of the taproot; internodes short, surpassing the sheaths at least above; nodes bearded, sheaths and blades puberulent and pilose in nodal areas. Panicle red, commonly with three racemes; pedicels short, capillary; spikelets two- to four-flowered; lemmas 3–4 mm long; the awn slightly longer than the rounded lobes. Caryopsis compressed, 2 mm long. Often in extensive colonies in coastal sands and sand hill associations, Fla., Tex., Central America.

TRIBE 11. Chlorideae

1. Inflorescence racemose on elogate spike
   rachis.
  2. Rachis terminating with the ultimate
     spikelets; fertile florets one or more.
    3. Rachilla, bearing a floral rudiment;
       spikes numerous, distant; spikelets
       appressed . . . .   23. *Gymnopogon*
    3. Rachilla not prolonged.
      4. Spikelets two- or more-flowered,
        spreading in maturity; racemes lax;
        grain falling without glumes . .   24. *Leptochloa*
      4. Spikelets one-flowered; racemes im-
        bricate, strongly unilateral; grain
        falling with glumes . . . . .   25. *Spartina*
  2. Rachis prolonged beyond the ultimate
     spikelet; with staminate floral rudi-
     ments; fertile floret one . . . . .   26. *Bouteloua*
1. Inflorescence an aggregate of digitate spikes.
  5. Spikelet with one floret, grain free be-
     tween lemma and palea.
    6. Rachilla produced back of the palea to
      a naked stipe or with one lemma;
      culms slender, sheaths not strongly
      compressed; colonial by leafy stolons   27. *Cynodon*
    6. Rachilla produced, bearing club-shaped
      floral rudiments; culms and sheaths
      strongly compressed, rarely stoloni-
      ferous . . . .   28. *Chloris*
  5. Spikelets with two or more florets, seed
     free from pericarp.
    7. Glumes pectinate, awned; culms de-
      cumbent, repent at base . . . .   29. *Dactyloctenium*
    7. Glumes smooth, awnless; not pectinate,
      culms geniculate, spreading at base .   30. *Eleusine*

23. GYMNOPÒGON Beauv. BEARD GRASSES

Perennials with short flat leaves to 15 mm wide, with imbricate sheaths. Culms short, overtopped by elongate wide-spreading diffuse panicle. Spikes slender, distant, spreading at right angles from the peduncle rachis; spikelets one- to three-flowered, remote in two rows along continuous rachis. Rachilla articulate above the glumes, prolonged behind the paleae to a naked tip, rarely with a rudimentary floret. Glumes one-veined, narrow, keeled, usually exceeding the uppermost floret. Lemmas bidentate, awned between the teeth. About ten species, mostly America.

1. G. floridanus Swallen—Plant to 4 dm tall in small tufts, commonly purple throughout, spikes floriferous to the base. Lemma awns 1–3 mm long, the body 2–2.2 mm long. Mature spikes 8–18 cm long, reflexed. Pinelands and sandy prairies, Fla., Gulf coast, endemic.

24. LEPTÓCHLOA Beauv. SPRANGLE TOP

Annuals or perennials with flat leaves and numerous spikes or racemes disposed on a common axis. Spikelets two- to several-flowered, compressed, alternating in two rows on the same side of rachis. Glumes unequal, one-veined; lemmas three-veined, glabrous or pubescent on the veins, obtuse, bifid, awned or mucronate. About 20 species in warm regions of both hemispheres. *Diplachne* Beauv.

1. Plants perennial.
  2. Lemmas bifid, lateral veins glabrous . .   1. *L. dubia*
  2. Lemmas acute, lateral veins pubescent .   2. *L. virgata*
1. Plants annual.
  3. Sheaths pilose; spikelets three- to four-
    flowered . . . .   3. *L. filiformis*
  3. Sheaths glabrous or scabrous; spikelets
    six- to twelve-flowered . . . . .   4. *L. fascicularis*

1. L. dubia (HBK) Nees, GREEN SPRANGLE TOP—Plants to 1 m tall, tufted, rhizomes knotty, culms wiry, glabrous, leafy. Blades 3–4 mm wide, long attenuate. Sheaths pilose at throat, ligule a fringe of hairs. Panicle flexuous, spikelets five- to eight-flowered, distant; lemmas lustrous, spreading from zigzag rachis. $2n = 40,80$. A polyploid species. Fla., U.S. to Mexico, South America. Summer, fall.

2. L. virgata (L.) Beauv.—Plants to 7 dm tall, branching above; internodes minutely scabrous; sheaths glabrous; ligule membranous. Blades elongate-attenuate, involute at tips. Panicle of numerous fili-

form spikes on zigzag rachis; spikelets imbricate, appressed, two- to four-flowered; glumes unequal, inner, the longer, 2 mm long; lemmas 1.5–2 mm long, awn pointed. Fla., Tex., tropical America.

3. L. filiformis (L.) Beauv. RED SPRANGLE TOP—Culms annual, 5–7 dm tall, geniculate, branching above the base; sheaths papillose-pilose, loose. Blades thin, flat to 10 mm wide, ligule lacerate, membranous. Panicle elongate, nodding; spikes distant, rachis filiform; spikelets 2 mm long, glumes acuminate, lemmas 1.5 mm long, awnless, grain amber, 0.8 mm long. Sandy soil, disturbed sites, tropical America, widely naturalized in U.S., Fla. to Calif.

4. L. fascicularis (Lam.) Gray—Annual, glabrous leafy plant to 8 dm tall, purplish at base; culms many from basal nodes. Sheaths loose, blades short, involute in young plants becoming elongate, flat in age; ligule hyaline, lacerate. Panicle stiffly erect with ascending spikes with overlapping spikelets; florets five, usually with rudiments at rachilla apex; lemmas keeled, 3–4 mm long, awn pointed. Frequent in swamps and mangrove shores, Fla., widely distributed in U.S., South America.

25. SPARTINA Schreb. CORDGRASS

Stout perennials creeping, spreading by scaly rhizomes; or rhizomes vertical in clump-forming plants. Spikelets one-flowered, laterally compressed, sessile, imbricated on one side of a continuous rachis, deciduous below the glumes; rachis apex rarely free; rachilla not produced beyond the ultimate floret; glumes unequal, keeled, pointed or short awned, the second longer than lemma; lemma membranous, only the midvein conspicuous; palea two-veined. About 14 species, largely America.

1. Plants without creeping rhizomes, tufted; spikes numerous.
    2. Inflorescence terete, spikes appressed, imbricate . . . . 1. *S. spartinae*
    2. Inflorescence not terete, spikes spreading.
        3. Blades less than 4 mm wide; second glume acute . . . . 2. *S. bakerii*
        3. Blades 5–15 mm wide; second glume obtuse . . . . 4. *S. alterniflora*
1. Plants with creeping rhizome; spikes few to several . . . . 3. *S. patens*

1. S. spartinae (Trin.) Merrill—Culms rigid to 2 m tall, polished, commonly in large tufts. Blades 3–4 mm wide, strongly involute. Inflorescence 8–30 cm long with twisting spikes; glumes keeled, glabrous, or hispid-ciliate. $2n = 40$. Marshy habitats, Gulf coast, Fla., Tex., Mexico.

2. **S. bakerii** Merrill—Culms to 2 m tall, thick, in large, dense tufts. Blades strongly involute, attenuate from a base 4–5 mm wide. Spikes several, overlapping, 3–5 cm long; spikelets densely appressed, twisted, 6–8 mm long; inner glume acute, longer than lemma, hispid on keel; outer glume half as long as lemma, scabrous on keel. Sandy soils on margins of swales, pinelands, Fla., Ga., S.C.

3. **S. patens** (Ait.) Muhl. SLENDER CORDGRASS—Culms commonly less than 1 m tall, in dense colonies in brackish sands; rhizomes strong, slender, remotely scaly. Leaves narrow, attenuate, involute, rigid. Spikes three to five, spreading, nearly sessile; spikelets 8–10 mm long; second glume acute, exceeding the lemma, scabrous on keel; first glume less than one-half the length of lemma. Saline marshes, Fla. to Quebec.

4. **S. alterniflora** Loisl. SMOOTH CORDGRASS—Culms 0.5–2.5 m tall, 1 cm or more in diameter at the spongy base; leaf blades flat, involute at tips, usually glabrous throughout. Spikelets 10–13 mm long, glabrous, glumes rarely hispid on keels at summit, alternate on rachis. Anthers and stigmas white in anthesis. In brackish waters of coastal lagoons and marshes. New England to Fla., Tex., Pacific Northwest. Summer.

26. BOUTELÒUA Lag. GRAMA GRASS

Perennial; spikelets with one functional flower, sessile, in two rows on the same side of a continuous rachis, produced to a sterile tip; glumes one-veined, unequal; lemmas commonly longer than the second glume, three-veined, the veins extending to narrow lobes as awns; plants with slender stems. About 38 species, mostly North America. *Triathera* Desv.

1. **B. hirsuta** Lag. HAIRY GRAMA GRASS—Culms 2–4 dm tall, tufted on knotty rhizomatous base, simple, erect or sometimes geniculate at nodes above. Leaves mostly basal with elongate blades, pustulate-hirtellous above the hairy lingule; sheaths imbricate, shorter than internodes. Spikelets one or two, pectinate, 1–2.5 cm long, terminal, distant; outer glume black tuberculate, pointed, sterile florets various. Caryopsis 1.2 mm long, gibbous trigonous, minutely mucronate. $2n = 20,36$. Fla. to Tex., Wis., Calif., Mexico.

27. CÝNODON Richard BERMUDA GRASS

Creeping perennial lawn grass, freely escaping. Spikelets one-flowered, sessile, appressed in two rows against the continuous rachis, glumes one-veined; lemmas three-veined, compressed, keeled, ciliolate on marginal veins. About six species of warmer regions. *Capriola* Adans.

1. **C. dactylon** Pers.—Culms 1–2 dm tall or taller from strong scaly rhizomes; stolons. Leaves mostly basal with imbricate sheaths, pilose at

throat, ligule of hairs; blades 3–6 cm long; cauline leaves few. Spikes terminal 2.5–5 cm long; glumes spreading in anthesis, lemmas cymbiform, paleae one-veined, a little longer than the prolonged rachilla rudiment. Apparently a polyploid complex with $2n = 18, 36, 40, 36 + 2B$ chromosomes. Southern U.S., naturalized from Europe.

28. CHLÓRIS Sw.   FINGER GRASS

Tufted perennials, or annuals with flat or folded scabrous blades and radially aggregated spikes at the culm summit. Spikelets with one bisexual floret, sessile, disposed in two rows along one side of a continuous rachis, the rachilla disarticulating above the glumes and produced beyond the bisexual floret, with one or more sterile florets. Glumes unequal, the outer shorter and narrower than the inner one, acute; lemma keeled, one- to five-veined, sometimes awned between the apical teeth, awns slender or reduced to a mucro. About 60 species in warmer regions of both hemispheres. *Eustachys* Desv.

1. Lemma mucronate; sheath orifice glabrous.
  2. Spikelets about 2 mm long.
    3. Plant stoloniferous; lemma short ciliate   1. *C. petraea*
    3. Plant tufted; lemma glabrous . . .   2. *C. glauca*
  2. Spikelets 3 mm long . . . . . .   3. *C. neglecta*
1. Lemma awned; sheath orifice pilose.
  4. Lemma copiously silky ciliate; awn about
     3 mm long . . . .   4. *C. polydactyla*
  4. Lemma not manifestly ciliate; awn to 18
     mm long . . . .   5. *C. radiata*

1. **C. petraea** Sw.—Culms 5 dm tall or taller, green or glaucous, decumbent; sheaths compressed, strongly keeled, distichous, usually reddish below. Spikes commonly four to six; mature spikelets shiny, blackish brown. $2n = 40$. Common throughout fields, sandy woods, waste areas, Fla., Tex., tropical America, coastal plain states to western Tex.

2. **C. glauca** (Chapm.) Wood—Plant in large tufts 8–15 dm tall, with glaucous foliage and culms; culms flat edged, 5 mm wide. Leaves many in basal tufts to 7 dm long, 5–10 mm wide; blades 3.5 dm long, equaling or exceeding the sheaths; successive internodes elongate with fascicles of fewer leaves. Spikes numerous; matured spikelets blackish brown. Marshy sites, low, flat pineland, Fla., coastal plain states.

3. **C. neglecta** Nash—Similar to **C. petraea** in habit; mature spikelets golden brown and slightly longer. Sandy woods and swamps, endemic to Fla.

4. C. polydactyla (L.) Sw.—Culms to 1 m long, compressed, not flat; sheaths loose, the lowermost with short blades; upper nodes geniculate. Cauline blades 7–9 mm wide, 3 dm long, attenuate at tips; spikelets many, flexuous; peduncles 2 mm long, pilose; mature spikelets stramineous. $2n = 72$. Sandy soil, south Fla., W.I., South America.

5. C. radiata (L.) Sw.—Tufted, decumbent, branching annual; sheaths distichous, keeled, compressed. Blades flat, attenuate, thin, 2–3 mm wide. Spikes ten or more, aggregate, outcurving; 3–6 cm long; spikelets silvery gray, appressed hairy on veins or glabrous; lemmas narrow, 2.5 mm long, awn straight to 10 mm long; floral rudiment slightly divergent, short awned. $2n = 40$. Waste places in Fla., U.S., naturalized from tropical America.

29. DACTYLOCTÉNIUM Willd. CROWFOOT GRASS

Culms soft, compressed with ascending branches forming radiate colonies. Spikelets three- to five-flowered, sessile, imbricated in two alternating rows on continuous three-sided rachis, dorsally angled, ventrally flattened, floret bearing. Glumes unequal, the outer persistent and keeled, one-veined, awned; the inner, deciduous. Lemmas cymbiform, awn pointed, three-veined; paleae sulcate with scabrous keels. Seeds free from pericarp, orange, minutely ridged. About three species, naturalized from Eurasia.

1. D. aegyptium (L.) Richt.—Culms to 6 dm long, decumbent creeping, rooting at base. Spikes three to six, terminal, spreading radially, distant, from the uppermost leaf; sheaths glabrous-pilose at throat; ligule membranous; blades glabrous or pustulate-hirtellous. $2n = 40$. Common in warm parts of both hemispheres; naturalized in America, Fla. to Me., Ariz., Mexico, W.I.

30. ELEUSÍNE Gaertn. YARD GRASS

Tufted annuals. Spikelets two- to three-flowered, sessile, compressed, awnless, crowded on one side of the rachis in two rows. Spikes two to seven, mostly subdigitate, 6–9 cm long; glumes unequal, acute. Lemma keeled, three-veined. Seed within pericarp, 1 mm long, bright brown, apex truncate, pointed at hilar end, sides three-angled, transversely ridged. About six species in Old World.

1. E. indica (L.) Gaertn.—Culms 1.5–4 dm tall, leafy, glabrous. Blades flat, sometimes overtopping the culms, ligule membranous; sheaths loose, overlapping, pilose at mouth, culms flattened, branching from the base, forming wide-spreading colonies. $2n = 18$. Fields and

waste places, naturalized in Fla., to Ariz., Calif., Kans., Mass., W.I.,
South America.

V. PANICOID GROUP

TRIBE 12. *Paniceae*

1. Fruiting spikelets subterranean; sterile
   spikelets aerial . . . .   31. *Amphicarpum*
1. Fruiting spikelets and sterile spikelets aerial.
   2. Spikelets within an involucre of free bris-
      tles or within a bur formed by free bris-
      tles or fused spinate segments.
      3. Involucre of bristles only, falling with
         spikelets . . . .   32. *Pennisetum*
      3. Involucre a bur, falling entire with
         spikelets . . . .   33. *Cenchrus*
   2. Spikelets without involucre.
      4. Spikelets falling free from one or more
         persistent, subtending rachial bristles   34. *Setaria*
      4. Spikelets not subtended by bristles.
         5. Spikelets partly embedded in cavities
            of a fleshy, articulate rachis   . .   35. *Stenotaphrum*
         5. Spikelets in inflorescences of slender
            axes, never fleshy.
            6. Sterile lemmas and glumes awned
               or awn pointed.
               7. Leaf blades linear, elongate
                  eight to twenty times longer
                  than wide.
                  8. Foliage glandular, sweet
                     scented; sterile lemma with
                     straight awn . . . . .   36. *Melinis*
                  8. Foliage not glandular.
                     9. Spikelets silky, fading red,
                        panicle open . . . .   37. *Rhynchelytrum*
                     9. Spikelets scabrous, green or
                        purplish, panicle con-
                        tracted . . . . . .   38. *Echinochloa*
               7. Leaf blades lanceolate, three to
                  four times as long as wide;
                  glumes awned . . . . .   39. *Oplismenus*
            6. Sterile lemmas and glumes awn-
               less.
               10. Mature lemmas chartaceous, in-
                   durate, margins not inrolled;
                   spikelets pubescent or silky;

inflorescence paniculate or
whorled . . . .   40. *Digitaria*
10. Mature lemmas rigid, polished,
chartaceous, with inrolled
margins.
11. Back of the fruit facing away
from the rachis spike.
12. Spikelets silky, fertile lemma
awned; callus annular . 41. *Eriochloa*
12. Spikelets glabrous; callus
wanting.
13. First glume present; spikes
racemose; culms uni-
formly leafy . . . 42. *Brachiaria*
13. First glume lacking; spikes
digitate; leaves basal . 43. *Axonopus*
11. Back of fruit facing the rachis
of spike.
14. First glume usually lacking,
rachis compressed, her-
baceous; second glume
and sterile lemma about
equal . . . . . .   44. *Paspalum*
14. First glume present.
15. Primary branches of panicle simple, spikelets in
two unilateral rows;
plants succulent aqua-
tics . . . . . .   45. *Paspalidium*
15. Primary branches of panicle rebranched; spike-
lets in symmetrical ra-
cemes; plants not suc-
culent aquatics.
16. Second glume saccate,
gibbous, many-veined   46. *Sacciolepis*
16. Second glume convex.
17. Culms woody, branch-
ing, high rambling,
viny; grain with
apical depression,
tufted with villous
hairs . . . .   47. *Lasiacis*
17. Culms herbaceous,
erect or creeping;
grain with tightly
inrolled apex with-
out hairs, branch-
ing basal or upper
nodes . . . .   48. *Panicum*

31. AMPHICÁRPUM Kunth BLUE MAIDEN CANE

Annual or perennial stoloniferous grasses with flat leaves. Panicles narrow, aerial and subterranean. Aerial spikelets bisexual but sterile; underground spikelets cleistogamous and fertile. Outer glume of the aerial spikelet small or lacking; inner glume and sterile lemma subequal, lemma and palea indurate. Fertile spikelet without outer glume, the inner glume and sterile lemma prominently nerved; fertile lemma and palea strongly indurate. Two species, restricted to southeastern U.S.

1. **A. muhlenbergianum** (Schultes) Hitchc.—Decumbent perennial to 8 dm tall; culms tufted, rooted at nodes of the widely creeping scaly stolons. Blades 3–7 cm long, linear-lanceolate, firm, white cartilaginous margins; sheaths sparsely hirtellous-pustulate, at least when young; ligule obsolete. Sterile spikelets 5–7 mm long; fertile spikelets larger on slender subterranean branches. Low, wet pinelands, prairies, central and south Fla. Fall. *Amphicarpum floridanum* Chapm.

32. PENNISÉTUM Richard NAPIER

Annual or perennial robust grasses in large clumps or colonies, foliage mostly pustulate-pilose. Panicles dense, cylindrical, rather small for a grass so large. Spikelets solitary or in groups of twos and threes, subtended by an involucre of bristles, only united at base, falling entire; outer glume short, inner glume equaling sterile lemma. About 30 species, tropical Africa and India.

1. **P. purpureum** Schum.—Culm to 4 m tall, copiously branching from the base. Blades 5–9 dm long, 2–3 cm wide, flat linear-attenuate with prominent white midrib, scabrous, margins white cartilaginous, pilose at auriculate base; ligule hairs 2–3 mm long; nodes hairy or glabrate. Mature spikes purplish or tawny, 17–18 cm long, 4 cm in diameter including bristles; spikelets reflexed, rachis spreading, hairy; short bristles scabrous, long bristles plumose; spikelets in group of threes, lateral pedicelled. Forage plant escaping to glades and roadsides, central and south Fla., W.I., Mexico, introduced from Africa. Fall.

33. CÉNCHRUS L. BURGRASS; SANDBUR

Annuals or perennials with compressed or terete culms, erect or prostrate; blades flat, sheaths sometimes inflated, nodes often swollen; ligules hairy fringes. Spikelets solitary or a few together within a spiny burlike involucre of connate segments or free bristles, articulate with peduncle, deciduous as a whole. Spines barbate, usually spreading, reflexed or erect, germination within the bur. About 25 species, temperate, tropical, and subtropical regions. *Cenchropsis* Nash

1. Spikelets subtended by basally connate bristles; ligneous or suffrutescent perennial .   1. *C. myosuroides*
1. Spikelets subtended by an involucre or coalesced, spinate segments with or without whorls of basal bristles; annual or perennial herbs.
  2. Burs with whorls of basal bristles.
    3. Burs globose; at least some of the basal bristles surpassing the tips of inner spines; burs stramineous . . .   2. *C. brownii*
    3. Burs truncate at base, wider than thick; none of the basal bristles surpassing the tips of inner spines; burs soon turning purplish . . . .   3. *C. echinatus*
  2. Burs without whorls of basal bristles.
    4. Leaves 2–3 mm wide; body of burs glabrous, to 5 mm wide . . . .   4. *C. gracillimus*
    4. Leaves to 7 mm wide or wider.
      5. Spines relatively few, twenty-five or less; bur body finely pubescent .   5. *C. incertus*
      5. Spines to thirty or more; bur body densely pubescent, or glabrate in maturity.
        6. Bur body with spines to 15 mm thick; spines slender; spikelets two or three . . . .   6. *C. longispinus*
        6. Bur body with spines to 17 mm thick, spines stout; spikelets one or two . . . . . .   7. *C. tribuloides*

1. **C. myosuroides** HBK—Reedlike perennial with culms to 1.5 m long; internodes polished, exceeding the sheaths; branches fasciculate, spreading, geniculate from the lower nodes. Blades 5–12 mm wide, retrorsely appressed, scabrous on margins. Racemes 10–25 cm long; bur with solitary spikelet 4–5 mm long; bristles free above the base, very scabrous. $2n = $ c.68. Fla., Caxambas Pass, Marco Island, Tex., La., Ga., tropical America. Fall. *Cenchropsis myosuroides* Nash

2. **C. brownii** Roem. & Schultes—Erect annual from base to 3–9 dm tall. Blades 0.8–3 dm long, to 12 mm wide, flat, sparingly pilose above or glabrous; sheaths exceeded by internode. Racemes 4–8 cm long, densely congested with contiguous burs 5–8 mm long; spines in mature burs of inner circle clasping across the summit between the

spikelets; spikelets one or two, 4–6 mm long. Rare in Fla., Key West and Key Largo; common in tropical America. *C. viridis* Spreng.

3. **C. echinatus** L.—Annual with compressed, geniculate culms branching from the base, to 7 dm long. Blades 4–8 mm wide, pilose on upper surface; sheath margins and orifice pilose or glabrate. Racemes 4–10 cm long; bur compressed, 4–7 mm long, 4–8 mm wide; spikelets three or four, bristles spreading to reflexed; tips of spines strongly clasping across the summit between the spikelets or sometimes only bending inward. Open sandy ground, widely distributed, Fla., La., Ga., tropical America. *C. insularis* Scribn. & Millsp.

4. **C. gracillimus** Nash—Tufted slender, wiry, perennial to 2–5 dm tall. Blades pilose or glabrous above at base, drying involute. Racemes 3–7 cm long; burs 4–6 mm long, lobes flat, erect, becoming reflexed, glabrous, spikelets usually two or three per bur. Dry open ground, disturbed pinelands, Fla., Ala., Miss., W.I. Summer.

5. **C. incertus** M. A. Curtis—Perennial with ascending leafy shoots from decumbent purplish base to 1 m long; culms often prostrate, spreading, rooting at nodes. Basal internodes short, often geniculate; blades pilose above the ligule. Racemes 3–8 cm long; bur 3–5 mm long, puberulent or glabrate, segments flat, ciliate or glabrous; spikelets one to three. Open sands, upper beaches or cleared areas, on sand hills, Fla., Tex., Va. Summer. *C. pauciflorus* Benth.

6. **C. longispinus** (Hack.) Fern.—Annuals in large tufts; culms terete to 7 dm long or less, branching, decumbent and geniculate at base; sheaths compressed, keeled, pilose along margins and at orifice; ligule hairs 1.6 mm long. Blades pilose or glabrate, harsh, to 18 cm long, 6 mm wide; spines grooved on margins. Coastal strand, south Fla., Ala. to Ky., N.C. Summer.

7. **C. tribuloides** L.—Similar to preceding taxon. Sheaths relatively loose, blades 1 cm wide, coriaceous, dorsally scabrous, overlapping. Burs 1–2 cm thick, lanate-villous, as are the flattened spines toward the base. Coastal plain, Fla. to N.J. Summer.

34. SETÀRIA Beauv.   Foxtail Grasses

Annual or perennial grasses with spikelike panicles, flat, linear leaves; ligule consisting of hairs. Spikelets awnless, subtended by one to several bristles, deciduous above them; first glume three- to five- nerved; second glume about equal to sterile lemma or shorter, five- to seven-nerved; fertile lemma indurate, hard, transversely rugose or glabrous. About 140 species, warm temperate, tropical, and subtropical regions. Most abundant in warm, tropical regions. *Chaetochloa* Scribn.

1. Subtending bristles five or more; fertile
   lemma rugose; rhizomatous perennials;
   spikelets 2.5 mm long, blades 5–7 mm
   wide . . . .   1. *S. geniculata*
1. Subtending bristles one to three; fertile
   lemma glabrous or rugulose.
   2. Panicle terete, 3–10 cm long; plant
      slender; blades 8–10 mm wide . . .   2. *S. corrugata*
   2. Panicle lobulate, 5 dm long or less; plant
      robust.
      3. Annual; bristles obscuring the spikelets;
         blades 2–3 cm wide . . . . .   3. *S. magna*
      3. Perennial; bristles not obscuring the
         spikelets.
         4. Rachis scabrous or sparingly inter-
            mixed with villous hairs; blades
            glabrous, 1–2 cm wide . . . .   4. *S. macrosperma*
         4. Rachis scabrous densely intermixed
            with long villous hairs; blades pu-
            bescent above and below, to 12 mm
            wide.
            5. Blades 12 mm wide . . . . .   5. *S. setosa*
            5. Blades less than 5 mm wide . .   6. *S. rariflora*

   1. S. geniculata (Lam.) Beauv.—Culms mostly erect, wiry from
knotty rhizomes to 8 dm tall. Blades flat, erect 7–14 cm long, 4–7 mm
wide, glabrous. Panicle 3–5 cm long, rachis puberulent; bristles yellowish or purple; lower floret neuter. Pinelands, shores, disturbed areas,
coastal plain states, Fla., Tex., Calif., tropical America.
   2. S. corrugata (Ell.) Schultes—Culms to 8 dm long, branching and
spreading at geniculate base. Blades scabrous and commonly pilose
above, to 1 cm wide, margin white, cartilaginous; midnerve often white;
sheaths pilose on margins and at throat; ligule of hairs. Panicle 3–5 cm
long, rachis hispid-villous; bristles to 2 cm long, purplish when mature;
spikelet 2 mm long, prominently rugose. Fields, waste ground, coastal
plain states, Fla., Cuba. Summer.
   3. S. magna Griseb.—Robust grass to 4 m tall, sparingly branched;
base of culm to 2 cm thick; sheaths shorter than the internodes. Blades
2–5 dm long, antrorsely scabrous, midnerve white, lateral nerves prominent. Panicle 2–3 dm long, densely lobulate, nodding; bristles green,
1–2 cm long; spikelets 2 mm long, fruit glabrous. Marshy lake shores,
ditches, coastal plain, Fla., Tex., Cuba. Summer.
   4. S. macrosperma (Scribn. & Merrill) Schum.—Culms to 1.5 m
tall, tufted, branching at base. Blades elongate, linear, 3–4 dm long,

antrorsely scabrous on margins; sheaths overlapping, at least below; pilose at throat. Panicle 15–23 cm long, lanceolate, branches spreading to 15 mm long; bristles green or purplish; spikelets 3 mm long, fertile lemma finely regulose. Hammocks, coral sands, Fla., Bahama Islands. Summer.

5. **S. setosa** (Sw.) Beauv.—Culms erect, wiry, commonly less than 10 dm tall, becoming decumbent and often suffrutescent below. Blades flat, stiffly spreading to 10 mm wide, pubescent. Panicle to 20 cm long, usually narrow or an open spike; spikelets 2 mm long, bristles 5–10 mm long, relatively short; fertile lemma acute, finely cross-ridged. Open, clear margins, adventive from W.I., on ballast in Camden, N.J., and Key West. Summer.

6. **S. rariflora** Mikan ex Trin.—Differs from the preceding species by narrower, more-interrupted or lobulate panicles with fewer spikelets, and narrower leaf blades. Collected at Key West, rare. *Chaetochloa rariflora* Hitchc. & Chase

S. glauca (L.) Beauv.—Annual; spikelets 3–3.5 mm long; lower floret staminate; culm decumbent, branching from the base. May occur in the area wastelands. Widely distributed in North America, Mexico, W.I. $2n = 36$. *S. lutescens* (Weig.) Hubb.

35. STENOTÀPHRUM Trin.   St. Augustine Grass

Creeping stoloniferous perennial with obtuse blades, axillary and terminal spikes. Spikelets embedded in two alternating rows on one side of thickened, compressed rachis, falling with disarticulating joints. Outer glume minute, inner glume and sterile lemma similar. About seven species, tropical and subtropical.

1. **S. secundatum** (Walt.) Kuntze—Culms geniculate, compressed, with internodes longer than the sheaths, prostrate; rooting at lower nodes; erect flowering shoots to 3 dm high. Propagated as lawn grass by its leafy stolons. Escaping from cultivation, Fla., Tex., Calif., tropical America. Fall.

36. MELÍNIS Beauv.   Molasses Grass

Viscid perennial, branching from the lower nodes. Spikelets dorsally compressed, one-flowered. Sterile lemma below the fertile floret, rachilla disarticulating below the glume. Outer glume minute, inner glume similar to sterile lemma, prominently nerved, somewhat longer than the floret; fertile lemma and the palea subhyaline above the middle. Known for its characteristic odor and oily foliage. One species of South America, 17 of tropical Africa.

1. M. minutiflora Beauv.—Culms to 1 m tall with decumbent, geniculate base; sheaths copiously spreading hirsute, blades 8–20 cm long, 10 mm wide, flat, with prominently white midrib, hirsute above and below; ligule a fringe of hair 1 mm long. Panicle tardily exserted, 10–14 cm long, with dense ascending branches; spikelets 2 mm long; sterile lemma two-lobed at apex; awn between the lobes to 10 mm long. Naturalized from Central and South America, south Fla., cultivated, becoming established.

37. RHYNCHELÝTRUM Nees NATAL GRASS

Perennial or annual grass with open panicle silky. Spikelets alike on slender pedicels; outer glume small, overtopped by long hairs; inner glume and sterile lemma equal, gibbous; hirsute, divergent at apex; fertile lemma navicular, surpassed by the glumes, cartilaginous, enclosing the margins of the palea, not tightly inrolled. About 37 species, tropical Africa, Madagascar, Arabia, to Indochina. *Tricholaena* Schrad.

1. R. repens (Willd.) C. E. Hubbard—Culms to 1 m tall, usually less, geniculate, decumbent at base, ascending rooting at lower nodes; blades flat linear-attenuate; sheaths shorter than internodes, pilose-papillose, ligule pilose. Panicle overtopping the leaves, white in early anthesis, fading brilliant red or purple. $2n = 36$. Established throughout in wastelands and cultivated sites, Fla., introduced from Africa. All year. *R. roseum* Stapf & C. E. Hubbard, *T. rosea* Nees

38. ECHINÓCHLOA Beauv. BARNYARD GRASS

Coarse annuals or perennials with fleshy or succulent culms, compressed sheaths, flat blades, obsolete ligules. Spikelets one-flowered, disposed irregularly more or less on the same side of stout raceme rachis; nodes of rachis usually with scattered bristly hairs. Glumes unequal, mucronate, hispid; second glume and sterile lemma equal, mucronate, or the lemma awned, enclosing a membranous palea with a staminate floret; fertile lemma shiny, polished margins inrolled. About 30 species throughout temperate, tropical, and subtropical regions.

1. Leaf sheaths copiously pustulate-hirsute;
   panicle lax . . . . 1. *E. walteri*
1. Leaf sheaths glabrous; panicle stiffly erect.
   2. Spikelets fasciculate, crowded on raceme
      branches.
      3. Sterile floret usually neuter; glumes
         usually without pustulate hairs . . 2. *E. crusgalli*
      3. Sterile floret staminate; glumes sparingly pustulate, hairy . . . . . 3. *E. paludigena*

2. Spikelets closely imbricate in two or more
     secund rows on rachis . . . . .   4. *E. colonum*

1. **E. walteri** (Pursh) Heller—Culms erect, to 1 m tall or taller from tufted leafy base, lower nodes thickened, often geniculate. Leaf blades 8–18 mm wide, antrorsely scabrous on margins; midvein white, prominent; lower sheaths strongly hispid-pustulate. Mature panicle 1–2.5 dm long, nodding, usually purple with branches 3–7 cm long, overlapping or distant. Spikelets 3.5–4 mm long, awns of variable length to 20 mm long. Fruit to 3.2 mm long, shiny, pointed. Shallow shores and brackish waters, Fla., Tex., Midwest, coastal plain states. Summer.

2. **E. crusgalli** (L.) Beauv.—Culms 6 dm tall or more, erect from decumbent base. Leaf blades elongate 5–15 mm wide, antrorsely scabrous above or glabrous. Panicle 8–15 cm long, green or purplish, with distant, spreading to appressed branches 2–5 cm long; spikelets 3–4 mm long, hispid on the veins, glumes awned or cuspidate. Fruit 1.8–2.5 mm long, beak inflexed, body shiny. $2n = 54$. Low ground, along ditches, waste sites and cultivated lands, Fla. to Calif., coastal plain states to New Brunswick. Summer.

Polymorphic species, including *E. pungens* (Poir.) Rydberg; *E. muricata* (Beauv.) Fern.; *E. crusgalli* (L.) Beauv. var. *mitis* (Pursh) Peterm.

3. **E. paludigena** Wieg.—Culms turgid at base, solitary to 1 m tall, erect; blades elongate 10–30 mm wide, with broad white midrib, scabrid on margins, interruptedly scabrous along the veins, sometimes sparingly pilose below. Panicle to 3 dm long, with ascending to spreading branches 3–10 cm long; spikelets pustulate, hispid-pectinate along the veins, purplish when mature; rachis three-angled, setose at nodes and at base, pectinate-scabrous along angles. Spikelets 3 mm long, with awns to 20 mm long. In shallow water, cypress ponds, shores of lakes, and rivers, endemic to central and south Fla. All year.

4. **E. colonum** (L.) Link—Plants essentially glabrous; culms stiff, 5–9 dm long, tufted, branching from the base; internodes two-edged, scabrous, sheaths commonly overlapping; blades flat, attenuate, 3–6 mm wide, green or purplish. Inflorescence narrow, 12–15 cm long, slightly spreading, green or purple. Mature spikelets 2 mm long, mucronate, scabrous on the veins; rachis nodes with a few long hairs. $2n = 54$. Open ground, in pinelands, Fla., Tex., N.J., Calif., tropical America. Summer, fall.

39. OPLISMENUS Beauv.

Panicles long peduncled; blades flat, spreading. Spikelets in pairs on one side of the rachis. Glumes entire or emarginate, awned between the lobes; lemmas entire exceeding the glumes, notched or mucronate; palea hyaline, closed in the fertile lemmas, not inrolled. Perennial, shade-loving grass, freely branching and creeping over damp soil in hammocks. About 15 species, tropical and subtropical regions.

1. **O. setarius** (Lam.) Roem. & Schultes—Culms slender, wiry, to 7 dm long, prostrate or ascending, freely rooting at nodes and loosely matted; lateral branches numerous; blades ovate or elliptic, 1–3 cm long, to 10 mm wide; sheaths usually overlapping ciliate on margins, pilose at throat; ligule a short, fringed membrane. Peduncle naked below, elongate with five or more distant racemes; spikelets conspicuous, awns 4–10 mm long. In shallow soil over oolitic outcrops, naturalized from tropical America, central and southern Fla., Tex., Ark., N.C. Summer.

40. DIGITÀRIA Heist.

Annual or perennial, erect or prostrate grasses, mostly with slender culms. Leaves flat or folded, elongate; ligules membranous. Spikelets paired, solitary, or in threes, alternating in two rows on one side of three-angled, winged or wingless rachis. Spikelets plano–convex, lanceolate or elliptic, short pedicelled in digitate or paniculate racemes. Outer glume lacking or minute; inner glume as long as the sterile lemma or shorter; fertile lemma cartilaginous. About 380 species in tropical and temperate regions. *Syntherisma* Walt.; *Trichachne* Nees

1. Rachis winged; margin flat at least as wide
     as the midvein.
  2. Rachis bearing interruptedly very fine
       long hairs; rachis wing a mere margin,
       spikelets 2 mm long . . . . . .   1. *D. horizontalis*
  2. Rachis not bearing fine hairs.
    3. Plant with leafy stolons; spikelets 1.5
         mm long . . . .   2. *D. longiflora*
    3. Plants without leafy stolons, spreading
         at base; spikelets 1.7–3.3 mm long.
      4. Culms repent, mat-forming, outer
           glume lacking . . . . . .   3. *D. serotina*
      4. Culms ascending, erect, outer glume
           small, rarely lacking.
        5. Inner glume one-half the length of
             the spikelet . . . . . .   4. *D. sanguinalis*
        5. Inner glume two-thirds the length
             of the spikelet . . . . . .   5. *D. adscendens*
1. Rachis essentially wingless.
  6. Racemes relatively few; spikelets short
       pubescent; hairs of sterile lemmas
       capitellate.

7. Spikelets 2–2.4 mm long; blades flat,
     straight . . . .   6. *D. villosa*
7. Spikelets 1.5 mm long; blades involute,
     flexuose . . . .   7. *D. dolichophylla*
6. Racemes numerous; spikelets silky vil-
     lous; hairs long, not capitellate . . .   8. *D. insularis*

1. **D. horizontalis** Willd.—Decumbent annual to 7 dm tall. Culms slender, glabrous, internodes elongate; sheaths pilose, blades linear-at-tenuate, 8–15 cm long, to 8 mm wide, glabrous, or sparingly pilose on surfaces; ligule membranous, the orifice pilose-tufted; blade margins white, minutely scabrous; midvein prominent, white; veins below minutely scabrous. Racemes 3–8 cm long, slender, approximate; rachis narrowly winged, beset with long delicate setae. Spikelets pubescent; first glume minute or lacking; second glume two-thirds as long as the fruit; margins sparingly villous. Waste ground, abandoned lots, hammocks, south Fla., W.I., Mexico, tropical America. Summer. *Syntherisma digitata* (Sw.) Hitchc.

2. **D. longiflora** (Retz.) Pers.—Culms glabrous to 4 dm high from leafy stolons, slender, nearly leafless above; blades 1–3 cm long, linear-lanceolate, margin white cartilaginous; sheaths glabrous, overlapping; ligule membranous; racemes two or more, long exserted; spikelets elliptic, minutely puberulent; fertile lemma pale. $2n = 18$. Low pinelands in depression and sides of sandy trails, central and south Fla., introduced from Old World. Summer.

3. **D. serotina** (Walt.) Michx.—Culms 3–4 dm tall, geniculate, from creeping, stem-rooting nodes; sheaths overlapping, pilose; blades 2–7 cm long; broadly linear, pilose above and below, margin white, scabrous. Racemes three to seven, long exserted, 2–7 cm long. Spikelets 1.7 mm long, minutely pubescent at summit; outer glume wanting, the inner less than half as long as the spikelet, fertile lemma pale. Low pineland, along ditches, lawns, coastal plain, Fla., Pa., Cuba. Summer.

4. **D. sanguinalis** (L.) Scop.—Culms ascending from decumbent rooting base, 3–9 dm tall. Blades 5–10 mm wide, sparingly pilose below, or glabrate. Sheaths strongly pilose, ligule membranous, 3–4 mm long; panicle long exserted. Racemes three to seven, 5–15 cm long, stiffly spreading; spikelets 3 mm long, villous along marginal veins, appearing fimbriate with spreading hairs; outer glume small, inner half the length of spikelet; grain elliptic-acute. $2n = 36$. A weedy grass widely distributed, naturalized from Old World. Summer. *D. fimbriata* Link

5. **D. adscendens** Henr.—Habit similar to **D. serotina** (Walt.) Michx.; sheaths pilose or glabrous. Blades sparingly pilose at base, above the ligule, or glabrate; spikelets 2.3 mm long; outer glume nearly obso-

lete; inner glume two-thirds to three-fourths the length of spikelet; grain narrowly elliptic-acuminate. $2n = 27$. Frequent in Fla., naturalized from Europe. Summer. *Panicum adscendens* HBK

6. D. villosa (Walt.) Pers.—Culms wiry in large tufts to 1 m tall. Leaves mostly basal, blades 2–4 mm wide, linear, strongly pilose above and below or glabrate. Sheaths purplish, copiously grayish villous or nearly glabrous; ligule membranous. Racemes three to five, ascending or spreading, 8–15 cm long, rachis slender, floriferous to the base; spikelets 2–2.5 mm long, numerous, copiously pubescent with capitellate hairs. Glades and coastal hammocks, Fla., Tex., Mexico, Cuba. Summer.

7. D. dolichophylla Henr.—Culms wiry, glabrous in large tufts 5–10 dm long. Blades elongate, narrow, commonly folded or involute, strongly flexuous. Racemes one to four, very slender, 10–20 cm long; spikelets 1.5 mm long, pubescence capitellate on veins; fruit brown. Dry pinelands southern Fla., Tex. to Md., Cuba, Mexico. Summer.

8. D. insularis (L.) Mez. ex Eckman, SOUR GRASS—Culms 5–15 dm tall, leafy in large clumps. Blades elongate, sheaths overlapping, loose pilose, at least when young; ligule membranous; 4–7 mm long. Spikelets tawny, silky; glume and sterile lemma three- to five-veined, 4 mm long excluding the hairs; grain brown, tapering to subulate apex, the lemma margins not enclosing the palea. $2n = 36$. Low pinelands and waste ground, Fla., Tex., Mexico, W.I., Australia. Summer. *Trichachne insularis* (L.) Nees

41. ERIÓCHLOA HBK CUP GRASS

Annual or perennial grasses, branching from the lower nodes, with elongate leaves and terminal inflorescences. Spikelets in pairs or solitary on the same side of the rachis; the outer glume reduced, sheathlike, adnate to an annular, thickened rachilla joint, subtending the inner glume and sterile lemma; sterile lemma sometimes with a hyaline palea, or with stamens, or empty; mature caryopsis minutely striate. About 20 species, tropics and subtropics.

1. E. michauxii (Poir.) Hitchc. var. **michauxii**—Culms tufted, erect, 6–8 dm tall, or trying to grow in overpopulated sites, proliferous from the upper nodes stolon fashion; creeping, reclining, layering over older growth; lower nodes thickened, bluish, commonly shorter than the blades. Sheaths loose, puberulent at the collar; ligule a dense fringe of white hairs. Blades flat or involute, 2–4 mm wide, long attenuate; inflorescence overtopping the leaves. Axes appressed silky pubescent throughout; spikelets 4–5 mm long, appressed villous, in distant racemes; sterile lemma usually with stamens. Brackish or fresh marshes, prairies, glades, Gulf strand, throughout Fla., Ga. Summer. *E. mollis* Kunth

E. michauxii var. **simpsonii** Hitchc. differs from var. **michauxii** in narrower leaves, fewer panicles, and empty sterile lemmas. Gulf dunes, in brackish sand, Bonita Springs to Cape Sable. Fall.

E. polystachya HBK, West Indian species, called Malojilla, a forage grass, is established in the area. Vegetatively it is similar to Para grass in having long, creeping runners; often an associate in a colony, Fla., Tex., Mexico, South America. All year.

42. BRACHIARIA (Trin.) Griseb. SIGNAL GRASS

Annuals or perennials with linear, flat blades and terminal racemose panicles. Spikelets in two rows on one side of three-angled, winged rachis. Outer glume adaxial, short or nearly as long as the floret. Inner glume and sterile lemma equal, five- to seven-veined, the latter with a reduced palea or stamens; fertile lemma indurate, papillose-rugose with inrolled margins, apex usually awnless, sometimes mucronate. About 50 species, warm regions of North and South America, Asia.

1. **B. subquadripara** (Trin.) Hitchc.—Leafy perennial, creeping and spreading by nodal roots. Culms geniculate, to 1 m long. Blades flat to 15 cm long, 4–10 mm wide; margins finely scabrous, pustulate-ciliate at cordate base; ligule a short fringe of hairs. Sheaths pilose along margins. Racemes three to five, distant, 3–5 cm long, pedicels short, puberulent. Spikelets 3–3.2 mm long, glabrous. Introduced as forage grass, Asia, established in southern Fla. Fall.

43. AXÓNOPUS Beauv.

Mainly perennials, creeping or tufted, mat-forming; foliage typically soft, of flat, obtuse, sheathed blades, spikelets depressed, biconvex, oblong-acute, sessile, alternate in two rows on one side of the rachis. The back of the fertile lemma turned from the three-angled rachis. First glume wanting; the second glume and sterile lemma equal, palea absent; fertile lemma and palea indurate, with inrolled margins of the lemma. About 35 species, South America.

1. Spikelets 2 mm long, minutely silky villous
    at summit; racemes two to four . . . 1. *A. affinis*
1. Spikelets 4–5 mm long, glabrous; racemes
    usually two . . . . 2. *A. furcatus*

1. **A. affinis** Chase—Tufted or stoloniferous with slender culms 3–5 dm tall, sheaths strongly keeled, thin with broad hyaline margins. Blades 0.6–2 dm long, 3–4 mm wide or wider on stolons, essentially glabrous, or pilose above the base; ligule membrane less than 1 mm long.

Spikelets biconvex, strongly compressed on each side, the midvein of the glume and sterile lemma not discernible. Stolons usually mat-forming. Established commonly in sandy, mucky soil, Fla., N.C., Okla., Tex., Cuba, Mexico; from Australia. Fall. *Axonopus compressus* (Sw.) Beauv. var. *affinis* (Chase) Henders.

2. **A. furcatus** (Fluegge) Hitchc.—Stoloniferous perennial more robust than the preceding one, culms to 8 dm tall, strongly compressed at the base, decumbent or erect, tufted. Blades 5–10 mm wide, ciliate, hirsute or glabrate. Racemes to 10 cm long; spikelets appressed, sterile lemma and the glume five-veined, produced to a beak, one-third the length of the spikelet; grain broadly ovate 3.2 mm long, flattened on each side. Hammock margins, moist banks, coastal plain states, Fla. to southeast Va. Summer, fall.

44. PÁSPALUM L.

Perennial or annual grasses with flat leaves, membranous ligules, spikelike racemes. Spikelets plano–convex, solitary or in pairs on one side of a narrow or dilated rachis. The back of the fertile lemma adaxial. Outer glume rarely present; inner glume and sterile lemma equal or the glume sometimes suppressed; fertile lemma strongly indurate; margins inrolled. About 250 species, warm regions, temperate America.

Species of this genus constitute a considerable proportion of the pasture or pampas grasses in tropical countries.

1. Rachis deciduous, foliaceous, winged, produced beyond the ultimate spikelet . . 1. *P. fluitans*
1. Rachis persistent branchlike with or without wing margin.
  2. Racemes two, or another, below, conjugate or approximate.
    3. Racemes conjugate, spikelets 1.7 mm long, silky; stolons leafy . . . . 2. *P. conjugatum*
    3. Racemes paired or approximate.
      4. Spikelets 3.5–4 mm long; ovate-acuminate; stolons or rhizomes remotely invested by sheaths.
        5. Sterile lemma five-veined, palea four-veined . . . . . . 3. *P. vaginatum*
        5. Sterile lemma and palea five-veined . . . . 4. *P. distichum*
      4. Spikelets 3–3.5 mm long; ovate or obovate; rhizome thick, invested by closely imbricate sheaths . . . 5. *P. notatum*
  2. Racemes one or several.
    6. Outer glume usually present; leaf

blades flat at base, terete toward
apex; culms strict . . . . . .   6. *P. monostachyum*
6. Outer glume lacking; leaf blades flat.
  7. Racemes terminal and axillary; leaves
    mostly below midstem . . . .   7. *P. setaceum*
  7. Racemes terminal.
    8. Spikelets lacerate winged; annual .   8. *P. fimbriatum*
    8. Spikelets wingless; perennials.
      9. Plants large, solitary or clump-
      forming.
        10. Spikelets silky fringed, pointed;
        spikes appressed . . . .   9. *P. urvillei*
        10. Spikelets silky fringed, pointed;
        spikes spreading, divaricate   10. *P. giganteum*
      9. Plants small to moderate in size,
      cespitose or tufted; racemes
      apparently axillary.
        11. Culms 1 mm or a little more in
        diameter.
          12. Spikelets 2 mm long, glume
          pubescent . . . . .   11. *P. laxum*
          12. Spikelets 2.5 mm long,
          glume glabrous . . .   12. *P. pleostachyum*
        11. Culms slender, less than 1 mm
        thick at base.
          13. Racemes two or more; plant
          cespitose; spikelets less
          than 1.5 mm long.
            14. Spikelet obovate, glandu-
            lar-pubescent; sheaths
            pilose . . . . .   13. *P. blodgettii*
            14. Spikelet oval, sparingly
            appressed pilose, glab-
            rate; sheaths glabrous .   14. *P. caespitosum*
          13. Racemes commonly solitary;
          plant tufted; spikelets 1.6
          mm long, minutely pubes-
          cent; sheaths pilose at
          throat . . . . . .   15. *P. saugetii*

1. **P. fluitans** (Ell.) Kunth—Aquatic annual with spongy culms;
rooting at nodes, branching, and creeping in shallow waters or floating
in deeper shores of lakes and rivers; sheaths inflated, papillose-pilose,
dotted with purple. Blades lanceolate to 25 cm long, 4–20 mm wide,
papillose-ciliate above the auriculate base; auricles erect 2.2 mm long.
Inflorescence emersed; racemes numerous, 5–6 cm long, rachis spathi-
form, scabrous on abaxial surface. Spikelets 1.5 mm long, solitary in alternating rows, minutely viscid-villosulus; glume and sterile lemma two-

veined. Alluvial soils, in water, Fla., Tex., Mo., Ill. Fall. *P. mucronatum* Muhl. included in *P. repens* Bergius

2. **P. conjugatum** Bergius—Culms erect, from leafy, creeping stolons to 5 dm tall. Blades glabrous, flat, thin, broadly linear, 3–10 cm long, 4–8 mm wide; sheath orifice pubescent within and without. Racemes conjugate, broadly divaricate, slender, 4–10 cm long; spikelets about 1.5 mm long, pale, villosulus; plant colonial, internodes of stolon villous. $2n = 40$. Often looping above the soil in dense populations, waste places, Fla., Tex., W.I., tropical America. Summer.

3. **P. vaginatum** Sw.—Widely creeping stoloniferous perennial, culms ascending, erect to 6 dm tall. Blades flat to 15 cm long, 3–5 mm wide, sheaths loose, overlapping, auriculate, pilose at throat. Racemes two, rarely another one below, on short peduncles; spikelets solitary, 3–4 mm long, compressed above and below, imbricate on very short pedicels; outer glume five-veined and sterile; lemma four-veined, the midvein reduced. $2n = 20$. Seacoast in brackish sand, colonial, Fla., Tex., N.C., tropics of Eastern Hemisphere. Fall.

Effective sand-binding grass on seacoasts around the world.

4. **P. distichum** L.—Similar to **P. vaginatum**. Freely stoloniferous, forming colonies, stolons pubescent at nodes. Culms erect, spikelets commonly 2–5mm long, incurved or somewhat spreading; outer glume often developed, the inner pubescent or glabrous, five-veined; mature fruit subconvex. $2n = 48,60$. Ditches and brackish waters, widely distributed Fla., Tex., N.J., South America. Summer, fall.

Valuable pasture grass on alluvial flats.

5. **P. notatum** Fluegge, BAHIA GRASS—Culms to 5 dm tall, ascending from a ligneous, rhizome imbricate with scales. Leaves 2–4 mm wide, linear-elongate; sheaths slightly inflated auriculate at summit, shorter or longer than the blades. Racemes approximately 3–7 cm long. Spikelets 2.5–2.8 mm long, obovate or ovate, lustrous green, smooth. $2n = 40$. Introduced as lawn and pasture grass, sandy and clay soils, Fla., southeastern states, Mexico, W.I., South America.

P. notatum var. **saurae** Parodi, Paraguay Strain—More hardy form planted in southern states.

6. **P. monostachyum** Vasey—Culms to 9 dm tall, close-set on scaly rhizomes, stiffly upright. Leaf blades to 6 dm long, rigid, involute, long tapering to apex. Raceme one or two, usually erect, 10–30 cm long, spikelets 3–3.2 mm long, paired, obovate in outline, on pubescent pedicels; anthers white in anthesis, tip of mature grain exposed. Conspicuous in seasonally wet pinelands and borders of glades, Fla., Tex. *P. solitarium* Nash

7. **P. setaceum** Michx.—Culms 3–9 dm tall, slender, erect or spread-

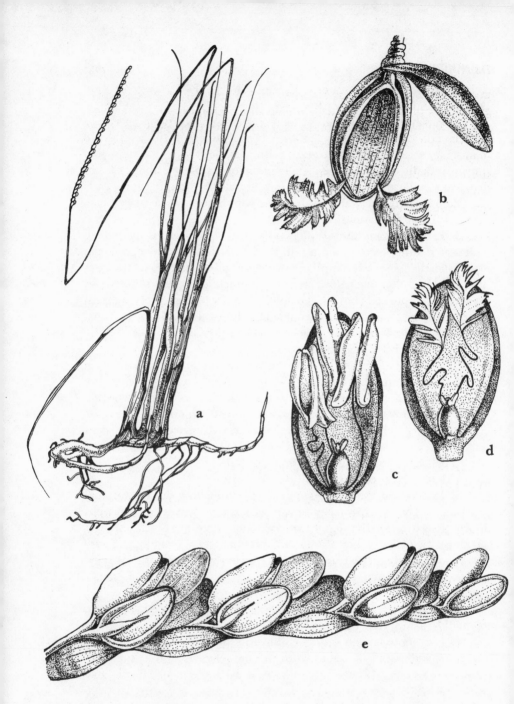

Family 32. **POACEAE.** *Paspalum monostachyum:* a, flowering plant, × ¼;
b, floret with exserted stigma, × 11; c, androecium and subtending fertile
lemma, × 19½; d, gynoecium and subtending fertile lemma, × 19½; e, fruiting raceme, × 6.

ing from short rhizomes. Leaf blades thin 5–30 cm long, 3–20 mm wide, glabrous or pilose on both surfaces with sheath margins often papillose-pilose; ligule a fringed membrane. Racemes solitary, or often two, 5–15 cm long, peduncled; spikelets to 2.7 mm long, to 2 mm wide, glabrous or puberulent, sometimes viscid, spotted, suborbicular, obovate or elliptic; sterile lemma veinless or prominently three-veined. Coastal sands, pinelands and oak scrub, Fla., Tex., Nebr. to N.H., Cuba. Represented in our area by six varieties.

1. Spikelets 2 mm long or less; midvein of
   sterile lemma lacking.
  2. Leaves cauline; blades erect to ascending   7a. var. *setaceum*
  2. Leaves radical; blades spreading.
    3. Blades 3–8 mm wide, essentially gla-
      brous on lamina . . . .   7b. var. *longependunculatum*
    3. Blades 3–10 mm wide, hirsute villous .   7c. var. *villosissimum*
1. Spikelets 2 mm long or more; midnerve of
   sterile lemma present or lacking.
  4. Culms lax, spreading; blades 3–15 mm
    wide, pilose on both surfaces . . .   7d. var. *supinum*
  4. Culms erect; leaf blades ascending to
    spreading.
    5. Blades 2–6 mm wide, rigid, puberulent 7e. var. *rigidifolium*
    5. Blades 3–20 mm wide, thin but firm,
      glabrous on surfaces . . . . .   7f. var. *ciliatifolium*

    7a. P. setaceum var. **setaceum**—Plants grayish green, usually erect from knotted base; leaves linear-attenuate mostly in line with culms. Racemes one to two, slender, arching, long peduncled; spikelets 1.4–1.9 mm long, elliptic, suborbicular, viscid-puberulent, spotted, or glabrate. Pinelands, sand hills, oakwoods, Fla., Tex., coastal plain states to Long Island west to Ohio, Mexico.

    7b. P. setaceum var. **longepedunculatum** (LeConte) Wood—Plants yellowish green in small tufts or solitary from knotty rhizomes, aggregate leaves with folded blades, pilose on margins. Racemes 3–5 cm long, on slender, elongate peduncles; spikelets 1.5–1.7 mm long, obovate, usually viscid-puberulent. Hammocks, open pinelands, south and central Fla., Tex., to Va. *P. longepedunculatum* LeConte

    7c. P. setaceum var. **villosissimum** (Nash) D. Banks—Grayish green, culms to 5 dm tall, coarsely villous hirsute throughout; leaves recurving, rhizome strongly knotty. Racemes commonly two, terminal and axillary; spikelets 1.5–1.9 long, elliptic-oblong, viscid, spotted with reddish brown. Coastal strand, sandy openings in pinelands, Fla., Cuba. *P. debile* Michx.; *P. villosissimum* Nash

    7d. P. setaceum var. **supinum** (Bosc) Trin.—Yellowish green, relatively stout, branching, spreading radially, prostrate, copiously hirsute throughout; blades 14–25 cm long. Racemes three or four, 3–7 cm long; spikelets elliptic-ovate, 2–2.6 mm long, puberulent. Glade borders, moist sands in lagoon embankments, Fla., Tex. to Va., Midwest. *P. dasyphyllum* Ell.; *P. supinum* Bosc ex Poir.

7e. **P. setaceum** var. **rigidifolium** (Nash) D. Banks—Plants greenish purple, to 7 dm tall; leaves to 3.5 dm long, pilose on margins or glabrate, sheath orifice pilose. Racemes one or several, spikelets 2–2.6 mm long, elliptic-obovate, puberulent or glabrous. Open pinelands, embankments, Fla., Ga., to N.C. Summer. *P. rigidifolium* Nash

7f. **P. setaceum** var. **ciliatifolium** (Michx.) Vasey—Plant dark green, to 9 dm tall, erect or spreading, branching at base, stout or slender; blades 2.5 dm long, smooth on surface; sheaths pustulate-pilose on margins. Racemes one to three, commonly congested, to 10 cm long; spikelets 1.4–2 mm long, suborbicular, puberulent or glabrate. Pinelands, canal banks in glades, sandhills, Fla., Tex., Midwest, W.I. Summer. *P. propinquum* Nash, *P. ciliatifolium* Michx., *P. ciliatifolium* Michx. var. *brevifolium* Vasey

8. **P. fimbriatum** HBK—Annuals. Culms 3–8 dm long, erect in clumps, separable into small tufts. Leaves to 2.3 dm long, 4–12 mm wide, flat, smooth on surface, pustulate-ciliate; sheaths loose, pustulate-ciliate, overlapping the internodes. Racemes terminal, spikes three or more with edged rachis, spikelet with wing about 5 mm wide and long, plano-concave, suborbicular, glume margin alate-lacerate; achene body about 2 mm long and wide. Established in subtropical Fla., W.I., naturalized from Bolivia. Summer.

9. **P. urvillei** Steud. VASEY GRASS—Culms erect, stout to 1 m tall or more, from massive rootstocks; blades to 3 dm long, flat, glabrous, lower sheaths pustulate-pilose. Inflorescence 3–4 dm long, of many stiffly erect racemes. Spikelets 2.5–3 mm long, abruptly pointed, plano–convex, copiously white silky fringed. $2n = 40,60$. In wastelands, canal banks, roadsides, Fla., Tex. to Calif., South America.

Introduced as pasture grass to many countries, including India.

10. **P. giganteum** Baldw. ex Vasey—Culms to 1 m tall or more, solitary or in small tufts from short scaly rhizomes. Leaves basal, elongate, linear-attenuate, 1–2 cm wide, glabrous, or in blades sparsely ciliate at base, above the ligule; lower sheaths hirtellous. Panicle of three or more racemes, 1–2 dm long, spreading or recurving at maturity. Rachis pilose-tufted at base, 1.5 mm wide, often flexuous, spikelets 3.2–3.5 mm long, broadly obovate, glabrous or minutely puberulent, mature grain russet, reticulate. Sandy soil, open pinelands, hammocks, south Fla., Ga. *P. longicilium* Nash

11. P. laxum Lam.—Culms tufted 5–7 dm tall, erect. Blades to 3 dm long, flat or involute, linear-attenuate, varyingly pilose on upper surface. Sheaths pilose at throat and margins or glabrate. Racemes solitary; if two or more, remote; 3–10 cm long, arcuate, spreading; spikelets obovate, plano–convex, glume puberulent or glabrescent. Sandy oolitic soils, Fla. Keys, W.I. *P. glabrum* Poir.

12. P. pleostachyum Doell—Culms to 1 m tall in large tufts, leafy up to inflorescence. Sheaths pilose at throat, pubescent around the collar, copiously pilose on margins. Blades 3–6 mm wide, varyingly pubescent on upper surface or glabrescent. Racemes three or more ascending, 14 cm long. Spikelets obovate, plano–convex; glume pubescent, glabrescent in age. South Fla., Marathon Key, Cuba.

13. P. blodgettii Chapm.—Cespitose; culms 4–10 dm tall, slender, wiry with swollen bases aggregated in hard tufts; basal scales and sheaths densely hirsute. Blades flat, 5–10 mm wide, mostly basal, pilose-papillose on margins, hairs commonly deciduous. Racemes three or more, remote, long exserted, 2–8 cm long; spikelets viscid-pubescent. Hammocks, oolitic crevices in thin soil, south Fla., Fla. Keys, Mexico, Bahama Islands, South America. *P. simpsoni* Nash; *P. gracillimum* Nash

14. P. caespitosum Fluegge—Cespitose; culms 3–6 dm tall, erect, relatively slender from thickened bases; sheaths and scales glabrous, at least in age. Blades flat or involute, 4 to 8 mm wide, essentially glabrous. Racemes three to five, ascending, remote, 3–5 cm long; spikelets ovalelliptic, minutely viscid-puberulent, becoming glabrous. Hammocks, crevices of oolitic rocks, in humus soil, south Fla., Mexico, Central America, W.I.

15. P. saugetii Chase—Culms to 4 dm tall, very slender from cespitose, tufted base. Blades basal flat, 3–15 cm long, glabrous, or pilose above the ligule; sheaths loosely pilose or glabrate. Racemes one or two, 2–4 cm long, arcuate; rachis pilose at base; spikelets, finely pubescent or becoming glabrous. Rocky, open hammocks, south Fla., Cuba, Greater Antilles.

45. PASPALIDÍUM Stapf

Subaquatic perennials with leafy culms to 2 m tall, succulent, decumbent, stoloniferous. Ligule a circle of hairs. Spikelets arranged singly in two rows on flattened rachis, the fertile lemma turned toward the rachis; the ultimate spikelet exceeded by the prolonged free apex of the rachis, sometimes bearing a rudimentary spikelet; fertile lemma and palea transversely rugose. About 20 species, warm parts of Old World, tropical Asia.

1. Culms tufted, rhizomatous; spikelets to 2.6
   mm long, five-nerved . . . . . .   1. *P. geminatum*
1. Culms from creeping, rooting rhizomes;
   spikelets to 3 mm long, three-nerved . .   2. *P. paludivagum*

1. P. geminatum (Forsk.) Stapf—Culms from rhizomatous tufts 2.5–8 dm tall, herbaceous, barely succulent. Blades 7–15 cm long, 3–6 mm wide, linear-attenuate, becoming involute. Panicles to 2.2 dm long, racemes ten to fifteen, in line with the rachis; spikelets in two alternating secund rows; the inner glume obovate, almost equal to the sterile lemma; grain sharply rugulose. In shallow fresh or brackish water, south Fla. to La., Tex., Okla., warm regions of both hemispheres.

2. P. paludivagum (Hitchc. & Chase) Parodi—Branching aquatic with stout, succulent culms to 2 m tall from creeping, rooting rhizomes. Leaf blades linear, elongate to 4 dm long, scabrous on surface, flat or folded, overtopping the panicles. Panicles 3–3.5 dm long; racemes distant; spikelets 2.8–3 mm long, appressed against the two-angled rachis; outer glume truncate, one-fifth as long as spikelet; inner glume one-half to two-thirds of the spikelet length, three-veined; sterile glume with staminate flower. Rooted and leaning against banks in canals, south Fla., Tex., Mexico, South America. Fall. *Panicum paludivagum* Hitchc. & Chase

46. SACCIÓLEPIS Nash

Annual or perennial grasses with contracted terminal panicle. Spikelets gibbous, tapering to the apex; first glume deltoid, short, five-veined; second glume inflated, eleven-veined. Sterile lemma five-veined. Fertile lemma not enclosing the palea at apex. About five species, one southeastern U.S., four tropical and subtropical Asiatic provinces, Australia, introduced to Africa and America.

1. S. striata (L.) Nash—Culms erect to 15 m tall from geniculate base, with forking branches below. Sheaths mostly hirsute, becoming glabrous above. Blades 15 cm long, 20 mm wide, soft, spreading, sparingly pubescent. Panicle to 2 dm long; spikelets 3–4 mm long, striate. Shores and marshes, Fla., Tex., Tenn., Va., W.I. Summer, fall.

47. LASIÀCIṢ (Griseb.) Hitchc.

Culm ligneous, branching, high clambering through crowns of hammock trees. Spikelets subglobose, obliquely articulate on pedicels. Outer glume about one-third the length of the spikelet, inflated. Inner glume and sterile lemma about equal, apiculate, many-veined, shiny, chartaceous. Sterile floret with or without stamens; fertile lemma and palea white, bony-indurate with a tuft of woolly hair at apex. About 30 species, American tropics.

1. **L. divaricata** (L.) Hitchc.—Essentially glabrous throughout; culms climbing to the height of 3–4 m; main stem to 6 mm in diameter at base; branching fasciculate, divaricate. Blades narrowly lanceolate, 5–20 cm long, 5–15 mm wide; panicles terminal, few-flowered; spikelets to 4 mm long, black in maturity. South Fla., Bahama Islands, W.I., South America.

48. PÁNICUM L. PANIC GRASS

Annual or perennial grasses, tufted or with repent rhizomes. Inflorescence usually a panicle, with symmetrical or one-sided branching. Spikelet with one bisexual, fertile floret above the reduced sterile floret with lemma and hyaline palea, rarely bearing stamen; first glume minute or half as long as the spikelet; second glume as long as the sterile lemma opposing it, these glumes enclosing the mature spikelet. Fertile lemma and palea, veinless, indurated, and polished, enclose the stamens and the pistil; the free grain maturing between them, becoming completely enclosed by inrolling lemma margins; embryo small, embedded in large starchy endosperm. Large cosmopolitan genus of 500 species, throughout tropical and warm temperate regions. A polyploidy complex based on $x = 9$.

Key to Series

1. Axes of the branchlets produced beyond the ultimate spikelets . . . .   1. *Distantiflora*
1. Axes of the branchlets not produced as free tips beyond the ultimate spikelets.
  2. Basal leaves different from cauline leaves; winter rosettes formed.
    3. Culms branching at the base; autumnal growth in copious tufts . . . .   2. *Laxiflora*
    3. Culms branching from the middle and upper nodes or remaining simple.
      4. Spikelets symmetrical.
        5. Blades 5 mm wide or less, moderately stiff, autumnal phase fasciculate . . . . . .   3. *Angustifolia*
        5. Blades 25 mm wide or less, pliable, autumnal phase freely branching.
          6. Ligule hairs conspicuous; plants hirsute . . . .   4. *Lanuginosa*

6. Ligule hairs inconspicuous;
    plants glabrous or moderately
    pubescent . . . . . .   5. *Dichotoma*
    7. Spikelets spheroidal or nearly
        so; panicle axes viscid,
        spotted . . . . .   6. *Sphaerocarpa*
    7. Spikelets obovoid, ellipsoid;
        panicle axes not viscid.
        8. Blades to 3 cm long; culms
            lax, slender . . . .   7. *Ensifolia*
        8. Blades to 15 cm long; culms
            moderately thick . . .   8. *Commutata*
    4. Spikelets pyriform, culms wiry, often
       geniculate . . . .   9. *Lancearia*
2. Basal and culm leaves alike, without win-
    ter rosettes; only primary panicles
    formed; all spikelets fertile.
    9. Plants annual.
      10. Panicle of several spiciform racemes;
         fruit transversely rugulose; inner
         glume and sterile lemma more or
         less reticulate . . . . . .   10. *Fasciculata*
      10. Panicle open or diffuse; fruit and
         glumes variable.
        11. Spikelets tuberculate, 2 mm long .   11. *Verrucosa*
        11. Spikelets smooth.
          12. Outer glume about one-fourth
            the length of spikelet, trun-
            cate or deltoid; spikelet 2 mm
            long . . . .   12. *Dichotomiflora*
          12. Outer glume about one-half the
            length of spikelet, acute;
            spikelet 4–5 mm long . . .   13. *Capillaria*
    9. Plants perennial.
      13. Fruit transversely rugose.
        14. Racemes unilateral, spreading;
          stolons elongate, creeping . .   14. *Purpurascentia*
        14. Racemes symmetrical, diffuse; rhi-
          zome short, bunchy . . . .   15. *Maxima*
      13. Fruit not rugose.
        15. Plant succulent with creeping base;
          spikelets 6–7 mm long . . .   16. *Gymnocarpa*
        15. Plant not succulent.
          16. Panicles narrow; racemes ap-
            pressed, few-flowered; aquat-
            ics, or swamp plants.
            17. Culms from elongate, creeping
              rhizomes; spikelets subsessile   17. *Hemitoma*
            17. Culms from knotted rhizomes,

tufted; spikelets short pedicelled . . . .   18. *Tenera*
16. Panicles open, diffuse or congested; racemes many-flowered.
18. Culms stout in large clumps,
or if scattered, outer glume
truncate; rhizomes some-
times strong and scaly . .   19. *Virgata*
18. Culms slender or moderately
thick.
19. Sterile lemma enlarged at
apex; panicle branches
floriferous at tips . . .   20. *Laxa*
19. Sterile lemma not enlarged;
panicle branches florifer-
ous to the base . . .   21. *Agrostoidea*

Key to Species

1. DISTANTIFLORA

Represented only by a single species . . .   1. *P. chapmanii*

2. LAXIFLORA

1. Sheaths retrorsely pilose; blades pilose on
surface; spikelets 1.7 mm long, pilose . .   2. *P. xalapense*
1. Sheaths glabrous; blades glabrous on sur-
face, margins ciliate.
2. Spikelets to 2 mm long, pilose or glabrescent . . . .   3. *P. ciliatum*
2. Spikelets 1.5 mm long, glabrous . . .   4. *P. polycaulon*

3. ANGUSTIFOLIA

1. Spikelets pointed beyond the fruit; sym-
metrically arranged.
2. Spikelets 2–2.5 mm long, twisted; sheaths
glabrous . . . .   5. *P. pinetorum*
2. Spikelets 3.3–3.5 mm long; sheaths villous   6. *P. fusiforme*
1. Spikelets obtuse about as long as the fruit;
unilaterally arranged . . . . . .   7. *P. neuranthum*

4. LANUGINOSA

Represented only by a single species . . .   8. *P. ovale*

5. DICHOTOMA

1. Nodes bearded; spikelets 2 mm long, pubescent . . . .   9. *P. nitidum*

1. Nodes glabrous; spikelets glabrous.
  2. Spikelets 2 mm long; leaf blades erect to
     ascending, usually olive . . . . .   10. *P. roanokense*
  2. Spikelets 1.5 mm long; leaf blades spread-
     ing, usually purplish . . . . . .   11. *P. caerulescens*

6. SPHAEROCARPA

Represented only by a single species . . .   12. *P. erectifolium*

7. ENSIFOLIA

1. Blades margined with white, aggregated
   toward the base.
  2. Secondary panicles from basal nodes; up-
     per cauline leaves much reduced . .   13. *P. albomarginatum*
  2. Secondary panicles from upper and mid-
     dle nodes; upper cauline leaves only
     slightly reduced . . . .   14. *P. trifolium*
1. Blade margins and surfaces all green;
   spikelets glabrous or puberulent.
  3. Spikelets glabrous; leaf blades 1–3 m wide.
    4. Spikelets 1–1.2 mm long; autumnal
      culms profusely branching, and re-
      branching from lower nodes . . .   16. *P. chamaelonche*
    4. Spikelets 1.4 mm long; autumnal culms
      branching from upper and middle
      nodes . . . .   17. *P. glabrifolium*
  3. Spikelets puberulent; leaf blades 1–4 mm
    wide.
    5. Spikelets 1.3–1.4 mm long; leaf blades
      outcurving, involute . . . . .   18. *P. breve*
    5. Spikelets 1.3–1.5 mm long; leaf blades
      flat, usually recurved . . . . .   15. *P. ensifolium*

8. COMMUTATA

1. Outer glume 1–1.2 mm long.
  2. Leaf blades lanceolate, cordate-ciliate at
    base; spikelets 1.4 mm wide . . . .   19. *P. commutatum*
  2. Leaf blades inequilateral, subcordate at
    base; spikelets 1.2 mm wide . . . .   20. *P. joorii*
1. Outer glume to 2 mm long; leaf blades
   linear; spikelets 3.2 mm long . . . .   21. *P. equilaterale*

9. LANCEARIA

1. Spikelets 1.6 mm long; blades 3–6 mm wide,
   glabrous . . . .   22. *P. portoricense*
1. Spikelets 2.1–2.6 mm long; blades to 8 mm
   wide, usually puberulent.

2. Spikelets 2.1 mm long; blades ascending,
    glabrous above, puberulent below . .   23. *P. lancearium*
2. Spikelets to 2.6 mm long; blades spread-
    ing, glabrous on both surfaces . . .   24. *P. patentifolium*

10. FASCICULATA

1. Spikelets 2.1–2.5 mm long, rarely 3 mm,
    strongly veined; pedicels bearing tri-
    chomes . . . .   25. *P. fasciculatum*
1. Spikelets 3–5 mm in length, transversely
    rugose . . . .   26. *P. adspersum*

11. VERRUCOSA

Represented only by a single species . . .   27. *P. verrucosum*

12. DICHOTOMIFLORA

1. Aquatic perennial; blades 2–3 mm wide;
    spikelets ellipsoid . . . .   28. *P. lacustre*
1. Terrestrial annuals; blades to 20 mm wide;
    spikelets narrowly oblong, acute.
    2. Sheaths papillose-hispid . . . . .   29. *P. bartowense*
    2. Sheaths glabrous . . . .   30. *P. dichotomiflorum*

13. CAPILLARIA

Represented only by a single species . . .   31. *P. miliaceum*

14. PURPURASCENTIA

Represented only by a single species . . .   32. *P. purpurascens*

15. MAXIMA

Represented only by a single species . . .   33. *P. maximum*

16. GYMNOCARPA

Represented only by a single species . . .   34. *P. gymnocarpon*

17. HEMITOMA

Represented only by a single species . . .   35. *P. hemitomon*

18. TENERA

Represented only by a single species . . .   36. *P. tenerum*

19. VIRGATA

1. Rhizome extensively creeping, relatively

slender; spikelets to 2.5 mm long, outer
  glume truncate, short . . . . . .   37. *P. repens*
1. Rhizome stout and scaly, bunch-forming.
  2. Panicle elongate or contracted; spikelets
    to 7 mm long, outer glume about one-
    half the length of the spikelet; blades
    involute; rhizome vertical . . . .   38. *P. amarulum*
  2. Panicle diffuse; mature spikelets to 3 mm
    long; blades flat; rhizomes horizontal .   39. *P. virgatum*

20. LAXA

Represented only by a single species . . .   40. *P. hians*

21. AGROSTOIDEA

1. Plant with scaly rhizomes; spikelets disposed
    obliquely on pedicels . . . . . .   41. *P. rhizomatum*
1. Plant tufted, without scaly rhizomes; spike-
    lets disposed in direct line . . . . .   42. *P. agrostoides*

1. P. chapmanii Vasey—Culms ascendent to 1 m tall, closely tufted.
Rhizomes knotted. Blades linear-attenuate, erect-elongate, to 4 dm long,
2–5 mm wide, flat or involute, pilose above the ligule; sheaths short
pilose toward the throat. Panicle elongate, narrow; racemes appressed,
distant; spikelets 2 mm long; outer glume one-third as long as the
spikelet, the inner glume as long as the spikelet. Sandy soil, south Fla.,
Bahama Islands, and Cuba. *Setaria chapmanii* (Vasey) Pilger

2. P. xalapense HBK—Vernal culms 1–2.5 dm tall, slender, from
matted winter rosettes. Cauline blades one to three; blades 7–12 cm long,
linear-attenuate, strongly ciliate; sheaths retrorsely pubescent. Panicle
diffuse, axes pilose; spikelets 2–2.2 mm long, pilose. Autumnal culms
scarcely surpassing the leaf tufts. Moist soil, hammocks, Fla., Tex., Me.,
Mo., Central America.

3. P. ciliatum Ell.—Vernal culms in dense tufts, 0.7–2.5 dm tall.
Blades 3–5 cm long, 3–8 mm wide, stiffly pustulate-ciliate, glabrous on
surface. Panicle diffuse, 3–4 cm long; axes pilose or glabrate; spikelets
2 mm long or less, pubescent or glabrescent. Autumnal growth matted;
culms somewhat exceeding the leaves. Cypress margins, moist soil, coastal
plain states, Fla. to N.C., Tex., Mexico.

4. P. polycaulon Nash—Vernal culms 1–2 dm tall, from dense rosettes. Blades flat, 3–6 cm long, 3–6 mm wide, strongly ciliate. Panicles
few-flowered; axes pilose, spikelets 1.5 mm long, glabrous or merely
pubescent. Autumnal culms mostly reclining among the leaves. Sandy
pineland, coastal plain, Fla. to Miss., W.I., Central America.

5. P. pinetorum Swallen—Vernal culms 5–9 dm long, wiry. Blades to 9 cm long, 2–3 mm wide, becoming involute, glabrous; sheaths glabrous, or, the lowermost, appressed-pubescent. Panicle narrow, 7–9 mm long, with ascending branches; spikelets pointed beyond the grain, 2.3–2.5 mm long or longer before maturing, minutely pubescent. Autumnal culms reclining, copiously branching; leaves involute, numerous, concealing the reduced spikelets. Pinelands, endemic to Collier and Lee Counties, Fla. Fall.

6. **P. fusiforme** Hitchc.—Vernal culms erect, 3–7 dm tall. Tufted basal blades flat, 3–6 dm long, 4–8 mm wide; pubescent on lower surface; lower sheaths pubescent. Panicle 4–10 cm long, spreading, reflexed; spikelets fusiform, 3–3.5 mm long, pointed beyond the grain. Autumnal culms, reclining branches fasciculate at upper nodes; blades becoming involute. Sandy pinelands, Fla., Ga., Miss., Central America.

7. **P. neuranthum** Griseb.—Glabrous vernal culms 3–6 dm tall, erect to ascending. Basal leaves few; cauline leaves 10–15 cm long, 3–5 mm wide. Panicle 5–9 cm long, with ascending branches; spikelets obtuse, 2 mm long, papillose-pubescent. Autumnal phase with shorter leaves, involute blades, and reduced panicles. Open pinelands, prairie, Fla., Tex., Miss., Central America. Fall.

8. **P. ovale** Ell.—Pubescent throughout, vernal culms 2–3 dm tall, erect, appressed pilose below. Blades elongate 5–10 mm wide, hirtellous beneath, sparsely pilose above; nodes bearded. Panicle diffuse, axes pilose; spikelets 2.5–2.9 mm long. Autumnal culms stiffly decumbent, branching from the middle and upper nodes. Sand hills, Fla. to Tex., on the coastal plains.

9. **P. nitidum** Lam.—Vernal culm, slender, upright 3–8 cm tall. Blades glabrous, 3–5 mm wide, ascending to deflexed; sheaths mottled between the veins. Panicle 4–6 cm long, many-flowered, ovoid-rounded. Moist pinelands, coastal plains, Fla. to Tex., Bahama Islands, Cuba.

10. **P. roanokense** Ashe—Vernal culms 5–6 dm tall, slender, wiry, glaucous. Sheaths essentially glabrous, surpassed by internodes; blades glabrous, linear-attenuate, 3–5 cm long, 2–4 mm wide, glabrous, usually erect, stiff. Mature panicle open, 2–5 cm long, branches ascending, spikelets purplish along margins; outer glume oblong-obtuse about one-third as long as the spikelet. Autumnal branching from middle and upper nodes, fasciculate, with reduced panicles. Pinelands, south Fla., Del. to Tex., W.I.

11. **P. caerulescens** Hack. ex Hitchc.—Vernal culms bluish green, erect 6 dm tall or more. Blades erect or spreading, 3–7 cm long, 4–8 mm wide, glabrous. Panicle 5 cm long or longer, open or diffuse; axes capillary rigid, spikelets obovate, glumes margined with purple; grain plump, lustrous, about 1 mm long; autumnal growth of fasciculate branches in

the middle and upper nodes. Glades with lime-rock substratum, marshy lands on coastal plain, Fla. to N.J., Cuba, Miss., W.I.

12. P. erectifolium Nash—Vernal culms 3–7 dm tall, erect, tufted. Leaves crowded at base; blades erect to spreading, 3–9 cm long, 4–12 mm wide, gradually reduced toward summit, purplish. Panicle exserted, 6–7 cm long, densely flowered; spikelets 1–1.2 mm long, spheroidal, puberulent. Autumnal culms branching from the middle nodes. A distinctive grass of moist pinelands and prairie depressions, Fla. to La., Cuba.

13. P. albomarginatum Nash—Vernal culms erect to 3 dm tall. Foliage grayish green or purplish; basal blades 5–7 cm long, with one to two smaller, remote, cauline blades, all with white cartilaginous margins. Panicle 3–6 cm long, open; spikelets 1.5 mm long, puberulent. Autumnal culms spreading and branching in dense tufts. Pine flatwoods, Fla. to Tex., Va., Tenn.

14. P. trifolium Nash—Vernal culms to 5 dm tall, slender, glabrous, ascending from crowded rosette leaves. Cauline leaves few, the lower blades 4–5 cm long, 3–5 mm wide, the upper blades nearly equal, or slightly reduced. Panicle diffuse, many-flowered, spikelets 1.5 mm long, puberulent. Autumnal phase, erect to reclining, from upper and middle nodes. Pine flatwoods in inclusion of hardwoods, Fla., Ga. to N.C. Spring.

15. P. ensifolium Baldw. ex Ell.—Vernal culms, slender, 1.5–4 dm tall, glabrous, leaning from basal leaf tufts. Blades 1–3 cm long, 1.5–4.5 mm wide, flat, thin, glabrous, cauline blades progressively smaller toward the summit. Panicle 1–3 cm long, diffuse, few-flowered; spikelets glabrous or puberulent; autumnal culms and foliage fasciculate in the upper axils, olive green. Low pinelands, cleared strands of lagoons; Fla. to N.J. Spring.

16. P. chamaelonche Trin.—Glabrous, vernal culms from dense tufts, 0.7–2 dm tall. Blades flat, firm, ascending, 2–6 cm long, 2–3 mm wide, glabrous or pilose at the base; sheaths glabrous. Panicle 2–3.5 cm long, diffuse, many-flowered, often barely exserted above the rosette leaves, green or purple. Autumnal tufts densely fasciculate at upper nodes concealing the reduced panicles; blades involute. Sandy soil, in openings and moist depressions of marginal woodlands, Fla. to La., N.C.

17. P. glabrifolium Nash—Vernal culms to 5 dm tall, geniculate at nodes, mostly erect. Blades flat, erect 3–10 cm long, distant; sheaths loose, shorter than the internodes; basal leaves few. Panicle 4–7 cm long, diffuse, branches spreading divaricate. Autumnal culms fasciculate in upper nodes; glades involute, panicles reduced, few-flowered. Sandy pinelands, endemic to peninsular Fla.

18. P. breve Hitchc. & Chase—Vernal culms 5–15 cm long, erect, rather stiff or wiry; tufts dense, aggregated into spreading mat. Blades flat, becoming strongly involute, erect 3–10 cm long, ciliate at base; sheaths short, imbricate; pubescent along margins; panicle 3–5 cm long, diffuse, branches divaricate. Autumnal culms concealed in profusely branching tufts, few-flowered panicles. Hammock margins, white sand oak scrub association, endemic to south and central Fla.

19. P. commutatum Schultes—Vernal culms 3–6 dm tall, erect from knotty crowns; internodes purple, glabrous, exceeding the sheaths. Leaf blades lanceolate, 10–25 mm wide near the middle, tapering to acute apex, distant, about uniform in size throughout; surface glabrous above, pubescent below, the cordate base ciliate when young. Panicle to 12 cm long, open, long exserted; spikelets to 2.9 mm long, obtuse, seven to nine-veined, pubescent. Autumnal culms from middle nodes, becoming upwardly dense, finally reclining. Moist hardwoods, cypress margins, hammocks, Fla., Tex., Okla., Mich. Summer.

20. P. joorii Schultes—Vernal culms erect, to 5.5 dm tall, internodes elongate. Leaf blades commonly asymmetrical, 5–15 cm long, 5–16 mm wide, broadly linear, narrowed to subcordate, ciliate base, thin glabrous on surface. Panicle diffuse, 5–9 cm long, spikelets 3–3.1 mm long, elliptic-pubescent. Autumnal culms short, fasciculate, branching and spreading from the lower nodes. Coastal plain, wet woods, hammocks, Fla. to Tex., Ark., Mexico. Summer.

21. P. equilaterale Scribn.—Vernal culm 2.5–7 dm tall, wiry, erect, internodes slender, elongate; sheaths glabrous; blades stiff, spreading horizontally, 0.6–1.7 dm long, 3–10 mm wide, linear-attenuate. Panicle 5–10 cm long, loosely flowered; spikelets elliptic, pubescent. Autumnal culms closely clustered, leafy in the upper nodes. Sandy pinelands, hammocks, coastal plain states, Fla., N.C., S.C. Summer.

22. P. portoricense Desv. ex Ham.—Vernal culms 1.5–3 dm tall, wiry puberulent or glabrate. Blades 2–5 cm long, 3–6 mm wide. Panicle diffuse, branches spreading; spikelets asymmetrical, pyriform, puberulent. Autumnal culms with fasciculate branches from all the uppermost node; blades reduced, pointed. Sand hills, prairies, moist habitats, Fla. to Tex., N.C., Cuba.

23. P. lancearium Trin.—Vernal culms 2–5 dm tall, grayish puberulent. Sheaths puberulent, blades 2–6 cm long, 3–7 mm wide, essentially glabrous. Panicle open, spikelets asymmetrical, pyriform, glabrous or puberulent. Autumnal culms geniculate, spreading with branches from the middle nodes. Sandy pinelands, cypress margins, coastal plain, Fla. to Tex., Cuba, South America.

24. P. patentifolium Nash—Vernal culms to 6 dm tall, decumbent or ascending, puberulent or glabrescent; internodes surpassing the sheaths. Blades 5–8 cm long, 3–5 mm wide, spreading, glabrous narrowed to base, usually sparsely pilose, glabrate. Panicle 2–7 cm long, narrow, few-flowered, spikelets strongly obovoid-puberulent, outer glume about one-third as long as the spikelet, obtuse. Autumnal culms, spreading and branching from upper and middle nodes. Plant, as a whole, in both phases purplish. Pine flatwoods, central and south Fla. to Ga.

25. **P. fasciculatum** Sw.—Culms erect or spreading, sometimes geniculate at base. Blades erect, 3–20 cm long, 5–15 mm wide, thin, pubescent or glabrate; sheaths papillose-hispid or glabrate, shorter than the internodes; ligule membrane fringed with hair. Mature spikelet prominently reticulate veined, abruptly pointed. Grain yellowish, plump, sharply rugulose, 2.2 mm long. $2n = 18$. Moist pinelands, cleared margins of lagoons, Tex., south Fla., tropical America.

26. **P. adspersum** Trin.—Glabrous; culms decumbent, leafy, 2–8 dm long, with creeping, rooting base. Blades 5–20 cm long, 6–18 mm wide, thin, ciliolate at base or glabrous; sheaths shorter than the internodes, pilose at throat or on margins. Panicle 7–15 cm long with ascending racemes; spikelets hispidulous at pointed apex, prominently veined, often obscurely reticulate veined. Grain rugulose, 2.9 mm long. $2n = 54$. Moist soil over lime rocks or sand and shell, south Fla. to Pa.

27. **P. verrucosum** Muhl.—Culms to 12 dm tall, erect, becoming diffusely spreading; blades soft, thin, lax, to 2 dm long, 3–7 mm wide. Panicle to 3 dm long, diffuse, spikelets to 2 mm long, elliptic, remote, outer glume less than 1 mm long. $2n = 36$. Moist banks along creeks and marshy borders of canals, Fla., Tex., to Mass., Mich., Ark. Fall.

28. **P. lacustre** Hitchc. & Ekman—Aquatic perennial, or terrestrial; culms of terrestrial plants simple to 1 m tall; innovations short. Blades to 10 cm long, 2–4 mm wide, pilose above; sheaths pilose, shorter than the internodes; ligules membranous, fringed with hairs. Panicle erect, 10–25 cm long, branches, ascending with appressed spikelets 2–2.2 mm long, glabrous. Cypress ponds, everglades, Fla., Cuba.

29. **P. bartowense** Scribn. & Merrill—Culms stout, erect, branching above 1 m or taller, 7 mm thick at base. Blades 1.8–3 dm long, sheaths loose, longer than the internodes; ligule 2–3 mm long. Panicle simple, open, spreading 1.5–6 dm long; spikelets 2.5 mm long, glabrous; grain shiny. Low grounds, Fla., Bahama Islands, Cuba.

30. **P. dichotomiflorum** Michx.—Culms stout, ascending, 5–10 dm tall, branching from the base. Blades 1–5 dm long, scaberulous; lower sheaths overlapping the internodes; ligule hairs 1–2 mm long. Panicle

terminal and axillary, 1–4 dm long, partly included in sheath; spikelets 2.5 mm long, acute, grain shiny. $2n = 54$. Along streams and ditches, in moist ground, Fla. and Tex., Nova Scotia to Minn.

31. P. miliaceum L. Broomcorn Millet—Culms to 1 m tall, stout, usually erect; blades to 3 dm long, 2 cm wide, hirsute above and below or glabrate. Spikelets to 5 mm long, seven- to nine-veined, glabrous, the outer glume clasping, the inner glume and sterile lemma pointed; grain lustrous, 2 mm wide. Disturbed sites, escaped from cultivation, naturalized from Old World throughout U.S.

32. P. purpurascens Raddi, Para Grass—Culms lax, 2–5 m long. Blades 1–3 dm long, glabrous or sparsely pilose; nodes hirsute; sheaths papillose-hispid. Panicle 8–15 cm long, racemes remote; few to many, ascending or spreading; spikelets glabrous, disposed in cluster on side of axis; outer glume, short deltoid. Grain obscurely rugulose. Widely cultivated as forage grass, freely escaping, marshy ditches, Fla., Ala., Tex., Oreg., and tropical America. *P. barbinode* Trin.

33. P. maximum Jacq. Guinea Grass—Culms to 2 m or more, tall, pale green, glabrous in large bunches. Blades flat, linear to 5 dm long or more, 1–3 cm wide, scabrous on margins, usually hirsute above at base, the midvein prominent white. Sheaths glandular-pustulate, hirsute, or glabrescent, commonly exceeding the nodes. Panicle to 5 dm long, diffuse, the primary branches in whorls, pendent, pilose in axils. Spikelets 4.5 mm long, first glume about one-third the length of the spikelets. $2n = 32$. Naturalized from Africa, escaped from cultivation in Fla., widely distributed in warmer parts of both hemispheres.

34. P. gymnocarpon Ell.—Creeping horizontal base to 2 m long; culms upright, commonly remote, to 1 m tall. Blades elongate, 10–35 mm wide, scabrous on margins. Panicle 2–4 dm long with ascending racemes; spikelets 6–7 mm long, outer glume slightly shorter than the sterile lemma; the glumes and sterile lemma acuminate, 2.1 mm long, pointed beyond the narrowly obovate, shiny brownish fruit. Cypress ponds, muddy banks of streams and lakes, Fla. to S.C., Ark.

35. P. hemitomon Schultes, Maidencane—Culms 0.5–1.5 m long, stiff; sterile shoots numerous. Blades 1–2.5 dm long, 3–10 mm wide, scabrous above. Panicle 10–17 cm long; racemes erect, distant at the base; outer glume less than half the length of the spikelet. Wet banks and fields, coastal plain, Fla. to Tex., N.J., Tenn., South America.

36. P. tenerum Beyr.—Culms to 9 dm tall, wiry, essentially glabrous, contiguous, numerous strong tufts; basal sheaths pilose, blades 3–5 cm long, 2–4 mm wide, involute at tips, pilose above at base. Panicle 3–8 cm long, racemes erect or spreading at maturity; spikelets greenish, pedicel hairs white; outer glumé one-half the length of the spikelet, fruit brownish ellipsoid. Wet soil, glades, swamp and canal margins, Fla., Tex., to N.C.

37. P. repens L.—Culms decumbent, rigid, 3–8 dm long, branching from the rhizome. Blades 6–12 cm long, flat or folded, 2–5 mm wide; sheaths sparingly pilose or glabrate. Panicle 7–12 cm long, open, branches stiff, ascending, spikelets to 2.2 mm long, outer glume truncate about one-fifth as long as the spikelet. $2n = 36, 45, 54$. Moist ditch banks along glade margins, south and central Fla., Tex., subtropical regions of both hemispheres.

38. P. amarulum Hitchc. & Chase, BEACH GRASS—Culms decumbent in large bunches, about 1 cm thick at base and 1–2 m tall, with bluish green foliage; rhizomes vertical in sand or ascending. Blades elongate to 5 dm long, 5–12 mm wide, involute, glabrous or pilose near the base; sheaths loose, glabrous. Panicle large, elongate, nodding. Coastal dunes, Fla., Tex., La., N.J., W. Va., Mexico, Bahama Islands, Cuba.

39. P. virgatum L. SWITCH GRASS—Culms 1 m tall, slender, solitary or a few together in bunches. Blades elongate, 4–7 mm wide, midrib white, prominent. Panicle terminal, broadly diffuse, with verticillate spreading branches; spikelets distant, in anthesis 2.2 mm long; mature spikelets 3 mm long, glabrous. Pinelands, south Fla., central Fla. to Miss., Mich., Tenn., Cuba.

40. P. hians Ell.—Culms to 6 dm tall, tufted, branching, decumbent base; sheaths sparingly pilose or glabrate. Blades 5–15 cm long, 2–5 mm wide, pilose at base on upper side. Panicle 5–15 cm long, branches slender, few, remote, naked below the floriferous tips, the obovate accrescent sterile lemma functional in opening the floret during pollination. Moist banks of lakes and ditches, hammock borders, and low pinelands, Fla., to Va., southwestern U.S. to Mexico. Summer.

41. P. rhizomatum Hitchc. & Chase—Culms to 8 dm tall with vertical and lateral scaly rhizomes and wide-spreading cordlike roots. Blades elongate, linear to 3.5 dm long, often glaucous, pilose, becoming glabrous; sheaths villous on margins and on abaxial side of ligule. Panicles terminal and axillary to 2.5 dm long, open with slender ascending branches. Spikelets 2.5–3 mm long, grain plump, ellipsoidal, pale brown. Hammocks, open pineland, central and south Fla., Tex., Tenn., coastal plain to Me.

42. P. agrostoides Spreng.—Culms compressed, to 9 dm tall in dense tufts from knotted rhizomatous crown, with numerous leaf innovations. Blades deep green to 5 dm long, 4–12 mm wide, with a broad white midvein, at least when young, scabrous above and below; sheaths often purple at base, loose, minutely puberulent or glabrous. Panicles to 2 dm long, racemes distant or congested, branchlets floret bearing to the base, and turned underside of the branches; pedicels apically pilose with delicate hairs, deciduous in age. Spikelets 2.2–2.5 mm long, glumes keeled. Represented in the area by two varieties:

1. Panicle branches remote, spreading, lax;
    spikelets to 2 mm long . . . . . .   42a. var. *agrostoides*
1. Panicle branches appressed, ascending, con-
    gested; spikelets 2.5 mm long . . . .   42b. var. *condensum*

42a. P. agrostoides var. **agrostoides**—Mature plants usually green. Pedicels about 1 mm long; branchlets few-flowered, often reflexed in maturity. Moist margins of hardwoods, QUERCUS, SALIX, SABAL in drier land, central and south Fla., Tex. to Me., Kans., northwest Pacific coast, South America. Fall.

42b. P. agrostoides var. **condensum** (Nash) Fern.—Mature plants usually purple throughout; spikelets sessile or nearly so, crowded in ascending branches, pedicels rarely pilose. Borders of canals, marshy shores of ponds, central and south Fla., Tex., coastal plain states to Ark., W.I. Fall.

TRIBE 13. *Andropogòneae*

1. Internodes of the spike rachis slender with-
    out cavities.
    2. Spikelets homogamous, all essentially
        alike.
        3. Rachis continuous; spikelets falling
            with the hairy callus . . . . .   49. *Imperata*
        3. Rachis articulate; spikelets falling with-
            out or with the rachis internode and
            the pedicel.
            4. Culms erect, robust, panicles plume-
                like.
                5. Spikelets awnless, hairs silky white   50. *Saccharum*
                5. Spikelets awned, hairs brown or
                    tawny . . . .   51. *Erianthus*
            4. Culms low, creeping, mat-forming,
                panicles not plumelike . . . .   52. *Polytrias*
    2. Spikelets heterogamous, the sessile bisex-
        ual, the pedicelled staminate rudimen-
        tary or lacking.
        6. Pedicelled spikelet represented only by
            pedicel.
            7. Racemes two- to three-jointed in con-
                tracted panicles . . . . . .   53. *Sorghastrum*

7. Racemes three- to many-jointed in
   twos or compound panicles . . 54. *Andropogon*
6. Pedicelled spikelets rudimentary or
   staminate.
8. Pedicelled spikelet rudimentary.
9. Outer glume simple; panicle
   branches terminating in solitary
   racemes . . . . 55. *Schizachyrium*
9. Outer glume winged at apex; inflorescence a terminal spike . . 56. *Eremochloa*
8. Pedicelled spikelets staminate.
10. Fertile lemma awned; the awn
    often readily deciduous.
11. Midvein of blades and spathes
    demarked with a row of aro-
    matic sessile glands . . . 57. *Heteropogon*
11. Midvein of blades and sheaths
    without glands.
12. Pedicelled spikelet and sessile
    spikelet about equal in size 58. *Dichanthium*
12. Pedicelled and sessile spikelets
    unequal in size.
13. Racemes golden brown, the
    lower spikelets sterile and
    awnless, the upper pairs
    fertile and sterile; sterile
    lemmas keeled toward the
    apex . . . . . . 59. *Hyparrhenia*
13. Racemes silver gray or yel-
    lowish.
14. Spikelets in threes; pani-
    cle few-flowered; rachis
    branches floriferous at
    tips . . . . . . 60. *Chrysopogon*
14. Spikelets in twos.
15. Spikelets awnless in
    pedicelled racemes;
    inflorescence panicu-
    late; lemmas entire . 61. *Sorghum*
15. Spikelets awned in ses-
    sile racemes; inflores-
    cence subdigitate;
    lemmas pitted . . 62. *Bothriochloa*
10. Fertile lemma awnless; outer
    glume margined with rows of
    aromatic glands . . . . . 63. *Elyonurus*
1. Internodes of the spike rachis thickened
   and modified with cavities.
16. Florets bisexual.
17. Pedicels flanked against the rachis

joints; outer glume transversely rugose	64. *Manisuris*
17. Pedicels adnate to rachis joints; outer glume minutely papillose	65. *Rottboellia*
16. Florets unisexual.	
18. Spikelets in separate inflorescences, staminate terminal, pistillate axillary; mature portion continuous . . .	66. *Euchlaena*
18. Spikelets in the same inflorescence, pistillate portion below the staminate portion; mature fertile portion fragmenting into joints	67. *Tripsacum*

49. IMPERÁTA Cyr. Satintail

Perennials from scaly rhizomes, with terminal, narrow panicles. Spikelets bisexual, awnless, all alike; pedicels unequal; rachis continuous, pilose, silky; glumes subequal, membranous; paleae sterile and fertile, lemmas thin, hyaline. About ten species in tropical and subtropical regions.

1. I. brasiliensis Trin.—Culms rhizomatous to 1 m tall. Leaves crowded toward the base; blades 3–8 mm wide, cauline leaves short, few, passing to bracts below inflorescence. Panicle cylindric, dense, silvery, 15 cm long, denuded pedicels funnel-like, imbricated on capillary axes; spikelets 3–3.8 mm long; inner glume and sterile lemma silky with hairs to 12 mm long. Coastal strand, pinelands and prairies, south Fla., Ala., Mexico, tropical America. Summer.

50. SACCHÁRUM L. Sugarcane

Tall, robust perennials. Spikelets awnless, in pairs, pedicelled and sessile, each bisexual, panicle decompound, rachis articulate, glumes indurate; sterile and fertile lemmas hyaline. About five species in tropical and subtropical regions of both hemispheres.

1. S. officinarum L.—Culms leafy, 2–5 m tall, 2–3 cm thick, solid, juice sweet; lower internodes short and tumid, sheaths overlapping, pilose at throat. Blades auriculate, 3–6 cm wide; midvein broad, white; ligule a fringed membrane. Panicle 3–6 dm long with a profusion of slender peduncled racemes; spikelets 3 mm long, encircled and surpassed by silky hairs. $2n = 80$. Escaping to fields and roadsides, cultivated in southern U.S. and central Fla., W.I. Spring.

51. ERIÁNTHUS Michx. Plume Grass

Tall reedlike grasses perennating by rhizomes. Spikelets alike in pairs, one sessile, the other pedicelled, both bisexual, rachis disarticulating below the spikelet, the pedicel falling attached to the sessile spikelet; glumes equal, coriaceous, subtended by long silky hairs; sterile and fertile lemmas hyaline, the midvein produced to an awn, panicle feathery, purple or tawny. About 28 species of tropical and temperate regions.

1. **E. giganteus** (Walt.) Muhl. SUGARCANE PLUME GRASS—Culms robust to 3 m tall, appressed-villous below the panicle; nodes appressed-villous, glabrescent in age. Blades harshly pubescent-elongate, 6–15 mm wide, linear, coriaceous, long attenuate, scabrous on margins. Panicle 1–4 dm long, elongate, arching; spikelets 5–7 mm long, with straight, terete awns to 25 mm long. Conspicuous in glades and on canal banks, coastal plain, Fla. to Tex., N.Y., Cuba. Summer. *Erianthus saccharoides* Michx.

E. ravennae (L.) Beauv.—Culms to 4 m tall, panicle to 6 dm long, probable garden escape, hardy as far north as New York City. Native of Europe.

52. POLÝTRIAS Hack.

Stoloniferous perennial, 3 dm tall. Spikelets bisexual, in threes, at the nodes of an articulate rachis. Sessile, spikelets two, back to back, one-pedicelled. Outer glume densely hairy. Sterile lemma wanting, fertile lemma hyaline, awn bearing. Awns bent, twisted. A single species, native of Java.

1. **P. racemosa** (Nees) Hack.—Characters of the genus. Near Miami, Fla., rare. Fall.

53. SORGHÀSTRUM Nash   INDIAN GRASS

Perennial grasses with or without rhizomes. Blades flat auricled; panicles terminal. Spikelets in pairs, one sessile, bisexual, the other wanting, only the pedicel present; glumes coriaceous, the outer hirsute, brownish; lemmas thin, hyaline, the fertile one awned. About 12 species in warm temperate regions.

1. **S. secundum** (Ell.) Nash, INDIAN GRASS—Culms tufted to 1 m tall or taller, bases glabrate in age. Blades elongate, 3–4 mm wide, involute at tips; ligule 1.5 mm long, membranous. Panicle unilateral, nodding, nodes of primary rachis bearded; branches fasciculate, flexuous; spikelets 7 mm long, pilose, yellowish brown, with bearded callus 1.5 mm long; awn to 3 cm long, twice geniculate. Pinelands, Fla. to Tex., S.C. Summer.

54. ANDROPÓGON L.   BEARD GRASS

Cespitose perennial grasses with solid stems, simple or compound panicles of spikelets in peduncled racemes from spathelike sheaths;

flowering culms simple or branching, commonly flattened, sheaths keeled, blades flat or folded. Spikelets in pairs at each node of an articulate rachis, one sessile and bisexual, the other pedicelled, staminate, neuter or reduced to a pedicel; the rachis and pedicels of the sterile florets commonly villous; glumes of the fertile florets coriaceous, awnless, the outer convex, plane or concave on the back; sterile lemma hyaline, empty; fertile lemma hyaline, entire or notched, awned from the apex or between the lobes; palea hyaline or sometimes wanting; stamens one or three; pedicelled spikelet awnless. About 113 tropical and subtropical species. *Schizachryium* Nees

1. Stamen one; outer glume nerveless between the keels.
    2. Inflorescence decompound; spathes aggregated in corymbiform masses of racemes 1. *A. glomeratus*
    2. Inflorescence simple, or, if compound, spathes loosely flabellate.
        3. Upper sheaths prominently enlarged; spathiform; racemes included or becoming exserted . . . . . . 2. *A. elliottii*
        3. Upper sheaths not prominently enlarged.
            4. Blades of the innovation filiform, permanently folded, without upper epidermis . . . . . . 3. *A. perangustatus*
            4. Blades flat.
                5. Peduncles less than 2 cm long; spikelets flexuous.
                    6. Villous hairs of racemes dense, obscuring the spikelets; spikelets 4–4.5 mm long . . . . 4. *A. longiberbis*
                    6. Villous hairs of racemes sparse, not obscuring spikelets; spikelets 3 mm long.
                        7. Rachis internodes longer than spikelets . . . . . 5. *A. virginicus*
                        7. Rachis internodes shorter than spikelets . . . . . 6. *A. capillipes*
                5. Peduncles more than 2 cm long.
                    8. Racemes 2.5 cm long; panicle branches capillary . . . . 7. *A. floridanus*
                    8. Racemes 1.5 cm long; panicle branches subfiliform . . . 8. *A. brachystachys*
1. Stamens three.
    9. Outer glume of sessile floret with two to three veins between keels; racemes tawny . . . . 9. *A. cabanisii*
    9. Outer glume of sessile floret obscurely one- to two-veined; racemes pure white 10. *A. ternarius*

1. **A. glomeratus** (Walt.) BSP—Culms stout, erect, to 15 dm tall, flattened; branching above, forming large clumps; sheaths loosely imbricate-pubescent at summit, pilose at throat, glabrescent. Blades elongate, 3–8 mm wide; scabrous on margins. Racemes to 3 cm long, paired, peduncled from prominent spathes; peduncles and ultimate branchlets long villous; sessile spikelets 3–4 mm long with straight awn to 15 mm long; rachis capillary, flexous, densely villous; sterile spikelet reduced or wanting. $2n = 20$. Glades, open pinelands, Fla. to southern Calif., W.I., Mexico, Central America.

2. **A. elliottii** Chapm.—Culms tufted, 3–8 dm tall; upper node strongly bearded; sheaths and bases of blades above the ligule diversely pilose, glabrescent. Blades flat, elongate 2–5 mm wide, 3–4 dm long, becoming curled or twisted in age; the upper sheaths subtending racemes becoming greatly enlarged, or inflated, successively in flabellate fascicles. Primary racemes long exserted; secondary racemes in twos, rarely exserted or included; sessile spikelet 4–5 mm long, with geniculate awn to 2.5 mm long; pedicelled spikelets rudimentary, white hairy. Pinelands, Fla. to N.J.

3. **A. perangustatus** Nash—Culms 3–7 dm tall, slender, wiry, in small tufts, branching above; blades folded, elongate, 1–3 dm long, 1–2 mm wide; ligule conspicuous, rounded or pointed; blades copiously pilose on lower side when young or glabrescent; sheaths keeled, glabrous in age. Racemes in twos on slender peduncles, with narrow spathes; racemes flexuous, silky white; sessile spikelet 4–5 mm long, with pilose tuft at base; awn to 2 cm long; pedicelled spikelet reduced, pedicels long pilose. Seasonally wet pineland, ditch banks, Fla. to Miss.

4. **A. longiberbis** Hack.—Culms wiry, branching above, to 8 dm tall; sheaths grayish villous, glabrate. Blades 1–5 mm wide, elongate, becoming flexuous in age; ligule truncate, less than 1 mm long. Panicle elongate with distant racemes in twos; subtending spathes narrow, 3–6 cm long; ultimate peduncles paired, capillary; racemes flexuous; sessile spikelet 3 mm long with awn 12 mm long, pedicels tufted, pilose, spikelet wanting. Similar to the following species, **A. virginicus**. Pinelands, Fla., Ga. Fall.

5. **A. virginicus** L. Broom Sedge—Culms to 10 dm tall, slender or moderately stout, branching above, in tufts of few to many. Leaves equitant, green, glaucous or purplish; sheaths glabrous or sparingly pilose on upper surface or glabrate. Inflorescence, simple, paniculate or corymbiform, spathes 2–6 cm long, usually shorter, racemes tardily exserted. Rachis slender, flexuous, long silky; sessile spikelets 3 or 4 mm long, awn straight, delicate, 2 cm long; pedicels long villous. A variable species represented in our area by two varieties:

1. Spikelets in simple racemes; leaves and
sheaths green, not pruinose . . . . . 5a. var. *virginicus*
1. Spikelets mainly in paniculate groups;
leaves glaucous, sheaths purplish, pruinose at base . . . . 5b. var. *glaucopsis*

5a. A. virginicus var. **virginicus**—Open pinelands, adjacent glades
and prairies, in sandy soil, Fla., Tex. to Mass., Calif., W.I., Central
America.

5b. A. virginicus var. **glaucopsis** (Ell.) Hitchc.—Marshy prairie,
glades, Fla. to Va. Summer.

6. A. capillipes Nash—Culms erect, 6–8 dm tall, tufted, rhizomatous, branching from upper nodes, prominently glaucous or purplish;
sheaths crowded at base, distichous, chalky white, yellowish green, or
purplish. Basal leaves elongate, 2–5 mm wide, flat or folded; ligule obsolescent. Racemes two or more, subtended by purplish or bronzed inflated spathes 2–3 cm long, reflexed or arcuate on capillary peduncles;
sessile spikelet 3 mm long, long villous, with straight awn to 10 mm
long; pedicels villous, sterile floret wanting. Sandy pineland, white sand
oak scrub, Fla. to Carolinas. Fall.

7. A. floridanus Scribn.—Culms usually stout 1–1.8 m tall, erect,
branching at the upper nodes. Blades elongate, 2–7 mm wide, glabrous;
ligule membrane short fringed, less than 1 mm long; sheaths loose, sometimes exceeding the internodes. Inflorescence of numerous pairs of silvery white to creamy racemes, exserted from slender spathes in loosely
corymbose flabellate fascicles; mature sessile spikelets 3 mm long with
awn 10 mm long, copiously soft villous throughout; sterile spikelet wanting, the pedicels villous. Low pinelands, endemic to central and south
Fla. Fall.

8. A. brachystachys Chapm.—Culms tufted, erect to 1.5 m tall, rhizomatous base, branching from the upper nodes; sheaths of tufted leaves
keeled, compressed, distichous, 15–25 cm long, pilose at throat; cauline
sheaths somewhat shorter than the internodes; blades folded, 4–6 mm
wide, glabrous; inflorescence elongate, open, branching somewhat flexuous with reflexed ultimate branches; spathes slightly dilated 2–3.5 cm
long, bronzed, subtending two racemes 1–2.5 cm long; sessile spikelet 3
mm long, with a straight awn 1 cm long; pedicel silky white, the spikelet
wanting. Seasonally wet pineland swales, shores of lakes, Ga., Fla.

9. A. cabanisii Hack.—Culms erect, slender, 8–15 dm tall, from rhizomatous base; branching about the middle; sheaths puberulent or
glabrate in age; blades 1–4 mm wide, flat or involute, pilose above the
ligule on the upper side, flexuous in age; ligule membranous, 0.6 mm
long, long-exserted peduncles; racemes two, grayish or tawny, 4–7 cm

long, spathes inconspicuous or slightly dilated; rachis straight; sessile spikelets 6–7 mm long, joints and pedicels sparsely villous; the outer glume scabrous, two-nerved between the keels; awn 1 cm long, twisted at base; pedicelled spikelet rudimentary. Sand hills and pinelands, central and south Fla.

10. A. ternarius Michx.—Culms to 1 m tall or less from tufted leafy base, rigid, branching above. Blades 1–2 dm long, 2–4 mm wide, glabrous, sheaths loose, ligule membranous about 1 mm long, innovations 1 mm wide or less, sheaths loose, shorter than the internodes. Racemes long peduncled, silvery, gray to white, 1–3 cm long; sessile spikelet, first glume channeled; lemma awn to 2 cm long, twisted, pedicelled spikelets reduced to a mere scale; fruit deciduous, attached to a villous joint and pedicel. Pine flatlands, sand hills, open woods, Fla., Tex. to Ky., Kans. Spring, fall.

55. SCHIZÁCHYRIUM Nees

Annual or perennial grasses, tufted or rhizomatous. Culms branched above. Leaf blades commonly narrow, flat or involute. Racemes terminal, solitary, pedunculate; spikelets in pairs, one sessile, the other pedicelled, a reduced spikelet, or only an awned glume; rachis articulate, hirtellous; sessile spikelets bisexual, one-flowered with awned lemma; glume dorsally compressed, rachis joints funnel shaped, conspicuous when denuded, falling attached to the spikelet and the pedicel. About 50 species, chiefly tropical and subtropical regions. *Andropogon* L.

1. Rachis straight; leaf blades flat or conduplicate.
  2. Raceme glabrous; sessile spikelet 4 mm long . . . .   1. *S. tenerum*
  2. Raceme pubescent, at least in lines; spikelets 6 mm long.
    3. Rachis internodes pubescent on outer edge; pedicels and the first glume of the sessile flower glabrous . . . .   2. *S. semiberbe*
    3. Rachis internodes, pedicels, and the first glume of the sessile flower pubescent . . . .   3. *S. hirtiflorum*
1. Rachis flexuous, leaf blades terete or flat; plant tufted or rhizomatous.
  4. Leaf blades terete, wiry; rachis hairs obscuring the spikelets; plant tufted . .   4. *S. gracile*
  4. Leaf blades flat; plant rhizomatous.
    5. Blades 1–3 mm wide; racemes strongly flexuous; sheaths convex or obscurely keeled . . . .   5. *S. rhizomatum*
    5. Blades 3–5 mm wide; racemes moderately flexuous; sheath strongly keeled   6. *S. stoloniferum*

1. **S. tenerum** Nees—Densely tufted grass in tight bunches from knotty crown; culms slender, 4–10 dm tall, spreading to reclining. Blades 1–2.5 mm wide, flat becoming involute, or involute on innovations; sheaths slightly shorter than the internodes, pilose at throat; ligule less than 1 mm long, membranous. Raceme slender, 2–5 cm long, nearly glabrous; sessile spikelet excluding awn 5–7 mm long. Pinelands and prairies, coastal plains to Ga., Fla., Tex., tropical America. *Andropogon tener* (Nees) Kunth

2. **S. semiberbe** Nees—Culms in small tufts, 6–12 dm tall, erect, rigid, pinkish, branching above; blades 2–4 mm wide, glabrous; racemes erect, 5–8 cm long, rachis joints white bearded at base; spikelets 6 mm long, awn to 15 mm long; sterile pedicel ciliate on outer margin. Pinelands, Fla. to tropical America. *Andropogon semiberbis* (Nees) Kunth

3. **S. hirtiflorum** Nees—Slender, rigid, tufted perennial 5–10 dm tall, with compressed culms. Blades flat, linear, elongate, 2–4 mm wide; ligule membranous 1 mm long, sheaths glabrous. Racemes 4–8 cm long, straight or curved; rachis joints and pedicels villous; sessile spikelets 5–6 mm long, outer glume villous on the convex back; awn 1 cm long, geniculate, twisted. $2n = $ c. 100. Sand hills, pinelands, Fla., Tex. to Ariz., Calif., Mexico. *Andropogon hirtiflorus* (Nees) Kunth

4. **S. gracile** (Spreng.) Nash—Culms 2–4 dm tall, slender, wiry, glabrous in dense tuft from knotty crowns. Blades terete, 4 cm long, filiform tipped with white hair. Racemes to 4 cm long, flexuous, copiously long villous, white; sessile spikelet 5 mm long, with awn 1–2 cm long. Crevice of oolite, pinelands, south Fla., W.I. Fall. *Andropogon gracilis* Spreng.

5. **S. rhizomatum** (Swallen) Gould—Culms 5–7 dm tall, in small tufts, erect from short rhizome, branching above the middle; sheaths rounded or slightly keeled, exceeding the internodes. Blades grabrous-elongate, 10–25 cm long, 1–3 mm wide, flat or somewhat involute. Raceme 2–3 cm long, spathe inconspicuous; peduncles 3–7 cm long; sessile spikelet 5–6 mm long, awn 8–10 mm long, geniculate, twisted; outer glume convex, obscurely keeled near the summit. Rocky pinelands, endemic to south Fla. *Andropogon rhizomatus* Swallen

6. **S. stoloniferum** Nash—Culms to 1.5 m tall, solitary, or few to several in a bunch, glabrous. Blades 2–5 mm wide, flat, elongate 4, dm long. Panicle large, with numerous peduncled racemes 3–4 cm long; rachis joints silky pubescent; pedicels silky villous; sessile spikelets 6–7 mm long, scabrous at summit; awns 10–12 mm long, bent and twisted. Pinelands and alluvial woods, Fla., Ga., Ala. *Andropogon stolonifer* (Nash) Hitchc.

56. EREMÓCHLOA Buese   Centipede Grass

Creeping perennial with stout stolons, short internodes with overlapping equitant sheaths. Spikelets two-flowered, flat, secund, sessile; glumes flat membranous, sometimes pectinate on margins. About eight species in India.

1. **E. ophiuroides** Hack. CENTIPEDE GRASS—Peduncles slender, terminal or axillary, solitary or fasciculate, subtending terminal spicate racemes 3–3.5 cm long. Spikelets 4 mm long, compressed, appearing as a single row along flattened rachis; outer glume alate at apex; fertile lemma and palea hyaline; grain lenticular, bright brown, 1.5 mm long, 1 mm wide. A coarse cover plant cultivated for lawns. Fla., southeastern U.S., Calif., Asia.

57. HETEROPÒGON Pers.   TANGLEHEAD GRASS

Robust annuals or perennials with flat blades, long leafy panicles of terminal racemes, rachis continuous at base, articulation below the sessile spikelets. Spikelets in pairs, one pedicelled, the outer sessile, the lower pairs sterile, staminate or neuter; the upper sessile spikelets bisexual, long awned; pedicelled spikelets staminate, awnless, glumes of the sessile spikelets equal, papery; fertile lemmas thin, narrow, produced into long, twisted, bent awns. About 12 tropical species.

1. **H. melanocarpus** (Ell.) Benth.—Annuals 1–2 m tall with numerous branching. Blades elongate, 5–10 mm wide; sheaths glabrous, the upper sheaths with concave glands on the midvein; sessile spikelets 9–10 mm long, with awn 10–15 mm long; callus sharp, copiously bearded. Emitting citronella fragrance. Pinelands, fields, prairies, Fla., Ga., Ala., Ariz. Fall.

58. DICHÁNTHIUM Willem.

Tall leafy grasses from decumbent base. Racemes mostly in twos and fours, sometimes solitary; sessile spikelets bisexual, about the size of the staminate, pedicelled spikelets; pedicelled spikelets awnless. About 15 species, Old World tropics. *Andropogon* L.

1. **D. aristatum** (Poir.) Nash—Culms to 1 m tall, stout, pubescent below the panicle. Leaf blades to 3 dm long, 4–6 mm wide, glabrous or pubescent at base; sheaths keeled, shorter than the internodes. Glumes of sessile spikelets fifteen- to nineteen-veined. $2n = 40$. Cultivated and naturalized in south Fla., Africa and Australia. *Andropogon nodosus* (Willem.) Nash. Palatable, excellent for fodder.

59. HYPARRHÈNIA Anderss. ex Stapf

Perennials with elongate leaves. Spikelets in pairs, the lower pairs alike and sterile; fertile spikelets one to few in each raceme, terete or flattened on the back; keeled toward the apex. Fertile lemma with geniculate awn; the sterile lemma awnless. Racemes digitate, paired. About 75 species, tropical Africa, Mediterranean, Arabia.

1. **H. rufa** (Nees) Stapf—Culms erect to 1 m or more tall; blades flat, elongate, 2–8 mm wide, scabrous. Panicles 2–6 dm long; peduncles filiform, flexuous, fertile spikelets pubescent with dark reddish brown hairs. Awn 15–20 mm long, twice geniculate. $2n = 30,40$. Naturalized from the tropics, escaping from cultivation, south Fla.

60. CHRYSOPÒGON Trin.

Decumbent annual grass. Spikelets in threes; the sessile bisexual, flanked by a staminate or rudimentary pedicelled spikelet on each side. Glumes coriaceous; lemmas thin, hyaline, the fertile one awned. About 25 tropical species, Old World.

1. **C. pauciflorus** (Chapm.) Benth. ex Vasey—Leaves and culms spreading radially from the crown of the primary root, culms 6 dm or more long, surpassing the leaves. Blades broadly linear-attenuate, 4–8 mm wide, prominently ciliate below the middle; sheaths overlapping. Panicle rachis to 10 cm long, with pairs of stalked, distant, reduced racemes; sessile spikelets to 12 mm long with golden brown silky callus 4–7 mm long, mature spikelets with awn to 15 cm long, twisted below; glumes dark brown, scaberulous at apex. Open sandy pineland, central and south Fla., Cuba. Fall. *Raphis pauciflora* (Chapm.) Nash

61. SÓRGHUM Moench JOHNSON GRASS

Annual or perennial grasses with broadly linear, flat blades, sheaths shorter than the internodes. Spikelets in pairs, one sessile, perfect; the other pedicelled, staminate. Terminal, bisexual, spikelet flanked by two pedicelled spikelets. Panicle overtopping the leaves. About 60 species in tropical and warm temperate regions. *Holcus* L.

1. **S. halepense** (L.) Pers.—Culms glabrous, to 1 m tall from scaly, creeping rhizomes. Blades auriculate, 0.8–1.6 dm wide, midvein white, prominent; ligule a fringed membrane 1–2 mm long. Panicle racemose, 1.5–3 dm long; bisexual spikelets 4–5 mm long, short-silky; awn geniculate, early deciduous. $2n = 20$. Roadsides and wasteland, forage grass escaping from cultivation, Fla., Tex., central U.S. *Holcus sorghum* L. S. VULGARE Pers.—A robust annual cultivated as Broomcorn, or as other races, Soudan grass, doubtfully persisting in waste areas without cultivation. Naturalized from Eurasia.

62. BOTHRIÓCHLOA Kuntze

Perennials; culms simple or branching. Racemes several to numerous in leafless panicles; rachis straight, joints and pedicels flat, bearded on margins, subhyaline at center. About 20 species of tropics, mostly Asiatic. *Andropogon* L.; *Amphilophis* Nash

1. **B. pertusa** (L.) A. Camus—Culms ascending, becoming prostrate, giving rise to nodal flowering branches to 3 dm tall, or erect in knotty tufts to 6 dm tall, nodes bearded, sheaths shorter than internodes. Blades 2–4 mm wide, flat linear-attenuate; ligule pilose. Flowering peduncles above the uppermost leaves to 2 dm long, panicle terminal; racemes aggregate or subdigitate, few to many, purplish; pedicels and internodes grayish hirsute, sessile spikelets awned; pedicelled spikelets awnless, pitted. $2n = 40,60$. Tenacious grass of waste areas; disturbed sites, Fla. Keys, naturalized from Old World. *Holcus pertusa* L.; *Andropogon pertusus* Willd.; *Amphilophis pertusa* Nash

63. ELYONÚRUS Humb. & Bonpl. ex Willd.

Perennials with spicate racemes, hirtellous rachis. Spikelets awnless in pairs, rachis joints and pedicels parallel, thickened; sessile spikelet bisexual, the pedicelled staminate, both falling with the rachis joint; outer glume coriaceous, dorsally compressed, ciliate above the middle and the glutinous aromatic margins incurving around the second glume; sterile and fertile lemmas similar, hyaline, paleae obsolete. About 15 species. Important forage grasses in tropical America.

1. **E. tripsacoides** Humb. & Bonpl. ex Willd.—Culms 0.6–1.2 m tall, rigid, glabrous branching from the rhizomatous base. Blades elongate involute, strongly pilose above the short ligule. Spikelets 6–8 mm long; pedicels pilose in lines. Glades and prairies, Fla. to Tex., Ga., Miss., Mexico, South America.

64. MANISÙRIS L. Necklace Grass

Tall grass with flat blades, involute toward tips; ligule short, hyaline. Spikelets sessile, awnless, paired with bisexual flowers, the pedicelled rudimentary at the joints to a thickened rachis, the pedicel appressed against the rachis or sometimes adnate to it; glumes coriaceous, covering the floret-containing cavity between the joint and the pedicel; racemes terete, several, short peduncled. About five species, pantropical. *Rottboellia* L. f.

1. **M. rugosa** (Nutt.) Kunze—Culms to 7 dm tall, branching above, glabrous; sheaths shorter than the internodes, keeled, compressed. Racemes 3–6 cm long; with bases included in sheaths, sessile spikelets 3.5–5 mm long, outer glume transversely ridged. Flat pinelands, Fla., Tex., Ark., on the coastal plain. Summer. *Rottboellia rugosa* Nutt.

65. ROTTBOÉLLIA L. f. ITCHGRASS

Robust, leafy branching annual with broad linear leaves from rhizomatous base. Rachis joints thickened, hollow, cylindrical, flanked by the adnate pedicel of sterile spikelet and by the sessile fertile spikelet; outer glume coriaceous, the inner glume thinner; sterile and fertile lemmas and paleae hyaline. Four species of tropical and subtropical regions.

1. **R. exaltata** L.—Culms to 3 m tall, glabrous, solid, at least above; sheaths loose, papillose-hirsute. Blades to 3 cm wide, elongate, with scabrous margins and a prominent white midrib. Racemes long peduncled to 10 cm long; bases concealed by spathes; apical spikelets abortive; sessile spikelets 5–7 mm long; outer glume papillose. Naturalized in Miami, Fla., from tropical Asia. *Manisuris exaltata* (L. f.) Kuntze

66. EUCHLAÈNA Schrad. TEOSINTE

Robust annual or perennial plant with flat, broad blades 3–8 cm wide, ligule a ciliolate membrane; sheaths coarsely hairy at least along margins. Inflorescence similar to that of *Zea*. Staminate spikelets terminal, pistillate axillary. Mexican forage grass, two species, Mexico.

1. **E. mexicana** Schrad.—Culms branching from the base 2–3 m tall or taller in large clumps. Pistillate spikelets enfolded in series of foliaceous spathes, with leaf sheaths, style elongate, florets alternating on slender rachis; staminate flowers in pairs, exposed, tassel fashion. $2n = 20$. Escaping from cultivation, rare in south Fla.

67. TRÌPSACUM L. GAMA GRASS

Coarse perennials with flat, mostly wide, blades and one to several terminal or axillary racemes, monoecious, staminate and pistillate spikelets in the same inflorescence; staminate spikelets two-flowered, in pairs at nodes of the terminal continuous portion of the rachis, one sessile, the other pedicelled; glumes equal, membranous, many-veined, outer glume two-keeled with sharply inflexed margins. Pistillate spikelets solitary on opposite sides on lower portion of the rachis, recessed within the thickened indurated disarticulating segments, composed of one fertile floret and sterile lemma; outer glume coriaceous, obtuse, the spikelet nearly enclosed by its margins; sterile lemma, fertile lemma, and palea hyaline, each successively diminished. Seven species, temperate and tropical America. Represented in our area by two species.

1. Blades 1–3 cm wide, culms 1–3 m long . . 1. *T. dactyloides*
1. Blades less than 8 mm wide, culms to 1 m
   long . . . . 2. *T. floridanum*

1. T. dactyloides L. EASTERN GAMA GRASS—Culms in large clumps from tough, hard rhizomes; blades 6 dm long or more, margins white cartilaginous, scabrous, midvein broad, white. Racemes 15–25 cm long, usually three together, or one if axillary; staminate spikelets with hyaline lemmas and paleae, floret of three stamens, anthers 3–4 mm long, golden colored, deciduous with the subtending rachis as a whole; styles plumose, purple, exserted beyond the tip of glumes 15–20 mm. Pond margins, hammocks, Fla., Tex., coastal plains states, Mexico, South America, W.I. Summer, fall.

2. T. floridana Porter ex Vasey, FLORIDA GAMA GRASS—Similar to preceding species. Culms relatively slender, blades to 2 dm long, racemes solitary with fewer pistillate spikelets. The plant as a whole usually purplish; inflorescence overtopping the leaves. $2n = 36$. In shallow soil of rock crevices, endemic to south Fla. Spring, fall.

33. CYPERÀCEAE SEDGE FAMILY

Perennial or annual herbs, tufted or from creeping rhizomes. Stems solid or rarely hollow, three-sided, sometimes four-sided to terete. Leaves alternate in three ranks; blades rarely altogether reduced; sheaths closed or rarely open. Flowers bisexual or unisexual, disposed in spikelets, aggregated into inflorescences; typical perianth lacking or reduced to scales, bristles, or hairs; spikelet scales distinct, or a saccate perigynium, as in the pistillate flower of CAREX; stamens one to three with free filaments; anther dehiscent lengthwise. Pistil two- or three-carpellate; style with two to three branches; ovary one-locular with a solitary basal ovule. Fruit an achene or drupe, lenticular or trigonous; embryo small, embedded in copious endosperm. Cosmopolitan distribution with 90 genera and 4000 species.

1. Flowers bisexual with stamens and pistils on the same receptacle.
  2. Scales of spikelets two-ranked; perianth lacking.
    3. Achene smooth, brownish, stalkless or nearly so . . . .   1. *Cyperus*
    3. Achene rugose, white, stipitate . . .   2. *Abildgaardia*
  2. Scales of spikelets spirally imbricate; perianth of bristles present or lacking.
    4. Flowers several to many in a spike.
      5. Base of style thickened.
        6. Bristles wanting.
          7. Style wholly deciduous . . .   3. *Fimbristylis*

7. Style deciduous above its tubercled base . . . . . .   4. *Bulbostylis*
6. Bristles present.
  8. Achene crowned by persistent,
tuberclelike base of style;
stems leafless, not branching   5. *Eleocharis*
  8. Achene without tubercle; base of
style deciduous; stems leafy,
branching . . . . . .   6. *Scirpus*
5. Base of style not thickened.
  9. Achene subtended by bristles or
bristles and scales.
    10. Flower surrounded only by bris-
tles (rarely by scales, alone,
Sect. *Kyllinga*) . . . . .   6. *Scirpus*
    10. Flower surrounded by stipitate
scales, alternating with bristles . . . .   7. *Fuirena*
  9. Achene only subtended by scales.
    11. Subtending scale one, minute;
bract one . . . . . .   8. *Hemicarpha*
    11. Subtending scales two, patent;
bracts two to three . . . .   9. *Lipocarpha*
4. Fertile flowers relatively few, one to
two; the lower scales flowerless.
  12. Flowers destitute of bristles.
    13. Stigmas two; achene lenticular, tu-
bercled.
      14. Spikelets white, in heads; sub-
tended by basally white in-
volucral leaves . . . . .   10. *Dichromena*
      14. Spikelets blackish, in large pani-
cles; involucral leaves lacking   11. *Psilocarya*
    13. Stigmas three; fruit without tuber-
cle.
      15. Spikelets aggregated in involu-
crate heads; fruit a trigonous
achene; stoloniferous plant of
sea beaches . . . . . .   12. *Remirea*
      15. Spikelets in panicles; fruit
drupelike, three-lobed endo-
carp; tall plants of glades . .   13. *Cladium*
  12. Flowers with bristles.
    16. Stigmas two, achene lenticular,
beaked; spikelets brownish . .   14. *Rhynchospora*
    16. Stigmas three, achene trigonous,
beakless; spikelets blackish . .   15. *Schoenus*
1. Flowers unisexual, pistillate and staminate
in the same or in distinct spikes.

17. Achene naked, white, crustaceous, usually subtended by disk . . . . . 16. *Scleria*
17. Achene concealed within perigynium . 17. *Carex*

1. CYPÈRUS L. GALINGALE

Annual or perennial, rhizomatous or cormose herbs with fibrous roots; culms three-sided, with basal leaves, and umbelliform terminal inflorescence spikelets or spikes subtended by involucral leaves. Flowers bisexual without perianth; stamens one to three. Stigmas one to three; ovary three-carpellate, one-locular, maturing into a lenticular or trigonous achene in axils of two-ranked scales. About 550 species in tropical and warm temperate regions.

Key to Subgenera

1. Stigmas two, achene lenticular.
  2. Spikelets one-flowered, falling entire
    from the rachis; rachilla articulate . . 1. *Kyllinga*
  2. Spikelets two- to several-flowered, only the
    scales falling from the rachis; rachilla
    continuous . . . . 2. *Pycreus*
1. Stigmas three, achene trigonous.
  3. Rachilla articulate at the base of each
    scale; scales distant on the same side of
    the rachilla . . . . 3. *Torulinium*
  3. Rachilla not articulate; internodes
    straight; scales imbricate on the same
    side of the rachilla.
    4. Spikelets deciduous from the rachis;
      scales mostly persistent . . . . 4. *Mariscus*
    4. Spikelets persistent on the rachis;
      scales usually deciduous from the
      rachilla . . . . 5. *Cyperus*

1. KYLLINGA

1. Tufted annual; keels of fruiting scales
  crested with pectinate wings . . . . 1. *C. metzii*
1. Stoloniferous perennial; keels of fruiting
  scales scabrous, wingless . . . . . 2. *C. brevifolius*

2. PYCREUS

1. Denuded rachis straight; scales of spikelets
  pointed or obtuse.
  2. Midvein of scales excurrent, sides hyaline,
    silvery . . . . 3. *C. pusillus*
  2. Midvein of scales not excurrent, sides hya-
    line, yellowish . . . . 4. *C. flavescens*

1. Denuded rachis geniculate; scales of spikelets mucronulate, sides of scales· stramineous or ferruginous . . . . 5. *C. polystachyos*
   var. *texensis*

3. TORULINIUM

1. Culms slender, leaves about 0.3 mm wide,
   spikes in sessile simple umbel . . . . 6. *C. filiformis* var. *densiceps*
1. Culms stout, leaves 10 mm wide or less,
   spikes in peduncled decompound umbel . 7. *C. odoratus*

4. MARISCUS

1. Spikes solitary, capitate.
   2. Involucral bracts spreading to reflexed;
      scales ovate, 2.5–3 mm long, greenish
      gray . . . . 8. *C. filiculmis*
   2. Involucral bracts ascending, erect; scales
      oblong-ovate, 1.8–2 mm long, maroon
      red . . . . 9. *C. fuligineus*
1. Spikes more than one, usually in simple umbelliform clusters.
   3. Plants relatively slender; umbel simple.
      4. Spikelets radiate in subspherical heads 10. *C. globulosus*
      4. Spikelets reflexed or ascending in cylin-
         dric, ovoid, ellipsoid heads.
         5. Spikes less than 1 cm long, short-
            ovoid; plants glaucous . . . . 11. *C. nashii*
         5. Spikes 1–3 cm long; plants usually
            bright green.
            6. Mature spikelets strongly retrorse,
               one- to two-flowered; spike sim-
               ple, clavate . . . . . . 12. *C. retrorsus*
            6. Mature spikelets mostly ascending,
               four- to seven-flowered; spike
               ovoid . . . . 13. *C. pollardii*
   3. Plants relatively robust; umbels com-
      pound.
      7. Blades and bracts conspicuously septate-
         nodulous with scabrous margins;
         spikelets 3–6 mm long . . . . 14. *C. ligularis*
      7. Blades and bracts not conspicuously
         nodulous, margins sometimes scaber-
         ulous.
         8. Umbels maroon brown, more or less
            contracted; spikelets 10–17 mm
            long . . . . 15. *C. planifolius*
         8. Umbels greenish brown, diffuse; rays
            elongate.
            9. Spikelets distant, 3–6 mm long . . 16. *C. tetragonus*

9. Spikelets congested, plumosely disposed, 7–19 mm long.
　10. Scales linear-elliptic, appressed . . 　17. *C. strigosus*
　10. Scales linear-lanceolate, spreading . 　18. *C. stenolepis*

5. CYPERUS

1. Wings of rachilla inconspicuous or wanting.
　2. Culms and foliage glutinous, viscid;
　　achene blackish, obpyramidal . . . 　19. *C. elegans*
　2. Culms and foliage not glutinous, viscid.
　　3. Spikelets with tips of scales over the
　　　base of next scale; spikes few in loose
　　　radiate clusters.
　　　4. Keel of scale produced into recurving
　　　　awn.
　　　　5. Scale with one raised vein on each
　　　　　side . . . . 　20. *C. cuspidatus*
　　　　5. Scale with four raised veins on each
　　　　　side . . . . 　21. *C. aristatus*
　　　4. Keel of scale not produced into awn,
　　　　tips only mucronate or cuspidate;
　　　　spikelets racemose or in digitate
　　　　clusters.
　　　　6. Scale 1.5 mm long; keel curving to
　　　　　a retuse scarious apex . . . 　22. *C. iria*
　　　　6. Scale 3–3.5 mm long; keel straight,
　　　　　ending in a cusp; margins
　　　　　broadly scarious . . . . . 　23. *C. compressus*
　　3. Spikelets with margins of scales in line
　　　with keels of adjacent scales; spikelets
　　　numerous in capitate or radiate clus-
　　　ters.
　　　7. Achene ellipsoid-fusiform, obtusely
　　　　angled . . . . 　24. *C. ochraceus*
　　　7. Achene linear, sharply trigonous.
　　　　8. Base of achene minutely stiped;
　　　　　culms roughened toward the
　　　　　summit . . . . 　25. *C. surinamensis*
　　　　8. Base of achene minutely bulbous;
　　　　　culms smooth . . . . . . 　26. *C. distinctus*
1. Wings of rachilla manifest.
　9. Wings deciduous; spikelets dense;
　　leaves present . . . . . . 　27. *C. erythrorhizos*
　9. Wings persistent.
　　10. Leaves reduced to basal sheaths, or
　　　rarely with blades.
　　　11. Culms septate; involucral leaves
　　　　obsolescent, very small . . . 　28. *C. articulatus*
　　　11. Culms continuous.

12. Plants 2–5 dm tall, sheaths rarely
  with blades . . . . 29. *C. haspan*
12. Plants 1–2 m tall.
  13. Involucral leaves three to ten,
    relatively small . . . . 30. *C. papyrus*
  13. Involucral leaves four to
    twenty, ample . . . . 31. *C. alternifolius*
10. Leaves present, linear-elongate.
  14. Scales cucullate at apex, commonly
    variegated with red . . . . 32. *C. lecontei*
  14. Scales ovate, acute at tips; elongate
    tuberiferous rhizomes manifest.
    15. Spikelets reddish purple . . . 33. *C. rotundus*
    15. Spikelets yellowish brown . . 34. *C. esculentus*

1. C. metzii Mattf. & Kukenth.—Plant annual, cespitose with fibrous roots; culms 0.5–4 dm long, slender. Blades flat, elongate, 2 mm wide; sheaths dilated toward summit. Involucral leaves, commonly three, dilated at base. Spike globose, sessile, to 10 mm in diameter, brown in maturity, crowded with scales; fruiting scales ascending on short rachis, 3.5 mm long, ovate, compressed; the crested keel rounding the base up to above the middle; mature achene 1.3 mm long, light brown, lenticular. Lawns in Everglades City, Collier County, also, farm land, Levy County, Fla., naturalized from tropical Asia and Africa. *Kyllinga squamulata* Thonn. ex Vahl

2. C. brevifolius (Rottb.) Hassk.—Perennial; culms to 2 dm long, ascending from scaly rhizomes. Lowermost sheaths bladeless, the upper with flat thin blades 2–4 wide; involucral leaves usually three, to 12 cm long, one often very short. Spike sessile, ovoid, solitary or two to three confluent at base; fruiting scales numerous, 2.5 mm long, mucronate, hyaline with green usually scabrous keel; stamens two; achene 1.2 mm long, lenticular bright brown. Margins of ponds and lawns among grasses, tropical America. Summer. *Kyllinga brevifolia* L.

3. C. pumilus L.—Cespitose annual with fibrous roots; culms 4–12 cm tall, sometimes surpassed by the leaves. Lowest sheaths bladeless; involucral leaves commonly three, of unequal length. Spikelets 4–9 mm long, numerous, crowded into glomerulate heads; achene 0.5 mm long, obovate in outline, truncate, mucronulate, dark brown. Frequent in wet sandy muck, central and south Fla., Ala., W.I., Old World.

4. C. flavescens L.—Tufted annual 0.6–3 dm tall, densely matted; sheaths loose, red striate, culms slender, leaves flat, 0.5–1.5 mm wide, attenuate, minutely serrulate at apices; involucral leaves similar, overtopping the spikes. Spikelets 15 mm long or longer, in radiate clusters, compressed, sessile, or on rays 1–4 cm long; fruiting scales thickened on margins, reticulate next to midvein at base. Mature achene suborbicular or obovate, strongly biconvex, 1 mm long with stalk and mucro, black, tardily becoming white verrucose. Pond margins, shores in wet sandy muck, Fla., tropical America, Mexico, Old World.

5. **C. polystachyos** Rottb. var. **texensis** (Torr.) Fern.—Annual; culms few together or matted in large tufts, 1–6 dm tall, overtopping the leaves. Blades 1–3 mm wide, sheaths stramineous or brownish, loose, the lowest bladeless, purplish; involucral leaves, similar, slightly dilated below. Umbel with one to six rays, rufous brown; spikelets linear, commonly ten to fifteen, 8–15 mm long, 1–1.2 mm wide; scales mucronulate, 1.5–2 mm long. Achenes 0.9–1 mm long. Glades, swamp margins, Va. to Fla., Mo., tropical America, Argentina. *C. paniculatis* Rottb.

6. **C. filiformis** Sw. var. **densiceps** Kukenth.—Plant wiry perennial from short, somewhat knotty rhizome, densely cespitose; culms to 2 dm tall or taller. Involucral leaves often longer than the culm, with basal leaves overtopping the inflorescence. Sheaths dilated at summit, maroon red, spikelets linear, 6–9 mm long, aggregated into sessile spikes, scales oblong, 3 mm long, midvein maroon, sides green, finely several-nerved. Achene linear, apiculate, maroon brown, 1.5 mm long. Chokoloskee, shell ridge, sandy soil, Everglades and Fla. Keys, W.I. Fall. *C. floridanus* Britt.

7. **C. odoratus** L.—Robust annual, few to several in clumps or solitary; culms 1–7 dm tall, stout at base, invested within leaf sheaths. Foliage conspicuous; the cauline blades 4–12 mm wide and the similar involucral leaves much overtopping the inflorescence. Umbel rays five to six, radiate with plumose spikes; spikelet axes pulvinate; scales 2.5–3 mm long, coriaceous, golden brown when mature, shortly imbricate or contiguous on the same side of the rachilla. Achene plano–convex, 1.5–2 mm long, dark brown, embraced by the rachilla wing. Brackish or saline shores, tropical America, Fla., Tex., Calif., Mass., N.Y. Summer. *C. ferax* Richard

8. **C. filiculmis** Vahl—Culms one to several, 1–7 cm long, from hard, cormose rhizome; leaves 1–2 mm wide, conduplicate or flat, grayish green. Involucral leaves similar. Umbel reduced to a solitary, subglobose head, with numerous radiating spikelets; scales 1.5–3 mm long, obtuse to acute, subcoriaceous, midvein green, sides hyaline; rachilla wingless or very narrowly winged. Achene 1.8–2.2 mm long. Sandy pineland, Fla. to Tex., north to N.Y., Mass. Fall. *C. martindalei* Britt.

9. **C. fuligineus** Chapm.—Plant perennial to 4 dm tall, densely tufted on knotty rhizomes. Leaves linear, barely 1 mm wide, obtuse and scabrous on margins at tips; sheaths maroon red. Involucral leaves three to four, similar, elongate; spikes maroon red or brown, spherical, sessile,

1–1.5 cm in diameter; spikelets linear, 4–11 mm long, 1.5 mm wide; scales ovate, obtuse, 2 mm long, mucronate, appressed. Achene black, 1–1.2 mm long, abruptly pointed. Oolitic soils, Fla., Key West, W.I. Fall.

10. C. globulosus Aubl.—Rhizome ascendant, knotted with corms, giving rise to culms 5–6.5 dm tall; sheaths purplish at base. Blades flat, soft, glaucous, 1–4 mm wide, attenuate and flexuous at apices; involucral bracts similar, commonly three to seven. Umbel simple; the sessile terminal spike overtopped by six to eleven rays 4–7 cm long; spikes nearly spherical to 2 cm in diameter when mature; scales with green keel, sides golden yellow, to 2.5 mm long, imbricate. Achene olive brown, obovate, trigonous, 1.7–2 mm long, short-mucronate, sharply three-angled. Seasonally wet pinelands, prairie, pond margins, Fla. to Tex., and Midwest, tropical America. Summer. *C. echinatus* (Ell.) Wood

11. C. nashii Britt.—Tufted, glaucous, slender perennial to 7 dm tall; culms wiry, sharply three-angled, exceeding the leaves. Blades 2–3.5 mm wide; involucral bracts similar, spreading. Spikes five to eighteen, subglobose or ellipsoid; rays capillary 1–5 cm long; spikelets subulate 2–3 mm long; scales acute. Persistent achenes linear, dark brown, 2 mm long. White sand scrub association, central and south Fla.

12. C. retrorsus Chapm. var. **retrorsus**—Rhizome cormlike, abbreviated; culms rigid to 9 dm tall, grayish green; acutely three-angled, exceeding the leaves; sheaths stramineous or purplish. Blades 2–5 mm wide. Involucral bracts similar, dilated at base; umbel simple, terminal spike sessile, 1–2 cm long, surpassed by five to eight rays; spikelets lanceolate, acute, one- to two-flowered; scales oblong, nine-veined, 2–3 mm long, keel green, yellowish, or drab. Achene trigonous, 1.7 mm long, mucronate, bright brown. Moist sandy soils, pineland, and hammocks, Fla. to Tex., north to N.J. Fall. Includes var. *cylindricus* (Ell.) Fern. & Griscom

13. C. pollardii Britt. ex Small—Perennial tufted plant, to 7 dm tall; sheath loose, purplish. Blades flat, 3–4 mm wide, flexuous. Involucral bracts exceeded in length by the cauline leaves; spikes ellipsoid to cylindric, 1.5 cm thick, densely flowered; spikelets 5–6 mm long, mucronate 1.5 mm wide. Sandy soil, prairies, pinelands, brackish shores, endemic to south Fla. *C. blodgettii* Torr.; *C. winkleri* Britt. ex Small; *C. deeringianus* (Britt.) ex Small; *C. litoreus* (Clarke) Britt.

14. C. ligularis L.—Plant perennial, commonly in large clumps, to 1 m tall, glaucous. Sheaths loose, purple striped below; blades flat or involute, attenuate; margin white cartilaginous, scabrous. Involucral bracts similar, flat or folded, surpassing the umbel. Umbel compound, congested, rays nine or more, 3–6 cm long; spikes ovoid, cylindric-ellipsoid; often lobulate or confluent; spikelets 3–6 mm long; scales ovate, 2–

2.5 mm long, obtuse or mucronate, nine-nerved, thin translucent, cellular-reticulate between the nerves. Achene 2 mm long; golden brown; sides inequilateral, sharply angled. Mangrove shores, hammocks, Gulf coast, Fla. to Ala., W.I., Central America, Old World.

15. C. planifolius Richard—Glaucous, loosely tufted perennial 0.8–1 m tall from cormlike base; sheaths maroon, purplish, becoming blackish with age. Blades flat membranous, 4–10 mm wide, usually scabrous on margins in the long-attenuate tips. Involucral bracts similar, prominently elongate; spike sessile, terminal or more often on elongate rays, maroon red, especially on the veins of scales; spikelets 8–15 mm long, 1.5–2.5 mm wide; scales spreading ovate, 3–4 mm long, thirteen-nerved, keel green, excurrent margins narrowly hyaline. Achenes 1–7 mm long, black, mucronate, embraced by conspicuous rachis wings. Brackish sands on beaches and dunes, Fla., W.I. *C. brunneus* Sw.

16. C. tetragonus Ell.—Plant perennial to 1 m tall from cormose rhizome, glabrous and glaucous; culms sharply triangular, leafy below the middle; sheaths loose, reddish at base. Blades linear, elongate, to 8 mm wide; involucral leaves similar, conspicuously surpassing the inflorescence. Umbel decompound, with nine or more rays, 5–16 cm long; spikes 1–4 cm long, oblong with somewhat distant spikelets; scales ovate, 2.5–3 mm long, fifteen-nerved, with green midvein, castaneous sides, hyaline margins. Achene castaneous to purplish brown, 2 mm long; adaxial face oblong-equilateral, the sides inequilateral. Coastal sands in openings of pinelands, Fla., Tex., N.C.

17. C. strigosus L.—Perennial with cormlike rhizome; culms one or more, glabrous, 5–10 dm tall, stout, leafy below the middle; sheaths loose, purple at base. Blades flat, 1–1.5 cm wide; involucral leaves two to seven, surpassing the inflorescence. Umbel decompound; rays nine to thirteen, 5–15 cm long, or longer; spikes open, spikelets linear-subulate, 10–15 mm long, spreading horizontally; scales overlapping, linear, appressed, golden brown with green keel, seven-nerved, 3–3.5 mm long. Achene lustrous brown, 2 mm long, mucronate, surface cells isodiametric, embossed. Glades, shores in damp soil, central Fla. to Miss., north to N.Y., N.J.

18. C. stenolepis Torr.—Glabrous perennial, 1 m tall or less from hard rhizomatous base, in large tufts; culms stout, sharply angular, longer than the loosely sheathed, glaucous leaves. Blades to 2 cm wide, flat, scabrous-attenuate at tips. Inflorescence 12–20 cm long, exceeded by the longest subtending bracts; spikes dense, compound, numerous, sessile to long peduncled; spikelets 10–12 mm long, fragile, stramineous; scales spreading 4–5 mm long, pointed. Achene linear, three-angled, minutely mucronate, 2–2.2 mm long, without embossment. Ditch or canal banks in damp soil, south Fla., Miss., N.J. Fall. *C. strigosus* L. var. *stenolepis* (Torr.) Kukenth.

19. C. elegans L.—Cespitose perennial with culms from short rhizomes, 3–6 dm tall, 2 mm thick at apex, sometimes septate-nodulous, roughened, obtusely trigonous. Leaves equaling the culms in length, 1–4 mm wide; sheaths brownish. Involucral leaves similar, three to five, scabrous on margins, 5–40 cm long, unequal; rays of the umbel three to eight, 3–5 cm long, unequal; spikelets flabellate, three to twenty in a cluster, 3–15 mm long, 3–4 mm wide, many-flowered; rachilla wingless; scales imbricate, viscid; cuspidate, reddish purple to stramineous, clasping the achene. Achenes 1.5 mm long, sharply angled, concave sides, obconic, blackish purple, lustrous. Sandy soils, W.I., Mexico, British Honduras, south Fla., Fla. Keys. *C. rubiginosus* Hook. f.

20. C. cuspidatus HBK—Tufted annual 3–12 cm tall; culms and foliage filiform; sheaths purple. Involucral bracts five or more, exceeding the culm. Umbels compound, spikelets in radiate heads 1.5 cm wide; scales 1.8–2 mm long, the blade often orange red, notched at the base of the subulate recurving awn; awn nearly as long as the blade. Achene obpyramidal, lustrous brown, colliculate, mucronate. Moist sandy hammock margins and pinelands, Fla. to Tex., tropical America. *C. squarrosus* sensu Small

21. C. aristatus Rottb.—Tufted, glabrous, fragrant annual with reddish roots. Leaves 1–2 mm wide, flat, linear, sheaths purplish orange. Rays three to four, surpassed by involucral leaves. Spikes to 12 mm wide, subglobose, congested, composed of radiately disposed spikelets. Scales 2.2 mm long, yellowish purple, midvein green, excurrent, awn shorter than the blade. Mature achene 0.8–1 mm long, obovoid, trigonous, mucronate, reddish brown, with lustrous colliculate surface. In dry-to-moist soil, shore sands, shell, Fla., coastal plain to Vt., southwestern U.S.

22. C. iria L.—Tufted annual to 3.5 dm tall; sheaths purplish, loose. Blades linear-flat, shorter than the culm, 2–4 mm wide. Involucral three-leaved, similar, unequal, shorter or longer than the inflorescence; rays five to six, spikelets compressed, disposed in plumose fashion, green to bronzed brown when mature; scales broadly oval in side view, 1.3 mm long, keel green, edged, flanked by two prominent ribs on each side. Achene yellowish brown, 1.2 mm long, angles obpyramidal, thickened. Moist ground, swampy borders, abandoned cultivated sites, Fla., southeastern U.S., native of Asia.

23. C. compressus L.—Tufted annual 0.5–3 dm tall, glabrous; culms slender, pale green, surpassing the flexuous leaves; sheaths purple, loose. Involucral three-leaved, unequal; inflorescence a simple umbel, the culm terminating in a sessile spike with usually three rays; spikelets three to five in a spike; scales closely imbricate, pale green, yellowish white; rachilla wings small. Achene obpyramidal, dark brown, 0.9 mm long, with thickened angles. Pinelands and gravelly roadside banks, Fla., tropical America and Old World tropics.

24. C. ochraceus Vahl—Glabrous; culms 7 dm tall or taller from ligneous rhizomes. Leaves 3–5 mm wide, sparingly septate-nodulose, about as long as the culms; sheaths broadly hyaline, dotted with brown, readily rupturing. Involucral bracts five to seven, foliose, 3–5 dm long. Inflorescence umbelliform of slender primary rays 3–9 cm long subtending secondary rays, or simple. Spikelets linear, 2–2.5 mm wide, 6–15 mm long, yellowish gray, radiate. Fruiting scales 2–2.3 mm long, 0.4 mm wide from margin to midvein at the truncate base, progressively deciduous from the wingless rachilla. Mature achene 1.3 mm long, 0.2–0.3 mm wide, tapering to base and apex, porose-punctate, lustrous on surface. Moist prairie, marshy openings in pineland, Fla. to La., Cuba, W.I. Fall.

25. C. surinamensis Rottb.—Tufted perennial 1–7 dm tall; culms thickened at base, smooth below, roughened on angles toward the summit, rigid, pale green. Leaves relatively few, blades flat, 4–5 mm wide; sheaths brownish, loose. Involucral bracts three to five, the longest exceeding the decompound, lobulate umbel; rays numerous, unequal; spikelets congested in small, compressed heads; scales dorsally convex; achenes brick red, apiculate, 0.7–0.9 mm long. Moist pinelands, shores, and drying ditches, coastal plain, Tex., Fla., W.I., tropical America.

26. C. distinctus Steud.—Tufted perennial from a thick short rhizome to 9 dm tall; culms glabrous, obtusely angled, leafy below; blades 5–8 mm wide, flat; sheaths loose, septate-nodose, splitting into hyaline margins. Involucral leaves elongate, unequal; umbel decompound, rays seven to eleven; spikes glomerulate, fuscous brown; spikelets in flabellate clusters; scales 2.2 mm long, inward curving, folded, truncate. Achene rufous, 1.5 mm long, apiculate, minutely knobby at base. Fla. to Carolinas, Brazil.

27. C. erythrorhizos Muhl.—Tufted annual from reddish fibrous roots, with culms to 7 dm tall, stout, bluntly angled, fistulose in age; sheaths loose, reddish. Blades elongate, 2–10 mm wide, upwardly scabrous on margins; involucral leaves similar, the longest much exceeding the inflorescence. Umbels decompound, the terminal sessile; rays five to six, spikes cylindric, congested, 1–2 cm long; scales 1.2–1.5 mm long, ovate, striate, reddish; keel green, excurrent as a mucro. Achene 0.7 mm long, elliptic, fuscous, glossy, upcurving at each end on ventral side. Swampy shores, ditches, Fla., Tex., Calif. to Ontario, Mass.

28. C. articulatus L.—Gregarious perennial with culms to 1.8 dm

long, from strong reddish rhizomes; culms septate irregularly, basal nodes with thin, loose, reddish, bladeless sheaths. Involucral bracts, 10 mm long or less; involucre decompound, rays one to twelve, with linear spikelets to 4 cm long, in lax clusters; scales 3–3.5 mm long, ovate-imbricate, appressed; keel green, sides brownish with hyaline margins. Achene 1.5 mm long, yellowish brown, smooth, apiculate. Canals in deep water; shores and swamps, coastal plain, Fla., Tex., W.I., South and Central America.

29. C. haspan L.—Tufted gregarious perennial from short rhizomes to 8 dm tall; culms spongy, sharply triangular, with loose basal, usually bladeless, sheaths. Involucral bracts few, at least one, overtopping the inflorescence; umbels decompound, rays ten or more, firm; spikes diffuse or congested of a few to many lanceolate spikelets 5–12 mm long; scales spreading, ovate-elliptic or lanceolate, 1.5 mm long, mucronulate, keel green, fading red or drab in age; rachilla glutinous, red. Mature achene white, 0.4 mm long, black mucronulate or light buff, minutely papillose, lustrous. Pinelands, swamps, coastal plain, Fla. to Tex. and Va., W.I., Mexico, Central and South America, Old World.

30. C. papyrus L.—Plants cespitose from ligneous rhizome to 2 m tall; culms smooth, erect, strong, three-sided; sheaths bladeless, spongy. Umbel terminal, large, subtended by a few involucral bracts; rays fifty to one hundred filiform, reflexed to 5 dm long; spikelets linear, squarrose, numerous, pale brown, on winged rachis. A native of Old World, introduced and persistent on old sites, naturalized in south Fla., Fla. Keys.

31. C. alternifolius L.—Cespitose from matted roots to 1 m tall or taller; culms stout, strong, erect, three-sided; sheaths basal, bladeless. Involucral leaves ample and numerous, 2.5–3.5 dm long, 1–2 cm wide, conspicuously three-ranked; rays numerous, 4–5 cm long; spikelets twelve- to twenty-four-flowered, oblong. Cultivated, persistent on old sites, native to Madagascar, Africa, South America.

32. C. lecontei Torr.—Gregarious perennial from elongate rhizomes; culms to 4 dm tall; blades linear, shorter than the culms; sheaths drab to brownish, fibrillose. Involucral leaves elongate, surpassing the umbel. Umbel decompound, loose or congested, rays ascending; spikelets in loose clusters, 1–3 cm long, 3–4 mm wide, yellowish, becoming blotched with red; scales ovate, 3 mm long, hooded at apices with appressed tips. Achene trigonous, mucronulate, brownish, 1 mm long. Pinelands, prairies, on the coastal plain to Fla., La.

33. C. rotundus L.—Plant glaucous, gregarious perennial from tuberous creeping rhizomes; culms to 5 dm tall. Leaves spreading, basal; blades linear, exceeded by the culms, 3–4 mm wide; sheaths loose, purplish. Involucral leaves four to five, usually one as long or longer than the umbel; rays two to three, with a few spikelets to 25 mm long, ovate, appressed. Achene 1.5–1.9 mm long, obovate. Fields, roadsides, lawns, Fla. to Tex., naturalized from tropical America.

34. C. esculentus L.—Gregarious perennials from tuberiferous elongate rhizomes; culms 2–6 dm tall, sharply three-edged, leafy at base. Blades spreading, elongate, 4–5 mm wide, glaucous; sheaths loose, drab or purplish at base. Involucral leaves longer or shorter than the rays; umbel simple, loose or congested; spikelets usually distant, 1–2 cm long, 1.5 mm wide, few to several in a head; scales yellowish brown, 2.8 mm long, ovate, with mucronulate, spreading tips. Achene 1.5 mm long, obovoid. Vacant lots, fields, roadsides, in sand, Fla., widely distributed in the U.S., W.I., Mexico, tropical America. *C. phymatoides* Muhl.; *C. fulvescens* Liebm.

2. ABILDGAÁRDIA Vahl

Glabrous perennial; culms from a hard bulbous base; leaves setaceous; spike terminal, solitary, rarely two; scales imbricate in two ranks; bristles lacking; stamens one to three; style wholly deciduous, pubescent, stigmas three; achene trigonous, tuberculate. About 15 species, widely distributed in warm regions.

1. A. ovata (Burm. f.) Kral—Culms tufted 1–4 dm tall, wiry, filiform, minutely sulcate. Blades less than 1 mm wide, involute 10–15 cm long, attenuate; sheaths short, hyaline at throat. Involucral bract one; spike 5–18 mm long, ovate, yellowish white; scales keeled, serrulate, pointed; achene 2–2.5 mm long, pearly white, verrucose, globose with stalk 0.5 mm long. Moist, grassy mats over oolitic rocks, central and south Fla., Mexico to Argentina, Old World tropics. All year. *Fimbristylis monostachya* Hassk., *A. monostachya* (L.) Vahl

3. FIMBRISTÝLIS Vahl

Annual or perennial sedges, commonly with fibrous roots, knotty or rarely elongate rhizomes; culms leafy below. Spikelets terete, many-flowered in cymose or capitate inflorescence subtended by ciliate bracts. Scales concave, spirally imbricate, all fertile, gradually deciduous on maturing; perianth lacking; stamens one to three, style two- to three-cleft, pubescent or glabrous, compressed, deciduous with the enlarged base. Achene lenticular or trigonous, verrucose or cancellate, with latticed, horizontally elongate cells. About 300 tropical and subtropical species, especially Indomalaya and Australia.

1. Styles three, achene trigonous; annuals.
    2. Spikelets linear, pointed; scales with excurrent midvein . . . .   1. *F. autumnalis*
    2. Spikelets subglobose, rounded; scales obtuse, midvein not excurrent . . . .   2. *F. miliacea*
1. Styles two, achene lenticular; perennials. (Annual No. 8.)
    3. Spikelets usually solitary, rarely two; achene stalked, white . . . . .   3. *F. schoenoides*
    3. Spikelets several to numerous.
        4. Umbel compacted; achene less than 1 mm long . . . .   4. *F. spathacea*
        4. Umbel open or diffuse; achene more than 1 mm long.
            5. Leaves scabrous on margins, or glabrate; plants in large tufts.
                6. Scales of spikelets, at least the lowermost, pubescent; sheaths brownish.
                    7. Plant stoloniferous; base of stem nonbulbous . . . . . .   5. *F. caroliniana*
                    7. Plant nonstoloniferous; base of stem bulbous . . . . .   6. *F. puberula*
                6. Scales of spikelets glabrous; sheaths castaneous . . . .   7. *F. castanea*
            5. Leaves ciliate, or glabrate in age; plants in small tufts . . . . .   8. *F. dichotoma*

1. F. autumnalis (L.) R & S—Tufted glabrous annual; culms 0.6–4 dm tall, soft, 0.5–1.2 mm wide at base. Blades 3–10 cm long to 2.5 mm wide, flat, scabrous at acute tips; sheaths 2–3 cm long. Panicle open, diffuse, subtended by two or more bracts; spikelets linear, 3–15 mm long, scales with prominent excurrent midveins, slightly spreading at tips, peduncles capillary; styles not fimbriate; stamens one to two. Achene 0.7 mm long, sharply trigonous. Mostly in wet glades, ditch banks, garden sites, south Fla. to Me., Minn., British Honduras, Cuba.

2. F. miliacea (L.)Vahl—Tufted annual with soft culms to 7 dm tall, angled toward the summit. Blades linear-attenuate, elongate 1–3 mm wide, flat; sheaths loose, 2–3 cm long. Involucral bracts short, filiform; panicle decompound, diffuse, spikelets distant, subglobose or oblong, 2–4 mm long; scales ovate, obtuse, mucronulate; style fimbriate; stamens one to two. Achene trigonous, 0.6 mm long, the four vertical bands on each face minutely granular. Wet sandy peat or marly soil, Fla., Calif., Mexico, tropical America.

3. F. schoenoides Vahl—Tufted glabrous perennial herbs with slender culms to 4 dm tall. Blades 4–12 cm long, less than 1 mm wide, involute, pale green like the stems; sheaths 3–6 cm long, somewhat dilated at the oblique summit. Involucral bracts small, inconspicuous; spikelets conoidal to ovoid, off-white or grayish, 5–10 mm long, solitary or rarely another above, or very rarely, a cluster of threes; scales ovate-orbicular, several-veined; achene 1.6 mm long, obovate in outline, lenticular, stipitate, prominently reticulate with isodiametric cells. Sometimes colonial in moist sands of disturbed pineland margins or cleared ditches, south and central Fla., southeast Asia, Australia.

4. **F. spathacea** Roth—Cespitose perennial from rhizomatous crown; culms 1–4 dm tall, subterete, stiff-sulcate, twisted below. Blades 3–7 cm long, linear, thick, channeled above or involute; sheaths dilated at base. Inflorescence capitate, or a dense compound umbel; involucrate; spikelets numerous, ellipsoid, 3–6 mm long, scales keeled, margins hyaline. Mature achene 0.7–0.8 mm long, dark brown, lenticular, obovate in face view; finely reticulate. Brackish soil along canals, mangrove borders, south Fla., Old World species of wide distribution in both hemispheres.

5. **F. caroliniana** (Lam.) Fern.—Tufted stoloniferous perennial; culms 2–7 dm tall, slender in relatively small tufts. Blades flat to 25 cm long, 2–3.5 mm wide, acute, scabrous on margins; sheaths pale brown, slightly dilated, 3–7 cm long. Involucral bracts lanceolate, long attenuate, puberulent; inflorescence umbelliform, with one to five rays; simple or compound, open; spikelets ellipsoid or ovoid, 5–10 mm long, becoming cylindric-oblong in age; scales ovate, 2.8–3 mm long, many-nerved, bright brown. Achene 1.2 mm long, fuscous brown, ribbed reticulate, obovate. Brackish sands of shores and prairies, Fla. to Tex., N.J. Fall.

6. **F. puberula** (Michx.) Vahl—Perennial to 7 dm tall in tussock from knotty crowns; culms slender, glaucous, glabrous. Sheaths light brown, somewhat dilated; blades linear, 7–12 cm long, 1 mm wide or less, crisply short pubescent or glabrate. Inflorescence umbelliform with three to six rays, involucral bracts manifest; scales of spikelets pubescent, 3–4 mm long, ovate, several-nerved; achene 1.3 mm long, obovate in outline, longitudinally lined. Dry pinelands, Fla. to Tex. and North America. *F. drummondii* Boeckler

7. **F. castanea** (Michx.) Vahl—Perennial with wiry culms 3–8 dm tall, from knotty rhizomes covered with leaf bases. Blades narrow, elongate, rigid, 2–4 dm long; sheaths dilated, castaneous or dark purple. Inflorescence umbelliform, open, three- to ten-rayed; peduncles unequal and ascending, or spreading; involucral bracts inconspicuous; mature spikelets cylindric; achene blackish, 1.4 mm long, obovate in outline; indistinctly reticulate. Brackish sands, marshes, Fla. on Atlantic coast to

Long Island, Gulf coast to Tex., Bahamas. *F. spadicea* (L.) Vahl var. *castanea* (Michx.) Gray

8. F. dichotoma (L.)Vahl—Annual or often perennial; culms to 5 dm tall, glaucous; blades flat, 1–3 mm wide, ciliate, sheaths brown adaxially, thin, becoming lacerate below the orifice; epidermal cells prominently cellular-recticulate, commonly red punctate. Spikelets 5–13 mm long, ovoid-ellipsoid, bright brown, arranged in peduncled cymules, raylike, with subtending bracts, on the primary axis, scales 2–3 mm long, keeled, mucronulate. Achene about 1 mm long, two-edged, obovate, with about ten rows of latticed cells separated by as many longitudinal, minutely tuberculate ribs. Low, wet or dry soil, open woodland, sandy margins of ponds, creeks, Fla., Tex., Mo., Panama, W.I., Old World. Summer. *F. laxum* (L.) Vahl

4. BULBOSTÝLIS Kunth

Annual or perennial, tufted, scapose herbs from fibrous roots. Leaves setaceous or linear, very narrowly; sheaths dilated below, thin and relatively loose. Inflorescence umbelliform, diffuse or dense, involucrate; spikelets small, many-flowered; scales prominently keeled, longer than the fruit; perianth wanting; stamens one to three. Stigmas two to three; achene three-edged, rarely two-edged; smooth or transversely wrinkled. About 100 species, warm temperate, tropical, and subtropical regions, Fla. to N.C., W.I., Mexico to South America.

1. Spikelets pedicelled, disposed in loose compound umbel; achenes bluish . . . .   1. *B. ciliatifolia*
1. Spikelets sessile, disposed in headlike umbels.
  2. Involucral bracts equal to or shorter than the umbel; achene semitranslucent, smooth, areolate . . . .   2. *B. barbata*
  2. Involucral bracts exceeding the umbels.
    3. Umbels subglobose, 1–2 cm in diameter; spikelets radially arranged; bracts spreading; culms to 6 dm tall, slender   3. *B. warei*
    3. Umbels ovoid, 5–7 mm in diameter; spikelets ascending; bracts erect elongate; culms 8–10 mm tall, filiform   4. *B. stenophylla*

1. B. ciliatifolia (Ell.) Fern.—Tufted glabrous annual to 3 dm tall; sheaths loose, thin fimbriate or glabrate at summit; abruptly passing to capillary setulose blades. Inflorescence an umbel, appearing lateral; the lowest involucral bract usually surpassing the rays; scales of spikelets minutely puberulent, chestnut brown, mucronulate. Achene 1 mm long, 0.8-1 mm wide, three-edged, colliculate, appearing white around the walls and sometimes the slightly convex center of each peridermal cell, becoming white over the entire surface. Pinelands, hammocks, coastal plains to Fla., Cuba. Summer. *Stenophyllus ciliatifolius* (Ell.) Mohr

2. **B. barbata** (Rottb.) Clarke—Culms numerous, erect to 2 dm tall, capillary, densely tufted; sheaths copiously pilose toward summit. Blades setaceous. Umbels chestnut brown; spikelets few- to several-flowered, of varying lengths; scales mucronulate, recurving at excurrent tip of the prominent midrib; achenes pale, three-edged, 0.6 mm long; periderm smooth, more or less. Moist sandy opening, shores, and flat pinelands, Fla. to Ga., Ala. to Tenn., N.C., South America, naturalized from Old World tropics. *Stenophyllus floridanus* Britt.

3. **B. warei** (Torr.) Clarke—Short-time perennial 1-5 dm tall in large tufts from hard crown. Culms slender, grooved or striate; sheaths pilose at summit, thin, split or lacerate in age. Blades linear-obtuse, spreading or flexuous; inflorescence contracted, capitate; spikelets subtended by pectinate bracts; scales to 5 mm long, thin, many-nerved, keeled toward the mucronulate apex. Mature achenes 0.5 mm long, transversely corrugate, white, obcordate, style base depressed; sides narrowed to base. Conspicuous in white coastal sands, of south and central Fla. to Ga. *Stenophyllus warei* (Torr.) Britt.

4. **B. stenophylla** (Ell.) Clarke—Tufted annual to 1.5 dm tall or less; culms capillary, sheaths pilose at throat. Blades setaceous. Involucral bracts conspicuously elongated, setaceous above the base. Spikelets compacted in small heads; scales 2.5 mm long, puberulent at base, midrib prominent, margins thin, apex aristate. Achene 1-1.2 mm long, three-edged, transversely corrugate, brownish blue variegated with white. Local, in moist sands of pineland, N.C. to Fla., Cuba. *Stenophyllus stenophyllus* (Ell.) Britt.

5. ELEÓCHARIS R. Brown CLUB RUSH; SPIKE RUSH

Glabrous annual or perennial sedges; culms simple, strict, terete, bluntly three-angled or compressed with leaves reduced to basal sheaths. Spikelets solitary, terminal without subtending involucral bracts; scales concave, spirally imbricate; perianth represented by retrorsely barbed bristles or rarely wanting; stamens two to three. Style two-cleft, fruit lenticular or biconvex or three-angled, base of the style persistent as a tubercle at the summit of the achene. About 200 species, cosmopolitan.

1. Perennials.
  2. Spikelets 2-5 cm long; culms stout, 3-5
     mm wide; never prolific.

3. Culms septate, terete; achene bristles
    retrorsely barbed . . . . . .   1. *E. interstincta*
3. Culms continuous, terete, rarely angled;
    achene bristles smooth   . . . .   2. *E. cellulosa*
2. Spikelets 4–5 mm long; culms less than
    1.4 mm in diameter; prolific.
    4. Fruiting spikelets usually perfected in
        basal sheath; culms soft   . . . .   3. *E. vivipara*
    4. Fruiting spikelets in apical inflores-
        cences; culms wiry . . . . . .   4. *E. baldwinii*
1. Annuals.
    5. Styles two; achenes lenticular, bristles longer
        or shorter than the achene, or some-
        times lacking.
      6. Achenes 0.7–1 mm long; spikelets few-
        flowered, obtuse . . . . . .   5. *E. caribaea*
      6. Achenes 0.5 mm long; spikelets many-
        flowered, acute . . . .   6. *E. atropurpurea*
    5. Styles three; achenes trigonous, bristles
      none . . . .   7. *E. nigrescens*

1. **E. interstincta** (Vahl) R & S—Perennial with strong creeping root-stocks; culms terete to 10 dm tall, nodose. Sheaths thin, sometimes foliose at apex. Spike cylindric, barely wider than the culm 3–4 cm long; scales imbricate, oblong; the deltoid, exposed apices hyaline margined. Achene plump, shiny, striate, with conoidal depressed tubercle, 2.5–3 mm long; styles two to three. Marshes, south Fla., eastern U.S., Bermuda, Cuba, Bahama Islands.

2. **E. cellulosa** Torr.—Rhizomatous colonial plants, with obscurely three-sided, compressible culms 5–8 dm tall. Sheaths reddish to 15 cm long, oblique-apiculate at summit. Spikes 3–4 cm long; thicker than the culm; scales coriaceous, oblong-ovate, hyaline on margins. Achene up to 3.2 mm long, cancellate, stramineous, bristles smooth, about as long as achene or longer; tubercle constricted, about 0.3 mm long, black, as long as the body. Glades and marshes, south Fla., Tex., tropical America.

3. **E. vivipara** Link—Culms 1–3 dm tall, filiform from a vertical rhizome with coarse, brown roots, lustrous green with reddish bases invested in yellowish sheaths. Spikelets linear, to 8 mm long, mostly proliferous; fruit-perfecting florets in basal sheaths. Scales appressed, oblong-obtuse, without prominent midvein, margins hyaline, sides reddish. Achene trigonous, obovate, reticulate, thickened on angles with pyramidal style base; body of achene grayish, bristles retrorsely barbed, shorter than the achene, changing to darker in aging; base of tubercle white ridged. In sandy soils, pool and ditch margins, Fla., Miss., Va. Summer, fall. *Eleocharis curtissii* Small; *E. prolifera* Chapm.

4. **E. baldwinii** (Torr.) Chapm.—Mat-forming perennial in loose

tufts; culms 3–10 cm long, wiry, capillary, commonly proliferous. Sheaths purple, apex acute. Spikelets compressed, 3–5 mm long, three- to eight-flowered; scales linear-acute, strongly keeled, red or brown, the lowest scale shortest. Style branches three; achene 1 mm long, sharply three-angled, olive brown, sometimes striate, tubercle pyramidal, subulate; bristles shorter than the achene. Exsiccating ditch beds, seasonally wet pinelands, Fla., Ga., Ala., La. to N.C., S.C. Summer.

5. E. caribaea (Rottb.) Blake—Tufts dense, large with numerous culms 0.6–3 dm tall; sheaths stramineous or reddish at base, acuminate orifice oblique. Spikes subglobose, 3.5–5 mm long; scales oblong to sub-orbicular, cartilaginous or membranous. Achene 1–1.2 mm long, pur-plish black, lustrous; bristles brownish, six to eight, usually longer than the achene. Moist sands on shores and roadsides, Fla. to Midwest, coastal plain states, tropics of both hemispheres. *E. geniculata* (L.) R & S

6. E. atropurpurea (Retz.) Kunth—Low, tufted annual with culms to 3–10 cm long, capillary often arcuate, or erect. Sheaths reddish brown, attenuate at apex. Spikelets oblong-ovoid, 2–8 mm long; scales ovate, membranous, with broad green midrib, brownish sides, and thin hyaline margins. Achene lustrous purplish black, 0.5–0.7 mm long; bristles six, delicate, translucent, minutely barbed, shorter than the achened body or sometimes obsolescent. Tubercle green before maturity; colorless, often flattened, when mature. Moist sand, shores, and pinelands, Fla., Mid-west, Wash., Tex., widely distributed in tropical regions of both hemis-pheres.

7. E. nigrescens (Nees) Steud.—Tufted annual with white fibrous roots or perennial with ascending rootstock. Culms 3–7 cm tall, three-ribbed, capillary, maroon red at base, invested by sheaths to 20 mm long, tapering to a short free apex with excurrent midvein. Spikelets 2–3 mm long or longer by progressive maturing of the fruits and elongation of the rachis. Scales imbricate, oblong, indented at apex, broadly scarious margined, the green midvein flanked by stripes of maroon. Mature achenes 0.3 mm long with thickened angles, grayish-smooth, style base short apiculate, tubercle margins slightly elevated, bristles none. Season-ally inundated prairies, flat pinelands, Fla. to S.C., Mexico, Brazil. Sum-mer, fall. *Heleocharis atropurpurea* Boeckler

6. SCÍRPUS L. Bulrush

Annual or perennial, caulescent or scapose herbs. Leafy, or leaves represented by bladeless sheaths; culms three-sided or terete with terminal or apparently lateral, mostly involucrate, inflorescences. Flower bisexual, subtended by a scale with one to six perianth bristles;

stamens two to three, style two- to three-cleft. Achene lenticular or trigonous. About 300 species, cosmopolitan.

1. Involucral bract one, apparently a continuation of the culm; spikelets 5–6 mm wide.
  2. Culm sharply trigonous; spikelets in sessile glomerules . . . .   1. *S. americanus*
  2. Culm terete; spikelets in open umbels .   2. *S. validus*
1. Involucral bracts several, foliose; spikelets 10–12 mm wide . . . .   3. *S. robustus*

1. S. americanus Pers.—Gregarious perennials from strong, elongate, deep-seated rhizomes. Culms 0.5–1.5 m tall, trigonous with concave sides; sheaths loose, becoming shreddy; orifice oblique. Blades linear-elongate, to 2 dm long. Involucre erect 5–10 cm long, firm pointed; spikelets ovoid in glomerules two to eight; scales ovate, ciliolate, cleft into rounded lobes and shortly awned by the excurrent midvein. Achene plano–convex, obovoid, tawny, 2.5–3 mm long, including the short beak. Swamps, sandy shores, widely distributed in U.S., Fla., tropics.

2. S. validus Vahl—Gregarious perennials from elongate, creeping rhizomes. Culms, terete 0.5–3 m tall, to 2 cm thick at base, spongy throughout. Sheaths 2–3 dm long, thin, loose, becoming lacerate in age. Involucral bract firm pointed, 0.8–5 cm long; spikelets in fascicles or in glomerules or nodding peduncles of varying lengths; spikelets rufous, ovate-elliptic; scales 3.5–4 mm long, minutely fimbriate-ciliate, shallowly cleft, the keel scabrous, excurrent. Mature achene castaneous, 3 mm long, plano–convex, smooth, short pointed; exceeded by eight retrorsely setulose bristles. Swamps, glades, local tropical America, extending north to Fla., S.C.

3. S. robustus Pursh—Culms trigonous, 0.5–1.5 m tall, leafy nearly to the summit from moniliform rhizomes. Sheaths imbricate; blades 3–10 mm wide, flat, linear-attenuate, overtopping the inflorescence. Spikelets in glomerules, 10–20 mm long or longer, rufescent; scales ovate, puberulent with recurved awns; achenes plano–convex, blackish brown, 3–3.5 mm long. Brackish waters Fla., north to Mass., Mexico, W.I. *Scirpus confervoides* Poir. of authors. WEBSTERIA SUBMERSA (Sauv.) Britt. may occur in our area. Fruiting specimens have not been found. Vegetatively similar plants have been authenticated as ELEOCHARIS VIVIPARA Link.

7. FUIRÈNA Rottb.   UMBRELLA GRASS

Mostly perennial caulescent herbs, leafy, or leaves represented by sheaths alone; ligules membranous. Spikelets ovoid, scales spirally imbricate awned; flowers bisexual; perianth of three stipitate scales alternating with three barbed bristles; stamens three. Stigmas three; achene stipitate, trigonous. About 40 species of warm temperate, tropical, and subtropical regions.

1. Culms leafy; only the lowest sheaths bladeless.
   2. Rootstock cormose, culms tufted . . . 1. *F. squarrosa*
   2. Rootstock not cormose, culms distant . . 3. *F. longa*
1. Culms leafless; blade reduced to a short
   point; plants with elongate rhizomes . . 2. *F. scirpoidea*

1. **F. squarrosa** Michx.—Culms 3–8 dm tall or more, cespitose on knotty or cormose rhizomes, compressible. Blades puberulent, attenuate, spreading or deflexed; lower sheaths papillose-hirtellous, the upper glabrate. Inflorescence terminal; spikelets sessile in clusters of four to six; scales obovate, pubescent, ciliate, with one to three prominent median nerves, ending in a recurved awn as long as the scale; perianth scale 2 mm long, blade suborbicular apiculate, stalked; the alternating bristles shorter than the achene stipe. Mature achene barely 1 mm long, short pointed, the three edges sharp, thin, sides concave, glossy, ivory white. Sandy mucky soil, moist pineland swales and shores, cypress margins, coastal plain, Fla., Tex. to N.C. Includes *F. hispida* Ell.; *F. breviseta* Cov.

2. **F. scirpoidea** Michx.—Culms to 6 dm long from strong, creeping rhizomes; internodes slender, sheaths loose. Spikelets 5–12 mm long, ovoid, in terminal bracted inflorescences, commonly two to three in a cluster; scales appressed, pubescent awns less than 2 mm long; perianth scales oblong. Mature achene 1.5 mm long with stipe and mucro; bristle a little longer than the stipe. Conspicuous colonial plant on shores, flatwoods, moist banks, Fla., Ga., La., Cuba.

3. **F. longa** Chapm.—Culms to 6 dm long, pubescent below inflorescence. Well-developed leaves, two to five in upper nodes; blades to 11 cm long, linear, 3–5 mm wide at base, pubescent on upper surface when young; ligule 0.8 mm long. Spikelets one to five, ovoid, scales 3–4 mm long with recurved scabrid awns 2–3 mm long. Perianth scale suborbicular, often wider than long. Mature achene with stipe and mucro 1.5 mm long, shiny brown. Wet soil, flatwoods, ditches, in openings of saw grass glades, central and western Fla. to Miss.

8. HEMICÁRPHA Nees & Arn.

Small, glabrous, scapose annuals in dense tufts; culms and leaves filiform. Spikelets one to three in involucrate heads, many-flowered; scales spirally imbricate, each subtending a bisexual floret in axil of one delicate perianth scale; stamens commonly one, stigmas two, achene subterete. About six species, temperate and tropical regions.

1. **H. micrantha** (Vahl) Pax—Culms mostly 5–8 cm long, with radical leaves, purplish sheaths. The longest involucral bract apparently a prolonged stem tip, overtopping the leaves. Spikelets one to three, sessile, conoidal, 4–7 mm long. Mature achenes to 0.3–0.4 mm long, obovoid, colliculate, iridescent, mucronulate. Pioneer plant on wet shores and cleared areas, widely distributed in North and South America. All year.

9. LIPOCÁRPHA R. Brown

Tufted annual herb with basal leaves. Terminal involucrate inflorescences; scales spirally imbricate, each subtending a pair of inner scalelets enclosing a flower; flowers bisexual without perianth bristles; stamens one to two. Stigmas two to three. Achene two- or three-edged. About 12 species, warm temperate, tropical, and subtropical America, Africa, Asia.

1. Scales spatulate, awnless . . . . 1. *L. maculata*
1. Scales obovate, awned . . . . 2. *L. microcephala*

1. **L. maculata** (Michx.) Torr.—Culms 5–20 cm long, slender, erect. Sheaths reddish, folded, narrowed to linear-attenuate blades. Spikes three to six, ovoid, subtended by foliose involucral bracts, commonly two, the longer one spreading horizontally or becoming deflexed; scales acute, 2 mm long, with green midvein, and white margins becoming spotted with purple. Mature achene 1 mm long, brown, enclosed between hyaline scales, obovoid, compressed-mucronate, colliculate. Moist sandy soil, often margins of landscaped lawn, Fla., coastal plain states to Va., adventive farther north. Summer, fall.

2. **L. microcephala** Kunth—Plants glabrous, tufted. Culms filiform, 8–13 cm tall. Leaves 2–4 cm long, narrowly linear; sheaths loose, purple-striate, abruptly narrowed to short blades. Spikes three, subtended by two basally dilated involucral bracts; the ascending bract 12–20 mm long, the descending one about one-third as long; floral scales numerous, imbricate around the rough rachis; the scale with awn 1.2 mm long; the body purple striate, concave toward the base, widest at the apex, abruptly narrowing toward the base of the recurving awn; the hyaline scales minute; stamens two; anther and filament equal in length, about as long as the pistil with three sessile style branches in anthesis. Mature nutlet brown, 0.3 mm long, three-angled, obovate, mucronulate, colliculate at surface. South Fla. near Immokalee a small colony was

located in white sand of ditch margin of pineland ranch, naturalized from Australia. Summer, fall.

10. DICHRÓMENA Michx.

Leafy herbs from perennial, creeping rhizomes. Involucral leaves white at base, subtending the terminal spike, spikelets white, scales imbricate, conduplicate or navicular; stamens three. Styles two-cleft; perianth lacking. Achene lenticular, transversely rugulose, tubercled. About 60 American, W.I. species.

1. Involucral bracts seven to ten, broadly lance-
    olate; body of achene blackish, tubercle
    decurrent . . . .   1. D. latifolia
1. Involucral bracts four to six, linear or nar-
    rowly lanceolate.
    2. Culms tufted, rhizomes knotty; achene
        black, tubercle depressed . . . .   2. D. floridensis
    2. Culms scattered on elongate rhizomes;
        achene yellowish, tubercle acute, flat . .   3. D. colorata

1. D. latifolia Baldw.—Culms 6–10 dm tall, bluntly angled to sub-terete from woody rhizomes. Sheaths loose, blades folded, broadly linear-attenuate to 2 dm long. Involucral leaves white to the attenuate apex; spikelets elliptic in heads 1–2 cm wide; scales 4–6 mm long, white, membranous. Achene obovoid or globose 1.6 mm long. Wet pineland, cypress margins, Fla. to Tex., Va.

2. D. floridensis Britt.—Culms 1–3 dm tall, capillary from knotty tufts. Leaves filiform above the sheathing bases, 5–10 cm long, flexuous. Spikelets elliptic in heads to 10 mm wide; scale 3–4 mm long, obtuse, white. Achene ovoid or globose, 1 mm long. Pinelands, Everglade Keys, endemic to south Fla.

3. D. colorata (L.) Hitchc.—Culms to 5 dm tall from slender, elon-gate rhizomes. Leaves to 3 dm long, 1–3 mm wide, flat or folded; sheaths loose, brownish. Bracts spreading to reflexed, to 7 cm long; spikelets 6–8 mm long. Achenes with beak 1.7 mm long, yellowish, rugulose, the body truncate at apex. Glades, moist hammock margins, Fla. to Tex., Va., W.I., Mexico, British Honduras.

11. PSILOCÁRYA Torr. BALD RUSH

Caulescent annual from fibrous roots; culms leafy. Inflorescence in terminal and axillary cymes; spikelets ovoid-acuminate, scales nu-merous, imbricate; stamens two. Style two-cleft, the base persistent as

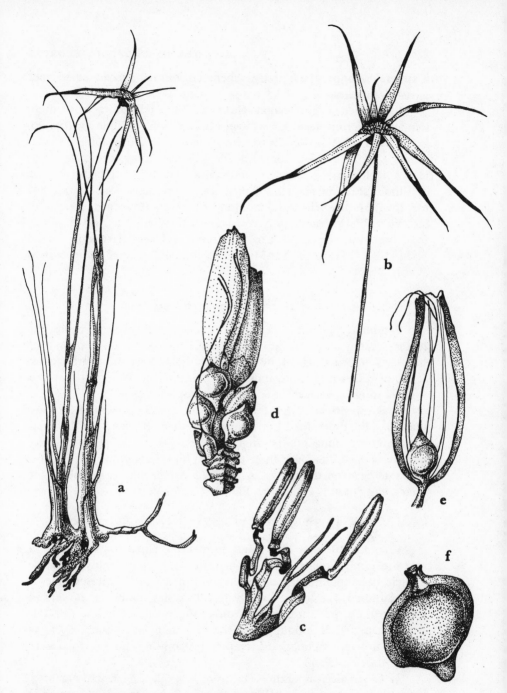

Family 33. **CYPERACEAE.** *Dichromena latifolia:* a, flowering plant, × ¼;
b, inflorescence, × ½; c, floret with androecium and gynoecium, × 12½; d,
developing fruits, × 10; e, developing fruit and subtending scales, × 12½; f,
mature fruit, × 15.

a tubercle, topping the shiny achene. About 60 species of warmer parts of America.

1. **P. nitens** (Vahl) Wood—Culms to 7 dm tall, solitary or tufted, slender, compressible; sheaths loose, dilated at the oblique hyaline orifice, readily splitting. Blades linear to 3 mm wide, flat or folded. Cymes simple or compound in umbelliform, wide-spreading, peduncled clusters; spikelets to 10 mm long, including the denuded rachis left by deciduous scales of maturing fruits; scales castaneous, one-nerved, 2–3 mm long. Mature achene 1.3 mm long, obscurely two-edged, obovate in face view, black, transversely corrugate; tubercle white, broad, decurrent, becoming depressed. Coastal strand, wet pine flatwoods, coastal plain, Fla., Tex. to N.Y., Midwest. Summer. *Rhynchospora nitens* (Vahl) Gray

12. REMÌREA Aubl.   BEACH STAR

Perennial caulescent herb with elongate, creeping, ligneous rhizomes rooting at nodes, at intervals, giving rise to fasciculate leafy branches terminating in flowering heads. Leaves imbricate, many. Scales of spikes numerous, imbricate; perianth lacking, stamens three, stigmas three, achene three-edged. One tropical species.

1. **R. maritima** Aubl.—Stems 1–3 dm tall. Leaves subdistichous; sheaths split, short, broadly hyaline margined. Blades 3–10 dm long, rigid, arching, pungently pointed. Spikelets 4–5 mm long; scales many-nerved, hooded, the lower three or more empty; the upper fertile scales punctate between the nerves. Achene 2.3–3 mm long, trigonous. Sand dunes, east coast of Fla., south Fla., W.I., Old World.

13. CLÀDIUM P. Browne   SAW GRASS

Coarse, caulescent perennials from thick rhizomes. Leaves rigid, antrorsely spinulose-scabrous on margins and on the midvein below. Panicle decompound, plumose, terminal or axillary. Spikelets one- to three-flowered, the lowest usually fertile; scales imbricate; flowers bisexual without perianth, stamens three, stigmas three, style continuous with the ovary, deciduous above the persistent base. About 50 to 60 species, nearly cosmopolitan, temperate, tropical, and subtropical regions. *Mariscus* Scop.

1. **C. jamaicensis** Crantz—Culms to 3 m tall, 2.5 cm in diameter at base, bluntly three-angled. Blades flat, 6–10 dm long, 6–14 mm wide, linear-attenuate to elongate capillary apices, erect or spreading; sheaths coriaceous, brownish red, orifice oblique. Panicle umbelliform, spikelets 4–5 mm long, in glomerules at the ends of branchlets, reddish, fad-

ing ferruginous; the lower several scales empty. Mature fruit 2.5–3 mm long, ovoid, beaked, lustrous brown, pericarp soft, endocarp black, indurated. Dominant sedge in glades. Fla., Tex., to Va., W.I. *Mariscus jamaicensis* (Crantz) Britt.; *C. effusum* (Sw.) Torr.

14. RHYNCHÓSPORA Vahl BEAK RUSH

Mainly perennial caulescent herbs; culms triangulate, leafy, rarely subterete; leaves narrowly linear, often filiform. Spikelets in cymes, variously arranged in panicles; the scales one-veined, flat or scarcely concave, spirally imbricate, the lower one to two empty, one or more succeeding ones subtending bisexual flowers; the uppermost scale bearing a staminate or rudimentary flower. Stamens usually three, pistil one, style cleft into two stigmas, or merely bilobed, crowning the achene by its tubercular base or persistent as the beak of achene. Achene body lenticular, or globular with characteristic surface markings. About 200 species, cosmopolitan, especially in tropical regions.

1. Achene 3–6 mm long; crowned by the style as a whole.
   2. Tubercle linear, much narrower than the achene summit; spikelets sessile, radially crowded into one or more spherical heads . . . .   1. *R. tracyi*
   2. Tubercle dilated at base, covering the rounded achene body; spikelets short pedicelled, arranged in paniculate glomerules.
      3. Bristles shorter than the achene body .   2. *R. corniculata*
      3. Bristles longer than the achene body .   3. *R. inundata*
1. Achene 3 mm long or less; crowned by the tubercled base of style.
   4. Bristles lacking, or minute; culms and leaves filiform tubercles depressed.
      5. Achene 1 mm long, transversely wrinkled . . . .   4. *R. intermixta*
      5. Achene 0.7–0.9 mm long, cancellate   .   5. *R. divergens*
   4. Bristles present.
      6. Bristles retrorsely barbed; inflorescences globose or loosely glomerulate; achene 1.1–1.6 mm wide, lenticular, shouldered at summit . . . . .   6. *R. cephalantha*
      6. Bristles antrorsely barbed, or plumose below.
         7. Perianth bristles silky plumose below serrulate tips; spikelets in congested fascicles; tubercles conoidal, depressed.

8. Leaves filiform to 1 mm wide; fascicles spiciform . . . . .   7. *R. plumosa*

8. Leaves narrowly linear to 2 mm wide; fascicles corymbiform . .   8. *R. intermedia*

7. Perianth bristles, barbed only, not plumose.

  9. Floral bracts ciliate; leaves crowded to a loose rosette . . . . .   9. *R. ciliaris*

  9. Floral bracts not ciliate.

    10. Achene surface smooth, disk commonly pale; floral bracts conspicuous . . . . . .   10. *R. fascicularis*

    10. Achene with shallow surface markings.

      11. Tubercle conical, confluent with the achene summit; achenes strongly biconvex.

        12. Achene 3–4 mm long and nearly as wide; surface silvery brown, shallowly alveolate pitted; leaves to 7 mm wide, tufts large . .   11. *R. megalocarpa*

        12. Achene 1.8–2.4 mm long and nearly as wide; surface patently alveolate, reddish brown in maturity; leaves 1.4 mm wide, tufts small .   12. *R. grayii*

      11. Tubercle mostly deltoid or conical, basal rim elevated or parted from achene summit.

        13. Plants relatively slender; leaves 4 mm wide or less.

          14. Perianth bristles shorter than the achene; spikelets in glomerules, mostly longer than pedicels . . . . .   13. *R. globularis*

          14. Perianth bristles equal to or exceeding the achene, or sometimes lacking; spikelets in glomerules, mostly shorter than pedicels.

            15. Bristles rudimentary, caducous or missing, achenes obovate, strongly transversely ridged, 1–1.3 mm long, spikelets 2.5

mm long, tubercle
deltoid . . . . 14. *R. perplexa*
15. Bristles well developed.
16. Plants stoloniferous;
cymes on divaricate
peduncles; achenes
3–4 mm long, rug-
ulose-striate, tuber-
cle conoidal, de-
pressed . . . 15. *R. miliacea*
16. Plants tufted; culms
flexuous at sum-
mit; tubercle del-
toid.
17. Cymes diffuse; ped-
icels spreading;
achene 1.3 mm
long, transverse-
ly rugulose, can-
cellate.
18. Leaves filiform . 16. *R. rariflora*
18. Leaves linear,
2–4 mm wide 17. *R. decurrens*
17. Cymes fasciculate
on ascendent
pedicels; achenes
1 mm long, al-
veolate pitted . 18. *R. microcarpa*
13. Plants relatively stout;
leaves 10 mm wide or less,
achenes stiped.
19. Mature achenes strongly
cross-striate, dark brown 19. *R. odorata*
19. Mature achene cross-
ridged, golden brown . 20. *R. caduca*

1. R. tracyi Britt.—Culms to 0.6–1 m tall, from spreading rhizomes,
glaucous, wiry, subequal to the involute elongate leaves. Spikes capitate,
spherical, terminal. Achene with beak to 6.5 mm long, the body clavate
transversely, striate; the beak scarcely dilated at base; bristles upwardly
barbulate nearly twice the length of the body. Wet pinelands, ditches,
coastal plains, Fla. to Ga., W.I. Summer.

2. R. corniculata (Lam.) Gray—Plant nonstoloniferous to 9 dm
tall or taller; culms leafy, stocky; leaves 9–18 mm wide, rigid. Cauline
sheaths elongate, tight. Inflorescence decompound, cymose; mature
spikelets golden brown; floral bracts aristate. Achene body 5 mm long;
tubercle 15 mm long, subulate above the dilated base, bristles three,

much shorter than the achene. Marshy habitats, coastal plains, and interior states, Fla. to Tex., Pa., Del. Summer.

3. **R. inundata** (Oakes) Fern.—Strongly stoloniferous plant to 9 dm tall; culms relatively stout, sharply three-angular. Sheaths loose; blades flat, membranous, 1–2 cm wide, flat, involute at tips. Involucral leaves scabrous, exceeded by the inflorescence; cymes decompound to 20 cm wide, diffuse; spikelets golden brown, 1–2 cm long; scales ovate, thin, the lower short-aristate. Achene body 4–4.5 mm long, beak 17–19 mm long, the dilated base nearly as wide as the achene summit; bristles six, slightly shorter than the body. Wet canal beds and ditches, south Fla. to Mass. Summer.

4. **R. intermixta** C. Wright—Culms tufted, 5–20 cm long, filiform. Blades shorter, filiform, flexuous; sheaths loose. Cymes 1.5 cm long and about as wide, subtended by setaceous involucral bracts; spikelets in diffuse cluster, bright brown. Achene 0.8–0.9 mm long, obovoid, transversely striate, crowned with minute tubercule, tipped with black or pale. Moist pinelands, often over oolitic rocks, coastal plains, Fla. to Tex. Summer.

5. **R. divergens** Chapm.—Culms tufted 2–3 dm tall, capillary, wiry; leaves filiform, exceeded by the culms; sheaths loose; cymes corymbiform, 3.5 cm long, acute; scales mucronulate, golden brown; achenes 0.7–0.8 mm long, smooth, obovate. Moist or wet ground, endemic to central Fla., extending into Collier County. Summer.

6. **R. cephalantha** Gray—Plants tufted 4–9 dm tall; culms obtuse, three-angled, rigid, rather thick or relatively slender. Leaves narrowly linear or to 4.5 mm wide; radical blades, elongate, to above the middle of the culm; cauline leaves erect and short. Involucral bracts foliaceous and prominent; inflorescence 2–3 dm long; glomerules remote, terminal and axillary on short branches or sometimes sessile; spikelets elliptic, 4–5 mm long, chestnut brown or blackish. Achene to 4 mm long with beak, the body glossy buff with truncate median summit, margined and rounded to a broad stipe; bristles six, bandlike, converging just below the subulate apex of the beak. Shady woodland, moist cypress borders, coastal plain states, Fla. to N.J.

7. **R. plumosa** Ell.—Culms 4–7 dm long, tufted, wiry, filiform, overtopping the usually flexuous filiform leaves. Sheaths short. Spikelets 2.5–3.5 mm long, elliptic-ovoid, in clusters of four or five, in more- or less-interrupted terminal inflorescence; scales acuminate-aristulate, 2.5 mm long, golden brown. Achenes plump, obovoid 1.5 mm long, strongly transversely rugulose; tubercle depressed, bristles silky white or buff plumose. Pinelands, Fla., Tex., Ga., coastal plain to N.C.

8. **R. intermedia** (Chapm.) Britt.—Culms tufted from knotty rhi-

zomes to 7 dm tall, obtusely angled, wiry, slender. Sheaths tight, firm, brownish; leaves numerous, usually flexuous, surpassed by the culms, narrowly linear, involute. Involucral leaves two to three, short, stiff; spikelets ellipsoid, ovoid, in glomerules of four or five gathered in loose semispherical heads; scales castaneous, acute or mucronate 3–5 mm long; perianth bristles longer than the achene body. Achene 2 mm long, transversely rugulose, obovoid, tubercle depressed. Frequent in sandy swales and pinelands, endemic to Fla.

9. **R. ciliaris** (Michx.) Mohr—Conspicuously tufted glaucous plants. Radical leaves flat, 4–6 mm wide, ciliate, obtuse; culms obscurely angled, erect to 9 dm tall, glabrous; cauline blades few erect, short-ciliate. Inflorescence commonly solitary, corymbiform, fascicles subtended by linear, subulate bracts; ciliate floral bracts; spikelets rufous brown, 4–6 mm long, usually two-fruited; scales acute or the lower short-aristulate. Achene biconvex, with tubercle 2.5 mm long, minutely pitted; bristles six, caducous. Moist pinelands, endemic to central and south Fla.

10. **R. fascicularis** (Michx.) Vahl—Plants to 1 m tall from close-set rhizomes; culms relatively thick, obtusely angled; leaves 1–4 mm wide, elongate, flexuous; cauline blades short. Spikelets 3–4 mm long, fasciculate in terminal corymbiform glomerules; lateral glomerules distant, pedunculate; scales ovate to lanceolate, short-aristate; bristles five to six, caducous. Achene biconvex, umbonate, 1.7–2 mm long with tubercle. Pinelands, pond margins, central and south Fla., Tex. to Va., coastal plain states.

Also in our area is:

R. fascicularis var. **distans** (Michx.) Chapm.—Differs from fascicularis chiefly by more-slender habit, darker and slightly smaller achene, and longer bristles. Similar habitats and distribution. Of doubtful taxonomic significance.

11. **R. megalocarpa** Gray—Rhizomatous perennial to 1 m tall with rufous scales at base; culms obtusely triangular. Sheaths loose, rufous brown; blades coriaceous, flat, 4–10 mm wide; nearly as long as the culms. Spikelets in distant corymbiform fascicles, on slender peduncles; spikelets 6–7 mm long, ovoid, one-flowered and one-fruited; scales mucronate, becoming castaneous in age; stamens ten to twelve, bristles six to eight, shorter than the achene body, caducous. Achene 3–4 mm long and nearly as wide; surface glossy, minutely pitted or shallowly alveolate; tubercle conoidal, apiculate, depressed above annular rim of the achene summit. White sand, Pinus, Ceratiola, Quercus scrub, coastal dunes, central Fla., Ga., N.C., S.C. *R. dodecandra* Baldw.; *R. pycnocarpa* Gray

12. **R. grayii** Kunth—Culms tufted, 2–7 dm tall, trigonous, sur-

passing the leaves. Blades 2–4 mm wide, flexuous; sheaths loose, blackish brown, becoming fibrous in age. Spikelets two to several, glomerulate, rufous brown, ovoid, 5–6 mm long, scales mucronate, 3–5 mm long; perianth bristles six, stamens commonly three, florets three, one-fruited. Achene obovoid, 2–2.4 mm long and nearly as wide, tubercle conoidal, apiculate; achene body compressed, rimmed at summit; surface alveolate, reddish brown. Moist depressions in pinelands, coastal plain states to Tex., Fla.

13. R. globularis (Chapm.) Small—Culms slender, sharply trigonous, wiry to 1 m tall, tufted from knotty rhizomes. Basal leaves numerous, 1–4 mm wide commonly below the midculm; cauline leaves erect, involute, distant above the middle. Spikelets one to several in fasciculate glomerules, subtended by bracts of varying length; lateral cymes long peduncled; spikelets 1.5–4 mm long, one- to three-fruited. Achene with beak, 2.1 mm long, transversely rugulose; tubercle conoidal, bristles four to six, antrorsely barbed, shorter than the achene body, deciduous. Peaty sands and moist depressions, pinelands and shores, coastal plain, Fla., Tex., northeastern states, west to Calif. Includes var. *recognita* Gale

Also in our area is:

R. globularis var. **pinetorum** (Small) Gale—Similar to **globularis**. Achene to 1.8 mm long, tubercle triangular; surface alveolate, or reticulate isodiametrically; culms leafy; blades flexuous, bright green. Similar distribution and habitats, Fla., Cuba, Jamaica.

14. R. perplexa Britt. ex Small—Plant strongly cespitose; culms relatively slender, to 1 m tall. Blades flat, linear, flexuous, minutely scabrous along margins; spikelets disposed in corymbiform cymes; subtending bracts shorter than the peduncles; spikelets ovoid, suborbicular, three- to five-flowered; two- to four-fruited, congested in numerous ultimate fascicles; scales mucronate, thin, ferrugineous, readily deciduous. Achenes obovate in face view, 1.3 mm long, transversely rugulose; tubercle deltoid, compressed; bristles obsolete, lacking. Pinelands, lake beds, and pond terraces, coastal plain, Va. to Fla., Tex., Cuba. Summer.

15. R. miliacea (Lam.) Gray—Stoloniferous with scaly rhizome; culms trigonous to 1 m tall, leafy. Blades 3–10 mm wide, membranous; sheaths short, brownish, loose. Inflorescence corymbose,* decompound with wide-spreading peduncles from the rachis, ultimately reflexed; spikelets ovoid 2.5–3·mm long, distant; pedicles slender, four- to several-flowered, three- to ten-fruited; scales aristulate, readily caducous, exposing the achenes. Achene 1.3 mm long, obovoid, the transverse ridges intersected by longitudinal striations; tubercle conoidal, depressed.

Hammocks and cypress margins, coastal plain to Fla., Ark., Tex., Cuba. Summer.

16. R. rariflora (Michx.) Ell.—Plant wiry, densely tufted from knotty crown; culms filiform, reclining, to 5 dm long. Blades filiform, flexuous, serrulate toward the tips. Inflorescence cymose, spikelets relatively few, remote, diffuse, long peduncled; spikelets two- to four-flowered, one- to three-fruited, 3–4 mm long; scales ovate, golden brown, thin; perianth bristles six. Achene obovate in outline, 1.9 mm long, biconvex, transversely rugulose; tubercle deltoid, compressed. Glades and grassy berm, exsiccating ditch beds, coastal plain, Fla. to N.J., Tex. to Cuba, Jamaica.

17. R. decurrens Chapm.—Plant cespitose to 1 m tall; culms obtusely trigonous, lax, leafy. Blades linear, flat 2–4 mm wide; sheaths thin, loose, spreading, flexuous. Inflorescence decompound; cymes diffuse, spreading, corymbiform; scales mucronate to 3 mm long. Achene cuneate, obovoid, 1.7 mm long, pitted, irregularly transverse-rugulose; tubercle deltoid, decurrent, hispid; perianth bristles equaling the achene in length. Shallow shore waters of lakes and ponds, Fla. to N.C.

18. R. microcarpa Baldw. ex Gray—Plants strongly cespitose or solitary; culms to 1 m tall, slender, rigid, leafy below the middle. Cauline leaves few; blades 2–4 mm wide; flexuous. Spikelets in decompound corymbiform cymes, three to four remotely disposed along upper third of the culm; spikelets ovoid, sessile, 3–4 mm long, three- to four-flowered, two- to three-fruited, aggregated into fascicles; scales aristulate, dark brown, bristles six, as long as or longer than the achene. Achene biconvex, alveolate, with tubercle 1.8 mm long. Moist soil, swamps, and pond margins, Fla. to N.C., Cuba, W.I., Jamaica.

19. R. odorata C. Wright ex Griseb.—Stout, leafy, glaucous, stoloniferous perennials to 1⁻ m tall or more. Blades elongate, linear, flat, 3–12 mm wide, minutely scabrous on margins, at least when young. Spikelets in four- to six-peduncled conspicuous corymbiform clusters; floral bracts subulate, scabrous at tips; scales mucronate, 3–8 mm long, golden brown, becoming castaneous in age, readily caducous; spikelets three- to eleven-flowered, one- to seven-fruited, 6–10 mm long. Achene obovate in outline, biconvex, with stipe and tubercle 2.6 mm long, transversely rugulose; perianth bristles six, exceeding the tubercle; tubercle deltoid, compressed, upwardly hispid, minutely scabrous on edges. Low pinelands, glades in hammock margins, coastal plain, Fla. to Miss., N.C., Cuba, Bahama Islands. *R. stipitata* Chapm.

20. R. caduca Ell.—Culms to 1 m tall, rhizomatous spreading by means of stolons; leaves flat, elongate to 8 mm wide, keeled, narrowing to triquetrous scabrous apex. Inflorescence elongate of remote cymes, each subtended by a foliose bract; cyme freely forking; spikelets fasciculate on slender pedicels, usually two- to five-flowered; achenes 1.4–1.6 mm long, with deltoid, setulous beak, exceeded by bristles. Swampy forest border, cypress margins, south Fla., Tex., Ark., coastal plain states, N.C., W.I.

15. SCHOÈNUS L.

Perennial tufted herbs from short rootstock. Leaves few, capillary, sheaths reddish black. Inflorescence terminal, subtended by involucral bract; spikelets in dense heads with three lower scales, empty, the fourth scale fertile; perianth three to six short bristles; stamens three, stigmas three. Achene three-angled, white with black style base. About 100 species mainly in Southern Hemisphere, Europe, Asia, Malaya, America.

1. S. nigricans L.—Culms 2–7 dm tall, wiry, slender, overtopping the leaves. Blades semiterete, stiff, sharp pointed, upwardly scabrous on margins; sheaths hyaline on margins with ligule forming a rounded membranous ligule in the folded base of the blade. The lower involucral bract exceeding the spike; the joints of the spike rachis yellowish, sulcate, subtend the flower; spikelets to fifteen in a head; scales glossy, conduplicate, keeled, blackish maroon under light. Achene glossy white, glabrous, 1.3 mm long, trigonous, with a black apical scar left by the deciduous style. Pinelands and prairies, south Fla., Calif., Mexico, W.I., Old World.

16. SCLÉRIA Bergius　NUT RUSH

Mostly perennial. Culms sharply trigonous, leafy, with overlapping sheaths, the lowest bladeless. Sheaths trigonous, the two sides of the keeled angle continuous with the blade. Blades linear, keeled, the midvein flanked by two prominent and many fine parallel veins on each side; the third side produced to a short free deltoid ligule beyond the orifice, or obsolete. Plants monoecious, flowers unisexual, without perianth. Pistillate spikelets one-flowered with empty scales below, contrary to staminate spikelets with several flowers, and empty scales above. Stamens one to three; pistil one, styles three-cleft, deciduous. Ovary subtended by hypogynium, a disk, or if lacking, only a stipiform conical base. Achene white or gray, body crustaceous, smooth or variously embellished by salient marks. About 200 tropical and subtropical species.

1. Hypogynium none.
　2. Achene surface alveolate-verrucose, plant
　　hirtellous at base, annual　.　.　.　.　1. S. verticillata

2. Achene surface glossy, smooth, perennials.
   3. Inflorescence paniculate, achene base
     with three cavities . . . . . .   2. *S. lithosperma*
   3. Inflorescence spicate.
     4. Glomerules of spikelets several, remote on rachis . . . . . .   3. *S. hirtella*
     4. Glomerules of spikelets solitary, terminal.
       5. Culms slender, wiry; achene base with three cavities that are two-pitted . . . .   4. *S. georgiana*
       5. Culms stout; achene with three cavities without pits . . . .   5. *S. baldwiniana*
1. Hypogynium present, with three entire lobes; plants perennial.
   6. Achene surface smooth, hypogynium incrusted with crystalline papillae. . .   6. *S. triglomerata*
   6. Achene surface variously demarked.
     7. Hypogynium tuberculate.
       8. Plants strongly pubescent; hypogynium with three tubercles, sometimes each lobed, achene surface papillose, verrucose . . . .   7. *S. ciliata*
       8. Plants glabrous, glabrescent, or puberulent.
         9. Achene papillose or verruciform, tubercles six . . . . . .   8. *S. pauciflora*
         9. Achene reticulate, tubercles six .   9. *S. curtissii*
     7. Hypogynium etuberculate; achene reticulate, pubescent . . . . .   10. *S. reticularis*

1. **S. verticillata** Muhl.—Annual from fibrous roots; culms to 6 dm tall, simple trigonous, slender, essentially glabrous. Leaves 0.5–3 mm wide, exceeded by the culms; sheaths pilose, ligule truncate, hirsute. Glomerules commonly four to nine, 4–8 mm long, the lower rarely a cymule of threes; spikelets 2–3 mm long, becoming castaneous. Achene 1.2–1.5 mm long, globose, verrucose-reticulate, apiculate, stiped, falling with or without the rudimentary hypogynium. Moist, sandy pinelands, and canal banks, Fla. to upper Midwest, Ontario, northern states, Mexico, W.I.

2. **S. lithosperma** (L.) Sw.—Perennial from nodulose rootstocks, forming leafy tufts; culms in dense clusters to 6 dm tall; sheaths overlapping, the lowest bladeless. Blades 1–3 mm wide, dark lustrous green between the nerves; reclining, keeled, essentially glabrous; ligule mostly deltoid, short pilose at summit. Panicle lax; subtending bracts setaceous exceeding the distant spikelets; pistillate scales ovate, abruptly awned,

center green. Achene 2–2.5 mm long, glossy, white, smooth. Shady hammocks, open woods, in shallow soil over lime rocks, south Fla., Bahama Islands, Cuba, Mexico, tropical regions of Old and New Worlds.

3. S. hirtella Sw.—Perennial; rhizome elongate, aromatic; culms trigonous, to 6 dm tall, glabrous or hirtellous above, exceeding the leaves. Blades 2–5 mm wide, linear, flat, pubescent or glabrate; ligule rounded, hirsute. Inflorescence to 12 cm long; glomerules three to nine, sessile, nodding; rachis three-keeled; floral bracts long ciliate; pistillate scales ovate, cuspidate; spikelets 4–5 mm long, becoming reddish brown or blackish. Achene glossy white, the body 1 mm long, globose, apiculate, abruptly narrowed to a three-sided, jointed, stipelike base, each side with a depression, sculptured with five or more minute pits between the folds. Coastal plain, Fla. to Ga., N.C., S.C., W.I., Mexico, Central and South America. Summer.

4. S. georgiana Core—Rhizome nodose, horizontal, hard, becoming densely compacted in age; culms wiry, to 5 dm tall, slender, trigonous, exceeding the leaves; sheaths purplish, orifice truncate, the lower bladeless. Blades linear, filiform 1–2 mm wide. Inflorescence a terminal fascicle, two to four spikelets; the primary involucral bracts appearing as an extension of the stem; spikelets ovate, 5 mm long; scales reddish brown; pistillate scales ovate, acuminate. Achene white, 2 mm long, longitudinally ridged, the base trigonous, each side with a depression and two pores. Low pinelands, Fla. to Tex., coastal plain to Carolinas, W.I., tropical South America. Summer. *Scleria gracilis* Ell. not Richard

5. S. baldwinii (Torr.) Steud.—Perennial from nodose rhizomes; culms stout, sharply angled, purplish below, to 9 dm tall, exceeding the leaves. Blades involute, 1–2 mm wide. Inflorescence solitary, floral bract to 4 mm wide, a continuation of the culm; bractlets glabrous; pistillate scales 5–7 mm long, awned. Achene 3–3.4 mm long, longitudinally grooved, apiculate, base three-sided with two pores each side. Wet pinelands, hammocks, Collier County, Fla., to Tex., Ga., Cuba. Summer. *Scleria costata* Small

6. S. triglomerata Michx.—Leafy perennial from hard knotted rhizome, in large tufts, 0.5–1 m tall, culms sharply trigonous. Blades flat, scabrous on the keel and margins, 3–7 mm wide, sheaths purple, pubescent on angles; ligules, obtuse or broadly deltoid, short. Inflorescence terminal; barely overtopping the leaves; axillary glomerules on elongate peduncles; spikelets 5–6 mm long; pistillate scales ovate to orbicular; keel long excurrent, scabrous. Achene 2–3 mm long, glossy white, ovoid-globose. Glade and hammock margins in sandy soil, Fla. to Tex. and, Midwest. Summer. *S. nitida* Willd.

7. S. ciliata Michx.—Plants from horizontal nodose rhizomes, becoming intricately layered in age. Culms 2–7 dm tall, erect, sharply triquetrous, slender or stout. Leaf blades 1–25 cm long, 1–6 mm wide, glabrous or ciliate, keeled, pubescent on surfaces above and below, or glabrous; sheaths mostly retrorsely hairy on angles, ligule 2–5 mm long, obtuse-ciliolate or glabrous. Spikelets in glomerules of three or more, short pedicelled; pistillate scales 4–6 mm long, ovate-acuminate, with ciliate or glabrous keel, greenish, becoming purple in fruit. Mature achene apiculate, body 2–3 mm long, 2.5 mm wide, verrucose-papillose, brown, hypogynia with bilobed, yellowish tubercles, disk brown; less cutinized achenes white with yellowish disk. Sandy pineland, crevices of oolitic rocks, Fla., Tex., to Va., Cuba. Spring, summer. *S. brittonii* Core; *S. elliottii* Chapm.; *S. glabra* Britt. ex Small

8. S. pauciflora (Muhl.) ex Willd.—Tufted perennial from congested nodulous rhizomes; culms erect 2–5 dm tall, numerous from close-set nodes, rigid, triangulate. Leaves 8–15 cm long, 2–2.5 mm wide, scabrous-ciliate toward apices or glabrate; basal sheaths purplish brown, minutely puberulent. Bracts essentially glabrous, or scaberulate on margins, surpassing the inflorescence, the outermost continuous with the culm. Spikelets three- to six-flowered in terminal fascicles or borne on axillary peduncle; pistillate scales ovate-acuminate, 3–5 mm long, midvein shortly excurrent, scaberulous, sides brownish red. Achene 2–2.3 mm long, globose, usually white, papillose, apiculate, tubercles six, distinct. Coastal strand, sandy peat, borders of cypress ponds, Fla., Tex. to Va., Midwest. Summer.

9. S. curtissii Britt.—Rhizomes knotted, reddish orange, tuft-forming; culms 4 dm tall or more, reclining, essentially glabrous; leaf blades 1–2 mm wide, involute; basal sheaths purplish, minutely grayish puberulent, ligule about 1 mm long, deltoid. Spikelets few in terminal sessile fascicles or often peduncled in axils; bracts stiffish, attenuate, glabrate; scales glabrous, long acuminate, yellow, variegated with red, the midvein often green. Achene 2–3 mm long, reticulate, white, hypogynium in mature fruits brownish, disklike, subtending three pairs of tubercles. Coastal strand in moist sand, open pineland, oak scrub, Fla., Cuba.

10. S. reticularis Michx. var. **pubescens** Britt.—Short-lived perennials or tufted annual with reddish fibrous roots; culms glabrous to 8 dm tall, slender, leafy, bladeless basal sheaths cleft; leaf blades 2–7 mm wide, smooth, ligules short, rounded. Inflorescence terminal and axillary, leafy bracted, borne on elongate, filiform, pendent peduncles; flowers few, pistillate spikelets 3–4 mm long, scales ovate-acuminate, glabrous. Achene 2 mm long, whitish, patently reticulate, with scattered, yellowish villous hairs. Pinelands, shrubby borders of lagoons, in moist soil, Fla., Tex., W.I. to South America. Summer, fall.

17. CÀREX L. Sedge

Perennial solitary or cespitose herbs with fibrous roots, often stoloniferous; culms three-sided, leaves three-ranked, sheaths closed, leaf blades linear. Flowers unisexual without perianth, staminate or pistillate, in different or in the same spike on the same plant, or rarely dioecious. Stamens three, pistil one, stigmas two or three, ovary one-locular, one-ovuled, contained in the perigynium, a modified saclike scale. Achene lenticular or trigonous. Cosmopolitan genus of 1500 to 2000 species, especially in temperate and subtropical regions.

1. Stigmas two; achene lenticular; perigynium
flat; basal scales of spikes staminate.
  2. Perigynium 4–5 mm long, pistillate scales
    rough awned . . . . 1. *C. alata*
  2. Perigynium 5 mm long or less; pistillate
    scales not rough awned.
    3. Perigynia concavo–convex, suborbicular . . . . 2. *C. vexans*
    3. Perigynia plano–convex, lanceolate . 3. *C. longii*
1. Stigmas three; achene trigonous; perigynia
inflated; staminate and pistillate spikes
distinct.
  4. Style persistent; beak of perigynia 3–13
    mm long.
    5. Plant cespitose; pistillate spikes aggregate.
      6. Mature perigynia spreading at right
        angles . . . . 4. *C. gigantea*
      6. Mature perigynia ascending . . . 5. *C. lupulina*
    5. Plant solitary; pistillate spikes distant . 6. *C. louisianica*
  4. Style deciduous above the base; beak of
    perigynium 1 mm long . . . . . 7. *C. joorii*

1. **C. alata** Torr.—Culms to 9 dm tall, sharply angled, densely tufted; leaves to 5 mm wide, scabrous margined, three to five or more per culm; sheaths green, strongly nerved, hyaline at the orifice, blades 7–15 cm long. Inflorescence of five to seven or more, ovoid or clavate spikes, subtended by setaceous bracts; scales narrow, aristate, shorter than perigynia; perigynia to 5 mm long and nearly as ascending, imbricated on rachis, abruptly narrowed to a serrulate beak, margins broadly winged. Achene ovate in face view, 2.2 mm long, shortly stiped and apiculate. Moist habitats, canal banks, cypress terraces, glades, Fla., Tex. to Midwest, Mass. Spring, summer.

2. **C. vexans** Hermann—Culms tufted, to 8 dm tall, obtusely angled,

scabrous at summit, commonly exceeding the leaves; sterile shoots numerous, leafy. Blades linear, flat, 4–15 cm long, 2.5–4 mm wide, scaberulous on margins and midvein; spikes four to several, ovoid or clavate, aggregated or the lowest distant, 8–12 mm long, 5–8 mm wide in maturity; subtending bracts setaceous, roughened, scales acute, thin, green veined, hyaline. Perigynia 4–4.5 mm long and nearly as wide, broadly winged, narrowed abruptly to a serrulate beak, achene elliptic-oblong, 2 mm long, apiculate and stiped. Open ground, hammock margins, grassy roadsides, moist banks, endemic to central and south Fla.

3. **C. longii** Mackenzie—Culms usually in large tussocks, bluntly angled, stiff, scabrous, to 8 dm long, or sometimes reclining, rooting stolon fashion. Blades linear, 2–4 mm wide, 1–2.5 dm long, grayish green, flat; sheaths tight, green-striate, prolonged at the orifice. Spikes usually four to seven, ovoid, aggregated or the lowest often distant; 3–10 mm long, 4–6 mm wide, grayish green; pistillate scales acute, silvery hyaline with green midvein stripe, gradually narrowed to apex. Perigynia 3–4.5 mm long, brownish green, finely nerved, narrowing to apex; broadly winged, appressed; achene 2 mm long, ovate in outline, stiped and apiculate. Low pinelands, berms, grassy roadsides and banks, Fla., Tex., to Nova Scotia, west to Mich., Saskatchewan, Mexico to South America. Spring, summer. *C. albolutescens* Schw. of authors

4. **C. gigantea** Rudge—Culms stout to 1 m tall, from loosely cespitose base, with stolons; leaves 2–5 dm long, 5–12 mm wide, septate-nodulose, flat; sheaths yellowish, ventrally hyaline, ligule longer than wide. Staminate spikes one to five, nearly sessile or peduncled, bearing sometimes a few pistillate florets at base; pistillate spikes densely flowered, cylindric, 2–7 cm long; perigynia rounded-globose at base tapering to a slender beak; achene 2.5 mm long, trigonous with concave sides and thickened angles, with continuous twice-bent style. Cypress swamps, wet pineland, Fla., Tex. to Del., Miss. Valley flood plains. Summer.

5. **C. lupulina** Muhl. ex Schk.—Culms stout to 1 m tall from short rhizomes; leaves 3–6 dm long, 5–12 mm wide, septate-nodulose, lax. Staminate spike one, peduncled, pistillate spikes two to six, 3–6 cm long, cylindric, approximate at the top of the culm; perigynia 12–15 mm long, strongly inflated, numerous, erect or ascending, simulating hoplike fruits. Achene 3–4 mm long, with rounded sides, contorted style. Wet woods and swamps, Nova Scotia to Minn., south to Fla., Tex. Summer.

6. **C. louisianica** Bailey—Culms to 6 dm tall, smooth, slender, stoloniferous in small tufts or solitary; leaves septate-nodulose, flat-spreading, yellowish. Staminate spike solitary, long peduncled; pistillate spikes 2–2.5 cm long, cylindric, relatively few-flowered; perigynia 9–13 mm long,

thin, bladdery. Achene trigonous, sides rounded, 2.5–3 mm long, style contorted. Cypress swamps, wet woods, Fla., Tex. to N.J., Miss. Valley, Ind. Summer.

7. **C. joorii** Bailey—Cespitose; rootstock creeping; culms leafy below to 1 m tall, rough above, glaucous. Leaf blades 3–12 dm long, 3–10 mm wide, on sterile shoots. Pistillate spikes three to five, cylindric; staminate spike terminal, long peduncled, lowest involucral bract equaling the inflorescence; ligule truncate short; perigynia fifteen to fifty per spike, obovoid 4 mm long, orifice entire. Achene 2.5 mm long, concave sides, thickened angles. Pine flatwoods, cypress margins, hammocks, central and south Fla., Tex., Mo., N.C. Summer.

34. ARECÀCEAE PALM FAMILY

Trunks erect, simple (rarely branched), distinctly crowned by leaves; crown sometimes with shag of persistent leaves. Petioles elongate, woody, armed or smooth, the sheathing base often covering the terminal bud. Blades pinnate, or palmately cleft into induplicate or reduplicate segments. Flowers bisexual or, when unisexual, staminate and pistillate borne on the same, or different, spadix of the same, or different, individual trees. Perianth trimerous, parts free or connate, valvate or imbricate; stamens six or more, filaments dilated at base, usually partly connate. Carpels three, free or united, ovary one- to three-locular, each one-ovuled. Fruit one-seeded, drupe or berry; endosperm uniform or ruminate. About 217 genera, 2500 species widely distributed in tropical and warm temperature regions. Nomen alt. PALMAE.

1. Leaves pinnate or costapalmate.
  2. Leaves pinnate, leaflets rising from the rachis.
    3. Ovary of three united carpels; fruit a drupe; lowest leaflets foliose.
      4. Spathes at summit of crownshaft or in leaf axils.
        5. Drupe fistulose, large; spathes in leaf axils . . . .   1. *Cocos*
        5. Drupe solid, less than 2 cm long; spathes at summit of crownshaft . . . .   2. *Pseudophoenix*
      4. Spathes at the base of crownshaft .   3. *Roystonea*
    3. Ovary of three free carpels; fruit a berry; lowest leaflets spinescent . .   4. *Phoenix*
  2. Leaves costapalmate, deeply cleft into segments, at least one to two pairs of

      basal leaflets rising from the extended
        rachis . . . .   5. *Sabal*
1. Leaves palmate.
  6. Petioles armed with marginal spines;
     ovary three-carpellate.
    7. Blades filiferous; crown with shag of
       persistent leaves . . . . . .   6. *Washingtonia*
    7. Blades nonfiliferous; crown without
       shag.
      8. Petiole armed with small spines;
         stem prostrate, tardily ascending;
         anthers not in groups of twos . .   7. *Serenoa*
      8. Petiole armed with large curving
         spines; stems erect; anthers didy-
         mous . . . .   8. *Acoelorraphe*
  6. Petioles unarmed; ovary one-carpellate.
    9. Leaves silvery beneath; perianth seg-
       ments six; endosperm uniform . .   9. *Thrinax*
    9. Leaves silvery scurfy beneath; perianth
       segments usually ten; endosperm
       ruminate . . . .   10. *Coccothrinax*

1. CÒCOS L.   Coconut Palm

Palms monoecious, unarmed, only needlelike fibers in webby petiolar sheaths; petioles channeled above, short, stout; blades with numerous leaflets. Flowers bracted in clusters on spadix, borne in spathes, crowded in leaf axils. Staminate flowers to 10 mm long; sepals and petals valvate; imbricate in pistillate flowers. Ovary with one functional locule; pericarp fibrous in fruit; endocarp indurate, three-porose, endosperm fistulose, within fluid-filled albuminous shell, the embedded embryo oriented centrally below the level of the pores. A single species widely distributed throughout the tropical regions of the world.

1. C. nucifera L.—Trees with shaft usually 10–20 m tall; to 5 dm in diameter at the enlarged base, usually inclined, diminished above to half the thickness; scarred by irregular rings and vertical cracks in age. Leaves 2–5 m long, arching, leaflets linear-lanceolate, acuminate, to 5 cm wide, lustrous green above, pale below. Drupe 1–3 dm long, obtusely three-sided or ellipsoid-ovoid; endosperm about 10 mm thick; embryo 9 mm long, 4 mm in diameter. $2n = 32$. Naturalized and cultivated as ornamental, south Fla., Fla. Keys, pantropical.

2. PSEUDOPHOÈNIX Wendl. ex Sarg.   Sargent's Palm

Palms solitary, erect, unarmed, trunk smooth, banded. Leaves spreading, blades pinnate, petioles channeled throughout. Spadix elongate, branched, spreading in anthesis; flowers bisexual, pistillate with valvate petals. Drupe pedicelled, globular, two- to three-lobed. Four species, south Fla., Caribbean shores.

1. **P. sargentii** Wendl. ex Sarg. SARGENT's CHERRY PALM—Trunk to 7 m tall, 2–3 dm in diameter. Leaves 1–2 m long, spreading, blade abruptly pinnate; leaflets narrowly linear-lanceolate, abruptly pinnate, 4–5 dm long in midsection of the rachis. Perianth 1 cm wide, calyx reduced to a mere rim on the thickened torus; petals yellowish, persistent in fruit. Drupe orange scarlet. $2n = 34$. Sand, Elliott, and Long Keys, south Fla., W.I. *P. vinifera* of various authors, not Becc.

3. ROYSTÒNEA O. F. Cook ROYAL PALM

Unarmed monoecious trees; trunk columnar. Leaf bases joined over the terminal bud, forming the crownshaft. Blades appearing terete in outline, leaflets disposed in four rows, each at a different angle along the rachis; petiole subterete above. Spadix with numerous pendulous branches below the crownshaft. Perianth six-parted, petals valvate, persistent; stamens six to twelve; staminate flowers with rudimentary pistils; pistillate flowers with staminodes. Fruit globular, violet blue. About 17 species, south Fla., tropical America, Cuba, W.I.

1. **R. elata** (Bartr.) F. Harper, FLORIDIAN ROYAL PALM—Stately palm with columnar bole to 40 m tall. Mature leaves to 3 m long or longer, poised over the lofty crownshaft. Leaf blades, as seen on seedling trees (Big Cypress Swamp), are 4–5 dm long, abruptly pinnate, with twenty-four alternate or subopposite, linear, long-tapering leaflets, to 3 dm long; an excurrent, free median filament to 4 cm long, produced beyond the apical pair of leaflets. Lamina spread 4–15 mm, with three to five prominent veins. $2n = 36$. Swamps and drier sites, Everglades National Park, Collier Seminole State Park, and widely cultivated, south Fla. *R. floridana* O. F. Cook

4. PHOÈNIX L. DATE PALM

Palms dioecious, armed, the trunk covered with stubblelike, long-persistent leaf bases. Leaves short petioled, blades spreading, recurved in a dense crown; spadices erect, ultimately drooping; racemes large, strands many, fasciculate outcurving among the leaves. Flowers unisexual, white, fragrant, with cuplike calyx, with reduced sepals; petals longer, free, stamens usually six. Pistillate flowers with three simple free pistils, maturing into terete one-seeded berry. About 17 species in warm temperate and tropical regions, introduced from Africa.

1. **P. dactylifera L.**—Trees to 30 m tall, giving rise to offsets, trunk

leaning to inclined, not altogether perpendicular. Leaves about twenty in a crown, to 6 m long; leaflets induplicate, to 4 dm long or less, glaucous, rigid, pungent at tips, in several ranks. Spadix with hanging strands of fruit, in profusion; berry 3–4 cm long, seed horny, grooved, semiterete. Various sites, south Fla., about dwellings, ornamental, doubtfully naturalized.

5. SÀBAL Adans. CABBAGE PALM

Trees or shrubs with contorted or simple roots. Leaves costaplamate, flabellate, bilaterally plaited into filiferous segments. Spadix paniculate, spreading, extending beyond the leaf blades. Flowers bisexual, calyx cup shaped, petals free, stamens six, filaments subulate, dilated at base. Drupe spheroidal or obovoid; seed depressed. About 25 species in warm regions of Western Hemisphere.

1. Shrubs usually trunkless above the ground;
   crowns at ground level.
   2. Leaves flat, rib short or lacking, drupe
      5–6 mm in diameter . . . . 1. *S. minor*
   2. Leaves recurved, midrib prominent,
      drupe to 15 mm in diameter . . . . 2. *S. etonia*
1. Trees; trunks erect, covered by crosses of
   persistent, split, petiole bases; crowns
   elevated . . . . 3. *S. palmetto*

1. S. minor (Jacq.) Pers. BLUE STEM—Leaves in a crown at ground level to 1 m tall or more. Blades commonly bluish green, flat, almost never filiferous; midrib short, segments cleft at tip, short plaited at petiole apex. Spadix slender, to 2 m long, branched at intervals. Flowers numerous, small bright orange in life. $2n = 36$. Rocky hammocks, river banks. Central and south Fla., N.C., Tex., Ark. *S. glabra* Sarg. not Miller; *S. acaulis* of various authors

2. S. etonia Swingle ex Nash, SCRUB PALMETTO—Leaves to 1 m long from S-shaped to spiral rootstock; blades dark green, filiferous, segments cleft at tip, plaited below the middle, toward the petiole. Spadix 5–8 dm long with numerous branches, prostrate with heavy fruits. Flowers numerous; petals 3–5 mm long; stamens with tapering filaments with introrse, sagittate anthers. Ripe fruit juicy, bluish black, seeds 10–12 mm wide. White sand scrub, endemic to central Fla., south Fla. No recent collections. (Reported by Small and Moldenke.) Summer.

3. S. palmetto (Walt.) Lodd. ex Schultes, CABBAGE PALM, STATE TREE OF FLORIDA—Trunk columnar to 20 m tall. Leaves to 3 m long, spreading all around from the heavy crown; blades commonly wider than long, strongly filiferous. Spadix copiously branched, to 2 m long with slender ultimate branches. Calyx lobes deltoid, petals to 3 mm long, yellowish white. Fruits 5–9 mm in diameter, globose, black. Hammocks, prairie, many sites and in cultivation throughout Fla. to N.C., *S. jamesiana* Small

6. WASHINGTÒNIA Wendl. WASHINGTON PALM

Trees with stout trunks. Leaves plicate, with filiferous blades, petioles short or long, armed with marginal spines. Spadix paniculate from among the leaves, commonly exceeding the leaves. Flowers bisexual, white, denticulate, stamens exserted, ovary three-locular, only one maturing. Fruit black, ellipsoid, with solid endosperm. $2n = 36$. Two species, southwest North America, introduced as ornamental in Fla. and widely planted.

1. **W. robusta** Wendl.—Trees to 30 m tall, trunk columnar, smooth, swollen at the base. First-formed leaves long persistent, forming a shaggy mane below the leafy crown or a skirt at the base when young. Petioles reddish brown; teeth hooked, yellow when young, the margins thickened with orange sclerenchyma. Gulf coast of Fla., southwest U.S., Mexico. Preferred ornamental cultivated nearly everywhere throughout Fla.

7. SERENÒA Hook. f. SAW PALMETTO

Armed shrubs with bright green or glaucous foliage. Above ground, stem tardily elongating, creeping. Leaves ascendant, palmate, longer than spadix. Flowers nectariferous, fragrant; sepals three-dentate; corolla lobes valvate; stamens six, anthers ovate. Carpels three, united. Drupe one-seeded. One species, southeastern U.S.

1. **S. repens** (Bartr.) Small—Leaves flabellate to 1 m wide, circular in outline, palmately cleft into nonfiliferous segments, shortly divided at tips. Petioles slender, three-sided, prickly toothed on margin, broadly fibrous winged at base; hastula brown, deciduous in age. Spadix branched, shorter than the petioles. Drupe to 2 cm long, subglobose, bluish black, juicy. Seed oblong, prominently grooved, contained with inner shell. $2n = 36$. Coastal dunes, interior sand hills, pinelands, prairie, widely distributed Fla. to N.C.

8. ACOELORRÀPHE Wendl.

Slender, clustered trees rising from a common subterranean rootstock. Leaves palmate, deeply cleft; petioles slender. Spadix paniculate; flowers greenish white, sessile in glomerules, sepals three, suborbicular, petals three, valvate; stamens six, united at dilated bases; anthers short, stigma terminal. Monotypic genus of tropical America, Central America, Cuba.

1. **A. wrightii** (Griseb. & Wendl.) Wendl. ex Becc. EVERGLADES PALM —Gregarious or colonial, trunks 5–8 m tall; 1–1.5 dm in diameter. Leaves 6–10 dm long, vernation reduplicate; segments plicate, cleft into attenuate tips; petioles ferruginous at base, webby sheathed, armed with hooked spines. Spadix to 9 dm long; flowers 2–3 mm wide. Fruits 5–8 mm in diameter, spherical, bluish black, seed globose, free from pericarp, the raphe not intruding endosperm. $2n = 36$. Moist sites and widely cultivated, Fla., Bahama Islands, Mexico, Cuba. *Paurotis wrightii* (Griseb. & Wendl.) Britt.

9. THRÌNAX L. f. ex Sw. THATCH PALM

Slender, small, unarmed trees. Leaves glabrous; petioles slender; blades suborbicular, flabellate, plicate; segments cleft to attenuate tips. Spadix elongate, ultimately branching into pendulous flower-bearing branchlets. Calyx and corolla cupuliform; filament bases dilated, united. Stigma funnel-like; ovary one-locular, one-seeded, endosperm divided through the center by a cavity. Mature fruits 4 mm long or more, pale white. About 5 species, south Fla., W.I., Cuba, Mexico, British Honduras.

1. Flowers and fruits sessile; calyx lobes obtuse . . . . 1. *T. microcarpa*
1. Flowers and fruits pedicelled; calyx lobes ,solescent . . . . 2. *T. parviflora*

1. **T. microcarpa** Sarg. BRITTLE THATCH—Tree to 12 m tall to 2.5 dm in diameter; the base of trunk roughened by matting roots. Leaves long petioled, effecting an open crown; blades 6–8 dm wide, pale green above, silvery white below, white tomentose when young, divided nearly to the base; hastula concave tomentose. Spadix ample with numerous flowers on disklike pedicel of nodding branchlets. Spathes hoary tomentose above, leaf sheaths tomentose, split below insertion to petiole. Anthers exserted, filaments barely connate at base, ovary orange, maturing into a globular fruit about 4 mm in diameter; seed brown. $2n = 36$. Dry pineland sand over oolitic rocks, hammocks and shores, south Fla. Keys, W.I. *T. keyensis* Sarg.

2. **T. floridana** Sarg. FLORIDA THATCH PALM—Slender tree to 8 m tall, to 2 dm in diameter, enlarged at base. Leaves to 1 m wide, green lustrous on upper surface, paler green beneath; hastula a wide rim,

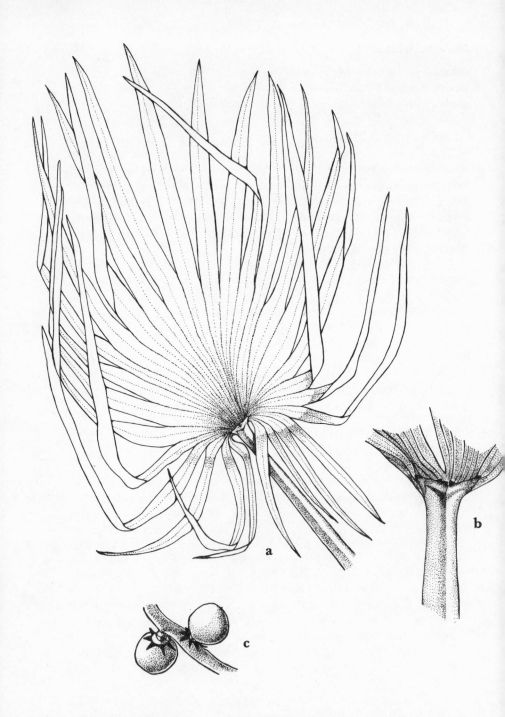

Family 34. **ARECACEAE.** *Thrinax microcarpa:* a, mature leaf, × ¼; b, blade base and part of petiole, × ½; c, mature fruit, × 2.

truncate or pointed. Spadix to 9 dm long, with separate alternate clusters of flowers. Fruit to 10 mm in diameter. $2n = 36$. Hammocks and shores, south Fla., Fla. Keys, W.I. *T. parviflora* Bailey and others

10. COCCOTHRÌNAX Sarg. SILVER THATCH PALM

Unarmed low or tall trees with slender trunks. Leaves densely silvery, scurfy below; blades flabellate, petioles webby-fibrous at base. Spadix short, many-flowered. Calyx and corolla united into cupuliform, six-lobed perianth; stamens eight to twelve, connate at dilated bases. Ovary one-locular with dilated style. Fruit with ruminate endosperm. About 10 species, south Fla., W.I.

1. **C. argentata** (Jacq.) Bailey, SILVER PALM—Tree to 8 m tall, 1.5 dm diameter, solitary. Leaves to sheath tubular tomentose 8 dm wide, deeply divided, bright green above; hastula short, 1–1.5 cm wide. Mature fruit 7–10 mm in diameter, black, short stalked, berrylike with thick pulp. $2n = 36$. Pinelands, Fla. Keys, W.I., Bahama Islands. *C. jucunda* Sarg., *C. garberi* Sarg., *Palma argentata* Jacq.

35. ARÀCEAE ARUM FAMILY

Perennials, mostly fleshy herbs from tubers, or creeping rhizomes; erect, prostrate or viny root climbers, rarely arborescent; or floating aquatics. Leaves simple or compound, sessile or petioled, sheathing at base. Inflorescence a spadix with or without subtending spathe. Flowers small, unisexual or bisexual, radially symmetrical, aggregated, wholly or partly covering the spadix below the appendage. Perianth segments four to six, present only in bisexual flowers, staminate flowers numerous, above the pistillate flowers, contiguous or separate, stamens one to six, with or without interposed staminodes. Ovary one- to four-locular with one or more ovules in each. Fruit a berry, in aggregate clusters. About 115 genera, 2000 species, mostly tropical.

1. Free-floating aquatics; leaves rosulate,
     scurfy . . . .   1. *Pistia*
1. Terrestrials or epiphytes.
    2. Leaves simple.
        3. Mature blades sagittate or ovate-sagittate; placentae basal or central.
            4. Blades parallel-veined, sagittate; placentae basal . . . .   2. *Peltandra*
            4. Blades netted-veined, ovate-sagittate; placentae central . . . . . .   5. *Xanthosomᵣ*

    5. Terrestrials; blades trifoliolate
       (rarely seven- to nine-foliolate);
       spathe hooded, incurved . . .   4. *Arisaema*
    5. Epiphytic vine; mature leaves di-
       chotomously pedisect; spathes
       not incurved . . . .   6. *Syngonium*
   3. Mature blades peltate; placentae parie-
     tal . . . .   3. *Colocasia*
 2. Leaves compound or pedately divided.

1. PÍSTIA L.   Water Bonnets

Plant spongy, stoloniferous with elongate, copiously branching roots, leaves spirally rosulate from abbreviated crown, petioles broad, parallel veined, dilating into flabellate, scurfy pubescent blades, 3–9 cm across the nearly truncate apex. Spathes foliose, greenish white, erect or spreading, copiously pilose without, adnate to spadix, flowers unisexual, naked, pistillate borne at base of spadix, staminate at the apex; floriferous parts not contiguous. Pistil one-carpellate, ovary with short style and fimbriate stigma. Seeds cylindric, truncate with buoyant endosperm, germinating in the surface film of water. One species.

   1. **P. stratiotes** L. Characters of the genus. $2n = 28$. In still waters of ditches, ponds, and creeks, Fla. to Tex., warm parts of America and Old World.

2. PELTÁNDRA Raf.

Monoecious colonial herbs perennating by cordlike roots. Leaves sagittate or hastate, long petioled, glabrous. Spathe white, spadix flower-bearing throughout, stamens six to ten, anthers sessile, imbedded in the fleshy spadix, each pair opening by pores. Pistillate flowers disposed in the lower quarter of the spadix, maturing into berrylike fruits; the staminate portion withering away. Four species, North America, south U.S.

   1. **P. virginica** (L.) Schott & Endl.—Leaves finely veined, to 5.5 dm long, about equaling the scapes, the blade apex and lobes acute or hastate; petioles sheathing at base; spathes 10–15 cm long, convolute, crisped along margins. Pistillate flowers with fleshy scalelike staminodes; ovary one-locular with 15 ovules; style evident with a small stigma. Fruit individually a red berry, aggregated into a globular body. $2n = 88$. Swamps, wet pinelands, Fla., La., Mich., northeast U.S.

3. CÓLOCÀSIA Schott   Elephant's Ear

Large herbs with ample foliage with tuberous rhizomes. Blades always peltate, often longer than the petioles. Spadix produced to a

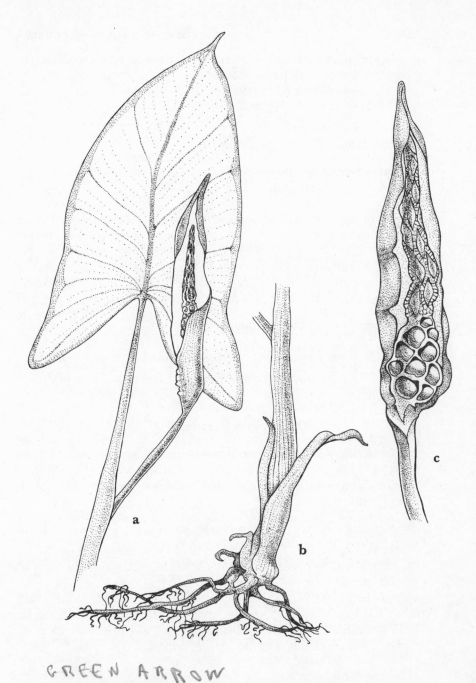

GREEN ARROW

Family 35. **ARACEAE.** *Peltandra virginica:* a, inflorescence and mature leaf, × ½; b, stem base and rhizome; c, spadix with staminate flowers on the upper part, pistillate flowers on the lower part, × 1.

sterile appendage, flowers unisexual, with stamens typically above the pistils, joined to a six-sided column. Ovary one-locular, placentation central. About eight species, Polynesian.

1. C. esculentum (L.) Schott, TARO—Leaves large, surpassing the scape; tubers large. The sterile appendage much shorter than the staminate portion of the spadix. $2n = 28,36,48$. Widely cultivated in warmer parts of the world, occasional in abandoned sites, in south Fla., not known as naturalized, probably native to Pacific Islands. *C. antiquorum* (Schott) var. *esculenta* Schott

4. ARISAÈMA Mart.   PARSON IN THE PULPIT

Herbs, monoecious or dioecious. Leaves long petioled, the outermost bladeless, sheathing the stem at base. Flowers unisexual, naked, crowded around the basal portion of the fleshy spadix, subtended by a hooded spathe; staminate flowers in clusters, each flower with two to four subsessile anthers opening by pores; pistillate flower with one-locular, several-ovuled ovary. About 150 species, North America, Mexico, Africa.

1. A. acuminatum Small—Herbs 3–6 dm tall arising from corm 3–4 cm in diameter. Sheaths purplish, 10–15 cm long, abruptly narrowed to petioles; leaflets ovate-acuminate to 15 cm long, and half as wide. Spathe long acuminate, green, striped with yellowish white, the convolute part nearly concealing the club-shaped, sterile appendage spadix. Mature fruiting cluster scarlet, 3–5 cm in diameter. Deep shade of hammocks, in rich, mucky soil, central and south Fla. Spring.

5. XANTHOSÒMA Schott

Terrestrial with milky latex, rootstocks thick; underground parts sometimes woody. Leaves long petioled, hastate, three- to several-lobed or pedatisect. Peduncles one or several, shorter than petioles; spathe tubular, convolute, constricted above the blade, persistent; spadix shorter than spathe; staminate portion twice the length of the pistillate portion. Plant monoecious, flowers unisexual; staminate with four to six stamens, anthers united into an angular column. Carpel one, style shieldlike, stigma discoid, yellow, ovary two- to four-locular. Fruit a berry, yellow, several-seeded. Large herbs cultivated for esculent tubers. About 45 species, tropical South America, W.I., Cuba, Mexico, introduced to south Fla.

1. X. hoffmannii Schott, AMERICAN TARO—Plants with tuberous underground caudex; leaves glabrous 2–3 dm long, blade pedatisect, reniform in outline; segments elliptic, 15 cm long, 7 cm wide, submar-

ginal veins two, venation reticulate, midvein prominent. Peduncles several within sheathing petiole base. Mature spathes not seen. South Fla., Mexico, Central America. Summer. *Xanthosoma wendlandii* Schott

6. SYNGÓNIUM Schott

Stems woody, climbing or creeping, with milky sap, with nodal leaves and roots; young leaf blades sagittate; petioles elongate, sheathing at base. Peduncles short, spathe yellowish white within, greenish without, longer than the spadix; staminate flowers four to six, crowded on the upper part of spadix, lacking the sterile appendage; ovaries connate at the basal part, fertile locule one, seed black, solitary. About 20 species, tropical America, Jamaica, Mexico.

1. **S. podophyllum** Schott—Plants soft and tender, the pinnatisect leaves 10–15 cm long with three to nine leaflets, the terminal one usually the longest, satiny to touch, venation pinnate-reticulate. Spathe with tube to 11 cm long, white within, green without. The cultivated plant is probably var. **albolineatum** Engelm., the leaves demarked with silvery white lateral veins. Escaped to hammocks in south Fla.; Mexico, Central America.

36. LEMNÀCEAE Duckweed Family

Small or minute floating herbs with or without roots, plant body discoid, globose or ligulate, without differentiation into stems or leaves. Flowers microscopic from spathes in marginal reproductive pouches, consisting only of the essential organs, staminate flowers one or two with delicate filaments and a single two-locellate or four-locellate anther. Pistillate flower solitary with a flask-shaped ovary, style and stigma simple, ovules one or several on a basal placenta. Fruit an utricle with ribbed or smooth seed. Fresh-water aquatics of temperate and tropical regions around the world. Four genera and 40 species, cosmopolitan.

1. Plant body discoid with roots; reproductive pouches lateral.
   2. Rootlets with vascular tissue, two or more per individual; anthers dehiscent lengthwise . . . . 1. *Spirodela*
   2. Rootlets without vascular tissue, only one per individual; anthers dehiscent transversely . . . . 2. *Lemna*
1. Plant body ligulate; roots lacking; reproductive pouches basal . . . . . . 3. *Wolffiella*

1. SPIRODÈLA Schleiden WATER FLAXSEED

Plants obovate, purplish below, veins prominent above, rootlets two to nine. Individuals oriented radially in groups of threes or fives by cordlike connective stipes. Utricle winged. Three species, one in Asia, two in America.

1. Plants 5 mm long or more; rootlets five to
   twelve, turion present . . . . . . 1. *S. polyrhiza*
1. Plants 2.5–4.5 mm long; rootlets two to five,
   turion lacking . . . . 2. *S. oligorhiza*

    1. S. polyrhiza (L.) Schleiden—Plants 3–10 mm long, broadly obovate; ventral side purple, dorsal side shiny green; veins five to twelve, prominent. Flowers and fruits not seen, vegetative propagation by turions. Widely distributed in the U.S. and Canada, Mexico, Central and South America, Old World.
    2. S. oligorhiza (Kurtz) Hegelm.—Plants flat, two to five or more together, 2.5–5 mm long, obovate or reniform in outline, with obtuse or acute apex; ventral surface strongly pigmented purple; dorsal surface yellowish green, nerves obscure, roots two to five, tips straight; turions not formed. Flowers and fruits not seen. Introduced to U.S. as aquarium plants, south Fla., La., Ill., Calif.

2. LÉMNA L. DUCKWEED

Monoecious plants, oval to obovate, subreniform, obscurely veined, green on both surfaces or pigmented ventrally; individuals often two to five together floating at the surface. Flowers in groups of threes, unisexual, two staminate, each consisting of a single stamen, and one pistillate with one pistil. Ovary one-locular with one or more ovules, style tubular, stigma a cupule containing a droplet of mucilage at maturity. Fruit one- to six-seeded, transversely striate. Reproduction by seeds only in warm regions, plants rarely flowering in the northern limits of distribution. About 15 species, cosmopolitan.

1. Plant green; root sheath appendaged . . 1. *L. perpusilla*
1. Plant purplish below and often around the
   margins on upper surface; root sheath
   not appendaged . . . . 2. *L. minor*

    1. L. perpusilla Torr.—Plants-in groups of two to five; individuals mostly ovate-obovate, 1.5–2.5 mm long, obtuse at apex, convex slightly

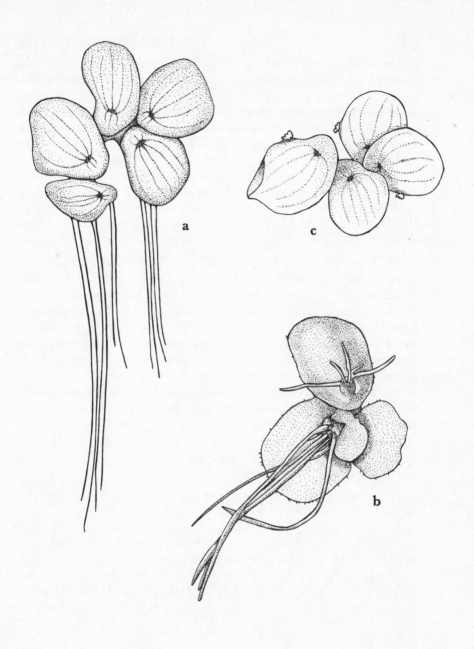

Family 36. **LEMNACEAE.** *Spirodela polyrhiza:* a, mature plant, dorsal side, × 6; b, mature plant, ventral side, × 10; c, apical portion, dorsal side, × 7.

on both surfaces, root curving at tip; veins obscure. Stamens exserted about 0.4 mm beyond the marginal cleft; anthers whitish, opening simultaneously, surpassing the rounded stigma. Seeds about 0.3 mm long, brownish, showing seven ribs in optical section, finely cross-lined. Often occurring with species of SPIRODELA, canals and lakes, Fla., coastal in warm regions of both hemispheres. Summer.

2. **L. minor** L.—Thallus 2.5–4 mm long, 2–3 mm wide, oval or orbicular, symmetrical, dark green above, purple beneath; slightly convex on both sides, indistinctly veined above. Stamens scarcely protruding from the reproductive cleft. Fruit clearly visible beyond the margin; seeds dark brown, 0.2 mm long, 0.1 mm wide. Canals, ponds, and shores of rivers, south Fla., widely distributed in North America, Old World. Summer.

3. WOLFFIÉLLA Hegelm.

Plant rootless, thin, acuminate, strap shaped, light green, punctate, solitary or commonly in twos, or numerous, forming dense colonies, floating at surface or below. Reproductive pouch basal; vegetative propagation by budding; sexual reproduction rarely seen. Seven species mostly in tropical regions.

1. **W. floridana** (Sm.) Thompson—Individual plants 5–9 mm long, less than 1 mm wide, falcate, attenuate. Reproductive pouch on one side of the median line; flowers spatheless, unisexual, represented by a single stamen and a single pistil. Fruit a compressed utricle; seed smooth. Coastal plains, Fla. to Tex., N.J. Summer.

37. XYRIDÀCEAE   YELLOW-EYED GRASS FAMILY

Scapose herbs, rhizomes with lateral buds and fibrous roots. Leaves equitant with sheathing bases. Scape ensheathed at base, inflorescence a terminal spike, flowers trimerous, sepals glumelike, petals yellow, stamens with extrorse anthers. Pistil three-carpellate, ovary one-locular; seeds numerous, fusiform longitudinally striate, sometimes farinose, opaque or translucent amber. Two genera, 250 species, temperate, tropical, and subtropical regions.

1. XYRIS L.   YELLOW-EYED GRASS

Flowers solitary in axils of closely imbricated, leathery bracts. Sepals three, unequal, the lateral navicular, persistent, the anterior larger, enfolding the petals in bud, withering in anthesis; petals equal, clawed, yellow or rarely white, early deciduous; functional stamens three, arising from the claws of the petals, alternating with three-bearded or fringed staminodes. Style three-cleft, capsule one-locular with three parietal placentae; dehiscence valvular.

1. Scape sheaths exceeding or equaling the leaf length.
  2. Leaves linear-attenuate; keels subentire; bracts lacerate . . . .   1. *X. brevifolia*
  2. Leaves linear-lanceolate; keels ciliate; bracts subentire . . . .   2. *X. flabelliformis*
1. Scape sheaths not exceeding the leaf length.
  3. Plant base hard, nodulose, bulbous or bulboid.
    4. Leaves broadly linear to 15 mm wide.
      5. Equitant portion of the leaves strongly indurate; leaves flexuous.
        6. Rhizome bulbous, terete; leaf bases spiraled, castaneous . .   3. *X. caroliniana*
        6. Rhizome compressed; leaf bases straight, brownish gray . . .   4. *X. platylepis*
      5. Equitant portion of the leaves soft, prominently distichous, tawny .   5. *X. ambigua*
    4. Leaves narrowly linear 1–3 mm wide; tufts compact, separable into nodulous clusters . . . .   6. *X. elliottii*
  3. Plant base soft, fibrous.
    7. Lateral sepals exserted; the keel lacerate; leaf bases pinkish purple, blades smooth on surface . . . . . .   7. *X. smalliana*
    7. Lateral sepals included.
      8. Leaves stramineous, tawny, toward the base; linear, smooth on surface   8. *X. jupicai*
      8. Leaves pinkish toward the base; linear, scabrid on surface . . .   9. *X. difformis*

1. X. brevifolia Michx.—Plant to 3.5 dm tall in small tufts, becoming aggregate in compact large tufts; scapes linear, capillary or filiform compressed, at least above, obscurely ribbed toward the base; leaves firm, linear-attenuate, often flexuous, sometimes sparingly scabrid, essentially smooth; spikes globose, scales firm, golden brown, becoming ragged-lacerate at apex in age; lateral sepals included, keels entire or remotely papillose; seeds ellipsoid, with numerous lines between the ends. $2n = 18$. Shores, ditches, sandy openings in glades, Fla. Keys to N.C., Caribbean Islands, South America. Summer, fall.

2. X. flabelliformis Chapm.—Plants in small, flabellate tufts to 7–25 cm tall; scapes filiform, twisted, glabrous, angled with low ribs toward the base. Leaves 2–3 cm long, shorter than the scape sheath; blades lanceolate, minutely scabrid on margins; base dilated; margins broadly hyaline. Spikes 4–5 mm long, scales firm, lustrous brown; lateral sepals included, the keel ciliate; seeds ellipsoidal, translucent, lustrous, barely 0.3 mm long. $2n = 18$. West central Fla., coastal plain Ga., rare in south Fla. Summer.

3. **X. caroliniana** Walt.—Plants in clusters, the bulbs covered with twisted, castaneous, polished, indurated scalelike leaf bases. Scapes rigid, twisted, to 7 dm tall; blades linear, 1–4 mm wide, flexuous; spike 1.5–3.5 cm long, ellipsoid, conical; lateral sepals linear, exserted; keel long fringed, petals white. Seeds 0.7 mm long, ellipsoid, longitudinally ridged. Pinelands, sand scrub, peaty soil, Fla., Tex., Ark., coastal plain to N.J. Summer, fall. *X. flexuosa* Muhl. ex Ell.

4. **X. platylepis** Chapm.—Plants tufted from nodulose base to 8 dm tall. Scapes two-edged above, becoming angular, ribbed below; bases commonly pinkish, becoming brownish, drab; blades linear to 5 dm long, 4–20 mm wide, glabrous, somewhat twisted; mature spikes ovoid, oblong 1–3 cm long; scales broadly rounded lateral sepals included; the keel narrow. Seeds translucent, ellipsoidal 0.5–0.6 mm long, longitudinally lined. $2n = 18$. Coastal plain, south Fla., central La., Va. Summer, fall.

5. **X. ambigua** Beyr. ex Kunth—Plant solitary or cespitose to 9 dm tall with a hard base covered with equitant sheathing bases of leaves, the outer persistent, marcescent, blackish, becoming fibrous in age; blades pliable 4–8 mm wide, papillose-serrulate on margins; spike ovoid 1–2.5 cm long; lateral sepals curvate, keel ciliate. Seeds translucent, 0.6 mm long, obovoid, pointed at base. $2n = 18$. Drier margins of ponds, canals in peaty sands, Ga. to south Fla., Tex., and northeast U.S. Summer.

6. **X. elliottii** Chapm.—Plants in extensive tufts to 7 dm tall, from lateral fleshy buds. Scapes slender, wiry, slightly twisted, compressed above, edged, minutely tuberculate; blades filiform to narrowly linear; bases ferruginous, loose; spikes 1–1.5 cm long; ovoid, bracts becoming lacerate in age; petals yellow; lateral sepals mostly included, the keel ascending, fibrillose. Seeds ellipsoid, 0.5–0.6 mm long, finely lined. Wet sandy soils, Fla. to S.C., west to Miss., on the coastal plain. Summer.

7. **X. smalliana** Nash—Plants to 1 m tall, lustrous throughout, from thickened rhizome. Scapes one- to two-edged above, becoming low ribbed toward the base; blades to 7 dm long, linear, sheaths broadly hyaline margined; mature spikes 1–2 cm long, ovate-ellipsoid; scales nearly entire on margins; lateral sepals exserted, the keel lacerate above the middle. Seeds ellipsoid, ribbed, and cross-walled, 0.6 mm long, translucent. $2n = 18$. In moist openings of coastal pinelands, peaty soils, south Fla., west to Miss., north to Me. Summer.

Family 37. **XYRIDACEAE.** *Xyris elliottii:* a, flowering plant, × ½; b, inflorescence, × 3½; c, flower, × 7½; d, mature fruit with two subtending persistent sepals, × 16.

8. X. jupicai Richard—Plants in small tufts from indurated, thickened base to 9 dm tall. Scapes compressed above, one- to two-edged, becoming several-ridged toward the base, the edges sometimes low tuberculate; leaves linear, 1–6 dm long, 4–8 mm wide, glabrous or remotely papillose; bases hyaline margined, sometimes reddish; mature spikes 0.5–1.5 mm long; bracts membranous, only becoming slightly lacerate on margins; lateral sepals included, the keel lacerate toward apex. Seeds translucent, ellipsoidal, many-lined, 0.3 mm long. $2n = 18$. Moist sand, swamps, and shores, or wet glades, adventive and naturalized from Central and South America; coastal plain Fla to N.J. Summer.

9. X. difformis Chapm.—Scapes 1.7 dm long, twisted below; leaves broadly linear 1–5 dm long, spreading fan fashion; 0.5–1.5 cm wide; the equitant part of leaf papillose; lateral sepal 4–5 mm long, the keel crested above the middle. Spikes 1 cm long or less. Seeds ellipsoidal, 0.5 mm long, translucent, finely papillose lined. $2n = 18$. Sandy peats, north Fla. to the northern limits of New England, Canada. Summer.

Three varieties occur in our area:

1. Scape apex two-edged; leaf surface smooth
   spike acute . . . . 9a. *X. difformis*
   var. *difformis*

1. Scape apex not compressed; at least the
   outermost leaves scabrid-papillose, spike
   obtuse.
   2. Leaves widest at base 3–5 mm wide; seeds
      farinose, opaque . . . . . . 9b. var. *floridana*
   2. Leaves widest at midblade 3–4 mm wide;
      seeds translucent, not farinose . . . 9c. var. *curtissii*

9a. X. difformis var. **difformis**—Range of the species.
9b. X. difformis var. **floridana** Kral—Plant to 7 dm tall in small tufts with nodulose base; scape twisted above, becoming terete, angular below with nine or more ridges, often strongly tuberculate or scabrid; leaves tuberculate-scabrid, the blade dilated above the middle; dark green to reddish at base mottled with brown. Spike 1–1.6 cm long, ovoid, conoidal; scales firm, the margins entire, scarcely lacerate; lateral sepals included, the keel lacerate. Seeds spindlelike about 0.5 mm long, amber yellow, ends opaque, dark, connected by fine lines. Openings in glades, grassy banks, Fla., La. Summer.
9c. X. difformis var. **curtissii** Kral—Plants less than 2 dm tall, in small tufts; the margins of the equitant part of leaves broad, pinkish, translucent. Surface of the leaves papillose; leaf bases pinkish, papillose; scapes terete, ridge, three to seven carinate. Mature spikes scarcely more than 5 mm long; lateral sepals included as long as the bract; keel lacerate

at tip. Seeds oblong 0.5 mm long, with twelve or more faint lines. Ditches, bogs, acid seepage areas. North Fla. to Tex., north to the Great Lakes and southeast Canada. Summer. *X. curtissii* Malme

38. ERIOCAULÀCEAE   PIPERWORT FAMILY

Scapose, monoecious or dioecious, aquatic or terrestrial herbs with rosulate leaves from branching rhizomes, the simple scape terminating in a capitate inflorescence. Leaves numerous, arising in a spiral from ascending rhizomes; internodes very short. Flowers unisexual with radial or bilateral symmetry, chaffy, perianth of two or three distinct connate sepals, two or three connate petals, two to six stamens, two to three carpels united into a three-locular ovary, with one style, three stigmas. Capsule two- to three-seeded, loculicidal; seeds with small apical embryo in ample endosperm. Thirteen genera, 1150 species, distributed in temperate, tropical, and subtropical regions, mostly in South America.

1. Stigmas two in our species; leaves cellular,
   pellucid; roots septate . . . . . .   1. *Eriocaulon*
1. Stigmas three, two-cleft; leaves opaque;
   roots continuous.
   2. Corolla replaced by tufts of hair; involu-
      cral bracts nondescript . . . . .   2. *Lachnocaulon*
   2. Corolla present; involucral bracts yellow-
      ish . . . .   3. *Syngonanthus*

1. ERIOCÀULON L.   HAT PINS

Leaves sessile with linear-attenuate blades, their dilated bases nonchlorophyllous. Scapes glabrous. Receptacle convex, imbricate-bracted; sepals and petals dorsally covered with opaline white multicellular hairs, staminate flowers with two cymbiform sepals, two- to three-lobed corolla tube, four to six exserted stamens; anthers black. Pistillate flowers with two sepals and two free petals, pistil with two-carpellate ovary borne on gynophore. About 400 subtropical and tropical species in the Western and Eastern Hemispheres.

1. Plant monoecious.
   2. Heads whitish, rigid, subspheroidal; scape
      stout . . . .   1. *E. decangulare*
   2. Heads nigrescent, globose to 4 mm in di-
      ameter; scape capillary . . . . .   2. *E. ravenelii*
1. Plant dioecious, heads powdery white, soft,
   hemispheroidal; scape slender   . . .   3. *E. compressum*

1. E. decangulare L.—Scapes to 1 m tall or taller, fluted, commonly ten-ribbed, sheath obliquely split, obtuse or often cleft at summit; leaves linear, obtuse exceeding the sheaths, in spirals on short branching rhizome with basal fringes of buff-colored soft trichomes, apparently rising from short internode. Heads 9–14 mm in diameter, white at a distance; involucral bracts, acute to acuminate, flavescent, stramineous, receptacle crowded with florets, trichomes copious, villous; sepals and petals cymbiform, clothed with white multicellular hairs. Capsule loculicidal; seed 0.5 mm long, amber, faintly reticulate. Pine flatwoods, around shallow ponds, glades, Fla., Tex., N.J. Summer.

2. E. ravenelii Chapm.—Plants small, delicate. Scapes filiform from short rhizomes; sheaths loose, shorter than the leaves; leaves tufted, linear-attenuate, 2–4 cm long; mature head 3–4 mm in diameter, black. Involucral bracts translucent, lustrous, oblong-obtuse; receptacle glabrous; sepals and petals oblanceolate, translucent, falcate in pistillate floret. Style two-cleft or sometimes simple; seeds 0.5 mm long, plump, grayish, reticulate. Prairies, flatwoods, Fla. Summer.

3. E. compressum Lam.—Scapes 2–6 dm tall, lax; sheaths 0.5–12 cm long, reticulate, dilated, often cleft at summit usually exceeding the leaves or in deep water, scape and leaves much elongate. Mature heads 10–15 mm in diameter with blackish ovate-obtuse involucral bracts; receptacle villous, crowded with florets; petal with a small black gland; seeds 0.5 mm long, obliquely ovoid, minutely reticulate. Pine flatwoods, Fla., Tex., N.J. Summer.

2. LACHNOCAÙLON Kunth BOG BUTTONS

Herb monoecious, florets trimerous. Leaves in loose tufts on lateral rhizome branches; roots firm, branching. Staminate floret with three sepals, petals lacking, stamens three, monadelphous, with stipelike base; anthers two-locular. Pistil two- to three-carpellate, three-locular, styles two- to three-cleft, seeds amber, longitudinally striate and finely cross-lined. About ten species, Fla., southeast U.S., W.I.

1. Perianth covered with white trichomes; seeds prominently striate; styles two-cleft; heads 4–5 mm in diameter . . . . . 1. *L. anceps*
1. Perianth parts covered with colorless clavate trichomes; styles three-cleft; heads 2.5–4 mm in diameter . . . . . . 2. *L. minus*

1. L. anceps (Walt.) Morong—Scapes to 4 dm long, hirtellous with upwardly divaricate trichomes; or glabrous; sheath 2–5 cm long. Leaves

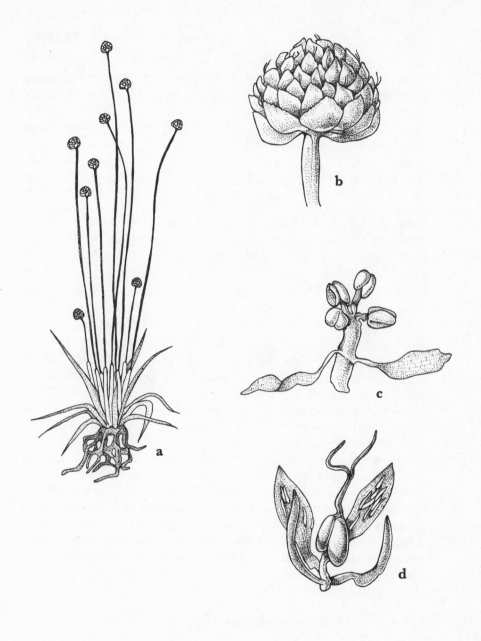

Family 38. **ERIOCAULACEAE.** *Eriocaulon ravenelii:* a, flowering plant, ×
½; b, inflorescence, × 6; c, staminate flower, × 15; d, pistillate flower, × 15.

2–7 cm long, 3–5 mm wide near the base, narrowed to an attenuate or acuminate apex, blades minutely puberulent on striae below, glabrous or sparingly villous; mature heads globose or cylindric, grayish brown. Flowers in small heads, inconspicuous among villous trichomes; seeds 0.6 mm long, conspicuously cancellate. Low pinelands, prairies, ponds, Fla., Tex., N.J. Summer. *Lachnocaulon floridanum* Small, *L. glabrum* Koern.

2. **L. minus** (Chapm.) Small—Scapes twisted, slender, 8–20 cm long, densely hirtellous; sheaths 3–5 cm long, foliose at apex. Leaves linear-attenuate, pubescent, 2–5 cm long; tufted on lateral branches of the rhizome. Mature heads 3–5 mm wide, 4–5 mm long, cylindrical or globose, brownish, with narrow involucral bracts inconspicuous in mature heads; florets small, obscured by copious trichomes; seeds amber, 0.4 mm long, pointed, cancellate. Shores and margins of pinelands, Fla., N.C. Summer. *Lachnocaulon eciliatum* Small

3. SYNGONÁNTHUS Ruhland   BANTUM BUTTONS

Leaves recurvate in dense rosettes, of short branching rhizomes; roots simple, spongy. Flowers in heads, subtending bracts conspicuous, staminate flowers with three free sepals, three united petals, three stamens. Pistillate flowers with three free sepals, the three petals connivent at tips, pistil three-carpellate with stiped three-locular ovary; capsule commonly three-seeded. About 195 species, temperate, subtropical, and tropical America, Africa.

1. **S. flavidulus** (Michx.) Ruhland—Scapes 3.5 dm tall or less, grooved, softly pubescent with ascending capitellate trichomes; sheaths pubescent, 3–4 cm long, dilated at apex; leaves 2–4 cm long, linear-subulate, copiously white floccose at base. Heads hemispherical, 3–6 mm in diameter; involucral bracts flavescent, lustrous shorter than the floral parts and trichomes. Fruiting heads globose; seeds 0.5 mm long, amber, vertically striate and finely cross-lined. Cypress margins and flatwoods, Fla. to N.C. Summer.

39. BROMELÌACEAE   PINEAPPLE FAMILY

Epiphytic perennial, offset-forming herbs, usually with roots only for anchorage. Stem leafy; the lower leaves crowded rosulate fashion, the upper gradually decrescent up to the inflorescence, or stem scapose, and sometimes filiform, elongate, pendulous. Leaves entire, dilated at base, peltate-scaly. Inflorescence simple or compound, spicate or racemose,

or flower solitary on naked peduncle, flowers bisexual, rarely functionally unisexual, trimerous; sepals and petals distinct or connate; stamens three to six, filaments simple or rarely spiral, spoonlike at apex.
Pistil one, three-carpellate, style undivided, stigmas three, ovary three-
locular; fruit capsular in our taxa, seeds comose or naked, embryo small
at the base of mealy endosperm. About 45 genera, 900 species, native of
tropical and subtropical America. Family includes many prized ornamentals from the tropics that are popular in cultivation.

1. Flowers in distichous spikes . . . . . 1. *Tillandsia*
1. Flowers in polystichous spikes.
  2. Petals free; anthers 1.4 mm long; fila-
    ments barely dilated at summit; flowers
    conspicuous . . . . 2. *Catopsis*
  2. Petals connate; anthers 4 mm long; fila-
    ments broadly dilated at summit; floral
    bracts conspicuous . . . . . . 3. *Guzmania*

1. TILLÁNDSIA L.

Leaves entire, saccate or dilated at base; flowers commonly purple,
white, or greenish yellow. Sepals usually free; petals clawed, free or connate; the three inner stamens opposite the petals; filaments capillary,
anthers usually linear-oblong. Ovary superior; style subulate with short
stigmas. Capsule dehiscing along the septa. Seeds erect, linear, funicle
elongate, splitting into fine threads (coma). About 400 species in warm
parts of America.

1. Leaves of uniform size; not dilated at base.
  2. Stems pendulous, filiform in branching
    strands; flower solitary, axillary . . 1. *T. usneoides*
  2. Stems radiate, outcurving from rooted
    crown; flowers usually two, terminal . 2. *T. recurvata*
1. Leaves not uniform in size; dilated, often
  saccate, at base.
  3. Flowering scape manifest.
    4. Leaves linear-subulate from abruptly
      dilated base, usually surpassing the
      scape; spike simple corolla, purple . 3. *T. setacea*
    4. Leaves lanceolate, tapering from di-
      lated base to an attenuate apex.
      5. Inflorescence paniculate; floral bracts
        ecarinate.
        6. Flowers spreading from the rachis;
          leaves banded and twisted at
          base; scape 7 dm high or higher. 4. *T. flexuosa*
        6. Flowers appressed against the .

rachis; leaf sheaths straight,
brownish, not banded; scape 1
m tall or taller . . . . .　　5. *T. utriculata*
5. Inflorescence pinnate, subpalmate or
simple; floral bracts carinate, at
least at tips.
7. Leaf sheaths not pseudobulbous;
leaves not contorted.
8. Spike usually simple; leaves pale
green, soft to touch . . .　　6. *T. valenzuelana*
8. Spike compound, pinnate or
subpalmate.
9. Leaves uniformly green, sur-
passed by the scape . . .　　7. *T. polystachia*
9. Leaves brown banded at base,
scape usually equal to or
rarely surpassed by the
leaves . . . .　　8. *T. fasciculata*
7. Leaf sheaths inflated; plant pseudobulbous; leaves contorted .　　9. *T. balbisiana*
3. Flowering scape obscured by elongate
pseudobulbs.
10. Leaves strongly scurfy-lepidote, surpassing the spike . . . . .　　10. *T. pruinosa*
10. Leaves moderately appressed-lepidote,
surpassed by the spike . . . . .　　11. *T. circinata*

1. **T. usneoides** L. SPANISH Moss—Profusely branching, rootless
herb; stems pendulous to 1 m long, becoming intricately aggregated on
tree branches. Leaves 3–5 cm long, 1–2 mm wide with narrow sheathing
bases, shorter than internodes, exposing the stem. Flowers greenish
yellow, fragrant; stamens included; capsule 15–20 mm long, seeds sub-
late, 2–3 mm long, comate hairs two-whorled. $2n = 32$. Hammocks,
usually on oaks, Fla., Tex., Va., Tenn., W.I., South America. Spring,
fall. *Dendropogon usneoides* (L.) Raf.

2. **T. recurvata** L. BALL Moss—Plant scapose, with distichous portion 12–18 cm tall, giving rise to curving stems clustering on the common rooted crown. Leaf sheaths contiguous or imbricate, completely
concealing the stem at very short internodes; the blades becoming inter-
laced, resulting in a ball-like growth. Scapes exserted; flowers usually
two to (five), petals purple violet; anthers included. Capsule to 2 cm
long, prismatic, linear; comate hairs two-whorled. Hammocks, usually
on broad-leaved trees, south Fla., W.I., tropical America. *Diaphoran-
thema recurvata* (L.) Beer

3. **T. setacea** Sw.—Plant to 3 dm long, strongly fasciculate, or an
aggregate of many small tufts. Leaves 4 dm long, channeled above,

abruptly dilated at base, becoming closely involute-attenuate, trigonous and often flexuous at tips. Flowering stem surpassed or at least equaled by the longer rosulate leaves; floral bracts reddish, longer than the sepals, flowers 2.5 cm long; sepals glabrous, connate posteriorly; petals erect, violet; stamens and pistils exserted. Capsule 2.5–3 cm long, valves twisted on dehiscence; endocarp shiny black; seeds slender, 2 mm long. On tree trunks and branches, in hammocks, and in swamps, Fla., W.I., Central and South America. Summer. *T. tenuifolia* L.

4. **T. flexuosa** Sw. Twisted Air Plant—Plants 2–7 dm long; flowering stem foliose-bracted, rosette leaves eight to sixteen, 0.7–2.5 dm long; the broad, banded, spirally curving sheaths, with overlapping margins gradually passing into linear-involute blades. Inflorescence racemose-paniculate, flexuous, remotely pinnately branched. Flowers spreading, sepals 2–3 cm long, petals 4 cm long, rose or purple; stamens exserted; capsule 5–6 cm long. Shell ridges or mounds, hammock trees and shrubs, south Fla., tropical America. Spring, summer. *T. aloifolia* Hook.

5. **T. utriculata** L.—Plant 0.5–2 m tall, pale green, erect. Leaves numerous, 3–8 dm long, large in urn-shaped rosettes; sheaths ovate; blades linear, trigonous-sulcate at tips, acuminate to 7 cm wide at the base. Inflorescence three-innate, rachis flexuous, branching, flowers 4 cm long, sepals obtuse 1.7 mm long, petals ivory white, 4 cm long, stamens exserted. Capsule cylindric 4 cm long. Plant dies after flowering. Large plants often fall to the ground and continue to grow, flower, and fruit normally. Hammocks, Fla., W.I., Central and South America. Late summer and fall.

6. **T. valenzuelana** A. Richard—Plant soft to touch, 3 dm long, outermost leaves 2–3 cm wide at the sheathing base, tapering to elongate, channeled, gray silvery blades, attenuate-involute apices. Cauline leaves flexuous with overlapping sheaths. Inflorescence a single spike or panicle. Floral bracts 3–3.5 cm long, crimson; sepals 6 mm long, white; petals 3.5 cm long, the exposed part violet purple, in contrast to the white basal part; stamens long exserted; anthers 1.3 mm long, versatile, green, pollen yellow. Stigmas three, rotate; ovary 6 mm long. Hammocks, Everglade Keys, south Fla., W.I., South and Central America. Fall.

7. **T. polystachia** (L.) L.—Rosettes large; flowering stems to 5.5 dm tall, overtopping the leaves; sheaths and blades concolor; lower sheaths bladeless, the inner spreading, abruptly narrowing to elongate, involute, membranous long-attenuate blades. Spikes upright, several, 1–15 cm long; floral bracts smooth, reddish, indurate, overlapping; apex acute, mucronate, often incurved; petals violet, to 3 cm long; stamens and pistils exserted. Capsules 3–4 cm long. Hammocks, cypress swamps,

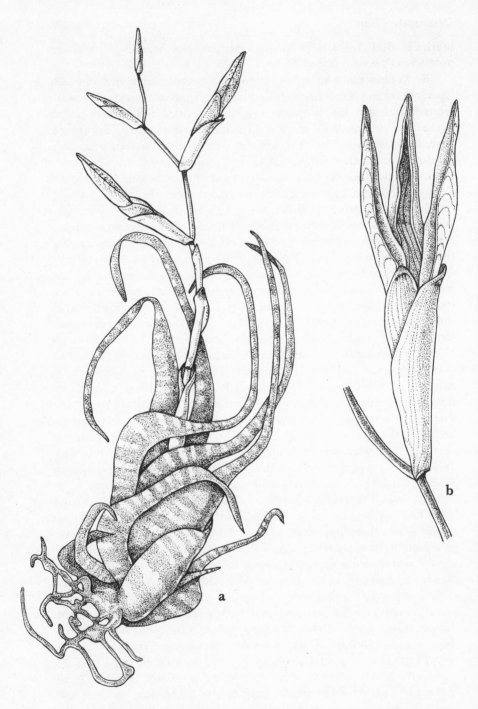

Family 39. **BROMELIACEAE.** *Tillandsia flexuosa:* a, flowering plant, × 1¾; b, flower, × 2.

south Fla., W.I., Cuba, Central and South America. Summer. *Renealmia polystachia* L.

8. T. fasciculata Sw.—Rosettes large; flowering stems to 6 dm tall, equaling or surpassed by the recurving leaves; sheaths membranous, dilated, brownish, bands with brown base; blades grayish green, tapering to involute tips; inflorescence fasciculate of numerous spikes 7–15 cm long; floral bracts yellow to rose purple, ovate-acute, loosely imbricate; 3–3.5 cm long; petals violet, 4.5 cm long; stamens and styles exserted. Capsules 3 cm long. Cypress swamps, hammocks; frequent; Fla., W.I., Central and South America. Fall. *T. hystricina* Small

9. T. balbisiana Schultes—Erect or pendent epiphytes 3–5 cm long; basal sheaths inflated, closely imbricate bulb fashion; blades flexuous, recurved, gradually narrowed to involute tips; scape leaves similar; the sheaths overlapping or in age shorter than the internodes. Inflorescence spicate, simple, becoming fasciculate; floral bracts 5–20 mm long, commonly reddish in anthesis, surpassing the sepals; petals violet, 3–3.5 cm long; stamens and stigma exserted. Capsule 4–4.5 cm long. Trees and shrubs in scrub and hammock, south Fla., Cuba, Panama, South America. Fall.

10. T. pruinosa Sw.—Plants 0.7–2.5 dm tall, clustered, stemless, silvery white throughout with copious spreading scales. Leaves borne on elongate pseudobulbs, sheaths suborbicular, inflated, abruptly narrowed into linear-triangular involute blades, recurving, filiform at tips. Spikes simple, rarely digitate, overtopped by uppermost leaves; floral bracts keeled, pinkish in anthesis, distichous; flowers five to many, sessile, in dense, compressed spikes to 7 cm long. Petals 3 cm long, violet, stamens and pistil exserted; capsule patently three-sided, pointed, dehisced valves 3.5–4.5 cm long, castaneous brown within. On trees, Big Cypress Swamp, Fla.; W.I., tropical America. *T. breviscapa* A. Rich.

11. T. circinata Schlecht. Plants 1.0–3.5 dm tall, clustered, stemless, silvery gray throughout with appressed scales. Leaves borne on elongate pseudobulbs, sheaths ovate, inflated, abruptly narrowed into subulate involute, acute blades, recurving or contorted. Spikes simple, linear, rarely digitate, overtopping the leaves; floral bracts imbricate, not keeled, spikes usually linear 5–8 cm long, two- to seven-flowered; flowers sessile, petals 3 cm long, bluish; mature capsule 3.5–4 cm long, grayish brown, pointed, the gaping valves sometimes filled with seedlings. On trees, coastal strand, Big Cypress Swamp, Everglades region, Fla.; Cuba to W.I., Mexico. *T. paucifolia* Baker; *T. bulbosa* Chapm. not Hook.

2. CATÓPSIS Griseb. AIR PLANT

Acaulescent epiphytes with rosulate strap shaped, sparingly lepidote leaves, inflorescence simple or bipinnate surpassing the leaves.

Scape bracts loose, urceolate, flowers bisexual or functionally unisexual; floral bracts small; sepals and petals free; stamens included, petals yellow or white, style short. Capsule septicidal, seeds with apical coma folded over and obtruding from the capsule. About 25 species, south Fla., W.I., Mexico, South America. Fall.

1. Sepals 15 mm long or longer; blades tapering to acute or acuminate tip; petals
   ovate or spatulate.
  2. Scape slender, recurved; petals larger
    than sepals . . . . 1. *C. nutans*
  2. Scape stout, erect; petals shorter than
    sepals . . . . 2. *C. berteroniana*
1. Sepals 9 mm long or less; blades long-attenuate from dilated bases; petals elliptic . . 3. *C. floribunda*

    1. **C. nutans** (Sw.) Griseb.—Rosettes relatively small with scapes 12 cm or more long; leaves outcurving, flexible, blades spatulate-acuminate; sheaths elliptic, dilated, at least the lowest, bladeless; scape deflexed, slender with remote bracts. Inflorescence usually simple, three-flowered or more; sepals strongly asymmetric, 1.5 cm long, finely nerved, petals 2 cm long, bright yellow, flaring at summit; stamens dimorphic, the outer three longer than the inner three; anthers sagittate. Seedlings forming on valves of dehisced capsule. Mature capsule ovoid, 2 cm long, beaked. South Fla., tropical America. Fall.

    2. **C. berteroniana** (Schultes) Mez ex D.C.—Plant in large rosettes to 7 dm tall; foliage sparingly lepidote; blades yellowish green; the sheaths pruinose-white, broadly dilate, thin. Scape rigid, thick; sheaths loose, imbricate, inflorescence bipinnate; floral bracts ovate, obtuse, equaling the petals; petals white to 12 mm long; stamens unequal, included. Dehisced capsule with beak 1.6 mm long; seed body constricted, brown in the middle, subcylindric, 1.5 mm long; coma abundant, seeds often germinating on capsule valves into rooted seedlings. South Fla., tropical America. Fall.

    3. **C. floribunda** (Brongn.) Sm.—Plants to 7 dm tall, with relatively large, rosulate leaves to scape, much exceeding the foliage; sheaths oval-elliptic, narrowed to attenuate blades. Scape sheaths long acuminate, overlapping the internodes; inflorescence compound, decurved, tripinnate, 25 cm long or longer; spikes stipitate, slender with many sessile flowers functionally unisexual, shorter or longer than the rachis internodes; pistillate flowers to 12 mm long; the staminate to 7 mm long; petals yellow, exceeding the sepals. Mature capsules to 14 mm long, beaked. South Fla., tropical America. Fall.

3. GUZMÁNIA R & S

Largely epiphytic acaulescent plants with erect, rooted rhizomes with numerous leaves in upright tufts. Inflorescence simple or compound, rachis internodes shorter than the bracts; sepals and petals partly united, anthers adherent around the stigma. Style slender, capsule cylindroid, seeds with straight basal coma. About 110 species in tropical America, W.I.

1. G. monostachia (L.) Rusby ex Mez—Plants pale green, to 4 dm tall, the scape and leaves about equal in length, forming cylindroid tufts in trees. Blades strap shaped, margins glabrous, parallel above the dilated flat sheaths, and slightly dilated acuminate apex. Surfaces sparingly lepidote; floral bracts membranous; sepals to 15 mm long, corolla white, the tube exceeding the lobes. Capsule 3–4 cm long. Tropical hammocks, south Fla., W.I., South America.

40. COMMELINÀCEAE SPIDERWORT FAMILY

Tender caulescent herbs with alternate leaves sheathing at the base, blades folded. Flowers ephemeral, bisexual, radially or bilaterally symmetrical, trimerous; sepals and petals differentiated as to texture and color; stamens six or three; ovary superior, usually three-carpellate, three-locular, style one, stigma capitate. Fruit a loculicidal capsule. About 30 genera, 500 species, chiefly tropical.

1. Floral bracts spathelike, strongly compressed, enclosing the cyme . . . . 1. *Commelina*
1. Floral bracts foliaceous, large, elongate or minute; cymes in pairs.
  2. Sepals petaloid, or greenish; stamen filaments barbate, hairs delicate.
    3. Inflorescence a cyme or a panicle, diffuse or moderately congested . . . 2. *Murdannia*
    3. Inflorescence cymose.
      4. Cymes sessile, subtending leaves large, similar to foliage leaves . . 3. *Tradescantia*
      4. Cymes pedunculate, subtending bracts minute . . . . . . 4. *Cuthbertia*
  2. Sepals paleaceous; stamen filaments glabrous . . . . 5. *Callisia*

1. COMMELÌNA L. DAYFLOWER

Herbs erect or spreading, creeping; stems slender, branching, tufted on short crowns of lanate roots. Flowers bilaterally symmetrical, corolla of two large blue petals and one small white one (rarely three blue petals of nearly the same size), exserted in anthesis above spathe margins. Capsule maturing within the spathe. About 230 temperate, tropical, and subtropical species.

1. Spathe margins free at base; petals all blue
    and nearly equal in size . . . . . . 1. *C. diffusa*
1. Spathe margins united at base; petals two
    blue, one white.
    2. Blades linear, 10 mm wide or less; plant
        erect . . . . 2. *C. erecta*
        var. *angustifolia*
    2. Blades ovate-lanceolate, 10 mm wide or
        more; plant decumbent, rooting at
        nodes . . . . 3. *C. elegans*

1. **C. diffusa** Burm. f.—Stems essentially glabrous up to 6 dm long, procumbent, mat-forming by nodal roots. Blades oblong-lanceolate, acute up to 6 cm long, 1.7 cm wide, abruptly narrowed to sheathing petiole; sheath long pilose about orifice or glabrate. Spathes acuminate 1–2 cm long, glabrous; flowers two to three, pedicelled, exserted above the spathes; corolla in anthesis 8–10 mm wide, evanescent. Mature seeds 2–2.5 mm long. $2n = 30$. Moist sandy margins of lawns, woods, widely distributed in eastern U.S., pantropical. Summer.

2. **C. erecta** L. var. **angustifolia** (Michx.) Fern.—Stems ascending or spreading radially from the crown of fasciculate, villous roots. Internodes and foliage puberulent or glabrate; blades 3–10 mm wide, narrowed to auriculate base, confluent with the sheath. Spathes puberulent, 16–24 mm long; corolla 16–25 mm wide in full anthesis; petals two large blue, one small white; anthers golden yellow. Capsule usually five-seeded; mature seed about 3 mm long, blackish, smooth; hilum linear, micropyle lateral; endosperm large. Widely distributed in central and south Fla., open pinelands, white sand scrub, glades, Fla. to Tex., N.C., on the coastal plain, W.I. Summer.

3. **C. elegans** HBK—Stems decumbent, branching, rooting from basal nodes up to 8 dm long. Internodes and foliage puberulous, becoming glabrous; blades narrowed to sheathing petioles. Spathes puberulous or essentially glabrous; corolla blue or white. Capsule obovoid, 4 mm long; seeds three, smooth. Hammocks and pineland, Fla. Keys, tropical America. Spring, summer.

2. MURDÀNNIA Royle

Annual herbs branching, rooting, and spreading from the base. Inflorescence terminal and lateral, flowers radially symmetrical, sepals concave with stamens attached at base; petals ephemeral; functional stamens three, alternating with staminodes. Capsule loculicidal, valves and the style base persistent, flowers protandrous. About 50 species of Asiatic tropics.

1. Leaves linear; flowers in condensed thyrsoid
    cymes . . . . 1. *M. nudiflora*
1. Leaves lanceolate; flowers in lax paniculate
    cymes . . . . 2. *M. spirata*

1. M. nudiflora (L.) Brenan—Plants 1–2 dm tall from fibrous roots. Leaves 5–10 cm long, 3–5 cm wide, firm, only sparsely ciliate at base and sheath margins, scaberulous at apex. Peduncle terminal or axillary from upper nodes; pedicels short; flowers few to several in congested cymules, sometimes cernuous; petals bluish lavender, 4–5 mm long, veined with darker colors; anthers white. Capsule lustrous, usually with transverse striae; seed usually two per locule, dark brown, angled, about 1 mm long and thick, verrucose. Fallow fields, hammock margins, central and south Fla., naturalized from Asiatic tropics. Late summer. *Aneilema nudiflorum* (L.) Kunth

2. M. spirata (L.) Brueckner—Stems slender, soft, up to 3 dm tall or less, diffusely branched in age, internodes pubescent only in adaxial groove into sheaths. Mature leaf blades 2–4 cm long, 5–12 mm wide, cordate, white margined. Peduncle forking at apex into lax, flexuous cymes, flowers in full anthesis 4–5 mm long; sepals broadly hyaline on margin, petals rose, veined in deeper shades, suborbicular with minute crenatures at apex; anthers blue, pollen white, filaments fringed all-around the thickened bases with moniliform hairs. Capsule ovoid, valves lustrous, chartaceous; seeds angled usually in tiers of four per locule, less than 1 mm wide and thick, sparsely verrucose, pale gray. $2n = 40$. Low prairie, glade margins, SABAL PALMETTO hammocks, naturalized from Asiatic tropics; known in North America, Fla., only from Immokalee and vicinity. Fall. *Aneilema spiratum* R. Brown

3. TRADESCÁNTIA L. SPIDERWORT

Perennial, upright, leafy-stemmed herbs from rhizomatous crown with essentially glabrous roots. Umbels sessile with two to three involucrate leaves; flower radially symmetrical; sepals herbaceous, petals blue, white, or rose, deliquescent; stamens six, fertile filaments pubescent or glabrous. Capsule sessile, one- to two-seeded, flowers protandrous. About 60 species in North American and South American tropics.

1. T. ohiensis Raf.—Plant to 6.5 dm tall, glaucous, leaves linear-lanceolate, 3–4 dm long, blades glabrous, sheaths pilose, usually tubular, dilated. Umbels congested, subtending bracts reflexed, pedicels 1–2 cm long, drooping, flowers deep blue to 2 cm long, sepals glabrous, hair tufted at apex; filaments fringed with moniliform hairs. $2n = 24$. Moist banks of lakes, canals, central and south Fla., Ga., La., Ark., Minn., to Mass. Spring. *T. canaliculata* Raf.

4. CUTHBÉRTIA Small Roselings

Tender, nearly glabrous herbs, solitary or tufted, with cluster of fistulous lanate or glabrous roots, with or without rhizomes. Stems erect or ascending, with three to four leaves above the base. Flowers born in paired cymes on paired peduncles from the axil of the uppermost sheath, or often another pedunculate cyme from a lower axil. Sepals three, green, suffused with rose; petals three, spreading, pink or rose with delicately crenate margins, filament hairs purple; ovary three-locular, style one, stigma tubular, white papillose, exserted ultimately beyond the closed sepals. In postanthesis each rachial cyme may proliferate, giving rise to rooted offshoots, two per peduncle. These mature to flower the following season. Three endemic species in southeastern U.S. *Tradescantia* L. authors, *Tripogandra* Raf. authors

1. Stems lax, as long as or shorter than the
     leaves . . . .   1. *C. graminea*
1. Stems firm, longer than the leaves . . .   2. *C. ornata*

1. **C. graminea** Small—Plants to 4 dm tall, strongly cespitose; underground parts becoming compacted, layered by vertical shoots; roots seemingly thickened at base, lanate, glabrous toward tips. Stems often geniculate, reclining, sheaths puberulent, pilose or glabrate at throat. Corolla 1.5–2.5 cm wide, sepals 4–5 mm long. Capsule normally six-seeded, seed 1.5 mm wide, corrugate, grayish, hilum punctiform. $2n = 12,24,36$. Oak, pine, and palmetto woods, sand hills, central and south Fla. to Va. Spring, fall. *Tradescantia rosea* Vent. var. *graminea* (Small) Anders & Woods.

2. **C. ornata** Small—Plants to 4.5 dm tall, solitary or in small tufts; roots copiously matted, villous, thick with adhering sand, upright, sheaths purple striate, puberulent on ribs, pilose hairs usually lacking. Flowers in full anthesis to 3.3 cm wide, petals spreading, essentially equal, margins prominently crenate; sepals 4–5 mm long. Capsule globose, seeds about 2 mm wide, corrugate. Coastal strand; oak, pine, and palmetto scrub, central and south Fla. Spring, fall. *Tradescantia rosea* Vent. var. *ornata* (Small) Anderson & Woodson

Rhoeo spathacea (Sw.) Stearn, Oyster Plant—A planted orna-

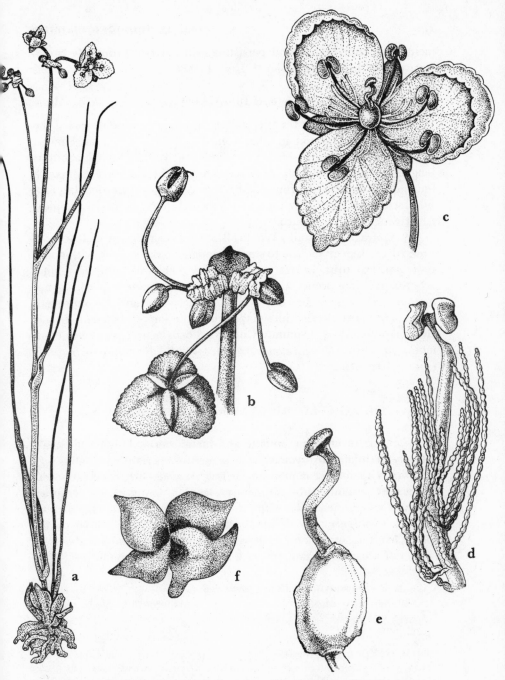

Family 40. **COMMELINACEAE.** *Cuthbertia ornata:* a, flowering plant, × ½;
b, inflorescence, × 1½; c, flower, × 3; d, stamen with moniliform hairs, × 15;
e, gynoecium, × 15; f, bracteoles, × 11.

mental, has not been found persisting without cultivation. $2n = 12$. Native of W.I. *Rhoeo discolor* (L'Her.) Hance

5. CALLÍSIA Loefl.

Creeping herbs; leaves alternate, progressively reduced toward the apex, becoming bractlike in inflorescence; leaf sheaths tubular. Flowers in paired cymes, sessile; sepals paleaceous, one-veined, keeled, about as long as the petals; stamens six, less commonly three, anther connective four-sided, filaments glabrous. Ovary three-carpellate, three-locular; capsule loculicidal, normally four- to six-seeded. Twelve species, tropical America, chiefly in Mexico.

1. **C. fragrans** (Lindl.) Woods, BASKET PLANT—Stems to 1 m long, upward arching, giving rise to strong, elongate basal stolons, with nodal roots and leaf tufts. Leaves to 15 cm long, 3 cm wide, elliptic-oblong, acuminate. Inflorescence a panicle with remote branches, bearing numerous, two-ranked, bracted cymes; branch internodes compressed. Flowers fragrant, petals white diaphanous, the smooth filaments crystal clear; capsule pilose at summit. Cultivated ornamental escaped to hammocks in south Fla. *Spironema fragrans* Lindl., *Rectanthera fragrans* (Lindl.) Degener

41. PONTEDERIÀCEAE PICKERELWEED FAMILY

Floating or creeping aquatic and marsh herbs. Flowers bisexual, radially or bilaterally symmetrical, trimerous; perianth petaloid, segments connate into a common tube below; stamens three or six, attached to base of perianth tube; anther two-locular, basifixed, or versatile. Ovary superior, one-locular with three parietal placentae or axile placentation with three, or one, fertile locules; style one, stigmas three- to six-lobed. Fruit a capsule or an achenelike body. Six genera and about 30 species of warm temperate, tropical, and subtropical regions.

1. Plants floating aquatics; fruit a three-valved
   capsule . . . . 1. *Eichhornia*
1. Plants rooted in soil; fruit a one-seeded
   achenelike body . . . . 2. *Pontederia*

1. EICHHÓRNIA Kunth WATER HYACINTH

Plants colonial, stoloniferous with inflated petioles and floating roots. Perianth bilabiate, lavender, the upper lip differentiated by a yellow spot; stamens three opposite the lower lip, exserted, upper three included. Six species of American and African tropics.

1. **Eichhornia crassipes** (Mart.) Solms—Leaves rosulate on horizontal stolons; roots pendent, finely branched. Petioles strongly thickened, aerenchymatous, contracting to a stiped blade usually much wider than long. Flowering stem with a basal leaf or merely a sheath terminating with two or more tubular bracts, subtending a contracted panicle. Flowers few to 5 cm wide with showy corolla of varying shades of purple, becoming marcescent around the ripening capsule. Prolific in waterways, from tropical America to south Fla., Va. Summer.

2. PONTEDÈRIA L. Pickerelweed

Emersed tufted aquatics from rhizomatous crown. Perianth bluish purple spotted with yellow; with spreading limb, the three inferior segments nearly free, the upper three connate to the middle; stamens six, three with short filaments inserted lower in the perianth tube than the longer stamen; anthers blue, versatile. Ovary three-locular, one locule fertile, with a single suspended ovule; fruit a utricle. Four species, America.

1. **P. lanceolata** Nutt.—Flowering stems to 8 dm tall, commonly with a solitary, short-petioled leaf above the middle; basal leaves long petioled, about equaling the stems; blades 12–20 cm long, 2–8 cm wide, deltoid, lanceolate, membranous, subtruncate to hastate with entire, white cartilaginous margin; sheaths loose, the basal with broad hyaline margins. Raceme 8–15 cm long, densely flowered, flowers up to 18 mm long, subsessile, puberulent with colored stipitate glands. Fruit 5–6 mm long, viscid, with four prominent, scalloped crests surmounted by the twisted, perianth limb. Fresh-water marsh and swampy ditches, moist soil, Fla., Tex., Mo., to Va., South America. Spring, summer.

42. JUNCÀCEAE Rush Family

Rhizomatous caulescent or acaulescent herbs; stems simple; leaves linear, flat, terete, or needlelike. Inflorescence open or congested cyme, often glomerulate. Flowers bisexual, rarely unisexual, trimerous, radially symmetrical; sepals and petals similar, stamens three to six, pistil three-carpellate, ovary superior, fruit loculicidal capsule, three-valved. Nine genera, 400 species, arctic to tropical regions.

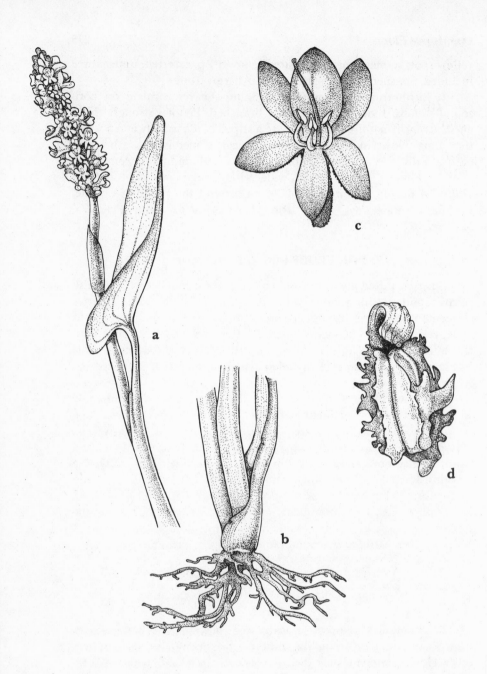

Family 41. **PONTEDERIACEAE.** *Pontederia lanceolata:* a, inflorescence and mature leaf, × ½; b, stem base and rhizomatous crown, × ½; c, flower, × 4; d, fruit, × 20.

1. JÚNCUS L. Rush

Glabrous perennial or annual plants, erect or creeping; leaf sheaths with free margins; blades often septate, channeled on adaxial side. Flowers subtended by one or two bractlets. Placentae parietal or often axile by intrusion with three-locular ovary; seeds numerous, minutely striate, pointed at each end or caudate. About 300 species, cosmopolitan. Coastal strand in moist sands, swampy margins of hammocks.

1. Inflorescence appearing lateral.
  2. Individual flowers subtended by two bracteoles; leaf sheath bladeless . .   1. *J. effusus*
  2. Individual flowers subtended by one bracteole; leaf sheath with blade . .   2. *J. roemerianus*
1. Inflorescence terminal.
  3. Leaves flat, nonseptate.
    4. Stems soft at the base, creeping, prostrate or floating, mat-forming; glomerules few, large . . . . .   3. *J. repens*
    4. Stems from knotty rhizome, erect, tufted; never floating; glomerules numerous, small . . . . .   4. *J. biflorus*
  3. Leaves terete, septate.
    5. Tepals linear, tapering; glomerules lobulate . . . .   5. *J. scirpoides*
    5. Tepals setaceous to narrowly lanceolate, glomerules spheroidal when many-flowered.
      6. Glomerules many-flowered, seeds pointed at each end.
        7. Fruiting glomerules contiguous; blade of uppermost sheath less than 3 cm long . . . .   6. *J. megacephalus*
        7. Fruiting glomerules distant; blade of uppermost sheath more than 3 cm long . . . . . .   7. *J. polycephalus*
      6. Glomerules two- to three-flowered, seeds winged at each end . . .   8. *J. trigonocarpus*

1. **J. effusus** L. Soft Rush—Stems to 1 m tall or more, 3–5 mm wide, in large tussocks, compressible, pale green, finely veined. Sheaths to 15 cm long or more, bladeless in varying shades of brown, loose, with filiform excurrent bristle at summit. Involucral leaf 10 cm long or more, gradually tapering to sharp apex. Inflorescence contracted or diffuse; flowers 2–3 mm long, greenish or brown, tepals lanceolate, about equal in length; stamens three. Capsule obovoid, rounded or umbonate at

beakless summit; seed amber 0.5 mm long, minutely ribbed, obliquely pointed at each end. $2n = 20$. Wet ground, cypress borders, shores of ponds, south U.S. to Fla. Spring.

2. **J. roemerianus** Scheele—Plants stiff, rigid to 1.5 m tall in large tufts; culms 3–5 mm thick, spaced in rows on horizontal scaly rhizomes. Lowermost short sheaths bladeless, succeeded by two to three sheaths with blades. Inflorescence compound, subtended by pungent involucral bract; flowers unisexual, sepals 3–4 mm long, scarious margined, center green with subaristate excurrent vein; petals shorter, obtuse; anthers longer than filaments, lacking in pistillate flowers; style evident, shorter than its branches. Mature capsule lustrous brown, 2.8 mm long; seeds 0.3–0.7 mm long, wedge shaped, obtuse at ends, fine striate. Salt flats, marshes, coastal areas on Gulf of Mexico and Atlantic Ocean, Fla., Tex., Md. Fall.

3. **J. repens** Michx.—Plant to 15 cm long, rosulate with numerous basal leaves; stems becoming prostrate, rooting at nodes, commonly floating or creeping in shallow waters; offshoots numerous, branching and rebranching, leaf blades 3.0–5.0 cm long, linear-attenuate; sheaths hyaline on margins, auricles often cucullate, united at the top margins. Inflorescence sessile or peduncled with each head five- to eight-flowered; sepals 5 mm long, petals twice as long; anthers small, oblong, shorter than filaments, stigmas sessile. Capsule beakless to 8 mm long; seeds numerous, amber, abruptly pointed. Coastal plain states, Fla., Tex., Okla., to Del., Cuba.

4. **J. biflorus** Michx.—Plants to 8 dm tall or taller from knotty protruding rootstocks. Blades flat 3–4 mm wide; sheaths loose, margins broadly hyaline. Inflorescence long peduncled, cymes diffuse, many-flowered, or often contracted; glomerules two- to four-flowered; flowers 3–4 mm long, sepals aristate, stamens persistent in fruit; style wanting or very short. Capsules golden brown, 2–2.8 mm long; seeds amber, about 0.6 mm long, ellipsoid, pointed, or short caudate. Moist sands, pine flatwoods, widely distributed, Fla., Tex., east U.S., Midwest. Summer.

5. **J. scirpoides** Lam.—Stems erect from stout rootstocks to 9 dm tall. Sheaths loose, margins hyaline; blades terete to 12 cm long with true partitions. Cymes ample, compound; glomerules globose, lobulate; flowers 3–4 mm long; tepals subulate, pungent in fruit; anthers small. Capsule one-locular, style base not separating on dehiscence; seeds fusiform, pointed at each end. Pine flatwoods, glades, Fla., Tex., N.J., Pa., Midwest, Va. Summer.

6. **J. megacephalus** M. A. Curtis—Stem to 1 m tall or less, purplish at the base and at the nodes; rhizome thickened, tuft-forming, giving

Family 42. **JUNCACEAE.** *Juncus biflorus:* a, inflorescence, × 1; b, stem base with basal leaves and rootstock, × ½; c, flower, × 15; d, cluster of mature fruits, × 8.

rise to several offshoots; stems to 6 dm tall or more, several, closely tufted on knotty crowns, basal sheaths, internodes and floral bracts purplish, especially in preanthesis. Internodes compressed, sheaths open, rather loose; basal blades subterete, septate 3–5 dm long, cauline blades short, the uppermost much reduced. Cymes contracted; ultimate peduncles short; glomerules spheroidal 10–13 mm thick, in fruit; flower 3–3.5 mm long, tepals attenuate, stamens included; stylar portion of capsules tardily separating. Wet sandy soil, river banks, coastal strand, Fla., La., Miss., Va. Fall.

7. J. polycephalus Michx.—Plants tufted, to 1 m tall or less, from thickened rhizomatous crown. Basal sheaths flat, ovate, bladeless, the upper three or four with blades, ligules 2–3 mm long; blades to 12 cm long, gladiate. Inflorescence decompound cyme, divergently branched, glomerules distant, 10–12 mm thick; tepals attenuate and rigid, recurving in fruits; stamens included, style exserted. Capsule dehiscent by three slits; style base persistent, entire; seeds 0.5 mm long, amber, ellipsoid, pointed at each end. Cypress swamps, in drier land, Fla., Okla., Tex., N.C., Va., Md. Summer.

8. J. trigonocarpus Steud.—Plants to 6 dm tall from rhizomatous base with stolons, leaves per culm two to three, blades terete, septate, pungent with auriculate sheaths; the basal sheaths loose, bladeless. Cyme erect, compound with rigid ascending branches, contracted; flowers two to four in cluster; tepals 3–3.5 mm long, three-veined, narrowly lanceolate, margins hyaline, usually evanescent; midvein mucronate, firm; stamens six or three. Capsules two, as long as tepals, dehisced valves lustrous maroon, outcurving, tips with style remnant; seed 2.3 mm long, with yellowish wing at each end. Hammock, Lignum Vitae Key, south Fla. to N.C., La.; rare. Fall.

43. LILIÀCEAE Lily Family

Plants from bulbs, corms, tubers, or rhizomes. Stems scapose or leafy. Leaves simple, alternate, verticillate, rosulate, sessile or petioled, fleshy or succulent, sometimes reduced to scales. Flowers bisexual, hypogynous, radially symmetrical, usually trimerous, with free sepals and petals; sepals petaloid or sometimes green, stamens six, in two alternating cycles. Pistil three-carpellate, ovary three-locular; ovules one to many in each locule. Fruit a loculicidal capsule or a berry. Cosmopolitan, about 250 genera, 3700 species.

1. Leaves typical, green, functional.
  2. Plants with cauline leaves; bulb of thickened, fleshy scales . . . . 1. *Lilium*

2. Plants with radical leaves.
   3. Leaves thick, juicy, margins prickly
     dentate . . . .   2. *Aloe*
   3. Leaves coriaceous, linear, elongate,
     margins smooth . . . . . .   3. *Schoenolirion*
1. Leaves reduced to scarious scales, sometimes
  spinose, functionally replaced by cladodes   4. *Asparagus*

1. LÍLIUM L.

Erect herbs with simple stems, or branching above. Leaves alternate, flat, sessile or petiolate. Flowers showy, large, solitary, or in racemose and umbelliform inflorescences, tepals free, recurring, deciduous; stamens hypogynous, filaments filiform, anthers oblong-linear, versatile or dorsifixed, ovary sessile. Capsule coriaceous, seeds compressed, two-seriate in each locule. About 80 species, north temperate regions, cosmopolitan.

1. **L. catesbaei** Walt. Pine Lily—Stem to 5 dm tall, leafy up to inflorescence without basal leaves from the bulb. Leaves 5–7 cm long, 3–7 mm wide, lanceolate, progressively shorter toward the summit. Flower solitary, rarely two, erect, terminating the stem; tepals 7–11 cm in length, long clawed, long acuminate, lamina dilated, brilliant red, yellowish at center, richly spotted with purplish brown; filaments 7 cm long, anthers to 4 mm long. Style with prominent stigmas overtopping the anthers. Capsule obovoid to 5 cm long, 1.6 cm in diameter. $2n = 24$. Pine flatwoods, south Fla., Ga., Ala., N.C. Summer, fall.

2. ALÒE L.

Succulent plants, sometimes arborescent. Leaves thick, in basal rosettes and at the ends of the branches, sharp pointed with hard, marginal teeth. Flowers in peduncled racemes, recurved in anthesis; perianth tubular, of basally connivent tepals; stamens six, anthers dorsifixed; stamens and the style about equaling the tepals in length. Fruit sometimes baccate. About 275 species of Old World tropics and subtropics.

1. **A. barbadensis** Mill.—Stoloniferous, nearly stemless succulent. Leaves to 5 dm long, narrowly lanceolate, long acuminate, juicy, yellowish green; in larger leaves spines prominent 1 cm distant. Scape to 1 m tall, scaly, inflorescence 1–3 dm long, flowers bright yellow to red, subtended by broad bracts, flowering perianth 2–3 cm long, arching, the segments linear, the free apical part longer than the tube. Fruit many-seeded. Commonly a greenhouse plant in cool climates; landscape ornamental in warmer regions, naturalized, south Fla., W.I., Bermuda.

Family 43. **LILIACEAE.** *Lilium catesbaei:* a, flowering plant with scaly bulb and roots, × 2⅓; b, flower, × ½; c, fruit, × ½.

3. SCHOENOLÍRION Torr. ex Durand

Wiry-stemmed scapose glabrous herbs from vertical rhizome with numerous elongate roots. Leaves rosulate. Inflorescence a raceme or panicle; tepals white; filaments subulate, connate at base, anthers versatile. Style simple, stigmas three, small, included. Native of North America, four species. *Oxytria* Raf.

1. **S. elliottii** Gray, SUNNYBELL—Scape to 6 dm tall. Leaves with shreddy, sheathing, distichous bases, hyaline on margins; blades rigid 3–5 mm wide. Inflorescence often with two to three distant branches toward summit; flowers 5–6 mm long, solitary on slender pedicels in axils of subulate bracts, tepals white or yellowish, anthers white, auriculate; ovary three-locular. Capsule 4–5 mm wide at summit, three-seeded; seed 3.5 mm long, 3 mm wide, plano–convex, obliquely cuneate at micropylar end, black, crustaceous, half-filled by the endosperm with embryo, buoyant, floating beetle fashion in water. Pine flatwoods, glades, south Fla., Ga., Ala. Summer.

4. ASPÁRAGUS L.

Woody vines or erect herbs with densely branched stems. Cladodes the ultimate branches, leaflike in texture and function, axillary to scale leaves. Pedicels in axils of cladodes, perianth six-parted, marcescent; stamens six. Ovary three-locular; fruit a saccate berry, usually two- to three-seeded. Old World genus, 300 species. Ornamental, escaping from cultivation. Spring.

1. Cladodes terete, filiform.
  2. Plant bushy, erect herb, from branched
      rootstock . . . . 1. *A. officinalis*
  2. Plant ramulose, woody climber . . . 2. *A. plumosus*
1. Cladodes flat, linear . . . . 3. *A. sprengeri*

1. **A. officinalis** L. GARDEN ASPARAGUS—The edible, scaly, spring shoots arising from cordlike fleshy roots; stems to 2 m tall, glaucous, diffusely branched above; ultimate branches threadlike, 5–8 mm long, functionally replacing the reduced leaves. Flowers campanulate, dioecious in axils of cladodes; staminate with abortive pistil; fertile flower with three exserted stigmas. Fruit berrylike, red, three-locular, two- to three-seeded. $2n = 20$. Escaping from cultivation, naturalized in Old World, introduced from Europe.

2. **A. plumosus** Baker, ASPARAGUS FERN—Vigorous woody climber from fleshy roots with numerous spreading, fernlike sprays and filiform

cladodes; leaf scales modified into spines. Flowers white; fruits purple. $2n = 20$. Naturalized from South Africa. Spring. Accessory ornamental in florist trade.

3. **A. sprengeri** Regel—Stems several to 2 m long, trailing from the crown of tuberous-fasciculate roots; cladodes linear 10–12 mm long, flat. Flowers fragrant, white or suffused with pink; fruit 5–6 mm in diameter, three-lobed, red, three-seeded. $2n = 60$. Garden ornamental, naturalized from South Africa, escaping from cultivation. Spring.

44. SMILACÀCEAE SMILAX FAMILY

Perennial, woody or herbaceous vines, smooth or spinose, scandent by stipule tendrils, dioecious. Leaves simple, evergreen, abscission layer lacking; petiole breaking off above its persistent base. Flowers unisexual, borne in umbelliform inflorescences on globose, sessile, or pedunculate receptacle, flowers trimerous, radially symmetrical, hypogynous; tepals and stamens six, each group in alternating series of threes; anthers versatile. Pistil one, three-carpellate, three-locular, three-seeded, sometimes functionally one-seeded. Fruit a berry. Four genera and about 275 species, tropical and subtropical America.

1. SMÌLAX L. GREENBRIERS

Stems prevailingly high climbing in woody species. Canes rising from tuberous rootstocks, with or without stolons; principal roots rising from internodes, lateral roots finely branched. Pistillate flowers with abortive stamens; staminate without pistillate rudiments. Stamens attached to base of tepals; anther locules linear, confluent in maturity. Fruit one- to three-seeded; seeds embraced in saccate, gelatinous endocarp; testa crusted, red or brown; the seed body globose, endosperm unbreakable. About 350 species tropical and subtropical.

The tips of growing shoots and the rootstocks were variously utilized as food by Indians. Crushing and washing the underground parts yield "red flour."

1. Leaf blades never dilated or lobed at the
     base.
     2. Margin of the blade rolled, consistently
          parallel with the submarginal vein;
          midvein keeled toward the base about
          two-thirds of its length; margins smooth   1. *S. laurifolia*
     2. Margin of the blade strongly ribbed,
          submarginal vein lacking; midvein

rounded throughout; margin com-
monly spiny or wholly entire . . .   2. *S. havanensis*
1. Leaf blades dilated, hastate, lobed at the
base.
3. Margin of the blade strongly ribbed, hispidulous, prickly, or smooth; high-
climbing spinose vine; rootstock short,
tubers with stolons . . . . . .   3. *S. bona-nox*
3. Margin of the blade rolled, never hispid
or prickly; high-climbing, sparingly
spinose vine; rootstock without stolons   4. *S. auriculata*

1. S. laurifolia L.—Bamboo Vine—Evergreen high-climbing vine
with terete stems 1.5 cm in diameter from tuberous, ligneous rootstocks;
spines to 10 mm long. Twigs geniculate; leaves, coriaceous, smooth,
blades lanceolate, oblong-ovate, 5–10 cm long, petioles to 8 mm long,
about as long as the axillary peduncles. Pedicels to 6 mm long, crowded
on the receptacle; flowers greenish white, tepals of staminate flower
4–5 mm long, about equaling the stamens, those of the pistillate flowers
to 4 mm long. Ovary three-carpellate, functionally one-locular, one-
seeded fruit; stigmas two to three, deciduous. Mature black berry 6–8
mm in diameter; seed maroon red, 5 mm long, 4 mm in diameter,
slightly flattened on one side. Fruit maturing the second year. Fla.,
Miss., Va. Fall.

2. S. havanensis Jacq.—Low- to high-climbing vine from nonstoloniferous rootstock. Stems and twigs angled, branches slender, geniculate,
internodes prominently papillose on ridges, punctate between scattered, dark-tipped, curved, flat spines. Leaves pale green, the blades
varying in shape from elliptic to suborbicular, 3–6 cm long, 1–6 cm
wide, spinose or smooth on margins. Flowers small; peduncle and pedicels 4–5 mm long; tepals 1.5 mm long, elliptic ovate. Mature fruit black,
5–6 mm in diameter, subglobose; seeds three, 3 mm long and wide, globose, maroon red, slightly flattened on one side. Pinelands and hammocks, Everglades, Fla. Keys, W.I. Fall.

3. S. bona-nox L.—High-climbing, strong vine from solitary tuber
or a group of ligneous tubers with stoloniferous rhizomes giving rise to
new canes by creeping and rooting at the nodes. Leaves prevailingly del-
toid-hastate or pandurate, 5–9 cm long, 1–5 cm wide, relatively thin,
with smooth or hispid-spinulose margins, often rounded, lobed; petioles to 15 mm long. Tepals white, 5–6 mm long; stigmas three; berry
6–8 mm in diameter, globular, black, one-seeded; seed 5 mm·long, 4.5
mm in diameter, maroon brown. $2n = 32$. Fla., Tex., Midwest, coastal
plain states, Mexico. Summer.

4. S. auriculata Walt.—Sparingly armed vine, trailing or high

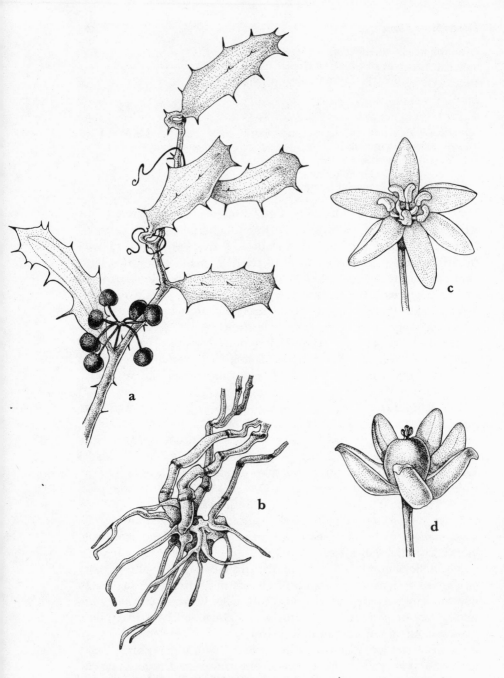

Family 44. **SMILACACEAE.** *Smilax havanensis:* a, fruiting branch with leaves
and tendrils, × ¾; b, stem base and rootstock, × ½; c, staminate flower, ×
12½; d, pistillate flower, × 11.

climbing from nonstoloniferous, woody, segmented rootstocks. Leaves 3–8 cm long, 1–5 cm wide, abruptly mucronate, dilated above the cuneate base; margin smooth, rolled, lamina veins usually three to five, rarely seven. Flowers yellowish green, fragrant; tepals in staminate flower 6–8 mm long, in pistillate flowers 3–4 mm long. Berry ultimately bluish black, 8 mm in diameter, 4.5 mm long, nearly plano–convex, brown. Hammocks, white sand scrub, Fla., Va., La., W.I. Fall.

45. AGAVÀCEAE Agave Family

Plants rhizomatous. Leaves basal or cauline, thick, fibrous with or without marginal prickles. Flowers bisexual, trimerous, essentially radially symmetrical, usually in large panicles, tepals free or connate; stamens six. Pistil three-carpellate, ovary inferior or superior, three-locular. Fruit loculicidal capsule, or berry; seeds numerous or solitary, compressed, endosperm fleshy. About 20 genera, 670 species, tropics and subtropics, desert regions, Western Hemisphere.

1. Ovary superior; leaf margins rough.
  2. Perianth segments free; fruit a capsule;
    stamens included . . . . 1. *Yucca*
  2. Perianth segments connate; fruit a berry;
    stamens exserted . . . . 2. *Sansevieria*
1. Ovary inferior; leaf margins prickly.
  3. Perianth funnel shaped; anthers with
    slender filaments exserted . . . . 3. *Agave*
  3. Perianth rotate; anthers with thickened
    filaments included . . . . . . 4. *Furcraea*

1. YÚCCA L. Spanish Bayonet

Large plants with woody, leafy stems or bracted scapes. Leaves often bearing marginal fibers, apex pungently spinose. Inflorescence a large panicle of nodding flowers, perianth campanulate, tepals distinct, stamens hypogynous. Ovary superior, sessile, three-locular; style columnar, stigma three-lobed. Fruit a capsule, indehiscent, seeds numerous, horizontally two-seriate in each locule. About 40 species, North America, W.I., Mexico.

1. **Y. aloifolia** L.—Trunk 1–15 dm or more tall, branching, stiffly inclining, colonial by rhizomes. Leaves to 4 dm long, the sheathing bases contracted to blades 2–3 cm wide, spreading at summit, reflexed toward the base. Panicle large, flowers many, to 5 cm long, glistening white throughout. Capsule to 8 cm long, 3 cm long, 3 cm in diameter;

seeds black, testa papery, rugulose. $2n = 42$. Coastal strand, sand dunes, shell mounds, south Fla., probably naturalized from W.I. and Mexico. Summer.

2. SANSEVIÈRIA Thunb. BOWSTRING HEMP

Succulent, fibrous herb with a thick creeping rhizome. Leaves upright, flat, stiff. Inflorescence a narrow panicle, terminating the scape, perianth white or greenish, lobes linear, one-veined, revolute, the base narrowly tubular. Pistil and stamens hypogynous; fruit berrylike, one- to three-seeded. About 60 tropical species, Old World. *Cordyline* Adans.

1. **S. thyrsiflora** Thunb.—Leaves to 1 m long, 5–8 cm wide, in tufts on the rhizome; lamina glabrous, with transverse color bands of white or yellow; margins entire, often with marginal stripe of yellow. Flowering scape surpassing the leaves, fruiting inflorescence to 6 dm long; flowers usually fasciculate, 3.5 cm long; anthers versatile, 3.5 mm long, filaments filiform, surpassing the tepals. Style to 4 cm long, filiform with capitate stigma, ovary 5 mm long, fruit spherical 8 mm in diameter, red. Ornamental, persisting in old sites, south Fla., naturalized from South Africa. *Cordyline guineensis* (L.) Britt.; *S. guineensis* Willd.

3. AGÁVE L. CENTURY PLANT

Leaves radical or on short caudex, succulent, fibrous and thick, margins with or without prickles or apical spine. Perianth funnelform; stamens inserted near the sinus level of segments or within the tubelike styles, long exserted, anthers versatile. Style curved, stigma capitate, shallowly three-lobed. Capsule cylindric, three-valved, dehiscence loculicidal; seeds numerous, thin and flat if matured. Bulbils usually formed after flowering, scape withering, suckers appear later. About 300 species, south U.S., South America.

1. Leaves sword shaped or linear; spines without marginal processes.
2. Blades concave; caudex becoming leafless   1. *A. decipiens*
2. Blades flat; caudex leafy . . . . .   2. *A. sisalana*
1. Leaves lanceolate, spines raised on fleshy
   marginal processes . . . .   3. *A. americana*

1. **A. decipiens** Baker, FALSE SISAL—Scape 3 m or more tall, woody, fibrous. Leaves to 1 m long and 5–10 cm wide in dense rosettes. Stamens attached near the middle of the tube. Capsule 3–4 cm long; bulbils abundant, rooting in clumps on floral axes before falling. Coastal

Family 45. **AGAVACEAE.** *Agave decipiens:* a, young plant with leaf rosette and rhizomes, × ½; b, part of inflorescence, × ½; c, flower, × 1⅛.

strands, hammocks, shell mounds, south Fla., introduced from Mexico. Summer.

2. A. sisalana Perrine, SISAL HEMP—Acaulescent or nearly so, with numerous leaves to 1.5 m long and 10 cm wide, usually glaucous, tapering to conical terminal spines; marginal spines weak or wanting. Filaments inserted above the middle of the perianth tube, flowers yellow; fruit seldom forms; bulbils abundant. Coastal strand, south Fla., introduced from Yucatan. Summer.

3. A. americana L. CENTURY PLANT—Scapes to 14 m or more tall from massive radical rosettes. Leaves to 1.5 m long, tapering to stout terminal spines; marginal spines uncinate, rising from mamillated outgrowths. Flowers erect in dense paniculate cymes on horizontal branches; perianth yellowish green; filaments inserted in the throat of the tube. Capsule ellipsoid, many-seeded; bulbils not formed. $2n = 120$. South Fla., tropical America. Summer.

4. FURCRAÈA Vent.

Large plants with subterranean caudex bearing dense rosettes. Leaves spinose-dentate on margins or rarely entire. Panicles large, the flowers in axils of bracts, solitary or fasciculate, or often replaced by bulblets. Perianth rotate, the segments shortly connate, stamens attached at the base; anthers linear. Style columnar, stigma obscurely three-lobed; ovules in two rows in each locule; seeds compressed. About 20 species, tropical America.

1. F. selloa K. Koch—Scape with panicle to 10 m tall; leaves succulent, ensiform to 9 dm long, narrowed at the base, contracted at the spine-tipped apex, thin, flexible, armed with curved spines along white margin. Flowers 4–5 cm long, 3–5 cm wide, white, nodding; perianth tube nearly obsolete. Style three-angled, the lower half-thickened and wing edged on the angles. Bulbils numerous, capsule not seen. Cultivated, ornamental, persisting on old sites, south Fla., South America. Summer.

F. macrophylla Baker, WILD SISAL—Another ornamental in cultivation; may persist in old sites, Bahamas, Jamaica, Cuba.

46. HAEMODORÀCEAE BLOODWORT FAMILY

Caulescent perennial herbs from fibrous roots. Leaves equitant with ensiform blades. Inflorescence cymose, hoary pubescent; flowers bisexual, trimerous, radially symmetrical, sepals and petals free, yellow within, hoary pubescent without; stamens three, anthers introrse. Ovary inferior, three-locular style, deciduous. Fruit a loculicidal capsule with many seeds. Fourteen genera, 75 species, Africa, Australia, and tropical America. Center of distribution in the Southern Hemisphere; one genus restricted to northeast North America.

1. LACHNÁNTHES Ell.   Red Root

Characters of the family

1. L. caroliniana (Lam.) Dandy—Plants to 8 dm tall from red slender rhizomes and roots with red juice. Flowering stem glabrous below, villous toward the summit. Leaves glabrous, with sheathing bases; cauline blades progressively shorter toward inflorescence. Sepals 4–5 mm long, petals 5–7 mm long, oblong, rounded. Capsule villous, spheroidal with marcescent floral remnants; seeds 2 mm wide, maroon black, centrally attached, orbicular or quadrate, narrowly winged, with a deep marginal sinus; endosperm 1.8 mm wide, 1.7 mm long, thin, with a small embryo. Pine flatwoods, bogs, hammock margins, coastal plain, Fla. to La., Mass. Summer.

47. AMARYLLIDÀCEAE   Amaryllis Family

Perennial acaulescent herbs from tunicate bulbs, corms, rarely rhizomes. Leaves usually radical, soft, fleshy, or stiff. Flowers bisexual, radially symmetrical, solitary or in umbelliform or racemose inflorescences, subtended by spathaceous bracts; perianth six-parted, six-lobed, tubular or campanulate; stamens six, filaments sometimes embraced in a membrane corona fashion; anthers introrse. Ovary inferior, or rarely only the base adnate to the perianth, three-locular, style elongate, stigma three-lobed. Fruit a capsule or a berry. About 85 genera, 1100 species, tropical and subtropical regions. *Leucojaceae;* including *Hypoxidaceae*

1. Plants from cormose or rhizomatous root-
    stocks.
  2. Perianth tube completely adherent to the
     ovary; flowers solitary, or a few in um-
     bel . . . .   1. *Hypoxis*
  2. Perianth tube adherent only to the base
     of ovary; flowers numerous in a spike .   2. *Aletris*
1. Plants from tunicate bulbs.
  3. Filament bases embraced in a mem-
     branous corona . . . .   3. *Hymenocallis*

Family 46. **HAEMODORACEAE.** *Lachnanthes caroliniana:* a, inflorescence, ×
½; b, stem base, basal leaves, and rhizomes, × ½; c, flower, × 5; d, mature
fruit, × 5.

3. Filaments free to the base.
  4. Corolla tube shorter than perianth
     segments . . . . 4. *Zephyranthes*
  4. Corolla tube longer than perianth
     segments . . . . 5. *Crinum*

1. HYPÓXIS L. STAR GRASS

Pilose or glabrous herbs from corms. Stems short with persistent fibers at base. Leaves three to many, linear, flat, sessile. Flowers few opening successively, perianth yellow within, tube short; anthers versatile or basifixed; filaments adnate at base of perianth segments. Seeds pebbled or muricate, beaked; hilum rostrate. About 80 species, tropical regions.

1. Mature seeds muricate; corm elongate . . 1. *H. juncea*
1. Mature seeds pebbled; corm subglobose . 2. *H. wrightii*

1. H. juncea Sm. YELLOW STAR GRASS—Perianth 1–1.5 cm long; gray pilose without; yellow, rarely white, within; stamens included, filaments adnate at the base of perianth. Leaves involute, three-nerved, appearing setaceous, overtopping the scapes. Ripe capsule 4–6 mm long, ellipsoid, capped by the marcescent perianth; seeds 1 mm long, black, crustaceous, muricate. Pine flatwoods, coastal plain Fla., W.I. Spring, summer.

1. H. wrightii (Baker) Brackett, FRINGED STAR GRASS—Corm 6–12 mm long, scales becoming fibrillose. Leaves linear, channeled to 3 mm wide. Capsule subglobose, 4–6 mm long; seeds lustrous, irregularly pebbled, about 1.1 mm wide. Pine flatwoods, Fla. Keys, W.I. Summer.

2. ÁLETRIS L. COLIC ROOT

Scapose herbs with radical leaves. Inflorescence spicate, distally floriferous; flowers bracted, perianth yellow or white, six-parted, cylindrical above the line of adnation with the ovary, glutinous, often mealy and varying papillose without, filaments adnate to the perianth about three-fourths of the length, anthers introrse, with stigmas included. Mature capsule dehiscent by splitting of the style and the free part of the valves from apex down to ovary summit; seeds 0.3–0.4 mm long, amber, sharply curved at one end. About 25 species, North America, eastern Asia.

1. Flowers yellow; perianth 8–10 mm long,
   segments narrowly oblong . . . . . 1. *A. lutea*
1. Flowers white; perianth 5–7 mm long, segments lanceolate . . . . 2. *A. farinosa*

1. A. lutea Small—Scapes 5–9 dm tall, erect, strict with five or more small bracts, diminishing toward the summit. Rosulate leaves numerous, 6–14 cm long, 6–20 mm wide, yellowish green, the veins converging at the shortly revolute apex, perianth lobes deltoid, ovate, or spatulate, spreading or recurving in fruit, free filament apices about as long as the anther. Capsule conoidal, long beaked, 3–5 mm long. Low pinelands, coastal plain, Fla. to La., Ga. Spring.

2. A. farinosa L.—Scape 2–6 dm tall, strict, erect with two to three subulate bracts toward the summit. Rosulate leaves broadly linear or elliptic 4–8 cm long, 5–10 mm wide, grayish green. Perianth 6–8 mm long, usually white cylindrical lobes deltoid, all roughened by papillose processes; free filament apex longer than the anther, exserting above the sinus level. Capsule 3–4 mm long, abruptly narrowing to a beak. Prairies, pinelands, south Fla., Fla. Keys, La., northeast U.S. Spring. *A. bracteata* Northrop

3. HYMENOCÁLLIS Salisb. SPIDER LILY

Herbs with tunicate bulbs, strap-shaped leaves, and solid scapes. Flowers solitary or in umbelliform clusters; hypanthium elongate, sepals three, petals three, stamens six, a connate stipular membrane embracing the basal half of filaments in a delicate, funnel-like corona; anthers versatile. Ovary inferior, style long exserted, stigma capitate. Capsule turgid, seeds one or two in each locule. About 50 species of tropical and subtropical America.

1. Flower solitary, terminal; leaves 10 mm
    wide or less . . . . 1. *H. palmeri*
1. Flowers sixteen or less in terminal umbels;
    leaves 7 cm wide or less . . . . . 2. *H. latifolia*

1. H. palmeri Wats.—Spreading leaves about 4 dm long; blade channeled, dorsally angled. Scape compressed, dilated at apex with three linear floral bracts; flowers to 5 cm wide; fragrant; hypanthium tube 7–10.5 cm long; sepals and petals linear 6–8 cm long, channeled, style filamentous; capsule globose, with several obovoid seeds about 10 mm long. Endemic to prairies, glades, cypress swamps, south Fla. Summer.

2. H. latifolia (Mill.) Roem.—Scape 7–8 dm long, bluish gray, compressed, sharp edged. Leaves broadly linear, acute, flat, veins straight-ribbed, interrupted at intervals. Hypanthium slender, bluntly angled, to 16 cm long; corona 3.5 cm in diameter, shallow, prominently six-lobed; sepals and petals linear, elongate, surpassing the stamens and the stye. Capsule ovoid, 4.5 cm long. Endemic to south Fla. and Fla. Keys. Summer, fall. *H. keyensis* Small, *H. collieri* Small.

4. ZEPHYRÀNTHES Herb. RAIN LILY

Acaulescent herb with coated bulbs and smooth foliage. Flower solitary, bisexual, radially symmetrical, subtended by a tubular spathe; tepals united into six-parted tubular corolla; stamens six, adnate to the perianth below sinus level; anthers dorsifixed. Qvary inferior, three-carpellate with axile placentation; stigmas three, spreading. Capsule loculicidal. About 35 to 40 species, tropical and subtropical America, W.I. *Atamosco* Adans.

1. **Z. simpsonii** Chapm.—Scapose plants to 3 dm tall; scape compressed, edged at the base. Leaves 3–5 dm long, 2–3 mm wide, semiterete fleshy, obtuse. Flowers 7–8 cm long, involucral spathe as long as the tube, limb white tinged with lavender or purple; lobes spreading to ascending at tips. Mature capsule 15–18 mm in diameter, subglobose, depressed, three-lobed, usually with persistent style remnant. Endemic to openings in low pineland, glade borders, central and south Fla., rare. Spring.

5. CRÌNUM L. STRING LILY

Leaves from columnar bulb apex. Flowers white, fragrant, radially symmetrical, bisexual, in umbels, subtended by two spathelike bracts. Hypanthium 7–11 cm long, slender; perianth salverform, 8–15 cm wide; segments oblanceolate; stamens inserted at the throat, filaments capillary, declinate, anthers red, versatile. Ovary inferior, three-locular with two or more ovules in locule; style slender, stigma small, capitate. About 100 to 110 species of the tropics and subtropics on seacoasts.

1. **C. americanum** L. STRING LILY—Distinctive showy plants from tunicate bulbs. Leaves 6–12 dm long, 4–7 cm wide, membranous, denticulate, not narrowing toward the base, scape 3–8 dm tall, with umbels of five or more flowers. Capsule 4 cm in diameter, subglobose, three-lobed; typical testa and integuments in seeds replaced by corky covering. Glades and marshy borders of creeks, ponds, coastal plains, Fla., Tex. Summer.

48. DIOSCOREÀCEAE YAM FAMILY

Chiefly herbaceous twining vines, dioecious, springing from underground tubers. Leaves alternate or opposite. Flowers bilaterally sym-

metrical, unisexual, staminate perianth campanulate, six-parted with
equal, rounded or lanceolate lobes; anther locules contiguous, or united.
Pistillate perianth six-parted, staminodia minute; ovary three-carpellate, three-locular styles three or bifid; ovules two in each locule. Capsule triquetrous, winged, loculicidal. Five genera, about 750 species,
chiefly tropical, rarely temperate regions.

1. DIOSCORÈA L.   YAM

Stem annual, twining, from tubers often regarded as rhizomes or
roots of varying morphological origin. Leaves cordate, principal veins
palmate. About 600 species, tropics and subtropics, represented by
one species in our area.

Some species cultivated for edible tuber propagated by sprouting
bud in tubers.

1. D. bulbifera L. AIR YAM—Robust, aggressive, high-twining vine
known for its aerial tubers. Leaves to 20 cm long, 25 cm wide or larger,
lamina thin, lustrous above, satiny to touch, prominently reticulate,
with thirteen veins rounding toward margin, four confluent with the
excurrent midvein at the apex; petioles tenuous, of varying length,
twisted at base. Tubers leathery to touch, tough, small or large, numerous. Flowers not produced in our area. $2n = 40,80$. Probably native of
tropical Asia, grown as ornamental; the tubers may be used as food
after cooking; an unwanted plant in central and south Fla.

49. IRIDÀCEAE   IRIS FAMILY

Scapose or leafy stemmed perennials from corms, bulbs, rhizomes,
or fibrous roots. Flowers trimerous, bisexual, radially or bilaterally symmetrical; sepals three, petaloid, petals three, basally connate into a
tube; stamens three, adnate to sepals, anthers extrorse. Ovary inferior,
three-carpellate, usually three-locular; style distally three-divided; capsule three- to six-angled, many-seeded. Plants generally herbs, rarely
shrubs, about 60 genera, 800 species of temperate, tropical, and subtropical regions. *Ixiaceae*

1. Sepals and petals unlike in form; style
    branches dilated, petaloid . . . . . .   1. *Iris*
1. Sepals and petals similar in form; style
    branches narrow, not petaloid . . . .   2. *Sisyrinchium*

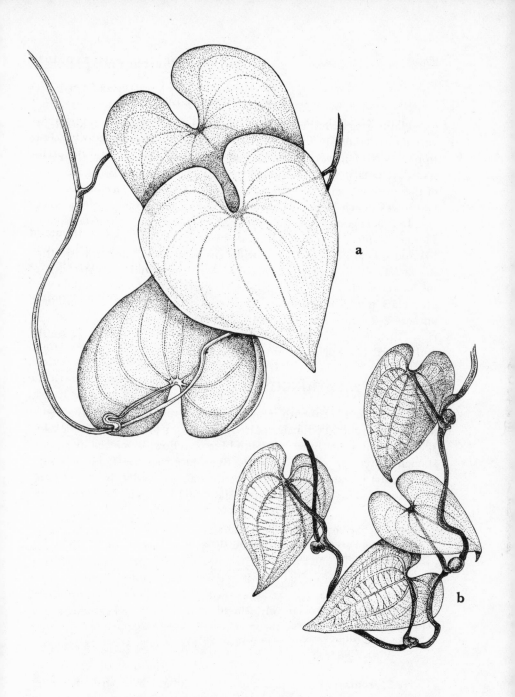

Family 48. **DIOSCOREACEAE.** *Dioscorea bulbifera:* a, mature leaves, dorsal side, × ½; b, mature leaves, ventral side, × ½.

1. ÌRIS L.   Fleur-de-Lis

Plants gregarious, often in extensive colonies. Flowers colorful, of varying shades of blue, white, or yellow, borne on peduncled spathes in upper axils of leaves; sepals clawed, recurving, larger than the erect spatulate petals; anthers situated beneath the stigmatic lip at the base of the terminal petaloid lobes of the style. Capsules coriaceous, seeds in two rows in each locule. About 300 species in north temperate regions.

1. **I. hexagona** Walt. var. **savannarum** (Small) Foster, Prairie Iris— Flowering stems 5–9 dm tall from fleshy sympodial rhizomes, with cord-like roots. Leaf blades 8–20 mm wide, ensiform, moderately stiff, sheathing at base, yellowish green. Sepals 7–8 cm long, oblanceolate with a median crest of yellow, variegated with white and blue toward the margins; claw greenish without; petals narrowly oblanceolate, 5–6 cm long, anthers 2–2.5 mm long, filaments shorter. Ovary in anthesis with six ribs. Capsule 4–5 cm long; seeds bouyant. Swamps, wet prairies, south and central Fla. Spring. *I. savannarum* Small

2. SISYRÍNCHIUM L.   Blue-Eyed Grass

Low, tufted herbs with compressed, wing-margined flowering stems and fibrous roots. Peduncles paired; flowers in umbels, subtended by spathe and an inner bract; perianth blue or yellow, six-parted, petaloid; stamens three, opposite the sepals. Style branches short, filiform surrounded by the anthers. Capsule loculicidal; seeds blackish, reticulate, glutinous or shiny before ripening. About 100 species in America.

1. Stem bases fibrillose from old leaf blades.
  2. Fibrils copious and long persistent; stems
    conspicuously winged.
    3. Leaves and spathes not rugose veined .   1. *S. arenicola*
    3. Leaves and spathes rugose veined . . 2. *S. solstitiale*
  2. Fibrils moderately conspicuous, tardily
    deciduous; stems narrowly winged at
    least along one edge . . . . . 3. *S. nashii*
1. Stem bases without persistent fibrils; scapes
  narrowly wing edged, stiff . . . . . 4. *S. atlanticum*

1. **S. arenicola** Bicknell—Plants 2–3 cm long; leaves 3.5–4 mm wide, green or often yellowish green at base, usually shorter than the stems. Spathes 8–10 mm long, scarious on margin; bracts subequal; perianth clear blue. Capsules 4 mm long; mature seeds globose 1–1.2 mm in diameter. Pine flatwoods, hammocks, Fla. to Miss. Spring.

Family 49. **IRIDACEAE.** *Iris hexagona* var. *savannarum:* a, mature plant, ×
⅙; b, flower, × ½; c, mature fruit and upper stem, × ½.

2. S. solstitiale Bicknell—Plants in large tufts to 7 dm tall from wiry brownish black roots. Leaf sheaths becoming fibrous, long persistent; mature blades rigid, linear-attenuate, 3–5 mm wide, often whitish on margins, nearly as long as the scapes, reclining. Scapes broadly winged, margins hyaline, minutely scabrous when young; spathes several, terminal, peduncled 1.5–2 cm long, margins hyaline, often purplish, perianth segments 3–4 mm wide, mucronate. Mature capsule black, subglobose; seeds black, reticulate 1.5 mm in diameter, spherical. Rosemary, sand pine, oak scrub, in white sand, central and south Fla., rarely coastal. Summer, fall.

3. S. nashii Bicknell—Stems to 4 dm tall, usually equaled by leaves, normally not blackened by drying. Spathes 10–15 mm long, green or purple; flowers and capsules exserted on capillary pedicels; perianth pale blue. Capsules drab, 3–4 mm long; immature seeds, glutinous, shiny. Pinelands, glades, sands, Fla. Spring, summer. *S. bicknellianum* Fern.; *S. floridanum* Bicknell

4. S. atlanticum Bicknell—Plants to 5 dm tall; tufted from short, hard rhizome with fibrous roots. Leaves 2–4 mm wide, about equaling the stems, pale green, lustrous brown at base on drying. Peduncles surpassing the subtending bract; spathes 15 mm long, subequal; perianth deep blue. Capsule 3–4 mm long, oblong, brown; seeds black, 1 mm in diameter. Glade margins, pine flatwoods, Fla. to Me. Summer.

50. MUSÀCEAE Banana Family

Treelike monoecious herbs; the aerial shoots rising from perennial stoloniferous rootstocks. Leaves spirally arranged, long petioled, with elongate, broad blade. Flowers unisexual by abortion, staminate flowers at the apex of the rachis above the pistillate flowers, perianth cycles differentiated into calyx and corolla; sepals mostly united, petals three or fewer, distinct or united, stamens five, polleniferous, accompanied by one petaloid staminode. Style simple, often three-lobed or more, ovary inferior, three-carpellate, ovules solitary or many. Fruit baccate, capsular or a berry, indehiscent or dehiscent. Two genera, 42 species, tropical Africa, Asia, Australia.

1. MÙSA L. Banana

Large herbs from perennial underground parts; stems commonly tall, chiefly consisting of elongate, tightly rolled, spirally arranged leaf sheaths developing successively; for each maturing blade at the apex another sheath is developing at the base. Blades oval-oblong, penninerved; midvein stout; margins entire, edged with white. Peduncle with a terminal inflorescence consisting of bracted spike appearing among leaf sheaths, basalmost flowers pistillate, the middle ones sometimes bisexual, the uppermost staminate with five fertile anthers, subtended by colorful bracts at the apex of elongating, drooping rachis. On the elongating pistillate portion of the rachis, flowers appear in clusters of five to seven or more, sepals with two anterior petals united into a tube, posterior petal free. Fruit a berry, three-locular, seedless in cultivated bananas. About 35 species, palaeotropical.

1. **M. sapientum** L.—Plant up to 15 m tall. Staminate flowers and bracts deciduous, mature fruit yellow, seedless, pollen grains granular. One of the most important foods of mankind, cultivated everywhere in the tropics and subtropics by propagation of the rhizome. Introduced from India, known in Fla. only on abandoned home sites and in other similar habitats. *M. paradisiaca* L. var. *sapientum* Kuntze

51. CANNÀCEAE CANNA FAMILY

Perennial caulescent herbs with simple stems arising from tuberiferous rhizomes. Leaves alternate, penninerved, with sheathing petioles. Inflorescence terminal, spicate, racemose or paniculate; flowers bisexual, asymmetrical, sepals three, free, persistent; petals three, partly united at base to the staminal tube, reflexed and deciduous; stamens five, the outer three modified into staminodia; blades dilated and colorful, of the two inner, one liplike, reflexed, the other fertile with marginal free anther, the filament adnate to the margin below. Ovary inferior, three-carpellate, placentation axile, each locule many-ovuled; style petaloid, opposite the anther-bearing segment; stigma marginal, embraced by dehiscing anther locule valves as seen in the convolute bud; pollen white. Surface cells of the ovary in bud green, bullate, becoming echinate flower and fruit. One genus, 55 species, tropical and subtropical America.

1. CÁNNA L. GOLDEN CANNA

Characters of the family

1. Flowers yellow, concolor . . . . . .   1. *C. flaccida*
1. Flowers red, varied with yellow and white .   2. *C.* × *generalis*

1. **C: flaccida** Small, BANDANA OF THE EVERGLADES—Plant up to 1.7 m tall, glabrous, green. Blades up to 6.5 dm long, 5–10 cm wide, elliptic,

Family 51. **CANNACEAE**. *Canna flaccida:* a, flowering branch and mature leaf, × ¾; b, basal leaves and rootstock, × ½; c, flower, × 1; d, fruiting raceme with mature capsules, × ½.

cuneate, margin entire, edged with white. Racemes simple or branching, flowers few, large, in anthesis up to 12 cm long, often in mirror-image pairs, bracted; sepals 2.5–3 cm long, erect, petals 4–6 cm long, becoming reflexed. Style petaloid, stigma dilated, marginal. Mature capsules 4–5 cm long, indehiscent, pericarp becoming denuded, disintegrating into a lattice of bundle fibers facilitating dispersal of seeds. Seeds 7 mm in diameter, spherical, black, sparsely pitted. Swamps and marshes, inundated pine flatwoods, southeast U.S., Fla., Ga., Miss., to N.C. Summer.

2. **C. X generalis** Bailey—Plants up to 1 m or more tall. Leaves 2–5 dm long, 6–15 cm wide with long-sheathing petioles. Flowers large, up to 15 cm long in spiciform racemes; staminodia in varying shades of red and yellow; bracts and floral parts often pruinose, sepals 2–2.5 cm long, petals to 7 cm long. Style bladelike with oblique stigma; anther white, 10 mm long. Young fruits dark brown, echinate; mature ones not seen. Gulf coast of Fla., escaped from gardens. Summer.

Believed to be of hybrid origin representing series of horticultural varieties, widely planted ornamental, naturalized from South American tropics. On shedding of pollen, the petaloid apex of the filament folds under and over the anther, covering it completely.

C. indica L. INDIAN SHOT—Plant as a whole smaller than the preceding ones. Raceme simple, flowers usually in pairs. Old favorite among garden ornamentals, cherished for its brilliant red orange variegated flowers and attractive foliage. Cultivated and escaped in south Fla., naturalized from India, Africa.

52. MARANTÀCEAE Arrowroot Family

Perennial herbs from underground rhizomes. Leaves alternate, chiefly basal. Flowers trimerous, bisexual, bilaterally symmetrical, spicate or paniculate, sepals three, free, petals three, free or united, unequal, stamens six, only half the anther fertile, the remaining reduced into petaloid staminodia. Ovary inferior, three-carpellate, three-locular, two-reduced; the remaining fertile, one-ovuled. Fruit fleshy or capsular, dehiscence irregular. About 30 genera and 400 species distributed throughout tropics and warm temperate regions.

1. THÀLIA L. Arrowroot

Tall herbs with long-petioled leaves. Flowers in open panicles, in pairs, mirror-image fashion, subtended by outer and inner bracts; sepals minute, corolla tube short or obsolescent; outer petaloid, staminodium

Family 52. **MARANTACEAE.** *Thalia geniculata:* a, flowering branch and mature leaf, × ½; b, stem base and rootstock × ½; c, flower, × 4; d, mature fruit with remnants of flower parts on the top, × 3; e, flower cluster, × 2.

one, dilated and liplike, larger than the petals, anther one-locular. Stigma two-lipped; fruit indehiscent. About 11 species, America, Africa.

1. T. geniculata L.—Plants up to 3 m tall with mostly basal leaves arising from strong rhizomes; internodes glabrous, angular; stem bases and sheaths soft with lacunar tissue. Petiole apex with a pulvinus of strongly indurated tissue confluent with the convex midvein, curving and subtending the blade at an angle of about 45°; blades to 5 dm long, 2 dm wide, ovate, asymmetrical; penninerved; margin entire, lamina parchmentlike to touch. Inflorescence branching and rebranching, lax, cernuous, rachises geniculate, hirtellous; flowers about 15 mm long, bluish purple, fruiting spathes to 22 mm long. Seed oblong-ovoid about 5 mm long, brown arillate; hypocotyl looped and coiled within horny endosperm emerging in germination through the ruptured, decaying pericarp. Marshy borders of swamps, lakes, and ditches, pine flatwoods, south Fla., W.I., Africa. Summer.

53. BURMANNIÀCEAE Burmannia Family

Small annual delicate herbs with alternate leaves. Flowers bisexual, radially symmetrical; sepals and petals similar, united into a three- to six-lobed perianth; stamens three to six, adnate to the perianth tube, sometimes dilated; anther connectives variable. Ovary three-carpellate, inferior, placentation axile or parietal, ovules numerous. Capsule dehiscence irregular, seeds small. Seventeen genera, 125 species, tropical, subtropical southeast U.S., Japan, Tasmania, New Zealand.

1. Flowers in racemes, ovary one-locular . .   1. *Apteria*
1. Flowers in capitate cymes, ovary three-
     locular . . . .   2. *Burmannia*

1. APTÉRIA Nutt. Nodding Nixie

Plants with simple or forking stems. Flowers purple white; stamens three, opposite inner perianth segments. Stigmas three, at even level with anthers. Withered perianth persistent over the capsule. Three species, warmer parts of America, W.I.

1. A. aphylla (Nutt.) Barnhart—Stems to 18 cm tall, capillary, rigid, purple above the soil. Scale leaves 1–3 mm long, purple. Perianth 10–12 mm long, white or purple; stamens three, filaments winged. Stigmas recurved; capsules 4–5 mm long, ovoid, dehiscent between placentae; seeds less than 0.5 mm long, light brown, striate. Hammocks in open shade south and central Fla., to Miss., W.I. Spring, fall. *A. setacea* Nutt.

2. BURMÁNNIA L.

Erect herb with filiform stems and threadlike rootlets. Leaves small, green, borne only in three or four nodes halfway up the stem. Flowers pedicelled, subtended by two or more bracts; perianth white, blue, or yellow, hypanthium winged or wingless; anthers three, nearly sessile, connective broad. Stigmas three, capsule three-locular, winged or wingless, overtopped by marcescent floral remnants. About 57 species, temperate, tropical, and subtropical regions.

1. Hypanthium wingless; flowers many in
    headlike clusters . . . . 1. *B. capitata*
1. Hypanthium winged; flowers two to three
    in cymose clusters.
    2. Wings broad, blue . . . . 2. *B. biflora*
    2. Wings yellow, narrow . . . . . . 3. *B. flava*

1. B. capitata (Walt.) Mart.—Stem simple 8–20 cm long. Lower cauline leaves 4–10 mm long, linear-attenuate, sheathing at base; rosette leaves apparently none. Floral bracts two or more, ovate, 3 mm long, flowers 2.5–4.5 mm long, white, tinged with blue, with sharply deltoid perianth lobes, the inner smaller than the outer; anther connective two-pointed, pollen shed by transverse clefts. Capsule 4 mm long, copper brown, seeds less than 0.5 mm long, crescentic, golden yellow; dehiscence by transverse rupturing of the pericarp. Moist soil of pond margins, dried-up water holes and ditches in pine flatwoods, central and south Fla., Tex. to Va., South America. Spring, fall.

2. B. biflora L.—Stem simple, 5–11 cm tall, rarely forking below inflorescence. Leaves very narrow, appressed 2–4 mm long. Floral bracts ovate, acute 3–4 mm long, flowers 4–6 mm long, pedicelled, two to five in forking cymes, clear blue, broadly winged above the ovary up to sinus level of the perianth. Fruiting specimens not seen. Pine flatwoods, moist soil around swampy spots, grassy, green terraces, central and south Fla., Tex., Miss. to Va., South America. Summer.

3. B. flava Mart.—Stem to 6 cm long or longer, slender, usually simple, erect, or sometimes forking. Inflorescence one-flowered or sometimes two- to nine-flowered; basal leaves linear, 7 mm long, 15 mm wide, obscurely three-veined; cauline .leaves scalelike, 2–4 mm long, floral bracts lanceolate, acute, 4–5 mm long. Flowers pale yellow, 9–11 mm long, outer perianth lobes erect 1–2 mm long, deltoid, the inner lobes to 2 mm long, spatulate, tube trigonous-cylindric, 3–4 mm long. Wings of perianth narrow, extending from the base of the limb to the base of

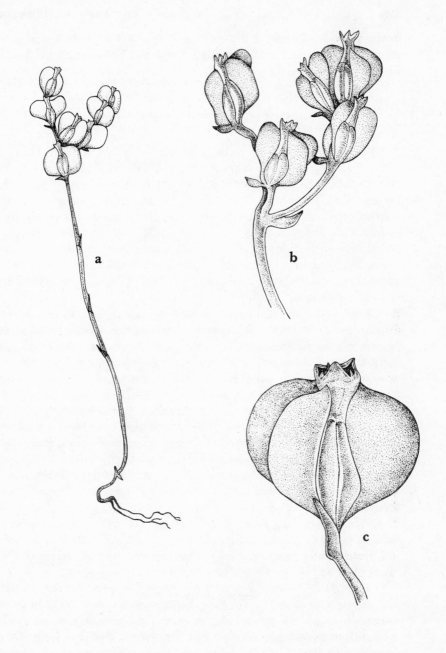

Family 53. **BURMANNIACEAE.** *Burmannia biflora:* a, mature plant, × 2;
b, inflorescence, × 4; c, flower, × 10.

ovary. Capsule obovoid 4–5 mm long, irregularly dehiscent, seeds numerous, minute, yellowish, slightly curved. Big Cypress Swamp, south Fla., Cuba, Guatemala.

54. ORCHIDÀCEAE ORCHID FAMILY

Perennial herbs with erect, prostrate, or scandent stems. Leaves simple, alternate, rarely opposite or whorled, basal sheaths usually closed around the stem. Inflorescence spicate, racemose, or paniculate; flowers bracted, bisexual, rarely unisexual, bilaterally symmetrical, trimerous, sepals three, petaloid in color, otherwise unlike the petals, two commonly connate usually below the middle; petals two, alike, lip (third petal) very different in form and color; stamens one or two, rarely more. Pistil one, three-carpellate, united with stamens to form the column, ovary with three parietal placentae. Fruit a capsule; seeds (ovules) countless with undifferentiated embryos.

The lip and the column constitute diagnostic features of orchids. The lip is the uppermost petal; such a flower is nonresupinate. In some orchids the lip is the lowermost petal, patently effected by twisting of the flower bud through a half-circle just before opening. The column, an extention of the floral axis, is formed by the union of the filaments and the style. The apical cavity on the inner surface lodges the anther, or, if two anthers, the attachment is marginal. Just below the anther is the stigmatic area, with lines of fusion of two or three stigmas. In orchids with one anther, the third stigma is modified into a rostellum, a beak facilitating pollination. The pollen, waxy or granular, in masses known as pollinia, is exposed or concealed in operculate clinandrium. A cosmopolitan family of more than 15,000 species, from tropical rain forests to temperate regions.

1. Plants rooted in soil, terrestrial.
  2. Leaves present in anthesis.
    3. Lip produced into a prominent spur.
      4. Spur slender; pollinia caudate with disk, exposed; sepals usually bipartite, the dorsal free, incurved cucullate . . . . 1. *Habenaria*
      4. Spur saccate; pollinia concealed in a clinandrium; lateral sepals simple, the dorsal connivent with petals, helmet fashion . . . . . . 2. *Erythrodes*
    3. Lip spurless.
      5. Leaves plicate, fleshy or membranous.

6. Lip the uppermost petal (flowers nonresupinate).
  7. Stems from short rhizome with fleshy, fasciculate, tuberous roots.
    8. Sepals free, or the lateral sometimes connate.
      9. Raceme lax, few-flowered; lip 5.5 mm wide, adnate to the column above the base; cauline bracts inconspicuous . . . .   3. *Ponthieva*
      9. Spike erect, many-flowered; lip 1.5 mm wide, adnate to the column at base; cauline bracts conspicuous . . . . . .   4. *Cranichis*
    8. Sepals united into a sepaline cup; spike slender, lip 2 mm wide . . . .   5. *Prescottia*
  7. Stems from corms; roots fibrous.
    10. Leaf basal; lip bearded; flowers showy, rose purple . .   6. *Calopogon*
    10. Leaves near midstem; lip glabrous; flowers yellow . .   7. *Malaxis*
6. Lip the lowermost petal (flowers resupinate).
  11. Inflorescence a spike.
    12. Rachis spirally twisted; lip base with two calli; stems from short rhizome, roots fasciculate, often strongly thickened . . . . .   8. *Spiranthes*
    12. Rachis straight; lip without calli; stem pseudobulbous with fibrous roots . . .   9. *Liparis*
  11. Inflorescence a panicle; stem leafy, branching from lanate roots . . . . . .   10. *Tropidia*
5. Leaves duplicate, thin or membranous.
  13. Plant caulescent, from fibrous and tuberous roots.
    14. Leaf sheaths inflated; blades loosely investing the stem; racemes spiciform, many-flowered . . . .   11. *Zeuxine*
    14. Leaf sheaths tight; blades cordate, clasping the stem; racemes lax, few-flowered . .   12. *Triphora*

13. Plant scapose, from thickened, tuberous roots.
  15. Leaf usually one, petioled; sepals united into spurlike mentum . . . . 13. *Centrogenium*
  15. Leaf one, sessile; sepals free, without mentum . . . . 14. *Basiphyllaea*
2. Leaves absent during anthesis.
  16. Raceme terminal.
    17. Stems pseudobulbous; roots fibrous.
      18. Lip arising from the column foot; lateral sepals connate with the lip, forming a mentum; flowers greenish white . . . . . 15. *Govenia*
      18. Lip arising from the base of the column, forming a spur; sepals free; flowers purple . . . . 16. *Galeandra*
    17. Stems rhizomatous; roots fasciculate, thickened; flower scarlet . . . 8. *Spiranthes*
  16. Raceme lateral from corms, distinct from leaf-bearing shoots.
    19. Lip articulated with the column foot; leaves sheathing the base, not petioled . . . . 17. *Eulophia*
    19. Lip articulated with base of column; leaves sheathing the corm, petioled 18. *Bletia*
1. Plants rooted on trees, rocks, not in soil (epiphytic).
  20. Plant scandent, viny; lip enfolding the column; capsule fleshy . . . . . 19. *Vanilla*
  20. Plants not viny.
    21. Plants essentially stemless and leafless; roots chlorophyllous.
      22. Lip the uppermost segment in a flower, flowers small.
        23. Inflorescence spicate; floral bracts prominent; lip with a saccate spur . . . . 20. *Campylocentrum*
        23. Inflorescence racemose; floral bracts obsolete; lip with spur, constricted above the base . . 21. *Harrisella*
      22. Lip the lowermost segment in a flower, flower solitary successively; lip with a filiform elongate spur; flower large . . . . 22. *Polyrrhiza*
    21. Plants with stems and leaves; roots not chlorophyllous.
      24. Peduncle terminal.
        25. Lip the uppermost segment in a flower; leaves duplicate.

26. Stems pseudobulbous; floral
    bracts vestigial . . . . . 23. *Polystachya*
26. Stems epseudobulbous; floral
    bracts manifest . . . . . 24. *Epidendrum*
25. Lip the lowermost segment in a
    flower.
  27. Leaf terminal, secondary stems
      from creeping rhizome.
    28. Cauline sheaths prominently
        funnel shaped . . . . 25. *Lepanthopsis*
    28. Cauline sheaths not funnel
        shaped; barely dilated
        above the base . . . . 26. *Pleurothallis*
  27. Leaves two or more on primary
      stems.
    29. Stems pseudobulbous; column
        partly adnate to the lip . . 27. *Encyclia*
    29. Stems epseudobulbous; col-
        umn wholly adnate to the
        lip . . . . 24. *Epidendrum*
24. Peduncle lateral.
  30. Inflorescence exceeding the leaves.
    31. Pseudobulb fusiform; hornlike;
        articulate; lip three-lobed or
        more, attached to the column
        foot . . . . 28. *Cyrtopodium*
    31. Pseudobulb pyramidal, small,
        sometimes compressed, not ar-
        ticulate.
      32. Flowers in a terminal spike;
          rachis slender or thickened;
          lip free, attached to column
          foot . . . . . . 29. *Bulbophyllum*
      32. Flowers in racemes or pani-
          cles.
        33. Sepals long caudate; flowers
            large and showy in a dis-
            tichous raceme . . . 30. *Brassia*
        33. Sepals ecaudate; flowers in
            panicles.
          34. Leaves from rhizomes; lip
              obcordate, with incon-
              spicuous calli; column
              wingless . . . . . 31. *Ionopsis*
          34. Leaves from the apex of
              pseudobulb; lip three-
              to five-lobed; disk with
              prominent calli; col-
              umn winged . . . 32. *Oncidium*

30. Inflorescence surpassed by leaves;
     pseudobulb one- to two-leaved;
     pollinia stiped.
     35. Peduncle erect; flower solitary;
         column foot short; pollinia
         without disk . . . . .   33. *Maxillaria*
     35. Peduncle pendent; flowers in
         racemes; column footless; pol-
         linia two with viscid disk . .   34. *Macradenia*

1. HABENÀRIA Willd.

Terrestrial herbs with thickened fleshy roots, or tubers, stem simple, erect, usually glabrous. Leaves cauline or basal, sheathing the internodes. Inflorescence raceme or a spike of white, greenish yellow, or orange red flowers; sepals free, the dorsal cucullate, forming a hood over the column with the converging petals; petals simple or bipartite, the anterior segment filiform, or essentially entire, lip tripartite, three-lobed, the divisions often fringed or toothed or entire, continuous with the column and produced to a spur, column footless; anther locules distant, pollinia two, granular, long stalked, each with a naked gland. Stigma appendages papillose or smooth. Capsule slender, ellipsoid. Polymorphic genus of some 600 species in temperate and tropical regions of both hemispheres. *Gymnadeniopsis* Rydberg; *Habenella* Small

1. Lip tripartite.
   2. Leaves basal, lateral lobes of lips upward
       arching . . . .   5. *H. distans*
   2. Leaves cauline, lateral lobes of lips pendent.
     3. Spur more than 4 cm long; stem base
        on ovoid tuber . . . .   1. *H. quinqueseta*
     3. Spur less than 4 cm long; stem base
        rhizomatous . . . .   2. *H. repens*
1. Lip entire.
   4. Leaves linear, elongate, few; raceme 5–15
       cm long; lip uppermost in flower . .   3. *H. nivea*
   4. Leaves lanceolate, several; raceme to 3.8
       dm long; lip lowermost in flower . .   4. *H. strictissima*

1. H. quinqueseta (Michx.) A. A. Eaton, LONG-HORNED ORCHID—
Herbs to 6 dm tall with leafy stems; roots fibrous. Leaves 5–8 cm long, ovate-lanceolate, acute, rounded to clasping bases, passing to floral bracts. Racemes to 15 cm long, few- to many-flowered, flowers white in full anthesis, 2–3 cm long, with the pedicellate ovary, spur clavate,

pendent, 5–10 cm long. Capsule with marcescent floral remnants. Hammocks, pinelands, Fla., Tex., S.C., Cuba, W.I., Central and South America. Fall. *H. michauxii* Nutt., *H. simpsonii* Small

2. **H. repens** Nutt. Creeping Orchid—Stem 1–9 dm tall, leafy, relatively stout. Leaves to 20 cm long, linear-lanceolate, with three primary veins, acute to acuminate. Racemes 8–20 cm long, densely flowered, conspicuously bracted; flowers greenish white, nearly erect, dorsal sepal ovate, short apiculate; the filiform anterior lobe of petals about equaling the falcate posterior lobe, lip 4–5 mm long, lateral lobes pendulous; spur to 10 mm long, incurved. Capsule 10 mm long. Seeds 1 mm long, curved at the ends. Swamps, wet ditch beds, south Fla., Tex., W.I., South America. Summer. *H. nuttallii* Small

3. **H. nivea** (Nutt.) Spreng. Bog Torch—Plant to 6 dm tall; the stem slender, from clustered tuberoid roots. Leaves two to three, linear-elongate. Raceme conical, crowded with snow-white flowers. Perianth 4–7 mm long; lateral sepals dilated, rounded, auriculate, upper sepal oblong, petals simple, falcate, spreading denticulate at apex, lip upcurving, spur slender flexuous, longer than the pedicellate ovary. Capsule to 12 mm long, straight or curving, six-ribbed. Glades, margins of hammocks, Fla., Tex., N.J. Spring, summer. *Gymnadeniopsis nivea* (Nutt.) Rydberg

4. **H. strictissima** Reichenb. f. var. **odontopetala** (Reichenb. f.) L. O. Wms.—Stem to 6 dm tall, stoutish, erect from fibrous roots often with tuberoids. Leaves to 15 cm long, 4 cm wide, lanceolate, elliptic. Racemes cylindric, open; flowers white, bracted with pedicellate ovaries to 25 mm long. Perianth 4–6 mm long; dorsal sepal cucullate; petals simple, oblong, auriculate on anterior margin; lip narrowly spatulate, pendent; stigmatic processes conspicuous, papillose, spur slender to 2.6 cm long. Mature capsule to 15 mm long; seeds 1.5 mm long, caudate, of hair thickness. Cypress margins, river banks, Sabal Palmetto hammocks, south Fla., Mexico, Central America. Fall, winter. *Habenella garberi* (Porter) Small, *Habenaria odontopetala* (Reichenb. f.) Small

5. **H. distans** Griseb.—Plant to 3 dm tall, cauline leaves reduced to bracts; basal leaves 4–9 cm long, 2–5 cm wide, acute, sheathing at base. Raceme narrow, loosely flowered, floral bracts lanceolate to 2 cm long; flowers greenish with pedicellate ovaries 1–15 mm long; petals bipartite, anterior lobe narrow, filiform, posterior linear, falcate, 8 mm long; midlobe of lip about 9 mm long, narrow; spur becoming clavate, 15 mm long; capsule 10–15 mm long, ellipsoid, strongly ribbed. Infrequent in hammocks, shady slopes, Fla., Mexico, Central America.

2. ERYTHRÒDES Blume

Caulescent terrestrial from fibrous roots. Stem simple, erect above, decumbent, repent at base with nodal roots. Leaves three-ranked, reticulate. Inflorescence spicate; perianth galeate; lip three-lobed, didymous, produced into a saccate spur; column short, pollinia two, granular. About 100 species of tropical and subtropical regions of both hemispheres. *Physurus* Richard

1. **E. querceticola** (Lindl.) Ames—Plant 1–20 cm long, glabrous throughout. Leaves with petioles 2.5–7 cm long; blades elliptic-lanceolate, thin, acute, narrowed to a loosely sheathing petiole. Spike dense or loose, 3–6 cm long; floral bracts ovate, long acuminate, shorter than the flowers; flowers white, 10–12 mm long; lateral sepals free, dorsal sepal adherent with petals; lip with spur to 7 mm long, spur prominently three-veined. Capsule ellipsoid 7 mm long, strongly ribbed. Moist woods and hammocks, central and south Fla., Tex., Mexico, Central and South America, W.I. Summer. *P. querceticola* Lindl.

3. PONTHIÈVA R. Brown

Terrestrial herbs with radical leaves from strong lanuginous roots. Scape simple, naked above, the lower cauline sheaths with dilated blades. Leaves petioled, blades reticulate veined. Racemes lax, flowers distant; sepals free, petals and the lip rising from the column; rostellum erect, anther broad, with four pollinia and granular pollen. About 60 species in tropical and subtropical America.

1. **P. racemosa** (Walt.) Mohr, SHADOW WITCH—Plants 2–5 dm tall, glandular pubescent above, the lower internodes and leaves essentially glabrous. Leaf blades chiefly oblong-elliptic or oblanceolate, abruptly narrowed at apex. Raceme 8–15 cm long, open pedicellate flowers 1.5–2 cm long, distant dorsal sepal oblong, rounded, lateral, ovate-oblique; corolla white, the lip 3–4 mm long, striped with green, terminal lobe apiculate, minutely papillose, limb dilated, saccate, outward folded, petals about 6 mm long, clawed, semicircular side by side, minutely ciliolate. Capsule ellipsoid to 12 mm long. Wooded lands with oolitic rocks, Fla., Tex., Ga., Va., W.I., South America. Fall, spring. *P. glandulosa* (Sims) R. Brown, *P. brittonae* Ames

4. CRÁNICHIS Sw.

Terrestrial herb from fleshy, fasciculate roots. Leaves radical. Scape with tubular bracts, often with reduced blades. Inflorescence spicate, floral bracts lance-attenuate, sepals unequal, connivent, petals free, lip erect, concave, column short; pollinia four, pollen granular. Capsule ellipsoid. About 34 species, hammocks, America, W.I.

1. C. muscosa Sw.—Plant to 3 dm tall, glabrous. Leaves petioled 5–15 cm long; blades elliptic-ovate, slightly inequilateral, reticulate veined. Spike 3–7 cm long; flowers white, 10 mm long; lip cymbiform with everted lateral folds, crested with three tuberculate, parallel veins to below the three-angled apex; column four-winged; rostellum pointed; anthers stalked. Capsule about 10 mm long. Hammocks, Fla., tropical America, W.I. Winter.

5. PRESCÓTTIA Lindl.

Terrestrial scapose herb from fasciculate, thickened roots. Leaves radical. Inflorescence spicate; sepals connate at base, cuplike; petals with the clawed lip and column adnate to the sepaline cup; pollinia four, granular; capsule nearly erect, ovoid. Tropical and subtropical genus of 22 species, Fla., Central America and tropical South America, W.I.

1. P. oligantha (Sw.) Lindl.—Stems to 3 dm tall with three or more leaves passing to progressively smaller, cauline bracts. Blades commonly oval, acute, 2–7 cm long, petiolate. Spike 3–7 cm long, densely flowered; flowers less than 3 mm long, white, pinkish, or greenish, perianth segments adherent at base; sepals ovate, recurved; lip saccate, three-nerved; column winged. Capsule six-keeled, less than 5 mm long. Hammocks, south Fla., Mexico, Central America, to Panama, W.I. and South America. Winter.

6. CALOPÒGON R. Brown GRASS PINK

Scapose terrestrial from corms. Flowers few to several in terminal raceme, showy rose magenta or white, sepals and petals free, lip uppermost, bearded, column free, anther terminal, operculate, pollinia four. Capsule erect. New World genus of four species, mostly in U.S., Canada, Cuba, W.I. *Limodorum* L.

1. Flowers 3–5 cm wide; anthesis sequential from the base toward the apex; lateral sepals falcate or obliquely obovate.
  2. Perianth rose magenta; leaf blades 10 mm wide . . . . 1. *C. pulchellus* var. *simpsonii*
  2. Perianth white or nearly so; blades 5 mm wide . . . . 2. *C. pallidus*
1. Flowers 3 cm wide or less; anthesis simul-

taneous; lateral sepals symmetrical or
inequilateral, perianth magenta.
3. Rachis short; anthesis simultaneous . . 3. *C. barbatus*
3. Rachis elongate; anthesis appearing sequential . . . . 4. *C. multiflorus*

1. **C. pulchellus** (Salisb.) R. Brown var. **simpsonii** (Small) Ames,
GRASS PINK—Plant to 9 dm tall; leaf solitary, linear to 4 dm long, less
than 8 mm wide, long acuminate, strongly ribbed, developing the first
year of growth, followed by a flowering scape the second year. Raceme
lax, few-flowered; floral bract lanceolate, acute, spreading, shorter than
the ovaries, flowers with pedicel to 3 cm long, sepals and petals free-
spreading, lateral sepals inequilateral, half-orbicular, obscurely three-
lobed; lateral lobes minute, middle lobe broadly emarginate, disk
bearded with clavate, multicolored hairs, or the flower wholly white in
f. *albiflorus* (Britt.) Fern.; column arcuate; apical wing to 9 mm wide.
Capsule 1.5–2 cm long. Pine flatwoods, south Fla., W.I. Spring. *Limodorum simpsonii* (Chapm.) Small
2. **C. pallidus** Chapm. PALE GRASS PINK—Plant to 7 dm tall, from
elongate corms. Leaf narrowly linear, long attenuate. Raceme flexuous
and lax; flowers magenta, to 1.5 cm long; sepals and petals subequal, to
14–20 mm long; lip 8–12 mm long, lateral lobes obsolescent; middle
lobe obcordate, bearded with clavellate hairs. Column to 9 mm long,
wing at apex 8 mm wide. Flat pinelands, Fla., La., Va. Spring. *L. pal-
lidum* (Chapm.) Mohr
3. **C. barbatus** (Walt.) Ames—Plant 2–3.5 dm tall from rounded
corms. Leaves narrowly linear, grasslike, to 1.5 dm long. Raceme flexuous, usually four- to five-flowered; flowers madder purple, opening simultaneously or nearly so, lip obscurely three-lobed, lateral lobes small
above the base, middle lobe suborbicular, disk bearded with clavate
hairs; column wing semicircular. Capsule 10 mm long, beaked. Prairie,
pineland, Fla., Ga., Ala., Miss., N.C. Spring. *L. parviflorum* (Lindl.)
Nash
4. **C. multiflorus** Lindl. MANY-FLOWERED GRASS PINK—Plant
slender, 3–4 dm tall from elongate horizontal corm. Leaves one to two,
narrowly linear, to 2.5 dm long. Inflorescence elongate, bearing several
(fifteen) bracted flowers opening more or less simultaneously in sequential groups. Flowers madder purple, lip cuneate, lateral lobes thin, winglike, middle lobe flabellate, nearly truncate at apex, disk densely
bearded with clavate hairs. Capsule 10–13 mm long. Prairie, pinelands,
Fla., Ga., Miss., Ala. Spring. *L. multiflorum* (Lindl.) Mohr, *L. pinetorum* Small

7. MALÁXIS Solander

Small terrestrial herbs from cormlike base. Leaves one to five, commonly near midstem. Inflorescence terminal, sepals and petals spreading, pollen waxy in four stalkless masses, two in each locule. Capsule ovoid-ellipsoid. About 300 species, cosmopolitan, attaining the greatest development in Asia and Oceania. Wide distribution in the Western Hemisphere. *Microstylis* (Nutt.) A. A. Eaton

1. M. spicata Sw.—Plant 3.5–4.5 dm tall, slender, slightly fleshy, stem base commonly invested with tubular sheaths. Leaves two, subopposite, the lower narrowed to a petiole; blades ovate or elliptic, reticulate veined. Racemes 10–20 cm long; many-flowered, floral bracts 4–5 mm long, lanceolate, persistent with pedicels. Flowers small with filiform pedicels to 8 mm long; sepals ovate or elliptic, greenish yellow, like the filiform recurved petals; lip entire, broadly cordate-ovate, auriculate, saccate, margins recurved; column 1 mm long, two-toothed at apex. Capsule 7 mm long, ellipsoid. Hammocks, pine flatwoods, coastal plain, Fla. to Va., W.I. Summer. *Malaxis floridana* (Chapm.) Kuntze, *Microstylis spicata* (Sw.) Lindl.

8. SPIRÁNTHES Richard Ladies' Tresses

Terrestrial herbs, roots fasciculate, fibrous or tuberous. Leaves cauline or radical or sometimes lacking during anthesis; blades suborbicular, ovate-lanceolate, elliptic or linear, sheathing at the nodes. Flowers secund, usually in spirally twisted spikes; sepals free, the dorsal adherent to the petals, the lateral decurrent on the ovary, forming a mentum. Column essentially footless; pollinia two, pollen granular, rostellum forked. Capsule erect. Cosmopolitan genus of 25 species of temperate to tropical regions. *Ibidium* Salisb., *Cyclopogon* Presl., *Mesadenus* Schlecht., *Stenorrhynchus* Richard, *Pelexia* Poit.

1. Leaves absent during anthesis; flower to 3
  cm long, scarlet . . . . 1. *S. orchioides*
1. Leaves present at least in early anthesis.
  2. Blades dilated, narrowed to petioles.
    3. Flowers green or white.
      4. Flowers greenish gray, lip lanceolate;
        callosities minute . . . . . 2. *S. polyantha*
      4. Flowers white, disk green.
        5. Perianth tubular, lip oblong, narrowed to deltoid apex; callosities terete, obtuse, 1 mm long . 3. *S. costaricensis*

5. Perianth campanulate, lip oblong,
crenulate; callosities conical . 4. *S. gracilis*
var. *brevilabris*
3. Flowers variegated with green, brown,
or purple.
6. Sepals maculate with madder purple;
lip purplish; leaves equilateral . 5. *S. cranichoides*
6. Sepals green; immaculate; lip brown-
ish green; leaves inequilateral . . 6. *S. elata*
2. Blades elongate, linear or linear-oblanceolate, epetiolate.
7. Spikes secund, rachis straight . . . 7. *S. longilabris*
7. Spikes spirally twisted.
8. Spirals in one to three rows; flowers
nodding . . . . 8. *S. cernua*
var. *odorata*

8. Spirals in one row.
9. Leaves linear, basal.
10. Blades filiform, terete; lip to 6
mm long, yellowish; callosities
mammillate, glabrous . . . 9. *S. tortilis*
10. Blades linear, 1–5 mm wide; lip
to 11 mm long, green veined;
callosities ciliate . . . . 10. *S. praecox*
9. Leaves basal and cauline; sheath-
ing at base.
11. Rachis and flowers copiously
pubescent with pointed hairs;
lip undulate . . . . . 11. *S. vernalis*
11. Rachis and flowers moderately
pubescent with capitate hairs;
lip irregularly lacerate . . 12. *S. laciniata*

1. **S. orchioides** (Sw.) A. Rich. SCARLET LADIES' TRESSES—Plant to 6 dm tall from fleshy, thickened roots viscid pubescent with gland-tipped hairs or glabrate in age. Leaves radical, appearing in postanthesis; blades 1–4 dm long, 2–5 cm wide, oblong-oblanceolate or lanceolate. Inflorescence 8–15 cm long, spicate; floral bracts 12–14 mm long, lance-attenuate, often ciliolate, flower 28–35 mm long, deep crimson; dorsal sepal erect, lanceolate, acuminate, adherent to the petals; lateral sepals forming a mentum, lip saccate, entire, ovate-lanceolate, 15–20 mm long, basal calli linear; disk pubescent; rostellum awnlike, column 10 mm long. Capsule ellipsoid, about 15 mm long. Low hammocks, prairie, pinelands, Fla., Mexico, Central and South America, W.I., Bahama Islands. Summer. *Stenorrhynchus orchioides* (Sw.) Richard, *Spiranthes jaliscanus* Wats.

2. **S. polyantha** Reichenb. f.—Plant 1.5–5 dm tall from fleshy, thick-

ened, fasciculate roots; stems slender, pubescent through the rachis, glabrous below. Radical leaves spreading, petioled. Inflorescence to 3 dm long, flexuous, flowers to 10 mm long, greenish purplish, sepals and petals linear, adherent to the dorsal sepal; lip broadly elliptic, prominently veined, short clawed; disk green, basal calli slender; column 2.5 mm long. Capsule 5.5–6 mm long, sessile, ellipsoid. Hammocks, south Fla. (not recently collected), Mexico, Guatemala, Bahama Islands, Puerto Rico. Summer. *Ibidium lucayanum* Britt., *Mesadenus lucayanus* (Britt.) Schltr.

3. **S. costaricensis** Reichenb. f.—Plant to 4.9 dm tall, glandular-pubescent above the base. Leaves 7–8 cm long, ovate-elliptic, narrowed to broad dilated petioles. Inflorescence lax, spicate, floral bracts conspicuous, lance-attenuate, flowers 12–15 mm long, greenish white; perianth parts with spreading, reflexed tips above the constricted throat; lip decurved, clawed, with two dorsally curved calli at base, narrowed abruptly to acute apex. Capsule ovoid about 9 mm long. Hammocks, south Fla., widespread in W.I., Mexico, Central America; rare. *Gyrostachys costaricensis* Kuntze

4. **S. gracilis** (Bigel.) Beck var. **brevilabris** (Lindl.) Correll—Plant to 1.8 dm tall, slender scapose from thickened, fasciculate roots; stems slender. Leaves basal with petioles 4 cm long, persistent. Inflorescence a simple spiral of flowers in a single row; sepals and petals white, glandular-puberulent without; lip quadrate-oblong, lacerate on margins, short clawed; calli two, mammillate, disk clearly veined with green at least in opening flowers, distally white villous; column 2–3 mm long. Capsule 4–5 mm long. Low, flat pinelands, central and south Fla., Ga., La., Tex. Spring. *Spiranthes brevilabris* Lindl.

5. **S. cranichoides** (Griseb.) Cogn.—Plant to 4.8 dm tall from tuberous, fasciculate roots. Scape, sheaths, and bracts spotted with white; glandular-puberulent above the middle. Leaves with petioles to 8 cm long; blades elliptic, inequilateral, brownish green, minutely punctate; spike with spreading pedicellate flowers 10–12 mm long; lip three-lobed, white, the lateral lobes upward curving, flanking the column; basal calli marginal, erect; column 3–4 mm long. Capsule ovoid, oblique, 8–9 mm long. Hammocks central and south Fla., Central America, British Honduras, W.I. Summer. *Pelexia cranichoides* Griseb., *Cyclopogon cranichoides* (Griseb.) Schltr.

6. **S. elata** (Sw.) Richard—Plant to 6 dm long, glandular-viscid above, glabrous below. Scape yellowish green or purplish, bracted. Leaves basal, petioles longer or shorter than the blades; blades 8–9 cm long, ovate-elliptic, thin, prominently reticulate veined. Floral bracts lance-attenuate; flowers 6–8 mm long, nodding, brownish green in a

loose spiral spike to 2 cm long or longer; lip short clawed, oblong, abruptly narrowed to an isthmus and dilated to a flabellate crenulate apex; column 2–3 mm long, rostellum pointed. Capsule 7–10 mm long. Hammocks, Fla., W.I., Central and South America, Mexico. Summer. *Cyclopogon elatus* (Sw.) Schltr.

7. **S. longilabris** Lindl.—Plant to 6 dm tall from fleshy, fasciculate roots; stem and cauline bracts essentially glabrous below, glandular-pubescent above. Leaves 3–10 cm long, basal, linear, narrowly lanceolate, dilated at base. Spike 5–10 cm long, secund or slightly spiraled. Floral bracts clasping, ovate, long attenuate. Flowers to 10 mm long, pedicellate, ringent, white; dorsal sepal erect, lateral sepals and petals spreading; lip to 10 mm long, arcuate, oblong-obovate with a short claw, two mammillate basal calli, narrowed toward apex with crenate, undulate margins; column 3–4 mm long, capsule about 5 mm long. Hammocks, Fla., Tex., La., Va. Spring. *S. brevifolia* Chapm., *Ibidium longilabris* (Lindl.) House

8. **S. cernua** (L.) Richard var. **odorata** (Nutt.) Correll—Plant erect to 1 m tall with cordlike roots giving rise to apical buds. Leaves deep green, membranous, to 4 dm long, 3–4 cm wide, broadly linear, acute, obtuse, narrowing to broad petiolar base diminishing to floral bracts. Spike to 1.8 dm long, to 3 cm in diameter in full anthesis. Floral bracts long acuminate, flowers 8–16 mm long, tubular, ringent, white, suffused with green; capitellate-pubescent, sepals to 12 mm long, acute acuminate, the lateral free; petals adherent with dorsal sepal, lip 6–14 mm long, obovate, striped with green, margin transparent, crisped, recurved, calli prominent, pubescent, retrorsely oriented; column to 5 mm long. $2n = 50$. Cypress swamps, central and south Fla., Tex., Me. Fall. *S. odorata* (Nutt.) Lindl.

9. **S. tortilis** (Sw.) Richard—Plants slender to 7 dm tall from fasciculate roots; stems glabrous below, glandular-puberulent above. Leaves to 2 dm long, basal, narrowly linear, sheathing at base; floral bracts ovate-acuminate; lip 4–5 mm long, ovate-quadrate, arcuate; claw short; basal calli prominently mammillate, disk yellowish green; column short. Capsule 4–5 mm long. Hammocks, south Fla., Central America, Bahama Islands, W.I. Summer. *Ibidium tortile* (Sw.) House

10. **S. praecox** (Walt.) Wats. GRASS-LEAVED LADIES' TRESSES—Plant to 7 dm tall from fasciculate, thick roots. Stems glabrous below, sparingly pubescent above; leaves 1–2 dm long, essentially basal, linear, acuminate, usually present in anthesis. Spike 3–12 cm long, densely flowered, spiral; floral bracts lanceolate-acuminate; flowers 10–12 mm long, white, the lip variegated with green on the veins; lateral sepals falcate, free, dorsal sepal oblong-elliptic, adherent to the petals; lip oblong, clawed, basal calli erect, compressed; column 3–4 mm long. Cap-

sule 8 mm long, obliquely ovoid. Moist soil, hammocks, prairie, canal banks, south Fla., Tex., Ark., N.J. Summer. *Ibidium praecox* (Walt.) House

11. **S. vernalis** Engelm. & Gray—Plant 1.5–7 dm long from elongate fasciculate roots. Stem densely pubescent above, especially through .the inflorescence. Leaves 1–3 dm long to 5 mm wide, basal and cauline, blades narrowly lanceolate. Inflorescence a dense spike, 8–25 cm long; floral bracts ovate or cuminate; flowers 8–10 mm long, yellowish white, fragment; perianth segments puberulent without. Ovary densely pubescent with viscid, articulate hairs; dorsal sepal with adherent petals protruding hood fashion over the arcuate lip and spreading lateral sepals; lip obovate, short clawed; broadest below the conical marginal calli, narrowed to a rounded, lacerate-sinuous apex. Low prairie, marshes, south Fla., Tex., N.J. Spring. *Ibidium vernale* (Engelm. & Gray) House

12. **S. laciniata** (Small) Ames, LACE-LIP SPIRAL ORCHID—Plant to 9 dm tall from elongate fasciculate roots; stem commonly glabrous below, sparingly to copiously pubescent above with capitate hairs. Leaves 9–30 cm long, basal and cauline, lanceolate or oblanceolate. Spike 5–15 cm long, strongly spiraled; floral bracts ovate-acuminate or lanceolate; flowers to 10 mm long, white, fragrant; perianth ringent, puberulent without; sepals and petals minutely denticulate; lateral sepals free; lip obovate-oblong, strongly arcuate, margins translucent, finely laciniate; basal calli curved, about as long as the claw of the lip; column 3–4 mm long. Capsule 6–7 mm long, obliquely ovoid. Glades, wet pinelands, south Fla., Tex., N.J. Spring. *Ibidium laciniatum* (Small) House

9. LÍPARIS Richard

Low herbs with pseudobulbs. Leaves plicate, basal. Raceme terminal, flowers remote, sepals and petals free, lip attached to the column, anther terminal, operculate; pollinia four. About 260 species with widest distribution in Asia, also Western Hemisphere, W.I. *Leptorchis* Thouars

1. **L. elata** Lindl. TWAYBLADE—Slender plants 4.5 dm tall or less, with sheathing bracts below the leaves, usually naked up to the inflorescence. Leaves three or more, blades 6–9 cm long, ovate-elliptic, acuminate, five- to seven-veined. Racemes 3–20 cm long, pedicellate flowers 9–13 mm long, lateral sepals obliquely ovate, greenish, veined with madder purple; petals oblanceolate with color pattern similar to the sepals, lip obcordate cuneate, madder purple, five-nerved; column incurved with lateral wings, toothed at apex. Capsule obovoid to 1.5 cm long. Hammocks, cypress swamps on decaying logs, widespread from central

and south Fla., W.I., Mexico, Central and South America. Summer. *Leptorchis elata* (Lindl.) Kuntze

10. TROPÌDIA Lindl.

Leafy branching terrestrial from a short rhizome with clavellate, fibrous roots. Leaves ovate-lanceolate, prominently veined, plicate. Flowers small, arranged in paniculate clusters, sepals connate at base with a small mentum, lip sessile, saccate, parallel to the short column, anther erect, about equal in length to the column, pollinia two, granulose. Capsule spreading. The genus includes about 22 species native of East Indies, Malaya, China, Japan; represented in Western Hemisphere by one species.

1. **T. polystachya** (Sw.) Ames—Plant to 5 dm tall, leaves 6–20 cm long, distichous. Inflorescence terminal, paniculate, flowers bracted to 6 mm long, reddish or greenish white; sepals and petals three-nerved, mainly ovate-oblong, rounded, or petals emarginate; lip cymbiform, concave, grooved, with thickened margins, constricted above the base, rounded at dilated apex; disk pubescent with converging crests; column cylindrical. Capsule six-ribbed, 10 mm long. In low hammocks, open woodlands, Fla., Cuba, Mexico, Galapagos Islands, W.I. Summer.

11. ZEUXÌNE Lindl.

Low terrestrial, glabrous, succulent, from elongate rhizome. Leaves linear or lanceolate, sessile. Inflorescence spicate; floral bracts exceeding the flower, perianth cucullate; sepals and petals broadly or wholly hyaline, lip panduriform adnate to the column, anther usually erect, pollinia four, pollen granular. Capsule erect, ovoid. A genus of some 76 species of tropical Asia and Africa; one species in Fla.

1. **Z. strateumatica** (L.) Schltr.—Stem simple, leafy, 2 dm tall or less with decumbent base tufted or often solitary. Blades keeled, 3–8 cm long, rachis short, to 4–5 mm long, sometimes overtopped by leaves. Flowers white, dorsal sepals with petals forming the hood, lateral sepals free, lip fleshy, basal portion deeply saccate, rostellum broad, deeply cleft. Capsule 7 mm long. Lawns, roadsides, hammocks, central and south Fla., warmest parts of Asia. All year. *Orchis strateumatica* L.

This is said to be India's most common orchid.

12. TRÍPHORA Nutt. NODDING CAPS

Fleshy terrestrial herbs with tuber-bearing stolons. Stems fleshy, often in clusters; upper sheaths bearing blades. Flowers pedicellate, subtended by foliose bracts, perianth white or magenta, sepals and petals similar; lip three-lobed, column free, anther erect, pollinia two, granular. Capsule pendent or erect, ellipsoid. About 13 species, North, Central, and South America, W.I.

1. **T. cubensis** (Reichenb. f.) Ames—Plants to 2 dm tall, from a tuber 6 cm long or less. Leaves ovate, apiculate, few and distant. Flowers ten or more, borne in the upper axils, sepals subequal, inequilateral, 6–11 mm long, 2 mm wide; petals linear 8–10 mm long, 1 mm wide, lip 8–10 mm long, three-lobed, terminal lobe crenulate on margins, disk with three parallel crests to lip apex, column 7 mm long. Capsule obovoid, 1–15 cm long. Pinelands, south and central Fla. Fall.

13. CENTROGÈNIUM Schltr. NEOTTIA

Fleshy herbs from fasciculate roots. Leaves plicate, petioled, rising from the crown. Scape bearing bladeless sheaths throughout. Raceme terminal, flowers large, few, noted for the spur formed by the column and the lateral sepals. A South American genus of about 11 species, subtropics and tropics of Western Hemisphere.

1. **C. setaceum** (Lindl.) Schltr.—Scape slender with a few-flowered open raceme; flowers 5.5 cm long, white; sepals and petals lanceolate, long acuminate, petals slightly shorter; lip arcuate, linear-oblong, narrowed below the disk, apical portion fringed on margins, apex prolonged, acuminate, column stout, tubular, continuous with the lip; anther erect, pollinia two. Capsule ellipsoid to 2.7 cm long. Leaf mold in hammocks, south Fla., Bahama Islands, W.I., South America. Winter. *Pelexia setacea* Lindl.

14. BASIPHÝLLAEA Schltr.

Terrestrial herbs from tuberous roots. Stems slender with one or more basal leaves. Raceme few-flowered, flexuous, sepals and petals free, porrect, lip three-lobed; column free. Capsule cylindroid, nodding. Three species in Fla., W.I. *Carteria* Small

1. **B. corallicola** (Small) Ames—Scapose, erect, glabrous plant to 3.5 dm tall; roots fasciculate. Leaves linear-oblanceolate; cauline bracts remote. Flowers distant with pedicellate ovaries to 15 mm long; sepals and petals yellowish green, flushed with magenta rose; dorsal sepal oblong; lateral sepals and petals linear-lanceolate, falcate; lip sessile, cuneate, with five parallel, crested veins ending submarginally in the broadly obtuse middle lobe. Capsule curved, 1–1.2 cm long. Hammocks, coral rocks, Fla., Bahama Islands, W.I. *C. corallicola* Small

15. GOVÈNIA Lindl.

Terrestrial herbs with thickened rhizome. Leaves plicate with elongate sheathing bases. Raceme terminal, sepals connivent, decurved with a small chin; lip articulate to the column foot, anther terminal, operculate; pollinia four, waxy. Capsule ovate, beakless. Genus of 20 species of tropical North and South America.

1. **G. utriculata** Lindl.—Plant 6 dm tall or less; cauline sheaths greatly inflated. Leaves commonly two, elliptic, tapering to petioles. Raceme to 15 cm long; floral bracts lanceolate; flowers to 15 cm long, white; sepals and petals 15 mm long, lip ovate, to 9 mm long. Capsule deflexed. Hammocks, south Fla., W.I., Mexico. Summer.

16. GALEÁNDRA Lindl.

Terrestrial or epiphytic pseudobulbous herb from thick roots. Stem leafless during anthesis. Leaves plicate, articulate. Raceme terminal, flowers nodding, relatively remote, sepals and petals free, lip produced to a spur, column two-winged at apex, foot short, anther terminal, operculate; pollinia four, waxy. Capsule ovoid. A genus of American tropics and subtropics comprising about 20 species.

1. **G. beyrichii** Reichenb. f.—Plant to 12 dm tall with distichous leaves in postanthesis; cauline sheaths, loosely imbricate. Leaf blades three- to five-nerved, ovate-elliptic, acuminate. Flowers yellowish green with pedicelled ovary 2.5 cm long, sepals and petals linear-oblong, one-nerved, spreading; lip orbicular, prominently nerved, margins undulate shirred;. disk pubescent, ornamented with four crests; column nearly 1 cm long, pubescent at apex. Capsule about 2 cm long. Woods, roadsides, south Fla., W.I., South America.

17. EULÒPHIA R. Brown

Tall terrestrial herbs from corms. Racemes on leafless scapes with sheathing bracts, rising from an axil of a basal scale embracing the leaf-bearing shoot; sepals and petals free, lateral sepals adnate to the base of the column, lip adnate to the column foot, saccate or spurlike at base, column short, winged, anther terminal, operculate; pollinia two to four, waxy. Capsule ellipsoid, recurved. About 200 species, tropical and warm regions of both hemispheres.

1. Floral bracts surpassed by the flowers; capsule pendent . . . . 1. *E. alta*
1. Floral bracts surpassing the flowers; capsules erect . . . . 2. *C. ecristata*

1. **E. alta** (L.) Fawc. & Rendle, WILD COCO—Plant to 15 dm tall, scapose; corm 4–6 cm in diameter. Leaves 10 dm long, long petioled; blade elliptic-acuminate, three- to eleven-nerved. Scapes erect, bearing four or more elongate bracts; raceme 3.5 dm long, simple; flowers conspicuously large, madder purple, sixty or more spreading on the rachis with pedicels 3.5 cm long; sepals erect, longer than the petals, lip three-lobed, veins crested-papillose, lateral lobes erect, terminal lobe crenate-undulate. Capsule 4–5 cm long, six sulcate. Hammocks, swampy ditches, south Fla., Mexico, W.I., Central and South America, Africa. Summer, fall.

2. **E. ecristata** (Fern.) Ames—Habit similar to the preceding species; scape to 13 dm tall or taller from a corm 8 cm long, 5 cm wide, juicy, and clear in content; cauline sheaths elongate. Leaves plicate, long petioled; blade elliptic, acuminate, three- to five-nerved. Raceme 10–15 cm long; floral bracts ascendent, variously colored; perianth yellowish flushed with magenta, lip deeply three-lobed, lateral lobes erect or like an open band over the limb; disk plain. Capsule 1–2 cm long, ovoid. Pinelands, sand hills, central and south Fla., Mexico. Fall.

18. BLÈTIA Ruiz & Pav. PINE PINK

Erect terrestrial herb. Leaves few from the upper node of globular corm, or pseudobulb. Flowering scape lateral, racemes simple or paniculate, sepals free, unequal, petals free, lip free, three-lobed, rising from base of the arcuate column, anther operculate; pollinia eight, waxy. Capsule ellipsoid. About 45 species, American tropics and subtropics.

1. **B. purpurea** (Lam.) DC.—Scape to 1 m tall, leafless. Blades on leaf-bearing shoots 2–9 dm long, 25 mm wide, elliptic, long attenuate at apex, petioled and dilated to a corm-sheathing base. Flowers 35 mm wide, purple, sepals and petals reticulate veined, lateral lobes of lip incurved, disk with lamellae or crests, only seven extending to the erose, apex, column clavate with lateral wings. Capsule 2–4 cm long, brown, erect; seed 1.3 mm long with tail at each end. Pinelands, south Fla., Mexico, South America. Summer. *Limodorum purpureum* Lam., *B. tuberosa* (L.) Ames

19. VANÍLLA Sw.

Scandent epiphytes with leafy or leafless branching stems. Sepals and petals similar, free, spreading, lip clawed, three-lobed, enveloping the column at base; anther arising from the margin of the clinandrium, pollen granular. Capsule elongate, fusiform, indehiscent. The genus includes about 90 species in tropical and subtropical regions around the world.

Cultivated for their attractive flowers and aromatic properties as well as for the capsule of one species that is the source of commercial vanilla.

1. Leaves present.
  2. Internodes surpassing the leaves; veins inconspicuous.
    3. Capsule 8 cm long; flowers greenish . 1. *V. phaeantha*
    3. Capsule 25 cm long; flowers greenish yellow . . . . 2. *V. planifolia*
  2. Internodes surpassed by the leaves; veins prominent . . . . 3. *V. inodora*
1. Leaves obsolescent or lacking.
    4. Capsule clavate, lip white and purple . 4. *V. dilloniana*
    4. Capsule subcylindric, lip green and red . 5. *V. barbellata*

1. **V. phaeantha** Reichenb. f. LEAFY VANILLA—Plant climbing by aerial roots; internodes to 17 cm long, 1.5 cm in diameter. Leaves to 14 cm long, alternate, sessile, fleshy, acute. Racemes axillary; floral bracts ovate to 1.5 cm long; flowers to 9 cm long, sepals and petals similar, oblanceolate, spreading; lip flabellate to 8 cm long. Claw adnate with the column except the apex within the upcurving ruffled. margins of the lateral lobes; disk crested with hair and papillae along the median veins. $2n = 30,32$. Hammocks, south Fla., W.I., Bahama Islands. Spring.

2. **V. planifolia** Andrews—Stems branching, leafy, climbing by aerial roots. Leaves to 23 cm long, sessile, acuminate, lanceolate, fleshy. Racemes many-flowered; floral bracts ovate-oblong, 10 mm long; sepals and petals similar, oblanceolate; lip 4–5 cm long, trumpet shaped, reflexed, limb flabellate, obscurely three-lobed; disk with hairy and verrucose parallel veins. $2n = 32$. Hammocks, swamps, south Fla., Mexico, Central America, W.I. Cultivated for commerce. All year. *V. vanilla* (L.) Britt.

3. **V. inodora** Schiede, SCENTLESS VANILLA—High-climbing vine; stems thick, internodes 5–15 cm long. Leaves 10–15 cm long, ellipticacuminate, short petioled; many-veined, thin and glossy on drying. Flowers odorless, greenish; lip three-lobed, white, margin orange, crenulate; column 2–3 cm long. Capsule 15–18 cm long, curved, compressed. South Fla., W.I., Cuba, Mexico, South America. Summer.

4. **V. dilloniana** Correll, LEAFLESS VANILLA—Stem branching and climbing by aerial roots; internodes to 14 cm long. Leaves reduced, becoming scarious scales. Racemes axillary, commonly eight-flowered, flowers 5–6 cm long; sepals and petals similar, obliquely oblong-obtuse;

lip trumpet shaped, attached to the column about one-third above its base; disk flabellate, plaited, three-lobed, tufted with hair, papillose in lines up to the fleshy apex, lateral lobes papillose, recurved over the column. $2n = 32$. Hammocks, oolitic rocks, south Fla., Cuba, Santo Domingo. Spring. *V. eggersii* of authors, not Rolfe

5. **V. barbellata** Reichenb. f. Link Vine—Stem succulent, articulate, climbing by aerial roots at the nodes; internodes to 3 dm long. Leaves abortive. Raceme axillary, floral bracts 5 mm long, ovate; flowers 3–4 cm long, sepals and petals similar, oblique or oblong-elliptic, lip flabellate, 3–4 cm long, greenish below, red above, white on margins, limb three-lobed, disk barbellate; apex of middle lobe thickened, lateral lobes plaited; column 2–3 cm long, glabrous. $2n = 32$. Hammocks, south Fla., Cuba, Bahama Islands, W.I. Summer. *V. articulata* Northrop

20. CAMPYLOCÉNTRUM Benth.

Small epiphytic leafless or leafy-stemmed plants from fasciculate, chlorophyllous roots. Leaves thick, distichous, fleshy, deciduous. Inflorescence radical or axillary. Flowers in bracted spikes; sepals free, connivent, spreading at tips; petals similar; lip sessile, spur obtuse; column short; anther terminal operculate; pollinia two. A genus of about 40 species, restricted to tropical and subtropical America.

1. **C. pachyrrhizum** (Reichenb. f.) Rolfe—Stem suppressed; roots 10 cm long or longer, green, thick, clustered. Spikes 2–4 cm long, sessile, springing forth directly from the crown; floral bracts 4–5 mm long, persistent, sessile, two-ranked, yellowish green; sepals and petals similar, oblanceolate; lip obscurely three-lobed, seven-veined, partly adnate to the column; spur saccate, slightly curved. Capsule 7–9 mm long, ovoid, six-ribbed. Big Cypress Swamp, south Fla., Cuba, W.I., South America. Spring.

21. HARRISÉLLA Fawc. & Rendle

Stemless and leafless epiphytic herbs. Roots 10 cm long or less, threadlike, chlorophyllous, appressed on slender twigs. From fasciculate crowns arise pedunculate racemes 1.5 cm long or longer. Flowers yellowish, less than 3 mm long, borne on jointed pedicels; sepals and petals similar, free, lip sessile at the base of the column, ovate, margins incurved, produced to a globose, constricted spur. Capsule 4–5 mm long, dehiscent at summit, six-valved. A monotypic genus restricted to tropical America and subtropical regions.

1. **H. porrecta** (Reichenb. f.) Fawc. & Rendle—Characters of the genus. The smallest orchid in North American flora. Germinating seeds

give rise to green leaves, transitory organs in developing plants. Epiphytic on shrubs and trees, in hammocks, cypress heads, mesophytic woods. Fall. *Harrisella amesiana* Cogn.

22. POLYRRHÌZẠ Pfitz.

Leafless epiphytic orchids with reduced stems. Roots elongate, thick, fasciculate, chlorophyllous. Inflorescence a bracted peduncle bearing a raceme of large flowers, usually one at a time; sepals and petals free, spreading. Lip attached to the column by sessile base; produced to an elongate spur. Anther terminal operculate; pollinia two, waxy. Capsule cylindric. A genus of four species of tropical and subtropical Fla., W.I.

1. **P. lindenii** (Lindl.) Cogn. ex Urban, PALM POLLY—Essentially stemless; roots extensive, creeping, flexuous. Flower white, large; dorsal sepal 2 cm long, narrowly lanceolate; lateral sepals to 3 cm long; lip three-lobed, cymbiform, lateral lobes deltoid, oblique, acute, erect to ascending, terminal lobe cleft into lance-attenuate tail, recurved lobes to 7 cm long; spur 11–17 cm long. Mature capsule to 9 cm long, 5 mm in diameter. Hammocks, swampy inclusions in cypress, Fla., Cuba. Summer. *Dendrophylax lindenii* (Lindl.) Benth. ex Rolfe

The flower is intriguingly beautiful, especially in its natural setting of deep cypress swamps.

23. POLYSTÀCHYA Hook.

Cespitose epiphyte with pseudobulbous leafy stems. Leaves distichous, articulate, with the sheath; inflorescence a bracted paniculate raceme. Floral bracts persistent, perianth bonnet shaped; dorsal sepal free; lateral sepals attached to the column foot, forming a mentum. Petals linear; lip three-lobed, disk velvety, column arcuate, with foot, wingless; anther erect, pollinia four, waxy. Capsule ellipsoid. About 200 species in tropical regions the world over, attaining greatest development in Africa.

1. **P. extinctoria** Reichenb. f. PALE-FLOWERED POLYSTACHYA—Clump-forming epiphyte usually less than 5 dm tall. Leaves 10–20 cm long, elliptic-oblanceolate, with prominent veins and white cartilaginous margins. Racemes many-flowered, sepals connivent, median lobe of lips emarginate, crenulate, disk velvety, anther terminal, operculate, pollinia four, waxy. Capsules ellipsoid, to 12 mm long, erect or nodding, copiously villous within, seeds ellipsoid, yellowish, 1 mm long. Hammocks, pinelands, south Fla., W.I., Mexico, Central and South America, Old World tropics. Summer. *P. minuta* Aubl.; *P. luteola* (Sw.) Hook.

24. EPIDÉNDRUM L.

Epiphytic herbs from rhizomes with fibrous roots, true pseudobulbs wanting. Blades duplicate, fleshy or coriaceous, articulate with the sheath, deciduous. Inflorescence terminal, sepals and petals free; lip adnate to the essentially footless column; anther terminal, opercular; pollinia four, waxy. Capsule ribbed and winged. One of the largest genera of the New World orchids with about 400 species in tropical America.

1. Lip entire, uppermost in the flower.
  2. Floral bracts distant; spikes elongate . .   1. *E. rigidum*
  2. Floral bracts imbricate; spikes congested. 2. *E. strobiliferum*
1. Lip lobed (entire in No. 3), lowermost in the flower.
  3. Stems flexuous; leaves 1–2.4 cm wide.
    4. Flowers in umbels; bracts linear, not enclosing ovary . . . . . . 3. *E. difforme*
    4. Flowers in racemes; bracts ovate, cucullate enclosing ovary . . . . . 4. *E. acunae*
  3. Stems not flexuous; leaves to 6 cm wide or wider.
    5. Flowers 4.5 cm long or longer, few in abbreviated racemes; midlobe of lip filiform at tip, with two lamellae at base . . . . 5. *E. nocturnum*
    5. Flowers less than 2 cm long, many in congested racemes; midlobe of lip retuse, apiculate . . . . . . 6. *E. anceps*

    1. E. rigidum Jacq.—Stem to 3 dm long, compressed, ascending from creeping rhizome. Blades articulated with inflated sheaths through internodes. Inflorescence spicate, flowers small, few, perianth to 8 mm long, sepals six- to seven-nerved, ovate; petals three-nerved, linear; lip 9 mm long, broadly obcordate. Capsule ellipsoidal, to 18 mm long. $2n = 40$. Hammocks, south Fla., W.I., Mexico, Panama. Summer. *Spathiger rigidus* (Jacq.) Small

    2. E. strobiliferum Reichenb. f.—Plants to 2 dm high, clustered on rooted rhizomes; stems compressed, internodes enveloped with leaf sheaths. Leaves distichous; blades 2–3 cm long, oblong-spatulate, spreading. Racemes few-flowered; flowers 4–9 mm long, sessile, yellowish white, striate with red; sepals and petals oblong, spatulate; lip ovate or deltoid-cordate, entire. Capsule ellipsoid, 6–9 mm long. Hammocks, south Fla., Mexico to South America. Spring. *Spathiger strobiliferus* (Reichenb. f.) Small

    3. E. difforme Jacq.—Plants cespitose with many ascending stems

4 dm long or longer. Leaves distichous to 10 cm long, oblong-lanceolate or ovate-elliptic, sheathing at base. Flowers 4 cm long, few in umbels, yellowish green; lip 8–10 mm long, 15 mm wide, with cordate winglike lateral lobes; median lobe a mere tooth in a retuse apex; disk greenish, lip margin crenulate; column 10 mm long, hoodlike over the rostellum; pollinia four, waxy. Capsule to 4.5 cm long, ellipsoid. $2n = 39,40$. Hammocks, south Fla., W.I., Mexico, Central and South America. Summer. *E. umbellatum* Sw.

   4. E. acunae Dressler—Plant to 9 dm tall, branching; internodes invested with leaf sheaths, nodal roots often present; leaves coriaceous, emarginate; racemes terminal, about as long as the leaves, rachis slender, geniculate; flowers greenish, less than 10 mm long, lip 3.5 mm wide at base, ovate-cordate, acute, with two calli, adnate to the column. Capsule 10 mm long, beaked. Epiphytic on trees, or epipetric. Rare in Fla.; Cuba, W.I., Mexico, Central and South America. *E. ramosum* Jacq.

   5. E. nocturnum Jacq.—Plant cespitose to 1 m tall; stems rigid, terete below, compressed toward summit, enveloped with cylindrical leaf bases; blades 10–16 cm long, linear-elliptic, articulate, finely many-veined. Racemes with short rachis; flowers with pedicelled ovary to 10 cm long, showy, commonly solitary, or two at a time; sepals and petals narrowly linear-attenuate, lip deeply three-lobed, white, spotted with purple at the base; column to 2 cm long, denticulate and dilated at summit. Capsule fusiform, to 5 cm long, attenuated at base and apex. $2n = 40,80,74,85$. Hammocks, south Fla., Mexico, Central America, W.I., Panama, South America. Summer, fall. *Amphiglottis nocturna* (L.) Britt.

   6. E. anceps Jacq.—Plant cespitose, to 1 m tall, stems upright, compressed, sharply margined, enveloped in leaf sheaths through internodes. Blades to 1.8 dm long, 3–4 cm wide, elliptic-lanceolate, obtuse and dorsally keeled. Inflorescence subcapitate, terminal peduncle to 2 dm long, flowers malodorous; sepals and petals 7 mm long; lip 9 mm long, 6.5 mm wide, three-lobed, yellowish green suffused with purple, claw connate to the stout column, 4 mm long, disk centrally ridged; rostellum yellow. Capsule 15 mm long, 10 mm in diameter, ovoid. Hammocks, south Fla., Mexico, Central and South America. All year. *Amphiglottis anceps* (Jacq.) Britt.

25. LEPANTHÓPSIS Ames

   Epiphytic herb with creeping rhizome from primary stem giving rise to erect leafy shoots. Inflorescence terminal, spicate, few to many flowers distichous, crimson purple, sepals free, margins incurved, lateral connate below the middle, uninerved, petals suborbicular, spreading, lip adnate to the base of the column, anther terminal, operculate; pollinia two, waxy. Capsule ellipsoid. About 15 species restricted to American tropical and subtropical regions.

1. **L. melanantha** (Reichenb. f.) Ames—Plant to 8 cm tall from short rhizomatous base with slender fasciculate nodal roots. Cauline bracts three to four, salverform, fringed throughout with coarse hairs, leaf solitary, in axil of the uppermost sheath, to 11 mm long, 3–4 mm wide, coriaceous. Lip clawed, cordate-rotund, three-veined with incurved entire margin. Capsule 3–4 mm long, six-ribbed, short pedicelled. Hammocks, Fla., Jamaica. Spring. *Lepanthes harrisii* Fawc. & Rendle

26. PLEUROTHÁLLIS R. Brown

Epiphytic herbs with cespitose or repent primary stems, secondary stems unifoliate at lower internodes. Inflorescence terminal, racemes one or more per axil, sepals nearly equal, erect or spreading, free or wholly connate, petals small, lip entire or three-lobed, often obscurely divided, column winged or wingless, with foot or footless. Anther terminal, operculate, incumbent; pollinia two to four, waxy. Capsule strongly three-angled. About 1000 species, chiefly of the tropical mountains of America, Cuba, W.I., Brazil.

1. **P. gelida** Lindl.—Plants glabrous to 4.5 dm tall, tufted on short rhizomes with wiry roots. Leaf 18–25 cm long, to 7 cm wide, elliptic, stems two-jointed below. Racemes 12–14 cm long, the flexuous rachis many-flowered; flowers with pedicellate ovaries 4–5 mm long, pale yellow, sepals 7 mm long, cupuliform, short pilose within; petals 3–5 mm long, ovate; lip 2.5 mm long, broadly rounded, disk crested, column short, two-toothed. Capsule 5–10 mm long, 3–5 mm wide. Hammocks, south Fla., Cuba, W.I. to Peru. Summer.

27. ENCÝCLIA Hook.

Epiphytic herbs with true pseudobulbs giving rise to terminal leaves and flowering stems. Leaves duplicate, membranous. Flowers in paniculate, spicate, or racemose inflorescences, lip three-lobed or entire, free or partly free from the column, anther one or more terminal, operculate. Capsule pendent. About 130 species in tropical and subtropical America. *Epidendrum* L.

1. Pseudobulbs aggregate, not on creeping rhizome.

2. Lip three-lobed, veins crimsom magenta,
   lobes erect . . . . 1. *E. tampensis*
2. Lip entire.
  3. Lip wider than long, suffused with
    purple, apiculate; perianth segments
    reflexed in anthesis . . . . . 2. *E. cochleata*
  3. Lip longer than wide, rhombic, angled,
    yellowish green; perianth segments
    erect in anthesis . . . . . . 3. *E. boothiana*
1. Pseudobulbs distant on a creeping rhizome;
  lateral lobes of lip broadly rounded,
  terminal lobe minute . . . . . . 4. *E. pygmaea*

1. **E. tampensis** (Lindl.) Small—Plants ascendent or pendent, glabrous, often in large clusters; pseudobulbs to 5 cm long. Leaves 0.5–2.0 dm long, 8–18 mm wide, linear-elliptic, usually one to three per pseudobulb. Inflorescence a flexuose panicle, perianth yellowish brown, spreading; sepals 1.5–2.2 cm long, oblong-spatulate; petals oblanceolate, acute, 1.2–1.8 mm long, lip porrect, white terminal lobe ovate, suborbicular, striate with magenta, terminal lobe ovate, suborbicular, crenulate; disk flabellately veined with crimson magenta; column 8 mm long, winged and auricled at summit. Capsule to 8 mm long. Hammocks, Fla., Bahama Islands to N.C. Spring, fall, winter. *Epidendrum tampense* Lindl.

2. **E. cochleata** (L.) Dressler, SHELL ORCHID—Erect plant, relatively stout, to 5.5 dm tall; pseudobulbs 5–7 cm long, stiped, ovate-elliptic, strongly compressed, giving rise to one to three leaves, 2.5 dm long, 2.5–3 cm wide. Flower with pedicellate ovary 5.5 cm long; sepals 3.7 cm long, yellow, reflexed; petals narrower, spatulate, 2.8 cm long; lip reniform-cordate, 2 cm long, 2.2 cm wide, madder purple; disk veined with white and magenta crimson, suffused with olive; column 5 mm long, upper half yellow, toothed; the lower flecked with maroon; anthers three. Capsule 2.5–3 cm long, winged. Hammocks, south Fla., Mexico, Bahama Islands, Central and South America, W.I. Fall, spring, and all year. *Epidendrum cochleatum* L.

3. **E. boothiana** (Lindl.) Dressler—Plant ascendent, 2–3 dm tall; pseudobulbs 2–5 cm long, fusiform, compressed in clusters. Leaves one to three, 7–15 cm long, slightly twisted at rounded apex; scape mostly simple. Flowers 2.6 cm long, with pedicellate ovary, orange, flecked with madder purple; lip 12 mm long, slightly adnate to the column base; disk with a medial groove, ending in a tooth at apex of the lip. Capsule 2–3 cm long, winged. Hammocks, south Fla., Bahama Islands, Cuba, South America. Fall. *Epicladium boothianum* (Lindl.) Small

4. **E. pygmaea** (Hook.) Dressler—Plants to 10 cm tall from a creeping, branching rhizome, pseudobulbs 2–3 cm long, numerous, fusiform,

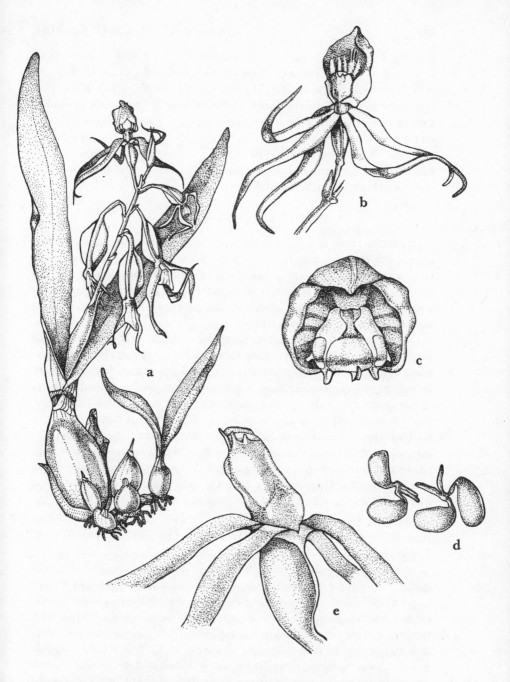

Family 54. **ORCHIDACEAE.** *Encyclia cochleata:* a, flowering plant, × ½;
b, flower, × 1; c, column and subtending petal, × 5; d, pollinia, × 7½; e,
developing fruit with persistent perianth parts, × 4½.

bearing at summit a pair of subopposite sessile leaves, one or more flowers in a reduced raceme. Leaves 2–5 cm long, elliptic-ovate. Flowers greenish white, the base of lip blotched with purple, sepals about 5–7 mm long; petals narrowly linear, about as long as the sepals, extended part of lip shorter than the sepals, the base shortly adnate to the column, median lobe deltoid, apiculate, shorter than the upcurving lateral lobes; column 2–3 mm long. Capsule 10 cm long, fusiform, with winged angles. Fla., Mexico, Central and South America. Summer. *Hormidium phygmaeum* (Hook.) Benth. & Hook.

28. CYRTOPÒDIUM R. Brown

Terrestrial or epiphytic herbs with fusiform, pseudobulbous stems, lateral flowering scapes. Leaves plicate. Flowers bracted in large panicles, sepals and petals free, spreading, lip three-lobed, articulate with column foot, anthers terminal, operculate; pollinia two to four, waxy. Capsules oblong or ellipsoid. About ten American species restricted to subtropical and tropical regions.

1. **C. punctatum** (L.) Lindl. COWHORN ORCHID—Flowering stems to 41 dm tall, branching above, pseudobulb 1.5–4 dm long, elongate, articulate with sheathing scales, leafy at summit when young. Blades to 9 dm long, narrowly lanceolate, long attenuate, three- to five-veined. Panicle diffusely branched above, axes purple punctate, flowers reddish brown, flushed with yellow; sepals obovate, greenish yellow, blotched with madder purple, petals and dorsal sepal, oblong-ovate, crisped, lip to 22 mm wide, about half as long; median lobe short, crenulate; disk with clusters of calli, the base tapering to a narrow claw; column clavellate, foot prominent. Capsule body to 8 cm long. Hammocks, south Fla., Mexico, South America. *Epidendrum punctatum* L.

29. BULBOPHÝLLUM Thouars

Repent epiphyte with rhizome, rooting at nodes, pseudobulbs two-leaved, sessile, in axils of scales. Flowering scape lateral with numerous sheaths, inflorescence a spike, the rachis slender or thickened, dorsal sepal free, lateral sepals adnate to the column foot; lip uppermost, articulate to the two-winged column foot; anther terminal, operculate; pollinia four, waxy. Capsule ellipsoidal. A large genus of 900 species of worldwide distribution, especially tropics and south temperate regions. A few species in Western Hemisphere. *Pleurothallis* R. Brown

1. **B. pachyrachis** (A. Rich.) Griseb.—Pseudobulbs about 2 cm long, ovoid-conical, four- to five-angled. Leaves to 18 cm long, oblong rounded at apex, leathery; scape to 4 dm tall, 8–20 mm thick, clavate,

remotely bracted. Flowers small, recessed in depression of the thick rachis, subtended by deltoid clasping bracts; sepals and petals navicular, white, spotted with purple, lip thick, clawed, entire, disk furrowed from base to apex. Capsule 8 mm long, six-keeled. Hammocks, south Fla., Cuba, Puerto Rico, Trinidad. *P. pachyrachis* A. Rich.

30. BRÁSSIA R. Brown SPIDER ORCHID

Plant terrestrial or epiphytic with elongate, creeping rhizomes giving rise to a short stem with sheathing scales or leaves, pseudobulb with a lateral flowering scape from its base. Inflorescence a raceme with remote, distichous, showy flowers, sepals and petals free, spreading, lip simple, column short, footless, anther terminal, operculate, pollinia two, waxy. Capsule ellipsoid. About 50 species native of tropical and subtropical America. Hammocks, south Fla., Mexico, W.I., Brazil, Peru.

1. **B. caudata** (L.) Lindl.—Plant to about 6 dm tall; pseudobulb 5–9 cm long, fusiform, compressed with two to three apical leaves. Blades to 5 dm long, 5 dm wide, strap shaped, apiculate, articulate at base. Scape to 7 dm tall; flowers large, greenish yellow, mottled with madder purple; bracts deltoid, spreading, sepals and petals long caudate; lip sessile, 4 cm long, 12 mm wide, oblong-elliptic, acuminate, spotted with purple near the base, crenulate. Capsule 4–5 cm long, stipitate. $2n = 60$. Hammocks, south Fla., Mexico, Central and South America, W.I. Spring, summer. *Epidendrum caudatum* L.

31. IONÓPSIS HBK

Plant epiphytic or sometimes terrestrial, rhizome usually short, leafy, bearing small pseudobulbs, with lateral peduncles, pseudobulbs unifoliate or leafless. Leaves from the rhizome rigid, distichous. Panicles lax, usually more than one per plant; flowers showy, dorsal sepal free, lateral united at base forming a saccate spur; petals free, lip clawed, attached to the winged, footless column; anther terminal, operculate; pollinia two, waxy. About nine species, tropical and subtropical America.

1. **I. utricularioides** Lindl.—DELICATE IONOPSIS—Scape to 5 dm tall with remote cauline bracts, pseudobulbs 2–3 cm long; rhizome leaves one to three, linear or oblong, apiculate, 7–15 cm long, 6–15 mm wide, coriaceous, often reddish ribs prominent on drying. Floral bracts small, persistent; rachises slender, flexuous, flowers pinkish lavender, lip broadly obcordate, veins flabellate, margin crenulate, claw crested at base; capsule ovoid, long beaked, about 16 mm long. $2n = 26$. Hammocks, central and south Fla., W.I., Mexico, South America.

32. ONCÍDIUM Sw.

Epiphytic or terrestrial rock-inhabiting herbs. Leaves from the apex of the pseudobulb or around the base. Flowering stems lateral and basal, inflorescence a large, usually many-flowered, widely branching panicle, sepals subequal, sessile, spreading, or reflexed; petals equal, lip panduriform, adnate at right angles to the base of the base of the footless column with apical wing; anther incumbent, terminal, operculate; pollinia two, waxy, Capsule ellipsoid, beaked. About 350 species, south Fla., W.I., temperate South America.

1. Leaves serrulate; lateral sepals connate;
   leaves less than 10 cm long . . . . .   1. O. bahamense
1. Leaves entire; lateral sepals free; leaves
   more than 10 cm long.
   2. Pseudobulb manifest; leaves linear, equi-
      tant . . . .   2. O. floridanum
   2. Pseudobulb abbreviated; leaves elliptic,
      solitary.
      3. Lip panduriform; flowers pale yellow
         spotted with lavender . . . . .   3. O. carthagenense
      3. Lip flabellate; flowers greenish yellow
         spotted with reddish brown . . .   4. O. luridum

1. **O. bahamense** Nash, Variegated Orchid—Pseudobulbs small, reduced, unifoliate, in basal leaf tufts. Leaves conduplicate, distichous. Flowering peduncle to 3 dm long, flowers commonly in racemes, few to many, white, variegated with yellow and magenta, with pedicellate ovaries 1–2 cm long, dorsal sepal 9 mm long, clawed, obovate; lateral sepals connate; petals to 9 mm long, with short claw, lamina obovate, broadly rounded, crenulate; lip three-lobed, 1–2 cm wide across the emarginate terminal lobe, disk ornamented with a yellow crest, two anterior and three posterior tubercles. Column short, conspicuously winged. Capsule ellipsoid to 2 cm long. Coastal dune forest and scrub, Fla., Palm Beach and Martin Counties, W.I. Winter, spring.

2. **O. floridanum** Ames, Florida Oncidium—Plant to 2.5 m tall, with slender rigid stem, pseudobulbs 8–10 cm long, compressed, narrowed to apex from a broad base; leaves to 6 dm long, 10–20 mm wide, coriaceous, linear, obtuse. Inflorescence branching, branchlets zigzag in fruit. Floral bracts 5–10 mm long, scarious, finely nerved, subtending one to two peduncles, flowers yellow mottled with brown, with pedicellate ovaries about 2 cm long and wide; sepals 10–12 mm long, elliptic, wavy on margins; petals similar; lip to 13 mm long, the disk ornamented

by five minute outgrowths; column 3–4 mm long, constricted above base with two apical, minutely scalloped wings; pollinia two. Dehisced capsules six-ribbed, fusiform, 2–2.3 cm long. Hammocks, endemic to south Fla. Summer, fall. *O. ensatum* Lindl. of authors

3. **O. carthagenense** (Jacq.) Sw. SPREAD-EAGLE ONCIDIUM—Plants to 2 m tall or taller, pseudobulbs 2.5 cm long, arising from short thick rhizomes. Leaf 4 dm long, solitary, borne at the apex of pseudobulb; blade oblong-elliptic, acute, thick, leathery. Inflorescence lateral, in axil of a scarious sheath; peduncle with panicle to 2 m tall, remotely bracted; flowers yellowish white mottled with lavender, magenta, red, to 2 cm long; sepals and petals clawed, limb oval, undulate, crisped; lip 9–12 mm long, three-lobed; disk with mammillate, lobulate calli; column 3–4 mm long, fleshy, with two-lobed wings. Capsule ellipsoid to 8 cm long. $2n = 28$. Hammocks, south Fla., Mexico, Central and South America. Spring, summer. *Epidendrum carthagenense* Jacq.

4. **O. luridum** Lindl.—DINGY-FLOWERED ONCIDIUM. Plant large from small pseudopodium with strong rhizome; roots numerous, clustered; leaf apical, solitary, commonly 3–5 dm long, 4–7 cm wide, leathery or coriaceous, elliptic-oblong; inflorescence lateral, with peduncle to 2 m tall arising from the base of the pseudobulb; nodal sheath scarious, 10–15 mm long; floral bracts deltoid-acute, 3–5 mm long, scarious. Flowers with distinctive color pattern in varying shades of green, yellow, brown, and red in blotches; sepals and petals similar, ovate-orbicular, clawed, crisped or undulate on margins; lip to 2 cm long, three-lobed, panduriform, midlobe semiorbicular, retuse, 2.5 cm wide at apex, margins crenulate, revolute; disk yellow at center, ornamented with five fleshy lobules with violet tubercles; column short, winged; capsule 3–4 cm long, 1.5 cm wide, ellipsoid. On trees and rocks, in hammocks, Fla., W.I., Mexico, tropical Americas. *O. undulatum* (Sw.) Salisb.

33. MAXILLÀRIA Ruiz & Pav.

Epiphytic herbs with abbreviated rhizome-bearing flattened stems, becoming thickened into pseudobulbs surrounded by leaf-bearing sheaths. Leaves fleshy or leathery, the midvein prominent. Peduncles one-flowered, arising from leaf axils or from the base of the pseudobulb; sepals equal, free, the lateral adnate at base to the column foot, forming a mentum, lip three-lobed, short clawed, adnate to the apex of the column foot, column arcuate, clavate; anther terminal, operculate; pollinia two, waxy. Capsule ovoid. A polymorphic genus of 300 species of tropical and subtropical regions of Western Hemisphere.

1. **M. crassifolia** (Lindl.) Reichenb. f.—Leaves to 4 dm long, 1–4

cm wide, elliptic-oblong, obtuse, conduplicate, articulate with sheaths. Flowers with pedicel 25 mm long, orange yellow; sepals 10 mm long, keeled, oblong, acute, stiffly fleshy; petals oblanceolate, slightly inequilateral, lip 15 mm long, 5 mm wide, denticulate, orange flushed with lavender, minutely red dotted, with a disk of densely massed farinose pubescence. Capsule 2.5–3.5 cm long. Hammocks, south Fla., W.I., Mexico, South America. Winter, summer. *Maxillaria sessilis* (Sw.) Fawc. & Rendle

34. MACRADÈNIA R. Brown

Small epiphytic herbs with short rhizomes, slender pseudobulbs with terminal leaves. Raceme pendent, lateral, sepals and petals similar, inequilateral; lip sessile, continuous with the base of the column; the basal half suborbicular, abruptly narrowed to a linear-oblanceolate terminal lobe; anther erect; pollinia two, attached to a viscid disk. Capsule ellipsoid, bluntly three-angled. A genus of eight species of Fla., tropical and subtropical America.

1. **M. lutescens** R. Brown—Plants glabrous; pseudobulbs terete, 2–4 cm long, unifoliate. Leaves 9–16 cm long, 1–2.7 cm wide; raceme lax; flowers whitish, coral pink variegated with purple; lip 7–10 mm long, three-lobed, lateral lobes erect, clasping the column; terminal lobe with revolute margins; column footless, ventrally sulcate, clavate, apex lacerate. Capsule 1.5–2 cm long. Hammocks, south Fla., Cuba, W.I., South America. Spring.

CLASS MAGNOLIATAE (DICOTYLEDONEAE)

Flower parts mostly in fours or fives; leaves mostly net veined; vascular bundles arranged in a circle in herbaceous forms, woody stems with a concentric cambium; embryo with two cotyledons.

55. CASUARINÀCEAE BEEFWOOD FAMILY

Monoecious or dioecious evergreen trees or shrubs with jointed branches generally resembling a conifer. Leaves reduced to toothed or scalelike whorls or sheaths surrounding the nodes of cylindrical branches. Staminate flowers spicate in sheaths toward the end of branches, stamen solitary and central. Pistillate flowers capitate, lateral or terminal on branches, ovary small, superior. Fruit a one-seeded nut, congested into a cone of persistent bracts opening and exposing the fruits. One genus, about 40 species of xerophytic appearance, chiefly Australia and southeast tropical Asia. Widely cultivated and naturalized in warmer parts of the world. The wood is very hard.

1. CASUARÌNA Adans.

Characters of the family

1. Scales (leaves) six to eight per whorl; seeds
   6–8 mm long . . . . 1. *C. equisetifolia*
1. Scales (leaves) twelve to sixteen per whorl;
   seeds 3–5 mm long . . . . 2. *C. glauca*

1. C. equisetifolia Forst. AUSTRALIAN PINE, BEEFWOOD, HORSETAIL TREE—Shrubs or tall trees up to 40 m with open, slender branches; leaves (scales) six to eight per whorl, 1–3 mm long on dark green "needles" that are jointed branches. Staminate spikes 1–5 cm long; pistillate spikes globular, 1–2 cm wide in fruit. Fruit conelike and woody. $2n = 18$. Sandy shores, pinelands, and Everglades area; south Fla.; native of Australia, naturalized in W.I., Mexico, tropical America. This species is spreading rapidly and seeds freely throughout our area.

2. C. glauca Sieb. BRAZILIAN OAK, SCALY-BARK BEEFWOOD—Shrubs or trees up to 20 m with dense, bushy branches; leaves (scales) twelve to sixteen per whorl, branches (needles) glossy green. Species apparently does not fruit in Fla., naturalized and widely planted inland but not salt tolerant, naturalized from Australia. *C. lepidophloia* misappl. Root sprouts grow up and form thick or coppice growth.

A third species **C. cunninghamiana,** with scales eight to ten per whorl, seeds 3–4 mm long, occurs in our area. It forms many root suckers and is said to be more cold tolerant than our other species.

56. SAURURÀCEAE LIZARD'S TAIL FAMILY

Succulent perennial herbs with jointed stems bearing simple, alternate, entire leaves that are strongly veined. Flowers hypogynous, bisexual in dense spikes or racemes; bracts especially or upper leaves conspicuous, perianth absent; stamens three, six, or eight free or fused to ovary. Ovary of three to four free or united carpels, styles free. Fruit dry or somewhat fleshy, indehiscent. Three genera and four species of Asia, North America, and subtropics.

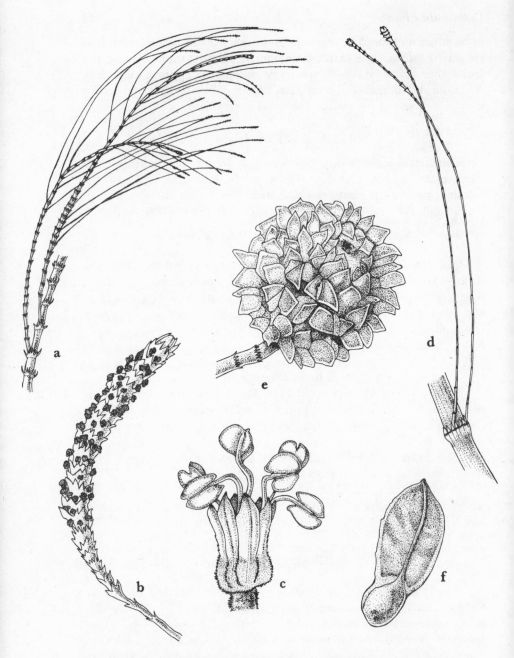

Family 55. **CASUARINACEAE.** *Casuarina equisetifolia:* a, mature branch bearing terminal staminate inflorescences, × ½; b, staminate inflorescence, × 3; c, staminate flower, × 7; d, mature branch bearing pistillate inflorescences, × 1½; e, pistillate inflorescence, × 3; f, fruit, × 8.

1. SAURÙRUS L. Lizard's Tail

Fresh-water aquatic or marsh herbs with distinctly petioled, thick, cordate leaves. Flowers in drooping racemes or spikes, minute bracts fused to pedicels or to ovaries. Fruit of three to four indehiscent carpels fused at base, wrinkled at maturity. Two species, the other in Asia.

1. S. cernuus L. Lizard's Tail—Erect stems 3–10 dm tall from creeping stoloniferous rhizomes. Leaves 8–16 cm long, ovate to ovate-lanceolate, cordate at base, acuminate, petioled, aromatic. Racemes nodding at tip 1–2 cm long, flowers whitish. Fresh-water marshes, swamps, moist soil, Fla. and Tex. north to New England, Quebec, and Minn.

57. PIPERÀCEAE Pepper Family

Herbs or shrubs, erect, reclining, or creeping. Leaves mostly entire, alternate, rarely opposite or whorled, with distinct petioles; stipules fused to petioles or lacking. Flowers minute, bisexual or unisexual, closely arranged and more or less imbedded in the fleshy axis of the spike, each borne in the axil of a bract; calyx lacking, stamens two to six, filaments free; stigmas short, one to five. Fruit baccate with fleshy, thin, or membranous pericarp. Leaves with a pungent flavor. About seven genera and 1150 species in the tropics.

1. PEPERÒMIA Ruiz & Pav.

Perennials with creeping, spreading, branching stems, often vine-like. Leaves leathery, alternate or opposite, evergreen, blades smooth. Flowers partially imbedded in pits of the honeycombed rachis; erect spikes congested and densely greenish flowered; subtending bracts ovate or shield shaped, stamens two. Ovary terminated by elongate beak bearing a minute stigma near the base. Fruit a berry partially buried in the rachis. Over 600 species of tropical distribution, ranging into subtropics. *Rhynchophorum* (Miq.) Small; *Micropiper* Miq. A number of species are popular as house plants and in cultivation.

1. Fruits distant and not crowded into a mass;
    stems pubescent, often reddish . . . . 1. *P. humilis*
1. Fruits crowded into a mass on the rachis;
    stems usually glabrous, not reddish.
    2. Spikes slender, nodding, branched; leaves
        cuneate to spatulate . . . . 2. *P. spathulifolia*
    2. Spikes stout, erect, simple; leaves obovate,
        ovate to suborbicular.

Family 56. **SAURURACEAE.** *Saururus cernuus:* a, flowering branch, × ½;
b, rootstalk, × ½; c, flower, × 15; d, mature fruit, × 25.

3. Leaf blades 10 cm or more long, narrowly obovate . . . . 3. *P. obtusifolia*
3. Leaf blades less than 10 cm long, ovate to suborbicular.
    4. Plants covered with black dots . 4. *P. glabella*
    4. Plants not covered with black dots.
        5. Fruit cylindric-ovoid, beak hooked 5. *P. floridana*
        5. Fruit globular, beak slightly curved 6. *P. simplex*

1. **P. humilis** Vahl—Stems usually up to 6–7 dm, terrestrial or epiphytic, bearing ovate, obovate to elliptical leathery leaves up to 6 cm long. Flowering spikes 6–12 cm long, about 2 mm thick, bracts 1 mm long or less. Fruit ovoid. Hammocks, mangrove belt, south Fla., W.I. *Micropiper humile* (Vahl) Small

2. **P. spathulifolia** Small, SPATULATE PEPEROMIA—Stems creeping, stout, with elongate branches up to 2 m, terrestrial in rich humus. Leaves glossy green 6–12 cm long, obovate to spatulate. Spikes slender, stalked, clustered two to five together, up to 15–20 cm long, nodding at tip. Fruit ovoid with a hooked beak. Hammocks, south Fla., W.I. *Rhynchophorum spathulifolium* Small

3. **P. obtusifolia** (L.) Dietr. FLORIDA PEPEROMIA—Stems stout, bearing elongate branches, often vinelike, epiphytic. Leaves 10–15 cm long, tapered to a short petiole, obovate to broadly spatulate. Spikes erect, 5–15 cm long, stalked, rachis up to 4 mm thick. Fruit elongate-ellipsoidal about 1 mm wide, beak hooked. $2n = 24$. Hammocks, epiphytic on trees and rooting logs, south Fla., W.I., tropical America. *Rhynchophorum obtusifolium* (Miq.) Small

4. **P. glabella** (Sw.) A. Dietr. CYPRESS PEPEROMIA—Stems climbing, succulent with divergent branches, epiphytic. Leaves ovate-lanceolate to ovate, acuminate, conspicuously three-nerved and black dotted. Fruit ovoid to spherical with short tip. Hammocks, south Fla., W.I.

5. **P. floridana** Small—Stems stout, branches elongated and often vinelike, epiphytic. Leaves ovate to orbicular, 5–10 cm long, narrowed to a short petiole. Spikes with short stalk mostly 6–10 cm long, rachis up to 5 mm thick. Fruit cylindrical-ovoid about 1 mm long, the beak hooked. Hammocks, on tree trunks and rotten logs, endemic to south Fla. *Rhynchophorum floridanum* Small; *Peperomia magnoliaefolia* Chapm.

6. **P. simplex** Ham. PALE GREEN PEPEROMIA—Stems stout, erect up to 50 cm tall, epiphytic or terrestrial. Leaves glabrous, pale green, ovate to elliptical 3–5 cm long. Spikes 5–10 cm long, erect, occurring singly at end of branch. Fruits globular up to 1 mm with a slight curved beak. Growing on trees and rotting logs, hammocks and swamps, south Fla., tropical America.

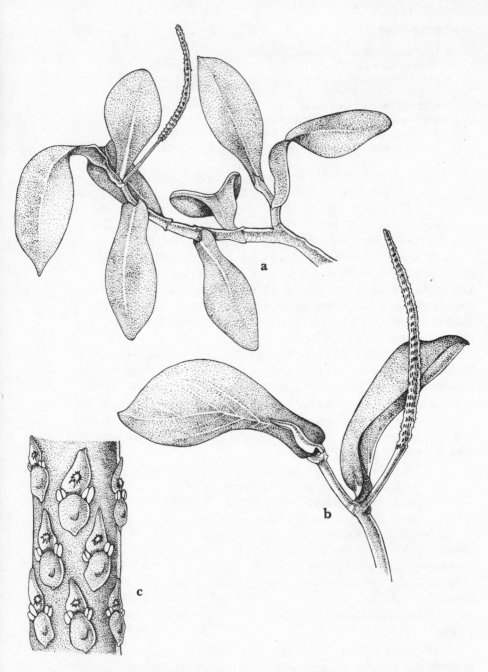

Family 57. **PIPERACEAE.** *Peperomia spathulifolia:* a, portion of flowering stem, × ⅓; b, inflorescence and mature leaves, × ½; c, detail of floral axis showing individual flowers with two stamens, brushlike stigma, and subtending bract, × 6.

58. SALICÀCEAE Willow Family

Trees or shrubs with alternate, simple, deciduous leaves, dioecious. Flowers borne in catkins, perianth absent; staminate flowers with two or more stamens. Pistillate flowers with ovary sessile, one-locular, style two-parted. Fruit a two- to four-valved capsule containing numerous small seeds bearing fine hairs. Two genera, about 180 species, chiefly north temperate zone but extending into Arctic and Alpine regions, few species in warmer parts of the world. Many natural hybrids.

1. SÀLIX L. Willow

Trees or shrubs bearing pinnately veined, mostly narrowly lanceolate leaves, lateral buds with a single bud scale. Staminate flowers with usually two to eight stamens subtended by one to two glands. Pistillate flowers with basal glands and two-branched stigma. About 200 species and many natural hybrids.

1. S. caroliniana Michx. Coastal-plain Willow—Shrub or trees to 10 m. Leaves 2–3 cm wide, lanceolate to lance-ovate, long tapering, glaucous or whitish below, dark green above, finely serrate. Catkins 7–10 cm long, somewhat drooping, bracts densely pubescent. Near ponds, swamps, sloughs, and in low ground, Fla., not common in the Fla. Keys, and Tex. north to Md. and Kans. Spring. *S. longipes* Anders., *S. amphibia* Small

59. MYRICÀCEAE Bayberry Family

Trees or shrubs with simple, alternate, resin-dotted leaves that are often aromatic, monoecious or dioecious. Flowers in dense, axillary spikes, with a perianth; sex of plant or shoot may vary from year to year; staminate flower subtended by a bract with two to twenty, or more often four, stamens. Pistillate flower with one-locular two-carpellate sessile ovary with short two-branched style. Fruit a small, often warty drupe with generally waxy exocarp. Three genera, 60 species of wide distribution in temperate and subtropical regions.

1. MYRÌCA L. Bayberry

Shrubs or low trees with evergreen leaves, blades entire or irregularly toothed, monoecious or dioecious. Staminate catkins with loosely

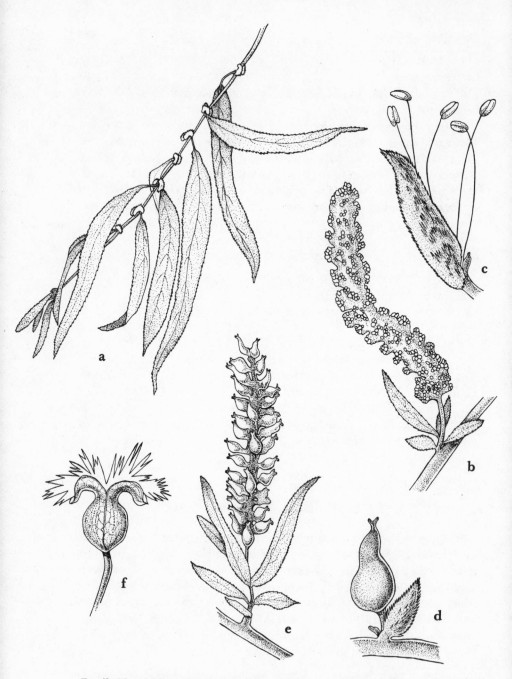

Family 58. **SALICACEAE.** *Salix caroliniana:* a, portion of stem with mature leaves and stipules, × ⅓; b, staminate inflorescence, × 1¼; c, staminate flower, × 10; d, pistillate flower, × 15; e, pistillate inflorescence, × ½; f, mature fruit, × 7.

imbricate bracts, flowers with two to sixteen stamens. Pistillate catkins not so conspicuously bracteate, the flowers with ovary subtended by two to four bracts. Fruit a globular drupe, in clusterlike spikes, waxy. *Cerothamnus* Tidestrom

1. Shrubs spreading, less than 1 m tall; leaf blades linear-spatulate to obovate, mostly less than 4 cm long . . . . *1. M. pusilla*
1. Tall shrubs or small trees over 1 m tall; leaf blades oblanceolate to elliptic-lanceolate, over 4 cm long . . . . *2. M. cerifera*

1. **M. pusilla** Raf. Dwarf Wax Myrtle—Shrub mostly 0.2–1 m tall, low spreading; leaves evergreen, 2–4 cm long, linear spatulate to obovate, usually entire, or with few small teeth near apex. Fruit 2–4 mm in diameter. Sandy soil, pinelands, Fla. and Tex. north to Va. on coastal plain. February, March. *M. pumila* (Michx.) Small; *Cerothamnus pumilus* Small; *M. cerifera* var. *pumila* Michx.

2. **M. cerifera** L. Wax Myrtle—Bushy shrub or tree to 12 m with oblanceolate to elliptic-lanceolate leaves 4–10 cm long but reduced toward tips of branches, toothed above the middle or entire. Staminate catkins 1–1.5 cm long, pistillate 0.5–1.0 cm. Fruit globose, 2–3 cm in diameter, waxy. Hammocks, wet sandy soil, borders of sink holes, swamps, Fla. and Tex. north to Ark., W.I. February, March. *Cerothamnus ceriferus* (L.) Small

60. JUGLANDÀCEAE Walnut Family

Trees with deciduous, alternate, pinnately compound leaves, monoecious. Staminate flowers in drooping elongate spikes, calyx three- to six-lobed, stamens three to forty. Pistillate flower in erect spikes, calyx one- to four-lobed, ovary inferior, one-locular. Fruit a drupe or nut covered by a dehiscent or indehiscent involucre. Six genera, about 40 species, north temperate and tropical Asia. Family includes pecan, walnut, hickory, and butternut trees.

1. CÀRYA Nutt. Hickory

Trees with hardwood, solid, continuous pith, and odd-pinnate leaves. Staminate catkins produced in groups of three, the calyx two- to three-lobed in axil of a larger bract, stamens two to ten, usually four. Pistillate spikes of two to ten flowers or flowers sometimes solitary, sub-

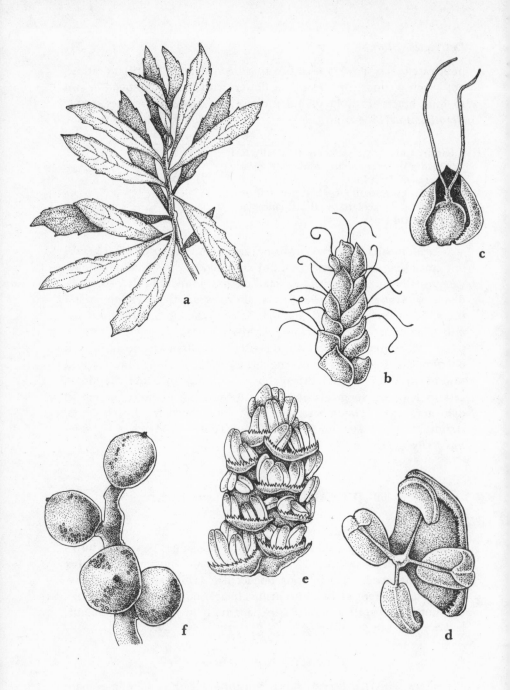

Family 59. **MYRICACEAE.** *Myrica cerifera:* a, portion of stem with mature leaves, × ¾; b, pistillate inflorescence, × 13½; c, pistillate flower, × 25; d, staminate flower, × 20; e, staminate inflorescence, × 7½; f, mature fruits, × 6.

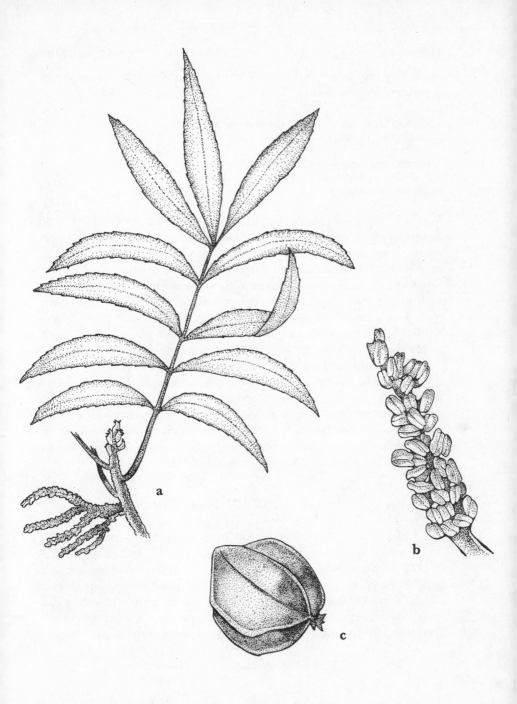

Family 60. **JUGLANDACEAE.** *Carya aquatica:* a, flowering branch with mature leaf, × ½; b, staminate inflorescence, × 6; c, mature fruit, × 1¼.

tended by four-lobed involucre. Fruit a nut enclosed by a four-valved involucre. *Hicoria* Raf.

1. Leaflets seven to thirteen, glabrous or tomentose on lower surface; bud scales valvate . . . .   1. *C. aquatica*
1. Leaflets five, rarely seven, rusty pubescent
   when young, becoming glabrous; bud
   scales imbricate . . . .   2. *C. floridana*

1. C. aquatica (Michx. f.) Nutt. WATER HICKORY—Trees up to 30 m or more tall, bark scaly. Leaflets seven to thirteen, ovate-lanceolate, rounded or cuneate-oblique at base, falcate, finely to coarsely serrate, covered with yellowish glandular dots, glabrous or tomentose on the lower surface. Catkins clustered or solitary; staminate flowers pedicelled, with six stamens. Pistillate flowers in clustered spikes, glandular-pubescent. Fruits usually clustered, compressed, conspicuously four-winged, wrinkled. River swamps and bottoms, low ground, Fla. and Tex. north to Miss. and Mo. $2n = 32$. Spring. *Hicoria aquatica* (Michx. f.) Britt.

2. C. floridana Sarg. SCRUB HICKORY—Shrubs or trees up to 25 m tall, bark close. Leaflets three to five, rarely seven, with rusty hairs especially on young petioles, lanceolate to oblanceolate, acuminate, rounded or cuneate-oblique at base, serrate. Staminate catkins scurfy pubescent, stamens four to five. Pistillate flowers in one- to two- flowered spikes. Fruit obovoid, variable, only slightly winged if at all. Sandy ridges and low hills, endemic to peninsula of Fla.

61. BATÀCEAE   SALTWORT FAMILY

Littoral, succulent shrubs with opposite, simple, fleshy, smooth leaves, dioecious. Flowers minute, in dense axillary spikes, calyx two-lipped, membranous, petals four, clawed, staminate flowers, stamens four, alternate with petals, filaments free. Ovary rudimentary or absent. Pistillate flowers four to twelve, perianth absent, joined in a conelike spike, ovaries eight to twelve, coherent, four-locular. One genus and one species widely distributed in tropical or warm temperate maritime areas.

1. BÀTIS P. Browne   SALTWORT

Characters of the family
1. B. maritima L. BATIS—Pale green, spreading, strong-scented shrub. Leaves 1–3 cm long, curved. Spikes up to 1 cm long, bracts suborbicular, often with an abrupt, pointed tip. Fruit 1–2 cm, oblong or

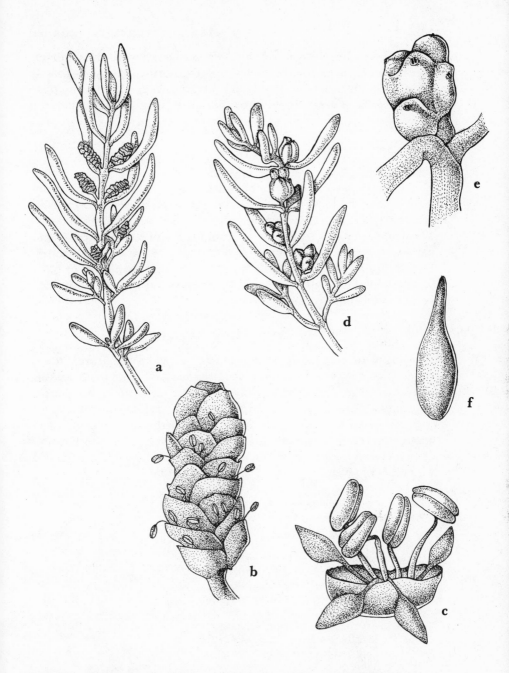

Family 61. **BATACEAE.** *Batis maritima:* a, flowering branch bearing axillary staminate inflorescences, × ¾; b, staminate inflorescence, × 4; c, staminate flower, × 7; d, flowering branch bearing axillary pistillate flowers, × ¾; e, pistillate inflorescence, × 3; f, mature fruit, × 6½.

obovoid, drooping. Shore habitats, salt marshes, low sandy seashores, rocky ground, mangrove belt, Fla. to Tex. north to N.C. on coastal plain, W.I:, Mexico, tropical America, Pacific Islands. Often forms pure stands over large areas, especially within small islands. Spring, summer.

62. FAGÀCEAE　Beech Family

Trees or shrubs with simple, alternate, deciduous or evergreen leaves, monoecious. Stipules deciduous. Staminate flowers numerous in erect or catkinlike spikes; calyx four- to six-lobed, stamens few- to many. Pistillate flower solitary within an involucre that becomes cupular and hardened in fruit, calyx four- to six-lobed, ovary inferior, three- to six-locular. Fruit a nut with a leathery or hard pericarp that is free or fused to involucre. Five genera, about 400 species, widely distributed in temperate and tropical regions.

1. QUÉRCUS L.　Oaks

Shrubs or trees with flowers appearing before the leaves in deciduous species. Leaves entire, lobed, or toothed. Staminate catkins slender, flowers with calyx six-lobed, stamens three to twelve, rarely sixteen. Pistillate flowers in short spikes or solitary, ovary three-locular, stigmas three. Nut partially enclosed by involucre (cupule) forming the acorn. About 275 species and many natural hybrids. The identification of oaks is complicated by widespread introgression and by the natural variability of several species.

1. Leaves or their lobes or teeth bristle tipped.
　2. Leaves deeply five- to seven-lobed, the sinuses more than one-half the distance to midrib . . . .　1. *Q. laevis*
　2. Leaves shallowly lobed or toothed, or entire, the sinuses if present less than one-half the distance to midrib.
　　3. Some or all the leaves oblanceolate or obovate, broadest above the middle.
　　　4. Leaves persistent, evergreen; margins slightly revolute; blades usually oblong-obovate; sandy pine woods or hammocks . . . .　2. *Q. myrtifolia*
　　　4. Leaves deciduous; margins flat; blades often with broadly dilated upper half, narrowed lower half; moist places . . . .　3. *Q. nigra*

3. Leaves narrowly lanceolate to ovate-lanceolate, widest at or below middle of blade.
  5. Low shrubs, never arborescent.
    6. Older leaves pubescent beneath; blades narrowly lanceolate to ovate-lanceolate, mostly entire . 4. *Q. pumila*
    6. Older leaves glabrous or puberulent beneath; blades ovate to elliptic-ovate, mostly toothed or lobed . . . . 5. *Q. minima*
  5. Trees . . . . 6. *Q. laurifolia*
1. Leaves or their lobes or teeth blunt or acute or only mucronate, not bristle tipped.
  7. Low shrubs; blades mostly toothed or lobed . . . . 5. *Q. minima*
  7. Trees or tall shrubs; blades entire or shallowly lobed, rarely toothed.
    8. Some or all the leaves widest above the middle with broadly dilated upper half, narrowed lower half, often two- to three-lobed at apex; leaves deciduous; moist places . . . . . . 3. *Q. nigra*
    8. Leaves widest at or near the middle; leaves persistent, evergreen; dry to moderate soils.
      9. Leaves ovate, oblong, or elliptic, entire, rarely lobed; acorn two times as long as wide or longer . . . 7. *Q. virginiana*
      9. Leaves oblong to obovate with slightly undulate margins or shallowly lobed or toothed above the middle of the blade; acorn less than two times as long as broad . 8. *Q. chapmanii*

1. **Q. laevis** Walt. TURKEY OAK—Trees often stunted, up to 20 m tall but usually smaller. Leaves mostly 10–20 cm long, oblong, obovate, gradually cuneate at base with deep sinuses forming three, five, or seven pinnate lobes, smooth except for pubescence of vein axils; leaves deciduous. Fruit (acorn) with white pubescence near the top, short pedunculate, 20–25 mm long, nut ovoid. Dry pinelands or sand ridges, Fla. and La. to Va. chiefly near coastal plain. February, March. *Q. catesbaei* Michx. Known to hybridize with Nos. 3 and 6.

2. **Q. myrtifolia** Willd. MYRTLE OAK—Shrubs or trees with short, spreading branches and smooth or shallowly furrowed barks. Leaves 2–6 cm long, variable, ovate or obovate, mostly dark green and shiny above, paler beneath, evergreen. Fruit sessile or short pedunculate; cup 10–12 mm wide, saucer shaped, puberulent within. Nut 10–15 mm long,

ellipsoidal. Dry sandy ridges, scrub vegetation, and coastal dunes, Fla. and Miss. north to S.C. February, March.

3. Q. nigra L. WATER OAK—Trees to 30 m tall, bark smooth or scaly, branches glabrous at maturity, buds puberulent. Leaves 5–14 cm long, obovate or cuneate, entire, or shallowly three- to five-lobed, often abruptly widened above, glabrous and dull green on both surfaces, evergreen or partially deciduous. Fruit 1–1.5 cm, cup saucer shaped, nearly sessile, nut 10–12 mm long. $2n = 24$. Moist sandy soil, swamps, river banks, hammocks, Fla. and Tex. to Del. and Mo. February, March.

4. Q. pumila Walt. RUNNING OAK—Low shrub with branches mostly 3–6 cm tall, stems underground. Leaves 6–12 cm long, variable, oblanceolate to elliptic, entire, finely pubescent beneath, deciduous. Fruits sessile or nearly so, cup 12–15 mm wide, involucre saucer shaped, nut ovoid 10–15 mm. Sandy soil, old fields, and pinelands, Fla. and Miss. north to N.C. on the coastal plain. February, March.

5. Q. minima (Sarg.) Small, DWARF LIVE OAK—Low shrub with branches 1 m tall or less, stems underground. Leaves 2–10 cm long, variable, repand toothed or entire, glabrous or finely pubescent beneath, evergreen or partially deciduous. Fruit pedunculate, clustered or solitary, cup (involucre) saucer shaped, mostly 12–15 mm wide; nut 15–17 mm long, ellipsoidal. Pinelands, old fields, and scrub vegetation, Fla. to Miss. north to N.C. on the coastal plain. February, March. *Q. virens* var. *dentata* Chapm.; *Q. virginiana* var. *minima* Sarg.; *Q. virginiana* var. *dentata* Sarg.

6. Q. laurifolia Michx. LAUREL OAK, WILLOW OAK—Trees up to 30 m tall with slender branches and dark scaly bark with deep furrows, branchlets glabrous, dark red when young. Leaves 4–12 cm long, elliptic; broadest above the middle, tapered at both ends, sometimes three-lobed at apex, shiny green above, duller green below, puberulent but with tufts of hairs in axils of veins, evergreen. Fruit pubescent, sessile or nearly so, broadly ovoid, cup (involucre) deeply saucer shaped, 9–15 mm wide. Nut ovoid, 10–15 mm. Sandy hammocks, stream banks, and swamps, Fla. and La. north to Va. February, March. *Q. phellos* var. *laurifolia* Chapm.; *Q. obtusa* (Willd.) Ashe; *Q. rhombica* Sarg. Hybridizes with No. 1.

7. Q. virginiana Mill. LIVE OAK—Trees up to 30 m with spreading crown, slender branchlets, or shrubby, bark slightly furrowed. Leaves 4–12 cm long, oblong, obovate, or elliptic, rounded at apex or tapered, base cuneate, usually entire, except on shoots, slightly or strongly revolute margins, glabrous to densely white pubescent beneath, dark green and shiny above, evergreen. Fruits in spikes of three to five or solitary,

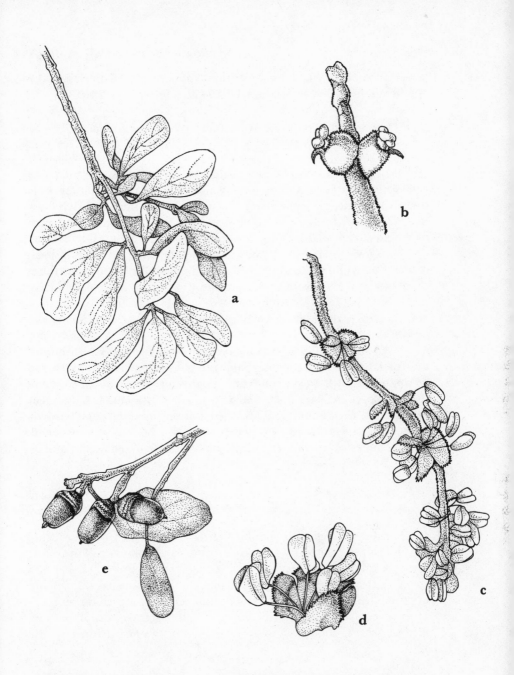

Family 62. **FAGACEAE.** *Quercus virginiana* var. *virginiana:* a, portion of stem with mature leaves, × ½; b, pistillate flowers, × 4½; c, staminate inflorescence, × 10; d, staminate flower, × 15; e, fruiting branch with mature fruits, × ¾.

pedunculate. Cup (involucre) turbinate, puberulent within, about 1.5 mm broad, nut ellipsoidal-ovoid, 20–25 mm long. Our varieties are:

1. Leaf margins flat or only slightly revolute . 　7a. var. *virginiana*
1. Leaf margins strongly revolute . . . . 　7b. var. *geminata*

    7a. Q. virginiana Mill. var. **virginiana,** LIVE OAK—Large trees with wide-spreading branches. Leaves oval-elliptic, margins flat or only slightly revolute, glabrous above, smooth below or only slightly pubescent. Hammocks, woods, into the mangrove belt, Fla. and Tex. north to Va. on coastal plain, Cuba, Mexico, Central America. March, April.

    7b. Q. virginiana Mill. var. **geminata** Sarg. SAND LIVE OAK—Shrub or small to medium-sized tree. Leaves oblong to elliptic, margins thickened and strongly revolute, conspicuously reticulate and hoary-tomentose below. Sandy soil, coastal dunes, oak scrub, Fla. and Miss. north to N.C., coastal plain. March, April. *Q. geminata* Small, *Q. virginiana* var. *maritima* (Michx.) Sarg.

    8. Q. chapmanii Sarg. SCRUB OAK—Shrub or tree to 15 m with stout branches. Leaves 5–10 cm long, oblong to obovate or spatulate, rounded at tip, entire with barely undulate margins, or obscurely lobed above the middle, thick, dark green above, pale or whitish below, puberulent or glabrous, deciduous. Fruit sessile, solitary or in pairs, cupule (involucre) deep, saucer shaped, 15–20 mm wide. Nut 1.5–2.5 cm long, ovoid. Coastal dunes, sandy soil, Fla. to S.C. on coastal plain. February, March.

63. ULMÀCEAE ELM FAMILY

    Shrubs or trees with simple, alternate leaves with oblique bases, stipules paired, monoecious, or flowers bisexual. Flowers small in cymose fascicles, calyx four- to eight-lobed, herbaceous, persistent; stamens inserted at the base of the calyx, the same number as the calyx lobes and opposite them; petals absent. Ovary composed of two fused carpels, one- to two-locular, styles two, divergent. Fruit a samara, drupe, or nut. Fifteen genera, about 150 to 200 species, mostly north temperate and tropical regions.

1. Flowers in branching cymes; fruits cymose;
    leaves densely pubescent beneath, mar-
    gins serrate . . . . 　1. *Trema*
1. Flowers solitary or in few-flowered clusters;
    fruit solitary; leaves glabrous or sparingly

pubescent beneath, margins entire or only
upper half serrate . . . .　2. *Celtis*

1. TRÈMA Lour.　NETTLE TREES

Trees and shrubs with smooth bark. Leaves oblique, serrate, persistent. Flowers polygamous in axillary cymes, calyx rotate, five-parted, the lobes exceeding the tube. Stigmas two. Fruit a drupe surrounded by persistent perianth. About 30 species in the tropics and subtropics.

1. Leaves 6–12 cm long . . . .　1. *T. micrantha*
1. Leaves 1–3 cm long . . . .　2. *T. lamarckiana*

1. Trema micrantha (L.) Blume, FLORIDA TREMA—Trees or shrubs up to 25 m with pubescent twigs. Leaves 6–12 cm long, oblong-lanceolate, cordate or obtuse at base, serrate, upper surface rough pubescent. Flowers greenish yellow in axillary cymes. Drupe about 3 mm long, orange or yellow, ovoid, smooth. In cleared or disturbed areas, edge of hammocks, south Fla., W.I., and tropical America. All year. *T. floridana* Britton

2. Trema lamarckiana (R & S) Blume, WEST INDIAN TREMA—Shrubs or small trees up to 6 m tall with thickly pubescent twigs. Leaves usually 1–3 cm long, lanceolate, elliptic or ovate, serrate, upper surface very rough pubescent. Flowers whitish or pinkish white in axillary cymes. Drupe pink, ovoid, about 3 mm long, smooth. Hammocks, disturbed areas, roadsides, south Fla., Fla. Keys, W.I. All year.

2. CÈLTIS L.　HACKBERRY

Shrubs or trees bearing toothed or entire, oblique, deciduous, alternate, simple leaves; leaf blades with three main veins. Flowers in cymes, racemose or solitary, polygamous, perianth usually five-parted, rotate, greenish colored. Fruit a drupe, subglobose. About 60 species in temperate and tropical areas.

1. C. laevigata Willd. HACKBERRY, SUGARBERRY—Trees up to 30 m tall with smooth bark or bearing corky outgrowths. Leaves 6–12 cm, variable, two-ranked, lance-ovate to ovate, entire· or serrate, often long tapering, slightly oblique to the base. Flowers small, greenish and unisexual, staminate with five to six stamens. Pistillate flowers two-carpellate, one-locular. Drupe 5–7 mm wide, ovoid-globose, reddish brown or orange red. $2n = 20$. Hammocks, river bottoms, disturbed areas, Fla. and Tex. north to Md. and ·Mo., W.I., Mexico. Spring. *C. mississippiensis* Bosc.

Family 63. **ULMACEAE.** *Trema micrantha:* a, flowering branch, × ½; b, flower, × 22½; c, mature fruit, × 7.

Celtis iguanaea (Jacq.) Sarg. [*Momisia iguanaea* (L.) Rose & Standl.] and **C. pallida** Torr. [*Momisia pallida* (Torr.) Planch.] are reported for peninsular Fla. and may occur in our area. Both are tropical spiny shrubs, although the latter species is smaller than the former and has scabrous leaves.

64. MORÀCEAE MULBERRY FAMILY

Trees or shrubs, rarely herbs, with milky juice, monoecious or dioecious. Leaves mostly alternate, simple or palmately lobed, stipules two, usually caducous. Flowers small, often in cymes, heads, catkins, umbels, or hollow receptacles; calyx lobes usually four or absent; staminate flowers with one to four stamens opposite the sepals; pistillate flowers with two carpels, two-locular. Fruit an achene or drupelike, more often a multiple fruit from fusion of fruits from several flowers. About 60 genera and 1000 species in tropical, subtropical, or temperate regions. *Artocarpaceae, Cannabinaceae*

1. Leaves serrate or dentate, often one- to three-lobed; flowers and fruits on outside of receptacle; stems and leaves without milky juice . . . .   1. *Morus*
1. Leaves entire or undulate, or palmately five- to seven-lobed; flowers and fruits within receptacle; stems and leaves with milky juice.
   2. Receptacle surrounding female flower, two-branched style exserted near the top; receptacle covered with male flowers; petioles less than 1 cm long . .   2. *Brosimum*
   2. Receptacle surrounding both female and male flowers, with a small opening near the top, styles within; receptacle not covered with male flowers; petioles over 1 cm long . . . .   3. *Ficus*

1. MÒRUS L. MULBERRY

Trees or shrubs with smooth or scaly bark, bearing simple, alternate leaves, mostly dioecious. Blades serrate, varying from ovate-oblong to two- to nine-lobed. Flowers unisexual, staminate flowers in elongate catkins, calyx four-lobed, stamens four. Pistillate flowers with four-lobed calyx and two-locular ovary. Fruit a fleshy syncarp. About ten species, chiefly north temperate.

1. Leaf blades glabrous beneath or nearly so;
   fruit black at maturity . . . . . .   1. *M. nigra*
1. Leaf blades pubescent beneath; fruit red or
   purplish at maturity . . . . . .   2. *M. rubra*

1. **M. nigra** L. BLACK MULBERRY—Shrubs or small trees to 10 m with pubescent young branches. Leaves 5–15 cm long, ovate, pubescent above or nearly glabrous. Staminate catkins 1–2 cm long; fruits black, about 1.5–2 cm long. Disturbed areas, roadsides, Fla. and Tex. north to N.Y., naturalized from Asia. Spring.

2. **M. rubra** L. RED MULBERRY—Shrub or tree to 20 m tall with glabrous or slightly pubescent young branches. Leaves 5–20 cm long, ovate and unlobed to two- to several-lobed, pubescent on both surfaces. Staminate catkins 5–8 cm long; fruit about 3–6 cm long, dark purple or red. $2n = 28$. Hammocks, woods, and pinelands, Fla. and Tex. to Vt. and Dakotas. Spring.

2. BRÓSIMUM Sw.

Trees with milky juice, bearing alternate, entire leaves, monoecious or sometimes dioecious. Flowers in globular inflorescence; staminate flowers with one to four stamens intermixed with numerous, peltate bracts. Pistillate flowers within the receptacle, stigma deeply two-lobed. Fruit globose, within a fleshy receptacle. About 24 species, south Mexico to Argentina.

1. **B. alicastrum** Sw. BREADNUT—Trees to 30 m tall, with oval to elliptic leaves, mostly 5–15 cm long, glabrous, obtuse to acuminate at apex, rounded at base. Inflorescence 3–6 mm wide. Fruit usually 1.5 cm in diameter. South Mexico to South America, W.I., introduced, cultivated, and possibly naturalized in south Fla.

3. FÌCUS L. FIG

Trees, shrubs, or vines with milky sap and smooth bark, monoecious. Leaves entire or lobed, conspicuously pinnately veined, usually persistent. Flowers sessile within the inside of hollow, globose, axillary receptacles, staminate with one to two stamens. Pistillate flowers many. Achenes enclosed within an elongate or globular multiple fruit. Over 600 species, mainly tropical regions.

1. Leaf blades entire, glabrous.
   2. Fruits sessile or subsessile; leaf blade
      tapering to petiole . . . . . .   1. *F. aurea*
   2. Fruits peduncled; leaf blade broad at
      base . . . .   2. *F. citrifolia*

1. Leaf blade palmately three- to five-lobed,
   pubescent . . . . 3. *F. carica*

1. F. aurea Nutt, STRANGLER FIG, GOLDEN FIG—Trees up to 20 m, often starting as an epiphyte, then a vine, or treelike, with glabrous, leathery, elliptical or ovate leaves usually 5–15 cm long, acute or acuminate at base. Fruits red, brown, or yellow, spheroidal, about 2 cm in diameter. Seedling plants develop often as epiphytes, growing in the branches of palms or other large trees, producing aerial roots that grow around host plant and eventually kill it. $2n = 26$. Hammocks, various sites, south Fla., W.I. All year.

2. F. citrifolia Mill. WILD BANYAN TREE—Trees 12 m tall bearing elliptical or ovate leaves 5–10 cm long, glabrous, cordate or rounded at base. Fruits bright red, subglobose, 1–1.5 cm in diameter, peduncles 1–2 cm. Hammocks, many sites. South Fla., Fla. Keys, W.I., to South America. Spring, summer. *F. populnea* Willd., *F. brevifolia* Nutt., *F. laevigata* Vahl

3. F. carica L. COMMON FIG—Small tree to about 10 m high. Leaves 7–15 cm long, thick, mostly three- to five-lobed, rough pubescent above, puberulent below, with long petioles, deciduous. Fruits axillary and subterminal, highly variable in size, usually pear shaped. $2n = 26$. Introduced from the Mediterranean region and possibly naturalized in disturbed areas. Spring, summer. This is the common fig of commercial use.

F. elastica Roxb. INDIA RUBBER PLANT with large, elliptic glossy leaves 10–30 cm long may persist for short times in disturbed areas and along roadsides, but it is doubtfully naturalized in south Fla. It is a native of Asia.

65. URTICÀCEAE NETTLE FAMILY

Low shrubs or herbs usually with stinging hairs and epidermal cells with prominent cystoliths, dioecious or monoecious. Leaves simple, opposite or alternate with or without stipules. Flowers small, unisexual, in cymes or crowded in an enlarged receptacle, perianth with four to five lobes, staminate flowers four to five and opposite the lobes of the perianth. Pistillate flowers with one-locular ovary, styles simple. Fruit a fleshy drupe or achene. About 41 genera and 480 species, in the tropics and the temperate zones.

1. Leaves alternate, deltoid-ovate to subrhombic . . . . 1. *Parietaria*
1. Leaves opposite, lance-ovate to elliptic.

Family 64. **MORACEAE.** *Ficus aurea:* a, flowering branch, × ½; b, inflorescence, × 3½; c, pistillate flower, × 20; d, staminate flower, × 20; e, mature fruits, × 4¼.

2. Leaf blades serrate, over 2 cm long; stems
   erect . . . . 2. *Boehmeria*
2. Leaf blades entire, less than 1 cm long;
   stems decumbent . . . . 3. *Pilea*

1. PARIETÀRIA L. PELLITORY

Succulent herbs without stinging hairs. Leaves alternate and entire without stipules. Flowers unisexual or polygamous in axillary clusters; staminate flowers with four sepals, four stamens. Pistillate flowers with four-lobed calyx, ovary ovoid. Fruit an achene. A small genus of about seven species.

1. P. floridana Nutt.—Annual or short-lived perennial with stems much branched, usually 30–50 cm long, often puberulent. Leaves ovate to rhombic-ovate, 0.5–5 cm long. Sepals acute. Fruit ovoid about 1 mm long. Hammocks, moist soils, shaded areas, disturbed sites, Fla. to N.C. on the coastal plain, W.I. Spring. *P. nummularia* Small

2. BOEHMÈRIA Jacq. FALSE NETTLES

Trees, shrubs, or perennial herbs without stinging hairs, monoecious or dioecious. Leaves opposite or alternate, three-nerved, toothed. Flowers unisexual, staminate flowers with four-parted calyx, three to four stamens; pistillate flowers tubular, two- to four-lobed calyx. Fruit enclosed in persistent perianth. Over 50 species, chiefly in the tropics. *Ramium* Rumph.

1. B. cylindrica (L.) Sw. BUTTON HEMP—Monoecious, erect herbs up to 1 m or more tall, lower stems becoming woody. Cauline leaves opposite, leaves up to 15 cm long, on branches alternate, lanceolate to ovate, acuminate at apex, coarsely serrate. Flower clusters sessile, spicate. Fla. to Canada, W.I., Central America, Brazil. Our two varieties are:

1. Stems glabrous; petioles over half as long as
   the blades . . . . 1a. var. *cylindrica*
1. Stems pubescent; petioles less than half as
   long as the blades . . . . . . 1b. var. *drummondiana*

1a. Boehmeria cylindrica (L.) Sw. var. **cylindrica**, BUTTON HEMP— Stem 0.2–1.5 m tall, glabrous. Leaves 3–15 cm long with long petioles, thin, lanceolate to ovate, eventually smooth above, glabrous below. Fruit about 1 mm wide. Woods, shaded sites, often on cypress stumps, Fla. and Tex. north to Canada. Spring, summer. *B. cylindrica* (L.) Willd.

1b. Boehmeria cylindrica (L.) Sw. var. **drummondiana** (Wedd.) Wedd. BOG HEMP—Stem 0.2–1.2 m tall, scabrous. Leaves 2–6 cm long

Family 65. **URTICACEAE.** *Boehmeria cylindrica* var. *cylindrica:* a, flowering branch, × ½; b, staminate flower, × 12½; c, pistillate flower, × 22½; d, fruiting branch, × 3¾; e, fruit, × 25.

with short petiole, firm, elliptic to ovate, pubescent above, glabrous below or pubescent along the veins. Fruit about 1.5 mm wide. Low ground, swamps in shaded areas, Fla. and Tex. north to Conn., Kans. Spring, summer. *B. drummondiana* Wedd.; *B. scabra* (Porter) Small

3. PÌLEA Lindl. CLEARWEED

Annual or perennial, often succulent, creeping herbs without stinging hairs, monoecious. Leaves opposite, entire or serrate, those of each pair equal or very unequal. Staminate flowers four-parted, mixed with pistillate, unequally three-parted flowers. Fruit an achene, compressed, orbicular or ovate. Over 100 species, chiefly tropical regions. *Adicea* Raf.

1. Leaf blades suborbicular to reniform-orbic-
    ular, about as wide as long . . . . . 1. *P. herniarioides*
1. Leaf blades oblanceolate to elliptic, over
    two times as long as wide . . . . . 2. *P. microphylla*

1. P. herniarioides (Sw.) Lindl.—Stems prostrate or creeping, filiform or very slender, glabrous up to 10 cm long. Leaves about 2–4 mm long, minute, orbicular. Cymes sessile in axils of upper leaves, staminate flowers 2 mm long, pistillate flowers about 0.7 mm. Achene oblong-ovoid, less than 1 mm long. Moist, shady soils, south Fla., Fla. Keys, W.I. All year.

2. P. microphylla (L.) Liebm.—Stems up to 30 cm long, usually prostrate or creeping, much branched, glabrous. Leaves up to 2 cm, mostly obovate, or smaller leaves orbicular. Cymes sessile in axils ɪof upper leaves, staminate flowers less than 1 mm long, pistillate flowers about 1 mm. Achene ovate about 0.4 mm long, moist, shady places, south Fla., Fla. Keys, American tropics. All year.

66. PROTEÀCEAE PROTEA FAMILY

Trees or shrubs with alternate, entire or pinnately divided leaves. Flowers bisexual or unisexual, in spikes, racemes, or heads, often very showy, radially or bilaterally symmetrical. Perianth in one series, colored, four-parted, variously split when fully open; stamens four, opposite the perianth segments, epipetalous. Ovary one-locular, style simple. Fruit a capsule, follicle, drupe, or nut. About 50 genera, 1000 species, chiefly in dry regions of Australia and south Africa, cultivated in many parts of the world.

1. GREVÍLLEA R. Brown

Trees or shrubs with evergreen leaves of various forms. Flowers bisexual, bilaterally symmetrical, usually terminal and showy, perianth tube slender, the lobes fused after the tube opens. Fruit a capsule. About 200 species, mostly from Australia.

1. **G. robusta** A. Cunn. SILK OAK—Trees up to 50 m with immature stems rusty tomentose. Leaves 15–30 cm long, bipinnatifid, secondary lobes usually entire, lanceolate, silky pubescent beneath. Flowers orange or reddish in short racemes from leafless branches. Styles long. Fruit about 2 cm long. From Australia, commonly planted in south Fla. and possibly naturalized. Other members of this family in south Fla. include MACADAMIA TERNIFOLIA F. Muell. QUEENLAND NUT and STENOCARPUS SINUATUS Endl., a species related to FLAMETREE.

67. LORANTHÀCEAE MISTLETOE FAMILY

Small semiparasitic shrubs attached to host trees by means of suckers or haustoria. Leaves opposite or whorled, simple, leathery, greenish or yellow, persistent or reduced to scales. Flowers radial, bisexual or unisexual. Inflorescence cymose or spicate; perianth usually composed of two similar two-parted or three-parted whorls; sepals fused to ovary, petals free or fused, stamens the same number as the petals, epipetalous. Ovary inferior, fruit berrylike or drupaceous. About 30 genera, 1000 species, tropical and temperate regions.

1. PHORADÉNDRON Nutt. MISTLETOE

Shrubby plants with succulent stems. Leaves opposite, entire, leathery or reduced to scales, dioecious. Flowers small, in short, solitary or clustered spikes, staminate flowers with deeply three-lobed calyx, anthers. Pistillate flowers with ovoid ovary, fruit a globose sessile berry. About 100 species, W.I. and tropical America, few in North America.

1. Young branches four-angled; fruit lemon
    yellow to orange colored . . . . . 1. *P. rubrum*
1. Young branches terete; fruit white or yel-
    lowish white . . . . 2. *P. serotinum*
    var. *macrotomum*

1. **P. rubrum** (L.) Griseb.—Stems four-angled. Leaves 4–9 cm long, near-lanceolate to oblanceolate, apex obtuse, base tapered, three to five

Family 66. **PROTEACEAE.** *Grevillea robusta:* a, flowering branch, × ½; b, flower, × 4; c, flower, post-anthesis, × 2; d, fruits, × 1.

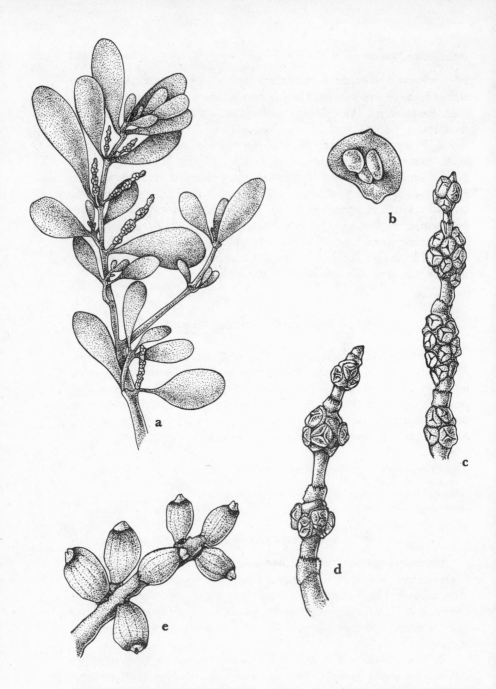

Family 67. **LORANTHACEAE.** *Phoradendron serotinum* var. *macrotomum:*
a, flowering branch, × ½; b, single petal with epipetalous stamen, × 20; c,
staminate inflorescence, × 5; d, pistillate inflorescence, × 6; e, mature fruits,
× 3½.

veins present. Spikes one to two axils, about as long as the leaves, conspicuously four- to five-jointed. Fruit lemon yellow to orange colored, globose. Parasitic on West Indian mahogany, south Fla., W.I., tropical America.

2. **P. serotinum** (Raf.) M. C. Johnston var. **macrotomum** (Trel.) M. C. Johnston—Stems terete, thick, brittle, diffusely branching. Leaves 2–6 cm long, oblanceolate to obovate, apex rounded, cuneate at base. Spikes usually solitary or two to three. Fruits white, globose. Parasitic on many kinds of trees, south Fla. and N. Mex., north to N.J., Mo. Fall, winter. *P. flavescens* (Pursh) Nutt., *P. eatoni* Trel.

P. trinervium (Lam.) Griseb., a West Indian species, has also been reported from Key Largo. It resembles **P. rubrum** except the leaves are obovate and generally smaller in **P. trinervium.**

68. OLACÀCEAE Ximenia Family

Trees, shrubs, or climbing plants with alternate, simple leaves. Flowers are small, radially symmetrical and usually bisexual, calyx lobes are imbricate, petals free or partially fused, stamens free or epipetalous opposite the petals or more numerous. Ovary superior or slightly inferior, one- to three-locular. Fruit usually a drupe or nut. About 25 genera and 125 species in the tropics and subtropics.

1. Stems unarmed; corolla reddish; fruit enclosed by disk . . . . 1. *Schoepfia*
1. Stems armed; corolla yellow or yellowish white; fruit naked . . . . 2. *Ximenia*

1. SCHOÈPFIA Schreb.

Unarmed, smooth trees or shrubs, reported as parasitic on roots. Leaves entire, alternate. Inflorescence axillary, usually a short raceme, calyx minute, petals three to six fused into a campanulate corolla, the lobes reflexed, stamens as many as the petals, fused to the corolla. Ovary half-immersed in hypanthium. Fruit a drupe. About 38 species, tropical regions.

1. **S. schreberi** J. F. Gmel. Whitewood—Shrubs or small trees up to 8 m with young branches pale or white, bark of older stems conspicuously grayish white. Leaves 2–8 cm long, lanceolate to ovate, shiny. Flowers clustered in axils, calyx about 1 mm long, corolla tube 2.0–2.5 mm long, red, the lobes deltoid. Ovary globose, papillate. Fruit 10–12 mm long, ellipsoidal, smooth. Hammocks, south Fla., W.I., Central and

South America. All year. *S. chrysophylloides* (A. Rich.) Planchon, *S. americana* Willd.

2. XIMÈNIA L. Spanish Plum

Armed shrubs or trees, the sharp spines axillary, parasitic on roots. Leaves elliptic to ovate, mucronate or acuminate at apex. Inflorescence, umbellate, simple or compound, calyx minute, petals four to five distinct with reddish brown pubescence, stamens eight to ten. Ovary conical shaped. Fruit an ellipsoidal drupe. About 15 species, pantropical.

1. **X. americana** L. Tallowwood—Shrub or small tree to 7 m. Leaves ovate, usually 3–10 cm long, mucronate or retuse at apex. Inflorescence with many small flowers, corolla yellow or pale yellow, densely pubescent within. Ovary 3–4 mm long, glabrous. Fruit ellipsoidal, pale yellow. Hammocks, south Fla., Fla. Keys, W.I. All year.

This species is especially common in the Fla. Keys where it may be found along roads and trails and down to the high-water level on the beach. The fruits are edible.

69. ARISTOLOCHIÀCEAE Birthwort Family

Twining vines, herbs, or shrubs with alternate leaves. Flowers bisexual, axillary, often brown, purple, or red, stamens six to twelve forming a column fused or connivent to style. Ovary inferior or superior with six fused carpels. Fruit a capsule. Over 200 species in temperate and tropical regions of the world.

1. ARISTOLÒCHIA L. Birthwort

Herbs, shrubs, or vines with broad, palmately veined leaves. Flowers axillary, clustered or solitary, strongly bilaterally symmetrical, often S-shaped or pipe shaped, or straight with a single terminal lobe, stamens six, anthers fused to stigma. Ovary inferior. Hypanthium usually ribbed. Fruit a six-lobed or six-angled capsule. Tropical and temperate regions.

1. **A. pentandra** Jacq.—Stems usually twining or decumbent; leaves 5–10 cm long, thick, broadly ovate. Perianth greenish or purplish, hypanthium puberulent. Fruit about 17–20 mm long, globular, drooping, wing angled. Hammocks, south Fla., W.I. All year.

A. maxima L. Dutchman's Pipe—With much larger S-shaped flowers of varied color, is cultivated widely, and has escaped locally. It is of tropical American origin.

Family 68. **OLACACEAE.** *Ximenia americana:* a, flowering branch, × ½; b, flower, × 5; c, fruit, × 1½.

70. POLYGONÀCEAE Buckwheat Family

Herbs, shrubs, vines, or trees with alternate, simple leaves. Base of the petiole expanded into a characteristic membranous sheath (ocrea); nodes are often swollen. Flowers bisexual or unisexual, small, radial, produced in large numbers in a racemose inflorescence. Sepals three to six, becoming membranous in fruit, usually persistent, petals absent. Stamens six to nine or fewer. Ovary superior, three-carpellate and unilocular, styles two to four. Fruit a three-sided or two-sided nut. About 40 genera and 800 species, chiefly north temperate regions, but cosmopolitan.

1. Trees, shrubs, or climbing vines with tendrils.
  2. Climbing vines; sepals rose to purplish .   1. *Antigonon*
  2. Trees or shrubs; sepals whitish.
    3. Lateral branches arising from nodes having sheathing ocreae . . . .   2. *Coccoloba*
    3. Some or all the lateral branches appearing to arise internodally between adjacent sheathing ocreae . . . .   3. *Polygonella*
1. Herbs, sometimes woody at base.
  4. Some or all the lateral branches appearing to arise internodally between adjacent sheathing ocreae . . . . .   3. *Polygonella*
  4. Lateral branches arising from nodes having sheathing ocreae.
    5. Perianth segments six in two series; fruit appearing wing margined from attached inner sepals; flowers in spikelike panicles . . . . . .   4. *Rumex*
    5. Perianth segments five; fruit not wing-margined; flowers in narrow spikelike racemes . . . .   5. *Polygonum*

1. ANTÍGONON Endl. Corallina

Climbing vines with woody bases. Leaves ovate to broadly sagittate; ocreae scalelike or much reduced. Flowers bisexual, in terminal and axillary racemes, peduncles ending in tendrils. Perianth five- to six-parted, colored. Stamens fused at the base, styles three. Fruit three-angled. About four species of tropical America. *Corculum* (Endl.) Stuntz

1. A. leptopus Hook. & Arn. Coral Vine—Leaves ovate to

broadly sagittate, 8–15 cm long, cordate at base. Sepals rose colored or rose purplish up to 2 cm long. Fruit 8–9 mm, beak three-angled. Disturbed areas, hammocks, roadsides throughout peninsular Fla. Escaped from cultivation and naturalized from Mexico. All year. *Corculum leptopus* (Hook & Arn.) Stuntz

2. COCCÓLOBA P. Browne ex L.   SEA GRAPE

Trees or shrubs with alternate, persistent, entire leaves, ocreae cylindrical and conspicuous. Flowers bisexual, or unisexual variously borne but often in spikelike racemes, terminal or axillary, perianth five-parted, becoming fleshy and enclosing the fruit, stamens eight, styles three. Achene three-angled, surrounded by fleshy hypanthium. About 150 species in tropical and subtropical America. *Coccolobis* P. Browne

1. Leaves orbicular, broader than long, cordate at base, thick and leathery . . . 1. *C. uvifera*
1. Leaves lance-ovate to ovate, longer than broad, tapering to base, thin . . . . 2. *C. diversifolia*

1. C. uvifera (L.) L. SEA GRAPE—Trees or shrubs up to 15 m tall, with stout branches. Leaves 10–27 cm long, orbicular to reniform, thick, fleshy, base rounded to cordate. Inflorescence racemose, stout, up to 30 cm long, staminate flowers clustered, pistillate flowers solitary. Fruit dark reddish in dense grapelike clusters. Coastal hammocks and dunes, south Fla., Fla. Keys, W.I., Central and South America. Spring, fall.

The species is frequently planted, and the edible fruits are often used in making jelly.

2. C. diversifolia Jacq. TIE TONGUE—Trees or shrubs with ovate to obovate leaves usually 6–10 cm long, cuneate or rounded at the base. Racemes 5–8 cm long, perianth parts up to 3.5 mm long. Fruit in drooping clusters, black. $2n = c. 400$. Hammocks, common in Fla. Keys, south Fla., W.I. Spring, fall. *C. floridana* Meissner; *C. laurifolia* Jacq.

3. POLYGONÉLLA Michx.   JOINTWEED

Annual or perennial herbs or low shrubs. Leaves small, numerous. Flowers in spikelike racemes, racemes paniculately arranged, white to red, bisexual or functionally unisexual. Perianth five-parted, persistent; stamens eight. Ovary three-angled, occasionally lenticular, styles three. Fruit a three-angled achene which the inner sepals tightly invest, forming winglike structures. About ten species in North America. *Dentoceras* Small; *Delopyrum* Small; *Thysanella* Gray

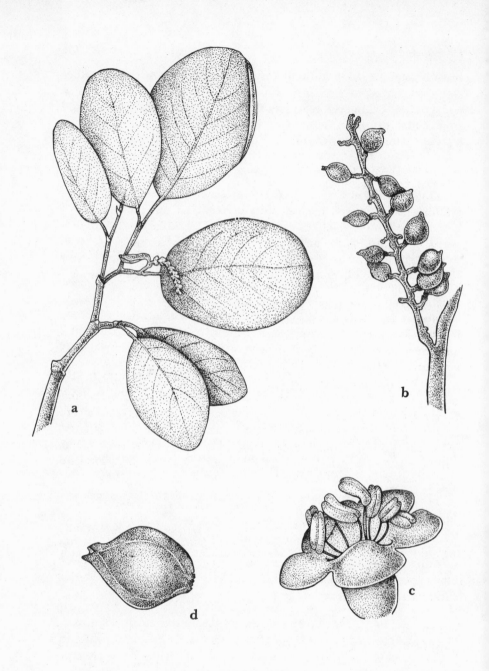

Family 70. **POLYGONACEAE.** *Coccoloba diversifolia:* a, flowering branch, ×
½; b, fruiting branch, × 1; c, flower, × 22½; d, fruit, × 2.

1. Leaves persistent, stems appearing leafy.
   2. Ocreae fringed with long bristles; inner
      sepals fimbriate . . . .   1. *P. fimbriata*
                                        var. *robusta*
   2. Ocreae not fringed with bristles or cilia;
      inner sepals not fimbriate . . . .   2. *P. polygama*
1. Leaves early deciduous, stems appearing
   leafless.
   3. Some or all the ocreae ciliate; inner
      sepals (pistillate) linear . . . . .   3. *P. ciliata*
   3. Ocreae not ciliate; inner sepals (pistillate)
      elliptic . . . .   4. *P. gracilis*

1. **P. fimbriata** (Ell.) Horton var. **robusta** (Small) Horton, SAND HILL POLYGONELLA—Perennial, becoming woody shrubs up to 1 m tall. Leaves linear, mostly 6 cm long; ocreae conspicuously fringed with long bristles. Racemes 2–5 cm long; inner sepals fimbriate. Achene about 2 mm wide, beaked. $2n = 26$. Pinelands, scrub vegetation and sand hills, Fla. Summer, fall. *Thysanella robusta* Small

2. **P. polygama** (Vent.) Engelm. & Gray—Slender perennials, becoming woody with stems up to 3–6 dm tall, often diffusely branching. Lower leaves 1–3 cm long, filiform, linear to narrowly spatulate or cuneate. Inflorescence racemose, 1–3 cm long. Sepals pinkish or white, ovate. Fruit ovoid, about 1–2 mm long. $2n = 24$. Sandy soils, pineland, scrub forests, coastal plain, Fla. to Va. Summer, fall. *Polygonella brachystachya* Meissner

Plants of this species are highly variable in height, leaf length and width, and raceme length.

3. **P. ciliata** Meissner—Annuals with stems slender up to 2 m tall. Leaves filiform or linear, about 2–4 cm long. Sepals oblong or nearly so, inner sepals of pistillate flowers much longer than outer. Fruit about 2 mm long. $2n = 20$. Coastal pinelands, Fla. Summer, fall. *Delopyrum ciliatum* (Meissner) Small

4. **P. gracilis** (Nutt.) Meissner, TALL POLYGONELLA—Stems up to 17 dm tall, stem leaves linear-spatulate or oblong-spatulate, mostly 2–3 cm long. Sepals of staminate flowers obovate; inner sepals of pistillate flowers longer than outer. Fruit about 2.5 mm long. $2n = 20$. Sand hills, sandy pinelands, Fla. to S.C. Summer, fall. *Delopyrum gracile* (Nutt.) Small

P. myriophylla (Small) Horton, a woody prostrate perennial, with smooth, needlelike leaves 5–10 mm long, and short racemes of white, pink, or yellow flowers, has been found in Dade County. It is endemic to the white sand ridges of the Florida lakes region. *Dentoceras myriophylla* Small

4. RÙMEX L.   Sheep Sorrel, Water Dock

Smooth, usually perennial herbs often becoming tall and shrubby, stems fleshy. Leaves alternate, entire or rarely toothed, ocreae hyaline. Flowers bisexual or unisexual, crowded in panicles, perianth six-parted, greenish, unequal, the outer three smaller and spreading or reflexed, the inner three cordate and becoming fused to the achene, stamens six. Ovary three-angled; fruit a three-angled achene enclosed by the fused calyx. About 170 species, cosmopolitan. *Acetosella* Raf.

1. Sepal wings serrate or spinulose; leaf blades
     elliptic to shallowly cordate . . . . . 1. *R. pulcher*
1. Sepal wings undulate; leaf blades linear-
     lanceolate to lanceolate . . . . . . 2. *R. verticillatus*

     1. R. pulcher L.—Stems up to 7 dm tall, slender or branching. Lower leaves about 8–20 cm long, elliptic-ovate, cordate or truncate at base. Inflorescence with many spikelike branches, flower clusters separate, conspicuously bracted, sepal wings of fruit with spinose teeth on each margin. Achene about 2 mm long. Disturbed areas, roadsides, low woods, Fla. and Tex., north to Del. and Okla., Pacific coast, W.I., and Mexico; naturalized in Europe. Spring, fall.
     2. R. verticillatus L. Water Dock—Stems up to 1 m tall, with numerous short branches. Leaves linear-lanceolate to lanceolate, tapered at the base. Flower clusters dense, sepal wings of fruit reniform, to deltoid, not spinose; fruiting pedicels strongly deflexed. Achene about 3 mm long. Inland swamps and wet ground, Fla. and Tex., north to Quebec, Wis. and Kans. Spring, summer. *R. floridanus* Meissner

5. POLÝGONUM L.   Knotweeds, Smartweed

Annual or perennial herbs, sometimes shrublike, ocreae conspicuous. Flowers perfect; calyx four- to six-parted, sometimes brightly colored, persistent and enclosing the three-angled achene, stamens three to eight. Ovary flat or three-angled with minute stigmas. About 300 species, mostly north temperate regions. *Persicaria* Adans., *Persicaria* Mill.

1. Ocreae not fringed with cilia or bristles . 1. *P. densiflorum*
1. Ocreae fringed with cilia or bristles.
     2. Stems hirsute . . . . 2. *P. hirsutum*

2. Stems glabrous or merely puberulent.
  3. Sepals and ocreae glandular-punctate .   3. *P. punctatum*
  3. Sepals and ocreae not glandular-punc-
    tate.
    4. Plants annual, roots fibrous . . .   4. *P. persicaria*
    4. Plants perennial, rhizomes woody.
      5. Ocreal bristles over 1 cm long . .   5. *P. setaceum*
      5. Ocreal bristles shorter, less than 1
        cm long . . . .   6. *P. hydropiperoides*

1. P. densiflorum Meissner—Perennial with stems up to 1.5 m. Leaves 5–25 cm long, lance-acuminate. Inflorescence racemose or paniculate, lax up to 6 cm, sepals punctate, white, about 3 mm long, up to 4 mm at maturity. Fruit 2–3 mm long, black. Inland swamps, shallow water, wet ground, Fla. and Tex., north to N.J., south Mo., W.I., South America. Summer, fall. *Persicaria portoricensis* (Bertero) Small

This species has been found in Brevard County and may be expected to occur within our range, although we have seen no records.

2. P. hirsutum Walt.—Perennial with stems hirsute. Leaves 5–10 cm long, lanceolate, cordate at base or nearly so; ocreae with coarse hairs. Inflorescence paniculate, erect, ocreolae barely fringed, cilia deciduous, sepals about 3 mm long at maturity, white or pinkish white. Fruit 2 mm long. Inland swamps, wet ground, Fla. and Ga., coastal plain. Spring, fall.

This species is probably rare in our area; it is more abundant in north Fla.

3. P. punctatum Ell. Water Smartweed—Annual with stems up to 1 m, simple or sometimes branching, glabrous or nearly so. Leaves up to 20 cm long, lance-linear to elliptic. Ocreae fringed, glabrous or pubescent. Racemes slender, up to 10 cm long, ocreolae fringed, sepals greenish or greenish white, up to 4 mm long at maturity. Achenes usually three-angled or lenticular, shiny, about 3 mm long. Inland swamps, margins of mangrove stands, shallow water, wet ground, Fla. to Canada, west to Pacific Ocean, tropical America. All year. *P. acre* HBK; *Persicaria punctata* (Ell.) Small

4. P. persicaria L. Heartweed—Erect annual with stems up to 8 dm, usually much branched, glabrous or nearly so. Leaf blades linear-lanceolate to lanceolate often with dark purplish blotch, ocreae thin, fringed. Panicles spicate, densely flowered, ocreolae short ciliate, perianth up to 3 mm long, sepals pink or rose white. Achenes mostly lenticular, sometimes three-angled, up to 2.5 mm long, shiny. $2n = 40,44$. Disturbed areas, roads and trails, low ground, Fla. and Tex., north to Canada, Alaska, Calif., naturalized from Europe. All year. *Persicaria persicaria* (L.) Small

5. **P. setaceum** Baldw. ex Ell.—Erect perennial from elongate rhizomes. Leaves up to 20 m long, linear-lanceolate to elliptic-lanceolate. Ocreae cylindrical, bearing marginal bristles up to 1.5 cm long. Panicles slender, ocreolae long-fringed, perianth up to 3 mm long, sepals pink or white. Achene 2–3 mm long, three-angled, shining. Everglades, swamps and moist places, margins of pools, Fla. and Tex., north to Mass., on coastal plain, Mo. All year. *Persicaria setacea* (Baldw.) Small; *Polygonum hydropiperoides* var. *setaceum* (Baldw.) Gl.

6. **P. hydropiperoides** Michx.—Perennial from rhizomes, erect up to 1 m tall. Leaves up to 13 cm long, narrowly lanceolate to lance-ovate, sometimes linear, often bearing stiff hairs. Ocreae always strigose and ciliate along the margins. Racemes slender, pedicels hairy, ocreolae short ciliate, calyx white, green, or pink. Achenes about 2.5 mm long, ovoid, black, smooth and shining, three-angled. Inland swamps, moist soil, low woods, Canada to South America. Spring, fall. *Persicaria hydropiperoides* (Michx.) Small

Polygonum lapathifolia L., an annual with glabrous ocreae, has been collected in the Miami area, but it apparently is rare. It is an introduced weed. *Persicaria lapathifolia* (L.) Small

71. CHENOPODIÀCEAE Goosefoot Family

Annual or perennial herbs or shrubs with stems sometimes jointed, often glaucous, and alternate or opposite leaves or apparently leafless. Flowers small and inconspicuous, bisexual or unisexual, often green, mostly radially symmetrical, calyx three- to five-lobed, petals absent, stamens as many as the calyx lobes and opposite them. Ovary superior or partially inferior, two- to three-carpellate, unilocular. Fruit a utricle or nut surrounded by a persistent perianth. About 75 genera and 500 species, nearly all halophytic, often weedy.

1. Stems apparently leafless, stems conspicuously jointed; flowers sunken in cavities of stems . . . . 1. *Salicornia*
1. Stems leafy, stems not jointed; flowers not sunken in cavities of stems.
  2. Leaves toothed, dissected or lobed, over 5 mm wide, not scalelike.
    3. Flowers bisexual; fruit not enclosed by bracts . . . . 2. *Chenopodium*
    3. Flowers unisexual; fruit enclosed by bracts . . . . 3. *Atriplex*
  2. Leaves entire or scalelike, mostly less than

5 mm wide.
4. Leaves linear, not spinescent . . . 4. *Suaeda*
4. Leaves scalelike, spinescent . . . . 5. *Salsola*

1. SALICÓRNIA L. GLASSWORT, SAMPHIRE

Annual or perennial succulent herbs or shrubs with jointed, opposite, branching stems. Leaves reduced to minute scales, opposite, stems appearing leafless. Flowers bisexual or unisexual sunken in depressions along the axis of the spike. Stamens one to two, styles two; utricle enclosed by fleshy calyx. About 15 species of saline soil, cosmopolitan.

1. Annual, main stem erect . . . . . . 1. *S. bigelovii*
1. Perennial, main stem trailing with ascending, slender branches . . . . . . 2. *S. virginica*

1. **S. bigelovii** Torr. ANNUAL GLASSWORT—Erect, stout annuals with green stems up to 6 dm tall. Spikes 2–10 cm, becoming bright red and succulent. Seeds black, about 2 mm long. Salt marshes, coastal areas, beaches, marl soils, Fla. and Fla. Keys and Tex., north to Nova Scotia, Calif., Mexico, W.I. All year.
2. **S. virginica** L. PERENNIAL GLASSWORT—Perennial with decumbent or prostrate stems and branches, often matted, greenish, turning lead colored or light brown and producing erect or ascending flowering stems. Spikes 1–4 cm long. Seeds about 1 mm long. $2n = 18$. Salt marshes, sea beaches, marl soils, Fla. and Tex., north to Mass., Alaska to Calif., W.I., Europe, Africa. All year. *S. perennis* sensu Standl., *S. ambigua* Michx.

2. CHENOPÒDIUM L. GOOSEFOOT

Annual or perennial herbs with entire or lobed, alternate, often scaly leaves. Flowers bractless, greenish, bisexual; calyx two- to five-parted, stamens one to five, styles two to three. Ovary subglobose. Fruit a utricle. Plants often with strong scent. About 100 species, many weeds, cosmopolitan. *Roubieva* Moq.; *Botrydium* Spach; *Ambrina* Spach

1. Leaves and fruits gland dotted . . . . 1. *C. ambrosioides*
1. Leaves and fruits not gland dotted . . . 2. *C. album*

1. **C. ambrosioides** L. MEXICAN TEA—Erect, coarse, aromatic perennials with stout stems up to 1.5 m tall. Leaves lanceolate to elliptic-oblong, irregularly dentate to entire, gland dotted. Spikes dense, calyx about 1 mm, glandular, enclosing the fruit. Seeds reddish brown, 0.6–

0.8 mm wide. $2n = 32$. Disturbed places, cultivated ground, a cosmopolitan weed. Summer.

2. **C. album** L. LAMB's QUARTERS, PIGWEED—Erect annual with ascending branches, stems up to 2 m tall. Leaves 3–9 cm long, green or whitish, mealy, waxy; rhombic or rhombic-ovate, larger ones always toothed or shallowly lobed. Flowers in dense clusters in paniculate spikes, sepals white, mealy, keeled, covering fruit. Seeds black, shining, mostly 1 mm wide. $2n = 18, 36, 54$. A highly variable species throughout much of U.S., Canada, Eurasia, polyploid races naturalized from Europe. Summer. *C. berlandieri* Moq.; *C. lanceolatum* Muhl.

3. ÁTRIPLEX L.   ORACHES

Annual or perennial herbs or shrubs, often appearing scaly, monoecious. Leaves alternate or sometimes opposite, toothed or irregularly lobed or entire. Staminate flowers without bracts, three to five sepals and stamens. Pistillate flowers bracteate, without perianth, styles two. Fruit a utricle. About 120 species, chiefly temperate and subtropical regions.

1. Plants dark green; leaf blades dentate . .   1. *A. pentandra*
1. Plants silvery or pale green; leaf blades entire to obscurely serrate   . . . . .   2. *A. arenaria*

1. **A. pentandra** (Jacq.) Standl. CRESTED ATRIPLEX—Annual or perennial with stems green, up to 6 dm tall. Leaves mostly 1–3 cm long, elliptic to elliptic-ovate, with entire to undulate margins. Fruiting bracts 2–3 mm wide, serrate, bearing two to four tubercles. Seeds brownish. Coastal dunes, Fla. to N.H., W.I. Summer.

2. **A. arenaria** Nutt. SAND ATRIPLEX—Annual with stems erect or prostrate, silvery scurfy up to 5 dm. Leaves 1–3 cm, alternate, oblong to obovate, with entire to undulate or sparsely toothed margins. Fruiting bracts 3–5 mm or wider with four to seven teeth, bearing two to four tubercles; seeds reddish brown. Coastal dunes, south Fla., Fla. Keys, W.I., South America. Summer.

4. SUAÈDA Forsk.   SEA BLITE

Fleshy herbs with many alternate, entire, linear leaves. Flowers minute, bisexual or unisexual, inflorescence spicate, calyx five-lobed, stamens five, styles two to five. Fruit enclosed by persistent calyx. About 50 species, widely distributed. *Dondia* Adams.

1. Stems glabrous; seeds about 1 mm wide . . 1. *S. linearis*
1. Stems glaucous; seeds about 2 mm wide . . 2. *S. maritima*

1. S. linearis (Ell.) Moq. SEA BLITE—Glabrous, erect or ascending stems, much branched, up to 1 m tall. Leaves linear, nearly terete, dark green, up to 5 cm long, upper shorter. Sepals keeled about 2 mm wide. Seeds 1–1.5 mm wide. Salt marshes, coastal beaches, marl soils, Fla. and Tex. to Me., W.I. Spring, fall.

2. S. maritima (L.) Dum.—Glabrous and glaucous erect or ascending stems up to 4 dm tall, usually much branched. Leaves linear or almost cylindric, usually glaucous. Sepals pale green, rounded, or obscurely carinate on the back. Salt marshes, margins of seashores, disturbed areas, Fla. to Quebec, naturalized from Europe. Summer.

5. SÁLSOLA L.

Herbs with diffusely branched stems, linear succulent leaves, usually spiny at the tip. Flowers axillary, bisexual, calyx five-lobed, enclosing the fruit, stamens five, styles two. About 50 species, found primarily in saline soils, cosmopolitan.

1. S. kali L. SALTWORT—Annual with stems glabrous or pubescent up to 6 dm tall. Leaves alternate, subulate-linear, the upper spine tipped and subtending a short flower spike or flower. Fruiting calyx leathery, up to 1 cm wide, wings 5–9 mm wide. Seeds 2–3 mm thick. Sea beaches, from Fla. to La., north to Newfoundland. Summer.

72. AMARANTHÀCEAE AMARANTH FAMILY

Annual or perennial herbs with opposite or alternate simple leaves. Flowers radially symmetrical, small, bisexual or unisexual, often with dry, scarious, persistent bracts, sepals three to five free or nearly so, dry or membranous, petals absent, stamens usually five, opposite the sepals. Ovary superior, one-locular. Fruit a utricle. About 50 genera and 500 species, chiefly tropical and subtropical regions, including many weedy herbs.

1. Leaves alternate.
    2. Flowers green or reddish, mostly unisexual; leaf blades linear-lanceolate to broadly ovate . . . . 1. *Amaranthus*

2. Flowers white or greenish white, bisexual; leaf blades deltoid-ovate . . . .   2. *Celosia*
1. Leaves opposite.
   3. Flowers in heads or short spikes.
     4. Flowers in heads or spicate heads less than one and one-half times as long as broad; perianth villous or not.
       5. Perianth villous; spikes subtended by two leafy bracts . . . . . .   3. *Gomphrena*
       5. Perianth not villous; spikes not regularly subtended by two leafy bracts.
         6. Staminodes present; leaves thin, lanceolate to lance-ovate; hammocks and pinelands . . .   4. *Alternanthera*
         6. Staminodes absent; often fleshy, linear to ovate; maritime plants or weeds.
           7. Leaves fleshy, linear-lanceolate to ovate; maritime plants . .   5. *Philoxerus*
           7. Leaves not fleshy, rhombic-ovate; weedy nonmaritime plants . . . .   6. *Centrostachys*
     4. Flowers in spikes over two times as long as broad on long peduncles; perianth villous . . . .   7. *Froelichia*
   3. Flowers in dense or spreading paniculate cluster . . . .   8. *Iresine*

1. AMARÁNTHUS L.   Amaranths

Erect annual herbs with alternate, mostly entire leaves. Flowers monoecious, dioecious, or polygamous in dense spikes or clustered, each subtended by one to two bracts, sepals three to five; styles one to three. Utricle circumscissile or indehiscent. About 50 species, temperate and tropical regions. *Acnida* L.

1. Stems spiny . . . .   1. *A. spinosus*
1. Stems unarmed.
   2. Plants dioecious.
     3. Seeds flattened; leaves broadly lanceolate to ovate, strongly pinnately nerved . . . .   2. *A. cannabinus*
     3. Seeds lenticular; leaves linear to narrowly lanceolate, not strongly pinnately nerved . . . .   3. *A. floridanus*
   2. Plants monoecious or polygamous.
     4. Upper stem leaves linear-spatulate, less

than 1.5 cm long; spikes narrow,
mostly 2–5 mm wide . . . . .   4. *A. australis*
4. Upper stem leaves lance-ovate to elliptic-ovate, over 1.5 cm long; spikes
over 5 mm wide.
  5. Fruit rugose . . . .   5. *A. viridis*
  5. Fruit smooth . . . .   6. *A. hybridus*

1. **A. spinosus** L. SPINY AMARANTH—Stems erect, glabrous, bushy branched, up to 1 m tall with two spines at most nodes. Leaves usually 4–6 cm long, narrowly ovate to rhombic-ovate. Flowers unisexual, in slender spikes, terminal chiefly staminate, basal pistillate. Fruit about 2 mm. Seeds about 1 mm, brown or black. $2n = 34$. Disturbed areas, fields, roadsides; a cosmopolitan weed. All year.

2. **A. cannabinus** (L.) Sauer, WATER HEMP—Stems erect, branched, up to 3 m or taller, dioecious. Leaves about 15 cm long, lanceolate to lance-ovate, tapering to the apex. Flowers in large terminal panicles composed of many-flowered spikes, staminate flowers with five-parted calyx 2 mm long, pistillate flowers without calyx; style branches three to five. Fruit 2–4 mm long, seed 2–3 mm, flat. Salt marshes, coastal areas, Fla. to Me. All year. *Acnida cannabina* L.; *Acnida cuspidata* Bertero ex Spreng.

3. **A. floridanus** (Wats.) Sauer—Stems erect, branched or simple up to 1.5 m or taller. Leaves about 12 cm long, linear to narrowly elliptic. Flowers unisexual in narrow panicles; staminate flowers with calyx 2.5 mm long, female flowers with calyx 1.5 mm long. Fruit 1 mm long. Marshy areas, swamps, endemic to Fla. All year. *Acnida floridana* Wats.

4. **A. australis** (Gray) Sauer—Stems erect, simple or branched, up to 3 m tall, dioecious. Upper stem leaves linear-spatulate, lower leaves lanceolate. Flowers in narrow spikes, usually less than 5 mm wide, calyx five-parted, segments about 2.5–3 mm long. Fruit 1–2 mm long, seeds flat. Moist soil, margins of hammocks, Fla. to Tex., W.I., Mexico. All year. *Acnida australis* Gray; *Acnida alabamensis* Standl.

5. **A. viridis** L. SLENDER AMARANTH—Stems erect, slender, usually branched from near the base, up to 1 m tall, monoecious. Leaves about 4–7 cm long, ovate to rhombic-ovate, with long petioles. Flowers in narrow spikes arranged in terminal racemes, calyx of pistillate flowers three-parted. Fruit about 1.5 mm, rugose on drying. Seeds less than 1 mm wide. $2n = 34$. Disturbed sites, margins of woods, pantropical weed. All year. *A. gracilis* Desf.

6. **A. hybridus** L. COMMON PIGWEED—Stems erect, stout, freely branching, often pubescent up to 2.5 m tall, monoecious. Leaf blades

up to 15 cm long, rhombic-ovate to deltoid or ovate-lanceolate. Flowers in dense spikes arranged in terminal racemes, staminate calyx 2.0–2.5 mm long, pistillate calyx 1.5–2.0 mm long. Fruit about 2.5 mm long, only slightly rugose on drying. Seeds 1.0–1.3 mm wide. $2n = 32$. Disturbed sites, beach ridges, old fields, throughout tropical and temperate Americas; adventive in Old World. All year. *A. chlorostachys* Willd.

2. CELÒSIA L. Cock's Comb

Annual or perennial herbs or shrubs with alternate, entire or lobed leaves. Flowers bisexual, sepals five, membranous; stamens five, fused at base. Ovary subglobose, style one. Fruit a circumscissile utricle. About 40 species, native of the tropics.

1. **C. nitida** Vahl, Cock's Comb—Smooth, straggling perennial, subshrub, up to 1 m long. Leaves up to 7 cm long, ovate to deltoid-ovate. Flowers spicate, the spikes 2–3 cm long or longer, lax. Sepals yellowish white, about 4 mm long. Seeds less than 1 mm long. Hammocks, coastal dunes, south Fla., south Tex., and Mexico, W.I., Central America. Summer.

3. GOMPHRÈNA L. Globe Amaranth

Erect or decumbent perennial or annual herbs with opposite, entire, usually pubescent leaves. Flowers bisexual in spikes or heads usually subtended by leaflike bracts, sepals five, stamens five, fused at base forming a stamen tube. Ovary globose, style one. Fruit an indehiscent utricle. About 100 species in the tropics.

1. **G. decumbens** Jacq. Globe Amaranth—Stems decumbent or prostrate-diffuse, pilose pubescent. Leaves up to 5 cm long, obovate to oblong. Bracts whitish or purplish red, up to 2 cm long, subtended by leafy bracts 1–2 cm long, sepals linear-lanceolate, densely pubescent. Fruit 1–2 mm; seeds usually reddish brown, 1–2 mm broad. Disturbed sites, roadsides, margins of hammocks, south Fla., Fla. Keys, W.I., South America. All year. *G. dispersa* Standl.

4. ALTERNANTHÈRA Forsk. Chaff Flowers

Erect or prostrate perennials or annuals with opposite, usually entire leaves. Flowers bisexual, sessile or nearly so in spikes or heads, sepals five, white or greenish white, stamens three to five, united at base. Ovary globose, style one. Fruit a membranous utricle. About 100 species, principally in the Western Hemisphere.

1. Spikes axillary and sessile . . . . .   1. *A. maritima*
1. Spikes terminal or long peduncled from up-
   per leaf axile.
   2. Flowers short pedicelled; leaves mostly
      ovate or elliptic . . . .   2. *A. ramosissima*
   2. Flowers sessile; leaves mostly linear to
      lance-ovate . . . .   3. *A. philoxeroides*

    1. A. maritima (Mart.) Standl.—Stems procumbent or scandent, much branched, up to 2 m long, glabrous except for leaf axils. Leaves about 3–5 cm long, obovate-elliptic. Sepals 4–5 mm, ovate. Utricle obovoid. Seed 1 mm long, brownish. Sandy sea beaches, hammocks, south Fla., W.I., South America. All year. *Achyranthes maritima* (Mart.) Standl.; *Telanthera maritima* Moq.

    2. A. ramosissima (Mart.) Chodat—Stems much branched, ascending or diffusely spreading, up to 4 m long, striate or glabrate. Leaves up to 8 cm long, lance-ovate to elliptic-ovate. Sepals lance-oblong, 4–5 mm long, pubescent. Seeds 1.5 mm long, red brown. Hammocks, coastal areas, south Fla., Mexico, W.I., and South America. All year. *A. floridana* Small; *Achyranthes ramosissima* (Mart.) Standl.

    3. A. philoxeroides (Mart.) Griseb. ALLIGATOR WEED—Ascending or decumbent stems up to 1 m long, glabrous except for leaf axils. Leaves elliptic-linear to obovate, 5–11 cm long. Sepals broadly ovate, 6 mm long. Disturbed areas, low ground, ponds, and ditches, Fla. and La. north to N.C. on coastal plain, South America. All year. *Achyranthes philoxeroides* (Mart.) Standl.

5. PHILÓXERUS R. Brown   SAMPHIRE

    Procumbent or creeping, branched, perennial herbs with opposite, narrow, entire leaves. Flowers bisexual in dense, white spikes, sepals five, unequal, stamens five, fused at base. Utricle ovoid, indehiscent. Seed lenticular. About three species, pantropical in distribution.

    1. P. vermicularis (L.) R. Brown, SAMPHIRE—Stems somewhat fleshy, much branched, up to 2 m long. Leaves about 2–6 cm long, sessile, linear to oblanceolate. Spikes up to 3 cm long, globose or cylindrical, flowers silvery white. Sepals 3–5 mm long, oblanceolate. Seeds about 1 mm, orbicular, dark brown. Sandy beaches, coastal dunes near water, Fla. to Tex., W.I., Central and South America, Africa. All year. *Iresine vermicularis* (L.) Moq.

6. CENTROSTÁCHYS Wall.

    Perennial or annual herbs erect or decumbent with glabrous or pubescent stems and opposite, entire leaves. Flowers bisexual, whitish or

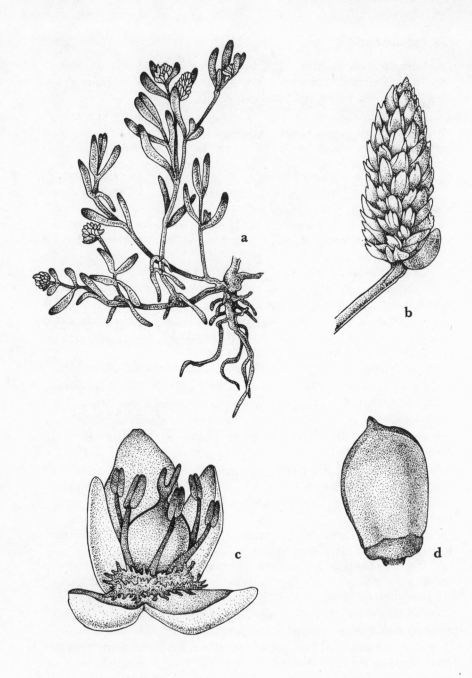

Family 72. **AMARANTHACEAE.** *Philoxerus vermicularis:* a, flowering plant, × ½; b, inflorescence, × 3½; c, flower, × 15; d, mature fruit, × 25.

greenish in slender spikes, sepals four to five, unequal; stamens two to five, united at base. Ovary oblong, glabrous. Utricle indehiscent. About ten species, tropical and subtropical regions.

1. **C. indica** (L.) Standl.—Erect annuals with stems up to 2 m, diffusely branching, puberulent. Leaves usually 3–8 cm long, obovate-orbicular, pubescent below. Flowers green, up to 4 mm long, midnerve of bract prolonged into a rigid spine somewhat longer than the bract. Sepals about 4 mm long, green; staminodia shorter than filaments. Fruit oblong, glabrous; seed about 2 mm long. Disturbed areas and old fields, Fla. to south Ala., W.I., South America, Africa, naturalized from Asia, Pacific Islands.

7. FROELÍCHIA Moench COTTONWEEDS

Perennial or annual pubescent herbs with opposite, entire leaves. Flowers bisexual, spicate, bracteate, calyx five-lobed, stamens five, fused at base to form a tube. Ovary ovoid, style elongate. Fruit a utricle, indehiscent, included within the calyx tube. About ten species, tropical and subtropical areas of the Western Hemisphere.

1. **F. floridana** (Nutt.) Moq. COTTONWEED—Erect, slender, pubescent stems up to 1.5 m tall, occasionally branched near the base. Leaves about 5–12 cm long, few, linear to spatulate, broadest below the middle, appressed pubescent or lanate below. Spikes all or mostly terminal up to 10 cm long, peduncles elongate, thickly pubescent, calyx tube greenish white, densely pubescent, up to 6 mm long at maturity with two lateral wings, one with two rows of tubercles or sharp ridges, the other with one. Seeds 1.5 mm long, brown. Sandy pinelands, old fields, Fla., and N. Mex., north to Md. and S. Dak. All year.

8. IRESÌNE R. Brown

Glabrous or pubescent herbs, shrubs, vines, or trees with opposite, entire or serrate, petiolate leaves. Flowers minute, unisexual or bisexual in spikes or headlike spikes, sepals five, subequal, stamens five, united at base into a tube. Ovary one-locular, stigmas two to three. Fruit an indehiscent utricle. About 40 species native to the Americas.

1. **I. celosia** L. BLOODLEAF—Erect or clambering annual or perennial up to 3 m tall. Leaves usually 5–15 cm long, ovate-rhombic or broadly lanceolate, apically acute or short acuminate, glabrous to densely pubescent. Flowers in paniculate spikes, unisexual, bracts subequal; pistillate flowers densely pubescent at base of calyx. Fruit 0.5–1 mm long, seeds reddish brown, about 0.5 mm. $2n = 34$. Hammocks, dis-

turbed sites, Fla. to La., N.C. on the coastal plain. All year. *I. celosioides*
L., *I. paniculata* (L.) Kuntze

73. NYCTAGINÀCEAE Four O'Clock Family

Herbs, shrubs, or trees with alternate or opposite leaves. Flowers
bisexual or unisexual, mostly cymose and often surrounded by brightly
colored bracts that simulate a calyx; perianth tubular, often petaloid;
stamens one to many, free or fused at the base. Ovary superior, one-
locular. Fruit indehiscent. About 20 genera and 160 species, mostly
tropical and subtropical, chiefly American in distribution. *Allionaceae,
Pisoniaceae*).

1. Perianth over 1 cm long.
  2. Flowers in cymes; stems glabrous . . . 1. *Mirabilis*
  2. Flowers solitary; stems pubescent . . . 2. *Okenia*
1. Perianth less than 1 cm long.
  3. Herbs, sometimes woody at base . . . 3. *Boerhavia*
  3. Trees, shrubs, or woody vines . . . . 4. *Pisonia*

1. MIRÁBILIS L.

Perennial herbs with opposite leaves and showy flowers. Involucre
resembling a five-parted calyx; calyx funnelform with long, slender tube,
stamens three to five, united at base. Style elongate, stigma capitate.
Fruit five-ribbed, ovoid. About 20 species, principally tropical America.
Allionia Loefl.; *Oxybaphus* L'Her.

1. M. jalapa L. Four O'clock, Marvel of Peru—Perennial herbs
with tuberous roots and opposite, ovate leaves. Stems and leaves glabrous, blades mostly 6–12 cm long. Involucres 6–8 mm long, lobes acute,
calyx deep red, purple, pink, yellow, white, or variegated, opening in
late afternoon. Fruit ovoid. Escaping from cultivation into disturbed
areas, roadsides, south Fla., W.I., Central and South America, naturalized from tropical America. Summer, fall.

Bougainvillea glabra Choisy, Bougainvillea—Shrubs or vines with
alternate, ovate leaves and small inconspicuous flowers enclosed by large,
showy, brightly colored bract, usually three together, calyx tube usually
rose or yellow. Popular as cultivated vines in south Fla.; may be found
near habitations and in waste areas but apparently is not naturalized.

2. OKÈNIA Schlecht. & Cham.

Decumbent or prostrate herbs densely pubescent with opposite,
fleshy, oblique leaves. Blades 3.0–4 cm long. Flowers bisexual, solitary in

three-bracted involucres, calyx deep violet or rose purple, funnelform, the lobes strongly emarginate, stamens fourteen to eighteen, united at the base. Fruit subterranean, 1–2 cm, elongated, bent. A single species in tropical America.

1. **Okenia hypogaea** Schlecht. & Cham.—Characters of the genus. Sea beaches and hammocks, south Fla., Mexico. Summer, fall.

3. BOERHÀVIA L.

Annual or perennial branching herbs with opposite or subopposite leaves. Blades sinuate or undulate. Flowers bisexual, small in terminal racemes or panicles, calyx tube five-lobed, stamens one to five. Fruit obovoid, glabrous or glandular, three- to five-angled. About 30 species in the tropics and subtropics.

1. Plants annual; ovary or fruit glabrous . .   1. *B. erecta*
1. Plants perennial; ovary or fruit glandular .   2. *B. diffusa*

1. **B. erecta** L. Spiderling—Annual herbs with erect or ascending branches often decumbent or spreading up to 1 m long. Leaves irregularly ovate-rhombic or deltoid-ovate, usually 2–9 cm long. Calyx white, pink, or purplish, tube glabrous, limb campanulate, about 1 mm long. Fruit obpyramidal, 3–4 mm long, grooves wrinkled. Weeds of fields, roadsides, beaches, and other disturbed sites, south U.S., W.I., Central and South America. All year.

2. **B. diffusa** L. Red Spiderling—Perennial herbs with decumbent or ascending stems up to 1 m long. Leaves orbicular to rhombic-ovate, undulate or sinuate, 2–6 cm long. Inflorescence paniculate, many-flowered, pink or purplish, calyx tube granular-pubescent; limb about 0.5 mm long, stamens three. Fruit obovoid, about 3–4 mm long, ribs glandular. All year. *B. coccinea* Mill.

4. PISÓNIA L. Cockspur, Beef Trees

Dioecious shrubs, vines, or trees, usually with stiff axillary spines or unarmed, and usually bearing opposite, entire leaves. Inflorescence umbellate, cymose, or corymbiform; flowers small, corolla absent, staminate flowers with six to ten unequal stems. Ovary one-locular, pistillate flowers tubular; fruit with longitudinal rows of glands along the angles, or drupaceous, red to black, cylindrical or subovoid. About 45 species, chiefly in the tropics. *Torrubia* Vell., *Guapira* Aubl.

1. Fruits dry, angular, bearing rows of glands.
   2. Stems armed with curved or hooked
      thorns, mostly climbing . . . . .   1. *P. aculeata*

2. Stems unarmed, erect . . . . . . 2. *P. rotundata*
1. Fruits drupaceous, without glands . . . 3. *P. discolor*

1. P. aculeata L. DEVIL's CLAWS, COCKSPUR—Woody, climbing vines with branched thorns, ultimate branches curved and very sharp. Leaves 2–8 cm long or longer, ovate or elliptical. Inflorescence 3–6 cm long, multiflowered; staminate flowers yellow green. Fruit 7–9 mm long, densely puberulent. Hammocks, moist thickets, disturbed sites, south Fla., tropical America, adventive in Africa, Asia. Spring, summer.

A number of varieties have been described for this species. **P. aculeata** var. **macranthocarpa** Donnell Smith with much larger fruits was collected near Miami many years ago and may still be in our area.

2. P. rotundata Griseb. COCKSPUR—Shrub or small tree with pale bark and bearing obovate or elliptic-ovate leaves usually 3–9 cm long. Inflorescence corymbiform, densely flowered; staminate flowers green or white, bearing pubescence along the edge. Fruit 0.5–6 mm long, sticky, obovoid. Hammocks and pinelands, south Fla., W.I. *Torrubia rotundata* (Griseb.) Sudw.

3. P. discolor Spreng. BLOLLY, BEEF TREE—Shrub or small tree with pale smooth bark and short internodes. Leaves opposite or alternate, often crowded, glabrous or sparingly puberulent, narrow lanceolate or ovate to elliptic-oblong or obovate, thin, rounded, obtuse or retuse at the apex, petioles slender. Flowers greenish yellow, racemose or paniculate. Fruit obovoid to oblong, 5–10 mm long, scarlet or red, juicy. Maritime shores and beaches, various sites. Two varieties occur in our area:

1. Leaf blades ovate or lance-ovate, broadest at
    or about the middle . . . . . . 3a. var. *discolor*
1. Leaf blades obovate or oblanceolate to nar-
    rowly spatulate, broadest above the mid-
    dle . . . . 3b. var. *longifolia*

3a. P. discolor var. **discolor**—Hammocks, thickets, especially along the east coast in south Fla., also in W.I. Spring, fall.

3b. P. discolor var. **longifolia** Heimerl—Hammocks, pinelands, especially in the Fla. Keys, south Fla., also in W.I. Spring, fall. Includes *Torrubia bracei* Britt. and *T. globosa* Small; *T. longifolia* (Heimerl) Britton. Considerable variation occurs in **P. discolor**, especially in leaf size, shape, and in fruit size. A pubescent leaf form from the Fla. Keys has been named TORRUBIA FLORIDANA Britt., but it apparently is very rare or has disappeared.

Family 73. **NYCTAGINACEAE.** *Pisonia aculeata:* a, flowering branch, × ½;
b, staminate flowers, × 7½; c, pistillate flower, × 15; d, mature fruit, × 3; e,
fruiting branch, × ½.

74. PHYTOLACCÀCEAE   Pokeweed Family

Herbs, shrubs, vines, seldom trees with simple, entire, alternate leaves. Flowers small, radially symmetrical, bisexual, rarely unisexual, in terminal or axillary spikes or racemes, calyx mostly four- to five-parted, green or colored; corolla absent; stamens three to many, inserted in fleshy disk, variable in number within the same species. Ovary globose, superior or inferior, gynoecium of one to sixteen carpels free or fused; styles as many as carpels. Fruit a berry, drupe, or achene. About 17 genera and 115 species, chiefly tropical and subtropical regions, especially in South America. *Petiveriaceae*

1. Perianth segments five . . . . . .   1. *Phytolacca*
1. Perianth segments four, rarely five.
   2. Ovary superior.
     3. Inflorescence spicate; fruits elongate,
       with hooks at the tip . . . . .   2. *Petiveria*
     3. Inflorescence racemose or narrowly pa-
       niculate; fruits spherical, without ap-
       pendages at the tip.
       4. Stamens four; fruit bright red . .   3. *Rivina*
       4. Stamens eight to sixteen; fruit pur-
         plish black . . . .   4. *Trichostigma*
   2. Ovary inferior . . . .   5. *Agdestis*

1. PHYTOLÁCCA L.   Inkberries, Pokeberry

Tall, perennial herbs or shrubs with simple, alternate, entire leaves. Flowers in terminal racemes, sepals five, petallike; petals absent, stamens eight to twenty-two, usually inserted on hypogynous disk. Ovary subglobose or ovoid, gynoecium five- to sixteen-carpellate, styles equal in number to carpels. Fruit a five- to sixteen-chambered fleshy berry. About 26 species, tropical and subtropical regions, chiefly American.

1. Fruit shorter than its pedicel; racemes nodding in fruit . . . .   1. *P. americana*
1. Fruit longer than its pedicel; raceme erect
   in fruit . . . .   2. *P. rigida*

1. P. americana L. Pokeweed, Pokeberry—Tall, smooth herb up to 3 m high; leaves about 1–3 dm long on long petioles, lance-ovate to ovate. Racemes pedunculate, lax, up to 15–20 cm long; flowers greenish white or pink, stamens and pistils usually ten. Fruit 1 cm in diameter, shorter than pedicel, dark purple. Disturbed sites, old fields, margins of hammocks, Fla. and Tex. north to Me. and Minn. All year. *P. decandra* L.

2. P. rigida Small, POKEWEED, INKBERRY—Tall, smooth herbs up to 3 m tall or taller and treelike. Leaves about 1–3 dm long on long petioles, lanceolate to oblanceolate. Racemes up to 20 cm long, erect. Fruit 10–12 mm wide usually, longer than its pedicel, dark purple. Disturbed sites, fields, margin of hammocks, Fla. to Tex. north on coastal plain. All year.

2. PETÍVERIA L.

Tall herbs usually woody at base with alternate, simple, entire leaves. Inflorescence at terminal or axillary spicate racemes, flowers small, sepals four, fused into a tube, petals absent, stamens eight, inserted on hypogynous disk. Ovary one-locular, gynoecium one-carpellate, tomentose, stigma brushlike. Fruit a linear achene. Two species in Western Hemisphere.

1. P. alliacea L. GUINEA-HEN WEED—Tall herbs often with angled stems, up to 1 m tall, producing a strong garlic odor when bruised. Leaves mostly 5–16 cm long, elliptic or obovate, puberulent. Inflorescence a slender spike up to 40 cm long, sepals white or greenish white or pale pink, about 4 mm long. Fruit linear about 8–10 mm long. Disturbed sites, moist woods, and hammocks, south Fla., south Tex., W.I., Central America and South America.

3. RIVÌNA L.

Erect or straggling herbs, often with woody base. Leaves alternate, simple, entire. Inflorescence a suberect raceme with numerous small flowers, sepals four, petals absent, stamens four. Ovary one-locular, gynoecium one-carpellate, stigma capitate. Fruit a globose red berry. Three species in the American tropics.

1. R. humilis L. ROUGE PLANT, BLOODBERRY—Stems often straggling, up to 1 m tall or more with vinelike stems. Leaves 6–12 cm long, lanceolate to ovate-elliptic, apex acuminate. Sepals white or pink, 2–3 mm long. Fruit 2–4 mm in diameter, scarlet or red. Disturbed sites, sandy soil, dunes, hammocks, Fla. to Okla., Tex., Mexico, W.I., Central America, South America. All year.

4. TRICHOSTÍGMA A. Richard

Woody vines or diffuse, reclining shrubs with alternate, simple leaves, entire. Flowers borne in terminal or axillary racemes. Sepals four,

Family 74. **PHYTOLACCACEAE.** *Rivina humilis:* a, flowering branch, × ½;
b, flower, × 15; c, fruiting branch, × 1; d, mature fruit, × 8½.

subequal, stamens eight to ten, on hypogynous disk. Ovary one-locular, gynoecium one-carpellate, subglobose. Fruit a globose drupe. Three species in tropical America.

1. **T. octandrum** (L.) H. Walt.—Decumbent or reclining shrubs or vines up to 10 m long. Leaves 12–15 cm long, oblong, apex acuminate, glabrous. Inflorescence a dense raceme about 5–7 cm long, sepals ovate, 2–3 mm, white or greenish white. Fruit black, 5–6 mm in diameter. Disturbed sites and moist woods, south Fla., W.I., Central America, South America.

5. AGDÉSTIS Moc. & Sessé ex DC.

Vines with turniplike roots and many branches. Leaves broad, ovate, alternate. Flowers in panicles; sepals four, spreading; petals absent, stamens numerous, mostly fifteen to twenty. Ovary inferior, gynoecium of four carpels, four-locular. Fruit a four- to five-winged achene. One species, native of Mexico and Guatemala.

1. **A. clematidea** Moc. & Sessé—Characters of the genus. Leaves 3–12 cm long, ovate-cordate, obtuse. Panicles mostly 1–1.5 dm long, sepals ovate, about 4–6 mm. Achene with wings 10–13 mm broad. Cultivated and established in disturbed sites, hammocks, south Fla., Tex., naturalized from Mexico; grown in south Fla. as an ornamental vine.

75. AIZOÀCEAE Carpetweed Family

Low shrubs or herbs either erect or prostrate with opposite or alternate, minute, filiform or fleshy leaves. Flowers bisexual or unisexual, bilaterally symmetrical, cymose calyx four- to five-parted or lobed; petals numerous or absent; stamens five to many, the outer forming series of petaloid structures. Ovary superior or inferior with one to several locules. Fruit a capsule, drupèlike or nutlike. About 23 genera and 1100 species, chiefly south Africa and Mediterranean regions. *Tetragoniaceae, Ficoideaceae*

1. Flowers in axillary or terminal cymes; leaves mostly whorled . . . . 1. *Mollugo*
1. Flowers usually solitary in upper leaf axils; leaves opposite.
　2. Leaves stipulate; ovary one- to two-locular; leaves elliptic to orbicular.
　　3. Stamens one to three; sepals greenish . 2. *Cypselea*
　　3. Stamens five to ten; sepals purplish within . . . . 3. *Trianthema*

2. Leaves not stipulate; ovary three- to five-
locular; leaves linear-lanceolate to
lanceolate . . . .   4. *Sesuvium*

1. MOLLÙGO L.   CARPETWEED

Annual or perennial, low, inconspicuous, slender herbs with numerous branches. Leaves basal, opposite or whorled. Flowers axillary
and white or greenish, bisexual, sepals five, stamens usually three to five.
Styles three to five, ovary three- to five-locular. Fruit a capsule. About 25
species, chiefly tropical regions.

1. **M. verticillata** L. INDIAN CHICKWEED—Prostrate annual, much
branched, forming mats up to 3–4 dm wide. Leaves 1–3 cm long, narrowly lanceolate or spatulate, in whorls of three to eight. Flowers usually
two to five on slender pedicels arising from the nodes, sepals green or
white, 4–5 mm wide, stamens three to four. Capsule 3 mm, ovoid, with
many small seeds. Disturbed sites, chiefly moist soil, cosmopolitan.
Spring, fall.

2. CYPSÈLEA Turp.

Annual or perennial succulent herbs with opposite, rather broad
leaves, forming mats up to 1 dm wide. Leaves 2–6 cm long, elliptic- ovate,
entire, obtuse. Flowers arising from axils on short pedicels, sepals four
to five, ovate, about 2 mm long, greenish white, stamens one to three,
styles two, short. Fruit about 2 mm wide, a capsule, subglobose. One
species.

1. **C. humifusa** Turp.—Characters of the genus. Sandy soil, pinelands, Fla. to Calif., W.I. Spring, fall.

3. TRIANTHÈMA L.

Annual or perennial succulent herbs or subshrubs with simple, opposite, unequal leaves. Flowers axillary, cymose or solitary, small, bisexual, calyx five-lobed, each lobe with an appendage near the tip, stamens five to ten or more, free or epipetalous. Ovary superior, one- to two-
locular, styles one to two. Fruit a circumscissile capsule. About 15 species in the tropics and subtropics.

1. **T. portulacastrum** L.—Annual herbs erect or prostrate with numerous branches up to 1 m. Leaves 1–4 cm long, of pairs unequal, obovate or oblanceolate. Flowers solitary or few, sessile, calyx tube campanulate, greenish outside, pink or purple inside. Stamens ten, adnate to
calyx. Ovary one-locular, style with lateral stigma. Capsule 4–5 mm long,

also enclosed in petiolar sheath. Disturbed sites, especially low ground, coastal areas. Pantropical weed, north to Okla., Mo. All year.

4. SESÙVIUM L. Sea Purslane

Succulent annual or perennial herbs or subshrubs with opposite leaves, erect or prostrate. Leaves simple, equal, fleshy, linear to elliptic. Flowers bisexual, sessile or pedicelled, calyx five-lobed, stamens five to many, free or adnate to calyx tube. Ovary superior, three- to five-locular, styles three to five. Capsule circumscissile. About six to eight species, maritime plants in tropics and subtropics.

1. Flowers or fruits sessile or subsessile in leaf
  axils . . . . 1. *S. maritimum*
1. Flowers or fruits on peduncles over 3 mm
  long . . . . 2. *S. portulacastrum*

 1. S. maritimum (Walt.) BSP, Sea Purslane—Smooth, prostrate or ascending annual with fleshy, much branched stems. Leaves about 1–2.5 cm long, oblong to spatulate-obovate. Flowers about 3–4 mm long, solitary, sessile, purplish within, sepals with minute appendages. Fruit about 4 mm long, ovoid. Sea beaches, coastal areas, Fla. to Tex., N.Y., W.I. All year. *S. pentandrum* Ell.

 2. S. portulacastrum (L.) L.—Smooth, prostrate or ascending perennial with few branches, these up to 1–2 m long. Leaves 1–5 cm long, linear to narrowly obovate. Flowers 2–10 mm long, solitary, pink on pedicels, flowers about 1 cm long, sepals with hornlike appendages. Fruit about 7 mm long, ovoid or obovoid. $2n = 16,36,48$. Sea beaches, coastal dunes, and low ground, tropics and subtropics, Fla. to N.C., W.I., Mexico, Central America. One of our most common seastrand plants. All year.

76. PORTULACÀCEAE Purslane Family

Shrubs or herbs often with smooth, succulent leaves. Leaves simple, alternate, opposite or basal, stipules tufted or scarious. Inflorescence axillary or terminal, the flowers solitary, cymose, racemose, or paniculate, flowers bisexual; sepals two, bractlike; petals four to five, fused at base or free, stamens opposite the petals, as many as the petals, fewer, or more, usually free. Pistil one, ovary superior to inferior, styles and stigmas one to nine. Fruit a capsule or sometimes nutlike, seeds one to many. About

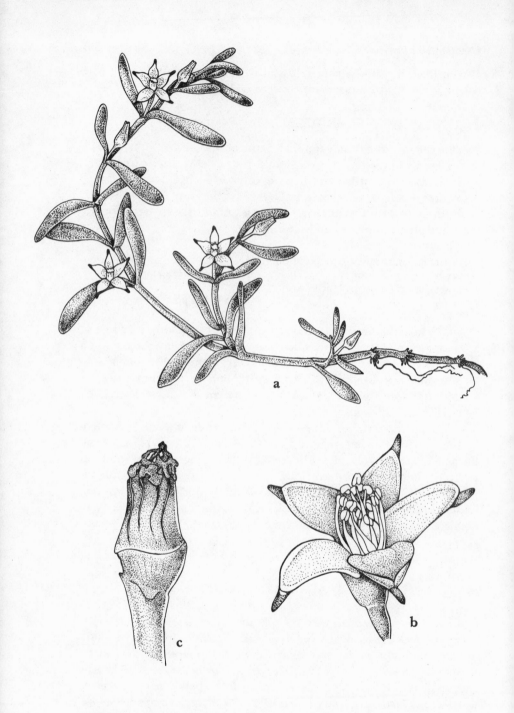

Family 75. **AIZOACEAE.** *Sesuvium portulacastrum:* a, flowering branch, × 1; b, flower, × 4; c, fruit, × 5.

19 genera, 500 species, widely distributed in tropical and subtropical America, Pacific Islands, including many weedy plants.

1. Ovary superior; flowers racemose or paniculate; stems erect . . . . 1. *Talinum*
1. Ovary half-inferior; flowers solitary or clustered; stems prostrate or suberect . . . 2. *Portulaca*

1. TALÌNUM Adans.   FLAME FLOWERS

Shrubs or herbs usually with succulent stems and leaves. Leaves alternate, fleshy, without stipules. Inflorescence terminal of racemes, panicles, cymose or sometimes solitary, sepals two, petals five, stamens five to thirty in clusters opposite the petals. Ovary superior, styles three, more or less united. Fruit a loculicidal capsule, three-chambered with numerous reniform seeds. About 50 species, tropical and subtropical, especially in Mexico.

1. T. paniculatum (Jacq.) Gaertn.—Erect, slender, smooth herbs from thick, fleshy roots. Leaves usually 5–12 cm long, alternate, broadly obovoid, elliptic or oblanceolate. Inflorescence a terminal, lax raceme of simple or compound cymes about 1–5 dm long, sepals ovate, petals elliptic, 3–5 mm long, reddish or yellow, stamens fifteen to many. Capsule ovoid 3–5 mm long. $2n = 24$. Sandy soil, disturbed sites, south Fla. to Tex., Ariz., naturalized from W.I. All year.

T. triangulare (Jacq.) Willd., a shrub with flowers corymbose, has also been reported for south Fla. but apparently is uncommon.

2. PORTULÁCA L.

Annual or perennial succulent herbs with alternate or approximate terete or flat leaves. Inflorescence near the end of stem, axillary, flowers solitary or small clusters, sepals two, fused at base, petals four to six; stamens four to many fused to base of petals. Ovary half to wholly inferior, styles three- to nine-parted. Fruit a circumscissile capsule, seeds numerous. Cosmopolitan genus of about 125 species in tropical and subtropical regions.

1. Leaf axils with conspicuous long hairs; leaf blades linear or linear-lanceolate.
  2. Corolla bright pinkish to rose purple; stems prostrate, axils villous pubescent   1. *P. pilosa*
  2. Corolla yellow; stems suberect, axils hirsute pubescent . . . . . . 2. *P. phaeosperma*
1. Leaf axils without conspicuous long hairs; leaf blades spathulate or obovate . . . 3. *P. oleracea*

1. **P. pilosa** L. PINK PURSLANE—Prostrate or diffusely branching annual with conspicuously hair-tufted leaf axils. Leaves about 1–2 cm long, linear-lanceolate, terete or nearly so. Flowers terminal, subtended by an involucre of six to ten bracts with brownish or white pubescence, sepals 2–3 mm wide, subglobose. $2n = 16,18$. Disturbed sites, sandy pinelands, Fla. to Tex., Ga. on coastal plain, tropical America. Spring, fall.

2. **P. phaeosperma** Urban—Stems suberect or ascending, diffusely branching, with tuberous roots and leaf axils merely hirsute. Leaves about 0.5–2 cm long, linear-lanceolate, nearly terete. Flowers terminal in involucre with five to eight bracts with whitish or brown pubescence, sepals 3–4 mm, suborbicular; petals yellow; stamens twelve to sixteen. Capsule about 3–4 mm wide, subglobose. Sea beaches, sand dunes, south Fla., W.I., tropical America. All year.

3. **P. oleracea** L. PURSLANE—Stems forming mats, prostrate and radially spreading, somewhat ascending, glabrous, with fibrous roots and leaf axils without pubescence. Leaves 1–3 cm long, elliptic to obovoid, or spatulate, flattened. Flowers sessile, crowded, sepals ovate 3–5 mm long, strongly keeled; petals yellow or orange yellow; stamens six to fifteen. Capsule ovoid, 6–9 mm long. $2n = 45$; c. 54. Disturbed sites, pinelands, sandy soil, cosmopolitan weed. All year.

77. BASELLÀCEAE   MADEIRA-VINE FAMILY

Vines with tuberous roots and somewhat fleshy, alternate, entire leaves. Flowers bisexual in narrow racemes; calyx with two sepals sometimes becoming winged in fruit; corolla of five petals; stamens five, opposite the petals. Ovary superior, one-locular, gynoecium three-carpellate. Fruit a utricle. About five genera and 20 species, chiefly tropical America.

1. ANRÈDERA Juss.   MADEIRA VINE

Vines with rather fleshy, broadly ovate leaves. Flowers borne in axillary or terminal racemes; sepals two, flat, not becoming winged; petals five. Ovary subglobose, stigma cleft. Two species of tropical America.

1. **A. leptostachys** (Moq.) Steenis—Smooth vine with slender stems. Leaves usually 3–7 cm long, elliptic to ovate, succulent, acute or acuminate, narrowed to 'a short petiole. Racemes slender, elongate; sepals about 1–2 mm long, ovate; petals ovate; stamens recurved. Styles three.

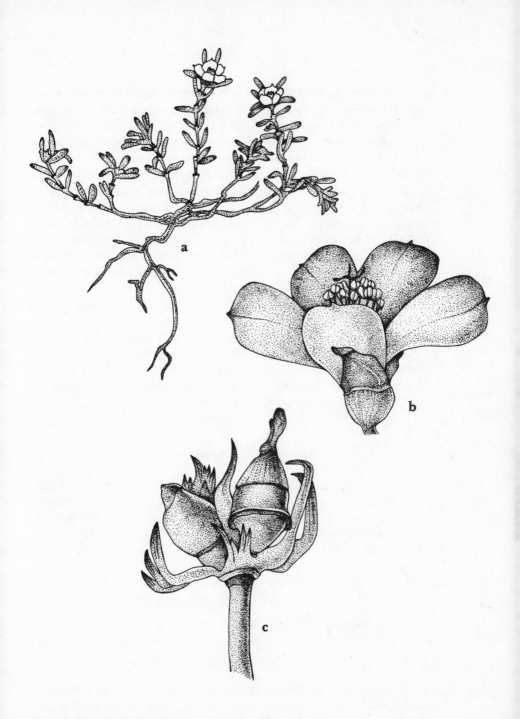

Family 76. **PORTULACACEAE.** *Portulaca pilosa:* a, flowering plant, × 1; b, flower, × 8; c, fruit, × 5.

Family 77. **BASELLACEAE.** *Anredera leptostachys:* a, flowering branch, ×
½; b, flower, × 15; c, inflorescence, × 1; d, mature fruit, × 32½.

Disturbed sites, hammocks, pineland, south Fla., tropical America, escaped from cultivation and naturalized. Summer. *Boussingaultia leptostachys* Moq.

78. CARYOPHYLLÀCEAE PINK FAMILY

Annual or perennial herbs or undershrubs with opposite, simple, entire leaves usually connected near the base by a transverse line, stipules present or not. Stems usually swollen at the nodes. Inflorescence cymose usually, flowers radially symmetrical, generally bisexual, solitary or often cymose, sepals four to five, free or united, sometimes scarious; petals four to five, sometimes smaller than sepals or absent; stamens eight to ten in two whorls or fewer. Ovary superior with free central placentation, carpels five or fewer; ovules numerous, fruit a capsule. About 80 genera and 1300 species, cosmopolitan. *Alsinaceae, Corrigiolaceae, Scleranthaceae*

1. Stem leaves reduced to scalelike bracts . .   1. *Stipulicida*
1. Stem leaves well developed with lanceolate to orbicular or reniform blades.
  2. Leaf blades orbicular, deltoid, or reniform, some at least as wide as long.
    3. Styles partially fused; petals 2–3 mm long . . . .   2. *Drymaria*
    3. Styles distinct; petals 3–6 mm long . .   3. *Stellaria*
  2. Leaf blades lanceolate, longer than wide.
    4. Flowers apetalous; styles two . . .   4. *Paronychia*
    4. Flowers with petals; styles three to four.
      5. Calyx of fused sepals . . . . .   5. *Silene*
      5. Calyx of distinct sepals . . . .   6. *Arenaria*

1. STIPULÍCIDA Michx.

Smooth perennial herbs with basal leaves broad, upper leaves much reduced, often scalelike. Flowers terminal in dense cymose or capitate clusters; sepals five, cleft or notched at the apex; petals five; stamens three to five. Style short, stigmas three. Two species, southeast U.S.

1. **S. setacea** Michx.—Stems 1–2 dm or less, erect, slender, with lower leaves spatulate or ovate-spatulate to suborbicular, 10–15 mm long; upper leaves much reduced, scalelike. Sepals elliptic, notched, about 1.5–2 mm long with a narrow transparent margin; petals clawed. Capsule about 1 mm long. Sandy soil, pinelands, Fla. to Miss., N.C. on the coastal plain. Spring, fall.

2. DRYMÀRIA Willd. ex Roem. & Schultes

Annual or perennial herbs, prostrate, spreading, or erect with opposite leaves and small stipules. Flowers in cymes or in small racemes, small
and inconspicuous, sepals five, free; petals five, white, usually deeply
notched; stamens five or fewer. Styles three, fused below. Capsule ovoid,
three-valved. About 60 species, tropical and subtropical America.

1. **D. cordata** (L.) Willd. ex Roem. & Schultes, WEST INDIAN
CHICKWEED—Annual with smooth or glandular-puberulent, erect or prostrate, often diffusely branching stems. Leaves smooth, orbicular, reniform or ovate, rounded to cordate at base, about 5–25 mm long, stipules
usually persistent. Flowers in terminal or axillary cymes; sepals acute,
3–4 mm long; petals 2–3 mm long; stamens two to three, sometimes five.
Capsule 1.5–3 mm long. Disturbed sites, lawn weed, old fields, gardens,
south Fla., W.I., tropical America. All year.

3. STELLÁRIA L.   CHICKWEED

Annual or perennial herbs with low, matted stems. Leaves rather
succulent, without stipules. Flowers in terminal cymes or solitary in axils,
sepals four to five, petals four to five, cleft or parted nearly to the base,
stamens usually ten, sometimes fewer. Ovary usually three-carpellate
with three styles. Fruit a capsule usually with six valves. About 100 species, chiefly in the temperate regions.

1. **S. cuspidata** Willd.—Annual with much-branched decumbent
stems, glabrous or glandular-puberulent. Leaves 1–2 cm long, ovate to
deltoid, cordate or truncate at the base. Flowers in open cymes, sepals
ovate 2–4 mm long, petals longer than the sepals. Capsule ovoid, somewhat longer than the calyx, seeds reddish brown. Moist soil, disturbed
sites, from Fla. east coast, presumably in our area, to Tex., Ga., Mexico. Spring, summer. *Stellaria prostrata* Baldw.; *Alsine baldwinii*
Small

S. media (L.) Cyr., the common chickweed, with pubescent stems
and smaller ovate leaves, a cosmopolitan weed, should occur in our area,
but no specimens have been seen.

4. PARONÝCHIA Mill.

Annual or perennial herbs with small opposite leaves and conspicuous hyaline stipules. Flowers bisexual, sepals free or barely fused at
the base; petals absent; stamens usually five, fused to the base of the

Family 78. **CARYOPHYLLACEAE.** *Drymaria cordata:* a, flowering plant, ×
½; b, inflorescence, × 3; c, flower, × 8½; d, mature fruit, × 10.

calyx. Ovary ovoid to globose becoming a one-seeded utricle, style two-branched. About 40 species, cosmopolitan. *Paronychia* Adans.; *Gastronychia* Small; *Anychiastrum* Small; *Siphonychia* T & G; *Anychia* Michx.; *Nyachia* Small).

1. **P. americana** (Nutt.) Fenzl ex Walp.—Stems and branches radially prostrate, about 8 dm long, minutely pubescent when young, becoming glabrate. Leaves linear-oblanceolate, about 5–20 mm long, minutely pubescent. Flowers small, barely 1.5 mm long; sepals obovate. Dry soil, disturbed sites, pinelands, Fla. to S.C. on coastal plain. Spring, summer.

5. SILÈNE L. CATCHFLY, CAMPION

Annual or perennial herbs, erect or decumbent. Flowers in cymes or solitary, pink, white, or red; calyx five-cleft, cylindrical and often inflated, petals five, narrowly clawed, stamens ten. Styles three, ovary with numerous ovules. Fruit a many-seeded capsule. About 400 species, widely distributed.

1. **S. antirrhina** L.—Annual with usually smooth or somewhat puberulent stems up to 8 dm tall, erect or decumbent. Basal leaves oblanceolate, stem leaves lanceolate to linear, 3–6 cm long. Flowers usually numerous, calyx 5–10 mm long, ten-veined, petals as long as or exceeding the calyx, sometimes absent. Capsule ovoid, 4–10 mm long, three-valved. Disturbed sites, sandy soil, seashores, a cosmopolitan weed throughout U.S., from Fla. west coast, presumably in our area, although no specimens have been seen. Summer.

6. ARENÀRIA L. SANDWORT

Annual or perennial herbs with prostrate or diffusely branching stems. Leaves without stipules, subulate to elliptic or ovate. Flowers in terminal cymes or axillary, or solitary. Sepals five, free or barely fused at the base, petals five, white, entire or shallowly notched, stamens ten, rarely fewer. Ovary usually three-carpellate with three free styles. Fruit a capsule with three valves. About 200 species, including many weeds, cosmopolitan. *Sabulina* Reichenb.

1. **A. lanuginosa** (Michx.) Rohrb. ssp. **lanuginosa**—Perennial with spreading puberulent stems. Leaves 5–20 mm long, lanceolate or narrowly elliptic, subsessile, barely to densely white puberulent. Flowers solitary in the leaf axils; sepals 2–5 mm long, attenuate, petals obovate, entire, 2–4.5 mm long, or absent. Ovary ovoid, styles two to four; capsule ovoid, 3–5 mm long. Fla. to Tex., N.C., tropical America. Spring, fall.

79. NYMPHAEÀCEAE　Water-Lily Family

Aquatic perennial herbs from rhizomes with peltate or cordate, floating or emersed leaves. Flowers bisexual, solitary, large and showy. Perianth parts free, six to many, usually differentiated into calyx and corolla, the petals often passing into stamens, stamens three to many. Ovary superior to inferior, carpels three to many, free or united in a many-locular ovary or sunken in the enlarged torus. Fruit a nut, pod, or berry, the seeds often arillate. Eight genera and about 100 species of water or marsh plants of cosmopolitan distribution in both tropical and temperate regions. *Cabombaceae, Nelumbonaceae*

1. Floating leaves inconspicuous, submerged ones palmately finely dissected . . . 　1. *Cabomba*
1. Floating or emergent leaves conspicuous, submerged leaves if present not finely dissected.
  2. Carpels distinct, nutlike fruits embedded in the top of receptacle; leaves peltate, the blade not cleft or slit to the petiole.
    3. Sepals and petals numerous; flowers over 10 cm wide, mostly yellow, pink, or white . . . . 　2. *Nelumbo*
    3. Sepals and petals in threes; flowers less than 5 cm wide, mostly purplish . . 　3. *Brasenia*
  2. Carpels united into compound pistil, fruit berrylike; leaves not peltate, the blade deeply cleft at base.
    4. Flowers deep yellow or orange to red or purplish; sepals petallike, petals small; stigmas on a disk . . . . 　4. *Nuphar*
    4. Flowers white, pale yellow, greenish yellow or blue; sepals green, petals conspicuous; styles and stigmas distinct . . . . 　5. *Nymphaea*

1. CABÓMBA Aubl.　Water Shield

Slender, mucilaginous, mostly submerged herbs with monomorphic to polymorphic leaves; submerged leaves dissected, the ultimate segments linear to filiform, opposite or whorled; floating leaves if present narrowly elliptic to ovate, alternate, peltate, bifurcate. Flowers trimerous, sepals three, petals three, stamens three to six. Carpels one to four, each with capitate stigma. Fruiting carpels diverging. Seven species in tropical and subtropical America.

1. **C. caroliniana** Gray, FANWORT—Stems green with dissected submerged leaves and peltate, shield shaped, oblanceolate, or ovate floating leaves 1–2 cm long, usually with a basal notch. Perianth parts obovate, 8–12 mm long, white or pinkish purple with yellow spots near the base, stamens six. Fruiting carpels beaked, about 5–7 mm long. Fresh-water pools, Fla. to Tex., N.C., Mo. Spring, summer. Stems and petals of **C. caroliniana** var. **pulcherrima** are purple or deep pink.

2. NELÚMBO Adans. LOTUS

Stout aquatic herbs from tuberous rhizomes and bearing erect, emersed leaves and flowers. Leaves large, orbicular, peltate, usually raised above the water with long, stout petioles. Sepals and petals numerous, barely differentiated; stamens many. Numerous ovaries sunken in the top of an obconical receptacle. Fruits large, nutlike, each embedded in the receptacle. Two species, America and Australia. *Nelumbium* Juss.

1. **N. lutea** (Willd.) Pers. LOTUS LILY—Leaves mostly 3–7 dm wide, raised above the water or floating, orbicular, margins slightly involute. Peduncles elongate, rising to 1 m or more above the water, flower about 15–25 cm wide, solitary, pale yellow. Fruiting receptacle 10 cm wide, individual fruits acornlike about 1 cm wide. $2n = 16$. Pools and quiet water, Fla. to Tex., N.Y., Minn., W.I. Summer.

N. nucifera Gaertn. SACRED LOTUS has escaped from cultivation and has become locally established. It has pink or white flowers.

3. BRASÈNIA Schreb. WATER SHIELD

Stout, long-stemmed aquatic herbs up to 2 m long with alternate, long-petioled, floating, peltate, entire leaves, about 5–12 cm long. Flowers with peduncles up to 10 cm long or more, trimerous, purplish, three sepals, three petals, free; stamens twelve to eighteen with filaments of two lengths. Carpels four to eight, free. Fruit nutlike, beaked. One species, cosmopolitan.

1. **B. schreberi** Gmel.—Characters of the genus. Ponds, quiet water, Fla. to Tex., Nova Scotia, Minn., Pacific slope, tropical America, Old World. Summer.

4. NÙPHAR Sm. YELLOW WATER LILIES

Aquatic or marsh plants with stout rhizomes and long-petioled, erect, floating or submerged broad leaves and bright yellow, purplish,

Family 79. **NYMPHAEACEAE.** *Nuphar luteum* ssp. *macrophyllum:* a, inflorescence and mature leaf, × ⅛; b, flower, × ½; c, mature fruit, × ½.

or red cuplike flowers on scapelike peduncles. Sepals five to six, concave and converging, petals many, smaller than sepals, stamens many. Ovary compound, prolonged into a thickish style and broad, disklike, many-rayed stigma. Fruit coriaceous. About 20 species, North America, Eurasia, temperate zones. *Nymphozanthus* Richard

1. N. luteum (L.) Sibth. & Sm. ssp. **macrophyllum** (Small) Beal, SPATTERDOCK—Leaves about 2–4 dm long, mostly erect or floating, elliptic-ovate, smooth, basally cordate or V-shaped, petioles and peduncles with conspicuous aerenchyma. Flowers spheroidal, about 3–5 cm wide, bright yellow within; sepals six, the outer three obovate, inner somewhat narrower; anthers larger than filaments. Fruit narrow-ovoid, about 3.5 cm long. Ponds, marshes, sluggish streams, Fla. to Tex., Mass., Minn., Nebr.; ssp. **macrophyllum** is endemic to Fla. Spring, summer. *Nymphaea macrophylla* Small; *Nuphar advena* Ait.

5. NYMPHÀEA L.   WATER LILY

Aquatic herbs with stout rhizomes and long-petioled, floating, orbicular or elliptic leaf blades that are cleft at the base. Flowers tetramerous, sepals four or eight; petals numerous in several series and grading into the stamens. Carpels three to many, fused into a compound ovary, superior or inferior, the carpellary styles radiating. Fruit baccate, many-seeded. About 35 species, almost cosmopolitan. *Castalia* Salisb.

1. Corolla bluish or violet . . . . . . 1. *N. elegans*
1. Corolla white, pink, or greenish yellow.
   2. Corolla white or pink . . . . . . 2. *N. odorata*
   2. Corolla bright yellow or greenish yellow . 3. *N. mexicana*

1. N. elegans Hook. BLUE WATER LILY—Leaves about 10–20 cm wide, polymorphic elliptic to ovate to orbicular, margins entire to undulate, often dark purplish red beneath. Flowers erect, raised 1–2 dm above the water; sepals about 4–5 cm long, narrowly lanceolate; petals six to ten, lanceolate, blue or pale violet, stamens many, styles fifteen to twenty, bluntish. Berrylike fruit about 2–3 cm wide, globose, depressed. Pools, quiet ponds, and ditches, Fla., south Tex. and Mexico. Spring, summer. *Castalia elegans* (Hook.) Greene

2. N. odorata Ait. WHITE WATER LILY—Rhizomes about 2–3 cm thick, elongate, stout. Petioles purplish green or red, leaf blades about 15–20 cm wide, orbicular, red tinged, basal cleft narrow. Flowers fragrant, expanding in the morning; sepals 5–6 mm long, elliptic, petals numerous, white or pinkish, elliptic-ovate, stamens very numerous, the outer petaloid. Styles fifteen to twenty. Berrylike fruit depressed-globose, about 2-3 cm wide. Ponds, lakes, and ditches, Fla. to Tex., Nova Scotia, Manitoba. Spring, summer. *Castalia odorata* (Ait.) Woodv. & Wood

N. minor Sims is a small-flowered extreme; *C. lekophylla* Small has larger flowers and leaf margins rolled upward. Neither appears to be worthy of taxonomic recognition.

3. N. mexicana Zucc. Banana Water Lily—Tuberlike rhizome erect, stoloniferous. Leaves about 8–12 cm wide, ovate, floating, obscurely and finely sinuate, dark green with brown blotches above, reddish brown with dark dots below; some of the leaves emersed when crowded. Flowers usually 10–12 cm wide, bright yellow, erect, about 10 cm above the water; petals many, stamens very numerous. Fruit subglobose. Quiet water, ponds, ditches, escaped from cultivation, naturalized from Mexico. Spring, summer.

80. CERATOPHYLLÀCEAE Hornwort Family

Aquatic monoecious herbs with rootless floating branches and dissected whorled leaves with threadlike or linear segments. Flowers unisexual, sessile, minute and axillary; staminate and pistillate flowers at different nodes, each subtended by an involucre of several bracts, perianth absent, stamens twelve to sixteen, clustered. Pistillate flowers with one simple pistil. A single genus with three species, cosmopolitan.

1. CERATOPHÝLLUM L. Hornwort

Characters of the family

1. C. demersum L. Coontails—Elongate stems up to 1 m or more long, freely branching and often forming large masses. Leaves whorled, about five to twelve per node, one to two forked and dissected into linear segments 5 mm wide. Segments of involucre elliptic-ovate. Achene about 4–6 mm long with two basal spines. Quiet fresh water, widespread, Canada to South America, Old World. The plant is highly variable in morphology of leaf segments. A second species, **C. submersum** L., distinguished by its unarmed achenes, has also been reported for south Fla.

81. MAGNOLIÀCEAE Magnolia Family

Trees or shrubs with alternate simple leaves and large, deciduous stipules. Flowers bisexual, terminal, large and solitary, sepals and petals often colored alike, in several series, free, stamens many, free. Carpels many, superior, ovary one-locular, arranged spirally on an elongated

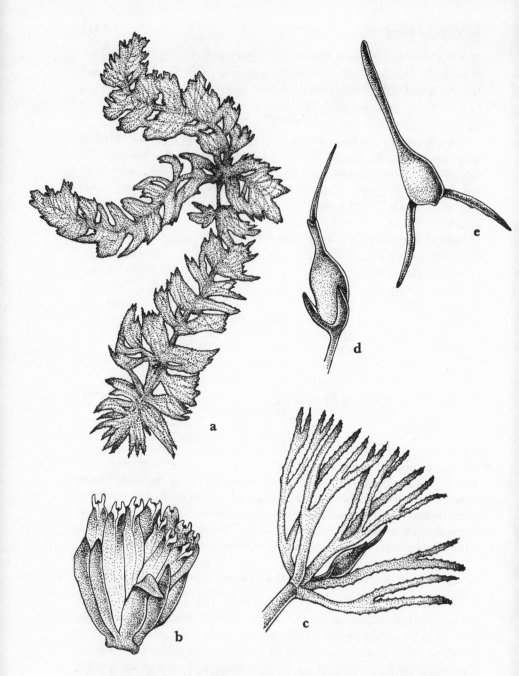

Family 80. **CERATOPHYLLACEAE.** *Ceratophyllum demersum:* a, leafy stem, × ½; .b, staminate flower, × 10; c, leaf whorl and pistillate flower, × 4; d, pistillate flower, × 10; e, mature fruit with persistent style and basal spurs, × 10.

torus. Fruiting carpels splitting or indehiscent, becoming a follicle, samara, or berry, rarely fused into a syncarp. About 18 genera, 220 species, mostly southeast Asia, America.

1. MAGNÓLIA L.

Evergreen or deciduous trees or shrubs with stipules free or fused to the petiole. Leaves often large, entire. Flowers at first enclosed by one to several bracts, often erect, white; sepals and petals about equal in length. Fruiting carpels dehiscing longitudinally, persistent in a cone-like cluster. About 20 species, southeast Asia, America. *Tulipastrum* Spach

1. **M virginiana** L. SWEET BAY, SWAMP BAY—Shrub or tree up to 20 m tall. Leaves usually 8–15 cm long, leathery, dark green above, pale, glaucous, and puberulent beneath, lanceolate to elliptic, apex obtuse; leaf buds silky pubescent. Flowers about 5 cm wide, white, subglobose, fragrant; petals ten to twelve, ovate about 3–5 cm long, stamens and pistils numerous on an elongated receptacle. Fruits ellipsoidal in cone-like structure. Bayheads, swamps, and low, wet woods, Fla. to Tex., to Mass. on the coastal plain.

82. RANUNCULÀCEAE BUTTERCUP FAMILY

Annual or perennial herbs with alternate leaves, or shrubs or climbing plants with opposite, often compound leaves. Flowers bisexual, rarely unisexual, radially symmetrical or rarely bilaterally symmetrical, solitary to paniculate, sometimes subtended by an involucre of leaves, perianth simple, petaloid; sepals five to eight, often petaloid; petals few to many or absent, free; stamens numerous, indefinite, free. Carpels one to many, free, one-seeded. Fruit a follicle or achene, rarely a berry. About 30 genera and 1200 species, chiefly in temperate zones, few in the tropics.

1. CLÉMATIS L.

Herbs or somewhat woody vines with opposite, entire to trifoliolate, pinnate, or biternately compound leaves. Leaflets entire to coarsely dentate, smooth or pubescent, rachis often twining. Flowers solitary, in cymes or umbels often on an elongate axis, unisexual or polygamous; sepals four to five, white, greenish, or purple, petaloid; petals absent; stamens many. Carpels numerous, free. Fruit an achene. About 300 species, cosmopolitan. *Viorna* Reichenb.

1. **C. baldwinii** T & G, PINE HYACINTH—Stems 2–6 dm tall with

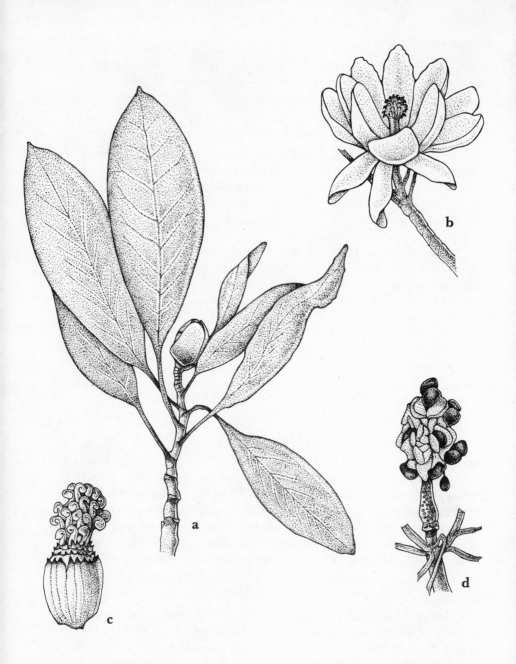

Family 81. **MAGNOLIACEAE.** *Magnolia virginiana:* a, leafy branch with floral bud, × ½; b, flower, × 1; c, androecium and gynoecium, × 2½; d, mature fruits, × 1.

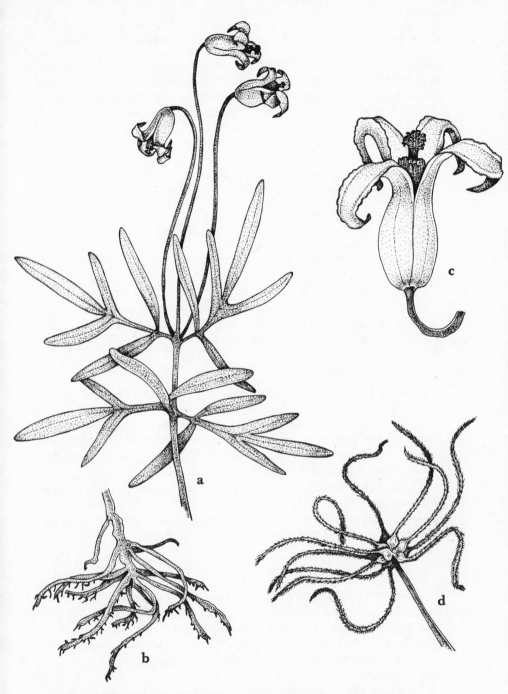

Family 82. **RANUNCULACEAE.** *Clematis baldwinii* var. *baldwinii*: a,
flowering branch, × ½; b, rootstock, × ½; c, flower, × 1¾; d, mature fruits,
× ¾.

leaves about 2–10 cm long, linear to elliptic-lanceolate, sometimes pinnately parted. Flowers usually solitary; sepals purplish or pale purple, about 3–5.5 cm long, each with a thinnish, crisped margin, acute. Style about 8–10 cm long. Achene 5 mm long. Endemic to peninsular Fla.

Two varieties occur in our area:

1. Sepals 3–4 cm long; leaves or leaf segments
   mostly linear to narrowly lanceolate . .   1a. var. *baldwinii*
1. Sepals 4–5.5 cm long; leaves or leaf segments
   mostly elliptic-lanceolate to narrowly
   ovate . . . .   1b. var. *latiuscula*

1a. C. baldwinii var. **baldwinii**—Pinelands, damp prairies. *Viorna baldwinii* (T & G) Small

1b. C. baldwinii var. **latiuscula** R. W. Long—Dry pineland.

C. crispa L. with climbing stems and **C. reticulata** Walt. with coarsely reticulate leaves may occur in our area, but they are more frequent north of the range of this manual.

83. MENISPERMÀCEAE Moonseed Family

Twining, mostly dioecious herbs or shrubs, rarely erect woody plants with alternate, usually simple leaves with palmate veins and lobes. Inflorescence cymose to paniculate; flowers unisexual, small, inconspicuous, staminate flowers; calyx mostly in two series with three sepals in each series; corolla similarly arranged or absent, often smaller than sepals; stamens three, six, or indefinite. Pistillate flowers with superior ovary with three or six fused carpels. Fruit a one-seeded drupe. About 70 genera and 300 species, chiefly in the tropics and subtropics.

1. CISSÁMPELOS L.

Twining woody vines with alternate, somewhat peltate or basifixed leaves. Staminate inflorescence axillary, fasciculate; flowers with sepals four, petals four, cuplike, stamens four, filaments fused. Pistillate flowers in axillary clusters, bilaterally symmetrical; sepal one, petal one, carpel one, free. Fruit a subglobose drupe. About 30 species in northeast tropics and subtropics.

1. C. pareira L.—Shrubby climbing vine with bright green, broadly ovate to suborbicular, entire, peltate or basifixed leaves 6–12 cm long, palmately three- to seven-veined, tomentose becoming pubescent, rarely

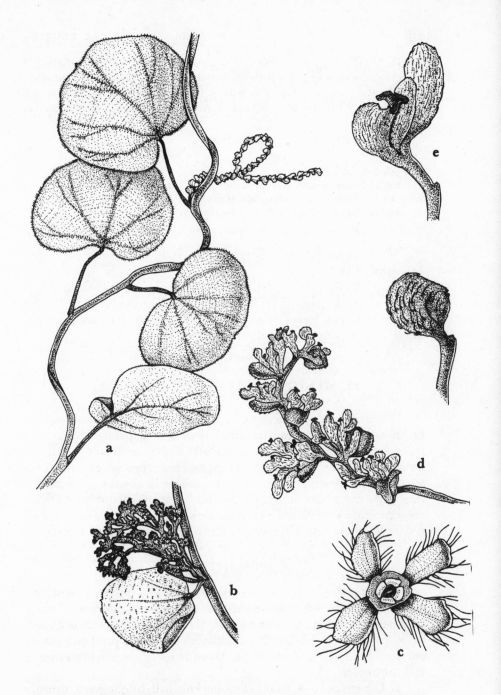

Family 83. **MENISPERMACEAE.** *Cissampelos pareira:* a, flowering branch, × ½; b, staminate inflorescence, × 2; c, staminate flower, × 20; d, pistillate inflorescence, × 5; e, pistillate flower, × 25; f, mature fruit, × 17½.

glabrous. Staminate flowers in branched cymes from the axils of reduced leaves; sepals 1–2 mm long, ovate; corolla campanulate, 1–2 mm long, petal obovate 0.5–1 mm long; carpel densely pubescent. Drupe 4–5 mm long, pubescent. $2n = 24$. South Fla., W.I., tropical America, Old World tropics.

84. ANNONÀCEAE Custard Apple Family

Trees, shrubs, or climbing plants with aromatic wood and leaves. Leaves alternate, entire, buds naked. Flowers bisexual, radially symmetrical, solitary or in various arrangements; sepals three, free or partially fused; petals six, thickish, in two series, stamens numerous, spirally arranged, with short filaments. Carpels few to numerous, free or rarely united with a one-locular ovary. Fruiting carpels free, often a fleshy berry, or united into an aggregate fruit. About 62 genera and 800 species, pantropical in distribution.

1. Fruit an aggregate from several confluent pistils; receptacle conical-elongate; leaves lanceolate to elliptic-lanceolate, apex acute . . . . 1. *Annona*
1. Fruit simple from a single pistil; receptacle hemispheric; leaves oblanceolate to obovate, usually rounded, obtuse, or truncate at apex . . . . 2. *Asimina*

1. ANNÒNA L. Alligator Apples

Trees or shrubs with simple or stellate hairs and pungent aromatic leaves. Leaves persistent, thick, pinnately veined. Flowers solitary or in few-flowered clusters; sepals three, petals six, or the inner three absent or rudimentary, white or yellow, stamens many; carpels many, ovules one in each carpel. Fruits aggregated and concrescent on the receptacle. About 110 species, chiefly tropical America.

1. Petals over 1 cm wide; fruit smooth; leaves elliptic-lanceolate . . . . 1. *A. glabra*
1. Petals less than 1 cm wide; fruit tuberculate; leaves lance-ovate . . . . 2. *A. squamosa*

1. A. glabra L. Pond Apple—Trees up to 10 m tall, often with buttressed trunks. Leaves 7–14 cm long, elliptic-lanceolate to oblong-elliptic or ovate, pale green, usually acute at the apex. Flowers solitary, from about the middle of the internodes below the leaves on often

Family 84. **ANNONACEAE.** *Annona glabra:* a, flowering branch, × ½; b, flower, × 1; c, androecium and gynoecium, × 2½; d, mature fruit, × ½.

drooping pedicels 1–2 cm long, sepals reniform or rotund, about 4–5 mm long, outer petals white, 2.5–3.0 cm long, stamens 3–4 mm long. Fruit globose, smooth, 7–12 cm long, yellowish blotched with brown. $2n = 28$. Mangrove swamps, ponds, low wet soil, south Fla., Fla. Keys, Mexico, tropical America, tropical Africa. Spring. *A. palustris* L.

2. **A. squamosa** L. Sugar Apple, Sweet Sop—Trees up to 10 m tall with smooth branches. Leaves mostly 10–15 cm long, lance-elliptic to elliptic-ovate, acute or sometimes obtuse at the apex. Flowers solitary; sepals deltoid-acute, about 2 mm long; outer petals lanceolate, 2.5–3.0 cm long, white. Fruit 6–10 cm long, ellipsoidal to subglobose, yellowish green, tuberculate. $2n = 14$. South Fla., Fla. Keys, naturalized from tropical America where it is cultivated for its edible fruits. Spring.

2. ASÍMINA Adans.

Shrubs or small trees deciduous or evergreen with pungent odor when bruised. Leaves usually large, alternate. Flowers axillary, solitary or in small clusters, short pedicelled, nodding; sepals three, soon deciduous, petals six, the inner smaller and erect. Pistils three to fifteen but only few maturing. Fruit one to few, oval or oblong fleshy berries. About ten species in east U.S.

1. Flowers arising from wood of previous year's growth, before or during emergence of current year's leaves.
2. Newly emergent leaves densely tomentose; outer petals white, inner yellowish white with yellow zone . . . . . 1. *A. speciosa*
2. Newly emergent leaves puberulent or only lower surface tomentose; outer petals white, inner yellowish with maroon or purple zone . . . . . 2. *A. reticulata*
1. Flowers arising after emergence of current year's leaves, axillary to new leaves or terminal to new shoot growth.
3. New shoots and peduncles with bright red pubescence; flower buds terminal on new growth; shrub or small tree over 2 m tall . . . . 3. *A. obovata*
3. New shoots and peduncles glabrous or sparsely pubescent; flower buds axillary to new leaves; dwarf shrub less than 1 m tall . . . . 4. *A. pygmaea*

1. **A. speciosa** Nash—Shrub up to 1.5 m tall, stiffly branching. Leaves 5–8 cm long, coriaceous, elliptical to obovate or ovoid, with whitish to-

mentum becoming sparse with maturity. Flowers one to four per node, nodding on densely pubescent peduncles; calyx 8–12 mm long; outer petals 3–7 cm long, oblong, white or yellowish white, inner petals about one-half as long; stamens pale green or pinkish. Carpels free, three to five, fusiform; fruit oblong, about 8 cm, yellow green. $2n = 18$. Sandy soil, pineland, central Fla. to Ga. Spring. *Pityothamnus incanus* (Bartr.) Small

2. **A. reticulata** Shuttlew. ex Chapm. Pawpaw—Shrub up to 1.5 m tall, stiffly branching. Leaves about 5–8 cm long, coriaceous, elliptic to oblong or cuneate, with sparse orange pubescence above, densely pubescent beneath, becoming glabrate with maturity; margins revolute. Flowers one to three per node nodding on orange pubescent peduncles; calyx 8–10 mm long; outer petals 3–7 cm long, oblong, white, inner petals about one-half as long, yellowish with deep purple zone; stamens pale green or pinkish. Carpels free, three to eight, fusiform. Fruit short, oblong, 4–7 cm, yellowish green. $2n = 18$. Sandy soil, low pinelands, endemic to Fla. Spring. *A. cuneata* Shuttlew.; *Pityothamnus reticulatus* (Chapm.) Small

3. **A. obovata** (Willd.) Nash—Shrub or small tree with coriaceous leaves about 4–10 cm long, oblanceolate to obovate with sparse reddish pubescence when young, becoming glabrate with age. Flowers white, 6–10 cm broad, subsessile or on densely red hairy peduncles less than 5 mm long; sepals 0.5–1.5 cm long; outer petals 4–6 cm long; inner petals less than one-half as long, white or occasionally with basal purple zone. Carpels free, three to eight, fusiform. Fruit 5–9 cm, yellow green. $2n = 18$. Dry sand ridges, coastal dunes, hammocks, endemic to Fla. Spring. *Pityothamnus obovatus* (Willd.) Small

4. **A. pygmaea** (Bartr.) Dunal, Dwarf Pawpaw, Gopherberry—Shrubs usually 2–3 dm tall with one to several shoots sparingly branched or unbranched, barely rusty hairy or glabrate towards the tip. Leaves 4–7 cm long, coriaceous, cuneate to obovate or oblanceolate, usually sparsely pubescent when young, becoming glabrous, strongly reticulate veined beneath. Flowers maroon, ill scented, nodding on peduncles 1.5–3 cm, solitary from upper axils. Outer petals 1.5–3 cm, pink with maroon streaks, inner petals about two-thirds as long, maroon. Carpels two to five, free, fusiform. Fruit 3–4 cm curved, cylindric, yellow green. Old fields, pineland, Fla. to Ga. Spring. *Pityothamnus pygmaeus* (Bartr.) Small

Deeringothamnus pulchellus Small, distinguished from Asimina by having six to twelve linear or linear-oblong petals, is reported from Lee and Charlotte Counties. Further work may demonstrate it properly belongs to Asimina.

85. LAURÀCEAE   Laurel Family

Trees or shrubs with aromatic, alternate, simple, evergreen leaves, or greenish yellow vines. Flowers small, bisexual or unisexual, numerous in axillary or terminal racemose or cymose inflorescence, calyx usually with four to six colored lobes, three to four whorls of stamens. Ovary superior with one carpel. Fruit a berry or drupe. About 50 genera and 1000 species, pantropical, especially in Brazil and southwest Asia. *Cassythaceae*

1. Greenish yellow herbaceous vines, parasitic; leaves scalelike or absent . . . . .   1. *Cassytha*
1. Trees or shrubs; leaves with broad blades.
   2. Stamens three, fused; anthers two-celled .   2. *Licaria*
   2. Stamens twelve, free, the inner present only as staminodes; anthers four-celled.
     3. Staminodes large, mostly sagittate.
       4. Calyx persistent, present under fruit   3. *Persea*
       4. Calyx deciduous, not present under fruit.
         5. Leaves pinnately nerved, not strongly aromatic . . . . .   3. *Persea*
         5. Leaves mostly trinerved, strongly aromatic . . . .   4. *Cinnamomum*
     3. Staminodes small, filamentous . . .   5. *Nectandra*

1. CASSÝTHA L.

Parasitic vines with yellowish or pale green stems and branches, holding on to herbs and shrubs. Leaves absent or reduced to mere scales. Flowers small, bisexual in racemes, spikes, or head, calyx with six sepals on short hypanthium; stamens nine, staminodia three. Fruit a drupe. About 15 species, pantropical.

1. **C. filiformis** L. Love Vine—Stems often thickly matted, yellowish green. Flowers three to six, spicate, inconspicuous; inner sepals deltoid, two to three times longer than outer ones. Drupe about 5–7 mm wide, globose. On many hosts and in great variety of habitats, pinelands, sand dunes, hammocks, south Fla., W.I. Closely resembling Cuscuta (dodder) of northern states in color and general habit. All year.

2. LICÁRIA Aubl.

Evergreen trees or shrubs with alternate or sometimes opposite, coriaceous, entire leaves. Inflorescence often of axillary or subterminal, few- to many-flowered, panicled cymes, the flowers small and inconspicuous. Flowers bisexual, sepals six, short, erect, outer stamens often petaloid or aborted, inner stamens fertile. Fruit a drupe, exserted. About 48 species, American tropics. *Misanteca* Cham. & Schlecht.

1. **L. triandra** (Sw.) Kosterm. GULF LICARIA—Trees with evergreen, coriaceous leaves and rather flaky bark. Leaves mostly 5–11 cm long, elliptic-ovate, entire, long pointed, acuminate, glabrous, dark green and lustrous above. Flowers minute, purplish or whitish, 2–3 mm wide in three- to five-flowered cymes on slender peduncles. Fruits about 2–3 cm long on elongated, thickened peduncles, ellipsoidal or acornlike. Hammocks, south Fla., especially near Miami, W.I.

3. PÉRSEA Mill.

Shrubs or trees with persistent, alternate leaves. Flowers in many axillary or subterminal panicles, bisexual, rather conspicuous, often greenish, calyx lobes six, unequal, pubescent, stamens in three series, often all fertile, staminodia usually large. Ovary ovoid, style filiform. Fruit small and globose or large and fleshy. About 145 species, chiefly tropical America. *Tamala* Raf.

1. Calyx deciduous; fruit over 5 cm long . . 1. *P. americana* var. *americana*
1. Calyx persistent; fruit less than 3 cm long.
  2. Leaves tomentose to appressed pubescent beneath; petioles and young branches pubescent . . . . 2. *P. palustris*
  2. Leaves glabrous beneath; petioles and branches smooth . . . . . . 3. *P. borbonia*

1. **P. americana** Mill. var. **americana** AVOCADO—Tree up to 20 m tall with elliptic or obovate-oblong, coriaceous, dark green leaves, about 10–30 cm long. Inflorescence an axillary panicle, densely pubescent, flowers small, about 6–7 mm long, with yellow green sepals, these about 4–5 mm long, deciduous. Ovary pubescent. Fruit large, pear shaped or globose, mostly 7–20 cm long, the pulp fleshy, oily. Widely cultivated in south Fla. and often found in deserted habitations, naturalized probably from Mexico and Central America. Spring.

2. **P. palustris** (Raf.) Sarg. SWAMP BAY—Shrub or tree up to 15 m tall with densely rusty tomentose young branches. Leaves mostly 10–15 cm long, lanceolate to oblong, thinly pubescent beneath, becoming glabrate; petiole 1–2 cm long, tomentose. Inflorescence a many-flowered panicle, peduncles tomentose. Fruits subglobose, 10–18 mm wide. Wet

soil, swamps, Fla. to Tex. and Va. on coastal plain, Bahama Islands. Spring. *Tamala pubescens* (Pursh) Sarg.

3. P. borbonia (L.) Spreng. RED BAY—Shrub or tree up to 20 m tall with young stems thinly pubescent or glabrous. Leaves mostly 10–20 cm long, lanceolate to oblong-lanceolate or elliptic, glossy green above, reticulate veined, paler beneath, tapering at both ends. Panicles few-flowered, sepals ovate, acute. Drupe subglobose, about 10–12 mm wide, dark blue or almost black, shiny. Two varieties occur in our area:

1. Leaf blade 4–13 cm long, sparsely strigose
   beneath . . . . 3a. var. *borbonia*
1. Leaf blade 3–8 cm long, densely rusty strigose beneath . . . . 3b. var. *humilis*

3a. P. borbonia var. **borbonia**—Leaf blades 1.5–6.5 cm wide. All soils, various sites as margins of swamps, hammocks, Fla. to Tex., N.C. Spring. The leaves are often deformed and swollen at the tips or edges. *Tamala borbonia* (L.) Raf.; *P. littoralis* Small.

3b. P. borbonia var. **humilis** (Nash) Kopp—Leaf blades 1–3 cm wide. Dry sand scrub, endemic to Fla. Spring. *P. humilis* Nash

4. CINNAMÒMUM (L.) Nees & Eberm.

Shrubs and trees with alternate or opposite evergreen leaves. Flowers small, bisexual or sometimes unisexual, in subterminal or axillary panicles or cymes, calyx short, with six segments, subequal, whitish or greenish; stamens usually five to nine in three series, fourth series of staminodes. Fruit a berry in cuplike perianth. About 50 species, Asia and Australia, widely cultivated. *Camphora* L.

1. C. camphora (L.) Nees & Eberm. CAMPHOR TREE—Shrub or tree up to 15 m tall. Leaves 5–12 cm long, alternate, long petioled, elliptic to ovate, acuminate, producing distinct camphor odor when bruised. Flowers in axillary panicles; sepals yellow, 1–1.5 mm long, deciduous. Fruits a globose or subglobose drupe, 6–9 mm in diameter. $2n = 24$. Drier sites, Fla., escaped from cultivation, naturalized from Asia. Spring.

5. NECTÁNDRA Roland.

Evergreen trees or shrubs with alternate or subopposite entire leaves. Flowers bisexual, pedicelled, in axillary or subterminal panicles, perianth present or absent; sepals six, equal, often reflexed, outer series of stamens usually petaloid and fleshy. Ovary globose or nearly so, style short, stigma capitate or discoid. Drupe ellipsoidal or subglobose, borne in a cupule. About 175 species in tropical America.

1. **N. coriacea** (Sw.) Griseb. LANCEWOOD—Shrub or small tree up to 10 m tall, often densely branching. Leaves 6–12 cm long, elliptic to narrowly ovate, glossy green above, paler beneath, aromatic. Calyx creamy white, 8–9 mm wide, obtuse, hairy pubescent within. Drupe 10–15 mm long, subglobose, dark blue or black in a reddish or yellow cupule. Hammocks, coastal woodlands, south Fla., Fla. Keys, W.I. All year. *Ocotea catesbyana* (Michx.) Sarg.

86. PAPAVERÀCEAE POPPY FAMILY

Perennial or annual herbs, rarely shrubs or trees, with clear or colored sap, alternate leaves or floral leaves opposite or whorled, often much divided. Flowers mostly showy, solitary, bisexual; sepals two to three, petals four to six or eight to twelve, free, radially symmetrical or less often bilateral; stamens numerous in two to three series. Ovary superior, one- to several-locular, gynoecium of two to many fused carpels, stigmas the same number as the placentas. Fruit a capsule opening by pores or valves; seeds numerous, small. About 28 genera and 700 species, chiefly north temperate and subtropical regions.

1. ARGÉMONE PRICKLY POPPY

Herbs, rarely shrubs, usually glaucous and with spines or prickles and yellowish sap. Leaves alternate, sessile, sharply pinnatifid, the lobes rigidly spine tipped. Flowers terminal, solitary, large, or on short axillary branches and appearing racemose; sepals two to three, early caducous; petals four to six, showy; stamens many; stigma lobes radiating from center, style short. Capsule oblong, often spiny, opening by three to six valves at top. About ten species in America.

1. **A. mexicana** L. MEXICAN POPPY—Stems glaucous, generally spiny, up to 1 m tall. Leaves 10–25 cm long, sessile-clasping, irregularly lobed or incised, pinnatifid, the lobes spine tipped and spiny along midrib, green but blotched with paler green. Flowers large, 3–6 cm wide, subtended by bracts of two to three reduced leaves; petals yellow or cream colored, broadly ovate, 2–4 cm long. Capsule ellipsoidal, about 3 cm long, conspicuously spiny. Disturbed sites, roadsides, cultivated fields, tropical America, north to Mass. $2n = 28$, $2n = 112$. The two forms occurring in our area are:

1. Inflorescence and fruit spiny	*A. mexicana* f. *mexicana*
1. Inflorescence and fruit without spine . .	*A. mexicana* f. *leiocarpa* (Greene) G. B. Ownbey

Family 85. **LAURACEAE.** *Nectandra coriacea:* a, flowering branch, × ½; b, stamens × 17½; c, flower, × 6¼; d, mature fruit, × 2¼.

Family 86. **PAPAVERACEAE.** *Argemone mexicana:* a, flowering branch, ×
½; b, flower, × 1; c, androecium and gynoecium, × 4½; d, mature fruit,
× 1½.

87. BRASSICÀCEAE   Mustard Family

Annual or perennial herbs with watery juice and alternate leaves. ⁙owers usually racemose, bisexual, radially symmetrical, sepals four, free, petals four, stamens two to six, free or fused in pairs. Ovary superior, gynoecium of two-fused carpels, one-locular, ovules many. Fruit a silicle or silique, dehiscent or rarely indehiscent. About 200 genera and 2000 species. Cosmopolitan but chiefly north temperate and Mediterranean regions. Nomen alt. Cruciferae.

1. Corolla yellow.
  2. Leaves, at least the lower ones, deeply serrate, lobed, pinnatifid, or pinnate.
    3. Petals 6–15 mm long . . . . . .   1. *Brassica*
    3. Petals smaller, less than 6 mm long.
      4. Leaves coarsely serrate or toothed .   2. *Diplotaxis*
      4. Leaves lobed or pinnatifid . . .   3. *Rorippa*
  2. Leaves simple, lanceolate, not lobed or pinnatifid . . . .   4. *Erysimum*
1. Corolla white, pale purple, or pink.
  5. Plants aquatic.
    6. Leaves dimorphic, submerged ones pinnate with linear segments, emergent leaves simple, entire or serrate . .   5. *Armoracia*
    6. Leaves all pinnate with rounded leaflets   6. *Nasturtium*
  5. Plants terrestrial.
    7. Ovary or fruit transversely jointed; leaves fleshy; maritime coastal beaches . . . .   7. *Cakile*
    7. Ovary or fruit not transversely jointed; leaves thin, not fleshy; inland plants.
      8. Cauline leaves simple, toothed or incised, entire or undulate.
        9. Leaves serrate or incised; ovary circular . . . .   8. *Lepidium*
        9. Leaves entire or undulate; ovary linear . . . .   9. *Warea*
      8. Cauline leaves deeply dissected, pinnate or pinnatifid . . . . .   10. *Cardamine*

1. BRÁSSICA L.   Mustard

Annual or biennial herbs, often weedy with simple stems branched above. Leaves simple, lobed, or divided. Inflorescence racemose; sepals oblong or obovate, petals spatulate, yellow; silique linear, usually with a tapering beak. Seeds globose, wingless. About 60 species, in Mediterranean region, Europe, Asia, introduced in America. Several species and varieties are grown as vegetables. *Sinapis* L.

1. **B. kaber** (DC.) L. Wheeler, Wild Mustard—Glabrous or rough pubescent annual up to 8 dm tall. Leaves 4–17 cm long, obovate, lower ones coarsely toothed, lobed, or serrate, upper progressively smaller and only serrate. Flowers 1–2 cm wide, bright yellow; petals 7–9 mm long, suborbicular or ovate narrowed to slender claws. Silique 1–2 cm long, glabrous or rarely pubescent, beak half as long as body of silique. Seeds smooth 1–1.5 mm. An introduced weed in disturbed sites, old fields, widely distributed in North America. Spring, fall. *Sinapis arvensis* L.; *Brassica arvensis* L.

Brassica rapa L. with upper leaves sessile-clasping and cordate-auriculate bases may occur as an occasional weed in our area. Also, Sisymbrium irio L., the London Rocket, with beakless siliques 3–5 cm long and deeply pinnatifid leaves has been collected as a weed in the Miami area.

2. DIPLOTÁXIS DC. Sand Rocket

Annual or perennial herbs usually with pinnately lobed or serrate leaves. Inflorescence racemose, sepals erect, petals clawed, yellow, stamens six, the short ones subtended by a small gland reniform or hemispheric in shape, the long ones subtended by a short prismatic gland. Ovary linear, style very short, stigma capitate. Silique somewhat compressed, short beaked, seeds smooth. About 35 species of central Europe, Mediterranean region, Asia.

1. **D. muralis** (L.) DC. Wall Rocket—Annual or biennial up to 5 dm tall, sparingly pubescent, the branches suberect or sometimes decumbent. Leaves malodorous, mostly basal, elliptic to spatulate, pinnately toothed to pinnatifid. Flowers in racemes, small; sepals 3–4 mm long, petals 7–8 mm long, yellow. Silique linear, not stipitate. Seeds less than 1 mm long. Disturbed sites, introduced from Europe, Fla. to La., north to N.C. on coastal plain. All year.

3. RORÍPPA Scop. Marsh Cress

Aquatic or terrestrial perennial or annual herbs with simple or deeply lobed pinnate or pinnatifid leaves. Inflorescence racemose, terminating all the upper branches. Flowers small, yellow, or sometimes white. Siliques short, slender to plump. Seeds turgid, numerous. About 50 species, cosmopolitan. *Radicula* Hill

1. **R. teres** (Michx.) Stuckey, Marsh Cress—Stems simple or dif-

fusely branched up to 5 dm tall. Leaves twice pinnatifid with the segments irregular, about 4–10 cm long. Sepals spreading, narrowly oyate; petals yellow, somewhat longer than the sepals. Silique long beaked. Disturbed sites, old fields, glades, hammocks, Fla. to Tex., S.C., Okla., Mex. Spring, summer. *Sisymbrium walteri* Ell.; *Radicula walteri* (Ell.) Small; *Rorippa walteri* (Ell.) Mohr

R. islandica (Oeder) Borbas var. **fernaldiana** Butt. & Abbe, with lower leaves merely pinnatifid or lobed and the upper serrate, also occurs in southern Fla., but it is very local in distribution.

4. ERÝSIMUM L.   TREACLE MUSTARD

Perennial or annual herbs with narrow pinnatifid to entire leaves with characteristic two- to four-pronged appressed hairs. Racemes terminating the branches; sepals usually erect, narrow; petals yellow to orange, obovate to spatulate narrowed to a claw. Ovary linear, pubescent, style short, stigma two-lobed. Silique hairy. About 80 species of the north temperate zone. *Cheirinia* Link; not *Erysimum* sensu Small

1. **E. cheiranthoides** L. WORMSEED MUSTARD—Erect annual with stems up to 1 m tall, leafy. Leaves linear to oblanceolate, entire or barely toothed covered with fine pubescence of branched hairs. Flowers small, racemes narrow with pedicels widely divergent, sepals 2–3 mm long; petals 3.5–6 mm long, bright yellow. Silique erect to ascending, 15–20 mm long, sparsely pubescent. $2n = 16$. Disturbed sites, cultivated fields, roadsides, circumboreal, Fla., Ark., Oreg., Colo. Spring, summer.

5. ARMORÀCIA Gaertn., Meyer, & Scherb.

Smooth herbs with erect or prostrate stems. Leaves oblong to pinnately dissected. Sepals elliptic to ovate, ascending; petals ovate, white, narrowed to a claw, short stamens subtended by a V-shaped gland, long stamens by a conic gland. Ovary ellipsoidal, style slender, stigma large. Three species of Europe and Asia. *Neobeckia* Greene

1. **A. aquatica** (A. A. Eaton) Wieg. LAKE CRESS—Aquatic, stems usually submerged, lax. Immersed leaves pinnately dissected into filiform segments, emersed leaves about 3–7 cm long, if present, narrowly oblong, serrate or dentate. Sepals about 3.5–4 mm long, petals spatulate, 6–8 mm long. Siliques about 5–8 mm long, ellipsoidal, often aborted, style persistent. In quiet water, wet soil, Fla. to Tex., Quebec, Minn. Summer. *Radicula aquatica* Robinson; *Neobeckia aquatica* Greene

Armoracia rusticana Gaertn. HORSERADISH is reported as an occasional escape in our area. It is terrestrial with lower leaves oblong, cordate at base, on long petioles.

6. NASTÚRTIUM R. Brown WATERCRESS

Smooth, diffuse, perennial aquatic herbs with diffuse stems and pinnately compound leaves. Racemes lax; sepals ascending, broad, with one pair saccate at base; petals ovate, white, narrowed to a claw, short stamens subtended by a pair of reniform glands. Ovary cylindrical, style short, stigma capitate and two-lobed. Silique linear, tipped with a short, persistent style. About 50 species, cosmopolitan. *Sisymbrium* sensu Small

1. **N. officinale** R. Brown, WATERCRESS—Stems submerged, floating, or prostrate, spreading on the mud, rooting freely from the nodes, up to 1 m or more long. Leaves with three to nine obtuse segments, the blades usually 10–20 cm long, the terminal ones much larger than the lateral blades. Flowers 5 mm wide, petals yellow, about twice as long as the sepals. Silique slender, about 1–2.5 cm long. $2n = 14$. Naturalized from Eurasia, widely established in quiet water in much of North America. Spring, summer. *Rorippa nasturtium* var. *aquaticum* (L.) Hayek

7. CAKÌLE Mill. SEA ROCKET

Succulent annual or biennial maritime herbs with branched, stout, decumbent stems. Leaves fleshy, entire to almost pinnate. Inflorescence racemose; sepals succulent, petals obovate, flowers white to purplish, pedicels thick. Silique of two joints which disarticulate freely at maturity, indehiscent. Seeds oblong. Four species, widely distributed.

1. Upper joint of pod prominently four-angled   1. *C. edentula*
1. Upper joint mostly six- to ten-ridged, not
    four-angled . . . . 2. *C. fusiformis*

1. **C. edentula** (Bigel.) Hook. SEA ROCKET—Stems much branched up to 5 dm tall. Leaves spatulate or ovate, coarsely lobed or toothed to subentire, narrowed at the base. Flowers pale purple, 5 mm wide; sepals 3–4 mm long, petals 5–6 mm long. Lower joint of silique obovoid or turbinate, up to 8 mm long, often sterile; upper joint four-angled, conspicuously larger and narrowed to articulation. $2n = 18$. Coastal beaches, Fla. to Labrador, along Great Lakes, Pacific coast. All year.

This species is reported for beach areas in north Fla. and presumably is in our area.

2. **C. fusiformis** Greene—Annual with rather coarse, decumbent stems up to 6 dm long, glabrous. Leaves mostly 3–6 cm long, narrowly elliptic to linear-oblanceolate, dentate or serrate to deeply cleft or dissected. Flowers about 5 mm wide, petals white. Siliques linear, the lower joint nearly cylindrical about 3–4 mm wide, upper joint lanceolate. Seeds

about 3.5 mm long. Coastal beach and dune vegetation, sandy shores, Fla. and Fla. Keys, to S.C., W.I., Central America. All year. Highly variable in leaf and fruit morphology.

8. LEPÍDIUM L.   Pepper Grasses

Annual or perennial herbs with entire to bipinnate or tripinnate leaves. Racemes terminating the upper branches, congested in flower and elongating in fruit, flowers small; petals absent or minute, white or pale yellow; stamens two to six; silique rounded to oblong. Seeds one in each locule. About 130 species, cosmopolitan.

1. **L. virginicum** L. Pepper Grass—Annual with stems freely branching, up to 6 dm tall, sparsely pubescent especially above. Lower leaves irregularly toothed or pinnately toothed or divided, upper leaves oblanceolate, irregularly toothed to almost entire, reduced upward. Racemes numerous, many-flowered; sepals about 1 mm long, petals white, slightly longer, stamens usually two. Siliques orbicular to slightly oblong, about 2.5–3.5 mm long, shallowly notched at apex. $2n = 18,32$. Disturbed sites, pineland, old fields, roadsides, a common weed throughout our area, Fla. to Newfoundland, west to Pacific Coast. All year.

9. WÁREA Nutt.

Glabrous, slender herbs with entire leaves. Sepals narrow, petals purple to white, narrowed to a slender claw. Stamens six, anthers elliptic to elliptic-sagittate. Style absent, silique linear-cylindrical with elongate stipe. About four species, southeastern U.S.

1. **W. carteri** Small—Stems up to 1.5 m tall. Lower leaves linear to linear-lanceolate, sometimes cuneate, about 1–3 cm long. Sepals 4–5 mm long; petals 5.5–6.5 mm long, white or nearly so, the claw densely glandular-pubescent. Silique 5–6 cm long, pedicel becoming about 1 cm long. Sand hills, pineland, endemic to Fla. from Brevard and Polk Counties to Dade County. Spring, summer. This species is closely related to and possibly conspecific with W. cuneifolia (Muhl.) Nutt.

10. CARDÁMINE L.   Bitter Cress

Perennial or annual herbs with leaves usually pinnately compound. Inflorescence racemose, without bracts or the lower flowers bracteate, sepals much shorter than petals; petals spatulate to obovate, white to pink, stamens fewer than six. Stigma usually two-lobed; siliques linear, dehiscent, often elastic. About 115 species, widely distributed.

1. **C. hirsuta** L.—Annual or biennial with hairy stems up to 3 dm tall. Basal leaves numerous, upper leaves few, plant appearing subscapose. Terminal leaflets orbicular to reniform, entire to few-lobed, petioles hirsute or at least ciliate. Petals white 2–3 mm, about two times as long as the sepals. Stamens four; silique erect, about 15–25 mm long. $2n = 18$. Disturbed areas, moist soil, pinelands, Fla. to N.Y., Ill., introduced from Old World. Spring.

88. CAPPARÀCEAE Caper Family

Trees, shrubs, less often herbs or vines with alternate simple or digitately compound leaves. Flowers mostly bisexual, radially symmetrical, terminal or axillary in various kinds of inflorescences, sepals mostly four, petals four, stamens few to many. Ovary superior, often supported on a gynophore, or sometimes the internode between petals and stamens develops into an androgynophore. About 40 genera and 45 species with many xerophytes, of tropical or warm temperate regions.

1. Herbs with palmately compound leaves; stems armed with thorns; fruit a dry silique . . . . 1. *Cleome*
1. Shrubs or trees with apparently simple leaves; stems unarmed; fruit fleshy . . 2. *Capparis*

1. CLÈOME L.

Herbs sometimes woody at the base and glandular-pubescent, or thorny, with usually palmately compound or occasionally simple leaves. Inflorescence racemose, few to many flowered, bracteate; calyx deeply four-parted; petals four, subequal; disk present; stamens six, rarely four on a short androgynophore. Ovary borne on an elongate gynophore. Fruit a capsule. About 200 species, chiefly in the tropical and subtropical regions. *Neocleome* Small

1. **C. spinosa** Jacq. SPIDERFLOWER—Stout herbs with stems up to 1.5 m tall, conspicuously glandular-pubescent and armed with prominent, paired thorns at the nodes. Leaves palmately compound with petioles 2–8 cm long, these also armed with stout yellow thorns. Leaflets 2–10 cm long, narrowly cuneate, pubescent. Inflorescence a many-flowered corymbose raceme, the peduncle greatly elongating in fruit; bracts foliaceous, cordate, sepals 5–8 mm long, densely glandular-pubescent; petals 1–2 cm, clawed, pink or white. Capsule fusiform, 5–12 cm long, the fruiting

gynophore 2–3 cm long. Disturbed sites, roadsides, hammocks, south Fla., Fla. Keys, escaped northward from cultivation; tropical America. All year. *Capparis spinosa* L.; *Neocleome spinosa* (L.) Small

2. CÁPPARIS L.

Shrubs or trees with simple or stellate hairs, lepidote, or glabrous. Leaves simple, entire. Inflorescence usually several- to many-flowered, bracteate; sepals four, free or fused at the base; petals four, equal or subequal; stamens few to over one hundred on an androphore. Ovary on a short or elongate gynophore. Fruit a fleshy capsule. About 350 species, pantropical in distribution.

1. Leaves and upper stems scaly (lepidote) . 1. *C. cynophallophora*
1. Leaves and upper stems glabrous, not scaly . 2. *C. flexuosa*

1. **C. cynophallophora** L. Jamaica Caper Tree—Shrub or small tree up to 6 m tall with branches densely lepidote to glabrate. Leaves mostly 5–20 cm long, elliptic to obovate, leathery, densely lepidote below. Inflorescence corymbose, terminal and axillary, sepals 7–12 mm long, petals ovate, 8–16 mm long, white to purplish or brownish, densely lepidote. Ovary elongating in fruit 15–40 cm long. Coastal hammocks, south Fla., Fla. Keys, W.I., Mexico, South America. Spring, summer.
Dwarfed forms may be found in hammocks and in bare rock in the upper Fla. Keys.

2. **C. flexuosa** L. Bay-leaved Caper Tree—Shrubs or small trees up to 8 m tall, glabrous. Leaves 4–16 cm long, very variable, linear to broadly obovate, glabrous and pale green. Inflorescence terminal, corymbose-paniculate, few-flowered, sepals 7–10 mm long, petals ovate 1–1.5 cm long, white or pale pink, glabrous. Ovary elongating into an irregular fleshy capsule up to 15 cm long, gynophore 1–9 cm long. Coastal areas, hammocks, and marl flats, south Fla., Fla. Keys, W.I., Mexico to South America. Summer.

89. DROSERÀCEAE Sundew Family

Herbs often stemless with rosettes of leaves bearing gland-tipped hairs on marginal bristles. Flowers radially symmetrical, bisexual usually in a simple cyme, sepals four to five, petals five, stamens five or four to many. Ovary superior, one-locular, styles three to five. Fruit a capsule.

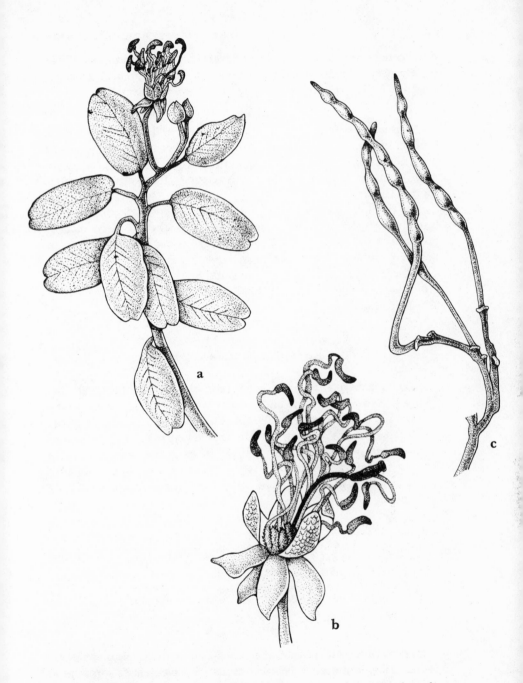

Family 88. **CAPPARACEAE.** *Capparis cynophallophora:* a, flowering branch,
× ½; b, flower, × 2; c, mature fruits, × 1.

Four genera and 93 species of carnivorous plants of sandy or boggy places, widely distributed. *Dionaeaceae*

1. DRÓSERA L. SUNDEW

Annual or perennial low scapose herbs covered with reddish, gland-bearing bristles. Leaves various, filiform to suborbicular, stipules scarious. Inflorescence a cyme, nodding at the tip; petals white, pink, or purplish; stamens as many as the petals. Capsule three-valved. About 90 species, chiefly America.

1. **D. capillaris** Poir. PINK SUNDEW—Stems up to 20 cm tall; leaves broadly spatulate, 5–10 mm long, shorter than the petioles, sepals 3–4 mm, petals 6–7 mm, pink. Capsule about as long as the sepals; seeds brown, papillose. $2n = 20$. Low pinelands, Fla. to Tex., Va., Tenn.

Large populations are often found in small areas covering the ground with reddish or pinkish small rosettes usually on moist or seasonally flooded soil.

90. CRASSULÀCEAE ORPINE FAMILY

Low shrubs or herbs usually with succulent leaves and stems. Flowers mostly cymose, bisexual, and radially symmetrical, sepals four to five, free or united into a tube, petals same number, free; stamens as many as the petals or twice as many. Ovary superior, gynoecium of carpels as many as the petals, free or sometimes fused at the base, one-locular. Fruit dry, membranous or leathery, dehiscent, opening down the ventral suture. About 30 genera, 1300 species, cosmopolitan, but chiefly of warm, dry regions. *Sedaceae*

1. Sepals fused for half or more their length,
    lobes short, calyx inflated; leaves opposite,
    petiolate, ovate flat . . . .   1. *Bryophyllum*
1. Sepals free or fused only at the base, not in-
    flated; leaves whorled or subopposite, ses-
    sile, linear-cylindric, or flat . . . .   2. *Kalanchoe*

1. BRYOPHÝLLUM Salisb.

Erect, succulent, perennial herbs or subshrubs, glabrous with opposite, fleshy, simple or pinnately compound leaves. Flowers in terminal panicles, bisexual, nodding; sepals four, fused in an inflated tube; petals four, fused, lobes shorter than tube; stamens epipetalous. Carpels dis-

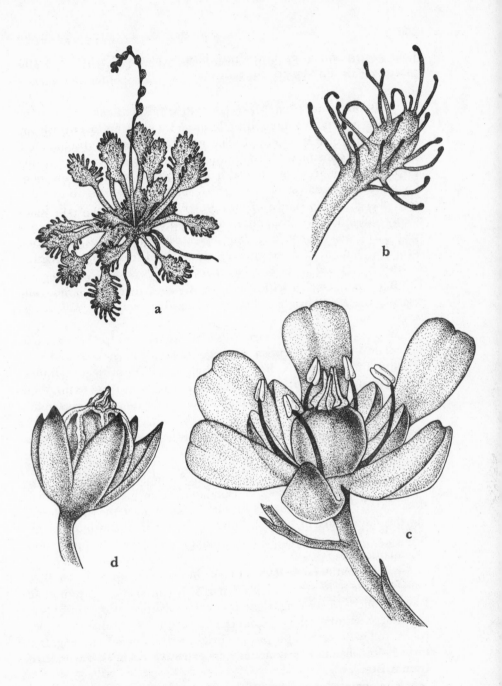

Family 89. **DROSERACEAE.** *Drosera capillaris:* a, flowering plant, × 1; b, detail of mature leaf, × 5; c, flower, × 10; d, mature fruit, × 11.

tinct or partly fused. Fruit of four follicles. About 20 species especially in the Old World, tropics, Madagascar, now cosmopolitan. *Kalanchoe* Hamet

1. **B. pinnatum** (Lam.) Kurz, LIFE PLANT, LIVE LEAF—Stems glaucous up to 1 m tall or more, little branched. Leaves up to 15 cm long, simple, fleshy, or three- to five-leaflets with both simple and compound leaves on the same plant; blade coarsely serrate, able to produce plantlets from indentations of the leaf margin; petiole up to 3–4 cm long, fleshy. Inflorescence a smooth terminal, compound panicle up to 5 dm long; flowers conspicuous; calyx tubular inflated, about 3 cm long, deeply four-lobed, greenish, often tinged with red; corolla tubular, contracted at base, 4–5 cm long, four-lobed, reddish; stamens eight. Disturbed sites, often in dense colonies, naturalized from tropical Africa. *Kalanchoe pinnata* Pers.; *B. calycinum* Salisb.

B. crenatum Baker with leaves usually auriculate, corolla not contracted at base, and bright red is also reported for south Fla. *Kalanchoe laxiflora* Baker

B. daigremontiana Hamet & Perrier, with leaves toothed but not auriculate and petals purplish, has been collected at Key West and may be naturalized in the Fla. Keys; **B. tubiflorum** Harv. with subcylindrical, mottled leaves that produce plantlets at the tips is naturalized on Marco Island, Fla., and may occur elsewhere in the range of this manual.

2. KALÀNCHOE Adans.

Erect perennial, tufted or branched succulent herbs sometimes woody at the base, with opposite or whorled, simple, fleshy leaves. Flowers borne in corymbose cymes, bisexual, often showy, calyx four-parted, not inflated, tube short; corolla four-lobed, salverform, the tube exceeding the calyx, stamens fused by their filaments. Follicles four, many-seeded, erect. About 100 species from the Old World tropics, now cosmopolitan.

1. **K. grandiflora** A. Rich.—Erect, simple stems up to 1 m tall or more, bearing opposite or whorled, narrowly lanceolate, fleshy, mottled leaves about 3–15 cm long, these producing plantlets at the tips. Flowers in large terminal trichotomous cymes, pendulous; corollas of various colors, yellowish brown, purplish to bright red, about 2–3 cm long. Disturbed sites, hammocks sometimes in extensive colonies, naturalized from Africa.

K. crenata Haw. is also reported for south Fla. It has ovate, flat, crenate leaves and cymes of yellow or orange flowers, the sepals are glandular-ciliate. Occurring in disturbed sites and hammocks, it has been

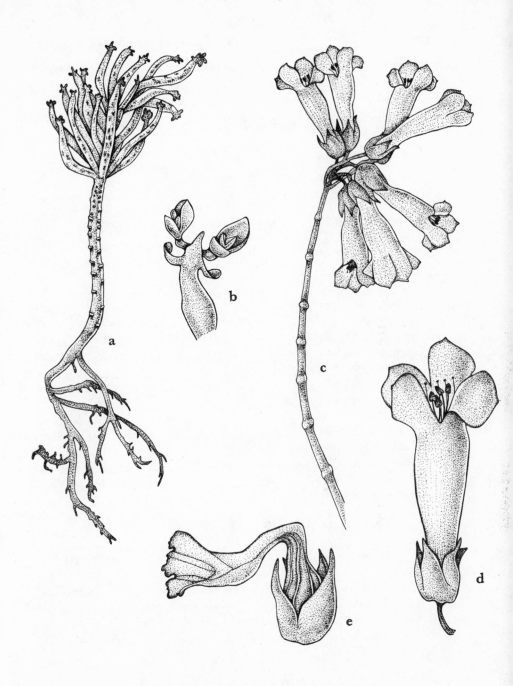

Family 90. **CRASSULACEAE.** *Kalanchoe grandiflora:* a, leafy plant, × ½;
b, leaf tip with plantlets, × 5; c, inflorescence, × 1; d, flower, × 2¼; e, fruit,
× 2½.

naturalized from Africa. No specimens, however, have been seen recently.

91. SAXIFRAGÀCEAE Saxifrage Family

Herbs or shrubs with alternate leaves and radially symmetrical, bisexual flowers. Sepals usually five, petals alternate with the sepals or absent; stamens five to ten free; receptacle polymorphic, varying from hypogyny to epigyny. Ovary one- to three-locular, free or fused to the tubular or flat receptacle. Fruit a capsule or berry, seeds numerous. About 80 genera, 1100 species, cosmopolitan but chiefly temperate regions. *Iteaceae* included.

1. ÌTEA L. Virginia Willow

Shrubs with alternate, pinnately veined, serrulate leaves and narrow, terminal racemes of small white flowers. Sepals small, petals oblong, larger; filaments pubescent. Ovary superior, two-locular, pubescent; stigma capitate. Fruit a capsule with persistent style, septicidal. About five species, chiefly Old World.

1. I. virginica L.—Shrub up to 3 m tall with pubescent stems and racemes. Leaves usually 3–8 cm long, elliptic to oblong-lanceolate or ovate, serrate, acuminate to the apex, base acute. Racemes 5–15 cm long; sepals lanceolate, 1–2 mm, petals white, linear-lanceolate, 5–6 mm long. Capsule slender, 8–10 mm long. $2n = 22$. Cypress heads and hammocks, along streams, swamps, low woods, Fla. to La., N.J., on the coastal plain; La. to Ill. in Miss. Valley.

92. HAMAMELIDÀCEAE Witch-Hazel Family

Shrubs or trees, deciduous or evergreen, usually with stellate pubescence. Leaves alternate, simple, toothed to palmately lobed, stipulate. Flowers bisexual or unisexual, radially symmetrical or sometimes bilaterally so, axillary, in spikes or heads; sepals four to five, petals four to five, or perianth absent, stamens two to eight. Ovary half-inferior to inferior, two-locular, styles and stigma two. Fruit a capsule. About 23 genera and 100 species, widely distributed.

1. LIQUIDÁMBAR L. Sweet Gum

Trees with palmately lobed leaves, monoecious. Flowers in globular heads or aments; stamens in conical clusters, very numerous, inter-

Family 91. **SAXIFRAGACEAE.** *Itea virginica:* a, flowering branch, × ½;
b, flower, × 6; c, mature fruits, × 2½.

mixed with small scales, pistillate inflorescence with numerous two-beaked, two-locular ovaries, cohering and becoming hard in fruit, forming a spherical ament. Seeds with wing-angled seed coat. Four species, Asia and North America.

1. **L. styraciflua** L.—Deciduous trees with grayish bark usually with corky ridges on the branchlets. Leaves rounded, deeply palmately five- to seven-lobed, appearing star shaped in outline, lobes acute pointed, glandular-serrate. Capsules woody, usually with numerous aborted seeds. Low ground, swamps, Fla. to Tex., Mexico, Mo. This species is not common in southern Fla. and appears to occur only in scattered sites at the northern edge of the range of this manual. Spring.

93. ROSÀCEAE ROSE FAMILY

Trees, shrubs, or herbs with alternate, stipulate, simple or compound leaves. Flowers generally bisexual, radially symmetrical and pentamerous, calyx free or fused to the ovary; corolla and stamens free or more or less perigynous; stamens numerous, often indefinite. Ovary superior or inferior; carpels one or more, distinct or often fused to the calyx tube or enlarged receptacle; styles free. Fruit various, dry or fleshy. Approximately 100 genera and 2000 species, cosmopolitan. A large and important family that includes several genera in which hybridization, polyploidy, and apomixis occur. *Malaceae, Amygdalaceae*

1. Stems armed with curved prickles; leaves
       compound.
     2. Leaves trifoliolate or palmately com-
          pound . . . .   1. *Rubus*
     2. Leaves pinnately compound with five to
          nine leaflets . . . .   2. *Rosa*
1. Stems unarmed; leaves simple   . . . .   3. *Prunus*

1. RÙBUS L. RASPBERRY, BLACKBERRY

Shrubs with erect, trailing or somewhat climbing, often prickly stems and usually glandular hairs. Leaves simple, or more often three- to seven-foliolate, serrated or lobed. Inflorescence racemose, cymose, paniculate, rarely solitary, axillary or terminal. Calyx five-lobed, petals five, distinct, white to purplish, stamens many. Carpels borne on a convex receptacle, closely packed. Fruit a syncarp of many coherent drupelets. About 700 species, especially in north temperate regions.

1. Stems erect or arching; leaflets white tomen-
       tose beneath   . . . .   1. *R. cuneifolius*

1. Stems prostrate, low, floral stems erect; leaflets not tomentose beneath   .   .   .   .   *2. R. trivialis*

   1. R. cuneifolius Pursh, Sand Blackberry—Stems erect or arching
and covered with hooked or straight prickles. Leaves three- to five-
foliolate on prickle-bearing petioles, densely white tomentose beneath,
the terminal leaflet cuneate from about midblade to the base, entire
margined, the upper half serrate. Inflorescence few-flowered, subtended
by normal leaves, petals white. Sandy soil, Fla. to Ala., Conn. on the
coastal plain.
   2. R. trivialis Michx. Southern Dewberry—Stems usually prostrate
or trailing, slender, with few, short, recurved prickles and also hispid
pubescent with reddish glandular bristles. Leaves somewhat leathery,
persistent, five-foliolate on prickle-bearing petioles, leaflets dentate-serrate, smooth. Flowers erect, mostly solitary on smooth pedicels; petals
white. Fruit black. Hammocks, sandy soil, disturbed sites as old fields,
Fla. to Tex., Va., Mo.

2. RÒSA L.

   Shrubs or woody vines mostly with pinnately compound leaves and
large stipules usually adnate to the petiole. Hypanthium urceolate to
globose with sepals foliaceous and attenuate, often persistent in fruit.
Petals five, white to red, or yellow, conspicuous; stamens numerous, inserted on the rim of the hypanthium. Pistils many in the bottom of the
hypanthium. Fruit an achene enclosed by the fleshy hypanthium.
About 100 or more species, chiefly in the temperate and subtropical regions of the Northern Hemisphere.
   1. R. bracteata Wendl. Macartney Rose—Stems much elongated,
trailing, climbing, or decumbent with paired, recurved prickles; younger
stem densely prickly and with stiff glandular hairs. Stipules about 4–5
mm long, adnate only at the base. Leaflets five to nine, elliptic-serrulate,
evergreen. Flowers usually solitary or few together, subtended by large
bracts; sepals tomentose, long acuminate; petals white, 2–3 cm long;
styles fused. $2n = 14$. Naturalized from China, well established in southeast U.S., in pineland, dry soil, Fla. to Tex., Va., Tenn.

3. PRÙNUS L.   Cherry, Plum

   Trees with alternate, simple, deciduous or evergreen leaves and
conspicuous flowers borne in umbels or solitary from axillary shoots, or
in terminal or axillary racemes. Calyx five-lobed, perigynous and forming, with the receptacular disk, an hypanthium; petals five, mostly
white, stamens fifteen to twenty, borne on the margin of the hypan-

thium. Carpel one with dilated stigma. Fruit a drupe. About 200 species, chiefly north temperate region. *Laurocerasus* Reichenb.

1. P. myrtifolia L. WEST-INDIAN CHERRY—Trees up to 15 m tall. Leaves evergreen, narrowly elliptic to oblong-ovate, often acuminate, glabrous, glossy in the upper surface, about 6–12 cm long. Flowers in axillary racemes, shorter than the leaves; sepals minute, petals yellowish white, about 2–3 mm long; stamens twelve to twenty. Fruit subglobose, glossy black purplish, about 1 cm wide. Hammocks, pinelands, south Fla., W.I. Fall, winter. *Laurocerasus myrtifolia* (L.) Britt. The leaves have a bitter cherry odor when crushed.

94. CHRYSOBALANÀCEAE   CHRYSOBALANUS FAMILY

Shrubs or trees with alternate, simple leaves, stipules entire. Flowers bisexual, rarely unisexual in simple racemes or in panicles; calyx five-parted, segments free or somewhat fused at the base; corolla of five petals, often unequal and more or less zygomorphic, inserted in the mouth of the calyx tube, stamens two to many, filiform, inserted with the corolla. Gynoecium with one carpel, ovules two, styles almost gynobasic, or lateral. Fruit a drupe. About 10 genera and 400 species in the tropics and subtropics.

1. Stamens fifteen or more; flowers in short
     axillary cymes . . . . 1. *Chrysobalanus*
1. Stamens three to ten; flowers in terminal
     panicles . . . . 2. *Licania*

1. CHRYSOBÁLANUS L.

Shrubs or small trees. Leaves leathery, stipules small. Flowers small, white, in axillary cymes; calyx lobes five, unequal or subequal; petals five, clawed; stamens fifteen or more, inserted with the petals, filaments free. Style basal, filiform ovules two, carpel one-locular. Fruit a drupe, eventually becoming dry. About 15 species in tropical Africa and America.

1. C. icaco L. COCO PLUM—Low shrub usually 1–2 m tall or trees up to 5–6 m tall, smooth except for conspicuous pale lenticels. Leaves variable, 2–8 cm long, elliptic-obovate to suborbicular, rounded or emarginate at the apex, usually dark green and glossy above, paler beneath. Flowers white in short-peduncled axillary cymes shorter than leaves; sepals 2–3 mm long, petals 3–5 mm long. Fruit globose or obovoid, 2–5 cm long, white to purple, edible. Coastal beaches, bayheads, swamps, Fla., W.I., Mexico and tropical America, tropical Africa. Two varieties occur in our area:

1. Fruit spherical; petals cuneate . . . . 1a. var. *icaco*
1. Fruit obovoid; petals spatulate . . . . 1b. var. *pellocarpus*

1a. C. icaco var. icaco—Shrub or tree often with radially creeping branches. Leaves mostly 5–8 cm long. Sepals 2–3 mm long, petals cuneate, drupe spherical, yellow, purple, or red, about 3–4 cm wide. Beaches, coastal hammocks, south Fla., W.I., tropical America.

1b. C. icaco var. pellocarpus (Meyer) DC. EVERGLADES COCO PLUM —Shrub up to 1 m or more tall. Leaves mostly 2–6 cm long. Sepals about 2 mm long, petals spatulate; drupe oblong-obovoid, purplish, 1–2 cm long. Bayheads, hammocks, south Fla., W.I. *C. pellocarpus* Meyer, *C. interior* Small

2. LICÀNIA Aubl.

Shrubs or trees. Leaves persistent, petiole often two-glandular at the apex, stipules subulate. Flowers small, in panicles or spicate, bracteolate; calyx tube globose, five-lobed, petals five, minute or absent; stamens three to ten, filaments often unequal. Ovary embedded in base of the calyx, one-locular, style basal, ovules two. Fruit coriaceous to woody, oneseeded. About 136 species, all in America except one species in Malaysia.

1. L. michauxii Prance, GOPHER APPLE—Low shrubs up to 3 dm tall with larger leaves oblanceolate or obovate, mostly 4–12 cm long, obtuse tipped or retuse, the leaf undersurface glabrous. Drupe ovoid, rarely seen owing to its use as food by small animals. Pinelands, oakscrub vegetation, sandy soil, and hammocks, Fla. to Miss., Ga. *Geobalanus oblongifolius* (Michx.) Small, *Chrysobalanus oblongifolius* Michx., *Geobalanus pallidus* Small

95. FABÀCEAE PEA FAMILY

Herbs, shrubs, vines, or trees, usually with alternate, prevailingly stipulate, compound leaves. Flowers typically pentamerous, synsepalous, petals free or sometimes sympetalous, hypogynous or perigynous with radial or bilateral symmetry, petals five or reduced to one; stamens ten or fewer or more numerous, monadelphous, diadelphous, or free; pistil one, simple; fruit a legume one- to several-seeded, dehiscent or indehiscent. Some 15,000 species distributed throughout the world, espe-

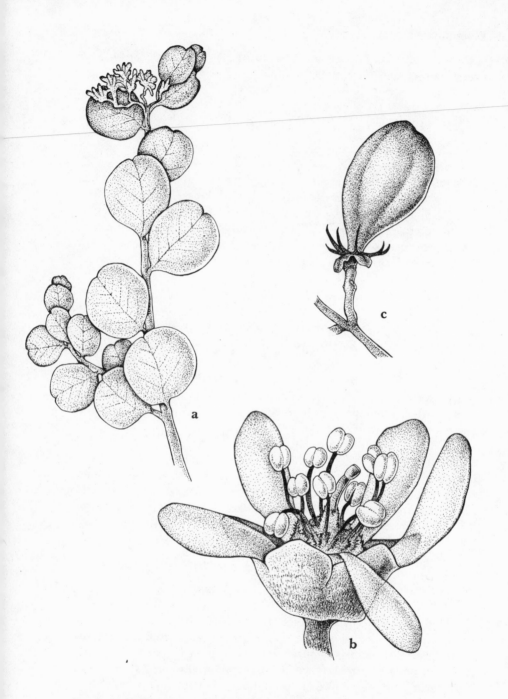

Family 94. **CHRYSOBALANACEAE.** *Chrysobalanus icaco* var. *icaco:* a, flowering branch, × ½; b, flower, × 10; c, mature fruit, × 1½.

cially numerous in warm temperate and tropical regions. Nomen alt. Leguminosae. *Mimosaceae, Caesalpinaceae, Cassiaceae, Krameriaceae*

1. Flowers radially symmetrical, hypogynous; petals valvate in bud, usually united below into a tubular base . . . . . . Subfamily I. Mimosoideae
1. Flowers bilaterally symmetrical, perigynous; petals imbricate in bud, usually distinct or merely connate.
  2. The odd petal innermost; stamens free . Subfamily II. Caesalpinioideae
  2. The odd petal outermost; stamens diadelphous (nine united, one free) or monadelphous (all united) or rarely all free . . . . Subfamily III. Faboideae

I. Mimosoideae

Leaves bipinnate. Flowers radially symmetrical, usually in capitate inflorescences; corolla valvate in aestivation, stamens numerous or at least twice as many as the petals.

1. Stamens numerous.
  2. Filaments united at or below the middle or slightly above the base; legume various, seeds in one row.
    3. Unarmed trees; stipules foliose.
      4. Legumes indehiscent, straight; valves not separating from marginal rib . 1. *Albizia*
      4. Legumes dehiscent; valves separating from the marginal rib . . . . 2. *Lysiloma*
    3. Armed trees or shrubs; stipules modified into spines.
      5. Legume contorted; seeds arillate . . 3. *Pithecellobium*
      5. Legume turgid; seeds usually not arillate . . . . 4. *Acacia*
  2. Filaments free; stems with stipular spines, rarely wanting; seeds in two rows . . 4. *Acacia*
1. Stamens ten, filaments distinct or nearly so.
  6. Trees.
    7. Leaflets 2–3 mm wide, minutely ciliolate; inflorescence capitate; axillary . 5. *Leucaena*
    7. Leaflets 7–18 mm wide, glabrous; inflorescence racemose; terminal . . 6. *Adenanthera*
  6. Shrubs or suffrutescent herbs.
    8. Plants briarlike, armed throughout with hooked prickles or spines; legume four-angled . . . . . . 7. *Schrankia*

8. Plants more or less prostrate or reclining, unarmed; legumes not angled.
9. Stems wiry, reclining; legume stipitate, broadly oblong . . . . 8. *Neptunia*
9. Stems suffrutescent at least below;
legume linear, sessile . . . . 9. *Desmanthus*

1. ALBÍZIA Durazz. WOMAN'S TONGUE

Trees of moderate size, unarmed. Leaves deciduous, even-pinnate
with numerous leaflets; petiolar gland adaxial near the base; stipules
setaceous or obsolescent. Inflorescence paniculate; flowers bisexual or
polygamous, pentamerous; calyx of united sepals; corolla funnelform;
stamens numerous, long exserted, filaments united below the middle.
Legume indehiscent, devoid of pulp, valves not separating from the marginal rib. About 100 to 150 species, Old World tropics or subtropics.

1. Leaflets less than 5 mm wide, strongly inequilateral, cuspidate . . . . . . 1. *A. julibrissin*
1. Leaflets 8 mm wide or wider, slightly inequilateral, ecuspidate . . . . . . 2. *A. lebbeck*

1. A. julibrissin (Willd.) Durazz.—Trees 10 m high or higher.
Leaves 3 dm long or shorter, with nine to twelve pairs of leaflets; leaflets
7–9 mm long, almost dimidiate, short ciliate, appressed puberulent at
least below; inflorescence a panicle of greenish yellow flowers in capitate
clusters; calyx 3 mm long, campanulate, lobes deltoid, serrulate; corolla
7–9 mm long, deeply cleft into strap-shaped lobes, surpassed by a staminal tube; filaments 2–2.5 cm long, anthers small. Legume 8–15 cm long,
2–3 cm wide, pointed, cuneate, usually eleven-seeded, valves reticulate
veined to the margin. $2n = 26,52$. Cultivated and naturalized in south
Fla. *Mimosa julibrissin* Scop.

2. A. lebbeck (L.) Benth.—Trees to 15 m tall with rounded, spreading crown, pale bark. Leaves 2–3.5 dm long, essentially glabrous in age;
leaflets oblique, broadly obtuse. Flowers in pedunculate, globose heads
arranged in panicles or umbelliform clusters; calyx 3–3.5 mm long, teeth
deltoid, puberulent; corolla 5–6 mm long, deeply cleft; stamens to 3 cm
long with bright yellow filaments and small anthers. Legume to 20 cm
long, cuneate, seven- to ten-seeded, valves reticulate veined over the
seeds. Plants in central and south Fla. and possibly naturalized. *Mimosa
lebbeck* L.

2. LYSILÒMA Benth. WILD TAMARIND

Trees or shrubs with bipinnate leaves, glabrous in age; leaflets usually numerous. Flowers in globose heads, pentamerous; stamens united

into a tube below; filaments long exserted; anthers shorter than wide, small, dorsifixed, ventrally dehiscent. Legume flat, elliptic, glabrous. About 33 species, subtropical and tropical America.

1. **L. bahamense** Benth.—Shrubs or trees to 20 m tall with wide-spreading crown; young twigs somewhat flexuous, zigzag. Leaves 10–18 cm long with four to eight pinnae; stipules early deciduous, conspicuous during leaf expansion. Floral bracts sagittate; heads 1–1.5 cm in diameter, cinereous in preanthesis; calyx 2 mm long, lobes deltoid; corolla greenish white, 4 mm long, with ovate reflexed lobes; stamens yellowish. Legume to 20 cm long, 2–3 cm wide, stipitate; valves thin, deciduous from marginal sutures; ovules transverse. $2n = 26$. Hammocks, Everglade Keys, W.I., tropical America.

3. PITHECELLÒBIUM Mart. CAT CLAW

Shrubs or trees with bipinnate leaves, with or without stipular spines. Inflorescence globose heads in terminal racemes or solitary on axillary peduncles; sepals five, united; stamens numerous with a short staminal tube; anthers small, dorsifixed, wider than long, ventrally dehiscent. Legumes red, contorted, dehiscent; seeds black, aril red. About 100 species, widely distributed in America and Old World tropics and subtropics.

1. Pinnae reduced to one to two pairs of leaf-
    lets; leaflets broadly obovate or elliptic.
  2. Leaflets chartaceous; stipular spines com-
    monly present.
    3. Calyx glabrous . . . .   1. *P. unguis-cati*
    3. Calyx puberulent . . . . . .   2. *P. dulce*
  2. Leaflets coriaceous; stipular spines rarely
    present . . . .   3. *P. keyense*
1. Pinnae five to six pairs of leaflets; leaflets
    1–3.5 cm long, oblong-lanceolate . . .   4. *P. graciliflorum*

1. **P. unguis-cati** (L.) Benth.—Shrub or a small tree to 16 m tall; stipular spines to 2 cm long. Leaves glabrous; petioles 5–20 mm long, dorsally sulcate with cuplike gland at summit; ultimate petiolules pulvinuslike with a short spinous tip between the leaflets. Inflorescence a terminal raceme of many peduncled heads; calyx 2 mm long, deltoid teeth; corolla 5–6 mm long; filaments pinkish or yellow, the tube included. Legume red, stipitate, coiled, constricted between the seeds. Coastal strand, shell mounds, Fla., W.I., South America. *Mimosa unguis-cati* L.

2. **P. dulce** (Roxb.) Benth.—Shrub or a small tree to 16 m tall. Leaflets somewhat inequilateral; stipular spines to 15 mm long. Heads

Family 95. **FABACEAE: MIMOSOIDEAE.** *Lysiloma latisiliqua:* a, flowering branch, × ½; b, inflorescence, × 6; c, flower, × 13; d, mature fruit, × ½.

small in panicles; peduncles pubescent; flowers whitish, densely canescent; calyx about 2 mm long, exceeded by the corolla; stamen tube short, included. Legume 8–12 mm wide, compressed, coiled. $2n = 26$. Naturalized in Fla., W.I., Mexico, South America. *Mimosa dulcis* Roxb.

3. **P. keyense** Britt. ex Coker—Shrub or small tree to 6 m tall, usually armed; bark gray, fissures shallow; twigs and foliage glabrous. Petioles 1–2 cm long, dorsally sulcate with a circular saucerlike gland below summit, terminating in a short spinous process; leaflets nearly sessile, subtended by pulvinuslike petiolule; blades obliquely obovate or often elliptic, mucronate; peduncles slender, exceeding the leaves in anthesis. Heads in full flower 2–3 cm in diameter; corolla 3.5 mm long, stamens long exserted, filaments roseate. Legumes 6–15 cm long, 8–10 mm wide, coiled; seeds black, lustrous. $2n = 26$. Hammocks, endemic to south Fla. *P. guadelupense* Chapm.

4. **P. graciliflorum** Blake—Unarmed shrub; young branches rufescent pilose. Leaves bipinnate with two to three pairs of pinnae; petiole 1–2 cm long, glandular; rachises densely rufous pilose, eglandular; secondary rachis with five to eight pairs of overlapping leaflets; leaflets oblong-lanceolate, 3–4 cm long; midrib pubescent below. Peduncles solitary in leaf axils, about as long as the leaflets; calyx tubular 3.4 mm long, lobes short, deltoid; corolla to 10 mm long; stamens many, united into tube at base. Ovary glabrous with fifteen ovules, fruit not seen. Naturalized from Honduras, hammocks, south Fla.

4. ACÀCIA Mill.

Trees or shrubs, usually armed with stipular spines. Leaves bipinnate; rachises tapering to a point. Flowers in spikes or racemes subtended by axillary peduncles, commonly bisexual; calyx and corolla four- to six-cleft; stamens numerous with free or slightly united filaments and small anthers. Legume cylindrical, compressed, dry or pulpy within, indehiscent or tardily dehiscent. About 750 to 800 tropical and subtropical species. *Vachellia* Wight & Arn., *Tauroceras* Britt. & Rose, *Poponax* Raf.

1. Armed shrubs; legumes not woody.
  2. Flowers in cylindric spikes; spines cornute, fistulous, much expanded at base   1. *A. cornigera*
  2. Flowers in globose heads; spines straight, solid, large or small.
    3. Legumes pubescent at least when young; spines thickened at base, pubescent or glabrescent.

4. Leaf pinnae eight pairs, or commonly
in numerous pairs; leaflet midvein
prominent . . . .   2. *A. macracantha*
4. Leaf pinnae commonly four pairs;
leaflet midvein not prominent . .   3. *A. tortuosa*
3. Legumes glabrous; spines slender, glabrous.
5. Legume beaked; corolla exceeding
the calyx by the length of its seg-
ments in full anthesis . . . .   4. *A. pinetorum*
5. Legume blunt or the tip incurved;
corolla twice the length of the
calyx . . . .   5. *A. farnesiana*
1. Unarmed trees; legume woody . . . .   6. *A. choriophylla*

1. A. cornigera (L.) Willd.—Shrub or small tree; spines 2.5–10 cm
long, paired, united at base, diverging horn fashion to attenuate sharp
tips. Leaflets 6–8 mm long, oblong, oblique at base, tipped with prominent nectar gland 2 mm long; flowers yellow. Legumes to 6 cm long; not
seen. Ornamental, Gulf Keys, beaches, shell mounds, south Fla., Mexico.
Tauroceras cornigerum (L.) Britt. & Rose, *Mimosa cornigera* L.

2. A. macracantha Humb. & Bonpl. ex Willd.—Small tree;
branches armed with solid, paired spines of varying sizes. Petioles and
rachises canaliculate, pubescent, or puberulent, a crateriform gland between the node and the lowest pair of pinnae, with similar rachial glands
between two ultimate pairs of pinnae. Leaflets numerous, 3–4 mm long,
0.9–1.2 mm wide, obtuse-oblong, oblique at base. Inflorescence of fasciculate peduncles with basal subtending bracts; receptacle subglobose 2.5
mm long; floral bracts 1.2 mm long, clawed, tips deltoid; flowers 3.5 mm
long at full expansion; calyx half the length of the corolla; segments incurved, tomentulous. Legume to 10 cm long, 12 mm wide, puberulous or
glandular. $2n = 26$. Hammocks, south Fla., W.I. *Poponax macracantha*
Humb. & Bonpl. ex Willd.

3. A. tortuosa (L.) Willd.—Arborescent shrub or small tree over
5 m tall with a trunk diameter of 10 cm or more; branches of crown
irregularly spreading and twisting. Leaves commonly with two to four
pairs of pinnae; petiole and rachis densely pubescent or glabrate, canaliculate; gland elevated near midpoint of the petiole, rachial glands between ultimate pairs of pinnae; leaflets 3–4 mm long, ciliolate or glabrate, oblong-obtuse. Peduncles solitary or a few together; receptacle
globose, flowers yellow 3.5 mm long at full anthesis; sweet-scented. Legume 5–7 cm long, slightly curved. $2n = 26$. Coastal strand, south Fla.,
W.I., South America. Summer. *Poponax tortuosa* (L.) Raf.

4. A. pinetorum (Small) Hermann—Shrub 3–4 m tall with flexuous branches; spines to 2 cm long, pale gray, glabrous. Leaves with three to four pairs of pinnae; petiole slender, gland-bearing near the midpoint; leaflets 3–4 mm long, less than 1 mm wide, essentially glabrous. Peduncles 2–3 cm long, slender; denuded receptacle cylindric clavate, shallowly alveolate, sparsely papillose; involucral bracts united, collarlike; flowers in full anthesis 4–4.5 mm long; floral bracts 1.5 mm long, segments deltoid, incurved; corolla yellow, funnelform, 2.4 mm long with incurved segments, papillose within and without; anthers versatile; filaments united half their length. Legume 3–6 cm long, turgid, fusiform, beaked, glabrous. Coastal strand, shell mounds, pinelands, endemic to Fla. Summer. *Vachellia peninsularis* Small

5. **A. farnesiana** (L.) Willd.—Arborescent shrub or small tree with flexuous branches; twigs and foliage pubescent when young; spines slender, glabrous, pale gray, 1–3 cm long. Leaves with two to six pairs of pinnae; leaflets ten to twenty pairs per pinna, 4–5.5 mm long, prominently veined on lower surface. Flowering peduncles 2 cm long, fasciculate; flowers yellow, fragrant; corolla 3 mm long, funnelform, segment lobes deltoid, papillose; stamens 5 mm long, free nearly to the base; anthers small, versatile. Pistil 5–6 mm long, ovary stipitate; ovules in two rows. Legume reddish brown, to 7 cm long; mature valves corky, tardily dehiscent. Seeds 5–6 mm long, olive brown, the flat sides clearly tracing the shape of the cotyledons to the hilum. Shell ridges, coastal strand, Fla., La., Mexico, W.I., Calif. to Argentina, Old World tropics. *Vachellia farnesiana* (L.) Wight & Arn.

6. **A. choriophylla** Benth. ex Hook.—Trees to 9 m tall or less, spineless; twigs and leaves glabrous; stipules minute, subulate, early deciduous. Leaves 5–6 cm long with one to three pairs of pinnae; petiole 1–8 mm long, sparsely glandular; an orbicular gland on the rachis between each distant pair of pinnae; leaflets sessile 12–16 mm long, 6–7 mm wide, oblong-cuneate, membranous, entire, reticulate veined; veins ending at margin. Peduncles 20–30 mm long, axillary, toward the tips of branches; inflorescence capitate, densely flowered; flowers yellow. Legumes woody, to 8 cm long, 2.5 cm wide, mostly straight, pointed, tardily dehiscent. North Key Largo, Fla., Bahama Islands, Cuba. Summer.

5. LEUCAÈNA Benth. LEAD TREE

Tree; leaves bipinnate, petiolar gland between the lowest pinnae or near the middle of the petiole. Flowers white or pinkish in globose heads; solitary or in racemes; calyx campanulate, dentate; corolla of free petals; stamens ten, free. Legume stipitate, linear, compressed, two-valved; embryo embedded in horny endosperm. About 50 species, tropical America, one pantropical, one Polynesian, warm regions of New World and Old World.

1. **L. leucocephala** (Lam.) de Wit—Shrub or small unarmed tree, 5–8 m tall. Leaves glaucous with four to eight pairs of pinnae, 8–15 cm long; leaflets falcate, inequilateral. Flower spikes axillary or in small terminal racemes; calyx 3 mm long, dentate, corolla 4–6 mm long; stamens 10–12 mm long. Legumes 10–15 cm long, pulpless, many-seeded. Seeds 6 mm long, stalked. $2n = 104$. Hammocks, coastal strand, Fla. Keys, Everglades, south Fla., W.I., South America. Summer. *Mimosa glauca* L., *Leucaena glauca* (L.) Benth.

6. ADENANTHÈRA L.

Trees; leaves to 5 dm long or longer, bipinnate. Racemes lateral, legumes curved, impressed between seeds, dehiscent. About eight species, tropical Asia, Australia, Pacific Islands, India, widely cultivated in warm regions.

1. **A. pavonina** L.—Foliage and axes minutely puberulent. Leaf pinnae two to five, paired; leaflets 2–4 cm long, oblong, elliptic, rounded at base and apex; petiolules 1 mm long. Flowers small in narrow racemes; calyx green, corolla golden yellow; stamens shortly exserted; fruiting peduncle stout, elongate. Legumes 23 cm long, 7 mm wide; seeds lenticular 4–5 mm wide, scarlet, adherent to impressions in coiled valves. Escaped from cultivation in south Fla., naturalized from India.

7. SCHRÁNKIA Willd. Sensitive Briars

Suffrutescent briarlike perennial with prostrate stems, strongly armed throughout. Leaves stipulate, bipinnate, sensitive; petioles eglandular; leaflets numerous. Flowers pentamerous; calyx minute; petals connate to the middle; filaments distinct or nearly so. Legume four-sided, beaked. About 30 species, mostly tropical and temperate America. *Leptoglottis* DC.

1. **S. microphylla** (Sm.) Macbride—Stems 1 m long or more, prostrate, armed with hooked prickles. Leaflets 3–4 mm long, linear, elliptic; petioles exceeding the peduncles; stipules setaceous. Legumes 6–7 mm long; seeds compressed, oval. Rocky pinelands, hammocks, south and central Fla. *Mimosa microphylla* Dryand. ex Sm.

8. NEPTÙNIA Lour.

Perennial, vinelike plants with prostrate branching stems; leaves bipinnate. Spikes elliptic, yellow; calyx tubular with short teeth; petals distinct or nearly so; stamens with free filaments. Ovary stipitate. Legumes oblong-oblique, commonly in clusters. About 15 species, America, Asia, Australia. *Desmanthus* Willd.

1. **N. pubescens** Benth. var. **pubescens**—Stems several from ligneous crown, essentially glabrous; taproot orange. Flowering branches ascending; leaves petiolate with two to four pairs of pinnae; stipules ovate, commonly attenuate at tip, oblique, striate; leaflets contiguous, triple-nerved at base, numerous, more or less ciliolate; rachises minutely edged, adaxially grooved, with a terminal mucro ca. 1 mm long. Inflorescence solitary, long peduncled; staminodes golden yellow; perfect anthers tipped with stalked glands. Mature legumes 3 cm long, 8 mm wide, strongly flattened; dehiscent; seed obliquely arranged on sutures. $2n = 28$. Rocky pinelands, hammocks, south Fla., La., W.I. *N. floridana* Small; *N. lindheimeri* Robinson; *N. pubescens* var. *floridana* (Small) Turner; *N. pubescence* var. *lindheimeri* (Robinson) Turner

9. DESMÁNTHUS Willd.

Perennial herbs or shrubs, erect, ascending or prostrate. Leaves bipinnate with a cupuliform gland below the lowest pair of pinnae; stipules subulate, setaceous; leaflets numerous. Inflorescence capitate; calyx campanulate, lobes short; petals clawed, distinct; stamens ten (five), free, exserted. Legume compressed, dehiscence marginal; seeds obliquely arranged. About 40 species in temperate, tropical, and subtropical America. *Acuan* Medicus

1. **D. virgatus** (L.) Willd. var. **depressus** (H & B) Turner—Stems 3–10 dm long, ascending to prostrate, branching, radiating from the crown of ligneous taproot. Internodes glabrous, angled along the four reddish decurrent lines; leaves 3–4.5 cm long, with two to four pairs of pinnae, petiole about 5 mm long, rachis terminating in a free tip; leaflets 2–2.5 mm long, only the midvein apparent; margins sparingly ciliolate, glabrescent, lamina linear, obtuse-oblique. Flowers white, fruiting peduncles 10 mm long or longer. Legumes 4–7 cm long, 2.8 mm wide, thickened on sutures, two-valved, linear, seeds nearly rhomboidal, about 2.2 mm long, 1.8 mm wide, crescentic mark in the middle of each side. $2n = 28$. Shell mounds, pinelands over oolitic rocks, south Fla., Fla. Keys, Everglades, W.I. All year. *D. depressus* H & B

II. Caesalpinioideae

Trees, shrubs, herbs. Leaves evenly once- or twice-pinnate. Petals imbricate, the uppermost internal; sepals usually distinct; stamens ten or fewer, free or somewhat united. About 500 species of the tropics and subtropics.

1. Leaves once-pinnate.
  2. Anthers dorsifixed; dehiscence lateral; legume pulpy indehiscent . . . . 10. *Tamarindus*
  2. Anthers basifixed; dehiscence terminal; legume dry dehiscent . . . . . 11. *Cassia*
1. Leaves twice-pinnate.
  3. Leaf rachis phyllodial with spinelike apex; petioles nearly obsolete; leaflets reduced in size or wanting; legume torulose . . . . 12. *Parkinsonia*
  3. Leaf rachis not phyllodial; petioles well developed; legumes not torulose; leaflets small.
    4. Shrubs, erect or straggling, prickly throughout or only at nodes; legumes dry . . . . 13. *Caesalpinia*
    4. Trees, unarmed; legumes pulpy . . 14. *Poinciana*

10. TAMARÍNDUS L.

Trees to 20 m high or higher; trunk massive to 1.5 m in diameter. Leaves even-pinnate, 6–12 cm long with a short petiole; leaflets 10–15 mm long, retuse reticulate, elliptic-oblique. Flowers in terminal racemes; calyx 8–10 mm long, exceeded by the petals and the stamens. Legume 5–12 cm long, fleshy; seeds 1 cm broad. One species, India.

  1. **T. indica** L.—Characters of the genus. $2n = 24$. Coastal strand, Fla. Keys, naturalized from India. Spring, summer.

11. CÁSSIA L.

Trees, shrubs, or herbs with evenly once-pinnate, stipulate leaves. Flowers bisexual, only slightly bilaterally symmetrical; inflorescence axillary or terminal; sepals united into a short tube; petals yellow or reddish, often unlike; stamens ten or fewer by abortion, unequal; anthers basifixed, opening by terminal pores. Ovary sessile or stipitate; legume dehiscent or indehiscent, one- or several-seeded; gland usually one. About 500 to 600 species, tropical and warm temperate regions except Europe. *Adipera* Raf., *Emelista* Raf., *Peiranisia* Raf., *Ditremexa* Raf., *Chamaecrista* (L.) Moench

1. Legumes elastically dehiscent; stipules persistent.
  2. Plants perennial; stems, roostocks ligneous; flowers showy, 2 cm wide or wider.

3. Plant gray hirsute throughout; petiolar gland 0.5 mm wide; leaflets not sensitive . . . .   1. *C. keyensis*

3. Plant with essentially glabrous stems and foliage; petiolar gland 1–2 mm wide; leaflets sensitive . . . .   2. *C. deeringiana*

2. Plants annual from taproot; flowers 1–1.5 cm wide or less.

   4. Glands depressed, saucerlike, or slightly raised, situated near mid-petiole . . . .   3. *C. brachiata*

   4. Glands elevated, stalked, situated toward distal end of petiole . . .   4. *C. aspera*

1. Legumes irregularly dehiscent; stipules deciduous.

5. Perfect stamens ten; legume flat; foliage reddish; petiolar glands slender, clavate   5. *C. surattensis*

5. Perfect stamens usually seven, with three staminodes; legumes convex or angular; foliage green; petiolar gland tubercular.

   6. Gland between the lowest pair of leaflets.

      7. Leaflets lanceolate, legume abruptly narrowed at apex . . . . .   6. *C. bahamensis*

      7. Leaflets obovate, legume tapering at apex.

         8. Margins glabrous, midvein not excurrent . . . .   7. *C. bicapsularis*

         8. Margins ciliolate, midvein excurrent . . . .   8. *C. obtusifolia*

   6. Gland at the base of the petiole; leaflets lanceolate or ovate-lanceolate.

      9. Gland cylindric . . . . . .   9. *C. ligustrina*

      9. Gland dome shaped . . . . .   10. *C. occidentalis*

    1. C. keyensis (Pennell) Macbride—Much branching ligneous stems from contorted rootstock; branches 1–5 dm long; internodes shorter than the leaves. Leaflets obovate 4–7 mm long, in four to nine pairs. Petals golden yellow, to 9 mm long, exceeding the sepals and the stamens; anthers reddish. Legume 4–5 cm long, 4–5 mm wide, gray; valves puberulent, marked with oblique impressions. Big Pine Key, endemic to south Fla. Summer, Fall. *Chamaecrista keyensis* Pennell

    2. C. deeringiana (Small & Pennell) Macbride—Stems several, rising from horizontal rootstocks; branches to 9 dm long, essentially glabrous. Stipules lance-acuminate; leaflets twenty pairs or fewer; blades 8–10 mm long, 2 mm wide, mucronate. Petals golden yellow, 2 cm long, exceeding the sepals and stamens; anthers usually reddish. Legume 5–

5.5 cm long, 5–6 mm wide; valves with oblique impressions. Showy shrub in glades and pinelands, endemic to south Fla., Everglade Keys, lower Fla. Keys. All year. *Chamaecrista deeringiana* Small & Pennell

3. C. brachiata (Pollard) Macbride—Stems to 2 m tall, widely branched, spreading from stout taproot; internodes and rachises puberulent with incurved hairs or glabrous. Leaves 3–6 cm long, leaflets ten to twenty-five pairs; petioles 3–5 mm long; stipules lance-acuminate, eciliate; petiolar gland 1 mm wide or less, margins slightly elevated. Flowers golden yellow; calyx 10–12 mm long, usually reddish, surpassed by the petals. Legume 5–9 cm long; glabrous, seeds flattened, usually rectangular, black pitted. Southwest Fla., Marco Island, south Fla. Fall.

4. C. aspera Muhl.—Densely branching shrub to 7 dm tall with three or more principal stems branching from the common crown of stout roots. Internodes short and rachises pilose, over incurved fine pubescence or glabrescent. Leaves 3–7 cm long, 10–20 mm wide; stipules oblique-lanceolate; leaflets seven to thirty-two pairs on hirtellous rachis, mucronate, inequilateral, the anterior margin linear, usually without veins, base oblique. Flowers less than 1 cm wide, calyx lobes hirsute on the excurrent midvein; corolla 5–8 mm long, odd petal the shortest. Legume 1–3 cm long, 3–4 mm wide, brushlike with white hairs when young, spreading hirsute in age. Seeds 2.5 mm long, nearly blackish, pustulate, wedge shaped. $2n = 28$. Pinelands, canal banks, sandy fields, Fla., Ga., S.C. Summer. *Chamaecrista aspera* (Muhl.) Greene, *C. aspera* Chapm., *C. aspera* var. *simpsonii* (Pollard) Macbride

5. C. surattensis Burman f.—Shrub or small tree with slender, pubescent branches and pubescent foliage. Petiole 3–5 cm long; rachis 3–6 cm long with three to five pairs of leaflets 2.5–3.5 cm long, oval in outline, reticulate veined; stipules linear, ciliate. Corymbs terminal or axillary. Legumes 7–9 cm long, glabrous, stipitate and beaked, flat, dehiscence valvate by two sutures. $2n = 32$. Hammocks, naturalized in south Fla. from tropical Australia.

6. C. bahamensis Mill.—Shrub 1–3 m tall with glabrous stem and branches. Leaves 3–5 cm long with three to five pairs of leaflets; leaflets 2–2.4 cm long, oblong-lanceolate, or elliptic, mucronate; base of midrib pubescent below. Inflorescence a corymbiform panicle; pedicels 2 cm long, slender; sepals and petals yellow, veined with red; anthers dark red. Legume 7–10 cm long, dehiscence by two sutures. Pinelands, coastal strand, south Fla., Bahama Islands, Cuba. Fall, winter. *Peiranisia bahamensis* (Mill.) Britt. & Rose

7. C. bicapsularis L.—Shrub 1.5 m tall or shorter. Leaflets three to four pairs 1.5–3 cm long, obovate-elliptic, glabrous, punctate on lower

Family 95. **FABACEAE: CAESALPINIOIDEAE.** *Cassia bahamensis:* a, flowering branch, × ½; b, flower, × 5; c, flower, postanthesis, × 6; d, mature fruit, × 1.

surface. Racemes paniculate in upper axils. Legumes turgid, stipitate, 6–15 cm long, 10–14 mm wide, curved, glabrous, obtuse at apex; dehiscence valvular by two sutures; mature seeds 4–5 mm long, compressed, oblique, oblong, narrowed to micropylar end, shiny olive brown, minutely pitted. $2n = 28$. Canal banks, mangroves, Everglades, south Fla., W.I. All year.

8. C. obtusifolia L.—Shrubby annual 4–10 dm tall, essentially glabrous. Leaflets three to six pairs, 1.5–4 cm long, rounded at apex, cuneate at base; stipules linear-setaceous. Flowers yellow, in terminal or axillary clusters; sepals 7–9 mm long; petals 13–15 mm long, spreading and surpassing the stamens. Legume 6–12 cm long, slender, pointed, falcate, somewhat four-sided. $2n = 26,28$. Pinelands, waste ground, Fla., W.I., Mexico, tropical America. Fall. *Emelista tora* (L.) Britt. & Rose

9. C. ligustrina L.—Suffrutescent shrub 2 m tall or less, essentially glabrous. Leaflets 3–6 cm long, short petioluled, lanceolate, oblique at base, in three to seven distant pairs on rachis; blades appressed puberulent below, becoming glabrous. Petals 10–15 mm long, exceeding the sepals, yellow with brownish obscure dots and darker veins; anterior pair of anthers overtopping the others. Legume 12 cm long or less, stipitate, falcate, compressed, valves obliquely impressed between the seeds; seed 3–3.2 mm long, biconvex, narrowed to micropylar end; a gray band of cutinized testa around the edges demarking cotyledonary facet on each side. Hammocks, central and south Fla., W.I. All year. *Ditremexa ligustrina* (L.) Britt. & Rose

10. C. occidentalis L.—Essentially glabrous annual with suffrutescent taproot 12 dm tall or shorter; stem angled, somewhat flexuous above. Leaflets six to ten, lanceolate, rounded at base, 3–7 cm long. Petals 12–17 mm long, yellow with darker veins, exceeding the sepals; two anterior stamens overtopping the others. Legume 12 cm long or shorter, compressed, valves obliquely impressed between seeds; seeds light brown, 2 mm long, lateral facets punctate. $2n = 28,26$. Pinelands, Fla., Tex., Kans., W.I., Mexico, South America, Old World. Summer. *Ditremexa occidentalis* (L.) Britt. & Rose

C. fistula L., C. grandis L., and **C. didymobotrya** Forsk. may persist in former homesites without cultivation.

12. PARKINSÒNIA L.

Armed trees or shrubs with smooth, thin bark and bipinnate leaves. Flowers in racemes, calyx five, petals five, and spreading unequal, stamens ten, legumes torulose. Widely cultivated. Two species naturalized from subtropical and tropical America and Africa.

1. P. aculeata L.—Tree to 10 m tall; branches slender, pendulous. Leaves nearly sessile with one or a few pairs of pinnae, pulvinuslike at base; rachis spinous; rachis of pinnae phyllodial, 2–4 dm long; leaflets numerous, 4–10 mm long, elliptic, obscurely veined; petiolule 1 mm long or less. Flowers perigynous; sepals 8–10 mm long, strap shaped, yellow; petals 9–17 mm long, clawed; blades oval or suborbicular, bright yellow, standard mottled with red, or fading orange red throughout; petal claws pubescent; anthers introrse, filaments connivent, hairy below. Pistil in anthesis 10 mm long; stigma small, capitate. Legume 15 cm long or less. $2n = 28$. Ornamental and possibly escaped in south Fla., tropical America, Africa. Summer.

13. CAESALPÌNIA L.

Erect, reclining or vinelike prickly shrubs with evenly bipinnate, usually stipulate leaves. Leaflets usually numerous and relatively large. Flowers in terminal racemes or panicles; stamens united at least below. Legume prickly or smooth, one- to two-seeded. About 100 species, tropical and subtropical. *Guilandina* L.

1. Plants devoid of fine pubescence; stamens
     long exserted; legume smooth   .   .   .   1. *C. pauciflora*
1. Plants pubescent; stamens about as long as
     the corolla; legumes prickly; stipules
     foliaceous.
  2. Leaflets ovate 2.5–5 cm long, 2–2.5 cm
       wide   .   .   .   .   2. *C. crista*
  2. Leaflets oblong-oval 2–4 cm long, 1–2 cm
       wide   .   .   .   .   3. *C. bonduc*

1. C. pauciflora (Griseb.) C. Wright—Shrub to 2 m tall or less, freely branching. Pinnae four pairs or more; stipules spinous; leaflets glabrous, oblong, oblique, retuse. Sepals 8–10 mm long; corolla yellow, somewhat longer than the calyx, filaments united above the anthers, two seriate. Legume oblong 3–4 cm, tapering to each end, Big Pine Key, south Fla.

2. C. crista L. Gray Nicker—Shrub straggly, reclining over vegetation, viscid pubescent and prickly with curved spines throughout. Leaves 4 dm long or more, with rachis and four or more pairs of pinnae; leaflets ovate, bristle-tipped, 5 cm long or less, 2.5 cm wide. Legume oval-elliptic 5–7 cm long. $2n = 24$. Coastal strand, hammocks, mangroves, south Fla. *Guilandina crista* (L.) Small

3. C. bonduc Roxb. Yellow Nicker—Similar to the preceding taxon; rachises and inflorescence with denser incurved indumentum. Floral bracts acuminate, subulate. Legume orange brown; mature seeds not seen. $2n = 24$. Key West, south Fla., widely distributed in the tropics. *Guilandina bonduc* L.

14. POINCIÀNA L.

Unarmed or armed shrubs or trees. Leaves estipulate, evenly bipinnate. Inflorescence terminal or axillary panicles. Flowers showy, pentamerous; stamens distinct, long exserted, or as long as the petals. Ovary sessile, compressed, ligneous, or chartaceous. Three species, tropical Africa, Madagascar. *Caesalpinia* L., *Delonix* Raf.

1. Leaflets oblanceolate, 8–10 mm wide, relatively few; filaments long exserted . . .   1. *P. pulcherrima*
1. Leaflets linear, 1–2 mm wide, relatively numerous; filaments about as long as the petals . . . .   2. *P. regia*

1. **P. pulcherrima** L.—Shrub to 3 m tall, glabrous; branches ascending, usually prickly. Leaves persistent, with twelve pinnae or more. Flowers orange yellow, long-pedicelled limb of petals flabellate. Legume linear 8–10 cm long. Ornamental, persisting near former habitations, south Fla., native in Africa, Madagascar.

2. **P. regia** Bojer—Trees to 10 dm tall with wide-spreading, flattened crown; trunk to 9 dm in diameter. Leaves 3–5 dm long, pinnae twenty or fewer. Calyx 1–2 cm long, sepals red within; petals clawed, 5 cm long or longer, blades reflexed, broadly flabellate, flame red; standard mottled. Legume to 6 dm long, compressed, solid between seeds. $2n = 24,28$. Spontaneous in hammocks, ornamental, native of Africa persisting around former habitations, south and peninsular Fla. *Delonix regia* (Bojer) Raf.

III. Faboideae

Trees, shrubs, herbs, and vines. Leaves simple, pinnately or palmately compound, stipulate; leaflets often stipellate, rachis sometimes produced beyond the ultimate leaflets. Flowers typically papilionaceous, with five petals (rarely reduced to one) consisting of a standard (the largest petal), which in bud is folded over, two wing petals, and two adherent keel petals enclosing the staminal sheath with the pistil; stamens ten or fewer with united filaments (monadelphous) or usually nine united and one free (diadelphous), rarely all free. Pistil one-carpellate with style and stigma, one-locular, one-ovuled or more. Fruit a legume, dehiscent along dorsal and ventral sutures, or a loment, indehiscent, constricted between seeds.

1. Leaves odd-pinnate.
  2. Filaments distinct; legume moniliform .   15. *Sophora*
  2. Filaments monadelphous or sometimes diadelphous.
    3. Leaf simple or unifoliolate by reduction.
      4. Anthers dimorphic.
        5. Legume inflated; staminal sheath split along upper side . . .   16. *Crotalaria*
        5. Legume compressed; staminal sheath a closed tube . . .   17. *Lupinus*
      4. Anthers uniform.
        6. Stamens diadelphous; anther locules dehiscent longitudinally.
          7. Leaves glandular dotted; legume valved, two-seeded . . .   18. *Rhynchosia*
          7. Leaves eglandular; loment partitioned, more than two-seeded   19. *Alysicarpus*
        6. Stamens monadelphous; anthers dehiscent by transverse splits; legume suborbicular, oblong .   20. *Dalbergia*
    3. Leaves compound.
      8. Leaflets several on rachis, more than three.
      9. Trees.
        10. Leaflets opposite on rachis.
          11. Legume four-winged . . .   21. *Piscidia*
          11. Legume wingless.
            12. Pedicels paired; fruit a legume . . . . .   22. *Lonchocarpus*
            12. Pedicels not paired; fruit drupaceous.
              13. Leaflets usually five to seven, estipellate . .   23. *Pongamia*
              13. Leaflets usually nine to thirteen, stipellate . .   24. *Andira*
        10. Leaflets alternate on rachis; legume complanate, one to two seeds . . . . . .   20. *Dalbergia*
      9. Shrubs and herbs.
        14. Foliage dotted with pellucid glands; stamens monadelphous; legume indehiscent.
          15. Corolla with five petals; stamens nine, or nine and one, tube open on upper side .   25. *Dalea*

15. Corolla represented by standard petal, wing and keel petals reduced or staminodial.

  16. Stamens ten, staminodia lacking . . . . .   26. *Amorpha*

  16. Stamens five, alternating with four petaloid staminodia . . . . . .   27. *Petalostemon*

14. Foliage without pellucid glands.

17. Plants erect or trailing; calyx not bracted.

  18. Fruit valved.

    19. Hairs short stalked, with one pair of appressed branches; fruit clusters reflexed . . . . .   28. *Indigofera*

    19. Hairs simple, spreading; fruit clusters not reflexed . . . . .   29. *Tephrosia*

  18. Fruit lomentaceous.

    20. Stipules neither bractlike nor produced beyond the point of attachment.

      21. Articles flattened, minutely uncinate puberulent . . . .   30. *Desmodium*

        22. Articles sparsely pustulate hairy . .   31. *Aeschynomene*

        22. Articles turgid, veiny   32. *Stylosanthes*

      21. Articles terete, prickly, hairs pustulate at base . . . . .   33. *Chapmannia*

    20. Stipules bractlike, produced beyond the point of attachment, subtending a flower between them . . . . .   34. *Zornia*

17. Vines or twining plants; calyx bracted; stamens diadelphous.

  23. Keel sigmoid or spiral; upper calyx lobes reduced, calyx bracts subulate, fugacious; flowers reddish .   35. *Apios*

  23. Keel small, incurved; calyx bracts persistent.

    24. Calyx five-lobed; bracts

striate; flowers mainly
blue; styles dilated.

25. Calyx bracts longer
than the tube; stan-
dard spurred at base;
leaves trifoliolate;
legume margins thick-
ened . . . . . 36. *Centrosema*

25. Calyx bracts shorter
than the tube; stan-
dard not spurred;
leaves five- to seven-
foliolate; legume mar-
gins thin . . . . 37. *Clitoria*

24. Calyx four-lobed; bracts
small ovate-deltoid;
flowers mainly pinkish,
white, or purplish; style
slender, glabrous . . 38. *Galactia*

8. Leaflets usually three, rarely reduced
to one.

26. Herbs.

27. Leaflets serrulate; veins ending
in teeth; sweet-scented herb.

28. Leaflets palmately arranged;
racemes capitate; legume oblique . . . . . . 39. *Trifolium*

28. Leaflets pinnately arranged.

29. Racemes elongate; legume
linear, centrally beaked . 40. *Melilotus*

29. Racemes spicate, abbreviated; legume coiled . . 41. *Medicago*

27. Leaflets entire; veins ending
submarginally.

30. Leaflets pinnately foliolate,
stipellate.

31. Stipules caducous; leaflets
stipellate . . . . . 30. *Desmodium*

31. Stipules persistent; leaflets
estipellate.

32. Stipular sheaths tubular
investing the stem; style
hooklike crowning the
fruit apex . . . . 32. *Stylosanthes*

32. Stipular sheath lacking;
style short, acute . . 18. *Rhynchosia*

30. Leaflets palmately trifoliolate,
estipellate . . . . . 16. *Crotalaria*

26. Shrubs.

33. Erect plants.
    34. Stems spine armed; leaflets
        eglandular; flowers scarlet;
        legume torulose, stipitate . 42. *Erythrina*
    34. Stems unarmed; leaflets gland-
        dotted; flowers yellow, le-
        gume linear, sessile . . . 43. *Cajanus*
33. Vines.
    35. Stems suffrutescent, high twin-
        ing; leaflets inequilateral;
        legumes coarsely hairy . . 44. *Mucuna*
    35. Stems herbaceous.
        36. Vines stout, prostrate to as-
            cending; leaflets equilat-
            eral; legume glabrous . 45. *Canavalia*
        36. Vines slender, twining or
            creeping.
            37. Style glabrous.
                38. Leaflets linear-elliptic
                    to ovate in outline;
                    flowers purplish or
                    white . . . . . 38. *Galactia*
                38. Leaflets rhomboidal or
                    obliquely deltoid in
                    outline; flowers yel-
                    low . . . . . 18. *Rhynchosia*
            37. Style pubescent.
                39. Corolla blue, much ex-
                    ceeding the calyx,
                    standard with spur . 36. *Centrosema*
                39. Corolla reddish, white,
                  or yellow.
                  40. Keel spiral, flattened;
                    legume with parti-
                    tions between the
                    seeds . . . . 46. *Phaseolus*
                40. Keel not spiral.
                  41. Flowers in ra-
                    cemes; legume
                    compressed; leaf-
                    lets ovate-lanceo-
                    late.
                  42. Corolla blue, leg-
                    ume.impressed
                    between seeds 47. *Pachyrrhizus*
                  42. Corolla white;
                    legume scimi-
                    tar shaped . 48. *Dolichos*

41. Flowers in close pe-
duncled heads;
corolla yellow,
legume subterete   49. *Vigna*

1. Leaves even-pinnate; rachis produced beyond ultimate pair of leaflets.
43. Plant viny.
44. Leaflets two to four; rachis terminated
by a tendril; stamens diadelphous .   50. *Vicia*
44. Leaflets numerous; rachis terminating
in a spine; stamens nine or ten, monadelphous . . . .   51. *Abrus*
43. Plant erect, shrubby, or ligneous at base;
rachis terminating in slender tip.
45. Legume two-seeded; seeds shed within
papery endocarp splitting off from
pericarp; flowers red . . . . .   52. *Glottidium*
45. Legume many-seeded; seeds separated
by partitions; flowers yellow . . .   53. *Sesbania*

15. SÓPHORA L.   Necklace Pod

Shrub with naked buds. Leaves odd-pinnate, leaflets opposite or nearly so. Racemes terminal with yellow flowers; calyx truncate, teeth obsolescent; petal clawed, standard oval anthers versatile. Legume indehiscent. Fifty species of tropical and warm temperate regions.

1. S. tomentosa L.—Shrub to 3 m tall, erect, tomentose throughout, glabrate in age. Branches numerous, stiff ascending. Leaves to 20 cm long or longer, estipulate; leaflets thirteen to twenty-one, oval or oblanceolate, cuneate, entire, midvein prominent, lateral, obscured. Racemes 10–20 cm long, flowers to 2 cm long, corolla bright yellow, pedicel slender, 5–10 mm long; legume 6–15 cm long, stipitate, four- to nine-seeded; seeds pealike, 7 mm in diameter, yellow. Coastal strand, hammocks, central and south Fla., Fla. Keys, Tex., W.I. All year.

16. CROTALÀRIA L.   Rattleboxes

Annual and perennial erect herbs and shrubs. Leaves punctate, unifoliolate or digitately three- to five-foliolate, stipules prominent or obsolescent. Flowers showy, yellow, sometimes brownish or purplish yellow, usually in racemes; calyx lobes nearly similar, standard, the largest petal; keel petals somewhat beaked and curved, stamens ten, dimorphic, the five smaller anthers with elongate filaments extended with the style to the keel apex. Stigma small, capitate, style sharply incurved, persistent in fruit. About 300 species, cosmopolitan.

1. Leaves unifoliolate.
  2. Stems 6–10 dm tall; stipules not decurrent
    through internodes.
    3. Stipules manifest; leaves apiculate;
      pedicel bract persistent; flowers yel-
      low . . . .   1. *C. spectabilis*
    3. Stipules obsolescent; leaves retuse;
      pedicel bract caducous; flowers yel-
      low, petals mostly reddish without .   2. *C. retusa*
  2. Stems 6 dm tall or shorter; at least some
    of the stipules decurrent through inter-
    nodes.
    4. Stems decumbent, reclining, spreading;
      leaf blades oval or linear-elliptic . .   3. *C. maritima*
                                              var. *maritima*
    4. Stems erect.
      5. Pubescence spreading; leaf blades
        oval-suborbicular . . . . .   4. *C. angulata*
      5. Pubescence appressed; leaves linear
        to narrowly elliptic . . . . .   5. *C. purshii*
1. Leaves trifoliolate.
  6. Plant decumbent; stems several, trailing;
    leaflets oblanceolate to suborbicular;
    flowers yellow, variegated with red . .   6. *C. pumila*
  6. Plant erect.
    7. Foliage dark green; overall pubescence
      loose, spreading, racemes short . .   7. *C. incana*
    7. Foliage bright green; overall pubes-
      cence appressed; racemes elongate .   8. *C. mucronata*

    1. **C. spectabilis** Roth—Erect annual to 1.5 m tall from stout tap-
root; stems leafy, branching above; internodes prominently ridged with
decurrent lines. Leaves 7–15 cm long; blades obovate, cuneate, rounded
mucronate, glabrous above, appressed puberulent below; stipules deltoid, herbaceous, flanking the swollen petiole base. Racemes terminal,
to 4 dm long; floral bracts lanceolate, cordate, flowers 2.2 cm long, nodding on bibracteate pedicels, calyx bilabiate, glabrous lobes subequal,
deltoid; standard bright yellow, retuse, variegated with dark lines to
the nearly clawless base with two bracketlike nectaries; wing petals concave; keel petals distinct below the beak, margins free ciliolate; staminal
sheath closed; longer anthers with an earlier dehiscence. Base of ovary
immersed in glandular disk. Legume 3–4 cm long with persistent geniculate style. Adventive, roadsides, waste ground, Fla. tropical and subtropical regions. All year.
    2. **C. retusa** L.—Annual 3–8 dm tall from ligneous branching roots;
stem appressed puberulent throughout, or glabrate; internodes ridged

6. Terminal petiolule 2–4 mm long; legume 1.5–1.7 mm long . . . . 6. *R. parvifolia*
5. Racemes many-flowered; peduncles
   elongate . . . . 7. *R. minima*

1. **R. reniformis** DC.—Plant to 2.5 dm tall from a slender crown; stem stiffish, upright branching from the base, sparingly short pubescent or glabrate. Leaflets long petioled, unifoliolate, 2.5–3 cm long, orbicular, or reniform-cordate, coarsely reticulate veined. Flowering branches in the upper axils; racemes terminal and axillary, nearly sessile; calyx 8–10 mm long, about as long as the golden yellow corolla. Legume 15 mm long, 5–7 mm wide, copiously glandular. Open pinelands, Fla., La. to Va. All year. *Rhynchosia simplicifolia* (Walt.) Wood, *Dolicholus simplicifolius* (Walt.) Vail

2. **R. michauxii** Vail—Prostrate plant from moniliform roots. Leaflets essentially unifoliolate, long petioled, distant, lamina oval to 6 cm wide, 4 cm long, or reniform, coarsely reticulate veined, glabrate above, pubescent on the veins below. Flowers in axillary clusters, bright orange; fruiting peduncles stout, up to 3 cm long; calyx to 16 mm long. Legume barely exserted; seed lenticular 4 mm long, brown, mottled with black. Open pinelands in sand, endemic to south and central Fla. Summer, fall.

3. **R. lewtonii** (Vail) Small—Woody vine trailing or twining with several stems arising radially from the stout crown, softly gray pubescent throughout, or glabrescent stipules setaceous, 3–4 mm long. Terminal leaflets 1–3 cm long, broadly obovate, on petiolule 8–10 mm long, lateral leaflets obliquely ovate, small on shorter petiolules. Calyx to 10 mm long, standard golden yellow, 11–12 mm long. Legume pubescent 13–15 mm long, seeds 3.5 mm long, brown, blotched with black. Coastal sands of pinelands, endemic to south and central Fla. Summer, fall.

4. **R. cinerea** Nash—Plant prostrate, cinereous pubescent throughout; stems several, radiating from the crown of the thickened root; leaflets to 1.5 cm long, 3 cm wide, appressed pubescent, short petioluled, ovate-oblique, to rhombic, obtuse, acute. Racemes short peduncled; calyx 10 mm long, lobes acuminate; corolla deep yellow, standard 6–8 mm long. Legume pubescent to 15 mm long, seed 3.5 mm long, brown, blotched with black. White sand oak scrub, endemic to south and central Fla. Summer, fall.

5. **R. swartzii** (Vail) Urban—Woody vine trailing or twining, softly pubescent or glabrate throughout; stipules 3–4 mm long, setaceous,

lanceolate; leaflets rhombic-ovate, acuminate at apex, rounded or trun-cate at base, pubescent about and below; terminal leaflets 4–6 cm long, 3–4 cm wide; lateral leaflets smaller, inequilateral. Racemes 1–2 cm long, pedicels filiform; calyx 2–3 mm long, teeth shorter than the tube, corolla yellow, standard 8 mm long, copiously resin dotted. Legume 2–3 cm long, 5–7 mm wide, falcate; mature seeds 5–6 mm wide, bright red. Hammocks, Fla. Keys, Bahama Islands, Cuba. *Dolicholus swartzii* Vail, *Rhynchosia caribaea* (Jacq.) DC.

6. R. parvifolia DC.—Plant low twining or trailing, grayish, vel-vety, tomentose throughout. Terminal leaflets elliptic-obovate, mucro-nate 1–3.5 cm long; lateral leaflets obliquely ovate 1–2.5 cm long. Calyx 5 mm long, lobes acuminate, longer than tube; standard 8–10 mm long, blade suborbicular, auricles rounded. Legume to 17 mm long, ellipsoid, pubescent, glandular-punctate. Pinelands, south Fla., Fla. Keys, W.I. Summer, fall. *Dolicholus parvifolius* Vail, *Leucopterum parvifolium* (DC.) Small

7. R. minima (L.) DC.—Intricately branching, clambering, or twin-ing plant; stems angled, retrorsely pubescent or glabrescent. Leaflets 1.5–3 cm long, tomentose becoming glabrous; terminal leaflet rhombic-orbicular; lateral leaflets obliquely ovate. Raceme rachis elongate with distant flowers; calyx 3–4 mm long; standard 5–6 mm long, striated with purple. Legume 17 mm long, 4 mm wide, valves slightly contorted; seeds 3 mm wide, compressed, oval in outline, brownish, mottled with black. Disturbed sites, Tex., S.C., Fla., temperate and tropical America, naturalized from Old World. Summer, fall. *Dolicholus minimus* (L.) Medicus

19. ALYSICÁRPUS Necker ex Desv.   FALSE MONEYWORT

Annual or biennial herbs with unifoliolate stipulate leaves. Flow-ers pedicellate in pairs on rachis nodes in a terminal raceme; calyx bi-labiate, lobes lance-acuminate, striate, corolla white or purplish; stand-dard suborbicular, wing and keel petals clawed, stamens diadelphous. Joints of pod separating into several one-seeded truncate articles. About 25 species, Africa, Australia.

1. Foliage and axes puberulent; pods exserted   1. *A. vaginalis*
1. Foliage and axes hirsute; pods included   .   2. *A. rugosus*

1. A. vaginalis (L.) DC.—Diffuse herbs with ligneous wide-spread-ing roots; stems to 1 m long, reclining, branching from the base, rooting at nodes. Leaflets 2–7 cm long, orbicular, oval or linear, of various shapes on the same plant; stipules 4–9 mm long, striate scarious; internodes strigillose and floral bracts with fine uncinate tomentum below. Flowers 5–7 mm long. Fruits mostly 2 cm long; joints reticulate uncinate tomentose; articles oval in transverse section; seeds 2 mm long, mottled with brown. Roadsides, wastelands, peninsular and south Fla., W.I., South America, Old World tropics. Summer and fall.

   2. A. rugosus (Willd.) DC.—Similar to preceding species; stems less diffusely branched. Leaflets oval, obovate, lanceolate, strongly reticulate; stipules and bracts tawny, scarious, copiously hirsute ciliate. (Flowers and fruits not seen.) Disturbed sites, naturalized south Fla., tropics of Old World.

20. DALBÉRGIA L. f.

   Shrubs or trees with odd-pinnate or unifoliolate leaves; stipules early deciduous. Inflorescence axillary; floral bractlets persistent; calyx campanulate, five-dentate; teeth unequal, the upper pair with shallow sinus, the odd one the longest; petals clawed, standard suborbicular, emarginate; wings somewhat falcate, slightly exceeding the keel; stamens nine to ten, didymous; filaments united below; anthers small, dehiscent by apical chinks. Pistil stipitate; ovary one- or several-ovuled. Legume stipitate, flat, orbicular or oblong; seeds compressed. About 180 species in tropical Asia, Africa, and America. *Ecastophyllum* L.

1. Leaves unifoliolate; stamens ten.
  2. Leaflets ovate; petiole 7–10 mm long, thick; legume usually circular, one-seeded . . . . 1. *D. ecastophyllum*
  2. Leaflets oblong-cordate; petiole 10–20 mm long, slender; legume oblong usually two-seeded . . . . 2. *D. amerimnon*
1. Leaves three- to five-foliolate; stamens nine; leaflets suborbicular, thin, abruptly acuminate; legume elongate, usually two-seeded or more . . . . 3. *D. sissoo*

   1. D. ecastophyllum (L.) Benth.—Shrub to 3 m tall with wide-spreading, reclining, or trailing branches; twigs stiff with prominent lenticels. Leaflets 5–15 cm long, including petiolules 8–10 mm long; blades puberulent, ovate, or elliptic, becoming thick coriaceous; stipules 1–15 mm long, linear-lanceolate, setaceous. Flowers in short, corymbiform, axillary clusters; calyx 3–4 mm long; segments broad; equal; corolla white or flushed with pink; standard to 8 mm long, obcordate, clawed. Legume flat, suborbicular or reniform, commonly one-seeded, 2–3 cm wide. Mangrove shores, coastal hammocks, central and south Fla., W.I., Mexico, South America, Old World. Spring. *Ecastophyllum ecastophyllum* (L.) Britt.

2. **D. brownei** (Jacq.) Urban–Arborescent shrub to 3 m tall with slender reclining branches. Leaflets with long slender petiolules 5–9 cm long, glabrous; blades ovate-truncate to subcordate, acute, thin, drying dark brown. Flowers axillary in corymbiform clusters; calyx bilabiate, lobes unequal; corolla white or pinkish, standard to 10 mm long, obcordate, short clawed. Legume compressed, oval or oblong, commonly two-seeded. Hammocks, Fla. Keys, W.I. tropical America. Summer. *Amerimnon brownei* Jacq.

3. **D. sissoo** Roxb.—Tree with full, rounded crown, gray bark; branches stout; season's twigs zigzag, finely striate, brownish, becoming gray. Leaflets with petiolules to 8 cm long or less; blades orbicular-ovate in outline, apex acuminate; rounded at base, firm in texture, finely reticulate veined. Flowers in short axillary panicles; calyx 5 mm long, bilabiate, cylindric, lobes unequal; corolla yellowish white, 10 mm long with flabellate, emarginate standard. Legume with stipe 7 cm long, 8–10 mm wide, oblong, compressed, one- to two-seeded. $2n = 20$. Introduced as an ornamental tree from India, naturalized in south Fla. Summer.

21. PISCÍDIA L. Jamaica Dogwood

Trees; leaves odd-pinnate with five to nine leaflets. Panicles congested; calyx lobes subequal, the upper lip short notched; corolla purplish, blue; petals clawed. Legume four-winged. About six species tropical and subtropical America. *Ichthyomethia* P. Browne

1. **P. piscipula** (L.) Sarg.—Tree to 10 m tall or more; trunk 5 dm in diameter, branches stiff. Leaves to 25 cm long, leaflets elliptic-ovate, estipellate, acute, cuneate, glabrous above, minutely puberulent below, petiolules 5 mm long. Flowers preceding the leaves; floral axes and calyx densely gray puberulent; pedicel 2–5 mm long, jointed; calyx campanulate, 4–5 mm long, corolla 5–6 mm long, cinereous; standard suborbicular, emarginate, keel and wing petals auriculate, rounded at apex; anthers versatile. Legume 3–7 cm long, wings thin, lobed and crisped; seeds oblong, oval, dark brown, 4–5 mm long. Coastal strand, tropics, Fla. Keys, south Fla. Spring. *Ichthyomethia piscipula* (L.) Hitchc., *Piscidia erythrina* L.

Powdered bark and other parts of trees used to inhibit alertness in fish.

22. LONCHOCÁRPUS HBK

Trees or shrubs. Leaves alternate, odd-pinnate, leaflets opposite. Flowers in racemes, with two-flowered pedicels. Calyx truncate, teeth obsolete; standard rounded, vexillary stamen adherent to staminal sheath above, free below. Stigma small, legume compressed, indehiscent; seeds one to three. About 60 species America and Africa tropics, one in Australia.

1. **L. violaceus** Kunth—Tree. Leaves 10–15 cm long, leaflets ovate 5 cm long, 2.5 cm wide, glabrous, membranous. Fruiting peduncle stout, five or more pairs of legumes. Legume stipitate 10–15 cm long, compressed, valves punctate, one- to four-seeded, margin sinuous, contracted between seeds; pedicels stout, jointed; style base persistent. Introduced from W.I., naturalized on Fla. Keys. *L. punctatus* HBK

23. PONGÁMIA Vent.

Tree to 10 m tall from spreading roots; bark smooth gray, demarked by yellowish lenticels. Leaves odd-pinnate, leaflets five to seven, ovate, thin, brown to bright green when young; lemon yellow, deciduous, when old. Racemes axillary, 7–9 cm long, axes puberulent, flowers purple, on slender, bracted pedicels. Calyx teeth obsolete, standard 15 mm wide, suborbicular, short clawed; stamens diadelphous, anthers uniform. Ovary sessile, style incurved, stigma small; legume one-seeded, 3–4 cm long, 2 cm wide, short beaked, nodding on short stipe; valves thick, woody, indehiscent. $2n = 20$. A single species, Ceylon.

1. **P. pinnata** (L.) Merrill—Characters of the genus. Planted ornamental in south Fla., doubtfully spontaneous. Summer.

24. ÁNDIRA Lam.

Trees; leaves odd-pinnate, leaflets opposite. Flowers purple, fragrant, in terminal panicles; calyx truncate, teeth nearly obsolete; uppermost stamen free; anthers versatile, small. Stigma capitate. Legume a dry, woody drupe. About 20 tropical species.

1. **A. inermis** (Wright) W/DC.—Trees to 10 m tall or more; crown wide-spreading. Leaves 2.5 dm long, with leaflets seven to thirteen, essentially glabrous, lanceolate, 5–7 cm long; petiolules 3–4 mm long. Panicle densely branched; flowers 9–10 mm long. Legume suborbicular, woody, indehiscent, one-seeded. $2n = 20$. Ornamental, naturalized from tropical America, rarely found on Fla. Keys. Summer.

25. DÀLEA Juss.

Shrubs or herbs. Leaves usually with numerous leaflets. Flowers in spicate inflorescences, calyx five-cleft, corolla imperfectly papilionaceous, standard arising from the base of the calyx, wings and keel from staminal sheath, anthers uniform in two series. Ovary one-ovuled. About 150 species in America. *Parosela* Cav.

1. **D. carthaginensis** (Jacq.) Macbride ssp. **domingensis** (DC.) Clausen—Shrub 8–30 dm tall, softly pubescent throughout. Foliage glandular-punctate, leaflets nine to fifteen, oval, cuneate or elliptic-obovate, 4–9 mm long. Floral bracts glandular; calyx campanulate, glandular dotted, lobes subulate, copiously villous; corolla white, fading pink; standard 4–5 mm long, blade oval, long clawed; wing and keel petals spatulate, long clawed. Legume dorsally gibbous, one-seeded, indehiscent, pubescent at summit; beak short curved. Hammocks and pinelands, Everglade Keys, peninsular Fla., W.I. All year. *D. domingensis* DC.; *Parosela floridana* Rydberg

26. AMÓRPHA L. Lead Plants

Shrubs, glabrous or pubescent, more or less glandular-punctate throughout. Leaves odd-pinnate; stipules setaceous. Racemes terminal or axillary; calyx obconic, persistent; corolla reduced to a single petal, the standard; stamen monadelphous. Ovary two-ovuled. Legume two-seeded, indehiscent. About 20 species in North America.

1. Leaflets crenulate, margins never contiguous   1. *A. crenulata*
1. Leaflets essentially entire, margins usually
    contiguous . . . . 2. *A. herbacea*

1. **A. crenulata** Rydberg—Shrub 0.9–1.5 m tall, rhizomatous; stems and foliage glabrous, branches slender, angled, purplish. Leaves to 17 cm long, ascending; leaflets eleven to twenty-nine, distant, spreading, oblong, cuneate, 5–11 mm long, mucronate, green above, paler beneath. Racemes two to three together or solitary, 15–20 cm long; flowers 8 mm long in relatively loose clusters, calyx dark green or purplish, 3.2 mm long, standard obcordate, showy white, 5.2 mm long, 4.2 mm wide; stamens long exserted. Style villous, 5 mm long, stigma capitate, ovary compressed, 1 mm long. Mature fruit 6–11 mm long, dorsally gibbous, straight on ventral suture, seed compressed, 5 mm long, often only one maturing. Pinelands over Miami oolitic rocks, endemic to south Fla. Spring.

2. **A. herbacea** Walt.—Shrubs colonial from stout rootstocks; stems

1. **I. endecaphylla** Jacq.—Annual or biennial prostrate herb to 9 dm long; branches ascending, leafy, internodes sparingly strigillose, glabrate. Leaves 3–4 cm long, spreading; stipules scarious, lanceolate-attenuate; leaflets seven to nine; blades 10 mm long, ovate-cuneate, strigillose above and below. Racemes axillary, calyx 3 mm long, tube shorter than the subulate lobes, corolla red purple, 7–8 mm long, the standard strigillose without; wings clawed, keel petals clawed and spurred. Legume 18 mm long, linear, angled, glabrate; seeds six to eight, truncate, four-angled, about 1.2 mm long. $2n = 32$. Naturalized from South Africa, hammocks, trail borders, established in south Fla. Summer, fall, or all year.

2. **I. hirsuta** Harv.—Plants hirsute, 3–7 dm tall from wide-spreading, cordlike, tenacious roots; basal branches procumbent, as long as or longer than the stem. Leaves 7–11 cm long, leaflets five to nine, terminal the largest, oblanceolate, cuneate. Inflorescence axillary, long peduncled; racemes 1–2 dm long; flowers short pedicled, numerous, bright orange red; calyx 4 mm long, standard 4.5 mm long, oblong, retuse, slightly exceeded by clawed wings and keel. Legume 1.5–1.8 cm long, four-angled, hirsute; seeds commonly six to eight, truncate, four-sided, 1.5 mm long, glandular dotted on all sides like the lining of the seed cavity. $2n = 16$. Wasteland, roadsides, introduced as a cover plant, naturalized from Africa, escaped cultivation, central and south Fla. All year.

3. **I. caroliniana** Mill.—Shrubby plant with several stems branching from the crown. Foliage gray green; branches slender; leaves 5–9 cm long; leaflets nine to fifteen, the terminal 7–12 mm long, elliptic, obovate, cuneate, mucronate, oblanceolate, glabrous above, glaucous, appressed pubescent below. Racemes axillary 5–6 cm long, peduncled 3–4 cm long, slender; flower 5–6 mm long, calyx strigillose, 2.5 mm long, lobes deltoid; corolla brownish yellow. Mature legume 12 mm long, stipitate, beaked, obscurely angled, sutures thickened, two-septate and two-seeded; seed 2.5 mm wide, compressed, quadrate. Hammocks, pinelands, coastal plain, Fla., La., N.C. Spring.

4. **I. suffruticosa** Mill.—Shrub 1.6–2 m tall, erect with spreading branches from the base; young twigs appressed strigose with decurrent lines. Leaves grayish green, leaflets seven to fifteen, elliptic, mucronate, cuneate, glabrous above, strigose below with petiolules 1.2 mm long. Racemes axillary, flowers orange, 4–5 mm long; calyx 2 mm long, lobes deltoid. Legumes 10–12 mm long, minutely puberulent; seeds 2 mm wide, laterally compressed. $2n = 32$. Vacant lots, hammocks, coastal plain, Fla., Tex., N.C., W.I., naturalized from Asia. All year. *I. anil* L.

5. I. tinctoria L.—Shrubby plants with branching stems to 2 m tall; young stems appressed strigillose. Leaves 4–7 cm long, rachis sparsely strigillose; leaflets 8–18 mm long, cuneate-spatulate, oblong-ovate, suborbicular, glabrous above, appressed strigillose below. Racemes axillary; flowers reddish, 6 mm long, calyx 2 mm long, lobes deltoid. Mature legume 2–3 cm long, linear, beaked, four-angled, sparingly puberulent; seeds 2.2 mm long, truncate, biconvex, twelve or less per pod. $2n = 16$. Pinelands, scrub oak, escaped from cultivation, Fla., La., N.C., naturalized from south Asia. All year.

This species was cultivated by early colonists for making indigo.

6. I. miniata Ortega—Plant 1.5–3 dm long; stems slender, from tenacious, elongate roots; internodes densely appressed strigillose. Leaves nearly sessile, 2–3 cm long; leaflets five to nine; blades linear-elliptic, narrowly oblanceolate, cuneate, mucronate; stipules setaceous. Inflorescence axillary, racemes long peduncled; calyx 3–5 mm long, tube shorter than the subulate lobes, corolla red, 8–10 mm long, standard 8–10 mm long, 7–9 mm wide, strigillose on dorsal side, wing petals clawed, keel petals clawed and prominently spurred. Legume elliptic-linear, 1.5–2 cm long, valves twisted in open pods. Pinelands, Fla. Keys, W.I., Mexico. All year. *I. argentata* Rydberg

7. I. keyensis Small—Plant 3–9 dm long with many decumbent stems from the crown of ligneous roots; internodes appressed strigillose; young stems angled. Leaflets 7–12 mm long, elliptic, mucronate, cuneate, prominently strigillose with two-pointed hairs above and below or the pubescence on the upper surface spreading and pilose; racemes few-flowered, flowers 5 mm long, reddish. Legume linear 3–3.5 cm long. Strand vegetation, pinelands, endemic to Fla. Keys. *I. subulata* Vahl

29. TEPHRÒSIA Pers. HOARY PEAS

Perennials with elongate, thickened, ligneous roots, stem monopodial or sympodial. Leaves odd-pinnate, stipulate. Inflorescence spicate or racemose, terminal or axillary, calyx slightly bilabiate, petals unguiculate; standard pubescent without, wings auriculate, coherent with the keel; stamens monadelphous; vexillar stamen free, or free only at base, anthers in two series, uniform. Ovary sessile, style barbate or glabrous. Legume dehiscent, five-seeded or more. About 300 species, tropical and subtropical, Southern Hemisphere. *Cracca* L.

1. Styles barbate.
  2. Inflorescence terminal and axillary; leaf-
    lets oblong to ovate-cuneate; pubescence
    coarse, spreading . . . . 1. *T. spicata*

usually several from perennial crown; overall pubescence uncinate, thin, glabrate in age. Leaflets 1–2.5 mm long, ovate, elliptic-lanceolate, or suborbicular, glabrate, longer than the petioles; stipules lanceolate, striate, deciduous. Panicle ample, branching, pedicels slender, 5–20 mm long; flowers purple, 5–6 mm long. Loment constricted; articles two to three, upper suture convex, lower rounded. Fla., Ga., Ala., Tenn., N.C., W. Va., to Mass. *M. marylandica* (L.) Kuntze

7. **D. paniculatum** (L.) DC.—Plant to 1 m tall with several stems radiating from a common crown; internodes angled, sparingly uncinate puberulent. Leaflets 3–6.5 cm long, lanceolate or the upper linear-lanceolate, sparingly pubescent above and below, or glabrate; stipules lance-attenuate, caducous. Inflorescence a diffuse panicle; flowers 5–6 mm long, pinkish lavender. Loments with three to five articles, with upper suture convex, the lower angled. South Fla., N.C., Kans., Mich., N.H. *M. paniculata* (L.) Kuntze

8. **D. tortuosum** (Sw.) DC.—Annual from taproot with wide-spreading branches; mature stem base terete, glabrate; younger branches angled or lanceolate, uncinate pubescent. Lateral leaflets 4–5 cm long, ovate; terminal leaflets 7–12 cm long, rhombic, all essentially glabrous except on the veins below, and margin minutely ciliolate at least when young; stipules to 5 mm long, broadly oblique to semicordate at base, ciliate, caudate. Floral bracts similar; panicles terminal and axillary; pedicels filiform to 2 cm long; flowers up to 4 mm long, purple. Loment four to six articles, equally constricted on both sutures. $2n = 22$. Introduced as a forage plant, escaped from cultivation in Fla., W.I., Mexico, Central and South America. Summer, fall. *M. purpurea* (Mill.) Vail, *M. tortuosa* (Sw.) Kuntze

31. AESCHYNÒMENE L. SHY LEAVES

Low, shrubby, often hispid perennials or annuals with pinnately many-foliolate leaves. Stipules produced beyond the level of insertion; leaflets usually respond to touch. Racemes axillary, calyx bilabiate, petals clawed, anthers uniform. Loment of several indehiscent articles, rounded on ventral suture, dorsal suture straight. About 55 species, worldwide distribution in tropics and subtropics. *Secula* Small

1. Annual shrub, tough, pithy, glabrous; flowers yellow streaked with red . . . . 1. *A. indica*
1. Perennial, woody plants.
   2. Leaflets linear.
      3. Blade three-veined, internodes crisply
         hairy . . . . 2. *A. americana*

3. Blade one-veined, internodes glabrous .   3. *A. pratensis*
2. Leaves obovate; blade reticulate veined;
    internodes viscid pubescent .   .   .   .   4. *A. viscidula*

1. **A. indica** L.—Plant shrublike about 1 m tall with stout taproot; stem base at soil line 18 mm in diameter, pithy, compressible solid, fistulose at upper internodes, branches stout, leafy; leaves 5–8 cm long; rachis pustulate, bearing thirty-eight or fewer pairs of leaflets, overlapping at tips; leaflets 3–10 mm long, 1–2 mm wide, entire; axes hispid to glabrous; floral bracteoles 2–4 mm long, subentire. Racemes axillary, rachis filiform three- to nine-flowered; sepals 4–8 mm long, standard 7–10 mm long, yellow; stamens about as long as the corolla. Fruit with five to nine articles, dorsal suture nearly straight, the ventral slightly constricted between the articles; young articles yellowish, spotted with dark glands, becoming verrucose, uniform dark brown. High sandy banks above lake shores, Immokalee, south Fla., Ga., India, Thailand. Summer. Stems and branches strongly tenacious, collectable only by cutting.

2. **A. americana** L. var. **americana**—Stems to 1.8 m tall, slender, erect to reclining, pustulate-hispid through internodes. Leaves 4–6 cm long, sixteen- to fifty-six-foliolate; leaflets 4–12 mm long, 1.5 mm wide, mucronate, dimidiate, obliquely cordate, serrulate on outer margin, glabrous on surface, rachis hispid. Inflorescence flexuose, rachis very slender, hispid, few-flowered. Flowers 6–10 mm long; standard short clawed, suborbicular, essentially entire, striate with purple and yellow, wings obliquely cordate, clawed; keel petals scythe shaped, two-lobed at apex. Fruit stipitate; five to eight articles, verrucose, semicircular on the ventral suture, straight dorsally. Pinelands, often disturbed soil, wastelands, peninsular and south Fla., Mexico, Cuba to South America. Summer.

3. **A. pratensis** Small var. **pratensis**—Stems to 1.5 m tall, woody, terete, glabrous branching above, appearing twiggy with a few remote leaves at a time. Leaves 4–7 cm long, with twenty-one to thirty-seven leaflets and entire stipules; leaflets to 5 mm long, rounded mucronate, dotted below, only the midvein prominent, bracteoles entire. Flowers 9–12 mm long; standard suborbicular, short clawed, purplish, variegated with yellow, essentially entire; calyx bilobed, the lower three-dentate, the upper two-dentate. Fruit articulate, constricted on both sutures, stipe to 15 mm long; articles four or more, verrucose. Endemic to south Fla. Winter.

4. **A. viscidula** Michx.—Plants prostrate, viscid pubescent throughout; stems slender. Leaves 1–3 cm long; stipules oblique, several-nerved; leaflets three to five, oblong-obovate, pale green, 4–10 mm long, hirtel-

lous below. Peduncles 3–4 cm long, flowers few, 4–5 mm long, distant on slender pedicels; calyx bilabiate with broad rounded segments, corolla yellow suffused with pink, standard broader than long, retuse, keel and wings auriculate. Legume one- to four, jointed, articles broad, 3 mm long, 4 mm wide, pubescent, indehiscent. South and central Fla., naturalized from tropical America. *Secula viscidula* (Michx.) Small

32. STYLOSÁNTHES Sw.   PENCIL FLOWERS

Tufted herbs with perennial suffrutescent base, stems erect or decumbent. Leaflets three, stipules adnate to the petiole. Flowers in terminal capitate or spicate inflorescences. Flowers of two kinds, sterile and fertile: the sterile ones complete, calyx tubular with two upper lobes connate, three free, or four connate, one free, corolla yellow, standard suborbicular, wing and keel petals with staminal sheath, attached to the throat of calyx; the fertile flower without perianth; pistil with long-exserted style, capitate stigma, ovary one- to two-ovuled. Loment two-jointed or one-aborted, style becoming hooklike or merely a curved beak. About 50 tropical and subtropical species, America, Africa, tropical Asia.

1. Pubescence of straight stiff hairs; style
    curved or nearly straight . . . . . 1. *S. calcicola*
1. Pubescence of fine silky hairs; style markedly
    hooked . . . . 2. *S. hamata*

1. S. calcicola Small—Stems low prostrate or decumbent; leaflets elliptic 5–7 mm long, veins conspicuous beneath, curving toward the margin and the mucronate apex; margin entire ciliolate, becoming glabrous. Legume joints cross-ribbed, style beak shorter than the body. Pinelands, endemic to Everglade Keys, south Fla. All year.
2. S. hamata (L.) Taubert—Stems with numerous ascending branches, relatively thick, ligneous from woody taproot; internodes villous at least in lines; ligules cleft, scarious, strongly veined, subulate. Leaflets elliptic, ciliolate, becoming glabrous; veins prominent on the lower surface, curving toward margins and the mucronate apex. Legume two-pointed, densely white pubescent, thick on dorsal suture, laterally reticulate; ripe seed black, plump, 2 mm long. Canal banks, roadsides. All year.

33. CHAPMÁNNIA T & G   ALICIA

Leaves odd-pinnate with three to seven leaflets. Flowers of two kinds, bisexual and unisexual; calyx bilabiate; stamens monadelphous, filaments free above the middle, anthers uniform. Legume a loment, terete. A single species in Fla.

1. **C. floridana** T & G—Herbs viscid from extensive elongate rhizomes with slender moniliform tubers; stems 5–8 dm tall, strict or virgately branched, pubescent with soft hairs and scattered pustular-based stiff hairs. Leaves 4–6 cm long, stipules setaceous; leaflets oblanceolate with excurrent midrib. Inflorescence terminal, lax corymbiform; calyx hirsute, teeth unequal, corolla golden yellow, standard 10–14 mm long, suborbicular, retuse, blades of wing and keel petal unequal. Pistils of bisexual flowers nonfunctional. Legume 1–3 cm long with strigose articles. Sandy woods, roadsides, white sand oak scrub, endemic to central and south Fla. Spring, summer.

34. ZÓRNIA J. F. Gmel.

Leaves palmately two- to four-foliolate, glabrate, punctate, stipules and bracts produced to a basal appendage below the point of attachment. Flowers of one kind, solitary or in spikes, enclosed by two bracts, calyx lobes four, subequal, corolla yellow, standard suborbicular, clawed, wings clawed, keel petals lunate; staminal sheath closed, persistent, anthers alternatingly long and short. Fruit a loment of three or more articles. North American genus of 12 species.

1. **Z. bracteata** (Walt.) Gmel.—Stem to 5 dm tall, slender, wiry, branching from ligneous base. Leaflets 12–17 mm long, 1–2 mm wide, linear-acuminate, or oblanceolate-cuneate to 14 mm long, 8 mm wide, sessile, midvein prominent, margins revolute, sparingly strigillose; petioles 8–12 mm long, slender, wiry. Floral bracts with appendage 7–12 mm long, elliptic-ovate, ciliolate or glabrate, flowers solitary, distant on slightly flexuous axis, calyx 3 mm long, corolla orange yellow, 10–12 mm long, petals pellucid-punctate. Loments 3 mm long, semilunate, retrorsely hispid prickly, compressed. $2n = 20$. Pinelands, south and central Fla. to N.C., tropical America, Brazil. Summer, fall.

35. ÀPIOS Medicus POTATO BEAN

Herbaceous vine with tuberous rootstock. Calyx campanulate, nearly entire or with shallow, rounded, lateral and dorsal˙segments, exceeded by the lanceolate ventral segment, standard broad, rounded, stamens diadelphous. Legume many-seeded. Five species of east North America and China. *Glycine* L.

1. **A. americana** Medicus—Nearly glabrous, high-climbing vine to 3 m long or longer. Leaves pinnately compound, 12–15 cm long, leaflets stipellate; stipules setaceous; leaflets five to seven, lanceolate or ovate,

perennial with ascending peduncles to 3 dm tall; petiole 13–15 cm long; leaflets to obcordate 10–25 mm long, sharply denticulate margin with excurrent tips of the lateral veins. Calyx tubular 3.5–4 mm long, lobes lanceolate; corolla white, 7–9 mm long, marcescent in fruit. Legume three- to four-seeded, yellowish brown. Cover plant in bare areas. Naturalized from Eurasia, now in temperate and subtropical areas, widespread in North America. Summer or all year.

40. MELILÒTUS Mill. SWEET CLOVER

Annual or biennial sweet-scented herbs. Leaves petiolate, pinnately trifoliolate. Racemes axillary; calyx lobes as long as the campanulate tube, corolla white or yellow. Legume one- to two-seeded, straight. Old World genus of about 20 species.

1. **M. alba** Desr.—Plant 7–8 dm long from elongate taproot, stem puberulent or glabrous. Leaves 5–6 cm long, puberulent; stipules adnate to petiole; leaflets 2–3 cm long, oblanceolate, oblong-obovate, denticulate. Racemes 10–15 cm long, stalked, in upper axils. Flowers becoming reflexed; petals white. Mature legume 3.5 mm long, ovoid, reticulate, exserted. $2n = 16,24,(32)$. Wastelands and margins of cultivated fields, south and central Fla., throughout U.S. and south Canada. Spring, fall.

41. MEDICÀGO L.

Annual or perennial herbs with several stems branching from the crown of deep-seated roots. Leaves pinnately trifoliolate. Flower blue or yellow, calyx campanulate with equal, lanceolate lobes, standard oblong, emarginate; keel and wing petals auriculate. Legume curved or spiraled, several-seeded. Old World genus of about 50 species.

1. Stems ascending; flowers blue; perennial . 1. *M. sativa*
1. Stems decumbent; flowers yellow; annual . 2. *M. lupulina*

1. **M. sativa** L. ALFALFA—Plants to 1 m tall, or shorter; leaves 4–5 cm long with adnate, linear stipules; leaflets cuneate, oblanceolate, 2–3 cm long, apex usually retuse; midrib excurrent with lateral veins ending in teeth. Racemes short peduncled, densely flowered; floral bracts setaceous, persistent; flowers 7–10 mm long, peduncles slender, 2 mm long. Legume 3–4 mm wide. $2n = 16,32,64$. Cultivated throughout U.S. and Canada, naturalized from Europe. Spring, summer.

2. **M. lupulina** L. BLACK MEDICK—Slender-stemmed trailing, prostrate, or creeping annual, from the crown of cordlike roots. Stems, leaflets, and axes viscid villous, intermixed with gland-tipped hairs. Leaves

short petioled, 2–2.5 cm long; leaflets obovate, cuneate, denticulate around the apex. Raceme spiciform on slender peduncles, flowers yellow; fruiting spikes cylindrical or subspherical 5–12 mm long. Mature legumes 2.5–3 mm wide, black, reticulate, curved. $2n = 16,32$. Wastelands, roadsides U.S. and south Canada, naturalized from Eurasia.

42. ERYTHRÌNA L. CORAL BEANS

Shrubs or trees; leaves trifoliolate, stipule spiny. Calyx obscurely lobed or nearly entire, corolla scarlet, standard erect, wing and keel petals reduced, barely protruding beyond the calyx, stamens diadelphous. Ovary stipitate. Legume torulose. About 100 species, tropical and subtropical regions. *Micropteryx* Walp.

1. **E. herbacea** L.—Shrubs becoming arborescent to 8 m tall; leaflets stipellate to 7 cm long, 7 cm wide or narrower, across the hastate lobes; petiole 8–15 cm long. Racemes 10–35 cm long; flowers many, opening in succession; calyx 6–9 mm long, oblique, standard folded, notched at apex 2–4 cm long, 11 mm wide; midvein prominent, lateral veins numerous, mostly parallel below the middle, submarginally reticulate at the apex, staminal sheath with maximum length of 26 mm with short and long free filaments at two different levels, all filaments arising above the middle, wing petals 11 mm long, broadly cordate, clawed; keel petal 10 mm long, clawed, flabellate. Pistil 5 cm long, style with small stigma; ovary slenderly fusiform with ten, semilunate, contiguous ovules. Legume stipitate, 7–10 cm long, black; seeds polished, crimson, 10–12 mm wide lengthwise through hilum. $2n = 42$. Hammocks, pinelands, coastal plains, Fla., Tex., N.C. Summer. *E. arborea* (Chapm.) Small

43. CAJÀNUS DC. PIGEON PEA

Erect shrub with velvety pubescence over glandular dots throughout; leaves petioled, pinnately trifoliolate. Upper calyx segments united almost entirely, corolla reddish purple and yellow, standard reddish without, striated with yellow and white; wings and keel yellow; style thickened below the stigma; ovules several. Legume linear, compressed, beaked, obliquely impressed between seeds. One species naturalized from Old World tropics.

1. **C. cajan** Millsp. PIGEON PEA—Shrub to 2 m tall, branching. Leaflets ovate or elliptic, prominently pubescent on principal veins below; terminal leaflet 3–8 cm long on petiolule 5–7 mm long; lateral leaflets smaller on shorter petiolules; stipules lanceolate, silky, persistent; internodes notably appressed hairy on ridges. Racemes few-flowered, axillary,

standard nearly orbicular with a green central spot above the claw at base, prominently crested on each side of the median line, the outer margins deflexed auricles; wing petals auriculate and coherent with keel; vexillar stamen geniculate at base. Legume 5–8 cm long, 10–14 mm wide, bronze purple, finely pubescent; ripe seeds orbicular, thickened on one side, brown, hilar area white. $2n = 22,44,66$. Naturalized from Old World tropics, cultivated in American tropics, escaping in south Fla., W.I. All year.

44. MUCÙNA Adans. Velvet Beans

Vines with twining, branching stems. Leaves trifoliolate. Racemes axillary, few- to many-flowered, calyx bilabiate, wing and keel petals twice as long as the standard, stamens diadelphous; anthers alternately short and long, versatile. Legume leathery, valves velvety with stinging hairs or transversely crested. About 45 species, mostly tropical. *Stizolobium* P. Browne

1. Vines herbaceous; legume turgid . . . 1. *M. deeringiana*
1. Vines woody.
  2. Legume with transverse crests; tip of keel
    not indurated . . . . 2. *M. sloanei*
  2. Legume without transverse crests; tip of
    keel indurated . . . . 3. *M. pruriens*

    1. **M. deeringiana** (Bort) Merrill—Leaflets 8–9 cm long or longer, terminal, rhomboidal, basal ones inequilateral. Inflorescence a raceme; flowers greenish yellow, upper calyx lip notched, lower three-lobed, standard 2–2.5 cm long, oval, nearly clawless, wings widest below apex, keel petals scythe shaped. Legume 5–9 cm long, dark brown, several-seeded. Old fields, cover plant, south Fla., naturalized from Asia. *S. deeringianum* Bort
    2. **M. sloanei** Fawc. & Rendle—Terminal leaflets ovate; the lateral ones inequilateral, thin 7–15 cm long, silvery pubescent below; inflorescence capitate, flowers red purple, calyx 1.5–2 cm long, the upper lip broad, rounded, the lower three-lobed, standard ovate, short clawed and auricles, keel petals scythe shaped, auricle like the wing petals. Legume inverted, oblong-ovate, irregularly crested, several-seeded. Hammocks, south Fla., W.I., South America. All year.
    3. **M. pruriens** (L.) DC.—Terminal leaflet 9 cm long, equilateral, lateral leaflets 9 cm long, 6 cm wide, inequilateral, thinly appressed pubescent with silvery straight hairs below, glabrous above. Flowers dark purple or blue, 3–3.5 cm long in elongate racemes, often two to

three together; calyx tube 5–6 mm long, as long as the three upper teeth, pubescence intermixed with bristlelike stinging hairs, especially on pedicels; standard about 18 mm long, wing petals twice as long, exceeded by cartilaginous apex of keel petals. Legume about 6 cm long, curved, covered with stinging hairs. Naturalized from tropical Asia, cultivated and escaping, marginal pinelands, south Fla., Bahama Islands, W.I., tropical America. Winter.

45. CANAVÀLIA Adans. BAY BEAN

Vine, fleshy, prostrate, from deep-seated roots. Leaves trifoliolate, petioled, coriaceous. Racemes peduncled, few-flowered; flowers pendulous, bracts deciduous; stamens diadelphous, anthers versatile. Ovary substipitate, several-ovuled. Legume oblong, 10 cm long. About 50 species, tropical and subtropical regions.

1. **C. maritima** (Aubl.) Thouars—Vine of ocean beaches and seaside dunes, usually several meters long; stems, floral axes strigillose with appressed retrorse hairs. Leaflets retuse or rounded at apex, obovate, oval, suborbicular, 7–9 cm long, margin entire, ciliolate, appressed pubescent on lamina when young. Raceme rachis thick, floriferous almost from the base; calyx campanulate, 10–12 mm long, 10 mm wide in anthesis, upper lobes rounded, the lower deltoid, corolla rose purple, 25–30 mm long, standard oblong-obovate, reflexed, auricles inflexed; wing and keel petals clawed. Legume prominently ribbed on each side of the upper suture, valves indurate, thick, minutely appressed puberulent. Coastal strand, Fla., pantropical shore plant. All year. *C. obtusifolia* (Lam.) DC.

C. gladiata DC. SWORD BEAN—Differs from the preceding species by its narrowed, usually apiculate leaf apex, by its longer legume to 25 cm in length, and by its tendency to climb rather than trail, at least on narrow beaches flanked by shrubs. Shore plant from Old World tropics, East Indies, cultivated and escaped in south Fla. All year.

46. PHASÈOLUS L. BEANS

Mostly vines, rarely erect herbs. Leaves pinnately trifoliolate. Flowers in racemes, few opening at the same time; calyx bracteate, bilabiate, corolla commonly purple or red of varying shades, standard the widest petal, keel coiled, stamens diadelphous. Legume valves convex, seeds large, usually reniform. About 470 species of warm and temperate regions. Widely cultivated for human consumption.

1. **P. lathyroides** L.—Stems branched above the base from perennial rhizomes (annual in cultivation), pilose, glabrescent. Leaves 4–12 cm

with alternate or radical, simple, or more often compound leaves. Flowers bisexual, solitary or in cymes or umbels; calyx five-parted, petals five, free or barely fused at the base; stamens usually ten, fused at the base. Ovary superior, five-locular, styles five, free and persistent. Fruit a capsule or berry. About 7 genera, 900 species, mostly tropical and subtropical in distribution. Generally low herbs with sourish watery sap and delicate trifoliolate leaves.

1. ÓXALIS L. Yellow Sorrel

Perennial or annual herbs with trifoliolate leaves and obcordate leaflets, usually relaxed at night, and small yellow, pink, purple, or white flowers. Sepals five, persistent, petals five, stamens ten, with long and short filaments. Fruit a capsule. About 750 species in warm regions, especially tropical America and south Africa. *Bolboxalis* Small, *Ionaxalis* Small, *Xanthoxalis* Small

1. Petals yellow.
  2. Stipules present; plants not rhizomatous;
    trichomes nonseptate.
    3. Stems erect or nearly so; peduncles over
      3 cm long . . . . 1. *O. dillenii*
    3. Stems reclining, creeping or trailing;
      peduncles less than 3 cm long . . 2. *O. corniculata*
  2. Stipules absent; plants rhizomatous;
    trichomes septate . . . . 3. *O. stricta*
1. Petals purple or violet . . . . 4. *O. violacea*

    1. O. dillenii Jacq.—Tufted perennials erect or becoming decumbent but not creeping, up to 4 dm tall, strigose with non-septate hairs; stipules present, 2–3 mm long. Leaflets 1–2 cm wide. Flowers yellow, in umbellike inflorescence; corolla 5–10 mm long. Fruit 2–3 cm long, strigose, seeds brown with whitish ridges. Disturbed sites, lawns, old fields, a cosmopolitan weed. Spring, summer. *O. filipes* Small, *O. florida* Salisb., *Xanthoxalis brittoniae* Small, *X. colorea* Small

    2. O. corniculata L. Lady's Sorrel—Tufted perennials with slender, trailing stems that root freely at the nodes, creeping. Stipules broad, brownish or purplish; leaflets 1–2 cm wide or smaller, tending to have a purplish or bronze cast. Flowers yellow in cymes; corolla 5–10 mm long, fruit 2–3 cm long, puberulent. Seeds brown without whitish ridges. Disturbed sites, lawns, gardens, old fields, south U.S. and tropics, widely distributed weed in warmer regions. $2n = 28,48$. Spring, fall. *O. repens* Thunb., *Xanthoxalis corniculata* (L.) Small, and *X. langloisii* Small

3. O. stricta L. Sour Grass—Tufted perennials with slender rhizomes, prostrate or erect up to 6 dm tall with pubescent stems bearing septate trichomes. Stipules absent; leaflets 1–2 cm wide, smooth. Flowers yellow or orange yellow in cymose or umbellate inflorescence; corolla 5–9 mm long. Fruit 8–15 mm long with flexuous hairs. Disturbed sites, lawns, old fields, cosmopolitan weed. Spring, fall. *X. cymosa* Small, *Xanthoxalis stricta* (L.) Small

4. O. violacea L. Violet Wood Sorrel—Bulbous perennials, glabrous, with scaly base, stemless. Leaves with smooth petioles, leaflets broadly obcordate. Flowers in umbels on scapes 1–2 dm tall; sepals bearing orange-colored callus tips, petals violet or purple or sometimes white. Capsule 4–5 mm long. Moist soil, disturbed sites, Fla. to N. Mex., Mass., Minn., Colo. All year. *Ionoxalis violacea* (L.) Small

98. TROPAEOLÀCEAE Nasturtium Family

Twining or prostrate herbs with succulent stems, watery juice, and alternate or opposite peltate leaves. Flowers showy, bisexual, bilaterally symmetrical, axillary and solitary, calyx two-lipped, the dorsal sepal prolonged into a spur; petals five, stamens eight, free. Ovary superior, three-locular, the carpels separating from a short, central axis. Fruit baccate. Two genera and about 80 species in tropical America.

1. TROPAÈOLUM L. Nasturtium

Characters of the family. Berry three-lobed. Nearly 80 species, tropical America.

1. T. majus L.—Stems diffusely branching or climbing, stems soft and succulent. Leaves orbicular, undulate, peltate. Corolla conspicuous bright red, orange, purplish, yellow; claws of petals toothed. Fruit a berry, about 1–2 cm wide. $2n = 28$. Disturbed sites, frequently cultivated and escaped, naturalized from Peru.

99. LINÀCEAE Flax Family

Shrubs or herbs with simple leaves and radially symmetrical, bisexual flowers in cymes. Sepals four to five, imbricate; petals four to five, usually clawed, stamens same number as the petals and alternate, fused at the base with them. Ovary two- to five-carpellate, styles two to five, filiform, free or fused. Fruit a capsule, rarely fleshy. Seeds flat. About 10 genera and 200 species, cosmopolitan.

1. LÌNUM L.   FLAX

Perennial or annual herbs with smooth or sometimes pubescent stems. Leaves alternate, opposite or whorled; blades entire, sessile, sometimes the upper leaves glandular-toothed. Flowers five-parted, in terminal scorpioid cymes; sepals five, petals five, usually free, blue, red, yellow, or white; stamens five, fused basally. Ovary superior, five-carpellate becoming ten-locular by false septa, styles five, stigmas filiform to capitate. About 150 species, cosmopolitan. *Cathartolinum* Reichenb.

1. Corolla blue; sepals without marginal glands — 1. *L. usitatissimum*
1. Corolla yellow; sepals with marginal glands.
  2. Styles distinct; corolla usually less than 1 cm wide.
    3. Dark, stipular glands present; staminodes present . . . . — 2. *L. arenicola*
    3. Stipular glands absent; staminodes absent.
      4. Capsule ovoid, 3 mm long or more . — 3. *L. floridanum*
      4. Capsule depressed-globose, less than 2.5 mm long . . . . — 4. *L. medium* var. *texanum*
  2. Styles united; corolla usually over 1 cm wide . . . . — 5. *L. carteri*

1. L. usitatissimum L. COMMON FLAX, BLUE FLAX—Erect, smooth annual up to 1 m tall, branched above. Leaves lance-linear, stipular glands none. Flowers paniculate on long erect pedicels; sepals short acuminate, about 7–9 mm at maturity; petals blue, 10–15 mm long. Capsule globose 6–8 mm wide, tardily dehiscent. $2n = 30$. Disturbed sites, old fields, roadsides throughout much of North America, introduced from Europe and escaped from cultivation. Summer.

2. L. arenicola (Small) Winkl. SAND FLAX—Erect perennial with one to several stems up to 7 dm tall, wiry, sparingly branched above. Leaves alternate except near the base, linear, mostly 7–10 mm long, acuminate; stipules glandular, reddish and becoming dark. Inflorescences with few, ascending branches; sepals glandular-toothed, 2–3 mm long; petals yellow, obovate, 4–6 mm long, staminodes present. Capsule 2–3 mm long, pyriform pointed, dehiscent into ten segments, false septa incomplete, ciliate margin. $2n = 36$. Pinelands, endemic to south Fla., Fla. Keys. All year. *Cathartolinum arenicola* Small

    This is the only species of LINUM found commonly in the Fla. Keys.

3. L. floridanum (Planchon) Trel.—Perennial, erect, smooth herb, solitary or sometimes several stems from the base, up to 1 m tall. Leaves linear-oblanceolate, 13–17 mm long, acute, opposite below and alternate above; stipular glands none. Inflorescence of few, short branches; sepals narrowly lanceolate, the inner conspicuously glandular-toothed; petals lemon yellow, obovate, 5–10 mm long, pubescent near the base inside; staminodes absent. Styles free, capsule pyriform or ovoid, 2–3 mm long or more, dehiscing into ten segments, false septa nearly complete, eciliate. Two varieties occur in south Fla.

1. Capsule pyriform, exposed portions purplish, 2–3 mm long . . . . 3a. var. *floridanum*
1. Capsule ovoid, yellowish throughout, 3–3.2 mm long . . . . 3b. var. *chrysocarpum*

3a. L. floridanum var. **floridanum**—Taller and somewhat more slender than var. **chrysocarpum**, with more numerous leaves. Capsule pyriform purplish, rather longer than wide, not as long as var. **chrysocarpum**, and the seeds somewhat smaller, usually less than 2 mm long. $2n = 36$. Open pine and pine–palmetto woodlands, Fla. to La., N.C. on the coastal plain. All year. *Cathartolinum floridanum* (Planchon) Small

3b. L. floridanum var. **chrysocarpum** C. M. Rogers—Generally less common than the typical variety, especially within our area. Capsule ovoid, over 3 mm long, yellowish, seeds somewhat larger than var. **floridanum**, over 2 mm long. $2n = 36$. Generally in pine–palmetto woodlands, Fla. to Miss., S.C. on coastal plain. All year.

Intergradation occurs between the varieties.

4. L. medium (Planchon) Britt. var. **texanum** (Planchon) Fern. YELLOW FLAX—Smooth perennial or sometimes annual herb with erect, often clustered, simple stems up to 6 dm tall. Leaves narrowly lanceolate, rounded or acute, entire, usually 10–15 mm long, stipular glands none. Inflorescence with spreading, ascending, stiff branches; sepals lanceolate, acute, 2–3 mm long, the inner somewhat shorter and glandular-toothed; petals lemon yellow, 5–8 mm long. Capsule depressed-globose, about 2 mm long, tardily dehiscent into ten segments, false septa nearly complete, eciliate. $2n = 36$. Open fields, low pineland, Fla. to Tex., Me., Ill., Mo., Bahama Islands. All year. *L. curtissii* Small, *Cathartolinum medium* (Planchon) Small var. *texanum* (Planchon) Moldenke

Highly variable, especially in south Fla.

5. L. carteri Small—Annual with erect stems up to 6 dm tall, rather scabrous near the base or on stem angles throughout. Leaves alternate, narrow, entire or the upper glandular-toothed; stipules present as dark glands. Inflorescence of spreading, striated branches, rather coarse, sepals lanceolate, glandular-toothed, 5–7 mm long; petals orange yellow,

moderately thick; petioles often winged.
2. Rind closely adherent to the fleshy pulp;
leaves lance-ovate to elliptic or ovate,
not long acuminate.
3. Flower buds white; fruit not mammillate.
4. Fruits oval, greenish yellow at maturity with a thin skin, conspicuously winged petioles, and strongly
acid pulp . . . . 1. *C. aurantiifolia*
4. Fruit globose, rarely ovoid, orange
colored, or if yellow, then large
and thick skinned.
5. Fruits very large, pale yellow; petioles broadly winged . . . . 2. *C. paradisi*
5. Fruits small or moderate sized,
orange or orange yellow; petioles
slightly to broadly winged.
6. Fruit pulp sweet with a solid
core; petioles slightly to moderately winged . . . . . 3. *C. sinensis*
6. Fruit pulp acid with a hollow
core; petioles broadly winged 4. *C. aurantium*
3. Flower buds tinged with red outside;
fruit ovoid and rather mammillate . 5. *C. limon*
2. Rind easily separating from the pulp;
leaves lance-oblong, acuminate . . . 6. *C. reticulata*
1. Leaves not articulate between the petiole
and blade; rind of fruit very thick, often
rough; petioles not winged . . . . . 7. *C. medica*

1. **C. aurantiifolia** (Christm.) Swingle, KEY LIME—Small tree or
shrub with uneven branches, armed with many stout spines. Leaves
mostly 5–7 cm long, elliptic-ovate, rounded at the tip, with crenate
margins; petioles narrowly winged. Flowers in axillary clusters; petals
white, stamens twenty to twenty-five. Fruit small, about 3–6 cm long,
rind very thin, with ten segments, pulp greenish yellow when ripe, very
acid. $2n = 18$–$21,27$. Naturalized in hammocks, coastal areas, Fla. Keys,
south Fla., tropical America. *C. lima* Lunan

2. **C. paradisi** Macf. GRAPEFRUIT—Trees with rounded crown and
even branching, spines slender, flexible or absent. Leaves large, elliptic
to ovate, acute, petiole broadly winged. Flowers solitary or in axillary
clusters; petals white; stamens twenty to twenty-five; fruit very large,
mostly 10–15 cm in diameter, globose, with eleven to fourteen segments,
rind thick, pale lemon yellow when ripe. Seeds numerous. $2n = 18,27,36$.
Found occasionally near former habitations, doubtfully naturalized. *C.
maxima* (Burm.) Merrill

3. **C. sinensis** (L.) Osbeck, SWEET ORANGE—Shrub to moderate-sized

tree with rounded crown and even branching, spines numerous, slender, flexible, or none; leaves elliptic to ovate, acute, petiole narrowly winged. Flowers in axillary clusters; petals white; stamens twenty to twenty-five. Fruit moderately large or smaller, globose or ovoid, with ten to thirteen segments, pulp sweet when ripe. $2n = 18,27,36,45$. Naturalized in hammocks, coastal areas, Fla. Keys, peninsular Fla., tropical America.

4. **C. aurantium** L. SOUR ORANGE, SEVILLE ORANGE—Trees with rounded crown and even branching, spines long and flexible; leaves elliptic to ovate, acute or acuminate, petioles broadly winged. Flowers axillary, in clusters or solitary; petals white; stamens twenty to twenty-four. Fruit 7–8 cm in diameter, globose, somewhat flattened at the apex with ten to twelve segments, pulp acid and bitter-tasting. $2n = 18$. Naturalized in hammocks, coastal areas, Fla. Keys, peninsular Fla., tropical America. *C. vulgaris* Risso

This species is much used in the U.S. as stock to which is grafted the sweet orange of commerce.

5. **C. limon** (L.) Burm. f. LEMON—Shrub or small tree with uneven branches, spines short, stout. Leaves elongate-ovate, barely serrulate, petioles not winged but often rather narrowly marginate. Flowers solitary or in small axillary clusters, red tinged in bud; petals white above, red purple below; stamens twenty to forty. Fruit ovoid or oblong with a nipplelike protuberance at the apex, lemon yellow at maturity, rind often rough and rather thick; pulp very acid. $2n = 18,36$. Naturalized in hammocks, coastal areas, apparently widely scattered by early Spanish settlers and Indians, south Fla., tropical America. *C. limonum* Risso

6. **C. reticulata** Blanco, TANGERINE—Low tree with slender, even branches. Leaves small, lanceolate, acuminate; petioles without wings or margins; flowers small; petals white; stamens eighteen to twenty-four. Fruit small, globose, orange yellow or reddish orange with a loose, easily separated rind from the pulp. Found occasionally near abandoned groves and habitations, doubtfully naturalized. *C. nobilis* Lour. var. *deliciosa* (Tenore) Swingle

7. **C. medica** L. CITRON—Shrub or low tree with uneven branches and short, rigid spines. Leaves broadly elliptic to ovate, obtuse or rounded at the apex, serrulate; petioles not winged. Flowers tinged with red in bud, in terminal panicles or clustered in axillary fascicles; petals white above, reddish underneath; stamens thirty to forty. Fruit very large, ovoid or oblong, often with a nipple at the apex, lemon yellow at maturity, the rind very thick, often rough, pulp rather small. $2n = 18$. Naturalized in pinelands, hammocks, south Fla., tropical America.

2. MURRÀYA L.

Shrubs or low trees without spines; leaves pinnately compound, alternate, the leaflets rather small and alternate or subopposite. Inflorescence a terminal or axillary cyme, or the flowers solitary; calyx of five partially fused sepals, petals four to five; stamens eight to ten, inserted on a disk. Ovary two- to five-locular; fruit baccate. Nine species, native of southeast Asia.

1. **M. paniculata** (L.) Jackson, ORANGE JASMINE—Shrub or tree with puberulent young branches. Leaves dark green, with three to nine ovate or obovate leaflets, these mostly 2–5 cm long. Flowers white, very fragrant; sepals deltoid, petals oblanceolate, about 1–2.5 cm long. Ovary smooth; fruit subglobose, bright red, 1–2 mm long. $2n = 18$. Introduced as an ornamental at Homestead, Fla., doubtfully established in our flora. *M. exotica* L.

3. GLYCÒSMIS CORREA

Shrubs or low trees without spines. Leaves pinnately compound, often unifoliolate to trifoliolate or apparently simple, persistent, alternate, with numerous pellucid dots. Inflorescence a terminal or axillary panicle, flowers small, bisexual, sepals five, partially fused; petals five, white, elongate, stamens ten, free, with numerous glands. Ovary two- to three-locular, on a cushion-shaped disk, style short. Fruit a berry with one to three seeds. About 35 species in southwest Asia.

1. **G. parviflora** (Sims) Little—Shrub or small tree up to 4 m tall. Leaves apparently simple or pinnately compound; leaflets elliptic, 8–18 cm long, glabrous. Calyx about 1 mm long, lobes ovate; petals white, elliptical, 3–4 mm long. Fruit a subglobular berry, 7–9 mm in diameter, pink or white. Hammocks, Key West, naturalized from Asia. Spring, summer. *G. citrifolia* (Willd.) Lindl.

4. TRIPHÀSIA Lour. LIMEBERRY

Shrubs armed with paired or solitary, sharp axillary spines. Leaves evergreen, alternate, trifoliolate, or occasionally with one to two leaflets, margins serrate. Flowers solitary or in axillary cymes, bisexual; calyx cupular, three- to five-lobed, lobes acute, petals three to five, white, linear to lance-oblong, stamens six, in two series, inserted on a fleshy ring-shaped disk, ovary three- to five-locular, styles fused, ovules solitary in each locule. Fruit a berry somewhat resembling a small orange. Three species of southeast Asia.

1. T. trifolia (Burm. f.) P. Wilson—Shrubs mostly 1–2 m tall, bearing usually paired or solitary spines. Leaves mostly trifoliolate, margins toothed, persistent. Flowers in two- to three-flowered axillary cymes or sometimes solitary, petals three to four, white, mostly 12–16 mm long. Berry small, dull orange or crimson, gland dotted. Key West, lower Fla. Keys, naturalized from southeast Asia.

5. ÁMYRIS L.   Torchwood

Shrubs or trees with resinous fragrant wood, without spines, branches smooth or pubescent. Leaves usually opposite, odd-pinnate, often three- to five-foliolate, or apparently simple (unifoliolate). Inflorescence of paniculate cymes with few to many flowers; flowers bisexual; calyx four-lobed, gland dotted, petals four, white, stamens eight in two series inserted at base of the disk. Ovary one-locular, style very short or absent. Fruit a drupe, one-seeded. About 20 species in tropical America.

1. Ovary pubescent, stipitate; drupe obovoid .   1. *A. balsamifera*
1. Ovary glabrous, sessile; drupe spherical .   2. *A. elemifera*

1. A. balsamifera L. Balsam Torchwood—Shrubs or trees up to 10 m tall or more. Leaves three- to five-foliolate, leaflets 5–12 cm long, lanceolate to ovate, acute to acuminate. Calyx about 1 mm long, lobes triangular; petals gland dotted about 3 mm long. Drupe obovoid, 6–15 mm long. Coastal hammocks, Fla. Keys, south Fla., W.I., tropical America.

2. A. elemifera L. Torchwood—Shrubs or trees up to 5 m tall. Leaves three- to five-foliolate, leaflets mostly 2–8 cm long, ovate, acute or acuminate. Calyx about 1 mm long, lobes ovate; petals narrowly ovate, about 3–4 mm long. Drupe globose, 5–8 mm long. Coastal hammocks, south Fla., W.I., Central America.

6. ZANTHÓXYLUM L.   Prickly Ash

Dioecious, monoecious, or polygamous trees or shrubs often armed with prickles on the branches and trunks, bark aromatic. Leaves deciduous or evergreen, alternate, odd- or even-pinnately compound, trifoliolate or apparently simple (unifoliolate). Flowers terminal or axillary, white, or greenish-yellowish white in spikes, cymes, or panicles; calyx absent, or sepals three to five, usually fused; petals three to five,

102. SIMAROUBÀCEAE QUASSIA FAMILY

Trees or shrubs with alternate, pinnately compound leaves. Flowers small, radially symmetrical, unisexual or bisexual, usually in axillary or terminal panicles or cymose racemes; calyx three- to seven-lobed; petals three to seven, stamens distinct, as many or twice the number of petals, usually inserted at the base of an intrastaminal disk. Gynoecium of two to six free to united carpels, locules one to three. Fruit a schizocarp with dry, sometimes winged, samaralike or drupelike mericarps, drupaceous, or a berry. About 30 genera and 200 species, chiefly tropical and subtropical.

1. Larger leaflets mostly less than 3 cm long; fruit a samara . . . . 1. *Alvaradoa*
1. Larger leaflets mostly over 3 cm long; fruit a drupe or berry.
  2. Stamens ten in staminate flowers; leaves even-pinnate, mostly twelve to eighteen leaflets . . . . 2. *Simarouba*
  2. Stamens five in staminate flowers; leaves odd-pinnate, mostly five to nine leaflets 3. *Picramnia*

1. ALVARADÒA Liebm.

Dioecious shrubs or trees with slender branches, bitter bark, and alternate, odd-pinnate, persistent leaves often crowded toward the ends of the branches. Flowers unisexual, minute, in dense spreading or drooping racemes; sepals five, distinct pistillate flower; petals five in staminate flowers, absent in pistillate flowers; stamens five, long exserted, intrastaminal disk present. Gynoecium three-carpellate, two- to three-locular with three recurved styles. Fruit samaralike, two- to three-winged capsule. Five species chiefly in tropical America.

1. **A. amorphoides** Liebm.—Shrubs or small trees up to 15 m tall with young stems puberulent. Leaflets numerous, nineteen to fifty-one, 1–2.5 cm long, ovate to obovate. Flowers green or yellowish white; staminate racemes up to 20 cm long, pistillate racemes 13 cm or less, very dense, sepals ovate, about 1–2 mm long; filaments pubescent. Samaras densely pubescent, lance-oblong, about 1–1.5 cm long. Hammocks, south Fla., Fla. Keys, Mexico, W.I., tropical America. All year.

2. SIMAROÙBA Aubl. PARADISE TREE

Dioecious or polygamodioecious trees or large shrubs with alternate, even-pinnate, persistent leaves and bitter bark. Flowers small, in large axillary or terminal panicles, calyx usually five-lobed, petals usually five; stamens usually ten; intrastaminal disk present; gynoecium five-carpellate, inserted in a disk. Ovary one-locular distinct. Fruit composed of a cluster of one to five drupes. About six species in tropical America, often cultivated in temperate zones.

1. **S. glauca** DC. PARADISE TREE—Small or moderate-sized tree up to 15 m tall, glabrous, bearing large leaves with ten to twenty coriaceous, narrowly oblong leaflets usually 5–10 cm long, entire, green above and pale beneath. Flowers in large panicles, open and somewhat lax, calyx 3–4 mm wide, lobes ovate; petals oblong, 4–6 mm long, whitish. Drupes ellipsoidal or 1–2 cm long, bright red and turning dark purple or black at maturity. Coastal hammocks, south Fla., south Mexico, Central America, W.I. Spring.

3. PICRÀMNIA Sw.

Dioecious or polygamous shrubs or trees, branches slender, curving, bark very bitter; alternate, odd-pinnate, persistent leaves. Flowers minute, in panicles, with slender spikelike or racemelike branches, or long racemes opposite the leaves, drooping; calyx three- to five-lobed, petals three to five, white or greenish, linear or absent; stamens three to five, opposite the petals, reduced to staminodes in pistillate flowers. Gynoecium two- to three-carpellate, styles short, ovary two- to three-locular. Fruit a berry, one- to two-locular. About 40 species in tropical America.

1. **P. pentandra** Sw. BITTER BUSH—Shrubs or small trees with leaves 20–36 cm long; leaflets five to nine, ovate-oblong or elliptic, 5–10 cm long, glabrous, acuminate. Flowers green in slender, pubescent racemes; calyx five-lobed; petals five, acuminate, hirsute; stamens five; stigma two- to three-lobed. Fruit a berry, ellipsoidal, red becoming black at maturity, 10–15 mm long. Coastal hammocks, south Fla., Fla. Keys, W.I., tropical America. Winter, spring.

The roots, bark, and leaves of this species have been used locally in the tropics to cure fever.

Ailanthus altissima (Mill.) Swingle, THE TREE OF HEAVEN—A small to medium-sized tree, with pinnately compound leaves usually having fifteen to forty-one lance-elliptic leaflets and samaras 4–5 cm long, is found along roadsides and in disturbed areas. It is native to China but has escaped and become naturalized throughout much of the U.S. and Canada and may occur in our area.

103. SURIANÀCEAE BAY CEDAR FAMILY

Shrubs or trees with grayish pubescent stems diffusely branching.

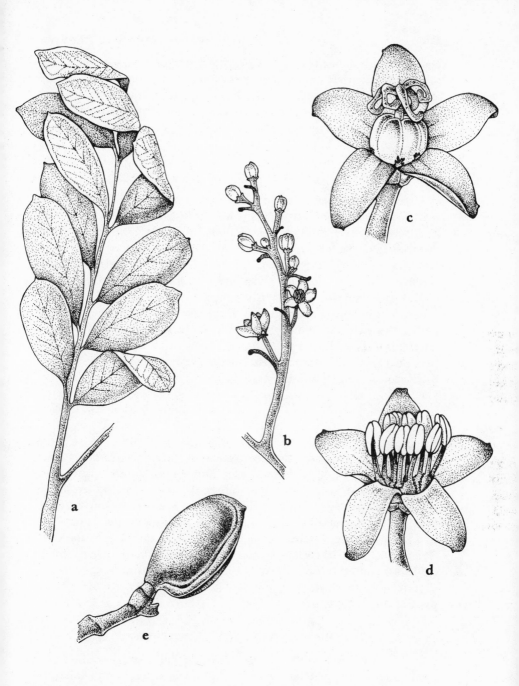

Family 102. **SIMAROUBACEAE.** *Simarouba glauca:* a, compound leaf, × ½;
b, staminate inflorescence, × 1; c, pistillate flower, × 5; d, staminate flower,
× 5; e, mature fruit.

Leaves simple, alternate, entire, without stipules. Flowers bisexual, radially symmetrical in axillary cymes subtended by conspicuous bracts; sepals five, fused at the base, petals five, imbricate, stamens ten, or fewer by abortion. Ovary five-carpellate with gynobasic style. Fruit of three to five, one-seeded drupaceous carpels. A single genus of tropical coastal areas.

1. SURIÀNA L.   Bay Cedar

Densely branched shrubs or small trees up to 8 m tall with simple, entire, linear-spatulate, sessile, fleshy leaves crowded on branches; blades mostly 1–4 cm long, 3–6 mm wide, densely pubescent. Flowers small and inconspicuous, yellow, solitary or in few-flowered clusters; sepals 6–10 mm long, petals 7–10 mm long, obovate, unequal, intrastaminal disk absent. Ovaries one-locular, styles five, filiform. Fruiting carpels 4–5 mm long, pubescent. A monotypic genus.

1. **S. maritima** L.—Characters of the genus. Coastal beaches, sand dunes, sandy thickets along seashores in south Fla., Fla. Bay Keys, and lower Fla. Keys. All year.

This species is a characteristic strand plant in many parts of the American tropics, especially W.I.

104. BURSERÀCEAE   Torchwood Family

Trees or shrubs mostly dioecious with alternate, pinnately compound, deciduous, or persistent leaves. Flowers racemose or paniculate, radially symmetrical, bisexual or unisexual, small and rather inconspicuous; sepals three to five, petals three to five, free or partly fused, deciduous; intrastaminal disk present, free or fused to the calyx tube; stamens twice the number of petals, inserted at the base of the disk. Gynoecium three-carpellate, syncarpous, ovary two- to five-locular, ovules two in each locule. Fruit usually drupaceous or dry, dehiscent or indehiscent with two to five stones. About 16 genera and 600 species, pantropical in distribution.

1. BÙRSERA Jacq. ex L.   Torchwood

Shrubs or trees with thin exfoliolating bark, strong-scented sap; producing resin or balsam. Leaves odd-pinnate, three- to nine-foliolate, or appearing simple, usually crowded near the end of branches, deciduous. Flowers very small, whitish, unisexual or bisexual; calyx three- to five-lobed; petals three to five; stamens six to ten nonfunctional in pistillate flowers. Stigma three- to five-lobed. Drupes ovoid, detaching into valves at maturity. About 100 species in tropical America.

1. **B. simaruba** (L.) Sarg. Gumbo Limbo—Dioecious tree up to 25 m tall with trunks often 1 m in diameter, with young bark characteristically greenish or greenish brown, older bark lustrous, light red to dark reddish brown that exfoliates in thin papery sheets. Leaves deciduous, usually with five to seven lance-oblong leaflets, these 6–12 cm long. Flowers in racemes or panicles 5–10 cm long; staminate flowers five-parted, greenish or brownish green; petals 2–3 mm long. Pistillate flowers three-parted; fruit variable in size, mostly 8–10 mm long, usually reddish tinged. Coastal hammocks, Fla. Keys, north to Brevard and Pinellas Counties, W.I., Mexico, Central America, tropical South America. Winter, spring. *Elaphrium simaruba* (L.) Rose

Various species of Bursera produce a pleasant fragrant resin used locally as a medicine. Resin of **B. simaruba** has been used as a cement substitute for mending glass and china.

105. MELIÀCEAE Mahogany Family

Trees and shrubs with alternate, mostly pinnately compound leaves. Flowers radially symmetrical, bisexual, in axillary cymes or panicles, calyx small, four- to five-parted; petals four to five, free or fused to the staminal tube; stamens mostly eight to ten, usually fused by their filaments. Ovary superior, three- to five-locular; stigma usually capitate or disk shaped. Fruit baccate or capsular, sometimes drupaceous. Almost exclusively restricted to the tropics with about 45 genera and 750 species. Family includes many valuable timber trees, such as the West Indian mahogany and the African mahogany.

1. Leaves once-pinnately compound, leaflets
   entire; fruit a capsule . . . . . .   1. *Swietenia*
1. Leaves twice-pinnately compound, leaflets
   usually serrate or dentate; fruit a drupe .   2. *Melia*

1. SWIETÈNIA Jacq. Mahogany

Large trees with alternate once-pinnate leaves. Flowers white; calyx small, five-lobed, petals, five, free; stamen tube with ten anthers; disk present, cupular or saucer shaped. Ovary five-locular, sessile on the disk. Fruit a capsule, large, woody, seeds ten to fourteen in each cell, winged. Three species in tropical America.

1. **S. mahagoni** (L.) Jacq. West Indian Mahogany—Trees up to

Family 104. **BURSERACEAE.** *Bursera simaruba:* a, flowering branch and mature leaf, × ½; b, pistillate flower containing nonfunctional stamens, × 20; c, staminate flower, × 13; d, mature fruit, × 3½.

Family 105. **MÉLIACEAE.** *Swietenia mahagoni:* a, flowering branch with mature leaves, × ½; b, flower, × 10; c, mature fruit, × ½; d, mature seed, × 1.

25 m tall. Leaves mostly four- to eight-foliolate, the leaflets elliptic to broadly ovate, 4–10 cm long, entire. Panicles 5–20 cm long, many-flowered; calyx 3 mm broad, lobes deltoid, petals obovate, white or greenish, 3–4 mm long. Ovary glabrous. Capsule ovoid, 6–12 cm long, erect. Coastal hammocks, extreme south Fla., in Fla. Keys, W.I. Spring, summer.

2. MÈLIA L.

Trees with alternate, pinnate or twice-pinnate leaves. Flowers conspicuous in axillary panicles, mostly purplish, sepals five to six, petals five to six, free, spreading, staminal tube with ten to twelve anthers, dentate, disk annular. Ovary three- to six-locular, stigmas five- to six-lobate. Fruit drupaceous with a one- to six-locular stone, seeds solitary within each cavity. About ten species in the tropics and warm temperate zones of the Eastern Hemisphere.

1. **M. azedarach** L. CHINA BERRY—Trees up to 15 m tall, usually less, with a broadly rounded crown, young stems stellate pubescent becoming glabrate. Leaves large, twice-pinnately compound with many leaflets; leaflets lance-ovate, 3–8 cm long, incised-serrate or lobed. Flowers fragrant in panicles 10–30 cm long, many-flowered, sepals 2–3 mm long, lanceolate-ovate; petals purplish or lilac or whitish, 8–12 mm long, staminal tube dark purple. Drupe globose, 1–2 cm wide, smooth and yellow. $2n = 28$. Disturbed sites, thickets, old fields, and in cultivation, naturalized from Asia in warmer parts of Western Hemisphere. Spring.

106. MALPIGHIÀCEAE MALPIGHIA FAMILY

Trees, shrubs, or woody vines usually with opposite, simple leaves and often with glands at the base or on the petiole. Inflorescence racemose with bisexual, mostly radially symmetrical flowers, sepals five, often with two glands, petals five, clawed, disk small; stamens mostly ten. Ovary usually of three carpels, rarely two or four. Fruiting carpels winged, or carpels fused into a fleshy or woody drupe. About 63 genera and 800 species, widely distributed in the tropics, especially the American tropics.

1. BYRSONÍMA Richard

Shrubs or trees with young stems and leaves tomentose or sericeous. Leaves entire, stipules intrapetiolar, usually fused. Inflorescence a simple, many-flowered raceme; sepals two-glandular, petals reddish, purple, yellow, or white, reflexed, the limb cordate or reniform, filaments barely fused. Styles subulate, fruit a drupe with one- to three-locular stone. About 100 species in tropical America.

1. **B. cuneata** (Turcz.) P. Wilson, LOCUSTBERRY—Shrubs or small trees up to 10 m tall, mostly smaller. Leaves obovate to spatulate, 2–5 cm long. Racemes 2–6 cm long with petals pink or white becoming rose or yellow, distinctly clawed, the limb reniform. Drupe 4–6 mm wide. Pinelands, especially on limestone soil, south Fla., W.I. *B. lucida* (Sw.) DC.

This is one of our most attractive native shrubs, conspicuous in the spring and early summer for its abundant, multicolored flowering.

107. POLYGALÀCEAE MILKWORT FAMILY

Herbs, shrubs, vines, or small trees with simple leaves. Flowers in racemes, spikes, or panicles with bracts present, bisexual, and bilaterally symmetrical; sepals five, free, the two inner larger, often petaloid, and winglike; petals three to five, upper two free, minute, and scalelike or absent; stamens eight, rarely fewer, fused by their filaments into a staminal tube. Ovary superior, two-locular. Fruit a capsule or drupe, seeds often pilose pubescent. About 10 genera and 680 species mostly in the genus POLYGALA, widely distributed.

1. POLÝGALA L. MILKWORT

Herbs with alternate, whorled, or opposite, usually entire leaves without stipules. Flowers in racemes, often rather small, calyx persistent, of five sepals, the three outer small, the two inner much larger (wings), often colored like the petals; petals three, fused and adnate to the staminal tube, the two upper ones similar, the lower one keel shaped or boat shaped (keel) with a fringed crest; stamens eight, sometimes six, fused by their filaments into a sheath split along the upper side. Ovary two-locular. Fruit a small capsule. About 500 species, worldwide except Australia. *Asemeia* Raf.; *Trichlisperma* Raf.; *Galypola* Nieuwl.; *Pilostaxis* Raf.

Taxonomically a very difficult group, in part, owing to the occurrence of hybridization and polyploidy.

1. Leaves whorled or opposite at least on lower
   nodes.
  2. Flowers rose or greenish; raceme sessile or

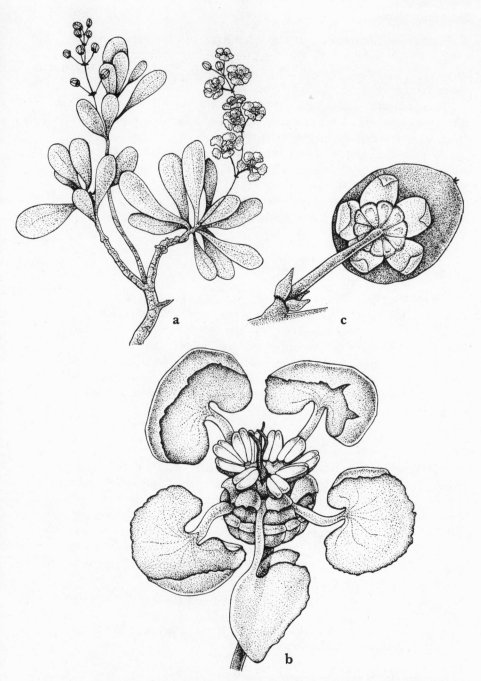

Family 106. **MALPIGHIACEAE.** *Byrsonima cuneata:* a, flowering branch, ×
½; b, flower, × 5; c, mature fruit, lower surface, × 5.

nearly so, densely flowered, bracts persistent . . . .   1. *P. cruciata*
2. Flowers white or greenish white; raceme pedunculate, bracts deciduous . . .   2. *P. boykinii*
1. Leaves all alternate.
  3. Flowers bright yellow, greenish yellow, or orange.
    4. Racemes in terminal compound cymes.
      5. Corolla greenish yellow, wings cuspidate; seeds glabrous . . . . .   3. *P. cymosa*
      5. Corolla bright yellow, wings acuminate; seeds pubescent . . . .   4. *P. ramosa*
    4. Racemes solitary, not in terminal cymes.
      6. Stems over 20 cm tall; wings mucronate or cuspidate.
        7. Wings 8–9 mm long; flowers lemon yellow . . . . . .   5. *P. rugelii*
        7. Wings 6–7 mm long; flowers orange yellow . . . .   6. *P. lutea*
      6. Stems short, less than 20 cm tall; wings acuminate . . . . .   7. *P. nana*
  3. Flowers purple, pink, green, or white.
    8. Racemes in terminal compound cymes   8. *P. baldwinii*
    8. Racemes solitary, not in terminal cymes.
      9. Leaves inconspicuous, subulate scales to linear-subulate leaves less than 2 mm wide.
        10. Corolla usually lavender or pink; wings less than one-half the length of keel . . . . .   9. *P. incarnata*
        10. Corolla usually white or greenish white; wings equaling or exceeding the keel . . . . .   10. *P. setacea*
      9. Leaves conspicuous, linear-lanceolate to oblanceolate, over 2 mm wide.
        11. Keel dissected or fringed; larger sepals two times longer than wide   11. *P. polygama*
        11. Keel truncate, not fringed; sepals as long as wide . . . . .   12. *P. grandiflora*

1. P. cruciata L.—Stems erect, mostly 0.6–3.6 dm tall, glabrous or nearly so. Leaves chiefly in whorls of three to four, lower ones spatulate or obovate, upper ones linear to oblanceolate. Racemes thick cylindrical 1–4 cm long, sessile or nearly so, flowers rose purple or greenish; bracts persistent; wings deltoid, 3–4 mm wide at the base, 4–6 mm long, terminating in a slender cusp. Capsule suborbicular, 2 mm long, seeds about 1 mm long. $2n = 36$. Lowlands, wet fields, Fla. to Tex., Me., Minn. *P. ramosior* (Nash) Small

2. **P. boykinii** Nutt.—Stems erect, mostly 2.5–5.2 dm tall. Leaves various, the lowermost subulate, linear to obovate or suborbicular to elliptic-obovate, the upper ones narrowly lanceolate to linear. Racemes cylindric, 4–25 cm long, 4.5–8 mm wide, bracts subulate, deciduous, flowers greenish white or white, upper sepals ovate, becoming 1.5 mm long; mature wings suborbicular, 2.2–3.1 mm long, keel 2–2.7 mm long. Capsule nearly spheroidal, about 3 mm long, seeds pilose, about 2 mm long. $2n = $ c.28. Pinelands, Fla. to La., Ga. Three well-marked varieties occur in our area:

1. Lowermost leaves elliptic-obovate to suborbicular, often mucronate.
   2. Leaf blades elliptic-obovate, upper blades
      linear . . . .   2a. *P. boykinii*
                       var. *boykinii*
   2. Leaf blades suborbicular, upper blades
      usually narrowly lanceolate . . . .   2b. *P. boykinii*
                       var. *suborbicularis*
1. Lowermost leaves subulate to linear or linear-lanceolate . . . .   2c. *P. boykinii*
                    var. *sparsifolia*

2a. **P. boykinii var. boykinii**—Stems erect, annual or biennial. Racemes mostly 6–25 cm long. Pinelands, Fla. to La., Ga.

2b. **P. boykinii var. suborbicularis** R. W. Long—Stems erect, annual. Racemes mostly 4–6 cm long. Pinelands, hammocks, endemic to peninsular Fla.

2c. **P. boykinii var. sparsifolia** Wheelock—Stems erect or decumbent, sparsely to sometimes densely leafy, upper leaves often scattered, few. Pinelands, endemic to south Fla. *P. praetervisa* Chodat, *P. flagellaris* Small

3. **P. cymosa** Walt.—Biennial 5–10 dm tall with stems simple. Leaves numerous, linear or linear-lanceolate, 3–12 cm long, reduced upward and scattered. Racemes several to many in a cymose panicle, yellow or suborbicular, greenish when dry, wings cuspidate. Seeds glabrous, suborbicular, less than 1 mm long. Pinelands, Fla. to La., Va. *Pilostaxis cymosa* (Walt.) Small

4. **P. ramosa** Ell.—Annual with simple or branched stems 1.5–4 dm tall. Lowermost leaves in a rosette, spatulate to elliptic or linear, obtuse, 7–24 mm long. Racemes numerous, forming a flattish cymose panicle, flowers lemon yellow; bracts persistent; wings 2.9–4 mm long, acuminate. Fruit suborbicular, seeds pilose, 0.6–0.7 mm long. $2n = 68$. Low ground, Fla. to Tex., Del. *Pilostaxis ramosa* (Ell.) Small

Family 107. **POLYGALACEAE.** *Polygala boykinii:* a, flowering plant, × ½;
b, flower, × 12½; c, mature fruit, × 13½.

5. **P. rugelii** Shuttlw.—Stems 2.5–7.5 dm tall. The lower leaves tufted, obovate, somewhat fleshy, the upper ones oblanceolate to lanceolate. Branch leaves narrowly elliptic to linear-lanceolate. Raceme lemon yellow, capitate, very densely flowered, loose at the base, upper sepals becoming 3 mm long, broadly ovate, mature wings oblong to elliptic, 6–6.8 mm long, mucronate. Capsule 1.4–1.7 mm wide, seed fully 1.5 mm long, ellipsoid. $2n = 68$. Moist low ground, endemic to south Fla. *Pilostaxis rugelii* (Shuttlw.) Small

6. **P. lutea** L.—Stems up to 4 dm tall, often clustered. Leaves clustered at the base, obovate to oblong, 2–5.8 cm long, somewhat fleshy. Racemes usually solitary, capitate or thick cylindrical, bright orange yellow, wings abruptly tipped to a sharp point. Capsule 1.5 mm long, seeds short pubescent, plump, 1.5 mm long. $2n = 68$. Pinelands, Fla. to La., N.Y. on coastal plain. *Pilostaxis lutea* (L.) Small

7. **P. nana** (Michx.) DC.—Stems 7–17.5 cm tall. Basal leaf blades tufted, spatulate to obovate, often broadly so, stem leaves few, alternate. Peduncles solitary, 2.3–7.5 cm long, raceme green or yellowish, conic-capitate; upper sepals elliptic-lanceolate to linear-subulate, 3–5.3 mm long, acuminate, wings elliptic-lanceolate to linear-subulate, 7–8 mm long, commonly mucronate, keel fringed, the fringe longer than the petals. Capsule about 1.5 mm wide. Seeds usually less than 1.5 mm long, the body ellipsoid, tipped with a curved beak. $2n = 68$. Low moist pinelands Fla. to La., S.C. on coastal plain. *Pilostaxis nana* (Michx.) Raf.; *P. arenicola* Small

8. **P. baldwinii** Nutt. WHITE BACHELOR'S BUTTON—Plant 2–6.5 dm tall. Leaves various, the lower ones spatulate to obovate, upper leaves lanceolate and acute to acuminate. Racemes numerous, capitate, blunt, white, bracts lanceolate, persistent; upper sepals subulate-ovate to lanceolate, 1.8–2.4 mm long, mature wings elliptic or oblong-elliptic, or broadened upward, 3 mm long, acuminate. Capsule plump, less than 1 mm wide, seed ovoid, sparsely pilose, 0.4–0.5 mm long. Pinelands, Fla. to Miss., Ga. on the coastal plain. *Pilostaxis baldwinii* (Nutt.) Small; *Polygala polycephala* Baldw.; *Polygala baldwinii* T & G; *P. carteri* Small

9. **P. incarnata** L. PROCESSION FLOWER—Smooth, glaucous annual or perennial herb, with stiffish, grooved, slender stems about 3.5 dm tall. Leaves about 5–10 mm long, linear, subulate, glaucous. Racemes nearly cylindric, pale purple or rose purple; claws of petal fused to form a long, slender tube surpassing the wings. Seeds plump, pilose, about 2 mm long. Glades, woodland, Fla. to Tex., N.Y., Kans. *Galypola incarnata* (L.) Nieuwl.

10. **P. setacea** Michx.—Stems slender and smooth up to 5 dm tall.

Leaves scalelike, scattered. Racemes rather small, cylindrical, pink or yellow; upper sepals 1–1.5 mm long, smooth, narrowly lanceolate to ovate, mature wings 1–2 mm long, obovate, acuminate, lateral petals 1–2 mm long, keel 2 mm long. Capsule 2 mm long, seeds plump, densely pilose. Pinelands, Fla. to Miss., N.C. on coastal plain.

11. **P. polygama** Walt.—Stems smooth, clustered, decumbent up to 5 dm tall. Leaves obovate or spatulate below, about 1 cm long, upper leaves linear-oblanceolate. Raceme rose purplish or white in an open, loose, cylindrical raceme, upper sepals 2–3 mm long, ovate; mature wings 4–6 mm long; keel 3–5 mm long, lateral petals 3–4 mm long. Cleistogamus flowers also produced in slender, subterranean racemes. Capsule ovoid, seeds 2–3 mm long. $2n = 56$. Sandy soil, Fla. to Tex., Me., Minn.

12. **P. grandiflora** Walt.—Stems 2–5 dm tall, glabrous to densely pubescent. Upper leaves linear to elliptic, loosely ascending. Flowers greenish to dark purple, upper sepals ovate, broadly egg shaped, becoming 2.5–3 mm long, mature wings 4–7 mm long. Capsule about 3–5 mm long, seeds cylindric, about 2.5 mm long. $2n = 28$. Pinelands, sandy woods, old fields, Fla. to Miss., S.C. *Asemeia grandiflora* (Walt.) Small. The following varieties have been described and occur in our area:

1. Wings greenish or purplish tinged.
  2. Leaf blades elliptic or elliptic-lanceolate   12a. var. *grandiflora*
  2. Leaf blades linear or linear-lanceolate .   12b. var. *angustifolia*
1. Wings dark purple . . . .   12c. var. *leiodes*

12a. **P. grandiflora** var. **grandiflora**—Stems pubescent, up to 5 dm tall. Upper leaves elliptic to elliptic-lanceolate. Flowers greenish or sometimes purple tinged, mature wings 6–7 mm long. Capsule about 5 mm long. Pinelands, Fla. to Miss., S.C. *P. cumulicola* Small, *P. miamiensis* Small

12b. **P. grandiflora** var. **angustifolia** T & G—Stems sparsely to densely pilose. Leaves linear to narrowly lanceolate, glabrous to pilose, about 2–5 cm long. Flowers similar to above, but usually smaller, often light purplish, wings 5–6 mm long, 4–5 mm wide. Pinelands, south Fla., W.I. *P. corallicola* Small

12c. **P. grandiflora** var. **leiodes** Blake—Stems 2–5 dm tall, mostly glabrous or the stem minutely blistered. Upper leaves linear or filiform-linear. Flowers purple; upper sepals at maturity less than 2.5 mm long, mature wings 4–5 mm long, dark purple. Capsule 3–4 mm long, seeds about 2 mm long. Pinelands, south Fla., W.I. *Asemeia leiodes* (Blake) Small

108. EUPHORBIÀCEAE Spurge Family

Trees, shrubs, or herbs sometimes vinelike and twining, monoecious or dioecious, stems and leaves often with milky juice. Leaves various, usually alternate, simple or palmately compound. Inflorescence extremely variable; flowers quite small or large, radially or sometimes bilaterally symmetrical; perianth usually small, of one or two whorls, often dissimilar in flowers of different sexes, or perianth absent, staminate flowers with one to many stamens, often as many as the lobes of the perianth, with disk present or represented by separate glands or lobes. Pistillate flower with usually three-locular ovary, or one to four or more, styles as many as the carpels, disk annular, cupular, glandlike or absent. Fruit usually a capsule or drupaceous and indehiscent.

One of the largest families of angiosperms with over 200 genera and 7000 species; cosmopolitan except in the Arctic, chiefly tropical and mostly woody, including many xerophytes, some of which resemble cacti in general appearance.

1. Trees and shrubs or large herbs over 1 m high.
  2. Some or all the leaves palmately, deeply or shallowly lobed.
    3. Leaf margins entire or irregularly incised.
      4. Stem sap colored, not milky; flowers with petals . . . . . .   1. *Jatropha*
      4. Stem sap milky; flowers apetalous .   2. *Manihot*
    3. Leaf margins finely to coarsely serrate or dentate.
      5. Leaves peltate; fruit echinate . .   3. *Ricinus*
      5. Leaves not peltate, petiole basal; fruit not echinate . . . . . .   4. *Croton*
  2. Leaves not palmately lobed, or leaves apparently absent.
    6. Stems succulent, flattened, or zigzag.
      7. Leaves present; flowers in branching cymes . . . .   5. *Pedilanthus*
      7. Leaves absent; flowers in clusters along margins of branches . . .   6. *Phyllanthus*
    6. Stems not succulent, woody at least below.
      8. Leaves stellate pubescent underneath, sometimes conspicuously so   4. *Croton*
      8. Leaves glabrous underneath, or essentially so.

9. Leaves sessile or nearly so; petiole
   less than 3 mm long . . . .   7. *Stillingia*
9. Leaves with definite petioles, these
   over 3 mm long.
  10. Stems spiny; leaves simple, cor-
      date . . . .   8. *Hura*
  10. Stems not spiny; leaves various,
      not simple cordate.
    11. Inflorescence a spike; leaf mar-
       gins distinctly to obscurely
       serrate.
      12. Petioles over 1 cm long, leaf
         base truncate; fruit a
         drupe . . . . . .   9. *Hippomane*
      12. Petioles less than 1 cm long,
         leaf base cuneate; fruit a
         capsule . . . . .   10. *Gymnanthes*
    11. Inflorescence not spicate; leaf
       margins entire to undulate
       or crenate-dentate.
      13. Petioles over 1 cm long and
         the leaves simple.
        14. Plants monoecious; leaves
            ovate-elliptical, less than
            8 cm long.
          15. Sap never milky; leaves
              without red blotches   6. *Phyllanthus*
          15. Sap milky; leaves espe-
              cially upper ones red
              or colored blotches .   11. *Poinsettia*
        14. Plant dioecious; leaves
            lance-ovate, over 8 cm
            long . . . . . .   12. *Drypetes*
      13. Petioles less than 1 cm long,
         or if more than 1 cm, the
         leaves trifoliolate.
        16. Inflorescence racemose or
            paniculate; leaf trifolio-
            late, apex acuminate .   13. *Bischofia*
        16. Flowers solitary or clus-
            tered; leaf simple, apex
            rounded or truncate .   14. *Savia*
1. Herbs usually less than 1 m high.
17. Leaves deeply three- to five- lobed . . .   1. *Jatropha*
17. Leaves not deeply lobed.
  18. Flowers in involucres (cyathium) that
      simulate a flower; sepals obsolete.
    19. Stems succulent, flattened, or zigzag;
       cyathium strongly zygomorphic .   5. *Pedilanthus*

19. Stems not succulent, flattened, or zigzag.
  20. Leaves all opposite . . . . .   15. *Chamaesyce*
  20. Leaves alternate, at least at the
    base.
    21. Leaves less than 2 cm long, linear
      to narrowly lance-ovate, en-
      tire; cyathium with four to
      five glands . . . . . .   16. *Euphorbia*
    21. Leaves over 2 cm long, lanceo-
      late to broadly ovate, often
      pinnately lobed or toothed,
      upper blades often red
      blotched near base; cyathium
      with a single gland . . .   11. *Poinsettia*
18. Flowers not in involucres; sepals present.
22. Stems and leaves bristly, often with
stinging hairs.
  23. Leaves palmately deeply lobed;
    flowers white, cymose . . . .   17. *Cnidoscolus*
  23. Leaves entire or toothed, not
    lobed; minute flowers green or
    purple, racemose . . . . .   18. *Tragia*
22. Stems and leaves glabrous or pubescent, not bristly or with stinging
hairs.
  24. Leaves distichous, margins entire,
    usually less than 2 cm long . .   6. *Phyllanthus*
  24. Leaves not distichous, margins ser-
    rate or serrulate.
    25. Pistillate flowers with conspicu-
      ous leafy bracts; inflorescence
      spicate or racemose.
      26. Corolla present in staminate
        flowers; inflorescence ter-
        minal . . . .   4. *Croton*
      26. Corolla absent in staminate
        flowers; inflorescence termi-
        nal or axillary.
        27. Stems prostrate or erect;
          leaves orbicular, often
          long petioled . . . .   19. *Acalypha*
        27. Stems erect; leaves lanceo-
          late or lance-ovate, sessile
          or nearly so . . . .   7. *Stillingia*
    25. Pistillate flowers without con-
      spicuous bracts; flowers in axil-
      lary clusters or interrupted
      spikelike racemes.

28. Stems decumbent, ridged:
styles several-branched . . 20. *Caperonia*
28. Stems erect, not ridged; styles
once-branched . . . . 21. *Argythamnia*

1. JATRÒPHA L.

Trees, shrubs, or tall herbs with alternate, simple, entire, toothed, or often deeply palmately lobed leaves, monoecious. Flowers small in dichotomous, often long-stalked cymes; staminate flower with five-fused sepals, petals five, free, stamens eight to ten, disk entire or dissected. Pistillate flower somewhat smaller, perianth similar to staminate flower; ovary two- to three-locular, styles fused near the base, disk cupular or ovoid. Fruit a capsule, seeds one in each locule. About 150 species, pantropical. *Adenoropium* Pohl

1. Leaves shallowly three- to five-lobed or unlobed, base broadly cordate . . . . 1. *J. curcas*
1. Leaves deeply cut below the middle into
   three to eight lobes.
   2. Leaf margins glandular-pubescent, three-
      to five-lobed . . . . 2. *J. gossypiifolia*
   2. Leaf margins smooth, not glandular-pu-
      bescent; seven- to eleven-lobed . . . 3. *J. multifida*

1. **J. curcas** L.—Shrub or small tree up to 8 m tall, usually smaller, with smooth, pale bark. Leaves mostly 8–16 cm long, elliptical to suborbicular, cordate at the base, shallowy three- to five-lobed or angled, with long, slender petioles about as long as the blade. Cymes small, densely flowered; sepals 4 mm long, ovate; petals yellowish green, obovate, densely pubescent within, stamens eight. Ovary glabrous. Capsule 2–4 cm long, somewhat fleshy, seeds pale with black lines. Disturbed sites, south Fla., native of Central America. All year.

This species, together with **J. integerrima** Jacq. with brilliant scarlet flowers and **J. podagrica** Hook. with small red flowers, are cultivated in south Fla. They may not be well established in our flora.

2. **J. gossypiifolia** L.—Herbs often 1 m tall or less, sometimes woody at the base, branched above. Leaves deeply palmately three- to five-lobed, the lobes acute-acuminate, denticulate, with long petioles bearing numerous gland-tipped branched hairs; leaf blades 8–15 cm long; stipules dissected with many gland-tipped filiform divisions. Flowers green, small, in peduncled cymes; sepals ovate, 5–7 mm long, petals obovate, purple; stamens eight. Ovary pubescent, capsule 1 cm wide, smooth. Seeds brown with prominent caruncles. Disturbed sites south Fla., Fla. Keys, naturalized from tropical America. All year.

Forms with glabrous leaves have been named var. **elegans** (Klotzsch) Muell. Arg.

3. J. multifida (L.) Pohl—Shrubs or small trees up to 2 m tall, branched above. Leaves palmately nine- to eleven-lobed, the lobes entire or denticulate, incised; blades up to 15–20 cm wide. Flowers corymbose, reddish purple or red; sepals five, ovate; petals five, up to three times longer than the sepals; stamens eight to ten. Capsule smooth, globose-ovoid, 2–3 cm long. Seeds brown, 2 mm long. Pinelands, disturbed sites, south Fla., naturalized from tropical America. All year.

2. MÀNIHOT Mill.   Cassava, Tapioca Plant

Tall herbs or shrubs usually with tuberous roots and stems producing milky juice, monoecious. Leaves alternate, long petioled, mostly palmately lobed. Flowers in terminal panicles, staminate flower with five-lobed calyx, stamens ten. Pistillate flower with three styles, three carpels, stigma dilated. Fruit capsular, sometimes winged. About 150 species, chiefly in Brazil and Mexico.

1. M. esculenta Crantz, Manioc—Shrub up to 2 m tall from fleshy, elongated, tuberous roots. Leaves deeply three- to seven-parted, lobes 6–15 cm long, linear-lanceolate, acuminate, smooth. Flowers small, 1–2 cm long in spreading panicles. Capsule globose, six-angled, 1 cm wide. $2n = 36,72$. In disturbed sites in south Fla., Fla. Keys, naturalized from Brazil. Spring.

Grown in tropical regions for its fleshy, starchy roots from which tapioca and other foods are made. Tubers must first be boiled; otherwise, they are poisonous.

Hevea brasiliensis (Willd. ex Juss.) Muell. Arg. Para Rubber Tree may be found planted in south Fla., but it is probably not naturalized. It may be recognized by its long-petioled, trifoliolate leaves, panicled, apetalous flowers, and large fruits.

Aleurites fordii Hemsl. Candlenut or Tung-oil Tree is widely cultivated in north and central peninsular Fla. and naturalized to some extent. However, it is not common in our area except in plantings. It produces showy flowers with petals 1.5 cm long, conspicuous three-lobed fruits, and entire, simple, long-petioled leaves.

3. RICÌNUS L.   Castor Bean

Tall, smooth herbs becoming somewhat woody, or shrublike, widely branched, up to 5 m tall, monoecious. Leaves alternate, large, usually 2–4 dm wide, long petioled, peltate, and palmately lobed; stipules large, fused. Flowers apetalous, in racemes or panicled clusters at the end of branches; the lower flowers staminate, the upper ones pistillate. Staminate flowers with three- to five-valvate calyx, stamens many, with filaments branched. Pistillate flower with cleft spathelike calyx. Ovary three-locular, style spreading and usually two-cleft. Fruit a large capsule, 1.5 cm wide, covered with soft spinelike structures, three-lobed; seeds about 1 cm long, lustrous and mottled, with a caruncle. A single species, native of Old World tropics, now pantropical.

1. **R. communis** L.—Characters of the genus. Found commonly in disturbed sites, old fields, along roadsides, canals, and embarkments. All year.

Large plants may be treelike in appearance. The seeds are attractively colored but are poisonous.

4. CRÒTON L.

Trees, shrubs, or herbs with pubescence usually of stellate hairs or scales, usually monoecious. Leaves alternate, often with two glands at the base or at the apex of the petiole; blades entire or toothed, seldom lobed. Flowers small, in dense spikelike racemes or clusters, strongly scented, the staminate in the upper part of the inflorescence, the pistillate flowers below, or the sexes mixed together, staminate flowers with usually five sepals, stamens five to many. Pistillate flowers with unequal sepals, ovary three-locular, styles one to many times forked or divided. Fruit a capsule, seeds smooth with an inconspicuous caruncle. A large genus with about 600 species, pantropical, only a few species extending into the temperate zones.

1. Leaf margins entire, subentire, or undulate.
   2. Leaf blades densely stellate pubescent below.
      3. Blades elliptic or ovate, over 1.2 cm wide; monoecious . . . . . . 1. *C. punctatus*
      3. Blades lanceolate, less than 1.2 cm wide; dioecious . . . . 2. *C. linearis*
   2. Leaf blades sparingly stellate pubescent below . . . . 3. *C. humilis*
1. Leaf margins serrate or crenate.
   4. Some or all the leaves three- to five-lobed 4. *C. lobatus*
   4. Leaves sharply dentate, not lobed . . . 5. *C. glandulosus*

1. **C. punctatus** Jacq.—Perennial herb often woody at the base up to 1 m high, usually less; erect, diffusely branched stems with densely appressed brown pubescence. Leaves 2–5 cm long, thick, on long petioles, elliptic to ovate, entire, densely stellate pubescent. Staminate

flowers few in racemes 1–2 cm long, stamens ten to twelve. Pistillate flowers one to three in racemes usually 1 cm long. Capsule subglobose, 5–8 mm long, seeds 6 mm long. Coastal dunes, beaches, Fla. to Tex., N.C. All year.

2. **C. linearis** Jacq. PINELAND CROTON—Shrubs up to 2 m tall, usually less, dioecious. Leaves 4–7 cm long, narrowly linear, obtuse with yellowish stellate pubescence beneath, smooth above. Racemes 4–10 cm long or longer; staminate flowers with five- to six-triangular sepals; petals longer than sepals, spatulate, pubescent; stamens about fifteen. Pistillate racemes shorter than staminate, mostly 4–5 cm long. Capsule subglobose, yellowish pubescent. Seeds about 3 mm long. Pineland, coastal areas, south Fla., W.I. All year.

3. **C. humilis** L.—Shrubs up to 1 m tall or less. Leaves usually 2.5–4 cm long, elliptic to ovate, rounded or subcordate at the base, margins glandular. Racemes of staminate flowers slender, flowers with fifteen to twenty stamens. Pistillate flowers at base of raceme, usually two to six, short pedicelled, sepals elliptic with marginal glands; styles three. Capsule subglobose, 4–5 mm wide. Hammocks, disturbed sites, south Fla., south Tex., Mexico. All year. *C. berlandieri* Torr.

4. **C. lobatus** L.—Herb with green stems only thinly stellate pubescent, especially above. Leaves deeply palmately three- to five-lobed, nearly glabrous, the lobes oblanceolate, acuminate, petioles 3–10 cm long without glands. Racemes terminal or axillary, ten to twelve; pistillate flowers subsessile. Ovary stellate pubescent. Capsule about 8 mm long, becoming glabrate. Seeds 5 mm long. $2n = 18$. Hammocks, disturbed sites, south Fla., tropical America, tropical Africa. All year.

5. **C. glandulosus** L.—Herbs with stellate pubescence, usually branched, up to 6 cm tall. Leaves 2–4 cm long, oblong-ovate to ovate, rounded or obtuse at the apex, coarsely dentate or crenate with long petioles having two saucer-shaped glands at the base of the stalk. Racemes usually 2 cm long or less, few-flowered and these subsessile; staminate flowers with ovate sepals 2 mm long, stamens ten. Pistillate flowers with spatulate, unequal sepals; ovary pubescent. Capsule 5–6 mm long, subglabrate. Seeds 3–4 mm long. $2n = 18$. Three varieties occur in south Fla:

1. Capsule and calyx pubescent.
    2. Leaves stellate pilose on both surfaces,
       margins serrate . . . .   5a. var. *glandulosus*
    2. Leaves densely stellate pubescent on both
       surfaces, margins crenate-serrate . .   5b. var. *simpsonii*
1. Capsule and calyx glabrate or nearly so . .   5c. var. *floridanus*

5a. C. glandulosus var. **glandulosus**—Widely distributed throughout the warmer parts of America, especially tropical America, highly variable. Found in various sites, pinelands, old fields, sandy soil, disturbed sites, southeast U.S., Mexico, W.I., tropical South America. All year. *C. arenicola* Small

5b. C. glandulosus var. **simpsonii** Ferguson—Endemic to south Fla., especially in dry pinelands. It is conspicuously densely pubescent. All year.

5c. C. glandulosus var. **floridanus** (Ferguson) R. W. Long—Endemic to south Fla. in dry pinelands. Essentially a smooth form with glabrous sepals and fruit. Spring, fall. *C. floridanus* Ferguson

Codiaeum variegatum Blume, the cultivated CROTON, is a popular shrub with yellow, red, or variegated linear to lance-ovate leaves. The blades may be either entire or lobed. Many forms are grown, but it is doubtful that any are truly naturalized in south Fla.

5. PEDILÁNTHUS Poit.

Trees or shrubs with milky sap, stems and leaves rather succulent. Leaves alternate, distichous, stipulate, entire, eglandular. Plants monoecious, cyathium bilaterally symmetrical, involucral tube elongated to a spur with four glands; pistillate flower solitary, central, staminate flowers in five bracteolate cymes each reduced to a single stamen; perianth absent. Pistillate flower pedicellate, carpels three, styles fused most of their length to form a slender column, style tips bifid. Ovary angled, ovules one in each locule. Fruit a capsule. About 14 species in the American tropics, chiefly Mexico. *Tithymalus sensu* Small

1. P. tithymaloides (L.) Poit. ssp. **smallii** (Millsp.) Dressler, JACOB'S LADDER—Erect branching shrub up to 2 m tall, usually smaller, with dark green, terete, succulent, zigzag branches. Leaves 4–8 cm long, broadly ovate, or ovate-oblong, succulent, subsessile, smooth, or the leaves absent and the plant appearing leafless. Partial inflorescences or cyathia in terminal dense cymes, bracts ovate-acuminate, caducous, involucre red, pink, or green, 10–12 mm long, appendage four-glandular. Capsule about 8 mm long, seeds ovoid, 5 mm long. $2n = 34,36$. Hammocks, pinelands, disturbed sites, south Fla., Cuba, naturalized from tropical America. All year. *Tithymalus smallii* (Millsp.) Small

In south Fla. ssp. **tithymaloides**, distinguished by its straight stems, may escape locally from cultivation. The species is popular as a hedge plant.

6. PHYLLÁNTHUS L.

Trees herbs or subshrubs with alternate, entire, often distichous, stipulate leaves, monoecious or dioecious, stems not succulent. Flowers small, apetalous in axillary glomerules, cymes or solitary; staminate sepals four to six, stamens mostly three to five. Pistillate flower with ovary three-locular, disk present. Fruit a capsule, explosively dehiscent, seeds usually three-angled. About 750 species, chiefly Old World tropical regions. *Xylophylla* L.; *Cicca* L.

1. Shrubs or trees.
  2. Stems flattened, leafless; flowers in marginal notches . . . . 1. *P. angustifolius*
  2. Stems with ovate leaves; flowers axillary, not in marginal notches . . . . . 2. *P. acidus*
1. Herbs.
  3. Main stem axes leafless and flowerless, branchlets bearing leaves and flowers (phyllanthoid).
    4. Perennial with numerous stems.
      5. Staminate calyx four-parted; stipules brown, lanceolate . . . . . 3. *P. abnormis*
      5. Staminate calyx five-parted; stipules black, auriculate . . . . . . 4. *P. pentaphyllus* var. *floridanus*
    4. Annual, mostly single-stemmed.
      6. Calyx five-parted; leaves not touch-sensitive . . . . 5. *P. amarus*
      6. Calyx six-parted; leaves sensitive, slowly closing when touched . . 6. *P. urinaria*
  3. Main stem axes bearing leaves and flowers (not phyllanthoid) . . . . . . 7. *P. caroliniensis*

1. **P. angustifolius** (Sw.) Sw.—Shrub up to 3 m tall with reddish brown or greyish branches. Branchlets bipinnate; phylloclades usually 3–10 cm long, elliptic to obovate-lanceolate, obtuse to bluntly acuminate at apex with ten to twenty-five nodes; stipules ovate, 1–2 mm long, ciliate when young. Inflorescence usually of two to five staminate flowers, one or two pistillate ones; calyx of staminate flowers, red or cream colored, stamens three to four. Pistillate calyx yellow green or pinkish; ovary rugulose. Capsule 3–4 mm wide. Hammocks, Key West, W.I. All year. *Xylophylla contorta* Britt.

This species was reported several years ago but has not been found recently.

2. **P. acidus** (L.) Skeels, GOOSEBERRY TREE—Monoecious shrub or

small tree up to 10 m tall with rough grey bark. Deciduous branchlets 25–50 cm long with twenty-five to forty leaves; blades 5–9 cm long, broadly ovate; stipules appressed, deltoid-acuminate, about 1 mm long, dark brown. Flowers unisexual in dense cymules, staminate flowers twenty-five to forty, pistillate one to nine; staminate calyx four-lobed, stamens four. Pistillate flowers up to 5 mm long at anthesis, ovary three- to four-lobed, brownish. Fruit drupaceous, greenish yellow to white. Disturbed sites, pinelands, south Fla., W.I., doubtfully naturalized, from South America. All year. *P. distichus* (L.) Muell. Arg.; *C. acida* (L.) Merrill; *Cicca disticha* L.

3. **P. abnormis** Baillon—Herbs up to 5 dm tall with branchlets up to 10 cm or more long. Blades elliptic to lance-elliptic or narrower, the main stem axis appearing leafless and flowerless, leaves borne only on lateral branches; stipules brown, lanceolate. Flowers in cymules; staminate calyx 2 mm wide, four-parted, the lobes ovate to orbicular. Pistillate calyx 3 mm wide, sepals ovate. Fruit a capsule 2–3 mm wide. Coastal sand dunes, south Fla., W.I., Tex., Mexico. All year. *P. garberi* Small

4. **P. pentaphyllus** C. Wright ex Griseb. var. **floridanus** Webster— Herbs up to 3 dm tall with stems often clustered on a caudex, dioecious. Leaf blades 3–7 cm long, elliptic to obovate or suborbicular, glabrous, dark green; stipules lanceolate, 0.5 mm long, acuminate, entire, black, auriculate. Staminate flowers five to twenty-five in narrow cymules, pistillate solitary in axils; staminate calyx five-lobed, stamens two. Pistillate calyx 1 mm long, disk five-lobed. Capsule ovoid 1–2 mm wide. $2n = 52$. A highly variable taxon. Limestone soils, pinelands, south Fla. All year.

5. **P. amarus** Schum.—Monoecious annual herb up to 5 dm tall. Leaf blades usually 5–11 mm long, elliptic-oblong, obtuse or rounded, light green above, grey or glaucous beneath; stipules ovate-lanceolate about 1 mm long, acuminate. Flowers in cymules of either one or two staminate flowers of one each, staminate and pistillate; staminate calyx five-lobed, subequal, segments ovate, stamens three. Pistillate flowers with deeply five-lobed disk. Capsule trigonous, 2 mm wide. $2n = 26, 52$. South Fla., W.I., tropical America, now a circumtropical weed. All year. *P. niruri sensu* Sw. and *sensu* Small, not L.

6. **P. urinaria** L.—Annual monoecious herb, erect or procumbent up to 5 dm tall. Leaf blades 8–20 mm long, oblong or oblong-obovate to narrower, bright green above, pallid or reddish tinged below; stipules 1–1.5 mm long, triangular-lanceolate, not auriculate, stramineous or brownish colored. Flowers with staminate calyx six-parted, lobes elliptic, yellowish white, stamens three, filaments united into a tube. Pistillate calyx yellowish, ovary spheroidal, verrucose capsule 2 mm wide.

$2n = 52$. Weed reported in various parts of Fla., doubtless in south Fla., W.I., naturalized from Asia. The leaves are touch-sensitive. All year.

7. **P. caroliniensis** Walt.—Annual herb, usually erect up to 5 dm tall with generally smooth stems. Leaf blades 1–2 cm long, obovate to oblong, distichous, entire, narrowed at the base; stipules triangular to lanceolate, pale brown or reddish brown, 1–2 mm long. Flowers paired in the axils, one staminate, the other pistillate; staminate calyx with usually six sepals, these spatulate, stamens three. Capsule subglobose, 1–2 mm wide, seeds sharply angled. $2n = 36$. Moist soil from Pa. and Ill., south to Argentina. Two subspecies are found in south Fla.:

1. Stems glabrous, styles erect . . . . . 7a. ssp. *caroliniensis*
1. Stems pubescent, styles appressed . . . 7b. ssp. *saxicola*

7a. P. caroliniensis ssp. **caroliniensis**—Branches smooth or furrowed, not pubescent, styles ascending or erect. Highly variable populations occurring in disturbed sites, old fields, reported for Key West, and generally found throughout the range of the species, but not common in south Fla. Spring, fall.

7b. P. caroliniensis ssp. **saxicola** (Small) Webster—Branches pubescent or with minute papillae in rows, styles appressed or horizontally spreading, not erect. This is the more frequently encountered form of the species in south Fla., occurring in pinelands, and W.I. Spring, fall. *P. pruinosus sensu* Small, not Poepp.

P. tenellus Roxb., herbaceous with five stamens, is also reported for south Fla. but apparently is not common in our area. It is a common greenhouse weed and is found in citrus groves of central Fla.

7. STILLÍNGIA Garden ex L.

Herbs or shrubs with smooth stems and glandular-serrate leaves, monoecious. Flowers apetalous in terminal spiciform racemes. Pistillate flowers solitary in the lower bracts of an otherwise densely staminate spike; staminate calyx two-lobed, stamens usually two; pistillate calyx three-lobed or none, ovary two- to three-locular. Fruit a capsule, three-lobed, seeds subglobose, caruncle sunk in a ventral depression. About 27 species, chiefly in American tropical regions.

1. **S. sylvatica** L. QUEEN'S DELIGHT—Herbs up to 1 m tall with narrowly lanceolate to lance-elliptic or oblanceolate, glandular-serrate leaves 3–8 cm long. Spikes stout, erect, up to 10 cm long or less; staminate flowers clustered in the axils of minute bracts, each bract with two conspicuous glands. Capsule about 1 cm wide. Two subspecies occur in south Fla.:

1. Leaf blades lance-elliptic, usually over 1.5
     cm wide . . . .   1a. ssp. *sylvatica*
1. Leaf blades narrowly lanceolate, usually
     less than 1.5 cm wide . . . . . .   1b. ssp. *tenuis*

   1a. S. sylvatica ssp. sylvatica—Spikes yellow; occurring in pinelands, sand hills, Fla. to Tex., Okla., Va. $2n = 36$. Summer, fall. *S. spathulata* (Muell. Arg.) Small

   1b. S. sylvatica ssp. tenuis (Small) Rogers—Spikes slender, appearing reddish; occurring in limestone soils, pinelands, endemic to south Fla. All year. *S. tenuis* Small, *S. angustifolia* (Torr.) Wats.

   S. aquatica Chapm. Corkwood has been reported for southwestern Fla. and may occur in our area. It is a shrub up to 2 m tall with narrow, serrulate leaf blades and yellow spikes, usually found in ponds or wet soil.

8. HÙRA L.

   Trees with alternate, petioled, broad simple leaves and often spiny stems, monoecious. Flowers apetalous, staminate in terminal spikes with cupuliform calyx, stamens eight to twenty, pistillate solitary at the base of the staminate spike, calyx broadly cupuliform, coriaceous. Ovary five- to twenty-locular, styles fused in a long fleshy column, branches divergent. Capsule large, cocci separating explosively. Two species in America, cultivated in Old World.

   1. H. crepitans L. Sand Box Tree—Tree up to 20 m with branches bearing sharp cylindric spines, or sometimes unarmed. Leaves 6–20 cm long, 4–8 cm wide, ovate, rounded or cordate at the base, entire or finely seriate, long petioled 6–15 cm long, stipules linear-lanceolate. Staminate spikes bright red, 5–6 cm long, pedicel of pistillate flower 1 cm long, elongating in fruit. Capsule 6 cm wide, deeply sulcate between the cocci. Cultivated and locally naturalized in Miami area from tropical America.

9. HIPPÓMANE L. Manchineel

   Low trees or shrubs with smooth bark and abundant milky poisonous juice, monoecious or dioecious. Leaves alternate, with petioles elongate, bearing a single gland at the apex; blades ovate, abruptly acute or short acuminate, entire or denticulate, pinnately veined. Flowers apetalous, in terminal greenish spikes; spikes 6–13 cm long, bearing usually one to two pistillate flowers sessile at the base and numerous staminate flowers in the upper bracts. Fruit drupaceous, applelike in appearance, yellow or reddish in color, six- to nine-locular. Two or three species in tropical America.

1. **H. mancinella** L.—Small tree with smooth stems. Leaves 5–10 cm long, more or less cordate, abruptly acute or short acuminate. Spikes 6–13 cm long, one to two pistillate flowers; staminate flowers with two- to three-lobed calyx, stamens two. $2n = 22$. Low ground and in hammocks in mangroves, near the coastline, south Fla., W.I., Central America and northern South America.

The milky sap is poisonous if taken internally and also can cause severe skin irritation. Smoke from burning the wood is dangerous to the eyes.

10. GYMNÁNTHES Sw. CRABWOOD

Trees or shrubs with smooth bark and alternate leaves, monoecious sap not especially milky. Flowers apetalous, greenish in terminal or axillary, solitary or clustered, bisexual spikes; staminate flowers in the upper axils, pistillate one to few in the lower bracts, solitary; stamens two to six, usually three. Ovary three-locular, stipitate, ovules one per locule. Seeds subglobose, carunculate, fruit a capsule. About 12 species, tropical America.

1. **G. lucida** Sw. CRABWOOD—Shrub or small tree up to 10 m tall with stiffish, slender, gray branches. Leaves 6–14 cm long, oblong-obovate, subentire or crenate-serrate, often two-glandular at the base. Spikes about 3–4 cm long, densely flowered, reddish brown, becoming yellowish green; staminate bracts three-flowered, stamens three to five in central flower. Ovary of pistillate flower on a rather long gynophore up to 1 cm long in fruit. Capsule 7 mm long, globose; seeds globose 4–5 mm in diameter. Hammocks, limestone soil, south Fla., W.I., Central America, Mexico.

11. POINSÉTTIA Graham

Herbs, shrubs, or small trees with milky sap, with entire, lobed, or serrate leaves, the upper ones often red blotched or discolored, monoecious. Flowers in cupuliform structure (cyathium) simulating a corolla with fused lobes, cyathia solitary, few-clustered or cymose; staminate flower consisting of a single stamen, pedicellate. Pistillate flower pedicellate, three-locular, both staminate and pistillate with one or sometimes few unappendaged glands. Capsule smooth, seeds tuberculate. About 12 species, mostly tropical America, often placed in EUPHORBIA L.

1. Leaves lanceolate to broadly ovate, over 2
   cm long, often pinnately lobed or

toothed; upper leaf blades often red
blotched near the base.
2. Floral bracts green or purple spotted,
never red at base; cyathial gland with a
circular opening . . . . 1. *P. heterophylla*
2. Floral bracts green or red at base; cyathial
gland bilabiate . . . . 2. *P. cyathophora*
1. Leaves linear to narrowly lance-ovate, less
than 2 cm long, entire . . . . . . 3. *P. pinetorum*

1. **P. heterophylla** (L.) Kl. & Gke. PAINTED LEAF—Erect, smooth, branched herb up to 1 m tall. Leaves alternate, highly variable, from linear to ovate, and from entire to serrate or lobed on the same individual; upper blades usually lobed, not red or white blotched near the base. Cyathia solitary, 2–3 mm long, campanulate, glands with circular opening. Capsule 3–4 mm long, seed 3–3.5 mm long. Pinelands, moist shaded soil, Fla. to Ariz., Wis., S. Dak., tropical America. All year. *Euphorbia heterophylla* L.

2. **P. cyathophora** (Murray) Kl. & Gke.—Erect, somewhat pubescent herbs up to 1.5 m tall. Leaves alternate, mostly lobed, varying from oblong to ovate, becoming narrower on the branches, about 5–20 cm long. Cyathia in reduced cymes, each about 3–4 mm long. Capsule 4–5 mm long, seed 3–3.5 mm long. $2n = 28,56$. Disturbed sites, hammocks, old fields, Fla., W.I., tropical America. All year.

3. **P. pinetorum** Small—Erect, smooth herbs up to 1 m tall. Leaves mostly 5–15 cm long, mostly less than 6 mm wide, alternate, linear-lanceolate, upper blades often red blotched or discolored near the base. Cyathia usually less than 3 mm long, capsule 2–3 mm long; seed 2–2.5 mm long. Pinelands, endemic to south Fla. All year.

The common POINSETTIA, **Poinsettia pulcherrima** (Willd.) Graham, is found in cultivation in many varieties and forms. The plants are shrubs with large, showy red or discolored upper leaves just below the inflorescences. They may persist in waste places or near former habitations but apparently are not naturalized.

12. DRYPÈTES Vahl WHITEWOOD, GUIANA PLUM

Trees or shrubs with alternate, usually leathery leaves, dioecious. Flowers apetalous, fascicled in dense axillary clusters. Calyx of staminate flowers four- to five-parted, ciliate, stamens three to twelve or more, disk central, pistillate flowers few in each cluster, calyx four- to five-parted. Ovary one- to three-locular, styles short, nearly obsolete. Fruit becoming drupaceous, indehiscent globose, seeds usually solitary in each locule, not carunculate. About 150 species, pantropical, chiefly Old World.

1. Calyx four-lobed, stamens four; fruit red,
   subglobose, less than 15 mm long . . .   1. *D. lateriflora*
1. Calyx five-lobed, stamens eight to ten; fruit
   white, elongate, over 15 mm long . . .   2. *D. diversifolia*

1. **D. lateriflora** (Sw.) Krug & Urban, GUIANA PLUM—Shrubs or trees up to 10 m tall, bearing leaves lanceolate to ovate, 8–10 cm long, coriaceous, acute, abruptly pointed; leaf bases often oblique, margins of blades entire. Flowers in dense clusters or fascicles; staminate flowers 3 mm wide, calyx four-lobed, stamens four, exserted. Pistillate flowers with two-locular pubescent ovary; drupe red, scarcely fleshy, subglobose, 1 cm long, densely tomentulose. Hammocks, south Fla., Fla. Keys, south Mexico, Central America, W.I. Fall.

2. **D. diversifolia** Krug & Urban, MILK BARK, WHITEWOOD—Shrubs or trees up to 12 m tall with smooth, milky white bark often marked with grayish or brownish patches. Leaves oblong-ovate, coriaceous, 8–12 cm long, rather rigid, entire margin, rounded to obtuse at the tip. Flowers in dense clusters, staminate many, pistillate often solitary or two- to three-flowered clusters; staminate calyx five-lobed, stamens eight. Pistillate flowers with pubescent, one-locular ovary. Drupe ovoid, 3 cm long, white at maturity. Hammocks, south Fla., Fla. Keys, Bahama Islands. Spring. *D. keyensis* Krug & Urban

13. BISCHÓFIA Blume

Trees with alternate, trifoliolate leaves with long petioles, dioecious. Leaf blades ovate, pinnately nerved with crenate-dentate margins. Flowers small, very numerous in axillary paniculate or racemose inflorescence; calyx deeply five-parted, stamens five, free. Ovary ovoid, three- to four-locular, fruit baccate. Two species of large trees of the East Indies and adjacent islands.

1. **B. javanica** Blume—Trees up to 10 m or more with smooth leaves and branches. Leaves trifoliolate, the leaflets ovate-elliptic, acuminate, base acute, about 15 cm long, margins crenate-dentate, terminal leaflet long stalked. Flowers minute, numerous in axillary racemose or panicled clusters. Fruit fleshy, globose, about 1 cm in diameter. Growing near Homestead, Fla., naturalized from Asia and possibly not well established locally in our flora. *B. trifoliata* Hook.

14. SÁVIA Willd.

Shrubs or small trees with alternate, entire, short-petioled, coriaceous leaves, dioecious. Flowers greenish in dense axillary clusters or few-flowered glomerules; staminate flowers with five-lobed calyx, petals five, smaller than sepals, stamens five, exserted. Pistillate flowers few or solitary, disk annular, calyx five-lobed, petals five, ovary three-lobed, ovules two in each locule. Fruit a capsule, seeds one per locule. About 20 to 25 species in the tropics and subtropics.

1. S. bahamensis Britt. MAIDEN BUSH—Shrubs or low trees with smooth, whitish or pale grayish bark. Leaves obovate or ovate up to 3 cm long, pale green and shining above, glabrous. Staminate flower with orbicular sepals 2 mm long; petals somewhat shorter, cuneate. Pistillate flowers with suborbicular sepals and petals, perianth about 2 mm long. Capsule spheroidal, 5–6 mm long. Coastal hammocks, low areas, south Fla., Fla. Keys, Bahama Islands, Jamaica, W.I. Spring.

15. CHAMAESỲCE S. F. Gray   SPURGE

Annual or perennial herbs or shrubs with opposite, oblique leaves and often with milky sap, monoecious. Flowers in cup-shaped involucres called cyathia, which are produced in axillary clusters or cymes; involucre campanulate with four to five lobes and with as many or fewer glands alternate with the lobes; petal-like appendages on lobes; staminate flower consisting of one naked stamen, usually in five groups surrounding a single terminal pistillate flower, without a calyx. Pistillate flower with or without a minute calyx, ovary sessile, three-locular, styles three. Capsule smooth or pubescent. About 250 species, widely distributed. Most species flower throughout the year in south Fla.

Many conservative authors prefer to treat all these species in the inclusive genus EUPHORBIA L.

1. Capsule glabrous.
  2. Stems erect or ascending.
    3. Leaves dentate or serrate, or if entire, then the leaves ligulate.
      4. Capsule less than 1.4 mm long; cyathia usually in peduncled leafless clusters; seeds wrinkled . . . .   1. *C. hypericifolia*
      4. Capsule more than 1.6 mm long; cyathia on leafy laterals; seeds with two to four lateral ridges on each face . . . .   2. *C. hyssopifolia*
    3. Leaves entire or obscurely serrate below the apex, leaves never ligulate.
      5. Leaves and young stems fleshy; stipules conspicuous, membranous, white, to 1 mm long . . . . .   3. *C. mesembryanthemifolia*
      5. Leaves membranous or coriaceous,

young stems not fleshy; stipules inconspicuous, brown, about 0.5 mm long . . . .   4. *C. porteriana*

2. Stems prostrate to decumbent.
   6. Stems stiff and wiry, usually many from a heavy rootstock; stem diameter usually less than 0.5 mm; leaves deltoid .   5. *C. deltoidea*
   6. Stems flexible, few to several from a rootstock; stem diameter up to 13 mm wide.
     7. Seeds terete or obscurely angled; stipules not fused, or barely so at the base, deeply parted or laciniate.
       8. Leaves are fleshy, dissimilar in size; seed 1–1.9 mm long . . . .   6. *C. bombensis*
       8. Leaves not fleshy, all similar in size; seed 1–1.4 mm long . . . .   7. *C. cumulicola*
     7. Seeds angular; stipules fused at least on upper or lower surface of tips of branches, the apex fringed or entire, never laciniate . . . .   8. *C. blodgettii*

1. Capsule pubescent.
   9. Leaves serrate.
   10. Cyathia solitary at leaf nodes, appearing clustered if on congested laterals but not in peduncled glomerules.
     11. Appendages of glands only slightly unequal in size, often much reduced; or, if two appendages much longer than other two, then capsule not fully exserted, splitting one side of cyathium at maturity.
       12. Ovary and capsule pubescent only along the angles.
         13. Stem short pubescent in lines at sides; seeds with deep transverse furrows . . . . .   9. *C. prostrata*
         13. Stem long hirsute, at least in lines at sides; seeds with rippled coat . . . . . .   10. *C. mendezii*
       12. Ovary and capsule pubescent over all.
         14. Capsule not completely exserted, splitting side of cyathium at maturity . . . . . .   11. *C. thymifolia*
         14. Capsule completely exserted at maturity . . . . . .   12. *C. maculata*
     11. Appendages of glands greatly unequal in size, one pair longer than the other pair; capsule fully exserted at maturity.

15. Stems long pilose on upper surfaces; leaf apex acute; cyathia
congested on short laterals . . 13. *C. conferta*
15. Stems short tomentose or strigose
on upper surfaces; leaf apex
rounded or obtuse; cyathia borne
singly or in small groups at upper nodes, or if in laterals, then
not congested . . . . . . 14. *C. adenoptera*
10. Cyathia in peduncled glomerules.
16. Cymules terminal and lateral on
leafless peduncles; stem branching at base; robust, ascending,
large-leaved plants . . . . . 15. *C. hirta*
16. Cymules terminal and on leafy laterals; stem branching freely; mostly
low, decumbent, small-leaved
plants . . . . 16. *C. ophthalmica*
9. Leaves entire or obscurely serrate.
17. Plants strongly suffrutescent, erect or
ascending, 3–10 dm tall . . . . 4. *C. porteriana*
17. Plants herbaceous throughout, or if
woody at the base, then plants of low
height rarely reaching 3 dm tall.
18. Plants robust; stems not wiry, 1–3 mm
in diameter, up to 3 dm long;
leaves 4–9 mm long . . . . . 17. *C. garberi*
18. Plants delicate; stems wiry, scarcely
reaching 1 mm in diameter, up to 2
dm long; leaves 2–5 mm long.
19. Plants closely appressed forming a
dense mat, sometimes becoming
diffuse with age; stems glabrous
or slightly pubescent . . . . 5. *C. deltoidea*
19. Plants erect or decumbent, at most
forming a loose mat; stems vil-
lous hirsute, tips canescent . . 18. *C. pinetorum*

1. **C. hypericifolia** (L.) Small—Herbs with erect or ascending stems
from slender rootstocks. Leaves serrate on few to several stems. Cyathia
1 mm long, numerous, in short-peduncled, leafless glomerules; appendages of the glands obovate to reniform, conspicuous. Seeds wrinkled,
capsule 1.4 mm long or less, smooth. $2n = 16,28$. Disturbed sites, pineland, Fla. to Tex. on coastal plain, Mexico, W.I., tropical America.

2. **C. hyssopifolia** (L.) Small—Herbs with erect or ascending smooth
stems from slender rootstocks. Leaves mostly serrate. Cyathia 1–1.5 mm
long, numerous, grouped few together on dichotomous leafy laterals;
appendages of the glands suborbicular or reniform, unequal. Seeds with
two to four lateral ridges on each face, capsule 1.6 mm or more long,

smooth. Disturbed sites, old fields, widely distributed in south U.S. and Canada, W.I., tropical America.

3. C. mesembryanthemifolia (Jacq.) Dugand—Herbs often woody at base, erect or ascending up to 1 m tall, glabrous, often purplish, and with somewhat fleshy stems. Leaves 8–12 mm long, ovate to elliptic, rather fleshy, entire, subsessile; stipules conspicuous, white, 1 mm long. Cyathia about 1.5 mm long, glands elliptic, appendages white. Capsule 2–3 mm wide, smooth. Coastal beaches, south Fla. and the Fla. Keys, W.I. *C. buxifolia* (Lam.) Small

4. C. porteriana Small—Herbs sometimes suffrutescent with erect or ascending, slender, smooth or pubescent, waxy stems. Leaves ovate to elliptic, surfaces with raised markings but not papillose, margins entire. Cyathia about 1 mm long, smooth; glands red or green, appendages minute to twice the width of the gland. Capsule about 2 mm wide, smooth or pubescent, seeds angled. Pinelands, endemic to south Fla., and Fla. Keys. The following varieties occur:

1. Capsule glabrous.
  2. Leaves ovate-elliptic nearly as wide as long, apex obtuse to rounded; rather sparingly branched . . . . . .   4a. var. *porteriana*
  2. Leaves linear-elliptic, all except the oldest much longer than wide, apex acute; freely branched, often strict and broom-like . . . .   4b. var. *scoparia*
1. Capsule pubescent . . . .   4c. var. *keyensis*

4a. C. porteriana var. **porteriana**—Pinelands, on limestone soils Dade and Monroe Counties, Fla.

4b. C. porteriana var. **scoparia** (Small) Burch—Pinelands, on limestone soils lower Fla. Keys. *C. scoparia* Small

4c. C. porteriana var. **keyensis** (Small) Burch—Coastal scrub, on limestone soils, sandy areas, lower Fla. Keys. *C. keyensis* Small

5. C. deltoidea (Engelm. ex Chapm.) Small—Prostrate perennial herbs, closely appressed and forming dense mats, stems wiry, usually less than 1 mm wide, delicate, puberulent to finely hirsutulous or canescent. Leaves 2–5 mm long, ovate-deltoid to triangular, obtuse entire. Cyathia 1 mm long, solitary at the nodes, appendages minute. Capsule about 1.5 mm wide, pubescent, seeds less than 1 mm long. Two endemic subspecies are recognized for south Fla.

1. Leaves about as long as wide; tight mat form maintained with age . . . . . .   5a. ssp. *deltoidea*
1. Leaves much longer than wide; mats becoming diffuse with age . . . .   5b. ssp. *serpyllum*

5a. C. deltoidea ssp. **deltoidea**—Occurring in pinelands, Dade County, Fla. *C. adhaerens* Small

5b. C. deltoidea ssp. **serpyllum** (Small) Burch—Pinelands only in the lower Fla. Keys. *C. serpyllum* Small

6. C. bombensis (Jacq.) Dugand.—Prostrate or decumbent herbs with smooth, much branched, flexible stems forming mats. Leaves somewhat fleshy, widely different on main stems and laterals, linear-lanceolate to broadly oblong, entire, about 1 cm long or less. Cyathia about 1 mm long, appendages minute or absent. Capsule about 2 mm long, seeds 1.4–1.9 mm long. Coastal beaches, Fla. to Tex., Va., tropical America. *Euphorbia ammannioides* HBK; *C. ammannioides* (HBK) Small

7. C. cumulicola Small—Prostrate or decumbent herbs with smooth, stringlike, flexible stems. Leaves all similar in size and shape, elliptic to elliptic-ovate. Cyathia about 1 mm long, appendages white, narrower than the glands or absent. Capsule about 1.5 mm wide, seeds 1–1.4 mm long. Coastal dunes, endemic to south Fla.

8. C. blodgettii (Engelm. ex Hitchc.) Small—Prostrate or decumbent herbs with smooth, flexible stems. Leaves lance-ovate, elliptic or spatulate. Cyathia about 1 mm long, appendages somewhat larger than the glands, whitish or pinkish. Capsule 1.5 mm wide, smooth, seed about 1 mm long. Widely distributed, Fla., W.I., occasionally in Central America. *Euphorbia blodgettii* Engel. ex Hitchc.

9. C. prostrata (Ait.) Small—Prostrate or decumbent herbs with lines of short pubescence on the stems. Leaves ovate, obovate, or elliptic, serrate. Cyathia about 1 mm long, usually solitary at leafy nodes; appendages subequal, very narrow. Capsule about 2 mm wide, pubescent along the angles, seeds about 1 mm long with deep, transverse furrows. Disturbed sites, sandy soil, Fla. to Tex., a widespread weed in warmer climates. *Euphorbia prostrata* Ait.

10. C. mendezii (Boissier) Millsp.—Prostrate or decumbent herbs with stems hirsute at least in lines or finely pubescent. Leaves ovate or elliptic, serrate. Cyathia solitary at leafy nodes, cyathia about 1 mm long, pubescent; appendages subequal, white or red, usually narrower than the glands. Ovary and capsule pubescent, along the angles. Seed about 1 mm long, surface ripples. Disturbed sites, sandy soil, Fla., Mexico, tropical America. *C. leucantha* (Kl. & Gke.) Millsp.

11. C. thymifolia (L.) Millsp.—Prostrate or decumbent annual herb, much branched and forming mats, generally pubescent. Leaves oblong, serrate, about 1 cm long or less. Cyathia usually solitary at leaf axils; glands small, appendages narrow or absent, inconspicuous. Capsule puberulent throughout, not completely exserted and slitting the side of cyathias at maturity; seeds tetragonus, oblong, with transverse

Family 108. **EUPHORBIACEAE.** *Chamaesyce mendezii:* a, flowering plant, × 1; b, flowering branch, × 5; c, pistillate flower, × 20; d, cyathia with pistillate flowers, × 15.

ridge. $2n = 18$. Common weed in disturbed sites, especially near salt water, Fla., Gulf coast, Mexico, pantropical. *Euphorbia thymifolia* L. This species may not occur commonly in south Fla.

12. **C. maculata** (L.) Small—Prostrate or decumbent herbs often forming mats, with excurrent, pubescent stems. Leaves serrate, ovate-oblong to linear-oblong, mostly 1–1.5 cm long, forming congested leafy laterals in upper portion of stem. Cyathia usually solitary, cleft on one side; appendages of glands white or red, narrow, subequal. Ovary strigose, capsule hirsute-strigose, about 1.5 mm long; seeds tetragonous, oblong, about 1 mm. $2n = 28$. Disturbed sites, hammocks, old fields, Fla. to Tex., Quebec, N. Dak., cosmopolitan weed. *Euphorbia maculata* L.

13. **C. conferta** Small—Prostrate or decumbent herbs with long-pilose stems. Leaves serrate, elliptic to ovate, acute. Cyathia villous, clustered on short laterals, appendages of glands very unequal in size, one pair longer than the other pair, red or magenta colored. Capsule fully exserted at maturity, about 1.5 mm wide, puberulent. Seeds less than 1 mm long. Pinelands, endemic to south Fla.

14. **C. adenoptera** (Bertoloni) Small ssp. **pergamena** (Small) Burch—Prostrate or decumbent compact herbs with tomentose or strigose freely branching stems. Leaves less than 5 mm long, elliptic or ovate-elliptic, serrate. Cyathia borne singly or in small groups in the upper nodes, about 1–1.5 mm long; appendages white or pinkish, petal-like, two much larger than the other pair. Capsule about 1.5 mm wide, densely pubescent, seed about 1 mm long, ridged. Pinelands, south Fla., W.I., tropical South America, Central America.

15. **C. hirta** (L.)Millsp.—Prostrate or decumbent herbs, freely branching near the base, less so above, robust, pubescent. Leaves 1–3 cm long, elliptic to elliptic-lanceolate, often blotched dark red, serrate. Cyathia about 1 mm long or less, in peduncled glomerules or headlike cymes. Appendages small, inconspicuous. Capsule pubescent, about 1 mm wide, seed ovoid, tetragonous. Disturbed sites, old fields, Fla., Mexico, tropical America. *Euphorbia hirta* L.

16. **C. ophthalmica** (Pers.) Burch—Decumbent or low, ascending herbs with freely branching stems. Leaves usually less than 1 cm long, elliptic or elliptic-ovate, sharply serrate, apex acute. Cyathia about 1 mm long or less in terminal, peduncled glomerules or cymes or on leafy laterals. Capsule about 1 mm wide; seeds 1 mm long. Disturbed sites, Fla., W.I., South America.

17. **C. garberi** (Engelm. ex Chapm.) Small—Prostrate, decumbent, or low, ascending, robust herbs with pubescent stems. Leaves ovate, 4–9 mm long, entire, or obscurely serrate. Cyathia about 1.5 mm long, solitary at nodes; appendages minute or absent. Capsule 1.5 mm wide, pu-

bescent, seeds smooth or with transverse ridges but not wrinkled. Pinelands, hammocks, endemic to south Fla.

18. C. pinetorum Small—Erect, decumbent, or ascending herbs with villous-hirsute stems, at most forming a loose mat, tips canescent. Leaves reniform, deltoid, ovate, to orbicular, conspicuously pubescent. Cyathia 1 mm long or more, pubescent, solitary at the nodes; appendages very narrow. Capsule fully 2 mm wide, pubescent, sharply three-lobed, the angles acute, seeds 1 mm long. Pinelands, Dade County, endemic to south Fla.

16. EUPHÓRBIA L. Spurge

Perennial herbs, shrubs, or small trees, sometimes cactuslike, more often with smooth stems bearing alternate leaves, sap milky and often poisonous, monoecious. Inflorescence umbel-like or cymose, flowers in a cupuliform structure (cyathium) simulating a corolla with fused lobes; staminate flowers few to many near the base of the cyathium, each represented by a single stamen with an anther at the end of a narrow stalk. Pistillate flower one, pedicellate, three-locular, the pedicel elongating and exserted, styles three, flowers surrounded by an involucre with five lobes, margin cyathium, bearing four to five nectariferous glands sometimes with petaloid marginal appendages. Seeds often carunculate. Over 1000 species, cosmopolitan in temperate and tropical areas. *Tithymalopsis* Kl. & Gke., *Galarhoeus* Haw.

1. Leaves very numerous, imbricate . . . 1. *E. polyphylla*
1. Leaves not imbricate, few to many but never
   very numerous.
  2. Leaves opposite; branches prostrate or
     spreading . . . . 2. *E. ipecacuanhǫe*
  2. Leaves alternate.
    3. Glands of the involucre with petal-like
      appendages . . . . 3. *E. corollata*
    3. Glands of the involucre without petal-
      like appendages . . . . . . 4. *E. trichotoma*

1. E. polyphylla Engelm.—Herbs with stout stems up to 3 dm tall, tufted and very leafy in appearance. Leaves mostly 5–15 mm long, blades linear to narrowly spatulate. Cyathia 1.5–2 mm long, appendages on glands white or pinkish white, oblong. Capsule 4–5 mm wide, seeds mostly 2.5 mm long. Pinelands, endemic to south Fla. All year. *Tithymalopsis polyphylla* (Englem.) Small

2. E. ipecacuanhae L. Wild Ipecac—Larger stems subterranean, aerial stems up to 3 dm long, diffusely and dichotomously branching.

Leaves 3–5 cm long, ovate to oblong or linear, green or purplish green. Cyathia solitary on elongate peduncles, cupules 1.5 mm tall, each of five glands with a narrow, colored appendage. Drier sites, pinelands, Fla. to N.J. on coastal plain. A highly variable species. Spring, fall. *Tithymalopsis ipecacuanhae* (L.) Small

3. **E. corollata** L.—Herbs up to 1 m from deep roots, stems smooth. Leaves sessile or subsessile, ovate or oblong, glabrous. Inflorescence umbellate, cyathia very numerous, 1.2 mm high with conspicuous, white, suborbicular, petaloid appendages up to 1 cm long. Capsule 3–4 mm wide. Drier sites, clearings, disturbed areas, Fla. to Tex., N.Y., Minn., Nebr. Summer. *Tithymalopsis corollata* (L.) Small

4. **E. trichotoma** HBK—Herbs with stems up to 4 dm tall. Leaves small, 0.5–1.2 cm long, elliptic or cuneate, entire. Cyathia in terminal cymose clusters or solitary, cupule 2 mm high. Capsule 4 mm wide becoming wrinkled, seeds subovoid or orbicular about 1.5 mm wide. Sandy soil, coastal areas, south Fla., Fla. Keys, W.I., Mexico. All year. *Galarhoeus trichotomus* (HBK) Small

A number of species of EUPHORBIA are cultivated in south Fla. and perhaps may escape locally or persist near former habitation. These include **E. lactea** Haw., cactuslike and bearing spines on branches prominently three- to five-angled, succulent; **E. milii** Des Moulins, CROWN OF THORNS, a low shrub or sometimes climbing with thick, woody stems and stout spines, the cyathia bear bright red bracts resembling petals; **E. tirucalli** L., PENCIL CACTUS or INDIAN TREE SPURGE, succulent, treelike, but with unarmed leafless branches produced in a brushlike crown.

17. CNIDÓSCOLUS Pohl

Trees, shrubs, or perennial herbs monoecious with milky sap, and with simple, alternate, coarsely serrate or palmately veined and lobed leaves; stems with stinging hairs. Flowers in branching terminal cymes, the staminate above, the pistillate below. Staminate flowers with five-lobed petaloid calyx, eight to ten or more stamens fused at the base. Pistillate flowers with three-locular ovary, disk annular. Fruit a three-seeded capsule. Seeds with a caruncle. About 50 species, chiefly tropical America. *Bivonea* Raf.

1. **C. stimulosus** (Michx.) Engelm. & Gray, TREAD SOFTLY— Plants armed with abundant, long, stiff, stinging hairs throughout, although somewhat less on the leaves; stems up to 1 m, usually much less, branched or simple. Leaves deeply three- to five-lobed, blades suborbicular in outline. Cymes few-flowered, calyx of staminate flowers white,

salverform, stamens ten, unequal. Fruit prismatic, seeds 6–8 mm long, subcylindric. $2n = 36$. Pinelands and beaches, disturbed sites Fla. to La., Va. chiefly on the coastal plain. All year. *Bivonea stimulosus* (Michx.) Raf.; *Jatropha stimulosus* Michx.

18. TRÀGIA L.

Perennial herbs or shrubs frequently with stinging hairs and often twining, monoecious. Flowers in slender racemes that arise opposite the leaves with many staminate flowers above, few pistillate ones near the base, staminate calyx three- to six-lobed, stamens five or many, petals absent. Pistillate calyx three- to eight-lobed, ovary three-locular, bearing stinging hairs, petals absent, styles three, connate at the base, recurved. Capsule three-lobed and three-seeded, seeds not carunculate. About 125 species, pantropical and in warm temperate North America.

1. Leaves orbicular or elliptic, margins dentate
    or serrate, bases cordate or truncate . .   1. *T. saxicola*
1. Leaves linear to elliptic-lanceolate, margins
    entire or irregularly serrate, bases cuneate.
    2. Leaf blades linear and entire or undulate   2. *T. linearifolia*
    2. Leaf blades elliptic and irregularly serrate   3. *T. urens*

    1. T. saxicola Small—Stems very slender and wiry, up to 2 dm tall or more. Leaves ovate to suborbicular or orbicular, 1–3 cm long, margins sharply dentate or serrate, leaf bases cordate or truncate. Staminate calyx 3–4 mm wide, sepals linear or linear-lanceolate. Capsule 7–8 mm wide, hirsute. Pineland, endemic to south Fla. and Fla. Keys. All year.

    2. T. linearifolia Ell.—Stems up to 5 dm tall. Leaves linear or linear-lanceolate, 3–12 cm long, entire or undulate. Staminate calyx 3–4 mm wide; pistillate calyx 5 mm wide, sepals ovate. Capsule 7–8 mm wide, pubescent. Sandy pineland, Fla., Ala. on coastal plain. All year.

    3. T. urens L.—Stems up to 4 dm tall, thinly pubescent. Leaves sessile or subsessile, elliptic to ovate or oblong, irregularly lobed or even subentire, variable, leaf base cuneate, blades 2–5 cm long. Inflorescence terminal on lateral branches, staminate calyx four-parted and with two stamens, pistillate calyx 4–5 mm wide. Capsule 8–10 mm wide, pubescent. Pinelands, dry soil, Fla. to Tex., Va. This species and **T. linearifolia** Ell. may be conspecific. All year.

19. ACALỲPHA L.   Three-Seeded Mercury

    Annual or perennial herbs, shrubs, or trees, monoecious with alternate, petiolate, mostly serrate leaves. Flowers minute in axillary or terminal, often conspicuous spikes, racemes, or panicles with each cluster of flowers or the flower subtended by a bract. Staminate flowers with four sepals and four to eight stamens. Pistillate flowers sessile, with three to five sepals, bracts often conspicuous, styles irregularly branched. Ovary smooth, three-locular. Fruit a three-locular capsule. About 400 species, chiefly America but widely distributed.

1. Stems prostrate, spreading; spikes bisexual . 1. *A. chamaedrifolia*
1. Stems erect; spikes unisexual.
  2. Stems and petioles finely pubescent; pistillate bracts cleft to about the middle . 2. *A. ostryifolia*
  2. Stems and petioles glabrous or nearly so; pistillate bracts cleft to base . . . . 3. *A. setosa*

    1. A. chamaedrifolia (Lam.) Muell. Arg.—Perennial herb with stems and branches prostrate or chiefly so, about 5–30 cm long. Leaves 8–20 mm long, ovate or narrower, elliptic-lanceolate, margins serrate-crenate. Bracts up to 7 mm long, serrate. Capsule about 2 mm wide, seeds small, 1 mm. Sandy soil, south Fla., W.I. All year.

    2. A. ostryifolia Riddell—Annual with stems erect, up to 6 dm tall, branched above, and finely pubescent. Leaves thin, widely spreading or drooping, ovate or ovate-oblong, long petioled, finely and regularly serrate or dentate. Staminate spikes short, axillary. Pistillate flowers in terminal spikes 3–8 cm long and a few in short axillary clusters, bracts cleft into thirteen to seventeen linear divisions, tending to enclose the capsule. Capsule deeply three-lobed, 3–4 mm wide, echinate. Disturbed sites, pineland, Fla. to Tex., N.J., Kans. All year.

    3. A. setosa A. Rich.—Annual with stems erect up to 8 dm tall, simple or somewhat branched, puberulent or glabrous. Leaves 3–10 cm long, ovate to suborbicular, finely serrate, acute. Staminate spikes axillary, short, about 1 cm long; pistillate spikes terminal, 3–6 cm long. Bracts of fruiting spikes 5–6 mm long, deeply parted, lobes seven to eight, filiform. Capsule 2 mm wide, pubescent. Disturbed sites, weedy areas, hammocks, south Fla., Mexico, tropical America. All year.

20. CAPERÒNIA St. Hil.

    Annual or perennial herbs with alternate, narrow, serrate leaves, monoecious. Flowers small, in pedunculate racemes or spikes, axillary, the individual flower greenish and solitary within bracts; staminate in the upper parts of the inflorescence, pistillate few in the lower part; sepals five, petals five, stamens ten. Ovary three-locular. Fruit a three-lobed capsule, hispid or echinate. Seeds subglobose, not carunculate.

About 35 species in the tropics of America and Africa, usually in wet soil or shallow water.

1. **C. palustris** (L.) St. Hil.—Erect annual up to 1 m tall, simple or with spreading or decumbent pubescent branches. Leaves petiolate, elliptic to ovate, crenate-dentate, the upper ones linear-lanceolate to lanceolate; stipules broadly ovate, acuminate, 5 mm long. Flowers in spikelike racemes, interrupted; staminate sepals up to 2.5 mm long, petals obovate, 3–4 mm long. Pistillate petals somewhat smaller, white, spatulate-ovate; ovary covered with fusiform glands. Seeds 2–3 mm wide. Capsule 7–8 mm wide, setulose. Low pinelands, ditches, glades, south Fla. to Tex., W.I., tropical America. All year.

C. castaneifolia (L.) St. Hil., a shorter plant with smooth stems, has also been collected in our area. It is a native of tropical America.

21. ARGYTHÁMNIA P. Browne

Shrubs or herbs, monoecious, usually abundantly pubescent and with alternate, short-petiolate leaves. Flowers in mostly bisexual racemes with staminate above and one to few pistillate flowers below; staminate calyx five-parted, petals five, disk dissected, stamens five to fifteen. Pistillate sepals five, petals five, ovary three-locular, capsule three-lobed, seeds subglobose. About 50 species in tropical and temperate America. *Ditaxis* Vahl

1. **A. blodgettii** (Torr.) Chapm.—Stems erect up to 6 dm tall with elliptic to ovate or spatulate leaves 2–5 cm long. Staminate calyx 7–8 mm wide, sepals lanceolate, petals ovate, somewhat shorter than the sepals; pistillate sepals linear-lanceolate to lanceolate, 5–6 mm long. Capsule 4–5 mm wide. Margins of hammocks, low moist soil, pineland, endemic to south Fla. and Fla. Keys. All year.

109. CALLITRICHÀCEAE WATER-STARWORT FAMILY

Aquatic or sometimes terrestrial herbs with slender, glabrous stems. Leaves opposite, entire, linear to suborbicular or spatulate. Flowers axillary, solitary, minute, bisexual or unisexual; perianth none, bracts two or none, stamen one. Ovary four-locular with one ovule in each locule, styles two, papillose. Fruit indehiscent, at maturity breaking into one-seeded nutlets. Family consisting of a single genus.

1. CALLÍTRICHE L. WATER STARWORT, WATER CHICKWEED

Characters of the family. About 20 species of wide distribution in both hemispheres but chiefly in the temperate zones. The genus appears to be poorly represented in south Fla., but the following species may be expected to occur in the swamps, shallow water of ponds and lakes, and on wet soil.

1. Fruit distinctly pedunculate; plants terrestrial in damp soil . . . . 1. *C. deflexa* var. *austini*
1. Fruit sessile or subsessile; plants aquatic or terrestrial in damp soil.
  2. Plants aquatic, submerged leaves linear, floating leaves spatulate or obovate . 2. *C. heterophylla*
  2. Plants terrestrial, all leaves essentially similar, broader than long . . . . 3. *C. peploides*

1. C. deflexa A. Braun var. **austini** (Engelm.) Hegelm.—Stems prostrate or ascending, often as dense tufts or in mats, 3–5 cm long. Leaves 3–5 mm long, oblanceolate or spatulate, three-veined, obtuse at the apex. Peduncles erect, half as long as or as long as the fruit. Fruit 0.5–0.7 mm long, notched at each end. Damp, shaded soil, wet banks, Fla. to Mexico, Mass., Ohio. A very small and inconspicuous plant easily overlooked when growing with other plants. *C. terrestris* Raf.

2. C. heterophylla Pursh—Stems submerged or floating about 1–2 dm long. Leaves of submerged stems linear, 5–10 mm long, floating leaves spatulate, tapering, sometimes crowded in a cluster or rosette. Fruit obovate to orbicular, about 1 mm long, margins rounded on each half and separated by a narrow shallow groove. Quiet water, widely distributed in North America and extending into South America.

3. C. peploides Nutt.—Stems prostrate, very slender and matted. Leaves of one kind, generally obovate, oblanceolate, or elliptic, mostly 2–3 mm long. Fruit orbicular, sessile or nearly so, margins rounded on each half. Low ground, moist soil, or even shallow water, Fla. to Tex., Ark.

110. EMPETRÀCEAE CROWBERRY FAMILY

Shrubs or subshrubs heathlike in appearance with small, crowded, linear and strongly revolute leaves. Flowers minute, unisexual, radially symmetrical in few-flowered, headlike, terminal inflorescence, sepals two to six, petals absent, stamens two to four, filaments free. Ovary two- to nine-locular, globose. Fruit drupaceous. A small family of three genera and four species in the arctic and temperate zones, ranging southward.

1. CERATÌOLA Michx. ROSEMARY

Shrubs 2–15 dm tall with numerous bushy branches and bearing crowded, evergreen, narrow leaves 8–12 mm long with revolute margins which cause the blades to appear tubular and needlelike. Flowers two to three in the axils, red or yellowish, with two sepals about 1 mm long, two stamens exserted. Ovary two-locular, fruit a drupe 2–3 mm wide with two nutlets. A single species in southeast U.S. coastal plain.

1. **C. ericoides** Michx.—Characters of the genus. $2n = 26$. Found in dry soil, scrub vegetation, and white sand associations, often in acid soil; widely distributed in Fla. All year.

A peculiar shrub resembling a gymnosperm in appearance because of its linear, revolute, tubelike, divergent leaves. It is not as common in our area as it is in central Fla., where it is a conspicuous element of the scrub vegetation.

111. ANACARDIÀCEAE CASHEW FAMILY

Trees or shrubs, rarely vines, with alternate, simple, trifoliolate or pinnately compound leaves. Flowers small, bisexual or unisexual, often polygamous, usually radially symmetrical in racemes or panicles; calyx usually five-parted; petals usually five-parted, free; disk annular; stamens mostly twice as many as the petals, inserted at the base of the disk: filaments free, reduced, sterile or absent in pistillate flowers. Ovary superior, ovoid, one-locular, or sometimes two- to five-locular, styles one to three. Fruit one- to five-locular, usually drupaceous. Plants often with poisonous or caustic oil or sap. About 70 genera and 600 species, chiefly pantropical. *Spondiaceae*

1. Leaves simple, blades lanceolate . . . .   1. *Mangifera*
1. Leaves trifoliolate or pinnately compound.
  2. Leaves trifoliolate . . . .   2. *Toxicodendron*
  2. Leaves chiefly pinnately compound.
    3. Leaflets mostly eleven to twenty-one,
      rachis winged . . . .   3. *Rhus*
    3. Leaflets mostly five to nine, rachis not
      conspicuously winged.
      4. Mature leaflets over 4 cm wide, bases
        truncate; fruits yellowish brown,
        elongate . . . .   4. *Metopium*
      4. Mature leaflets less than 4 cm wide,
        bases tapering or cuneate; fruits
        reddish, spherical.
        5. Leaflets unequal-sided (oblique),
          obovate to oblong-elliptical; ra-
          cemes or panicles few-flowered,
          corollas red or purplish . . .   5. *Spondias*
        5. Leaflets equal-sided, lanceolate to

lance-ovate;   panicles   densely
flowered, corollas white .   .   .   6. *Schinus*

1. MANGÍFERA L.

Trees polygamodioecious with alternate, petiolate, simple, ever-green leaves. Flowers in terminal, branching panicles, bracteate; calyx four- to five- parted, petals four to five, spreading, disk lobate; stamens one or four to five inserted on the margin of the disk. Ovary one-locular, compressed, style lateral and curved. Fruit drupaceous, reniform or ovoid, fleshy, often quite large. About 40 species, tropical Asia.

1. M. indica L. MANGO—Trees up to 15 m tall, often with dense spreading crown. Leaves 10–20 cm long or more, linear-lanceolate to oblong-lanceolate or narrowly elliptic, subcoriaceous, glabrous, acute or acuminate. Flowers whitish green or yellowish green in large panicles, sepals ovate, 2–3 mm long, petals elliptic, 5 mm long, fertile stamens one to two and three to four staminodia usually present. Drupe varying greatly in size and color, usually red or pink tinged on green or yellow. $2n = 40$. Widely cultivated and naturalized in hammocks, south Fla. and the Fla. Keys, introduced from south Asia. Summer.

The fruits are edible; the flowers, seeds, and bark have medicinal use in India; the leaves are a source of yellow dye; and the wood is used in carpentry.

2. TOXICODÉNDRON Mill.

Trees, shrubs, or woody vines. Leaves compound, trifoliolate or pinnate. Flowers bisexual or unisexual, often polygamous, in usually dense panicles; calyx five-parted, petals five, spreading, disk annular, stamens five or ten, free. Ovary globose, one-locular, styles free, short. Fruit a drupe, small and somewhat fleshy to dry, seeds ribbed. About 25 species, Asia and North America.

The resinous sap is poisonous and can cause skin irritation.

1. T. radicans (L.) Kuntze ssp. **radicans**, POISON IVY—Vines or low shrubs climbing by aerial roots. Leaves trifoliolate, blades thin mem-branous, coarsely toothed, lobed, or entire variable, 3–20 cm long. Pani-cles axillary or terminal sepals 1–2 mm long, ovate, petals white or greenish white, 3–4 mm long. Drupes greenish white, 3–6 mm in di-ameter. $2n = 30$. Many sites, pinelands, hammocks, swamps, disturbed areas, Fla. to La., Nova Scotia, Minn., W.I. Spring, supper. *Rhus radicans* L.

3. RHÙS L.

Trees or shrubs with pinnately compound leaves, dioecious or polygamous. Flowers in dense terminal panicles, sepals mostly five, petals five, white or greenish white. Ovary globose, pubescent, one-locular, styles short. Fruit a pubescent drupe, seeds smooth. About 125 species in warm temperate and temperate regions. *Schmaltzia* Desv.

1. **R. copallina** L. var. **leucantha** (Jacq.) DC. SOUTHERN SUMAC—Shrub or small tree up to 8 m tall, with young stems closely pubescent. Leaves pinnately compound; with fifteen to thirty-three leaflets, oblong to lanceolate or linear lanceolate, 3–8 cm long, entire or with a few teeth; leaf rachis winged, pubescent, interrupted at each pair of leaflets. Inflorescence a panicle up to 15 cm long; fruits red, pilose. Drier soil, hammocks, south Fla., W.I. All year. *R. leucantha* Jacq.

4. METÒPIUM P. Browne

Shrubs or trees polygamodioecious or dioecious with stout-petiolate, odd-pinnate leaves, usually clustered near the tips of branches, persistent; sap caustic. Flowers in open axillary panicles; sepals five, fused, petals five, twice the length of the sepals, disk annular; stamens five. Ovary one-locular, stigma three-lobed. Fruit drupaceous, oblong to obovoid, glabrous, shining. Three species in south Fla., W.I., Mexico, Central America.

1. **M. toxiferum** (L.) Krug & Urban, POISONWOOD—Shrub or tree up to 14 m tall with resinous sap. Leaves with three to seven ovate, coriaceous leaflets, these 3–9 cm long. Flowers in axillary panicles 1–2 dm long; sepals reniform or suborbicular, petals elliptic or ovate, greenish yellow. Drupe 10–15 mm long. Hammocks, pinelands, sand dunes, widely distributed in Fla. Keys and south Fla., north to Martin County; also in W.I. All year.

All parts of the plant and the sap can produce a painful irritation similar to poison ivy and should be avoided.

5. SPÓNDIAS L.

Trees with alternate, odd-pinnate, smooth leaves. Flowers polygamous in terminal or lateral panicles; calyx four- to five-cleft, small; petals four to five, spreading; disk crenate; stamens nine to ten. Ovary four- to five-locular, styles four to five. Fruit drupaceous, fleshy. About eight species in tropical regions.

1. **S. purpurea** L. HOG PLUM—Shrub or tree up to 12 m tall with smooth, grayish or whitish bark. Leaves with five to twelve pairs of sub-sessile or very short-petiolate leaflets highly variable in shape, usually oblong to obovate. Panicles or racemes small, narrow, bright red or reddish purple; petals 3 mm long, drupe obovoid, red, purple, or yellow,

Family 111. **ANACARDIACEAE**. *Metopium toxiferum:* a, fruiting stem, ×
½; b, staminate flower, × 8½; c, pistillate flower, × 10; d, mature fruit, ×
4.

3–3.5 cm long. Shell mounds, disturbed sites, naturalized in south Fla., introduced from tropical America.

6. SCHÌNUS L. PEPPER TREE

Trees or shrubs usually dioecious with alternate, odd-pinnate or rarely even-pinnate compound leaves. Flowers unisexual, bracteate, in terminal or axillary panicles. Calyx five-parted, persistent petals five, longer than the sepals, disk annular; stamens ten, in two series, inserted on the lobate disk. Ovary one-locular, styles three. Fruit drupaceous, globose, rather small and usually oily. About 28 species in tropical and temperate South America.

1. **S. terebinthifolius** Raddi, BRAZILIAN PEPPER TREE—Shrub up to 3 m tall with arching branches. Leaves pinnately compound with three to eleven membranaceous, lanceolate to elliptic, sessile or subsessile leaflets about 2–4.5 cm long. Flowers white, usually forming large, terminal panicles; sepals deltoid, 1 mm long, glabrous; petals 1.5 mm long; drupe red or orange red, about 6 mm wide. Introduced from tropical America, widely established and spreading in most soil types of peninsular Fla. *Rhus terebinthifolia* Schlecht. & Cham.

This species now occupies many sites formerly taken by native plants. It is an aggressive invader, spreading rapidly; it can be found in mangrove associations, hammocks, pineland, and very commonly in old fields and other disturbed areas in peninsular Fla.

112. AQUIFOLIÀCEAE HOLLY FAMILY

Trees or shrubs with alternate, simple, mostly evergreen leaves. Flowers bisexual or unisexual, radially symmetrical, often in axillary cymes or sometimes solitary; sepals four to five, small, imbricate, petals four to five, inconspicuous; stamens equal in number and alternate with the petals. Ovary often three-locular, superior, globular or ovoid. Fruit drupaceous. Three genera and 300 species in temperate and tropical regions.

1. ÌLEX L. HOLLIES

Shrubs or trees with usually coriaceous, persistent, entire, dentate or spinose-dentate leaves. Flowers bisexual or unisexual, fascicled, umbellate, or racemose inflorescence, whitish or yellowish, calyx persistent, four- to five-lobed, corolla four- to six-lobed, rotate; stamens four to six. Ovary four- to six-locular, subglobose; stigmas equal the number of locule. Drupe globose with four to eight smooth or ribbed nutlets. Almost 300 species, chiefly in South America.

1. Leaves evergreen, persistent, lanceolate to
   elliptic or ovate to orbicular.
  2. Larger leaves crenate toothed or teeth
     spreading.
    3. Leaf margins crenate toothed . . .   1. *I. vomitoria*
    3. Leaf margins deeply sinuate, teeth
       spreading . . . .   2. *I. opaca*
  2. Larger leaves entire to obscurely or distinctly crenate or toothed, especially
     above the middle of the blade.
    4. Leaves usually 3–6 cm long, distantly
      serrate above the middle of the blade.
      5. Drupe less than 6 mm in diameter .   3. *I. glabra*
      5. Drupe 6–7 mm in diameter . . .   4. *I. ambigua*
    4. Leaves usually 5–10 cm long, entire or
      distantly serrate.
      6. Drupe red or yellow; blades oblanceolate, petioles usually less than 7
        mm long . . . .   5. *I. cassine*
      6. Drupe black or purple; blades elliptic-ovate, petioles usually over 7
        mm long . . . .   6. *I. krugiana*
1. Leaves deciduous, narrowly lanceolate,
   spatulate . . . .   7. *I. decidua*

1. **I. vomitoria** Ait. YAUPON—Shrubs or small trees up to 8 m tall with leathery, elliptic-lanceolate, elliptic, or ovate leaves about 2–4 cm long, obtuse, smooth, dark green, finely crenate-serrate, dark green and shiny above and pale green below. Staminate flowers few to several in cymes, pistillate cymes subsessile. Drupe red, about 5 mm wide, nutlets grooved. Cultivated and occasional in low ground near the coastal plain, Fla. to Tex., Ark., Va. Spring.

2. **I. opaca** Ait. AMERICAN HOLLY—Shrubs or trees up to 20 m tall with leathery, evergreen leaves. Leaves oval to broadly lanceolate with remote spiny teeth, or sometimes subentire, mostly 5–10 cm long. Flowers fascicled in axils or along bases of young stems; calyx four-parted, acute-ciliate. Drupes 7–10 mm in diameter, red or yellow. Moist woodlands, Fla. to Tex., Mass., Mo., Okla. Spring. Commonly planted, especially north of the range of this manual, but apparently not common in our area.

3. **I. glabra** (L.) Gray, GALLBERRY, INKBERRY—Shrubs up to 2 m tall with finely pubescent stems. Leaves about 2–5 cm long, leathery, evergreen, oblanceolate, obtuse, punctate below and having one to few ob-

Family 112. **AQUIFOLIACEAE.** *Ilex glabra:* a, flowering branch, × ½; b, flower, × 8½; c, fruiting branch, × ½; d, mature fruit, × 2½.

scure teeth on each side of the blade near the apex. Staminate flowers clustered, pistillate flowers solitary. Drupe black, 4–5 mm wide, nutlets smooth. Sandy soil, low ground, Fla. to La., Nova Scotia on the coastal plain. Spring, summer.

4. I.ambigua (Michx.) Chapm.–Shrub or small tree up to 6 m tall, branches smooth. Leaves 5–7 cm long, ovate, elliptic-ovate to suborbicular, margins crenate-serrulate at least above the middle. Drupe globose, oval, 6–7 mm in diameter, red. Scrub, hammocks, hills, Fla. to Tex., N.C. on the coastal plain, Ark. Spring. *I. caroliniana* (Walt.) Trel.

5. I. cassine L. DAHOON—Shrub or tree up to 12 m tall with pubescent stems. Leaves 5–10 cm long, leathery, elliptic to oblanceolate, dark green and glabrous above, pale and pubescent below or occasionally smooth, somewhat revolute, entire or obscurely and distantly serrate. Flowers about 4–5 mm broad; drupe red or almost yellowish; about 6–8 mm wide, globose. Bayheads, low ground, Fla. to La., south Va. on the coastal plain. All year.

6. I. krugiana Loes.—Shrub or tree up to 10 m tall with glabrous stems. Leaves 5–7 cm long, elliptic or ovate, entire or almost so, acuminate, dark green and shiny above, on long slender petioles. Staminate corollas about 5–6 mm broad. Drupe black or purplish. Hammocks, pineland, south Fla., W.I. Winter.

7. I. decidua Walt. POSSUM HAW—Shrub or small tree up to 10 m tall. Stems glabrous or nearly so. Leaves cuneate-oblong, oblanceolate to narrowly obovate, membranous but becoming leathery, margins crenate or blunt toothed. Flowers four or five-parted, peduncles of staminate flowers longer than petioles. Drupe 6–8 mm in diameter, red or scarlet, shiny, usually clustered. Low ground, bottom land, thickets, Fla. to Tex., Md., Mo., Kans. Spring. *I. buswellii* Small

Reported for the lower Gulf coast and presumably in our area.

113. CELASTRÀCEAE BITTERSWEET FAMILY

Trees or shrubs or climbing vines with simple leaves. Flowers usually cymose or fasciculate, bisexual, radially symmetrical, calyx four- to five-lobed, petals five, stamens four to five, alternate with the petals and inserted on or below the margin of a well-marked disk which is often flat and fleshy. Ovary superior, one- to five-locular, fruit various, an indehiscent dry fruit, drupe, or a berry, seeds usually with a brightly colored aril. About 45 genera and 450 species, widely distributed in tropical and temperate regions.

1. Some or all the leaves opposite.
　2. Flowers bisexual; leaves spine toothed,
　　crenate or entire margin with obscure
　　serration near the apex . . . . . 1. *Crossopetalum*
　2. Flowers unisexual; leaves entire . . . 2. *Gyminda*
1. Leaves alternate.
　3. Fruit a capsule; leaf tips rounded or blunt 3. *Maytenus*
　3. Fruit a drupe; some or all the leaf tips
　　acute or tapered . . . . 4. *Schaefferia*

1. CROSSOPÉTALUM P. Browne

Shrubs or small trees with usually opposite or verticillate coriaceous leaves. Flowers small, cymose or solitary; calyx short, urceolate, petals four to five, reflexed; stamens five. Ovary confluent with the disk, three- to four-locular, stigma three- to four-lobed. Fruit dry or fleshy, small, coriaceous or drupaceous. About 20 species in tropical America. *Myginda* Jacq., *Rhacoma* L.

1. Leaf margins spiny . . . . 1. *C. ilicifolia*
1. Leaf margins crenate or mostly entire . . 2. *C. rhacoma*

1. C. ilicifolium (Poir.) Kuntze, CHRISTMAS BERRY—Low shrubs with spreading, pubescent stems. Leaves elliptical to ovate, 1–1.5 cm long, coarsely spiny toothed. Flowers in short-peduncled cymes; petals orbicular, 1 mm long, red. Fruit a subglobose drupe, 3–4 mm long, red. Pinelands, south Fla., W.I. All year.

2. C. rhacoma Hitchc. RHACOMA—Erect shrub or low tree with smooth stems. Leaves elliptic or more commonly ovate, 1–4 cm long, margins shallowly crenate or entire. Flowers in long-peduncled cymes; petals ovate to suborbicular, 1 mm long, reddish or purplish. Fruit an obovoid drupe, 5–6 mm long, red. Pinelands, hammocks, south Fla., W.I. All year.

2. GYMÍNDA Sarg.

Shrubs or trees with four-angled stems and opposite, evergreen leaves. Flowers unisexual in axillary cymes; sepals four, petals four, sometimes three, stamens four, disk fleshy. Fruit a drupe. Two species, one in North America, the other in Central America.

1. G. latifolia (Sw.) Urban, FALSE BOXWOOD—Shrub or tree up to 8 m tall with oblong-ovate to elliptical, short-petiolate leaves, margins entire or obscurely crenulate-serrate, especially above the middle, mostly 2–4 cm long. Petals elliptic or obovate, white, much longer than the calyx. Ovary two-locular with a two-lobed stigma. Drupe one- to two-seeded, black or dark blue, ovoid. Hammocks, south Fla., W.I., Mexico.

3. MAYTÉNUS Molina

Shrubs or small trees with smooth branches. Leaves persistent, coriaceous, alternate with minute stipules. Flowers small, polygamous, axillary, cymose or solitary; calyx five-parted, petals five, spreading, stamens five, inserted below the disk. Ovary immersed in the disk and confluent with it, two- to four-locular, stigma two- to four-lobed. Fruit a capsule, seeds arillate. About 70 species in tropical America.

1. **M. phyllanthoides** Benth.—Shrub or low trees up to 7 m tall with slender pale gray stems. Leaves 2–4 cm long, elliptic to oblong-obovate or oblanceolate, entire, cuneate at the base. Flowers mostly solitary or in dense clusters, very small, calyx lobes reddish, rounded, petals white. Ovary three- to four-locular, capsule four-angled, bright red, 8–12 mm long, three- to four-valved, valves opening to the base exposing the ellipsoidal seeds that are surrounded at the base by a bright red aril. Hammocks, dunes, south Fla., south Tex., Mexico, and southern Calif.

4. SCHAEFFÈRIA Jacq.

Dioecious trees or shrubs with terete stems and alternate, persistent, entire leaves. Flowers in axillary clusters; calyx four-lobed, persistent; petals four, stamens four, inserted under the margins of the small disk, absent in pistillate flowers. Ovary two-locular, style short with a large two-lobed stigma with spreading lobes. Fruit a small fleshy drupe. About five species in tropical America.

1. **S. frutescens** Jacq. Yellowwood—Shrub or tree up to 12 m tall with light gray older stems marked by the persistent wartlike clusters of bud scales. Leaves 4–6 cm long, alternate, not clustered, elliptical or ovate or oblanceolate, shining, bright yellow green with thickish revolute margins. Flowers pedicelled, greenish; staminate three to five together, pistillate solitary or two to three together. Drupe bright red or scarlet, about 5 mm wide. Hammocks, south Fla., Fla. Keys, Bahama Islands, W.I., tropical America.

114. HIPPOCRATEÀCEAE Hippocratea Family

Small trees, shrubs, or more commonly vines with opposite or alternate, simple leaves. Flowers usually quite small; bisexual, radially symmetrical, in cymes, panicles, or fascicles; calyx five-parted, petals five, stamens three, or sometimes two to five, inserted on a disk and alternate with the petals. Ovary superior, three-locular, fruit a capsule or berry. Three genera and about 150 species in tropical and subtropical regions.

1. HIPPOCRÁTEA L.

Trees or mostly woody vines with terete stems and opposite, coriaceous leaves. Flowers very small in axillary cymes or panicles with the pedicels two-bracteate near the base; petals larger than the sepals, spreading; disk conic, style very short. Fruit a two-valved, coriaceous capsule. About 80 species, pantropical.

1. **H. volubilis** L.—Large woody vine or climber up to 20 m long, the stems, petioles, and inflorescence brownish tomentulose or puberulent. Leaves mostly 5–14 cm long, elliptic to ovate, crenate-serrate or merely undulate, thinly coriaceous or membranous. Inflorescence 4–12 cm long, long pedunculate, branched; flowers 4–8 mm broad; sepals deltoid or ovate; petals white, often densely puberulent. Capsule 4–8 cm long, obovoid or narrowly oblong. Hammocks, climbing on trees, south Fla., Mexico, W.I., tropical America. Winter, summer.

Thick networks of these often much branched vines can be found in mangrove swamps, tropical hardwood hammocks, and similar habitats.

115. ACERÁCEAE Maple Family

Trees or shrubs with opposite, palmately lobed, simple, or pinnately compound leaves. Flowers borne in fascicles, bisexual or unisexual, radially symmetrical; sepals four to five, petals four to five, disk present; stamens four to ten, usually eight. Ovary superior, two-locular. Fruit a samara. Two genera and about 150 species mostly in Acer, widely distributed but chiefly in north temperate regions.

1. ÀCER L. Maple

Deciduous trees or shrubs with usually palmately three- to nine-lobed leaves or pinnately compound with three to five leaflets. Flowers polygamous or unisexual, small, in terminal or axillary racemes, corymbs or fascicles, sepals usually spreading, five, free, petals five, free, disk annular, thick stamens four to ten, usually eight, inserted within the disk. Ovary two-lobed, ovules two in each locule, stigmas two, shorter than style. *Rufacer* Small

1. **A. rubrum** L. var. **tridens** Wood, SOUTHERN RED MAPLE—

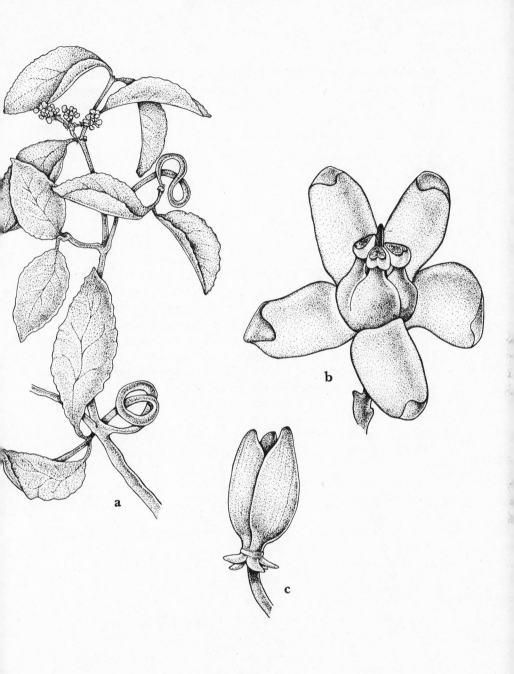

Family 114. **HIPPOCRATEACEAE.** *Hippocratea volubilis:* a, flowering branch, × ½; b, flower, × 13; c, mature fruit, × 1½.

Medium or tall trees up to 30 m tall with smooth branches and young stems reddish. Leaves elliptic to ovate in outline, 3–10 cm wide with three short, terminal lobes, the middle one 1–4 cm long, laterals somewhat shorter, these unevenly serrate. Flower in umbel-like clusters, reddish or rarely yellow, petals and sepals about equal. Ovary smooth, samara 1–3 cm long. $2n = 78, 104$. Stream banks and moist soils, swamps, south Fla., Tex., N.J., on the coastal plain. *Acer rubrum* var. *trilobum* K. Koch; *Rufacer carolinianum* (Walt.) Small

A. rubrum ranges from Canada southward in a wide variety of habitats, and variation is common. The taxonomy of these populations is poorly understood.

116. SAPINDÀCEAE Soapberry Family

Monoecious, dioecious, or polygamodioecious trees, shrubs, or woody climbers, rarely herbaceous, with alternate, simple or compound leaves. Flowers bisexual or unisexual, often very small and inconspicuous, in racemes or panicles; sepals four to five, free or partially fused, petals four to five or absent, disk usually present with the six to twelve stamens usually inserted within or on the disk. Ovary superior, lobed or divided nearly to the base or entire, two- to four-locular, often three-locular. Fruit generally large, capsule, nut, schizocarp berry, or drupe, seeds sometimes arillate, usually one per locule. About 150 genera and 2000 species, chiefly in the tropics and subtropics. *Dodonaeaceae*

1. Climbing vines with tendrils; fruits globular bladdery capsule . . . . 1. *Cardiospermum*
1. Trees or shrubs; fruits never bladdery capsules.
  2. Leaves simple; fruits winged . . . . 2. *Dodonaea*
  2. Leaves trifoliolate or pinnately compound; fruits not winged.
    3. Leaves mostly trifoliolate, leaflets up to 6 cm long . . . . 3. *Hypelate*
    3. Leaves mostly even-pinnately compound, leaflets mostly over 6 cm long.
      4. Rachis conspicuously wing margined.
        5. Drupe about 3 cm wide, green; leaflets 7–15 cm long . . . . 4. *Melicoccus*
        5. Drupe 1.5–2 cm wide, yellow or yellowish brown; leaflets usually smaller, 3–10 cm long . . . 5. *Sapindus*
      4. Rachis not wing margined.
        6. Leaflets mostly eight to fourteen.
          7. Margins of blades serrate, crenate, or crenate-serrate . . 6. *Cupania*

7. Margins of blades entire to
    sharply serrate . . . . .   7. *Koelreuteria*
6. Leaflets mostly two to six.
  8. Stems glabrous; seeds two in each
     cell of fruit . . . . . .   8. *Exothea*
  8. Stems puberulent; seeds one in
     each of fruit . . . . .   9. *Talisia*

1. CARDIOSPÉRMUM L.   Balloon Vine

Herbaceous or semiwoody vines with slender stems and bearing axillary tendrils. Leaves biternate or twice compound, sometimes trifoliolate leaflets coarsely serrate or crenate often with pellucid dots. Flowers unisexual or occasionally bisexual in axillary racemes or corymbs, white, bilaterally symmetrical; sepals four to five, petals four, stamens eight, disk with two glands. Ovary three-locular, stigmas three. Fruit a large, inflated, bladdery capsule. About 12 species, pantropical.

1. Sepals pubescent; leaflets finely pubescent
  beneath . . . .   1. *C. corindum*
1. Sepals glabrous or nearly so; leaflets glabrous
  or sparingly pubescent beneath on veins
  and margins . . . .   2. *C. halicacabum*

    1. C. corindum L.—Stems semiwoody, covered with fine, pale, pubescence. Leaflets ovate, acute or tapered, somewhat lobed or incised, margins crenate-dentate or crenate-serrate. Sepals up to 2 mm long, obtuse, pubescent; petals white, 3–4 mm long. Capsule subglobose to ovoid, 3–4 cm wide, rather longer than broad with a long stipe, seeds with a semicircular hilum. Hammocks, near coastal areas, south Fla., Fla. Keys, W.I. All year. *C. keyense* Small
    2. C. halicacabum L. Heart Seeds—Stems herbaceous, usually annual, smooth or puberulent, and much branched. Leaflets lanceolate or ovate, acuminate or obtuse, variously serrate or lobed or dentate, pubescent or glabrate. Sepals about 2 mm long, obtuse; petals white, obovate, 3–5 mm long. Capsule subglobose or turbinate globose, pubescent, 2–4 cm long, seeds with a heart-shaped or reniform hilum. $2n = 22$. New and Old World tropics, naturalized northward in U.S. to N.J., Mo. All year. *C. microcarpum* HBK
    This species of balloon vines is also commonly cultivated.

2. DODONAÈA Mill.   Varnish Leaf

Shrubs or trees usually viscid with alternate, evergreen leaves having many parallel lateral veins. Flowers small, in axillary or terminal panicles; sepals three to seven, petals none, stamens six to ten, filaments short. Ovary two- to four-locular, covered with glands, angled with style filiform, two- to four-parted. Fruit a coriaceous or membranous capsule, three- to six-angulate. About 54 species, all but five in Australia.

1. **D. viscosa** (L.) Jacq.—Dioecious shrub or small tree up to 3 m tall with very viscid leaves and dark, slender stems. Leaves 6–12 cm long, variable, mostly linear-oblanceolate or oblong-lanceolate, sessile or nearly so, acute to rounded at the apex, glabrous or puberulent beneath. Flowers pale yellow, in small, axillary clusters; sepals 3 mm long. Ovary mostly three-locular. Capsule 1.5–2.5 cm wide, three-locular and three-winged. $2n = 28,32$. Pinelands and hammocks, south Fla., Fla. Keys and pantropical. All year. *D. jamaicensis* DC., *D. microcarya* Small

This is a highly variable species with numerous forms and varieties of doubtful taxonomic status.

3. HYPELÀTE P. Browne   WHITE IRONWOOD

Monoecious evergreen trees or shrubs with alternate, trifoliolate leaves. Leaflets rather small, 3–6 cm long, obovate or oblanceolate, sessile with close, parallel veins. Flowers unisexual, few in long-peduncled axillary or subterminal panicles; sepals four to five, reddish, unequal, petals four to five, white, disk present, annular, lobed and fleshy, stamens seven to eight inserted on the disk. Ovary three-locular, stigma three-lobed, ovules two per locule. Fruit a small drupe, black, about 8–9 mm in diameter, subglobular, one-seeded. A monotypic genus of Fla. Keys and W.I.

1. **H. trifoliata** Sw.—Characters of the genus. Hammocks and rarely pinelands, lower Fla. Keys, Big Pine Key. One of the rarest species in our area. The wood is very hard and heavy. Spring, summer.

4. MELICÓCCUS P. Browne

Trees with smooth stems and alternate, pinnately compound, evergreen leaves. Flowers bisexual or unisexual in long, terminal, slender racemes or paniculate with many small flowers, sepals four, nearly free, petals four, oblong, rounded, or obovate; disk four- to five-lobed, large, stamens eight. Ovary superior, one- to two-locular, stigma large, peltate, two- to three-lobed, style short. Fruit one-locular, one-seeded drupe. Two species in tropical America.

1. **M. bijugatus** Jacq. SPANISH LIME—Trees up to 10 m tall or more, deciduous, usually with two-paired, pinnately compound leaves. Leaflets 8–11 cm long, ovate-elliptical or elliptical, apex acute or obtuse. Flowers 6–8 mm wide, fragrant in branching inflorescence. Drupe 3 cm wide, greenish with a fibrous pulp. $2n = 32$. Occasional on shell mounds and

Indian Middens in Fla. Keys and in cultivation south Fla., naturalized from tropical America. *Melicocca bijuga* L.

Grown as a shade tree and ornamental; fruit and roasted seeds are edible.

5. SAPÍNDUS L. SOAPBERRY

Monoecious or dioecious trees with alternate, pinnately compound leaves or appearing simple. Flowers in terminal racemes or panicles, sepals four to five, unequal, petals four to five, equal, white or greenish. Disk annular, fleshy, lobed, stamens eight to ten, inserted in the disk. Ovary two- to four-locular, entire or lobed, stigma small, three-lobed, style short. Fruit drupelike, one-locular, one-seeded, resinous. About 13 species in tropics and subtropics.

1. **S. saponaria** L. SOUTHERN SOAPBERRY—Trees up to 15 m tall with gray flaking bark and broad dense crowns, or sometimes low trees, or even shrublike. Leaves pinnate, leaflets usually six to twelve, narrowly lanceolate to oblong, 6–18 cm long, obtuse to long acuminate, asymmetric, entire, rachis wing margined. Flowers white or whitish, 4 mm wide in much branched panicles. Fruit globose, 1–2 cm wide, very fleshy, seeds pale, about 1 cm in diameter. Hammocks in coastal areas, Fla. Keys, south Fla., Mexico, tropical America, widely dispersed, also in the Old World. Winter, spring.

When mascerated in water, fruits yield abundant suds and have been used as a substitute for soap in tropical countries.

6. CUPÀNIA L.

Dioecious or polygamodioecious trees or shrubs with alternate, pinnately compound, evergreen leaves. Flowers small, in axillary racemes or panicles, sepals five, free, in two series, petals five or absent, disk present, annular, lobed or inserted within the disk, stamens eight, filaments short. Ovary two- to four-locular, pubescent, stigmas three. Fruit a three-lobed stipitate capsule, seeds ellipsoidal, arillate. About 45 species in tropical America.

1. **C. glabra** Sw.—Large shrub or tree up to 15 m tall. Leaflets usually seven to fourteen, obovate to narrowly oblong, 6–20 cm long, subentire or denticulate. Panicles equal to or shorter than leaves, puberulent, in the upper axils or subterminal, with many small white flowers; sepals about 2 mm long. Capsule turbinate, about 2 cm long, lobed. Hammocks of lower Fla. Keys, especially Big Pine Key, southern Mexico, Central America, W.I. Spring.

7. KOELREUTÈRIA Radlk.

Deciduous trees with alternate, pinnate or twice-pinnately compound leaves. Flowers polygamous, bilaterally symmetrical, in large terminal panicles, calyx five-lobed, petals three to four, clawed, disk present, stamens eight or fewer with long filaments. Ovary three-locular, style three-cleft. Fruit an inflated capsule, three-valved, with ovoid, black seeds. Four species in China and Japan.

1. **K. paniculata** Laxm. GOLDEN RAIN TREE—Trees up to 10 m tall with large, bipinnately compound leaves, leaflets seven to fifteen, 3–8 cm long, ovate to oblong-ovate, irregularly crenate-serrate and often incisely lobed near the base, dark green and smooth above, pale and pubescent along the veins beneath. Flowers about 1 cm long, numerous in broad panicles 4–5 dm long. Capsule 5 cm long, papery, the valves narrowed into a pointed tip. $2n = 22$. Naturalized from Asia in disturbed sites and widely cultivated in east U.S.

8. EXÓTHEA Macfadyen

Polygamodioecious or dioecious, tall, evergreen shrubs or trees with alternate, pinnately compound leaves or leaves appearing simple. Flowers small, in axillary panicles, calyx five-parted, persistent, petals five, white, stamens seven to ten, mostly eight, inserted on a complete fleshy disk. Ovary globose, two-locular, ovules two in each locule. Fruit a berry, globose, usually one-seeded. Three species in tropical America.

1. **E. paniculata** (Juss.) Radlk. INKWOOD—Shrubs or mostly trees up to 12 m tall with smooth branches and bearing two- to four-foliolate leaves, or some leaves unifoliolate and appearing simple; blades oblong, 5–8 cm long, entire, lustrous. Flowers in panicles clustered at the ends of branches, minutely tomentulose, sepals ovate, 3–4 mm long, petals white. Fruit subglobose, glabrous, about 1 cm wide, orange, turning dark purplish. Hammocks, shell mounds, calcareous soils, south Fla., Fla. Keys and in to Volusia County, Central America, W.I.

9. TALÍSIA Aubl.

Polygamodioecious trees or tall shrubs with alternate, pinnately compound leaves. Flowers small, paniculate, sepals five, petals five, disk present; stamens eight. Ovary villous, three-lobed and three-locular, stigma three-lobed. Fruit baccate, not lobed. About 40 species in tropical America.

1. T. pedicellaris Radlk.—Trees up to 15 m, usually less, with pubescent stems. Leaves four- to six-foliolate, blades ovate or elliptic-ovate, 5–10 cm long, tapering to an acute apex. Panicles small with puberulent branches, flowers small, greenish or yellowish; petals ovate or lanceolate, 4–5 mm long. Fruit ovoid, 1.5–2 cm long, papillose. Reported from Brickell near Miami, hammocks, doubtfully still present in our flora, naturalized from South America.

117. RHAMNÀCEAE Buckthorn Family

Trees, shrubs, or climbing vines, often with thorny branches and bearing simple leaves, stipules present, though often quickly deciduous. Flowers bisexual or rarely polygamous or unisexual, small, generally in cymose inflorescence, calyx four- to five-lobed, tubular or often cup shaped, petals four to five, usually cupped or hooded or absent; stamens four to five, alternate to the sepals and the anthers fitting into the hood of the petals; disk present. Ovary two- to four-locular, free or sunken in the disk. Fruit various, drupelike or dry and opening suddenly into dehiscent or indehiscent segments. About 50 genera and 500 species in temperate and tropical parts of the world. *Frangulaceae*

1. Climbing vines.
    2. Leaf margins irregularly dentate-serrate; ovary inferior, fruit dry, winged or crested . . . .   1. *Gouania*
    2. Leaf margins entire to undulate-crenate; ovary superior, fruit an elongate drupe.   2. *Berchemia*
1. Trees or shrubs.
    3. Some or all the leaves notched or obscurely mucronate or both at apex; fruit a fleshy drupe with a single stone.
        4. Leaves opposite on lower branches, conspicuously broadly notched at apex; sepal lobes not keeled within . .   3. *Reynosia*
        4. Leaves alternate on lower branches; sepal lobes keeled within . . . .   4. *Krugiodendron*
    3. Leaves acute or acuminate at apex, rarely notched; fruit dry or, if fleshy, then with more than one stone.
        5. Branches with thorns; flowers 1–2 mm broad in branching spikes . . . .   5. *Sageretia*
        5. Branches unarmed; flowers larger, in axillary clusters, flowers on short peduncles, not spicate . . . . .   6. *Colubrina*

1. GOUÀNIA Jacq.

Vinelike and bearing tendrils. Leaves alternate, deciduous with persistent or deciduous stipules. Flowers bisexual or polygamous, whitish in terminal and axillary racemes or paniculate spikes, the branches often terminated by a tendril, calyx tube short, five-lobed, petals five, stamens five, covered within the petals, disk within the calyx tube. Ovary inferior, three-locular, within the disk, style three-parted. Fruit coriaceous, usually three-winged. About 40 species, chiefly tropical America.

1. **G. lupuloides** (L.) Urban, CHEW STICK—Arching or climbing vinelike plants with smooth stems and membranaceous, ovate to elliptic leaves usually 5–10 cm long, blades crenate-serrate, generally smooth. Flowers small, white or yellowish in slender racemes or in large terminal panicles, calyx 1–2 mm long, petals 1 mm long. Fruit 8–12 mm wide, smooth. Coastal hammocks, mangrove associations, south Fla., especially in the Fla. Keys, north to Brevard and Manatee Counties, Mexico, Central America, W.I. Spring.

In the W.I. pieces of stems are chewed to cleanse the teeth and to harden and heal the gums.

2. BERCHÈMIA (Necker) DC.

Shrubs or vines with alternate leaves. Flowers bisexual, unisexual by abortion in our taxa, small, greenish white, in axillary or terminal inflorescence, calyx tube five-lobed, petals five, stamens five, disk covering the calyx tube. Ovary two-locular, fruit an ovoid drupe. About 20 species, one in North America, the rest in tropical Africa and Asia.

1. **B. scandens** (Hill) K. Koch, RATTAN VINE—Slender, flexible, smooth vine with tough, round stems. Leaves ovate or oblong-ovate, mostly 3–6 cm long, dark green and shiny above, margins entire or undulate, lateral nerves conspicuous and ascending to the tip parallel. Flowers 3 mm wide, mostly in small terminal panicles. Drupe 6–8 mm long, black. Hammocks, bay tree forests, swamps, Fla. to Tex., Va., Mo., Mexico, Central America, especially in low ground on the coastal plain. Spring.

3. REYNÒSIA Griseb.

Evergreen trees or shrubs with opposite, entire, coriaceous leaves. Inflorescence axillary, sessile, cymose, or reduced to a single flower. Flowers bisexual, small, greenish; sepals five, petals absent, stamens five, inserted on the upper margin of the disk. Ovary two-locular, superior style

Family 117. **RHAMNACEAE.** *Gouania lupuloides:* a, fruiting branch, × ½;
b, flower, × 12; c, mature fruit, × 4½.

bifid, short. Fruit a drupe, endosperm ruminate. About 16 species in the W.I., one in Fla.

1. **R. septentrionalis** Urban, DARLING PLUM—Shrub or small tree up to 10 m tall. Leaves ovate or ovate-elliptic. Sepals deltoid or acute, petals alternate with sepals. Drupe ovoid or obovoid, about 1–2 cm long, purplish or nearly black. Found in a variety of communities, scrub vegetation, cactus hammocks, hammock forests, margins of mangrove associations south Fla., especially the Fla. Keys, and Bahama Islands. Spring, summer.

Fruits of this species are edible and are pleasantly flavored.

4. KRUGIODÉNDRON Urban

Trees or shrubs with persistent, opposite, entire leaves up to 10 m tall. Leaves ovate or elliptical about 3–7 cm long, dark green above, paler beneath. Flowers bisexual, small, yellowish green in axillary cymose inflorescence, calyx five-parted, lobes longer than tube, petals absent, stamens five, disk annular. Ovary two-locular. Fruit a small ovoid drupe, black, about 5–8 mm long. One species in south Fla., Mexico, Central America, W.I.

1. **K. ferreum** (Vahl) Urban, BLACK IRONWOOD—Characters of the genus. Tropical hammock forests. Spring. *Rhamnidium ferreum* (Vahl) Sarg.

The very close-grained hardwood is the heaviest hardwood occurring in the U.S.

5. SAGERÈTIA Brongn.

Trees or shrubs often with spiny branches and opposite or sub-opposite leaves. Flowers bisexual, very small, in panicles, calyx five-parted, petals five, disk present, lining the calyx tube, five-lobed. Ovary ovoid, immersed in disk, three-locular, and with three stigmas. Fruit a drupe with three coriaceous or indurate nutlets. About 24 species in Asia, Africa, and America.

1. **S. minutiflora** (Michx.) Mohr, BUCKTHORN—Shrubs with pubescent, spiny stems. Leaves ovate or orbicular mostly 2–5 cm long. Flowers white, in spikes 1–4 cm long. Calyx 2–2.5 mm wide, petals orbicular. Drupe subglobose, 7–9 mm wide. Hammocks, sandy soils, Fla. to Miss., N.C. on the coastal plain. Spring.

A polymorphic species with very fragrant flowers reported from Lee County, presumably also in our area.

6. COLUBRÌNA Richard ex Brongn.

Trees or shrubs, sometimes bearing spines, with alternate leaves. Flowers bisexual, small, in axillary clusters, calyx tube five-lobed, spreading, disk angulate or lobed, petals five, stamens five. Ovary three-locular, immersed in the disk; styles three; fruit drupelike, becoming dry and schizocarplike, somewhat three-lobed, separating into three dehiscent parts. About 30 species, chiefly tropical America, Asia, Madagascar, Oceania.

1. Leaves tomentose beneath, usually less than
    7 cm long . . . . 1. *C. cubensis*
    var. *floridana*
1. Leaves glabrous to very sparsely pubescent
    beneath.
   2. Leaves strongly three-nerved at base, mar-
       gins serrate . . . . 2. *C. asiatica*
   2. Leaves pinnately nerved, margin entire.
      3. Leaves thick and with a few black
          glands scattered over the undersur-
          face; young stems and veins on lower
          surface of leaves rusty pubescent . . 3. *C. arborescens*
      3. Leaves thin and stems sparsely pubes-
          cent; glands very small, confined to
          lower part of the margin of leaf . . 4. *C. elliptica*

1. C. cubensis (Jacq.) Brongn. var. **floridana** M. C. Johnston—Shrub or small tree with stems closely pubescent. Leaves narrowly elliptic or elliptic-ovate, 5–10 cm long, strongly veined, finely tomentose. Flowers very small, sepals 1.5 mm long, petals 7 mm long; 5–6 mm wide. Hammocks and pinelands, south Fla., Bahama Islands. All year.

2. C. asiatica (L.) Brongn.—Shrub up to 3 m tall with smooth, trailing or spreading, weak, prostrate branches. Leaves 4–9 cm long, ovate or elliptic-ovate, serrate or crenate-serrate, dark green and lustrous above, paler beneath. Sepals 2 mm long, petals 1–2 mm long, greenish. Fruit subglobose, 7–10 mm wide. Coastal beach and dune vegetation, coastal hammocks, naturalized in Fla. Keys, south Fla., and the W.I.; native to Old World beaches from east Africa to India, Malaysia, Pacific Islands. All year.

3. C. arborescens (Mill.) Sarg. WILD COFFEE—Shrub or tree up to 7 m tall with young stems densely rusty tomentose. Leaves elliptic to ovate-lanceolate, 5–15 cm long, acuminate, entire, dark green and lustrous above, pale and with some rusty pubescence beneath and with scattered black glands. Flowers small in axillary cymes, covered with rusty pubescence; sepals 2.5 mm long, petals somewhat longer, white or nearly so. Fruit about 1 cm long, subglobose, dark purplish or black. Hammock forests, south Fla., Fla. Keys, W.I., Mexico, Central America. All year. *C. colubrina* (Jacq.) Millsp.

4. **C. elliptica** (Sw.) Briz. & Stern, NAKEDWOOD—Shrub or small tree up to 6 m tall or sometimes up to 10 m tall with young stems finely pubescent. Leaves elliptic to ovate, 3–9 cm long, entire, with two to four minute glands near the base of the margin. Flowers in densely pilose cymes, sepals 2 mm long, petals somewhat shorter, yellowish or greenish yellow. Fruit globose, orange red to brownish, 7–9 mm wide. Hammock forests, Fla. Keys, Mexico, tropical America, W.I. All year. *C. reclinata* (L'Her.) Brongn.

CEANOTHUS AMERICANUS L., NEW JERSEY TEA, with flowers in terminal, umbel-like cymes on elongate, leafless peduncles 4–6 cm long, may occur locally in our area. However, we have seen no specimens from south Fla.

118. VITÀCEAE GRAPE FAMILY

Climbing shrubs rarely erect or vines with tendrils, leaves alternate. Flowers bisexual or unisexual, small, radially symmetrical, usually in cymose inflorescence, calyx four- to five-parted, petals four to five, free or fused, stamens four to five opposite the petals and inserted at the base of a disk. Ovary two- to six-locular, fruit a watery berry. About 11 genera and 450 species, widely distributed in tropical and temperate regions.

1. Leaves pinnately twice compound, leaflets
   coarsely toothed or lobed . . . . .   1. *Ampelopsis*
1. Leaves simple, trifoliolate, or palmately
   compound.
   2. Leaves palmately compound, leaflets
      mostly four to five . . . . . .   2. *Parthenocissus*
   2. Leaves simple or trifoliolate.
      3. Leaves trifoliolate or, if simple, then
         ₗleaf base truncate and margins dis-
         tantly serrate . . . .   3. *Cissus*
      3. Leaves simple, leaf base cordate or
         V-shaped and margins coarsely den-
         tate or lobed . . . .   4. *Vitis*

1. AMPELÓPSIS Michx.

Shrubs or woody vines with or without tendrils. Leaves simple or twice-pinnately compound. Flowers unisexual, small, greenish, in cymose

Family 118. **VITACEAE.** *Ampelopsis arborea:* a, leafy branch, × ½; b, flower, × 15; c, mature fruit, × 4.

clusters; calyx shallowly lobed, petals free, disk cup shaped, stamens erect. Ovary surrounded by a disk. Berry with two to four seeds, flesh thin. About 20 species in subtropical Asia and North America.

1. **A. arborea** (L.) Rusby, PEPPER VINE—Erect bushy shrub or climbing vine with smooth stems. Leaves about 15–20 cm long, twice-pinnately compound, leaflets few to many, ovate, coarsely toothed or lobed, 2–5 cm long. Small, greenish flowers cymose, berry dark purplish, about 1 cm wide. $2n = 40$. Hammocks, low ground, Fla. to Tex., Va., Ill., W.I.

2. PARTHENOCÍSSUS Planch. WOODBINE

Woody vines climbing or trailing with tendrils bearing disks, leaves palmately lobed or compound. Flowers in compound cymes opposite the leaves, small, bisexual or unisexual; calyx five-toothed, petals free, spreading, stamens short. Berries with one to four seeds, flesh thin. About ten species in North America and Asia.

1. **P. quinquefolia** (L.) Planchon, VIRGINIA CREEPER—Climbing vine adhering to support with many disks on the ends of many-branched tendrils. Leaves with long petioles, palmately compound with five elliptic or obovate leaflets, these 6–12 cm long, serrate at least along the upper half. Inflorescence terminal or from the upper axils forming panicles of small, greenish flowers. Berries dark blue or almost black, 4–6 mm wide. Hammocks, low ground, Fla. to Tex., Me., Kans., W.I., Mexico, Central America.

Also occurring within the range of this manual is **P. quinquefolia** f. **hirsuta** (Donn) Fern. HAIRY VIRGINIA CREEPER, with leaves and tendrils conspicuously pubescent, generally found in drier sites. *P. hirsuta* (Donn) Small

3. CÍSSUS L. POSSUM GRAPE

Herbaceous or more commonly woody vines often with tendrils and bearing simple or trifoliate leaves. Flowers bisexual, greenish, in small cymose or corymbose inflorescence. Sepals four, petals four, spreading, disk four-lobed and fused to the base of the two-locular ovary. Fruit a one- to four-seeded berry. About 200 species, widely distributed, chiefly in the tropics.

1. Leaves simple . . . .   1. *C. sicyoides*
1. Leaves trifoliate.
  2. Leaflets usually 3–10 cm long; flowers in
     branching cymes . . . .   2. *C. incisa*

2. Leaflets smaller, usually 1–3 cm long;
flowers in umbellate cymes . . . . 3. *C. trifoliata*

1. C. sicyoides L. Possum Grape—Woody vine with thick but flexible stems. Leaves simple, 5–16 cm long, often asymmetric, oblong to ovate, coarsely to finely serrate, mostly densely pubescent or occasionally glabrous. Cymes pedunculate, flowers green or yellowish green; berries globose, black, one-seeded, about 6 mm wide. Hammocks, low ground, south Fla., Mexico, W.I., tropical America.

2. C. incisa (Nutt.) Des Moulins, Marine Ivy—Woody vine with stout stems. Leaves trifoliolate, leaflets ovate or elliptic or somewhat wedge shaped, 3–10 cm long, irregularly and coarsely toothed. Cymes branching, flowers greenish. Berries obovoid or globose, black, 10–12 mm long, mucronate. Sand dunes, dry soil, and coastal hammocks, Fla. to Tex., Kans., Mo.

3. C. trifoliata L. Sorrel Vine—Vines with somewhat fleshy, smooth, flexible stems. Leaves trifoliolate, fleshy, leaflets obovate or wedge shaped, 1–3 cm long, deeply toothed above the middle. Flowers in umbellate clusters, greenish yellow; berries globose or obovoid, black, mucronate. Coastal areas, south Fla., Fla. Keys., W.I., Mexico, Central and South America.

4. VÌTIS L. Grape

Woody polygamodioecious vines climbing with tendrils opposite the leaves or sometimes coming from peduncles. Leaves simple, usually rounded or cordate. Flowers in racemes or panicles, small, greenish, calyx cupular, very small, petals five, fused at their tips forming a deciduous cup; stamens five, disk of five glands fused to the ovary. Ovary two-locular, fruit a fleshy berry, usually edible. About 60 species, chiefly north temperate regions. *Muscadinia* Small

1. Bark not shredding, pith without diaphragms at the nodes; tendrils simple . 1. *V. rotundifolia*
1. Bark of older stems shredding, pith with diaphragms at the nodes; tendrils usually forking.
   2. Mature leaves densely and evenly tomentose beneath . . . . 2. *V. shuttleworthii*
   2. Mature leaves thinly cobwebby pubescent beneath or merely sparingly short pubescent.
      3. Leaves rusty cobwebby pubescent beneath; some leaves lobed with rounded sinuses . . . . . . 3. *V. aestivalis*

3. Leaves pale, greenish, or tan colored beneath, lightly pubescent; leaves usually not lobed . . . .　4. *V. vulpina*

1. V. rotundifolia Michx. MUSCADINE GRAPE—Vines with stout stems
without nodal diaphragms. Leaves leathery, glossy, orbicular to cordate-
ovate, 6–12 cm long, coarsely or irregularly serrate, subglabrous at maturity, although there may be small patches of hairs on veins. Inflorescence short, 2–5 cm long, densely flowered. Berries few, subglobose,
purplish, 1.5–2.5 cm wide. $2n = 40$. Hammocks, low ground, Fla. to Tex.,
Del., Mo. *Muscadinia rotundifolia* (Michx.) Small; *M. munsoniana*
(Simpson) Small

This is the wild form from which the Scuppernong grape was originated.

2. V. shuttleworthii House, CALUSA GRAPE—Vines with stems having nodal diaphragms. Leaves reniform or suborbicular, shallowly
toothed or lobed, or those of young new shoots often deeply lobed 4–10
cm wide, densely white to rusty tomentose beneath, becoming smooth
above. Inflorescence elongate, 6–12 cm long. Berries subglobose, about
8–15 mm in diameter. Hammocks, south Fla., W.I. *V. coriacea* Shuttlew.

3. V. aestivalis Michx. SUMMER GRAPE—Vines with stems having
nodal diaphragms. Leaves broadly ovate, suborbicular, or cordate-ovate,
usually shallowly to deeply three- to five-lobed, reddish, or cobwebby
pubescent or tomentose beneath when young, light colored, becoming
thinly cobwebby pubescent. Inflorescence elongate, 5–15 cm, slender.
Berries black or dark purplish, 5–10 mm wide. $2n = 38$. Various sites,
disturbed areas, Fla. to Tex., Mass., Minn. *V. rufotomentosa* Small, *V.
simpsonii* sensu Small

4. V. vulpina L. FROST GRAPE—Vines with flexible stems having
nodal diaphragms. Leaves broadly obovate or oblong, suborbicular, usually 5–15 cm wide, shallowly or distinctly toothed, or sometimes angularly three-lobed, or entire, glabrate, rusty or scurfy brown puberulent to tomentose beneath, glabrate or cobwebby above. Inflorescence an
irregular, lax panicle mostly 10–13 cm long when fully developed, fruits
crowded. Berries subglobose, 5–8 mm in diameter, black, sometimes
lightly glaucous. $2n = 38$. Hammocks and borders of hammocks, low
ground, Fla. to Tex., N.Y., Mo. *V. simpsonii* sensu authors, *V. baileyana*
Munson, *V. cinerea* Engelm. ex Millardet var. *floridana* Munson

119. ELAEOCARPÀCEAE　FALSE OLIVE FAMILY

Trees and shrubs with alternate or opposite leaves. Flowers bisexual, in racemes, panicles, or cymes, sepals four to five, free, petals four to five or absent, stamens many, adnate to the disk. Ovary superior with two to many locules. Fruit a capsule or drupe. About 10 genera and 400 species in the tropics and subtropics. Although many authors place this family in the Tiliaceae, the Elaeocarpaceae differs from the Tiliaceae in certain anatomical details.

1. MUNTÍNGIA L.

Trees or shrubs up to 12 m tall with stellate pubescence. Leaves two-ranked, 6–14 cm long, oblong-lanceolate, acuminate, base oblique, irregularly or coarsely serrate, smooth and green above, densely gray or white stellate pubescent below. Flowers rather large, one-flowered on axillary peduncles, or in sessile clusters of two to three, sepals five, lanceolate petals five, distinct white, about 1 cm long; stamens many, free, inserted on an annular disk. Ovary with glandular hairs, five- to seven-locular. Fruit a berry, subglobose, about 1 cm wide, smooth, red or yellow, seeds numerous. A single species in tropical America.

1. **M. calabura** L. Strawberry Tree—Characters of the genus. Hammocks and pinelands, and in cultivation, south Fla., Mexico, W.I., naturalized from tropical America.

120. TILIÀCEAE Basswood Family

Trees, shrubs, or herbs with alternate or rarely opposite simple leaves. Flowers bisexual, radially symmetrical, cymose; sepals five, petals five, free; stamens many, free or barely fused at the base. Ovary superior, two- to ten-locular. Fruit baccate, drupaceous capsule, or variously dehiscent. About 41 genera and 400 species, tropical and temperate.

1. Leaves ovate, usually palmately three-lobed,
   tomentulose beneath; fruit covered with
   hooked prickles . . . . 1. *Triumfetta*
1. Leaves lanceolate, not lobed, glabrous beneath; fruit smooth, unarmed . . . . 2. *Corchorus*

1. TRIUMFÈTTA L. Burbush

Subshrubs or herbs with stellate pubescence and membranaceous alternate leaves. Blades serrate, undivided or lobed. Flowers pedicelled, axillary or opposite the leaves, yellow or red, sepals five, distinct, petals five or absent, glandularly thickened or foveolate at the base, yellow, stamens many or five. Ovary two- to five-locular, densely covered with minute spines, style long, filiform. Fruit capsular or nutlike with prickles. About 100 species, pantropical.

1. T. semitriloba Jacq. BURWEED—Perennial herb or subshrub up to 2 m tall, usually much branched. Leaves long petioled, ovate to rhombic, acute, serrate or irregularly dentate, often shallowly lobed, 4–8 cm long, with stellate pilose pubescence above and densely tomentose or sparsely stellate pubescent below. Sepals linear, 5–7 mm long, petals yellow, stamens fifteen to twenty. Capsule 6–8 mm wide with numerous slender spines. $2n = 32$. Hammocks, pinelands, south Fla., Mexico, W.I., tropical America. All year.

T. pentandra A. Rich., an herb with five to eight stamens, has also been reported for Fla., but no recent collections have been seen.

2. CÓRCHORUS L. Jutes

Herbs, subshrubs, or small shrubs with simple or stellate pubescence. Leaves small, alternate serrate, membranaceous. Flowers axillary or opposite the leaves on very short peduncles, one- to few-flowered, small, sepals five, deciduous petals five, stamens numerous, free. Ovary two- to five-locular, style short. Fruit a capsule, usually elongate. About 40 species in the tropics.

1. Leaf blades with pair of bristle-tipped appendages on the lowermost pair of teeth; stout, three-winged capsule less than 2 cm long . . . . 1. *C. aestuans*
1. Leaf blades without basal appendages, teeth not bristle-tipped; capsule 4 cm long or more.
2. Capsule with four short points at apex; stems with few short hairs on one side . 2. *C. siliquosus*
2. Capsule with single beak; stems densely puberulous on one side . . . . . 3. *C. orinocensis*

1. C. aestuans L.—Annual herbs up to 1 m tall, widely branching, with finely pubescent stems. Leaves elliptic, ovate or orbicular-ovate, about 2–6 cm long, acute, crenate or crenate-serrate. Flowers few or small clustered, subsessile, sepals linear to linear-lanceolate 4–5 mm long; petals spatulate, 4–5 mm long. Capsule 1–3 cm long, narrowly wing angled with spreading beaks. $2n = 14$. Hammocks, disturbed sites, Fla., Ala., naturalized from Asia. Spring, fall. *C. acutangulus* Lam.

2. C. siliquosus L.—Shrub up to 1 m tall, usually less, often densely branched. Leaves ovate to oblong-lanceolate, mostly 1–3 cm long, crenate, thin, short petiolate. Flowers solitary or two together; sepals

linear, 6 mm long, petals obovate, 4–5 mm long. Capsule linear, cylindrical-compressed, 5–8 cm long. $2n = 28$. Hammocks, disturbed sites, south Fla., Mexico, W.I., tropical America. All year.

3. **C. orinocensis** HBK, JUTES—Annual slender herbs up to 6 dm tall, usually smooth, sparingly branched. Leaves 4–10 cm long, lanceolate to lance-ovate, acute to acuminate, crenate, glabrate. Peduncles solitary, sepals linear, 5 mm long, green, petals spatulate, 4–6 mm long. Capsule slightly flattened, linear, 4–7 cm long, attenuate short beaked, glabrate. Hammocks, disturbed sites, Fla. to Tex., Ariz., Mexico, W.I., Central America, South America. All year.

121. MALVÀCEAE MALLOW FAMILY

Herbs, shrubs, or trees with alternate, usually stipulate leaves. Flowers bisexual or rarely unisexual, radially symmetrical or nearly so, often solitary. Calyx mostly five-lobed and persistent in fruit subtended by an involucel of two to many bracts, petals mostly five, free but fused at the base to the staminal tube; stamens, monadelphous. Ovary usually three- to many-locular, often five superior, style branched above the staminal tube, the branches as many as or twice as many as the carpels. Fruit dry, rarely fleshy, capsular, drupelike, or breaking into cocci. About 92 genera and 1500 species of cosmopolitan distribution.

1. Involucel immediately beneath calyx of two to many bracts, bracts sometimes minute and inconspicuous or partially fused into a cup.
   2. Involucel of two to five bracts.
      3. Leaves and upper stems hirsute; corolla orange . . . .   1. *Malvastrum*
      3. Leaves and upper stems glabrous or nearly so; corolla pink, yellow, or white.
         4. Some or all the leaves three-lobed; bracts toothed; herbs or subshrubs   2. *Gossypium*
         4. Leaves entire, not lobed; bracts entire; trees or shrubs . . . . .   3. *Thespesia*
   2. Involucel of five to many distinct bracts partially united into a cup.
      5. Involucel of five partially fused bracts; mature carpels with spines . . .   4. *Urena*
      5. Involucel of five to many distinct bracts; carpels without spines.
         6. Bracts five to eight . . . . . .   5. *Pavonia*

6. Bracts many, indefinite.
  7. Styles undivided, stigma not
    spreading; corolla yellow . .   6. *Cienfuegosia*
  7. Styles divided, stigmas spreading;
    corolla yellowish green, rose, red,
    pink, or white.
    8. Corolla tending to remain
      closed; carpels forming a
      drupelike fruit; flowers nod-
      ding . . . .   7. *Malvaviscus*
    8. Corolla open; carpels forming a
      five-valved capsule; flowers
      erect or horizontal.
      9. Leaves hastate or sagittate,
        rough hirsute; locules one-
        seeded . . . . .   8. *Kosteletzkya*
      9. Leaves usually cordate or
        lobed, entire to serrate,
        smooth or tomentose; lo-
        cules many-seeded.
        10. Calyx not spathaceous . .   9. *Hibiscus*
        10. Calyx spathaceous, splitting
          on one side at flowering . 10. *Abelmoschus*
1. Involucel not present, no bracts immediately subtending the calyx (involucre of
  leafy bracts present in MALACHRA).
11. Upper stem and leaves hispid with long
  hairs; flowers in involucrate heads . . 11. *Malachra*
11. Upper stem and leaves glabrous to soft
  tomentose; flowers in cymose clusters or
  solitary but not in heads.
  12. Flowers 2–3 cm in diameter . . . 12. *Abutilon*
  12. Flowers smaller, less than 2 cm in di-
    ameter.
    13. Seeds two to many in each bladder-
      like carpel, carpels beakless; flowers
      solitary on long, delicate peduncles,
      1–4 cm long . . . . 13. *Herissantia*
    13. Seeds one in each leathery carpel, car-
      pels beaked; flowers mostly cymose
      or solitary on short peduncles, 1–2
      cm long . . . . 14. *Sida*

1. MALVÁSTRUM Gray   FALSE MALLOW

Annual or perennial herbs or subshrubs, pubescence usually of stellate hairs. Leaves unlobed, stipules linear to lanceolate. Flowers subsessile or short peduncled, axillary or in short spikes; involucel of one to three linear bractlets; calyx companulate, five-lobed, petals five.

Ovary five- to many-locular, style branches equal in number to the carpels. Fruit a schizocarp. About 50 species, chiefly tropical America.

1. **M. corchorifolium** (Desc.) Britt. ex Small, FALSE MALLOW—Herbs with stems up to 9 dm tall. Leaves ovate to lance-elliptic, 2–6 cm long, coarsely dentate. Calyx lobes deltoid-ovate, acuminate, petals orange yellow. Ovary bristly pubescent. $2n = 48$. Mature carpels beakless. Coastal hammocks, disturbed sites, south Fla., W.I. All year.

M. coromandelianum (L.) Garcke, with strigose stems with four-armed hairs, and subulate, beaked, mature carpels, should also occur in south Fla. $2n = 24$. It is a native of tropical America and is often a weed. All year.

2. GOSSÝPIUM L. COTTON

Tall herbs, shrubs, or small trees with long-petioled, stipulate, three- to nine-lobed leaves, black dotted with oil glands throughout. Flowers large, axillary, solitary; calyx five-toothed or truncate, rather small, subtended by three large, usually foliaceous bracts; petals five, white, yellow, or red, often dark colored at the base. Ovary three- to five-locular, style unbranched, stigmas five. Fruit a loculicidal capsule, seeds usually covered with long hairs or short tomentum. About 20 species in tropical and subtropical regions.

1. **G. hirsutum** L. WILD COTTON—Herbs or shrubs up to 4 m tall, or branches spreading widely, stems coarsely pubescent with long, spreading, simple and stellate hairs. Leaves 5–15 cm long, mostly three-lobed, smooth above, hirsute or villous below. Involucel bractlets 3–6 cm long, cordate-ovate, dissected into ten to thirteen segments. Petals pale yellow, whitish fading into pink, with purplish base. Capsule ovoid, rough, seeds bearing long white or brownish cotton. $2n = 52$. Coastal hammocks, south Fla., Fla. Keys, W.I., Mexico, tropical America. All year.

3. THESPÈSIA Solander ex Correa

Trees or shrubs with long-petioled, entire or lobed leaves. Flowers large, conspicuous; involucel of three to five small, deciduous bractlets, calyx truncate or five-lobed. Ovary five-locular, style five-branched; fruit a loculicidal capsule, or coriaceous and indehiscent. About seven species, pantropical.

1. **T. populnea** (L.) Solander ex Correa, SEASIDE MAHOE, PORTIA TREE—Shrub or tree with young stems bearing peltate scales, becoming glabrate. Leaves ovate-cordate, 5–20 cm long, cordate at base, entire. Flowers solitary, axillary; calyx cupular, 7–9 mm long, petals 5–6 cm

long, yellow becoming purplish. Capsule subglobose, 3 cm wide. $2n =$ 26. Coastal areas in south Fla., Fla. Keys, W.I., pantropical. All year.

4. URÈNA L.

Herbs or shrubs, the pubescence of stellate hairs. Leaves lobed or angulate, midvein beneath with a long slitlike gland. Flowers small, axillary or in terminal spikes. Involucel bractlets five, fused, calyx five-lobed, petals five, staminal tube about equal to the petals in length. Ovary five-locular, fruit style of ten branches of five mericarps, indehiscent and covered with barbed spines. About six species, pantropical.

1. **U. lobata** L. CAESAR WEED—Herbs or shrubs up to 3 m tall, branching, stellate tomentose or almost glabrate. Leaves broadly ovate to suborbicular, 5–10 cm long, shallowly lobed or angulate, dentate or serrulate, densely stellate tomentose beneath. Involucel bractlet 5–7 mm long, petals pink or rose, 1–2 cm long. Mature carpels 6 mm long, bearing stiff spines. Disturbed sites, various soils, south Fla., W.I., tropical America, tropical Asia. All year.

5. PAVÒNIA Cav.

Herbs or shrubs usually with stellate pubescence. Leaves various, stipules subulate. Flowers usually solitary in upper leaf axils; involucel bractlets four to many, calyx five-lobed, petals often colored and showy. Style with eight to many branches, fruit capsular or indehiscent, often with one to three spines at the tip. About 100 species, chiefly in tropical America. *Malache* B. Vogel

1. Corolla pink; leaves hastate . . . . . 1. *P. hastata*
1. Corolla white or greenish yellow; leaves
  ovate . . . . 2. *P. spicata*

1. **P. hastata** Cav.—Shrubs or subshrubs up to 2 m tall with gray, coarse pubescence. Leaves hastate or narrowly sagittate, about 3–5 cm long. Involucel bracts ovate, 4–5 mm long, petals reddish, about 15–30 mm long. Fruiting carpels 4 mm long, spines. $2n = 56$. Coastal areas, sandy soil, Fla., Ga. on coastal plain, naturalized from South America. Spring, fall.

2. **P. spicata** Cav. MANGROVE MALLOW—Coarse shrubs up to 3 m tall, puberulent. Leaves ovate to elliptic-lanceolate or elliptic-cordate, 6–15 cm long, subentire. Involucel bracts six to eight, elliptic-lanceolate; calyx lobes ovate, petals greenish yellow, 2–3 cm long. Mature carpels about 1 cm long, keeled, rostrate, the two beaks very short. Low coastal

mangrove forests, seashore marshes, south Fla., W.I., tropical America. All year. *Malache scabra* B. Vogel, *P. racemosa* Sw.

6. CIENFUEGÒSIA Cav.

Shrubs or subshrubs with erect, branching stems and alternate, entire or lobed leaves. Flowers conspicuous, usually solitary; involucel of three to many bracts, sepals five, fused below; petals five, styles fused. Fruit a three- to five-valved capsule. About 20 species, chiefly American.

1. **C. yucatanensis** Millsp.—Woody herbs or shrubs mostly up to 1 m tall or more. Leaves 2–6 cm long, broadly ovate or elliptic, becoming lance-ovate above. Involucel bracts very small, subulate; petals bright yellow. Capsule about as long as the calyx. $2n = 20$. Coastal hammocks, Fla. Keys. This species can be mistaken for HIBISCUS ssp. All year.

C. heterophylla (Vent.) Garcke is a related species occurring in South America.

7. MALVAVÍSCUS Adans. WAXMALLOW

Shrubs or low trees with dentate or lobed, petioled leaves and linear stipules. Flowers mostly solitary on distinct peduncles from the upper leaf axils; involucel bracts five to many, linear; calyx campanulate, five-lobed; petals five, usually bright red, and convolute forming a tubular-campanulate corolla. Ovary five-locular, styles ten, stigmas capitate. Fruit fleshy and berrylike, the mature carpels separating. Three species, chiefly American.

1. **M. arboreus** Cav.—Bushy shrubs up to 2 m tall. Leaves narrowly elliptic to suborbicular or broadly ovate, coarsely toothed to three- to five-lobed, glabrous to densely tomentose underneath, usually 3–6 cm long. Involucel bracts mostly six to ten, linear to narrowly spatulate, 0.5–1 cm long; calyx tubular about 1–1.5 cm long; petals forming a tubular corolla, dark red, remaining closed. Fruits fleshy, drooping, about 1–2 cm wide, bright red. Represented by two varieties in south Fla.:

1. Leaves cordate at base; usually densely tomentose underneath . . . . . . la. var. *drummondii*
1. Leaves truncate at base, glabrous beneath,
   or only sparingly pubescent . . . . lb. var. *mexicanus*

1a. M. arboreus var. **drummondii** (T & G) Schery, WAXMALLOW—Involucel bracts six to nine, linear to linear-lanceolate. Coastal areas, south Fla., Tex., Mexico, W.I. *M. drummondii* T & G

1b. M. arboreus var. **mexicanus** Schlecht.—Involucel bracts six to ten, narrowly spatulate or oblong. Disturbed sites, roadsides, introduced and locally naturalized. *M. grandiflorus* HBK

8. KOSTELÉTZKYA Presl FEN ROSE, SALT MARSH MALLOW

Shrubs or herbs with coarse stellate pubescence and leaves angulate-lobed or sagittate. Flowers solitary to many, axillary, or in terminal racemes or panicles, involucel bracts seven to ten or obsolete; calyx five-lobed, petals pinkish, yellowish, white, or purplish. Ovary three-locular, style branches five. Fruit a capsule. About 12 species in America, Mediterranean area, and Africa.

1. Corolla white; stems and leaves bristly with stinging hairs . . . . 1. *K. pentasperma*
1. Corolla pink; stems and leaves densely pubescent.
   2. Lower leaves three-lobed, cordate; calyx densely hirsute . . . . . . 2. *K. althaeifolia*
   2. Lower leaves ovate-hastate; calyx canescent . . . . 3. *K. virginica*

1. **K. pentasperma** (Bertero ex DC.) Griseb.—Slender herbs up to 1 m tall, usually less, often much branched, stems stellate pubescent. Leaves variable in shape, narrowly oblong to deltoid or ovate, mostly 3–7 cm long, crenate-serrate or dentate, sparsely pubescent. Flowers solitary, involucel bracts linear, calyx with bristly pubescence, 4.5–5 mm long; petals white, about 1 cm long. Capsule about 1 cm wide, deeply five-lobed. Coastal hammocks, south Fla., W.I. The leaves and stems have stinging hairs. All year.

2. **K. althaeifolia** (Chapm.) Rusby, FEN ROSE—Herbs with branching stems, about 1 m tall and covered with stellate pubescence. Leaves lanceolate to ovate, acuminate, 5–15 cm long, coarsely toothed, larger ones three-lobed and cordate. Calyx five-lobed, the segments lanceolate, involucel bracts subulate; petals pink, about 4 cm long. Capsule about 1 cm wide, deeply five-lobed. Coastal hammocks, south Fla., W.I. The leaves and stems have stinging hairs. All year.

3. **K. virginica** (L.) Presl ex Gray, SALT MARSH MALLOW—Herbs or subshrubs with stems mostly 1 m tall or more. Leaves coarsely stellate pubescent, about 5–15 cm long, ovate or hastate, dentate or crenate-serrate. Flowers solitary in upper, axils or in terminal panicles; calyx five-lobed, canescent, involucel bracts subulate or obsolete; petals pink, about 3–4 cm long. Capsule 1 cm wide. Low ground, marshes, Fla. to La., N.Y. on coastal plain. All year.

Possibly **K. althaeifolia** is only a variety of **K. virginica** rather than a separate species.

9. HIBÍSCUS L.

Shrubs, herbs, or trees with variable, stipulate, often lobed leaves. Flowers usually large and showy, solitary or clusters, racemes or panicles, involucel bracts usually many, or sometimes only four to five, free or fused, calyx campanulate, five-lobed or toothed, corolla rotate, petals usually five. Ovary five-locular, style branches five, the stigmas capitate or flattened. Fruit a loculicidal capsule, five-valved. About 250 species, chiefly tropical. *Pariti* Adans.

Hybrids between a number of species occur both in cultivation and in the wild. Many horticultural forms are found in cultivation.

1. Leaf blades scabrous or tomentose beneath.
  2. Larger leaves less than 6 cm long; corolla cylindric, nodding . . . . . .   1. *H. pilosus*
  2. Larger leaves over 6 cm long; corolla funnel shaped, erect or horizontal.
    3. Leaves elliptic-cordate to elliptic-ovate, entire, not lobed; corolla yellow, becoming pink or red . . . . .   2. *H. tiliaceus*
    3. Leaves orbicular-ovate to ovate-lanceolate, irregularly serrate, shallowly three- to five-lobed; corolla pink or purplish.
      4. Bractlets forked at tip; petals 6–10 cm long; sepals bristly . . . .   3. *H. furcellatus*
      4. Bractlets entire, not forked; petals 12–15 cm long; sepals tomentose .   4. *H. grandiflorus*
1. Leaf blades glabrous or nearly so beneath.
  5. Leaves deeply three- to five-lobed, serrate; bractlets strigose . . . . . .   5. *H. acetosella*
  5. Leaves ovate or elliptical, dentate; bractlets glabrous.
    6. Petals recurved, cut into narrow lobes .   6. *H. schizopetalus*
    6. Petals not recurved, entire . . . .   7. *H. rosa-sinensis*

1. **H. pilosus** (Sw.) Fawc. & Rendle—Herb or subshrub up to 18 dm tall with adpressed stellate pubescence often in dense lines. Leaves deltoid-ovate, mostly 2–5 cm long, irregularly crenate-dentate and sometimes with obscure angulate three-lobed margins; stipules 4–6 mm long. Flowers solitary in upper axils; involucel bracts linear, 7–9 mm long; calyx about 12 mm long, petals bright red, 2–2.6 cm long. Capsule 1 cm long or longer, with stellate pubescence. Hammocks, south Fla., Mexico, W.I. All year.

2. **H. tiliaceus** L. MAHOE—Evergreen shrub or small tree up to 6 m tall. Leaves long petioled, rounded cordate or ovate-cordate, 8–30 cm

long, deeply cordate at base, whitish beneath and covered with dense tomentum of stellate hairs. Flowers large, solitary, on short peduncles; involucel bracts fused for more than half their length to form an eight- to fourteen-toothed cup; sepals 1.5–3 cm long, petals 4–8 cm long, yellow turning red with age or green when dried. Capsule 1–3 cm long, ovoid, stellate pubescent. $2n = 80,92,96$. Coastal areas, south Fla., Mexico, pantropical. All year. *Pariti tiliaceum* (L.) St. Hil.

The species is unknown in the wild state and is found as an escape from local cultivation.

3. H. furcellatus Desr.—Coarse herb or shrub up to 2 m tall with young branches densely stellate tomentose. Leaves broadly ovate to suborbicular, mostly 8–12 cm long, cordate at base, undulate-dentate and often shallowly lobed, closely stellate tomentose on both surfaces. Flowers axillary, involucel bracts linear, ten to fourteen, clearly forked at apex, hirsute; calyx lobes triangular, pubescent; petals rose or pinkish, 6–8 cm long. Capsule about 2.5 cm long, ovoid, strigose pubescent. $2n = 72$. Sandy soils, south Fla., W.I., South America. All year.

4. H. grandiflorus Michx. SWAMP HIBISCUS—Woody herbs or subshrubs up to 2 m tall or more with adpressed stellate pubescence over stems. Leaves sagittate or hastate, three-lobed, these irregularly toothed. Flowers solitary in upper axils; involucel bracts linear, about 1.5–2 cm long; calyx longer, up to 3–4 cm long; petals pinkish or purple, about 12–15 cm long. Capsule 2–3 cm long, coarsely pubescent, ovoid. $2n = 38$. Low ground, gladeland, Fla. to Miss., Ga., on coastal plain. Spring, summer.

5. H. acetosella Welw. ex Hieron.—Slender shrubs up to 6 dm tall or more, leaves and stems often appearing entirely red. Leaves long petioled, deeply three- to five-lobed, serrate, reddish colored, broadly ovate to suborbicular in outline. Involucel bracts linear, eight to ten; calyx deeply lobed, the lobes deltoid with scattered long hairs, median vein with a gland; petals red, spreading. Capsule about 2 cm long. Old fields, disturbed sites, introduced and locally established in south Fla. Spring, summer. *H. eetveldeanus* de Wild. & Dur.

6. H. schizopetalus (Mast.) Hook. f. ROSE MALLOW—Slender shrubs with nearly smooth stems. Leaves elliptic-oblong or oblong-ovate, acute, serrate-dentate, smooth. Flowers solitary in upper axils, peduncles usually pendent, several times longer than subtending leaves. Involucel bracts minute less than 2 mm long, or absent; calyx tubular-campanulate, about 2 cm long, unequally divided; petals reddish, 4–5 cm long, reflexed, dissected into narrow segments. Widely distributed on shell mounds, near old habitations in south Fla. and Fla. Keys, naturalized from east Africa. All year.

7. H. rosa-sinensis L.—Woody herb or shrub up to 3 m tall, nearly glabrous. Leaves ovate-acuminate, crenate-dentate, with rather short petioles. Flowers solitary in upper axils; involucel bracts six to seven, more than 5 mm long, linear; calyx campanulate, lobes lanceolate; petals rose, red, purplish, white, or of other colors, variable also in size, sometimes "double," staminal column conspicuously exserted, equal to or greater in length than the petals. Capsule 2–2.5 cm long, usually not maturing on cultivated plants. $2n = 72,92$. Old fields, disturbed sites, near old habitations, and in widespread cultivation, naturalized from tropical Asia. All year.

H. coccineus (Medicus) Walt. with leaves deeply palmately parted or compound and **H. militaris** Cav. with hastate three- to five-lobed leaves have both been reported for peninsular Fla. They may occur in south Fla.

10. ABELMÓSCHUS Medicus

Herbs or subshrubs usually hispid pubescent. Leaves lobed or parted, stipulate. Flowers axillary, solitary or in racemes; involucel bracts four to sixteen, distinct or slightly fused at the base; calyx spathaceous; five-toothed at the tip, slitting at one side at flowering; corolla usually yellow with a dark purple center, petals five. Ovary five-locular, ovules many, style five-branched. Fruit a loculicidal capsule. About 15 species in Asia and Australia.

1. A. esculentus (L.) Moench, OKRA, GUMBO—Annual coarse herbs up to 1 m tall or more with dense, hirsute pubescence. Leaves long petioled, about 10–15 cm long, ovate or ovate-lanceolate, three- to five-lobed, deeply cordate at base, margins dentate-serrate. Flowers long peduncled, solitary, involucel bracts eight to ten, linear, hirsute, petals 4–8 cm long, bright yellow. Capsule lance-ovoid, 4–7 cm long. $2n = $ c. 130,132. Escaped from cultivation, disturbed sites, naturalized from Asia. Spring, fall. *Hibiscus esculentus* L.

11. MÁLACHRA L.

Coarse herbs or shrubs often with hispid pubescence. Leaves simple, lobed or angulate, long petioled, stipulate. Flowers small, in dense axillary or terminal heads subtended by large leaflike, often scarious involucral bracts; bractlets often irregularly scattered among flowers; calyx five-lobed, campanulate, petals five, reddish, yellow, or white, staminal tube short. Ovary five-locular; style branches ten. Mature carpels separating from the axils, obovoid, indehiscent or semidehiscent. About eight species from tropical America.

1. M. alceifolia Jacq.—Coarse herb or subshrub up to 2.5 m tall, erect or procumbent, variously branched, hispid overall with long, stiff, simple, branched or stellate hairs. Leaves mostly 6–15 cm long, rounded or broadly ovate three- to seven-angled or lobed, dentate, stipules filiform, up to 2 cm long. Heads in the upper axils, sessile or on short peduncles; involucral bracts 1–2.5 cm long, broadly ovate or deltoid, dentate or sinuate; calyx tubular-campanulate, whitish, petals yellow, about 1–2 cm long. Mature carpels 3–4 mm long. $2n = 56$. Disturbed sites, roadsides, south Fla., Mexico, W.I., tropical America. All year.

M. urens Poit., less hispid hirsute, with flowers red, is also reported for the Fla. Keys. It is probably only a form of **M. alceifolia.**

12. ABÙTILON Mill.

Herbs, shrubs, or small trees usually pubesent with mostly cordate, angulate, or lobed, stipulate, simple leaves. Flowers axillary, solitary, or in cymes, small or conspicuous. Involucel absent; calyx five-lobed, campanulate, petals five, staminal tube divided at the apex into slender filaments. Ovary five- to many-locular; style branches equal in number to locules. Mature carpels free or fused at the base, bivalved. About 150 species, chiefly in tropical and subtropical regions.

1. A. permolle (Willd.) Sw. INDIAN MALLOW—Coarse herb or shrub up to 1.5 m tall, stellate pubescent overall. Leaves suborbicular-cordate, mostly 5–10 cm long, deeply cordate at base, irregularly crenate, stellate tomentose below. Flowers solitary, axillary on long peduncles, calyx 8–10 mm long, densely stellate pubescent, petals 1.5 cm long, yellow. Ovary seven to ten-locular, mature carpels 1 cm long, beaked. Coastal areas, disturbed sites, south Fla., Fla. Keys, Mexico, Central America, W.I. All year.

13. HERISSÀNTIA Medicus

Subshrubs diffusely branched, erect or procumbent and vinelike with stems stellate pilose. Leaves ovate or rounded-ovate, 2–10 cm long, acute or acuminate, deeply cordate at base, crenate. Flowers solitary on thin 1–4 cm long peduncles, axillary; calyx 4–6 mm long, pubescent, campanulate, involucel absent; petals five, 6–9 mm long, free, spreading. Ovary of eight to fourteen carpels, style branches as many as the carpels. Fruit depressed-globose, 10–20 mm wide, carpels papery, thin, inflated, whitish or light green with stellate pubescence. *Gayoides* Small, *Bogenhardia* Reichenb.

A single polymorphic species native to tropical and warm temperate regions of America.

1. H. crispa (L.) Briz. BLADDER MALLOW—Characters of the genus. $2n = 14$. Coastal areas and hammocks, disturbed sites, Fla., Tex. on the coastal plain, W.I., tropical America, Old World tropical regions. All year. *Gayoides crispum* (L.) Small, *Bogenhardia crispa* (L.) Kearney

14. SÌDA L.

Low shrubs or herbs. Leaves variable, usually serrate or lobed, stipules linear or lanceolate. Flowers axillary or terminal, small, involucel absent, calyx five-toothed or lobed; petals usually yellowish, orange, white, or red. Ovary with five or more locules; mature carpels usually two-valved, often beaked. About 125 species in tropical and subtropical areas.

1. Stems decumbent or prostrate . . . . 1. *S. procumbens*
1. Stems erect, not decumbent.
  2. Leaves short tomentose beneath.
    3. Leaves ovate-cordate, cordate, or truncate at the base; stems and branches tomentose . . . . 2. *S. cordifolia*
    3. Leaves lanceolate, more or less rhombic, cuneate at the base; stem and branches not tomentose . . . . 3. *S. rhombifolia*
  2. Leaves glabrous or merely stellate pubescent beneath, not tomentose.
    4. Petals usually purplish red 6–10 mm long; flowers often in involucrate clusters . . . . 4. *S. ciliaris*
    4. Petals yellow, orange, or white; flowers not in involucrate clusters.
      5. Petioles mostly over 1 cm long, often with spinose tubercles; leaf base truncate . . . . 5. *S. spinosa*
      5. Petioles mostly less than 1 cm long, tubercles lacking; leaf base cuneate 6. *S. acuta*.
        6. Upper leaves narrowly linear; petals yellow . . . . . 7. *S. elliottii*
        6. Upper leaves lanceolate or narrowly obovate; petals orange . 8. *S. rubromarginata*

1. S. procumbens Sw.—Annual or perennial with much branched prostrate stems and woody roots; stems minutely stellate pubescent and pilose with simple hairs. Leaves ovate or rounded-ovate, 6–20 mm long, cordate or truncate at base, coarsely crenate, long petioled. Flowers axillary, solitary on slender peduncles, calyx 5–6 mm long, puberulent to long hirsute, stellate pubescent; petals pale yellow or whitish; mature carpels five, 3–4 mm long, smooth, short beaked. Disturbed sites, south

Family 121. **MALVACEAE.** *Herissantia crispa:* a, flowering branch, × ½; b, flower, × 7½; c, fruit, × 2½.

Fla., Fla. Keys, Tex., Mexico, tropical America, naturalized from tropical America. All year.

2. **S. cordifolia L.**—Annual or sometimes perennial velutinous herbs up to 1 m tall, usually less; stems covered with dense stellate tomentum. Leaves ovate, suborbicular, to lance-oblong, 3–8 cm long, cordate or obtuse at the base, dentate or serrate. Flowers axillary and terminal, racemous, panicled or fascicled on short peduncles, many flowered; calyx 6–7 mm, densely tomentose; petals about 1 cm long, yellow or yellow orange. Mature carpels seven to twelve, beaked, stellate tomentose. $2n = 28$. Pinelands, disturbed sites south Fla., Fla. Keys, Mexico, pantropical, naturalized from tropical America. All year.

3. **S. rhombifolia L.**—Herb or subshrub, usually annual, up to 1 m tall or more, branching, with glabrate or finely stellate-pubescent stems. Leaves lanceolate or rhombic-oblong, three-veined, finely serrate, pale tomentose or glabrate beneath. Flowers mostly solitary in axils on long peduncles; calyx 6–7 mm long, lobes triangular; petals yellow or whitish, sometimes purplish at the base. Mature carpels ten to fourteen, with one to two short beaks. $2n = 14,28$. Disturbed sites, roadsides, Fla. to Ariz., N.C., Mexico, W.I., pantropical. All year.

4. **S. ciliaris L.**—Annual or perennial herb, spreading or procumbent to ascending, diffusely branched from the base, up to 30 cm long, covered with appressed stellate pubescence. Leaves narrowly oblong to obovate or suborbicular, 1–3 cm long, serrate, stellate pubescent. Inflorescence terminal, headlike, calyx 4–5 mm long, hirsute, lobes triangular, petals usually dark purplish red, 6–10 mm long. Mature carpels seven to eight, about 2 mm long, with short murications. Hammocks, open places, Fla. Keys, Tex., Baja, Calif., Central America, South America. All year.

5. **S. spinosa L.** INDIAN MALLOW—Annual erect herbs usually less than 1 m tall, few branches, with stems minutely stellate pubescent. Leaves linear-lanceolate to ovate-elliptical, 1–4 cm long, often with one to two minute, spinelike, infrapetiolar tubercles below the leaf node, margins crenate-serrate, stellate pubescent beneath. Flowers solitary, sometimes glomerate; calyx 5–7 mm long, tomentose; petals normally yellow or lighter colored. Mature carpels five with two short spines at the tip. $2n = 14$. Disturbed sites, Fla. to Tex., Mass., Mich., W.I., pantropics. Summer.

6. **S. acuta Burm.**—Herbs up to 1 m tall, usually less, woody at the base, with erect, pubescent or glabrate stems. Stipules lanceolate to subulate, often falcate, persistent and conspicuous, prominently three-veined, larger, 1–1.5 cm long, leaves narrowly oblong-lanceolate to ovate, somewhat asymmetric, rhombic, distichous, serrate. Flowers solitary in

leaf axils, subsessile; calyx 6–8 mm long, lobes acute, petals buff yellow. Mature carpels seven to twelve, 3–4 mm long at maturity, with two short beaks at the apex. $2n = 14,28$. Disturbed sites, pinelands, Fla., southeast U.S., Mexico, pantropical. All year. *S. carpinifolia* L.f.

7. **S. elliottii** T & G—Perennial up to 0.5 m tall, branched. Leaves narrowly linear to lance-oblong, 2–6 cm long, subentire to sharply serrate. Flowers solitary in the leaf axils, pedicels often much longer than calyx, calyx glabrate or thinly puberulent, lobes triangular, often red margined, petals bright yellow. Mature carpels eight to twelve, each with two erect beaks. Old fields, disturbed sites, Fla. to Tex., Va., Mo. Summer.

8. **S. rubromarginata** Nash—Subshrubs up to 1 m tall, branching. Leaves variable, oblong-lanceolate to elliptic-ovate or narrowly obovate mostly 2–6 cm long, coarsely serrate. Calyx campanulate, lobes triangular; petals light orange. Mature carpels eight to ten, mostly 3–4 mm long, pitted on the back, usually aristate. $2n = 28$. Disturbed sites, sandy pineland, endemic to Fla. Spring, fall.

122. BYTTNERIÀCEAE Byttneria Family

Trees, shrubs, vines, or herbs with alternate, simple or palmately compound, stipulate leaves. Flowers bisexual, variously arranged; sepals five, petals five, stamens free, five to many, arranged singly or in groups or monadelphous. Ovary of usually five fused carpels, or sometimes reduced to one. Fruit usually dry, capsular indehiscent or dehiscent. About 50 genera and 600 species, principally in tropical regions. *Buettneriaceae* sensu Small

Some authors place these species in the inclusive family Sterculiaceae.

1. Ovary five-locular.
   2. Petals recurved with hoods; style one . . 1. *Ayenia*
   2. Petals flat without hoods; styles five . . 2. *Melochia*
1. Ovary one-locular; styles one . . . . . 3. *Waltheria*

1. AYÈNIA L.

Herbs or shrubs, pubescence chiefly of branched hairs, with membranaceous, toothed leaves. Flowers small, pedicelled, fascicled or in subsessile, axillary cymes; calyx five-parted; petals five, clawed, the expanded apex of each petal inrolled and fused to the staminal tube; staminal tube short, lobed, anthers five, solitary in the sinuses. Ovary five-locular; fruit a muricate two-valved capsule. About 20 species in tropical America.

1. A. euphrasiifolia Griseb.—Herbs or subshrubs with taproots and prostrate or decumbent stems up to 35 cm long, sparingly pubescent. Leaves broadly ovate, reniform, to suborbicular or orbicular, about 1 cm long, serrate, stipules 0.5–1 mm long, subulate. Flowers small, axillary; calyx 2–3 mm long; petals 6–6.5 mm long, reniform, or hastate, reddish; androgynophore 1–2 mm long. Capsule about 3 mm long, 4–5 mm wide, seeds tuberculate. Pinelands, Fla. Keys, W.I.

A. pusilla L. with petals only 4–4.5 mm long has been reported from Key West. It is a native of tropical America. No recent collections, however, have been examined.

2. MELÒCHIA L.

Herbs or shrubs with serrate leaves, pubescence chiefly of stellate hairs. Flowers small in various arrangements, headlike, glomerate spicate, cymose, or paniculate; calyx five-lobed, campanulate sometimes inflated, petals five, spatulate or oblong; stamens five, opposite the petals, fused near the base. Ovary five-locular, styles five. Fruit a capsule, five-valved. About 60 species, pantropical but chiefly American. *Riedlea* Vent.

1. Stems and leaves thickly pubescent; inflorescence headlike or glomerate . . . .   1. *M. hirsuta*
1. Stems and leaves glabrous or barely pubescent; inflorescence a loose, interrupted spike or spikelike panicle.
  2. Capsule strongly five-angled.
    3. Leaves green, glabrous or barely pubescent . . . .   2. *M. pyramidata*
    3. Leaves pale green, densely and closely stellate tomentose . . . . . .   3. *M. tomentosa*
  2. Capsule five-lobed but not angled . . .   4. *M. corchorifolia*

1. M. hirsuta Cav.—Herbs or subshrubs up to 6 dm tall or more with hirsute stems. Leaves subsessile or short petioled, ovate-oblong to rhombic-ovate, 2–7 cm long, serrate, usually densely pubescent. Flowers in headlike clusters, sessile, forming an interrupted spike; petals purplish, violet, or pink, 10–12 mm long. Capsule subglobose, about 3 mm wide, within the calyx. Disturbed sites, pinelands, south Fla., W.I., Mexico, tropical America. *Riedlea hirsuta* (Cav.) DC., *R. serrata* authors

M. hirsuta var. **glabrescens** Gray with thin, sparingly pubescent leaves has been described from south Fla. in low pinelands.

2. **M. pyramidata** L.—Herbs or subshrubs up to 1 m tall, usually less, glabrous. Leaves slender petioled, ovate-oblong to oblong-lanceolate, 3–7 cm long, serrate, usually glabrous but occasionally stellate pubescent. Flowers pedicelled in axillary cymes or forming interrupted spikes, calyx 3–4 mm long, petals rose or violet or reddish. Capsule five-angled, broadly pyramidal, about 6 mm long, glabrous or stellate pubescent. Pinelands, south Fla., W.I., tropical America.

3. **M. tomentosa** L.—Herbs or subshrubs up to 2.5 m tall, branches densely stellate tomentose, the plant appearing pale green or grayish. Leaves thick, oblong to rhombic-ovate, mostly 3–8 cm long, densely stellate tomentose on both sides, serrate to dentate. Flowers mostly pedicelled in axillary or terminal cymes; calyx 6 mm long, petals 8–18 mm, pink to violet. Capsule pyramidal, pubescent. Pinelands, south Fla., Mexico, W.I., Central America, tropical America. *Moluchia tomentosa* (L.) Britt.

4. **M. corchorifolia** L.—Herbs or subshrubs up to 1.5 m tall, glabrous or sparingly pubescent, branching. Leaves ovate to ovate-lanceolate, serrate, sometimes obscurely three-lobed; leaves in dense clusters, which often form interrupted spikes or spikelike racemes; calyx 3–4 mm long, campanulate; petals purplish. Capsule ovoid or subglobose, 4–5 mm wide. Disturbed sites, Fla. to Tex., S.C.

3. WALTHÈRIA L.

Herbs or shrubs with petiolate, serrate, stipulate leaves with pubescence chiefly of stellate hairs. Flowers variously arranged, small; calyx five-parted, petals five, spatulate; stamens five, fused at the base and opposite the petals. Ovary one-carpellate and one-locular. Fruit a smooth or pubescent capsule. About 30 species, chiefly tropical America.

1. **W. indica** L.—Herbs or subshrubs erect or decumbent, up to 1 m tall but usually less, often much branched. Leaves oblong to orbicular, mostly 4–6 cm long, obtuse, crenate-dentate, densely stellate tomentose on both sides. Flowers sessile in dense axillary headlike clusters, the inflorescence sessile or pedunculate; calyx 4–5 mm long, the segments lance-subulate; petals 4–6 mm long, bright yellow. Capsule 2 mm long, pubescent. Drier soils, disturbed sites, pinelands, hammocks, south Fla., Fla. Keys, tropical America, naturalized in Old World tropics.

DOMBEYA WALLICHII D. Jackson and D. BURGESSIAE Gerrard, small trees with palmately veined, cordate, toothed leaves and showy reddish or pink, pendulous flowers in loose umbels or heads, have been introduced into southern Fla., but they are doubtfully naturalized.

123. HYPERICÀCEAE St. John's-Wort Family

Trees, shrubs, or herbs with opposite, simple leaves. Flowers often showy, bisexual or unisexual, usually terminal in cymes, panicles, or solitary, sepals two to six, free, petals two to six, free, yellow or white; stamens numerous to few, often united into bundles. Ovary one-locular or three- to five-locular, styles free or connate. Fruit various, capsule, berry, drupe, or dry and indehiscent. About 50 genera and 900 species, chiefly tropical. The leaves are usually gland dotted and the sap resinous. Nomen alt. Guttiferae. *Clusiaceae*

1. Herbs or shrubs; leaves less than 6 cm long   1. *Hypericum*
1. Trees; leaves over 6 cm long.
  2. Stems and leaves with yellowish sap; fruit
     four to ten cells, many-seeded . . .   2. *Clusia*
  2. Stems and leaves without yellowish sap;
     fruit one to four cells, few-seeded.
    3. Ovary one-locular; sepals four . . .   3. *Calophyllum*
    3. Two- to four-locular; sepals two   . .   4. *Mammea*

1. HYPERÌCUM L. St. John's Wort

Herbs or shrubs generally smooth throughout with rather small, opposite, entire, usually gland-dotted leaves. Flowers solitary or cymose; sepals four to five, corolla four to five, petals unequal or equal, often black punctate; stamens many, united at the base to form a shallow ring or narrow band or definite bundles. Ovary three- to five-locular, styles three to five. Fruit a thin-walled or coriaceous capsule. About 300 species in tropical and temperate regions. *Sarothra* L., *Sanidophyllum* Small, *Triadenum* Raf., *Ascyrum* L., *Crookea* Small

1. Sepals four, the outer pair usually much
  longer than the inner; petals four.
  2. Styles two; leaves sessile, not clasping . .   1. *H. hypericoides*
  2. Styles three to four; leaves sessile and
    strongly clasping.
    • 3. Sepals and leaves similar in shape;
      leaves strongly clasping . . . .   2. *H. tetrapetalum*
    3. Sepals and leaves distinctly unlike in
      shape; leaves sessile, not strongly
      clasping . . . .   3. *H. stans*

1. Sepals five, approximately equal; petals five.
  4. Stems herbaceous.
    5. Leaves ovate, clasping, three- to five-
      veined . . . .   4. *H. mutilum*
    5. Leaves much reduced appearing as erect
      or appressed, minute scales . . .   5. *H. gentianoides*
  4. Stems woody.
    6. Mature leaf with expanded blade, not
      needlelike, usually over 2 mm wide;
      sepals usually broadened, linear-ellip-
      tic, oblong-ovate, or spathulate.
      7. Leaves linear-lanceolate to lance-
        ovate, acute or acuminate, sessile or
        partially clasping.
        8. Inflorescence appearing naked ow-
          ing to the much reduced bracts;
          leaves and sepals without a basal
          articulation or groove . . .   6. *H. cistifolium*
        8. Inflorescence appearing leafy ow-
          ing to the foliaceous bracts;
          leaves and sepals with a basal ar-
          ticulation or groove . . . .   7. *H. galioides*
      7. Leaves ovate to elliptic-ovate, clasp-
        ing . . . .   8. *H. myrtifolium*
    6. Mature leaves and sepals linear-subu-
      late or needlelike.
      9. Largest leaves usually over 13 mm
        long; sepals usually over 4.5 mm
        long . . . .   9. *H. fasciculatum*
      9. Largest leaves usually less than 11
        mm long; sepals usually 4.5 mm
        long or less.
        10. Mature capsules 6–10 mm long;
          stems often decumbent . . .   10. *H. reductum*
        10. Mature capsules 4–6 mm long;
          stems usually erect . . . .   11. *H. brachyphyllum*

1. **H. hypericoides** (L.) Crantz var. **hypericoides**, St. Andrew's
Cross—Perennial herbs often woody at the base, or subshrubs. Leaves
linear to oblanceolate 5–20 mm long, sessile, punctate. Flowers terminal
and axillary, yellow, about 12–18 mm broad; calyx in two whorls; outer
two sepals ovate, often cordate at base, inner ones narrower, shorter;
petals oblong-linear, about as long as the outer sepals, styles two. Capsule ovoid, about 4 mm long. $2n = 18$. Low pinelands, Fla. to Mexico,
N.J., Mo., Central America, W.I., Bermuda. Spring, fall. *Ascyrum hy-
percoides* L.

2. **H. tetrapetalum** Lam.—Perennial herbs woody at the base or
shrubs up to 1 m tall. Leaves ovate to elliptic-ovate, sessile, punctate.

Flowers terminal and axillary, solitary or few-clustered; calyx in two whorls; outer two sepals foliaceous, inner ones lanceolate to elliptic; petals bright yellow. Capsule ovoid, about 3–4 mm long. $2n = 18$. Low pinelands, Fla. to Ga. on the coastal plain, Cuba. Spring, fall. *Ascyrum tetrapetalum* (Lm.) Vail

3. **H. stans** (Michx.) P. Adams & N. Robson—Shrub with simple or branching stems mostly 3–8 dm tall. Leaves mostly 2–3 cm long, elliptic-oblong, thickly coriaceous, sessile but not strongly clasping. Outer sepals about 1–2 cm long, broadly ovate, inner sepals less than 1 cm long, lanceolate, petals showy, yellow; styles three, rarely four. Capsule ovoid, about 1 cm long. Pine flatwoods, scrub vegetation, dry-to-moist soil, Fla. and Tex. north to N.J., Ky. Summer. *Ascyrum stans* Michx.

4. **H. mutilum** L. DWARF ST. JOHN'S WORT—Perennial or annual herb up to 8 dm tall, often much branched above. Leaves lanceolate to elliptic or ovate, 1–4 cm long, three- to five-veined. Flowers usually in branching inflorescence; sepals linear-lanceolate, acute; petals spreading, 3–4 mm wide. Capsule ellipsoid, 3–4 mm long. Low ground, Fla. to Tex., Newfoundland to Me. Summer, fall.

5. **H. gentianoides** (L.) BSP, PINEWEED—Annual herb up to 5 dm tall with wiry, scaly-leaved stems. Leaves much reduced, appressed, often subulate. Sepals linear 2–3 mm long, corolla yellow or orange, 4–8 mm wide, stamens five to ten. Capsule conic, 4–5 mm long. Disturbed sites, sandy or drier soil, Fla. to Tex., Me., Ontario. Summer, fall. Reported from Broward and Lee Counties, and presumably in our area. *Sarothra gentianoides* L.

6. **H. cistifolium** Lam.—Shrubs up to 1 m tall. Leaves lanceolate to narrowly elliptic, mostly 2–8 cm long. Flowers in branching inflorescence, corymbose; sepals ovate or lanceolate, 3–4 mm long; petals cuneate, 5–8 mm long. Ovary three-lobed; capsule globose, 4–6 mm long. $2n = 18$. Coastal areas, Fla. to La., S.C. Summer. *H. opacum* T & G

7. **H. galioides** Lam.—Shrubs up to 2 m tall, usually less, with evergreen leaves. Blades linear to somewhat broader above, about 5–20 mm long. Sepals linear or linear-spatulate, 3–4 mm long; petals bright yellow, 4–7 mm long, cuneate. Capsule ovoid-conic, 5–6 mm long. Low pinelands, Fla. to Tex., N.C. Summer, fall.

8. **H. myrtifolium** Lam.—Shrub up to 1 m tall, usually less, with evergreen leaves. Blades ovate to elliptic-ovate or lance-ovate, 1–3 cm long, clasping. Sepals foliaceous, ovate, 5–7 mm long; petals obovate, 1–1.5 cm long, bright yellow. Capsule ovoid, 5–6 mm long. Pine flatwoods, low ground, margins of inland swamps, Fla. to Miss. Spring, summer.

9. **H. fasciculatum** Lam.—Shrubs or small trees up to 3 m tall, usu-

ally less, branched, with evergreen leaves. Leaves opposite, linear, 1–2 cm long, punctate. Flowers solitary, sessile, sepals linear, 4–6 mm long, resembling the leaves; petals cuneate to the base, 6–9 mm long. Capsule oblong-cylindric, 6–7 mm long. $2n = 18$. Low ground, Fla. to Miss., S.C., on the coastal plain. Spring, fall.

10. H. reductum P. Adams—Herbs or subshrubs often with decumbent stems forming a low matted plant usually less than 5 dm tall; nodes of youngest stems with auriculate structures, one on each side of leaf base. Leaves linear-subulate, usually less than 11 mm long, dull green above when fresh. Sepals linear-subulate, 4.5 mm long or less, resembling the leaves; petals yellow, cuneate. Ovary three-carpellate, styles three; capsule 6–10 mm long. $2n = 18$. Sandy soil, coastal dunes, Fla. to N.C.

11. H. brachyphyllum (Spach) Steud.—Herbs or subshrubs erect up to 1.5 m tall; stems without auriculate structures at leaf base. Leaves linear-subulate, usually less than 11 mm long, glossary green above when fresh. Sepals linear-subulate, 4.5 mm long or less, resembling the leaves; petals bright yellow. Ovary three-carpellate, styles three; capsule 3.5–5.5 mm long. Low pinelands, ditches, Fla. to Miss., Ga.

2. CLÙSIA L.

Trees, shrubs, dioecious or polygamodioecious, often epiphytic when young, or woody vines with viscid, resinous sap. Leaves opposite, coriaceous with parallel lateral veins. Flowers in branching inflorescences or solitary, small; sepals four to six, petals four to ten, stamens many. Ovary four- to many-locular. Fruit a fleshy or leathery capsule. About 145 species, chiefly tropical America.

1. C. rosea Jacq. BALSAM APPLE—Shrub or tree up to 18 m tall. Leaves short petioled, rounded-obovate, 7–15 cm long, somewhat cuneate-narrowed at base, dark green, thickly coriaceous. Flowers in few-flowered cymes, pedicels thick; sepals four, coriaceous, inner ones up to 1.5 cm long; petals six, rose white, waxy. Capsule eight- to twelve-locular, 5–8 cm in width. Hammocks, lower Fla. Keys, Mexico, W.I., tropical America. Summer.

Found at scattered locations on Little Torch, Big Pine, Sugarloaf, Cudjoe, and Bahia Honda Keys. **Clusia flava** Jacq. erroneously reported for south Fla.; it appears to be restricted to Jamaica and the Grand Cayman Islands.

3. CALOPHÝLLUM L.

Trees producing a colored sap and bearing opposite leaves that have very many close, parallel, lateral veins. Flowers polygamous, small, in axillary or terminal racemes or panicles; sepals four, petals four, fewer, or none; stamens many, free. Ovary one-locular, stigma peltate; fruit drupaceous. About 100 species, chiefly Old World tropics.

1. C. inophyllum L.—Trees up to 30 m tall, usually less, with smooth, four-angled young stems. Leaves 5–10 cm long, coriaceous, broadly elliptic-oblong to obovate, often with undulate margins. Flowers racemose, in the axils of upper leaves; sepals four, reflexed, the outer ones 7–8 mm long, orbicular; petals four, 9–12 mm long, white, obovoid; stamens in four fascicles, many. Ovary globose; drupe 2.5–4 cm in width, green, smooth. Summer.

Introduced into south Fla. from tropical Africa, widely planted, especially in the Lower Fla. Keys and Miami area; possibly naturalized locally, although we have seen no specimens to confirm this. It is included as an aid in identification.

4. MÁMMEA L. Mamey

Trees with resinous sap and thick, penninerved, coriaceous, opposite leaves. Flowers polygamous, axillary, sessile, one to three in a cluster; calyx splitting open into two valves, petals four to six, stamens many. Ovary two-locular, style thickish, stigma peltate. Fruit large, baccate. Four species, one tropical America, three Africa.

1. M. americana L.—Tree up to 15 m tall, usually smaller, with a dense crown and dark green, coriaceous, persistent leaves. Leaves 8–15 cm long, broadly elliptic to obovate, densely pellucid-punctate. Flowers solitary or clustered; sepals concave, 1–1.5 cm long; petals four to six, white, orbicular, 1.5–2.5 cm long. Fruit subglobose, 10–15 cm long with thick, russet colored exocarp and yellowish, juicy sweet mesocarp.

Reported by Moldenke as occurring along the Tamiami Trail canal, Dade County, and naturalized from tropical America, W.I. This species probably is not established in our flora.

124. TAMARICÀCEAE Tamarisk Family

Trees or shrubs with slender, irregular branches and scalelike, alternate, clasping leaves. Flowers very small, usually bisexual, in slender plumelike spikes, racemes, or panicles; sepals four to six, free, petals four to six; disk present, stamens five to ten, free. Ovary superior, one-locular; styles three to four. Fruit a capsule; seeds with tufts or leaves near the tip or all-around. About 5 genera and 100 species, Europe and Asia.

1. TÀMARIX L. Tamarisk

Trees or shrubs with small, scalelike, often imbricate leaves. Numerous small flowers in racemes arising from older wood or in terminal panicles, sepals four to five, petals four to five, stamens four to five. Ovary conic, styles three to four. Capsule separating into three to four valves. Three species from Mediterranean region.

1. **T. gallica** L.—Shrubs or small trees with irregular branches, the younger stems bearing imbricate scalelike leaves. Flowers minute in branching spikes or panicles; petals pinkish or white, deciduous; disk five-lobed. Capsule many seeded, 1 cm long. Disturbed sites, old fields, escaped locally from cultivation, introduced from Europe.

125. CISTÀCEAE Rock Rose Family

Shrubs or herbs with simple, mostly opposite leaves. Flowers bisexual, radially symmetrical, solitary to clustered; sepals three to five, petals five or absent, stamens many. Ovary superior, one-locular; style simple, with three to five fused or free stigmas. Fruit a three- to five-valved capsule. About 7 genera and 160 species, chiefly North America and Mediterranean regions.

1. Flowers minute, 1–2 mm wide; petals three,
    greenish or purplish, persistent . . .   1. *Lechea*
1. Flowers larger, 10–20 mm wide; petals five,
    yellow or wanting . . . .   2. *Helianthemum*

1. LÉCHEA L. Pinweed

Perennial herbs with erect or ascending stems. Leaves small, alternate, ovate to subulate. Flowers very small, numerous; sepals five, biseriate, petals three, usually shorter than sepals; stamens three to twenty-five. Style very short or none, stigmas dark red, plumose. Capsule three-valvate. About 17 species, chiefly eastern U.S., south to W.I., Central America.

1. Exterior sepals as long as or longer than the
    interior ones.
    2. Cauline leaves broadly lanceolate to ellip-
        tic; capsule not longer than calyx . .   1. *L. villosa*
    2. Cauline leaves narrowly lanceolate to
        lance-oblong; capsule much longer than
        calyx . . . .   2. *L. patula*

1. Exterior sepals shorter than interior ones.
  3. Capsule much longer than calyx.
    4. Cauline leaves linear-oblong, abruptly
      acute or obtuse; calyx nearly glabrous   3. *L. deckertii*
    4. Cauline leaves lance-elliptic, sharply
      acute; calyx subappressed pilose . .   4. *L. divaricata*
  3. Capsule shorter or about equal or slightly
    longer than the calyx.
    5. Basal leaves lance-elliptic to broadly
      elliptic-ovate, less than three times as
      long as broad; pedicels exceeding
      calyx; seeds one to two   . . . .   5. *L. cernua*
    5. Basal leaves, when present, elliptic-
      lanceolate to narrowly lanceolate or
      linear, more than three times as
      long as broad; cauline leaves nar-
      rowly lanceolate or oblanceolate to
      narrowly linear . . . .   6. *L. torreyi*

1. **L. villosa** Ell.—Stem simple to strongly branching, up to 9 dm tall, suberect, axis densely villous. Basal leaves 1–1.5 cm long, villous beneath, broadly elliptic-ovate, upper leaves oblanceolate to elliptic-lanceolate. Flowers in spreading many-flowered panicles, calyx 1.4–2 mm long, inner sepals broadly lanceolate, 2 mm long, almost equal to the outer ones. Capsule about 1 mm long. Sandy soil, Fla. to Tex., N.H., Mich. Summer.

2. **L. patula** Leggett—Stems simple or divergently spreading, up to 4 dm tall. Leaves linear, 4–10 mm long, 1–2 mm wide, glabrous; basal shoots with linear-elliptic or spatulate leaves. Flowers minute, clustered, obovoid, internal sepals 1–1.5 mm long, external sepals somewhat longer. Capsule ovoid, 1.5 mm long. Dry sandy pineland, Fla. to S.C. on the coastal plain. All year. *L. exserta* Small, *L. prismatica* Small

3. **L. deckertii** Small—Stems suffruticose, up to 3 dm tall, branches wiry. Leaves 1.5–3.5 mm long, linear, sparingly subappressed pilose. Flowers scattered, few to many, inner sepals about 1 mm long. Capsule subglobose, about 1 mm wide. Sand hills, scrub vegetation, south Fla. to Ga. Summer, fall. *L. myriophylla* Small

4. **L. divaricata** Shuttlew. ex Britt.—Stems nearly procumbent, densely villous, subshrubby. Basal leaves 2–3 mm long, few, lance-ovate to elliptic. Flowering stems numerous, divaricately branched, divergent, leaves abundant on branches, calyx pyriform, pilose to villose, outer sepals somewhat longer than inner ones. Capsule exserted, exceeding the calyx, ovoid about 2 mm long. Endemic to dry sandy soil, Fla. Summer, fall.

5. **L. cernua** Small—Stems branched near the base, less commonly

simple, up to 6 dm tall, often much branched above, silvery strigose pubescent. Leaves elliptic to ovate, about 1 cm long, 4–5 mm wide, strigose; basal shoots with broadly ovate, densely pale pubescent leaves. Flowers clustered, minute; calyx turbinate, 2 mm long at maturity. Capsule obovoid, 2 mm long. In sand scrub vegetation, endemic to central Fla., south on East Coast. Summer, fall.

   6. L. torreyi Leggett ex Britt. var. **congesta** Hodgdon—Stems simple or branching up to 4 dm tall, usually less. Leaves linear to linear-lanceolate, 1–2 mm wide, 5–20 mm long, glabrous above, sparingly pubescent below. Flowers in compact panicles; internal sepals copiously appressed pubescent, obovate, about 2 mm long, external sepals shorter. Pinelands, Fla. to N.C., Miss., Central America. All year.

2. HELIÁNTHEMUM Mill. Frostweed

Perennial herbs or subshrubs from rhizomes. Leaves opposite, narrow, entire with stellate pubescence. Flowers in racemes, cymes, or panicles; sepals three to five, if five then the external two much narrower than the inner; petals five, yellow, later flowers without petals; stamens many. Style short with capitate stigma. About 80 species, chiefly Europe. *Helianthemum* Adans., *Crocanthemum* Spach

1. Flowers in terminal dense cymes; leaves pale
     green beneath, dark green above . . .   1. *H. corymbosum*
1. Flowers in elongate panicles or racemes;
     leaves pale green both upper and lower
     surfaces . . . .   2. *H. nashii*

   1. H. corymbosum Michx.—Stems erect, finely and densely canescent pubescent, up to 2 dm tall. Leaves narrowly elliptic to ovate, or the lowermost obovate, mostly 1–3 cm long, pale green beneath, dark green above. Flowers in terminal dense cymose clusters; petals bright yellow, corolla 15–20 mm wide. Capsule many-seeded, 4–5 mm wide, from petaliferous flower, smaller from apetaliferous ones. Pinelands, dry soil, Fla. to S.C., coastal plain. All year. *Crocanthemum corymbosum* (Michx.) Britt.

   2. H. nashii Britt.—Stems erect, somewhat canescent pubescent or glabrate, up to 4 dm tall. Leaves linear-elliptic to elliptic, 1–3 cm long, midvein very prominent below. Flowers in elongate panicles or racemes; sepals canescent pubescent; petals bright yellow, with fifteen stamens or apetalous with five stamens; corolla 15–20 mm wide. Capsule broadly ovoid, 3–4 mm long. Endemic to scrub vegetation, south peninsular Fla. *Crocanthemum nashii* (Britt.) Barnhart

126. BIXÀCEAE ARNOTTO FAMILY

Trees or shrubs bearing alternate, simple, long-petioled leaves, with colored sap. Flowers bisexual, large, in terminal panicles, the pedicels bearing five glands subtending the calyx; sepals five, petals five, stamens indefinite on a thick receptacle. Ovary superior, one-locular, style simple, stigma two-lobed. Fruit a valved capsule often covered with long spines. A single genus in tropical America.

1. BÍXA L.

Characters of the family

1. B. orellana L. LIPSTICK TREE—Shrub or tree up to 8 m tall with a dense, rounded crown. Leaves 10–20 cm long, ovate to broadly deltoid-ovate, acuminate, palmately five-veined, smooth and green above, paler beneath, rather densely lepidote. Flowers in small, rather few-flowered panicles; sepals 12–15 mm long, brown lepidote; petals 2.5 cm long, variable, pinkish or white. Capsule 2.5–4.5 cm long ovoid, densely covered with reddish brown, soft spines; seeds covered with abundant reddish orange pulp. Introduced from tropical America, cultivated and possibly locally naturalized in south Fla.

127. CANELLÀCEAE CANELLA FAMILY

Trees with simple, alternate, gland-dotted, aromatic leaves. Flowers bisexual, radially symmetrical, in cymose clusters; sepals four to five, free, thickish; petals four to five, thin, imbricate; stamens five to twenty, filaments fused in a tube. Ovary superior, one-locular. Fruit a berry. Four genera and seven species of marked geographical discontinuity, tropical America, South America, east Africa, Madagascar.

1. CANÉLLA P. Browne CINNAMON BARK

Trees up to 10 m tall with gray bark and stems having large, orbicular leaf scars. Leaves 5–10 cm long, coriaceous, obovate, narrowed to a cuneate base. Flowers small, in terminal, subterminal, or axillary panicles or cymes; sepals three, petals five, fleshy, white or purplish rose colored, stamens about twenty. Ovary one-locular, stigma two- to three-lobed. Berry red, or very dark globose, fleshy. A single species in tropical America.

1. C. alba Murray, Wɪʟᴅ Cɪɴɴᴀᴍᴏɴ—Characters of the genus. Hammocks, south Fla., Fla. Keys, W.I., tropical America. Summer, fall. *C. winterana* (L.) Gaertn.

This is one of our most attractive flowering trees. The leaves and bark are strongly but pleasantly aromatic, and they have been used locally in medicinal preparations. Both the flowers and the dark colored berries are conspicuous.

128. VIOLÀCEAE Vɪᴏʟᴇᴛ Fᴀᴍɪʟʏ

Herbs or shrubs with alternate, simple, usually stipulate leaves. Flowers mostly bilaterally symmetrical, solitary or paniculate, bisexual, sometimes cleistogamous; sepals five, persistent; petals five, usually unequal, with the lowermost larger and often spurred or saccate; stamens five. Ovary superior, one-locular, style simple, stigma simple. Fruit a three-valved, loculicidal capsule. About 18 genera and 800 species of cosmopolitan distribution.

1. VIÒLA L. Vɪᴏʟᴇᴛs

Annual or perennial herbs, low, often with apparently all basal or alternate leaves and usually conspicuous, often foliaceous stipules. Flowers usually solitary on distinct axillary peduncles; early flowers complete and chasmogamous, later flowers apetalous and fertile or cleistogamous; sepals unequal, auricled at base, petals unequal, the lowest larger, usually connivent into a sheath enclosing the ovary; anthers subsessile, capsule three-valved, seeds ovoid or globular. A large and complex genus of 400 species, cosmopolitan, chiefly north temperate zone.

1. Corolla pale blue; leaves cordate . . .   1. *V. floridana*
1. Corolla white, purple veined on lower
    petals; leaves narrowly lanceolate to elliptic lanceolate . . . .   2. *V. lanceolata*

1. V. floridana Brainerd—Low herbs with leaves basal. Leaf blades about 3–4 cm long, 2–3 cm wide, cordate-acute, finely serrate, somewhat pubescent above; leaf blades produced after chasmogamous flowering, twice as long as wide on long erect petioles. Flowers on long peduncles that surpass the leaves; corolla pale blue or whitish; apetalous flowers produced underneath the soil or duff, sepals broadly lanceolate. Capsule

Family 127. **CANELLACEAE.** *Canella alba:* a, flowering branch, $\times$ ½; b, flower, $\times$ 7; c, mature fruit, $\times$ 3.

reddish brown, about 1.5 cm long. Sandy soil, southeast, central, and north Fla. Spring.

2. **V. lanceolata** L. Bog White Violet—Low herbs with leaves basal and producing leafy, slender stolons usually bearing apetalous flowers. Leaves mostly 8–15 cm long, narrowly lanceolate to elliptic, obscurely serrate. Flowers solitary on scapes up to 8 cm long, sepals broadly lanceolate, acute. Capsules produced by cleistogamous flowers 6–12 mm long; corolla white, purple veined on lower petals. $2n = 24$. Sandy soil, Fla. to Tex., Nova Scotia, Minn.

V. lanceolata ssp. **vittata** (Greene) Russell has been described as leaves linear-lanceolate up to 10 mm wide, lower surfaces villous pubescent; it also may occur in our area.

129. FLACOURTIÀCEAE Flacourtia Family

Trees or shrubs with simple, alternate, leathery, often pelluciddotted leaves. Flowers small, radially symmetrical, bisexual or unisexual; sepals two to fifteen, free, petals two to fifteen or absent; stamens many, indefinite. Ovary of two to ten fused carpels, one-locular. Fruit a berry, capsule, or indehiscent. About 70 genera and 800 species in tropical and subtropical regions.

1. FLACOÚRTIA L'Her.

Shrubs or small trees bearing short-petioled, sharply dentate leaves. Flowers small, bisexual or unisexual, mostly in axillary clusters or racemes; sepals four to five, scalelike, petals absent, stamens many. Ovary superior, surrounded by a lobed disk, styles two to six or many. Fruit berrylike, indehiscent. About 15 species in tropical Africa, Asia. *Xylosma* Forst.

1. **F. indica** (Burm. f.) Merrill, Governor's Plum—Smooth shrub or tree up to 7 m tall, often with axillary spines. Leaves ovate to elliptic, 5–8 cm long, crenate-dentate, acuminate. Flowers yellowish, in axillary clusters. Berry globose, 1–2 cm wide, purplish, with small, thin seeds. Hammocks, disturbed sites, south Fla., W.I. Introduced from south Asia, Madagascar, and probably naturalized.

130. TURNERÀCEAE Turnera Family

Shrubs or herbs usually pubescent with alternate, entire or lobed leaves. Flowers usually solitary in leaf axils, bisexual, radially symmetrical; calyx tubular, sepals five, imbricate, petals five, free, inserted on the calyx tube; stamens five, alternate with the petals. Ovary superior, one-locular, styles three, slender. Fruit a loculicidal capsule, three- to many-seeded. About 8 genera and 120 species, chiefly in tropical America and tropical Africa.

1. Low shrubs; flower sessile or subsessile, pe-
    duncle fused to the petiole . . . .   1. *Turnera*
1. Herbs; flower pedicelled, peduncle distinct   2. *Piriqueta*

1. TURNÈRA L. TURNERA

Perennial or annual herbs or low shrubs, usually pubescent with simple hairs, with toothed leaves. Flowers solitary, axillary in the upper leaves, mostly yellow and often showy, sessile or subsessile, two-bracteate at base, peduncles often fused to the petioles of subtending leaves; petals obovate. Styles three, the stigmas cleft to form a brushlike structure. Capsule globose, three-valved, seeds curved, numerous. About 60 species, chiefly tropical America.

1. **T. ulmifolia** L. var. **ulmifolia**—Erect herb or shrub up to 7 dm tall, usually less, closely pilose pubescent with pale, spreading or appressed hairs. Leaves lanceolate to oblong-ovate, mostly 3–10 cm long, acute or acuminate, two-glandular at the base, serrate. Peduncles fused with the petioles, flowers thus appearing sessile on the petiole; calyx deeply five-lobed, petals 1–3 cm long, yellow. Capsule 6–10 mm long, densely pubescent. $2n = 20,28,30,32$. Coastal areas, Fla., and Fla. Keys to La., naturalized from tropical America. All year. *T. ulmifolia* var. *angustifolia* Willd.

A highly variable species with about 12 varieties distributed throughout the range for the genus.

2. PIRIQUÈTA Aubl.

Erect herbs usually with abundant, stellate pubescence. Leaves alternate, serrate or dentate, rarely entire, often two-glandular at the base. Flowers with distinct pedicels, yellow, axillary, cymose or in reduced panicles or solitary; hypanthium ten-ribbed, sepals five, imbricate in bud; petals five, corona membranaceous, often inconspicuous. Stigmas two- or three-cleft. Capsule three-valved, seeds numerous. About 30 species, chiefly in tropical America.

1. **P. caroliniana** (Walt.) Urban—Stems erect, up to 4 dm tall, glabrous to stellate tomentose or hirsute. Leaves narrowly linear or lanceolate to ovate-lanceolate or elliptic, 2–7 cm long, entire to somewhat ir-

.

regularly dentate. Petals deep to light yellow, capsule 5–7 mm long. Pineland, Fla. to N.C. on the coastal plain, South America, W.I. A polytypic species with five varieties, two of which occur in our area:

1. Leaves and stems hirsute to tomentose; cauline leaves narrowly lanceolate to ovate-lanceolate . . . . 1a. var. *tomentosa*
1. Leaves and stems glabrous or nearly so; cauline leaves linear to linear-lanceolate . . 1b. var. *glabra*

1a. P. caroliniana var. **tomentosa** Urban—Leaves narrowly lanceolate to ovate-lanceolate, stems and leaves hirsute to tomentose. Drier sites, sandy soil, Fla. to N.C. All year. *P. tomentosa* sensu Small, not HBK

1b. P. caroliniana var. **glabra** (DC.) Urban—Leaves linear to linear-lanceolate, stems and leaves glabrous or nearly so. Low ground, south Fla., Fla. Keys. All year. *P. glabrescens* Small, *P. viridis* Small

131. PASSIFLORÀCEAE PASSIONFLOWER FAMILY

Trees, shrubs, or herbaceous climbing plants with axillary tendrils. Leaves alternate, simple, or compound, entire or lobed, often with glands on the petioles; stipules mostly small. Flowers bisexual or rarely unisexual, frequently large, showy, with the receptacles cuplike; sepals five or more, corona of one or more threadlike filaments or scales; stamens five or more, often fused in bundles. Ovary one-locular, stalked, styles free or fused, stigma capitate or discoid. Fruit a capsule or berry. About 12 genera and 580 species, mostly in PASSIFLORA, chiefly tropical.

1. PASSIFLÒRA L. PASSIONFLOWER

Herbaceous or woody vines, rarely erect shrubs or trees, with alternate, simple or compound, often lobed leaves, predominately three- to five-veined, stipules present. Inflorescence axillary, subtending bracts forming an involucre or sometimes small and remote; flowers bisexual, calyx tube cylindric to campanulate; sepals five, fleshy or thin, petals five or absent; corona of one to several series of free or more*or less fused filaments, operculum within the corona membranous, flat or plicate, entire or dissected, stamens five, fused to form a tube, usually raised on an androgynophore. Ovary borne on an elongate gynophore, styles three. Fruit berrylike, indehiscent, with a musculaginous pulp, seeds several to many. About 400 species, chiefly in tropical America.

1. Leaf margins finely serrate; leaf blade
   deeply three-lobed . . . .   1. *P. incarnata*
1. Leaf margins entire; leaf blades entire or
   lobed.
   2. Stems and leaves white tomentose; leaves
   thick . . . .   2. *P. multiflora*
   2. Stems and leaves glabrous to pubescent,
   not white tomentose.
    3. Petioles with stalked glands near base
   of leaf blade; stems glabrous or sparingly pubescent.
     4. Stipules large, ovate-foliaceous . .   3. *P. pallens*
     4. Stipules small, acicular . . . .   4. *P. suberosa*
    3. Petioles without glands; stems pubes-
   cent . . . .   5. *P. sexflora*

1. **P. incarnata** L. APRICOT VINE—Vines glabrate or finely pubescent
on the younger stems. Leaves three-lobed, blades 8–12 cm long,
finely serrate. Flowers conspicuous, sepals about 3 cm long; petals five,
pale lavender or bluish, corona purplish or lavender. Berry ovoid, 5–10
cm long, yellowish at maturity. $2n = 18$. Dry soils, disturbed sites, old
fields, Fla. to Tex., Va., Mo. Spring, summer.

The fruits, called maypops, are edible.

2. **P. multiflora** L.—Vines white tomentose. Leaves 6–12 cm long,
thick, elliptic or ovate, entire. Flowers white; sepals 5–6 mm long, yellowish green; petals linear or linear-lanceolate, white. Berry subglobose,
6–8 mm wide, purplish black at maturity. Hammocks, south Fla. and
Fla. Keys, W.I. All year.

3. **P. pallens** Poepp. ex Mast. PASSIONFLOWER—Vines glabrous and
glaucous. Leaves equally three-lobed, the blades 4–8 cm long, entire;
stipules conspicuous, foliaceous, ovate to reniform. Flowers conspicuous,
greenish white, sepals with a foliaceous appendage on the back just below the apex, petals absent. Berry ovoid, 4–5 cm long, yellow at maturity.
Coastal and interior hammocks, south Fla., W.I. All year.

4. **P. suberosa** L. CORKY-STEMMED PASSIONFLOWER—Vines glabrous
or sparingly pubescent. Leaves usually 4–10 cm long, entire to three-
lobed, stipules small, inconspicuous. Flowers greenish; sepals linear or
lanceolate or broader, 7–9 mm long; petals absent. Berry 5–10 mm wide,
globose, purplish black at maturity. $2n = 24$. Hammocks, south Fla.,
Tex., Mexico, W.I., tropical America, Hawaii, naturalized in the Old
World tropics. All year. *P. pallida* L. Older vines often develop broad,
corky outgrowths. A highly polymorphic species.

5. **P. sexflora** Juss.—Vines herbaceous or woody with densely pilose

Family 131. **PASSIFLORACEAE.** *Passiflora suberosa:* a, fruiting branch, ×
½; b, flower, × 6; c, mature fruit, × 2¼.

or tomentulose stems. Leaves three-lobed, about 4–8 cm long, entire, softly pubescent beneath. Flowers in two- to ten-flowered cymes; sepals 9–10 mm long, petals greenish white. Ovary densely brown pubescent. Berry globose, 6–10 mm wide, densely pubescent. Hammocks, Everglade Keys, south Fla., Mexico, W.I., tropical America. All year.

P. lutea L., with three-lobed leaves, entire margined, glandless petioles, greenish white flowers, and berries over 1 cm wide, has been reported for our area. It is common in the East and Midwest and in other areas of the southeast U.S.

132. CARICÀCEAE Papaya Family

Small trees, shrubs, or sometimes herbs with a terminal cluster of leaves and milky juice. Leaves alternate, simple, and variously lobed or divided. Flowers bisexual or unisexual, racemose; staminate flowers with five-lobed calyx, small; petals fused; stamens ten, epipetalous. Pistillate flower with five-lobed calyx, petals eventually free, ovary superior, one-locular. Fruit a pulpy berry, many-seeded. Four genera and about 40 species in tropical America and Africa. *Papayaceae*

1. CÁRICA L.

Small dioecious trees, shrubs, or large herbs with usually simple stems and long-petioled, simple, deeply lobed leaves. Flower axillary, cymose or racemose, staminate corolla with slender tube, lobes contorted in bud, stamens biseriate in the throat of the corolla. Style short, five stigmas linear or cleft. Berry small or quite large, many-seeded. About 30 species in tropical America.

1. C. papaya L. Papaya—Stems usually simple, columnar, up to 6 m tall, usually less, bearing conspicuous petiolar scars and a leafy top. Leaves very large on long petioles, palmately seven-lobed, the lobes pinnate lobed. Staminate flowers white in pedunculate panicles 10–30 cm long, corolla 2–3 cm long. Pistillate corolla longer than staminate flowers, petals distinct. Berry very variable, the wild plants producing rather small, obovoid fruits, the cultivated plants producing large, oblong, yellow or orange fruits. Hammocks and in cultivation, south Fla., pantropical. All year.

Naturalized in areas removed from cultivation and probably disseminated by birds. Wild plants have been found far up in peninsular Fla.

133. LOASÀCEAE Loasa Family

Herbs, subshrubs, or climbers usually covered with bristly pubescence. Flowers bisexual, radially symmetrical, solitary, cymose, or capitate; calyx tube four- to five-lobed, fused to the ovary, petals four to five, inserted on the calyx; stamens many, free or in bundles opposite the petals. Ovary inferior, one- to three-locular, style one. Fruit a capsule, one- to several-valved. About 14 genera and 250 species, widely distributed but chiefly in tropical America and southwest Africa.

1. MENTZÈLIA L.

Annual or perennial herbs, sometimes subshrubs, with brittle diffusely branching stems that become white or yellowish with age; covered with rigid, barbed pubescence. Leaves alternate, lobed or variously divided. Flowers terminal, cymose or axillary; calyx tube cylindric or ovoid, petals five to ten, stamens ten to many. Style three-cleft at the apex, ovary one-locular. Capsule three- to seven-valved. About 60 species, America.

1. **M. floridana** Nutt. Poor Man's Patch—Perennial with erect, much branched stems up to 1 m tall. Leaves 2–10 cm long, ovate or deltoid-ovate, three-lobed, dentate. Flowers mostly solitary or in few-flowered cymose clusters; calyx lobes lanceolate; petals 15–20 cm long, deep yellow. Capsule 1–1.5 cm long. $2n = 20$. Coastal hammocks, sand dunes, south Fla., Fla. Keys, W.I. All year.

134. BEGONIÀCEAE Begonia Family

Monoecious herbs or subshrubs with thick rhizomes or tubers usually with succulent, jointed stems. Leaves alternate, simple, mostly oblique. Flowers radially or bilaterally symmetrical; staminate flowers with two sepals, petals two to five or absent; stamens many. Pistillate flower with similar perianth, ovary inferior, two- to four-locular, rarely one-locular, angled or winged. Fruit a capsule or berry. Five genera and 900 species, of which about 800 are in Begonia, pantropical.

1. BEGÒNIA L.

Annual or perennial succulent plants with stems frequently woody below. Leaf blades strongly oblique. Perianth parts free, four in two pairs in staminate flowers, five parts in pistillate flowers, styles short, mostly three. Fruit an unequally three-winged capsule. About 800 species, pantropical, including many ornamental and cultivated forms.

1. **B. semperflorens** Link & Otto, EVERFLOWERING BEGONIA— Stems succulent up to 1 m tall, usually less. Leaves 5–7 cm wide with teeth ending in abrupt tips. Bracts of inflorescence fringed with marginal hairs; corolla pale white, petals 6–14 mm long in staminate flowers. Capsule lax. Low ground, south Fla., naturalized from South America. All year.

Numerous species and horticultural forms are cultivated in south Fla., some with very large leaves and showy, often highly colored flowers. that have a waxy appearance.

135. CACTÀCEAE   CACTUS FAMILY

Succulent herbs or shrubs with the fleshy stems often spiny and the leaves usually reduced to scales or apparently absent. Flowers solitary, bisexual, radially symmetrical or rarely becoming bilaterally symmetrical; calyx and corolla not clearly distinguishable, inner sepals petaloid, petals numerous, usually in several series, stamens many. Ovary inferior, one-locular; fruit a berry, many-seeded. About 25 genera and 1000 or more species, chiefly in warm, arid regions of Western Hemisphere, especially Mexico and South America.

Identical chromosome numbers of $2n = 22$ have been reported in numerous diverse genera. *Opuntiaceae*

1. Mature stems leafless or leaves rudimentary and highly modified; succulent shrubs.
   2. Stems conspicuously jointed and flattened or the main stems cylindrical and lateral and the terminal shoots jointed and flattened . . . .   1. *Opuntia*
   2. Stems and branches cylindrical with ridges or grooves, not jointed nor flattened.
      3. Epiphytic, pendent shrubs, diffusely branching; flowers minute, about 1 mm long . . . .   2. *Rhipsalis*
      3. Terrestrial shrubs, trees, or coarse vines, erect or reclining; flowers large, mostly over 3 cm long . . . . .   3. *Cereus*
1. Mature stems with broad leaves; vines . .   4. *Pereskia*

1. OPÚNTIA Mill.   PRICKLY PEAR

Low plants, branching from the bases or sometimes erect and

shrublike or treelike, branches (joints) compressed and flattened, succulent. Leaves very small, scalelike, deciduous, areoles axillary and bearing spines (or spines absent in the mature plant), bristles (glochids), pubescence, glands, or flowers. Flowers large, solitary; sepals green or colored, grading into the petals, petals colored, usually red or yellow, spreading; stamens shorter than the petals. Ovary bearing scalelike leaves, styles thick. Berry juicy or dry, spiny or naked. About 240 species, widely distributed in Americas, especially Mexico and South America.

1. Spines strongly barbed; stem joints attached
    loosely and separated readily.
    2. Spines circular to broadly elliptic in cross
        section; joints 4–8 cm long, 3–4 cm wide   1. *O. triacantha*
    2. Spines flattened at least basally and very
        narrowly elliptic in cross section; joints
        10–30 cm long, 5–6 cm wide . . . .   2. *O. cubensis*
1. Spines (if present) not strongly barbed; stem
    joints attached firmly.
    3. Main stems flattened, obviously jointed;
        all stems determinate; spines present or
        not.
        4. Joints elongate to obovate or elliptic,
            4–16 cm long, 4–12 cm broad; spines
            if present gray, white, or brown, not
            yellow at maturity . . . . . .   3. *O. compressa*
        4. Joints narrowly obovate, spatulate to el-
            liptic, 17–25 cm long, 7.5–12.5 cm
            broad; spines if present yellow or
            slightly reddish or brownish . . .   4. *O. stricta*
    3. Main stems cylindrical, not conspicuously
        jointed, indeterminate.
        5. Smaller branches all flattened and rela-
            tively thick . . . .   5. *O. spinosissima*
        5. Smaller branches of two kinds, some
            cylindrical and some flat and very
            thin . . . .   6. *O. brasiliensis*

1. **O. triacantha** (Willd.) Sw.—Stems nearly prostrate, irregularly branched; joints 4–8 cm long, 3–4 cm broad, loosely attached. Leaves (scales on new joints) 2–3 mm long, purplish or greenish, areoles inconspicuous; spines one to six, mostly 4 cm long. Flowers commonly solitary on the joints; sepals green; petals pale yellow, corolla about 2.5–3.5 cm in diameter. Berry urn shaped, 2–3 cm long, tuberculate, red or reddish purple. $2n = 22$. Hammocks, Big Pine Key, Fla. Keys, W.I. Spring, summer. *O. abjecta* Small

2. **O. cubensis** Britt. & Rose—Stems erect, up to 1 m tall, usually

less, much branched and loosely spreading; joints 1–3 dm long, 5–6 cm wide, bright green. Leaves (scales) 2–4 mm long, pinkish, ovoid, areoles conspicuous, armed, spines in clusters of usually two to three or up to five to six, these 4–5 cm long. Flowers few, sepals about five, greenish or yellowish, purple tinged; petals bright yellow, corolla 6–9 cm in diameter. Berry obovoid, 2–3.5 cm long, red or reddish purple, seeds many. Endemic to hammocks, Big Pine Key, Fla. Keys. All year. *O. ochrocentra* Small

3. **O. compressa** (Salisb.) Macbride—Stems erect or prostrate and spreading, irregularly branching and forming clumps or large mats. Leaves (scales) subulate, areoles conspicuous, spineless or with one or sometimes a few two-clustered, gray to nearly white or brown. Sepals green, narrow, petals yellow, corolla 4–8 cm wide. Berry obovoid, reddish or purplish, seeds many. Two varieties are found in our area:

1. Plant usually three or four joints high; joints elongate, 4–16 cm long, 4–6 cm broad . . . .   3a. var. *ammophila*
1. Plant creeping or ascending to about two joints high; joints obovate or sometimes elliptic, 7.5–10 or 14 cm long, 5–9 or 12 cm broad . ' . . . .   3b. var. *austrina*

3a. **O. compressa** var. **ammophila** (Small) L. Benson—Stems erect, spines solitary or sometimes two-clustered. Sepals 7–14 mm long, petals light yellow, corolla 4–7 cm in diameter. Berry 2.5–3.5 cm long, purplish. Pinelands, endemic to Dade County, Fla., north to Duval and Dixie Counties. Spring, summer. *O. austrina* Small

3b. **O. compressa** var. **austrina** (Small) L. Benson—Stems prostrate or nearly so, spreading and forming clumps or large mats, spineless or with one to two spines per cluster. Sepals 8–15 mm long, petals bright yellow, corolla 5–8 cm in diameter, usually with a reddish center. Berry reddish or purple, 3–5 cm long. Coastal areas from Fla. to La., S.C. Spring, summer. *O. compressa* sensu Small, *O. polycarpa* Small

4. **O. stricta** Haw.—Stems erect or suberect, much branched, at maturity up to 2 m tall and six or more joints high; joints narrowly obovate, spatulate to elliptic, thick, 17–25 cm long, 7.5–12.5 cm broad. Leaves (scales) subulate or cone shaped, 2–7 mm long, areoles distant but conspicuous, spines none or solitary, or three- to six-clustered, mostly yellow at maturity. Flowers few to many per joint, sepals green, deltoid to ovate or reniform, petals light yellow to reddish, corolla 5–10 cm wide. Berry purplish. Two varieties are found in our area:

1. Joints narrowly elliptic or narrowly obovate,
   mostly 1.7–2.5 dm long, 0.7–1.2 dm broad;
   spines none or few and solitary in the upper areoles of the joint . . . . . .   4a. var. *stricta*
1. Joints obovate, mostly 1–3 (or 4) dm long,
   1.2–1.7 (or 2.5) dm broad, spines one to
   eleven in each areole . . . . . .   4b. var. *dillenii*

4a. **O. stricta** var. **stricta**—Stems erect or suberect, leaves subulate 5–7 mm long. Spines none or solitary, flowers often many per joint, petals light yellow, most eight to ten, corolla usually 5–6 cm or up to 10 cm in diameter. Berry pyriform or obovoid, depressed. Shell mounds and fields, coastal areas, Fla. to Tex., Va., W.I. Spring, summer.

4b. **O. stricta** var. **dillenii** (Ker) L. Benson—Stems suberect, shrub-like, branches loosely spreading, leaves cone shaped, 2–5 mm long. Spines stout, three- to six-clustered, curved, pale to deep yellow, flowers few to many per joint, petals few, yellow, reddish, or salmon colored, corolla 6–8 cm in diameter. Berry compressed with a depressed center, seeds many. $2n = 22$. Hammocks, coastal sand dunes, Fla. to Tex., S.C., Mexico, W.I., South America. Spring, summer. *O. atrocapensis* Small, *O. zebrina* Small, *O. keyensis* Britt, *O. dillenii* (Ker) Haw.

These plants in the U.S. tend to be intergrades with var. **stricta,** and hybrid swarms abound.

5. **O. spinosissima** (Martyn) Mill. SEMAPHORE CACTUS—Stems up to 2 m tall, treelike, copiously spiny, and from fibrous roots; main stem ovate or elliptic in cross section. Areoles abundant, each with a cluster of five to nine spines, these salmon colored, becoming light gray, the longer ones 6–12 cm long; joints arising at the summit of the main stem, these elliptic, curved, thin, 2–3 cm long, abundantly spiny. Flowers many per joint, sepals green, deltoid to reniform, acute; petals bright red, 1 cm long or less, ovate. Berry obovoid, 3–5 cm long, yellow, seeds few, irregular. Endemic to hammocks, Key Largo and Big Pine Key, Fla. Keys. All year. *Consolea corallicola* Small

6. **O. brasiliensis** (Willd.) Haw.—Stems erect, treelike, up to 5 m tall, the main stem cylindrical, supporting cylindrical branches bearing both cylindrical and flattened joints. Leaves (scales) flat, oblong, 1–3 mm long, areoles small, minutely white woolly, spines slender, usually solitary on the flat joints, 1–4 cm long, reddish or brown tipped, becoming gray. Flowers several on terminal joints, sepals ovate, petals lemon yellow, corolla 3–4 cm wide. Berry subglobose, light yellow, slightly con-cave in the middle. Hammocks, disturbed sites, shell mounts, south Fla.,

naturalized from South America. All year. *Brasiliopuntia brasiliensis* (Willd.) Haw.

2. RHÍPSALIS Gaertn.

Slender, succulent epiphytes, usually hanging from the branches of trees, much branched, sometimes forming dense clumps; stems terete or leaflike. Leaves absent or represented by mere scales; areoles of the mature stems small, marginal, bearing pubescence, bristles, or flowers but never spines. Flowers small, solitary; perianth parts few, white, stamens few to many, in two rows on the outer margin of the disk. Ovary small, stigma with three or more lobes. Berry globose, white or colored. About 50 species, mostly in American tropics.

1. **R. baccifera** (J. Miller) Stearn, MISTLETOE CACTUS—Stems forming dense masses or clusters 1 m or more long, much branched, fleshy, terete, rather pale green; branches usually in pairs or verticillate. Flowers lateral, sessile, solitary; sepals 1 mm long; petals white, 2–3 mm long, elliptic to ovate; stamens 1–2 mm long. Berry naked, white or pinkish, 4–5 mm wide. $2n = 22$. An epiphyte on mangroves and buttonwood in hammocks, rare in south peninsular Fla., Mexico, tropical America, W.I., tropical Africa, Ceylon. All year. *R. cassutha* Gaertn.

3. CÈREUS (L.) Mill.

Larger stems cylindrical to prismatic, 15 to 100 times as long as their diameter, 0.3–16 m long, 6 mm to 7.5 dm in diameter; areoles on the ribs, bearing hairs and spines; areoles usually circular to elliptic. Leaves not discernible on the mature plant. Flowers and fruits on the old growth of preceding seasons, each within a spine-bearing areole or at its margin. Flower 2.5–22 cm in diameter, diurnal or nocturnal; floral tube above its junction with the ovary (superior floral tube) almost obsolete to funnelform or long and tubular. Fruit fleshy, usually pulpy, often edible, with or without surface scales, spines, hairs or bristles, usually orbicular to ovoid or ellipsoid, usually 1.2–7.5 cm long; seeds black, longer than broad. About 25 species in the American tropics.

1. Stems and branches strongly three- or four-
     angled at maturity (with up to six angles
     in immature stems).
  2. Stems suberect, not rooting . . . . 1. *C. pentagonus*
  2. Stems vinelike, climbing, bearing aerial
          roots at the nodes . . . . . . 2. *C. undatus*
1. Stems and branches fluted or grooved at

maturity, but not strongly angled.
3. Superior floral tube slender, at least 4 cm
long; stems mostly reclining or vinelike;
petals 6–7.5 cm long, 1.2–2 cm broad.
4. Hairs of the areoles on the floral tube
10–15 mm or more in length, white,
flexible, forming conspicuous tufts;
petals entire, narrowly lanceolate or
oblanceolate, the. upper margins not
expanded . . . .　3. *C. eriophorus*
var. *fragrans*
4. Hairs of the areoles on the floral tube
6–8 mm long, white or tawny brown,
flexible or stiff, not forming conspicuous tufts; petals erose-denticulate
toward the apices, oblanceolate, expanded above . . . .　4. *C. gracilis*
3. Superior floral tube broad, not more than
1.5 cm long; stems erect; petals 3.8–4.5
long, about 1.2 cm broad　. . . .　5. *C. robinii*

1. **C. pentagonus** (L.) Haw. DILDOE CACTUS—Stems and branches
reclining and spreading, up to 10 m long, stout, older ones usually three-
angled and dark green, the younger four- to six-angled and light green.
Areoles distant, with four or usually seven to eight slender spines, the
central ones 1–4 cm long; floral tube trumpet shaped, 8–10 cm long;
calyx green or purple tinged; sepals deltoid-triangular, acuminate, up to
4 cm long; corolla 8–12 cm in diameter; petals linear-acuminate, white.
Berry ovoid, 4–6 cm long, bright red and shiny. Coastal hammocks,
beaches, higher marl ground, Tex., south Fla. and Fla. Keys, to north
South America. *Acanthocereus floridanus* Small

2. **C. undatus** Haw. NIGHT-BLOOMING CEREUS—Stems (when terrestrial) arching or recurving, or (when epiphytic) vinelike and producing aerial roots, light green, often in dense, compact clusters; stems
usually three-ribbed, the margins undulate; areoles distant, with one to
four spines, in the stem sinuses. Flowers large, up to 30 cm long, the
outer perianth segments whitish, yellowish green, or pink tinged, the
inner segments white, lanceolate-acute. Style elongated up to 25 cm long,
lobes of the stigma up to twenty-five. Berry ovoid, 6–12 cm long, deep
red when mature, covered with large foliaceous scales. Hammocks,
around old habitations, south Fla., tropical America and widely cultivated. *Hylocereus undatus* (Haw.) Britt. & Rose

3. **C. eriophorus** Pfeiffer var. **fragrans** (Small) L. Benson—Stems
canelike, several to many, sprawling, 1–5 m long, 2.5–5 cm in diameter;
ribs eight to twelve, prominent, the intervening grooves obvious but

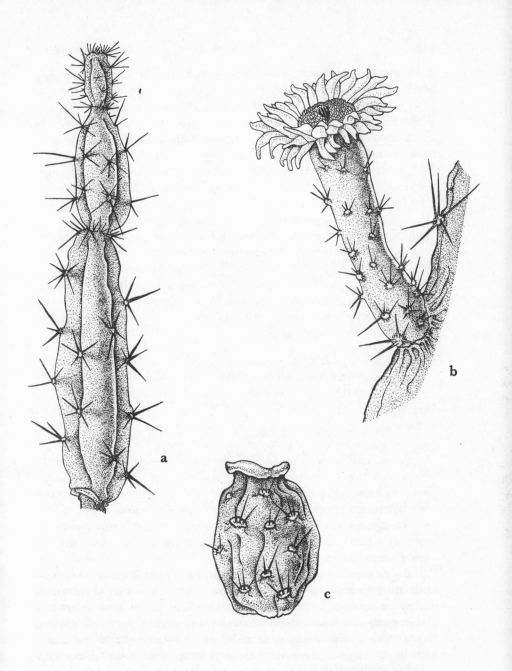

Family 135. **CACTACEAE.** *Cereus pentagonus:* a, stem portion, × ½; b, flower, × ½; c, mature fruit, × 1.

shallow; spines numerous, more or less obscuring the stem, brown with black tips or grayish tan with yellow tips, 1–4 cm long. Flower nocturnal, faintly scented, 7.5–10 cm in diameter, 15–17 cm long; hairs of the areoles of the floral tube 10–15 cm or more long, white, flexible, in conspicuous tufts; sepals purplish brown with paler or white margins, up to 5.5 cm long; petals pale pink or white, up to 7.5 cm long, entire; fruit orange red, fleshy, smooth, with some wool-like spines, obovoid, 5–6 cm long. Endemic to east coast of Fla. south to Cape Sable and Big Pine Key, Fla. Keys.

4. **C. gracilis** Mill.—Stems erect, reclining, often vinelike, terrestrial or epiphytic, simple or branched, often forming dense clusters up to 6 m long, nine- to eleven-ridged. Spines seven to nine on each areole, slender, 1–3 cm long, young buds often covered with white or brown pubescence. Swollen base of floral tube covered with lanceolate, imbricate scales, sepals linear, petals narrow-spatulate to oblanceolate, stamens erect or nearly so. Berry globular, mostly 6–7 cm in diameter. Two distinct varieties are found in our area:

1. Young buds covered with short brown pubescence; berry globular, 6–7.5 cm in diameter, dull yellow at maturity . . . 4a. var. *aboriginum*
1. Young buds covered with white pubescence; berry depressed, globose, 6 cm in diameter, dull red at maturity . . . . . . 4b. var. *simpsonii*

4a. **C. gracilis** var. **aboriginum** (Small) L. Benson, PRICKLY APPLES —Stems nine- to eleven-ridged, erect or reclining, spines up to 1 cm long, pink, becoming gray at maturity. Swollen base of floral tube covered with lanceolate scales, inner sepals narrowly linear-acuminate, petals oblanceolate, stamens erect. Shore hammocks and shell mound, endemic to the Ten Thousand Islands area, southwest Fla. coastal areas and Fla. Keys. *Harrisia aboriginum* Small

4b. **C. gracilis** var. **simpsonii** (Small) L. Benson, PRICKLY APPLES— Stems nine- to ten-ridged, simple or branched, often vinelike or terrestrial spines slender 1–2.5 cm long. Swollen base of floral tube covered with broadly lanceolate, imbricate scales with slightly protruding white pubescence, sepals linear, acuminate, petals white, narrow-spatulate, stamens nearly erect. Endemic to coastal areas, Tampa Bay to the Ten Thousand Islands area, Fla. *Harrisia simpsonii* Small

5. **C. robinii** (Lemaire) L. Benson—Shrub or small tree up to 10 m tall with erect, simple or fastigiate branches, branches sometimes numerous, dark green to pale green in color, nine- to thirteen-ribbed,

areoles pubescent, spines thirteen to twenty per areole, up to 2.5 cm long, usually less. Flowers about 5–6 cm long, campanulate, sepals ovate, petals 1–1.5 cm long. Berry depressed, reddish. Two varieties occur in our area:

1. Branches blue green, numerous, ten- to
  thirteen-ribbed; petals about 1 cm long .   5a. var. *robinii*
1. Branches gray green, not profuse, nine- to
  ten-ribbed; petals about 1.5 cm long, the
  outer ones clawed . . . .   5b. var. *keyensis*

    5a. C. robinii var. robinii—Shrub or small tree up to 10 cm tall with simple or fastigate branches forming a narrow head, branches numerous. Spines thirteen to twenty per areole, up to 1.0–1.5 or 2.5 cm long. Flowers diurnal about 5–6 cm long, sepals ovate, petals ovate-obtuse, about 1 cm long. Berry 3.5 cm in diameter, dark red. Rocky hammocks, Fla. Keys, Cuba. *Cephalocereus deeringii* Small
    5b. C. robinii var. keyensis (Britt. & Rose) L. Benson, TREE CACTUS—Shrub o'r small tree up to 6 m tall with erect, ultimately branching stems, the branching not profuse. Spines sixteen to twenty per cluster. Flowers about 5 cm long, sepals ovate-acute; petals about 1.5 cm long, the outer ones clawed, ovate to elliptic. Berry depressed-globose, about 3.5 cm wide, reddish. Endemic to rocky hammocks, Matecumbe Key, Fla. Keys. *Cephalocereus keyensis* Britt. & Rose

4. PERÉSKIA (Plum.) Mill.

    Shrubs, trees, or succulent vines, branched, bearing normal, green, fleshy, entire leaves. Spines in areoles in the leaf axils, leaves alternate, deciduous with age. Flowers solitary, corymbose or paniculate, rotate; sepals narrow, stamens many; stigma lobes linear. Berry globose, red or yellow, often bearing small leaves; seeds black. About 20 species in tropical America.
    1. P. aculeata Mill. LEMON VINE—Large, slender vine up to 10 m long; spines solitary or two to three together on older stems, axillary spines 2–4 mm long, strongly recurved. Leaves lanceolate to oblong-elliptic, short petioled, 4–9 cm long, acute or acuminate. Flowers paniculate or corymbose, 2.5–4.5 cm broad, white, yellowish, or pinkish. Ovary bearing small leaves, sometimes spiny. Berry subglobose, yellowish, 1.5–2 cm wide, smooth at maturity. Hammocks, Fla., W.I.
    Reported for peninsular Fla., presumably in our area, but no specimens have been seen.

136. LYTHRÀCEAE Loosestrife Family

Herbs, shrubs, or small trees mostly with opposite or verticillate, entire, simple leaves. Flowers usually radially symmetric, bisexual, in racemes, panicles, cymes, or solitary; sepals fused, forming a tube, petals six or four, present or absent; stamens inserted on the calyx tube. Ovary superior, two- to six-locular. Fruit a capsule opening by a transverse slit. About 23 genera and 450 species, chiefly tropical in distribution.

1. Shrubs or small trees; flowers in terminal
    panicles . . . . 1. *Lagerstroemia*
1. Herbs or low shrubs; flowers axillary or in
    terminal spikes or racemes.
    2. Hypanthium elongate, longer than wide.
        3. Upper stem glabrous or nearly so; flow-
           ers irregular . . . . 2. *Cuphea*
        3. Upper stem glabrous or nearly so; flow-
           ers regular . . . . 3. *Lythrum*
    2. Hypanthium ovoid almost as wide as long.
        4. Low, prostrate or diffuse herbs; fruit
           2–3 mm in diameter; intersepaline
           appendages triangular . . . . 4. *Rotala*
        4. Erect branching herbs; fruit 3–6 mm
           in diameter; intersepaline append-
           ages subulate . . . . 5. *Ammannia*

1. LAGERSTROÈMIA L. Crape Myrtle

Shrubs or trees with mostly alternate leaves. Flowers usually five- to eight-parted, often large and showy in terminal panicles; calyx sub-globose or turbinate; petals clawed, stamens many in a single series. Ovary globose, three- to six-locular. Capsule oblong or ellipsoid, three- to six-valved. About 30 species from Asia and Australia.

1. **L. indica** L.—Shrub or small tree with smooth bark and four-angled young stems. Leaves 2–7 cm long, oblong to rounded, sessile or with short petioles. Flowers showy, usually white, pink, or purple, mostly six-parted, in panicles 5–25 cm long; calyx tube 7–10 mm long, petals 10–20 mm long, stamens many. Capsule 9–14 mm long, about half included within calyx. Cultivated widely throughout Fla. and other parts of the Southeast, and in most warm regions, doubtfully naturalized, but often found around former habitations or waste places; introduced from Asia. Summer.

2. CÚPHEA P. Browne

Annual or perennial herbs, rarely subshrubs, with opposite leaves. Flowers bilaterally symmetrical, six-parted in leafy racemes causing the flowers to appear to be axillary; calyx tubular; petals six, rarely two or absent; stamens eleven or fewer. Ovary usually with a cupulelike disk. Capsule eventually dehiscent, seeds narrowly winged. *Cuphea* Adans; *Parsonsia* P. Browne

1. C. carthagenensis (Jacq.) Macbride—Stems erect, annual, up to 50 cm tall or less, usually much branched and generally puberulent or more or less glandular-pilose. Leaves 2–5 cm long, obovate to ovate, lance-oblong, scabrous on both surfaces. Flowers small and inconspicuous and usually subtended by large leaves; calyx light green, 4–6 mm long, petals pale purple, small, stamens eleven. Capsule with four to eight seeds. Disturbed sites in south Fla., naturalized weed from Central America. Spring, fall.

3. LÝTHRUM L. LOOSESTRIFE

Perennial or annual herbs with slender stems. Leaves opposite, verticillate or alternate, small, entire. Flowers small, solitary in leaf axils or in terminal spikes or racemes; calyx tubular, four- to six-lobed, petals four to six, stamens four to twelve. Ovary sessile, fruit a two-valved capsule. About 24 species, widely distributed in warm and temperate regions. Some species show floral dimorphism.

1. Leaves mostly alternate	1. *L. alatum* var. *lanceolatum*
1. Leaves mostly opposite.	
2. Stems prostrate; leaves ovate-lanceolate, usually less than 1.5 cm long . . .	2. *L. flagellare*
2. Stems erect; leaves linear-lanceolate, usually over 1.5 cm long	3. *L. lineare*

1. L. alatum Pursh var. **lanceolatum** (Ell.) T & G—Stems erect, smooth, branched, up to 1 m or more tall, branches four-angled in cross section. Leaves lance-ovate or linear-oblong, narrowed at the base, those on the branches tending to be more crowded together and narrower. Flowers axillary, dimorphic with either stamens or styles exserted, peduncles bracteolate near the top; hypanthium 5–7 mm long, narrowly twelve-winged; petals purplish, 5 mm long. Capsule 4–5 mm long. Low sandy ground, swamps, Fla. to Tex., Va., chiefly on the coastal plain. Summer.

2. L. flagellare Shuttlew.—Stems prostrate or creeping. Leaves about 5–10 mm long, elliptic to orbicular or cuneate, those on the branches not conspicuously reduced. Flowers axillary, hypanthium about 7–8 mm long at maturity; petals purplish, 4–5 mm long. Capsule 3–4 mm long. Low ground, endemic to south Fla. All year.

3. L. lineare L. LOOSESTRIFE—Erect, glabrous, perennial herbs up to 1 m tall or more with creeping basal offshoot. Leaves mostly opposite, mostly 2–4 cm long, linear or the lower ones linear-oblong, the upper reduced. Flowers axillary, dimorphic with either stamens or styles exserted; hypanthium 4 mm long. Petals pale purple to white, 4–5 mm long. Capsule 3–4 mm long. Low ground, sand dunes, marshes near fresh, brackish, or saline water. Fla. to Tex., N.J. Summer.

4. ROTÀLA L.

Smooth perennial or annual herbs growing in water or in moist soil. Leaves opposite or verticillate, rarely alternate. Flowers small, solitary in leaf axils or sometimes in terminal spikes or racemes, each with two bracts; calyx campanulate, three- to six-lobed, petals three to six or none, stamens one to six. Ovary two- to four-locular, style elongate or none. Capsule two- to four-valved, the walls densely striated. About 20 species, pantropical but chiefly in Asia and Africa.

1. R. ramosior (L.) Koehne, TOOTH CUP—Annual, erect or procumbent, much branched herbs up to 30 cm long. Leaves 1–4 cm long, opposite, oblanceolate or linear-oblanceolate, sessile or subsessile. Bracts about as long as the calyx, 2–3 mm long at flowering; sepals four, alternating with triangular appendages; petals four, pink or white, minute; capsule 2–4 mm long, three- to four-valved. Low ground, wet sandy soil, Fla. to Tex., Mass., on the coastal plain, Mo., Wash., Oreg. Summer.

5. AMMÁNNIA L.

Annual, smooth, succulent herbs with generally four-angled stems. Leaves opposite, narrow, entire, sessile. Flowers very small, axillary or cymose, greenish or reddish tinged; calyx ovoid to campanulate, four-angled, usually with small appendages in the spaces between sepals; petals four, stamens four to eight. Ovary enclosed within the calyx tube, subglobose, two- to four-locular, stigma capitate. Capsule two- to four-valved. About 20 species, pantropical.

1. Corolla present.
   2. Leaves mostly auriculate-cordate at base;
       style 1.5–3 mm long, filiform   .   .   .   1. *A. coccinea*

2. Leaves mostly tapering to the base; style
    0.5 mm long, stout . . . . . . 2. *A. teres*
1. Corolla absent . . . . 3. *A. latifolia*

1. A. coccinea Rottb. SCARLET AMMANNIA—Stems up to 1 m tall, much branched. Leaves linear to oblong-lanceolate, about 3–6 cm long, sessile-clasping or auriculate at the base. Flowers cymose with three to five flowers, subsessile; petals spatulate, pink or purple; calyx enclosing the mature capsule. Style filiform. Capsule 3–4 mm in diameter. Low ground, swamps, throughout most of east North America. Summer, fall.

2. A. teres Raf. TOOTH CUPS—Stems erect, simple or branched from near the base, fleshy, up to 1 m tall, usually less. Leaves linear-oblong to spatulate, only upper and midcauline auriculate at the base, blades up to 6–8 cm long; cymes closely three- to seven-flowered, sessile or subsessile; calyx lobed, short, broad; petals minute, pink, deciduous; style thick and very short. Capsule 5–7 mm in diameter. Wet ground, Fla. to Miss., N.J. on coastal plain. Summer, fall. *A. koehnei* Britt.

3. A. latifolia L.—Stems erect, simple or sometimes branched, up to 1 m tall or more. Leaves mostly 4–8 cm long, narrowly elliptic or linear-lanceolate, somewhat auricled at the base and clasping. Corolla absent. Fruit about 5 mm in diameter. Moist soil, hammocks, swamps, south Fla., Fla. Keys, W.I., tropical America.

137. PUNICÀCEAE POMEGRANATE FAMILY

Trees or shrubs, sometimes spiny, with opposite, subopposite, or fascicled simple leaves. Flowers bisexual, terminal, solitary or clustered; calyx tubular, five- to seven-lobed, colored, fused to the ovary; petals five to seven; stamens many, free. Ovary inferior, many-locular, the locules superimposed in two series. Fruit a spherical berry crowned with a persistent calyx, seeds numerous. Only two species in a single genus, PUNICA, native of Eurasia.

1. PÙNICA L. POMEGRANATE

Shrub or small tree up to 6 m tall, usually much branched from the ground. Leaves 2–6 cm long, elliptic to oblong or oblanceolate, smooth, on short petioles. Flowers conspicuous; petals obovate to suborbicular, bright red. Berry 5–10 cm in diameter with a pink or pinkish white pulp.

1. P. granatum L.—Characters of the genus. $2n = 165$. Occurring in disturbed sites, old fields, and in cultivation. Many of our plants apparently have persisted long after habitations have disappeared and therefore appear "wild." It is doubtful that the plant is truly naturalized.

138. RHIZOPHORÀCEAE Mangrove Family

Trees or shrubs with smooth leaves and branches. Leaves opposite, stipulate, coriaceous and petioled, mostly entire or sinuate-crenate or serrulate; stipules interpetiolar, caducous. Flowers bisexual in various kinds of inflorescences; calyx tube fused to the ovary, three- to fourteen-cleft, valvate and persistent; petals as many as the sepals, inserted on a disk. Ovary usually inferior, two- to five-locular or septa obscure and ovary appearing one-locular. Fruit leathery, crowned with a persistent calyx. About 15 genera, pantropical.

1. RHIZÓPHORA L. Mangrove

Smooth, evergreen trees with thick stems and opposite leaves, these petioled, thickly coriaceous, opposite; stipules conspicuous, interpetiolar, caducous. Flowers in clusters on elongate axillary peduncles, nodding; calyx subtended by two bracts, calyx tube short, four-parted; petals four, stamens eight to twelve, inserted on the petals. Ovary semi-inferior, two-locular. Fruit coriaceous, ovoid. About three species distributed along tropical seashores throughout the world.

1. **R. mangle** L. Red Mangrove—Evergreen trees up to 25 m tall but usually smaller with thin gray bark that is red within and with conspicuous aerial "prop" roots arising from the trunk and branches. Leaves 5–15 cm long, very thick, leathery, obovate or elliptic, obtuse, entire, deep green above, pale beneath; stipules 2.5–4 cm long. Flowers two to three-clustered; calyx about 1 cm long; petals 7–8 mm long, yellow, villous pubescent within; stamens eight. Fruit 2.5–3.5 cm long. Coastal areas and embayments, south Fla., tropical marine coastlines in tropical America. All year.

The long, cylindrical fruit contains the young seedling that begins elongating while still on the parent tree. If the fruit falls on muddy soil it begins to root and grow at once in salt or brackish water, less commonly in fresh water.

139. COMBRETÀCEAE Combretum Family

Trees, shrubs, or woody vines with simple leaves. Flowers small, bisexual or rarely unisexual in spikes or racemes; calyx tube fused to

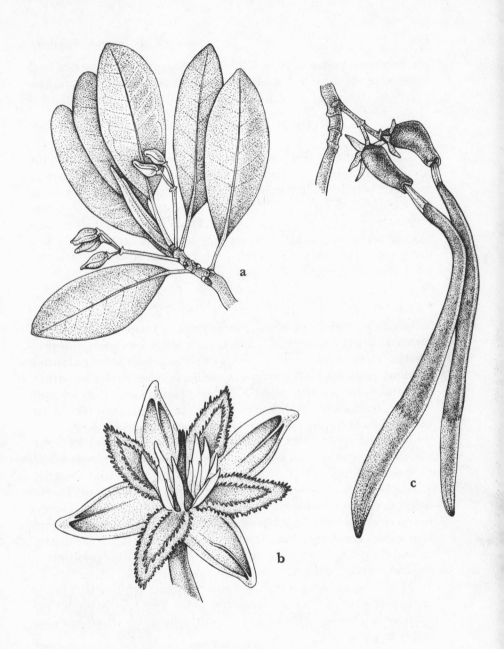

Family 138. **RHIZOPHORACEAE.** *Rhizophora mangle:* a, leafy branch with flower buds, × ½; b, flower, × 3; c, developing seedlings with elongated radicle, × ½.

the ovary, with four to eight lobes or divisions; petals four to five or none; stamens four to ten, disk present. Ovary inferior, one-locular. Fruit usually winged, mostly indehiscent. About 17 genera and 500 species in the tropics. *Terminaliaceae*

1. Leaves alternate, often clustered near ends
   of branches.
  2. Flowers in dense spherical or oblong
     heads; maritime trees or shrubs . . .   1. *Conocarpus*
  2. Flowers in simple or branched spikes or
     solitary; pinelands, hammocks.
    3. Larger leaves ovate, 1.5–10 cm long .   2. *Bucida*
    3. Larger leaves obovate, 10–20 cm long .   3. *Terminalia*
1. Leaves opposite . . . .   4. *Laguncularia*

1. CONOCÁRPUS Gaertn.

Shrubs or trees with smooth or sericeous, evergreen, alternate leaves. Flowers small in dense, conelike heads, these in terminal panicles; calyx tube five-toothed, petals none, stamens five, anthers cordate. Ovary one-locular, style villous. Fruit small, angulate, densely clustered together. Two species in mangrove swamps of America and west Africa.

1. C. erecta L. Buttonwood—Shrubs or trees up to 20 m tall with smooth or silky foliage. Leaves 2–10 cm long, obovate, ovate or elliptic, entire, obtuse or acute at the tip. Flower heads 1 cm or less wide, greenish. Fruits purplish green, conelike. Many soil types, brackish water, coastal areas, and sandy shores, south Fla., tropical America, west Africa. All year.

C. erecta L. var. **sericea** Forst. ex DC. Silver Buttonwood is distinguished from the typical variety by its densely silky-pubescent leaves. It occurs throughout the range of the species, and it is not uncommon in our area. It is also popular in cultivation.

2. BÙCIDA L.

Trees often with spines and bearing evergreen, alternate leaves that are usually crowded at the swollen tips of branches. Flowers bisexual in simple or branched spikes; calyx campanulate, shallowly five-toothed; petals absent, stamens ten in two rows. Fruit a coriaceous drupe crowned with a persistent calyx. Four species in Mexico, South America, and W.I.

1. Flowers in short spikes; leaves less than 2
   cm long; stems with small thorns . . .   1. *B. spinosa*
1. Flowers in elongate spikes; leaves over 2 cm
   long; stems mostly unarmed . . . .   2. *B. bucera*

1. **B. spinosa** Jennings, SPINY BUCIDA—Trees up to 8 m tall with a dense crown, young stems spiny and pubescent becoming glabrate with age. Leaves obovate-elliptic to elliptic, blades 1.5–2 cm long. Flowers whitish, in short axillary spikes, these numerous and densely flowered; calyx broadly campanulate. Fruit 3 mm wide or more, ovoid, turbinate. South Fla., Cuba, W.I. Reported from south Dade County and possibly naturalized in our area.

2. **B. bucera** L. BLACK OLIVE—Trees up to 15 m tall with young stems sericeous becoming glabrate. Leaves spatulate to obovate, ovate, or elliptic, crowded at the ends of branches, blades 3–9 cm long. Flowers whitish in pedunculate, slender spikes 5–10 cm long; calyx lobes triangular-acute. Fruit 8 mm long, ovoid, pubescent. Hammocks, south Fla., Fla. Keys, W.I., Mexico, Central America.

Popular as an ornamental and shade tree in south Fla. and widely planted.

3. TERMINÁLIA L.

Trees with leaves alternate or subopposite, crowded at the ends of branches. Flowers small, sessile, bisexual or unisexual, four- to five-parted, in spikes; calyx tube ovoid or cylindrical; petals absent, stamens ten in two rows. Fruit dry or drupelike. About 200 species, widely distributed in tropical regions.

1. **T. catappa** L. INDIAN ALMOND—Tree up to 25 m tall but usually smaller, with branches conspicuously whorled and spreading. Leaves mostly clustered at the ends of branches, blades 10–30 cm long, obovate, mostly rounded at the apex, narrowed to the obtuse or subcordate base. Flowers in spikes 8–15 cm long with pistillate flowers on the lower part of the spike, greenish. Fruit 4–7 cm long, a coriaceous or woody drupe, ellipsoidal. $2n = 24$. Pinelands and hammocks, introduced and naturalized from tropical Asia.

4. LAGUNCULÁRIA Gaertn. f.

Trees or shrubs with opposite, thickly coriaceous succulent leaves that are conspicuously two-glandular at the base. Flowers polygamous in elongate axillary, pubescent spikes; calyx tube five-toothed, urceolate, persistent; petals five, minute; stamens ten, in two rows. Ovary one-locular, stigma bilobed. Fruit coriaceous, one-seeded. Two species in tropical America, Africa.

1. **L. racemosa** Gaertn. f. WHITE MANGROVE—Shrub or tree up to 20 m tall. Leaves mostly 4–7 cm long, oblong to oval, rounded or very ob-

tuse at the end, smooth, the blade with two glands at the base. Flower spikes lax, often curved, calyx pubescent, 2–3 mm long. Fruit 1.5 cm long, drupaceous, reddish. Many different soils and near brackish water, coastal areas, tropical America, Africa. Spring.

The white mangrove is usually found on higher ground than the other mangrove species and is often associated with Conocarpus erecta, the Buttonwood. Both species are often on the landward side of mangrove associations.

140. MYRTÀCEAE Myrtle Family

Trees or shrubs with simple, mostly entire, evergreen, opposite or rarely alternate leaves. Flowers radially symmetrical, bisexual; calyx fused to the ovary, with three to five lobes or more; petals four to five or none, inserted on the margin of the disk, stamens many, often conspicuously colored, filaments free or fused at the base into a short tube or in bundles opposite the petals. Ovary inferior, one- to many-locular. Fruit baccate, berrylike or capsular. About 80 genera and 3000 species in tropical regions, Australia.

1. Corolla rose pink; buds and undersurface of
mature leaf blade tomentose . . . . 1. *Rhodomyrtus*
1. Corolla white or yellowish white or absent;
buds and undersurface of mature leaf glabrous to pubescent, not tomentose.
  2. Leaves alternate, blades with three to
eight parallel veins; fruit dry . . . 2. *Melaleuca*
  2. Leaves opposite, blades pinnately veined;
fruit fleshy.
    3. Flowers apetalous; calyx lidlike, deciduous at flowering . . . . . . 3. *Calyptranthes*
    3. Flowers with petals; calyx persistent
and not lidlike.
      4. Mature fruits 3–5 cm in diameter;
leaf blades conspicuously pinnately
veined . . . . 4. *Psidium*
      4. Mature fruits mostly less than 3 cm
in diameter; leaf blades not conspicuously pinnately veined.
        5. Seed coat roughened, adhering to
the pericarp of the fruit; fruit
greenish, yellowish, or purplish
red . . . . 5. *Syzygium*
        5. Seed coat smooth, free from the

pericarp of the fruit; fruit black,
red, or bluish black.
6. Inflorescence racemose or the
flowers glomerate or fascicled,
or if solitary, then arising in
lowermost axils of leafy
branches, fruit one- to two-
seeded . . . . *6. Eugenia*
6. Inflorescence cymose or the
flowers solitary.
7. Flowers solitary, berry many-
seeded . . . . . . *7. Myrtus*
7. Flowers cymose, berry one- to
two-seeded . . . . . *8. Myrcianthes*

1. RHODOMYRTUS (DC.) Reichenb. DOWNY MYRTLE

Shrubs or trees with opposite, tomentose, three-veined leaves. Flowers solitary, axillary, or in three-flowered clusters, calyx lobes four to five, petals four to five, rose colored and spreading, stamens many, inserted on the hypanthium, filament pinkish. Ovary one- to three-locular, appearing six-locular. Fruit a blue-black berry crowned by persistent calyx. About 25 species, native to Asia and Australia.

1. **R. tomentosus** (Ait.) Hassk.—Shrub up to 2 m tall, evergreen, with leaves and stems bearing short, soft pubescence. Leaves 5–7 cm long, ovate-elliptic, three-veined, entire, obtuse to acute. Flowers axillary, one to three or more on short peduncles; calyx with two small bracts at the base, tomentose, five-lobed; petals pinkish or rose colored, ovate, 1–1.5 cm long. Fruit ovoid, 1–1.5 cm long, purplish at maturity, and crowned with the persistent calyx. Seeds small, many. Naturalized from east Asia and Australia, escaped from cultivation.

2. MELALEUCA L.

Trees with alternate, one- to many-veined coriaceous leaves. Flowers sessile, bracteate, solitary or in dense or elongate spikes, the axis of which grows into a leafy shoot during or after flowering; calyx lobes five, deciduous; petals five, white, orbicular, spreading; stamens many, united in five bundles opposite the petals. Ovary three-locular. Fruit a loculicidal capsule crowned by the hypanthium. About four species, Australia.

1. **M. quinquenervia** (Cav.) Blake, CAJEPUT TREE, PUNK-TREE, BOTTLEBUSH TREE—Trees with often drooping, irregular branches, bark whitish, thick, soft. Leaves 5–12 cm long, elliptic to lanceolate-elliptic,

bright green, acute, subsessile. Flowers in many-flowered spikes; sepals about 2 mm long, obtuse; petals 3–4 mm long, white, obovate. Capsule 3–5 mm long, short-cylindrical. Low moist soil, cypress swamps, naturalized from Australia, introduced and naturalized in south Fla. All year. *M. leucodendron* misapplied

    Callistemon sp. R. Brown is the popular, cultivated BOTTLEBRUSH, distinguished from MELALEUCA by its bright red stamens.

3. CALYPTRÁNTHES Sw.   SPICEWOOD

    Trees and shrubs with opposite leaves. Inflorescence paniclelike, the branches terminating in single flowers or cymes; calyx undivided, white, petals absent; stamens many, inserted on the margin of the hypanthium. Ovary two-locular. Fruit a one- to two-seeded berry crowned by the hypanthium. About 100 species, native to tropical America and W.I.

1. Inflorescence glabrous; leaves subsessile . .   1. *C. zuzygium*
1. Inflorescence pubescent; leaves with distinct
    petioles . . . .   2. *C. pallens*
                                  var. *pallens*

    1. C. zuzygium (L.) Sw. MYRTLE OF THE RIVER—Trees or shrubs up to 12 m tall, young stems rounded, not winged. Leaves elliptic-ovate, 3–6 cm long, obtuse to acuminate, sessile or subsessile, obtuse tipped. Panicles glabrous, flowers conspicuously pedicelled and solitary, hypanthium about 2–4 mm wide after opening, stamens fifty to sixty. Berry dark red, subglobose or depressed, 8–10 mm wide. Hammocks, south Fla., W.I.

    2. C. pallens Griseb. var. **pallens**, PALE LIDFLOWER—Trees or shrubs with young stems prominently two-winged, generally pubescent, up to 8 m tall. Leaves coriaceous, elliptic to ovate-elliptic, often 5–10 cm long, petioled, tapered at both ends, with lateral veins close and parallel, usually glandular-punctate above. Panicles pubescent, coppery to yellow or gray white, three to four times compound, flowers in clusters near the tips of branches; hypanthium campanulate after opening, stamens fifty to sixty. Berry dark red, globose, 5–8 mm wide. Hammocks, south Fla., Mex., W.I. All year.

4. PSÍDIUM L.   GUAVA

    Trees or shrubs with opposite, pinnately veined leaves. Flowers axillary, solitary, or in few-flowered cymes. Calyx in the species in our area undivided when young, separating irregularly, petals four to five, white, spreading. Stamens many, free, inserted on the hypanthium. Ovary in the species in our area four- to five-locular with many ovules. Fruit a many-seeded berry crowned by persistent segments of the calyx. About 150 species in tropical America.

1. **P. guajava** L. GUAVA—Shrub or tree up to 10 m tall, young stems pubescent. Leaves 8–14 cm long, elliptic or oblong, rounded or narrowed at the tip, veins prominent beneath and strongly impressed above with twelve to twenty pairs, finely appressed pubescent below or sometimes glabrous. Inflorescence axillary, one-flowered; hypanthium 5–7.5 mm long, calyx splitting irregularly into four to five lobes, petals white, elliptic, 10–12 mm long, yellow or pinkish. $2n = 21,22,30,33$. Many sites in south Fla., roadsides, hammocks, old fields, waste places, naturalized from tropical America, also naturalized in Old World. Summer.

5. SYZÝGIUM Gaertn.

Trees and shrubs with evergreen, opposite leaves. Flowers solitary or in few-flowered terminal clusters or in the axils of lowermost leaves; calyx four- to five-lobed, hypanthium not produced beyond the ovary; petals four, white, stamens numerous. Ovary two- to three-locular; fruit a fleshy berry. Seed coat roughened and adhering to the fruit. About 74 species in the Old World tropics. *Eugenia* by various authors.

1. Leaves less than 10 cm long; styles less than
  1 cm long . . . .   1. *S. cumini*
1. Leaves 15–25 cm long; styles 3–4 cm long .   2. *S. jambos*

1. **S. cumini** (L.) Skeels, JAMBOLAN, JAMBOLAN PLUM—Trees up to 25 m tall with smooth, often whitish stems. Leaves broadly oblong to ovate, 5–10 cm long, bluntly acuminate on petioles 1–3 cm long. Flowers in branched cymes; petals white, fused into a calyptra or cap. Berry 1–2 cm long, ovoid, purplish red, edible. Introduced in south Fla. from Asia, doubtfully naturalized. All year. *Eugenia cumini* Druce

2. **S. jambos**(L.) Alston, ROSE APPLE—Tree up to 10 m tall with smooth stems. Leaves oblong-lanceolate, 10–25 cm long, acuminate, thick and lustrous. Flowers in short corymbs, petals greenish white, not fused into a cap. Berry 2.5–5 cm long, spheroidal or ovoid, greenish or yellowish, edible. Introduced in south Fla. from Asia, possibly naturalized locally. All year. *Eugenia jambos* L.

6. EUGÈNIA L. Stopper

Trees or shrubs with opposite leaves. Flowers in racemes; or the axis sometimes very short and inflorescences resembling axillary fascicles, umbels, or glomerules; or, if the flowers solitary, then they arise from the lowermost axils of leafy stems. Petals four, white, ovate to orbicular, stamens many. Ovary two-locular; berry one- to two-seeded, crowned by the persistent lobes of the calyx. About 500 species, pantropical.

1. Flowers mostly solitary in axils of bracts;
   mature fruit up to 3 cm in diameter,
   bright orange red . . . .   1. *E. uniflora*
1. Flowers in axillary clusters or racemes.
   2. Pedicels of flowers or fruits short and
      stout, less than 5 mm long.
      3. Leaves usually rounded or obtuse at the
         apex . . . .   2. *E. foetida*
      3. Leaves usually acute or pointed at the
         apex.
         4. Pedicels stout, shorter than the flow-
            ers; flowers in clusters or dense
            fascicles . . . .   3. *E. axillaris*
         4. Pedicels usually longer than the flow-
            ers, slender; flowers usually in cy-
            mose, few-flowered clusters . . .   4. *E. rhombea*
   2. Pedicels of flowers or fruits long and slen-
      der, over 5 mm long.
      5. Leaves narrowed at the apex into a pro-
         longed tip; fruits 5–8 mm in diameter   5. *E. confusa*
      5. Leaves acute at the apex, not narrowed
         into a prolonged tip; fruits 9–16 mm
         in diameter . . . .   4. *E. rhombea*

1. E. uniflora L. Surinam Cherry—Shrub or small tree up to 9 m tall, glabrous or with few reddish hairs on young stems and leaves. Leaves 3–7 cm long, ovate, bluntly acuminate at the apex, rounded at the base, dark green, shiny above, pale beneath. Inflorescence a very short axillary raceme with one to three pairs of flowers, or flowers solitary from the lower leaf axils; hypanthium 1 mm long, calyx lobes leaf-like, reflexed, petals white, 7–8 mm long, stamens fifty to sixty. Ovary two-locular. Berry orange red, very juicy, 2–3 cm wide, depressed-globose, edible. $2n = 22$. Cultivated and locally naturalized in south Fla., native of South America and now grown in many tropical regions. All year.

2. **E. foetida** Pers. Spanish Stopper—Small tree up to 12 m tall; young stems, inflorescence and young leaves puberulent. Leaves 3–6 cm long, obovate, rounded at the apex, dark green above, paler beneath. Inflorescence an axillary raceme with three to six pairs of flowers; hypanthium cup shaped, calyx lobes four, scarious; petals smooth, about 1.5 mm long, stamens about forty. Ovary two-locular. Berry globose, dark red to black, about 5–7 mm wide. Hammocks near coastal areas, south Fla., W.I., Mexico, Central America. All year. *E. buxifolia* (Sw.) Willd. *E. myrtoides* Poir, *E. monticola* DC.

3. **E. axillaris** (Sw.) Willd. White Stopper—Small tree or shrub with scaly bark up to 7 m tall. Leaves 3–7 cm long, ovate, elliptic, or rhombic-ovate, coriaceous, acute or cuneate at base, bluntly acuminate at the tip, dark green and somewhat lustrous above, paler beneath. Inflorescence a short axillary raceme mostly with five to seven pairs of flowers; hypanthium cup shaped, calyx lobes in two unequal pairs, petals white, 2–3 mm long, stamens about forty. Ovary two-locular; berry bluish black, globose, about 7–8 mm wide. Hammocks, south Fla., W.I., Mexico, Central America. All year.

4. **E. rhombea** (Berg) Krug ex Urban, Red Stopper—Shrub or small tree up to 3 m tall, with smooth bark. Leaves 3–6 dm long, ovate, coriaceous, narrowed to a rounded tip, dull green above, paler beneath. Inflorescence axillary, racemose but appearing clustered, the two to four flowers seeming to arise from the axil; hypanthium hemispheric, calyx lobes in unequal pairs, ciliate, petals ciliate, white, 4 mm long. Ovary two-locular, many ovules. Berry black, globose, 7–8 mm wide. Hammocks, lower Fla. Keys, W.I., Mexico, Central America. All year. *E. anthera* Small

5. **E. confusa** DC. Ironwood—Trees or shrubs up to 6 m tall with scaly bark. Leaves 3–5 cm long, elliptic, elliptic-ovate, or ovate, narrowed at the apex into a prolonged tip. Inflorescence a short raceme; hypanthium very short, sepals four, scarious; petals white, corolla about 5–6 mm wide; stamens about forty. Ovary two-locular. Berry scarlet or red, subglobose or obovoid, about 5–8 mm wide. Coastal hammocks, south Fla., W.I.

This is one of our most common stoppers in the hammock vegetation of south Fla.

7. MÝRTUS L. Stopper

Shrubs or small trees with opposite leaves. Flowers solitary, peduncled, appearing in the axils of leaves or bracts in the lower nodes of leafy branches; calyx four-lobed, persistent, petals four, white, spreading; stamens many. Ovary two- to three-locular with many ovules. Fruit a many-seeded berry crowned by a persistent calyx. About 16 species, chiefly in W.I. and in Fla.

1. Calyx less than 1 cm wide; branches decumbent or prostrate . . . . 1. *M. verrucosa*
1. Calyx over 1 cm wide; branches erect . . 2. *M. bahamensis*

1. M. verrucosa Berg—Small tree or shrub with narrow stems, often decumbent or lax. Leaves 1-3 cm long, ovate or elliptic, dull green above, somewhat paler beneath. Flowers solitary on distinct peduncles; sepals about 2 mm long, petals creamy white, 4-5 mm long, corolla about 1 cm wide; berry globose, 6-9 mm wide, black. All year. *Mosiera longipes* (Berg) Small, *Eugenia longipes* Berg

2. M. bahamensis (Kiaersk.) Urban—Shrub or small tree with erect stems. Leaves usually 3-5 cm long, elliptic to ovate or orbicular, shiny or lustrous above, duller beneath. Flowers solitary, sepals about 3 mm long, petals white, about 2-3 mm long, corolla about 1.5 cm wide. Berry 1 cm wide, subglobose, black. Hammocks, Fla. Keys, south Fla., W.I. All year. *Mosiera bahamensis* (Kiaersk.) Small, *Eugenia bahamensis* Kiaersk.

This species may best be considered as a variety of **M. verrucosa** Berg; future study may require a reduction in its status.

8. MYRCIÁNTHES Berg NAKEDWOOD

Tall or moderate-sized trees or shrubs with opposite or ternate leaves. Flowers one to seven in an axillary dichasium with the central flowers sessile in the fork of the axils, or flowers numerous and forming a compound inflorescence; calyx four-lobed, occasionally five-lobed persistent; petals four, white, spreading; stamens many. Ovary two-locular, each locule with many ovules. Fruit a one- or two-seeded berry crowned by the persistent calyx. About 50 species, tropical America extending into south Fla. *Anamomis* Griseb., *Eugenia* L. by various authors

1. M. fragrans (Sw.) McVaugh—Shrubs or trees up to 20 m tall, usually less, with reddish brown or light brown bark. Leaves 1-8 cm long, narrowly obovate to elliptic-cuneate, acute or obtuse, pale green above, slightly paler beneath. Flowers in cymes with three to fourteen flowers; sepals about 1-2 mm long, corolla about 1 cm wide. Berry globose or ellipsoidal. Hammocks, south Fla., tropical America. Two varieties occur in our area:

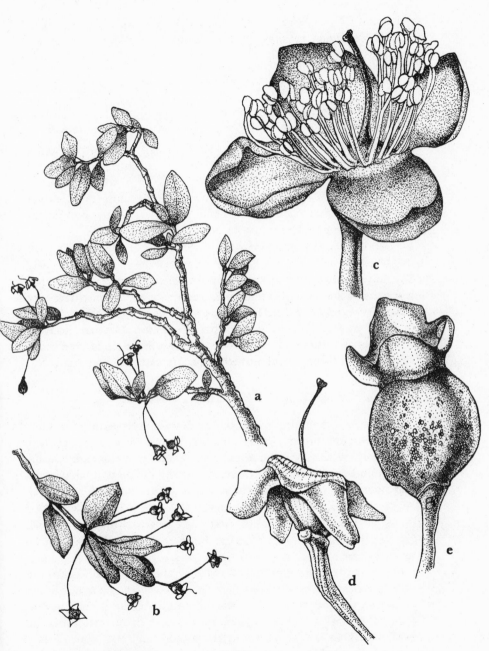

Family 140. **MYRTACEAE.** *Myrtus verrucosa:* a, flowering branch, × ½; b, inflorescence, × ¾; c, flower, × 7½; d, flower postanthesis, × 5; e, fruit with persistent calyx, × 7½.

1. Corolla less than 1 cm wide, cymes usually
   three- to seven-flowered . . . . . .   1a. var. *fragrans*
1. Corolla over 1 cm wide, cymes several-flow-
   ered, usually ten- to fourteen-flowered . 1b. var. *simpsonii*

1a. **M. fragrans** var. **fragrans**, Twinberry—Shrub or small tree up
to 8 m tall in Fla., calyx 4–5 mm wide. All year. Includes *Eugenia
dicrana* Berg.

1b. **M. fragrans** var. **simpsonii** (Small) R. W. Long—Shrub or tree
sometimes up to 20 m tall with strongly buttressed base, calyx 6–7 mm
wide. Endemic to south Fla. All year. *Anamomis simpsonii* Small

141. NYSSÀCEAE Sour-Gum Family

Monoecious trees or shrubs with alternate simple leaves. Flowers
unisexual, staminate in racemes, umbels, or heads, the pistillate usually
solitary; staminate flowers numerous, stamens five to twelve, petals five
or more; pistillate flowers larger than staminate, calyx fused to inferior
ovary, petals five or more. Fruit drupaceous or dry, one- to five-cham-
bered. About three genera and ten species in east Asia and east U.S.

1. NÝSSA L. Tupelo

Dioecious or polygamodioecious deciduous trees or shrubs with
terete stems and broad, simple, alternate leaves. Flowers greenish; staminate flowers usually five-parted, petals small; pistillate flowers five-
parted and with a one-locular ovary. Fruit a drupe with thin exocarp
and ridged endocarp. About eight species in east U.S. and Asia.

1. **N. sylvatica** Marsh. var. **biflora** (Walt.) Sarg. Black Gum, Swamp
Tupelo—Trees up to 30 m tall with trunk greatly swollen at base when
growing in water. Leaves 5–15 cm long, oblanceolate, often crowded
toward the tips of stems, entire or with a few coarse teeth, obtuse or
rounded at the apex. Staminate flowers in umbels or umbel-like racemes
on peduncles 1–3 cm long; pistillate flowers two to eight, sessile at tip of
peduncles that become 4–6 cm long. Drupe 1–1.5 cm wide, ellipsoidal,
dark blue black, stone with shallow grooves. $2n = $ c.44. Fresh-water
swamps, Fla. to Ga., Del. on coastal plain. Spring. *N. biflora* Walt.

A single collection from Lignum Vitae Key by J. K. Small would
extend the range of this species into the Fla. Keys. No recent collections
have been seen, and its occurrence in our area is doubtful.

142. MELASTOMATÀCEAE   MELASTOMA FAMILY

Herbs, shrubs, or trees with simple, opposite or verticillate leaves; veins characteristically three to nine, diverging at the base and running longitudinally before converging at the apex. Flowers bisexual, showy; calyx tubular; petals free, four to five, with corona usually present between petals and stamens; stamens the same number or double the number of petals; anthers opening by terminal pores. Ovary mostly inferior, two- to many-locular. Fruit a capsule or berrylike. About 200 genera and 4500 species, chiefly in tropical regions.

1. Herbs; petals purplish or white; fruit a
   capsule . . . .   1. *Rhexia*
1. Shrubs or small trees; petals white; fruit a
   berry . . . .   2. *Tetrazygia*

1. RHÉXIA L.   MEADOW BEAUTY

Erect, branched, perennial herbs with opposite, sessile or short-petioled leaves, the blades conspicuously three- to five-veined, serrulate to serrate, often ciliate margined. Flowers in terminal cymes or solitary, bracts resembling leaves, perigynous; hypanthium tubular or urceolate, prolonged beyond the ovary; stamens eight, petals four, showy, purplish or white, fugacious. Ovary four-locular. Fruit a capsule. About 12 species in east North America and W.I.

1. Anther straight, 1–3 mm long; corolla rose .   1. *R. nuttallii*
1. Anther curved, 5–11 mm long; corolla white,
   rose, or purple.
   2. Leaves linear to elliptic-obovate, blade
      one-veined . . . .   2. *R. cubensis*
   2. Leaves ovate-lanceolate to narrowly el-
      liptic, blade three- to five-veined.
      3. Hypanthium usually glabrous at matur-
         ity; sepals deltoid; roots often tu-
         berous . . . .   3. *R. nashii*
      3. Hypanthium with scattered hairs at ma-
         turity; sepals lanceolate; roots not
         tuberous . . . .   4. *R. mariana*
                                        var. *mariana*

1. **R. nuttallii** C. M. James—Small, often dwarfish plants, stems up to 35 cm tall, sometimes prostrate, unbranched. Leaves 4–12 mm long,

three-veined, subsessile, glabrous, the lower surface sometimes reddish. Inflorescence cymose, few-flowered, bracts persistent; hypanthium densely glandular-hispid, 5–8 mm long, calyx lobes 1–2 mm long; petals purplish rose. Capsule glabrous, 3–4 mm long, seeds curved. $2n = 22$. Sandy pineland, Fla. to Ga. Spring, summer. *R. serrulata* Nutt.

2. **R. cubensis** Griseb.—Stems 2–5 dm tall from extensive spreading roots. Leaves linear, obovate, 2–4 cm long, serrulate to shallowly serrate, one-veined. Flowers few to numerous, cymose; hypanthium glandular-hirsute, 10–13 mm long, calyx lobes deltoid; petals purple or dark rose. Capsule glabrous, 5–8 mm long, seeds cochleate. $2n = 22,44,66$. Low pineland, swamps, ditches, Fla. to N.C., Tenn., La., W.I. Spring, summer or all year. *R. floridana* Nash

3. **R. nashii** Small—Stems 3–8 dm tall from extensive spreading roots or from tuberous roots. Leaves lanceolate or elliptic, 4–5 cm long, three-veined, sessile or nearly so. Flowers few to many in open cymes, hypanthium glabrous at maturity, 10–12 mm long, calyx lobes deltoid; corolla deep purplish to dark rose. Capsule pyriform, glabrous, seeds cochleate. $2n = 44,66$. Organic soils, moist ground, ditches, Fla. to Va., and La. on the coastal plain. Spring, fall.

4. **R. mariana** L. var. **mariana**—Stems 1–5 dm tall from extensive, spreading, nontuberous roots, glandular-pubescent. Leaves 3–6 cm long, elliptic or sometimes lanceolate, three-veined, margins serrate. Flowers few to many in open cymes, hypanthium glandular-hirsute at maturity, calyx lobes deltoid or more often lanceolate; corolla commonly white, sometimes dark purple or rose. Capsule 4–6 mm long, seeds cochleate. $2n = 22$. Low ground, prairies, ditches, low pineland, Fla. to Mass., Mo., Tex. Spring, fall.

2. TETRAZYGIA Richard ex DC.

Shrubs or trees with conspicuously three- to five-veined leaves. Blades usually scurfy pubescent or covered with appressed hairs beneath. Flowers showy; sepals four to six, petals four to six. Ovary four- to five-locular. Fruit a berry. About 16 species in W.I. and Fla.

1. **T. bicolor** (Mill.) Cogn. TETRAZYGIA—Shrub or small tree bearing lanceolate or lance-ovate leaves, blades 8–20 cm long, silvery below. Flowers conspicuous in panicles 1–2 dm long; calyx 1 mm long; petals 6–8 mm long, white, cuneate. Berry purplish or purplish black, about 1 cm long. Hammocks, pinelands, south Fla., W.I. Summer.

This is one of our most attractive and characteristic W.I. plants in south Fla. It is very common in the Everglades, especially on the Everglade Keys.

Family 142. **MELASTOMATACEAE.** *Tetrazygia bicolor:* a, fruiting branch, × ½; b, flower, × 5; c, mature fruit, × 5½.

143. ONAGRÀCEAE   Evening Primrose Family

Herbs and shrubs, sometimes aquatic, with simple leaves alternate or opposite, and axillary or racemose flowers. Flowers radially or somewhat bilaterally symmetrical, bisexual, often solitary; calyx fused to the ovary, four- to five-lobed; corolla with four to five free petals, these often showy; stamens as many or twice the number of calyx lobes. Ovary inferior, four- to five-locular; stigma three- to four-lobed or capitate. Fruit a capsule or nut. About 18 genera and 650 species, chiefly in the temperate and arid subtropical regions. *Epilobiaceae*

1. Corolla white, fading pink; fruit a nutlike,
   indehiscent, one- to four-seeded capsule .   1. *Gaura*
1. Corolla yellow or flowers apetalous; fruit an
   eventually dehiscent, many-seeded capsule.
   2. Leaves and fruit silky tomentose; flowers
      opening near sunset . . . . . .   2. *Oenothera*
   2. Leaves and fruit glabrous to hirsute, not
      silky tomentose; flowers opening
      throughout the day . . . . . .   3. *Ludwigia*

1. GAÙRA L.

Annual or perennial herbs with alternate leaves. Flowers in dense racemes; calyx tube cylindrical, three- to four-parted, petals three to four, pinkish or white, stamens six or eight. Ovary three- to four-locular, or appearing one-locular, style filiform, stigma three- or four-lobed, surrounded by a cup-shaped border. Fruit nutlike, three- to four-angled. About 21 species, chiefly in Tex. and surrounding areas.

1. **G. angustifolia** Michx. Southern Gaura—Stems up to 2 m tall, slender, often radiating from the base, puberulent below. Leaves 4–12 cm long, oblanceolate or narrowly lanceolate, sharply serrate, dentate, or pinnatifid. Flowers small, about 5 mm long, opening near sunset and fading the next day, white, becoming pinkish. Nutlike capsule 8–10 mm long, smooth. $2n = 14$. Drier soils and dunes, roadsides, Fla., Ga., S.C. and N.C. on the coastal plain. Spring, fall. *G. simulans* Small

2. OENOTHÉRA L.   Evening Primrose

Annual, biennial, or perennial herbs. Leaves alternate, often forming basal rosettes, entire or more commonly dentate, pinnatifid, or lobed. Flowers axillary, solitary, nocturnal; hypanthium four-angled, cylindric; sepals four, petals four, obovate or obcordate, stamens eight. Ovary four-locular, stigma four-parted or entire. Fruit a four-valved capsule of various forms. About 85 species, chiefly in warm temperate and temperate regions of the U.S., South America. *Raimannia* Rose

1. O. humifusa Nutt. SEASIDE EVENING PRIMROSE—Perennial herbs often with rather woody stems, decumbent or ascending, spreading, branched from the base, 3–5 dm long, with silky appressed pubescence. Leaves 2–3 cm long, oblanceolate, entire to sinuate, or denticulate, covered with silky pubescence. Flowers sessile in the upper leaf axils or occasionally forming a short spike; sepals 10–12 mm, reflexed; petals bright yellow, 1 cm long, hypanthium and ovary hirsute. Capsule linear pubescent, about 3 cm long, slightly curved, seeds dark brown. $2n = 14$. Beaches, sandy soil, drier sites, disturbed areas, Fla. to La., N.C., W.I. along the coast, apparently not found in the Fla. Keys. *Raimannia humifusa* (Nutt.) Rose

3. LUDWÍGIA L.

Perennial herbs or shrubs, often aquatic, erect, spreading, or procumbent with alternate or opposite leaves. Flowers axillary, solitary, four- to five-parted; hypanthium cylindrical, often four-angled or winged; sepals persistent; petals yellow or absent, sometimes very small; stamens four, eight, or ten. Ovary inferior, four- to five-locular; fruit a many-seeded capsule. About 75 species in both temperate and tropical regions of both hemispheres. *Jussiaea* L.; *Isnardia* L.; *Ludwigiantha* Small

1. Stamens eight or ten; herbs or shrubs.
  2. Seeds with an enlarged raphe as big as the
    body of the seed; capsule terete, narrowly cylindrical, not winged.
    3. Sepals four . . . .   1. *L. octovalvis*
    3. Sepals five . . . .   2. *L. leptocarpa*
  2. Seeds with raphe not more than one-fifth
    the diameter of the seed; capsules more
    or less four-angled or winged.
    4. Shrubs; petals 1–3 cm long; plants usually densely pubescent . . . . .   3. *L. peruviana*
    4. Herbs; petals about 0.5 cm long; plants
    almost glabrous . . . . . .   4. *L. erecta*
1. Stamens four; herbs.
  5. Leaves opposite; stem prostrate.
    6. Flowers sessile; leaves obovate or spatulate, petioled . . . .   5. *L. repens*
    6. Flowers on elongate peduncles; leaves

linear to narrowly oblanceolate, sessile . . . .   6. *L. arcuata*
5. Cauline leaves alternate; stem erect or reclining. .
  7. Petals present, conspicuous.
    8. Leaves linear to linear-lanceolate, less than 3 mm wide; capsules indehiscent . . . .   7. *L. linifolia*
    8. Leaves wide; capsules dehiscent by a terminal pore.
      9. Style more than 6 mm long, the stigma held above the anthers at anthesis; lower leaves subglabrous . . . .   8. *L. virgata*
      9. Style less than 6 mm long, the stigma surrounded by the anthers at anthesis; lower leaves pubescent.
        10. Sepals ovate . . . . . .   9. *L. maritima*
        10. Sepals narrowly deltoid . . .   10. *L. hirtella*
  7. Petals absent or very small.
    11. Flowers in dense spikes . . . . .   11. *L. suffruticosa*
    11. Flowers in axils of leaves or leaflike bracts.
      12. Larger leaves narrowly ovate or ovate-spatulate, mostly less than 2 cm long; capsule 1–4.5 mm long.
        13. Upper leaves entire but with marginal glands; blades 5–10 mm long . . . . . .   12. *L. microcarpa*
        13. Upper leaves obscurely serrate near apex, eglandular; blades 10–20 mm long . . . . .   13. *L. curtissii*
      12. Larger leaves linear-lanceolate to lanceolate, mostly over 2 cm long; capsule 3–7 mm long . .   14. *L. alata*

1. **L. octovalvis** (Jacq.) Raven ssp. **octovalvis**—Stems erect, herbaceous, often woody below, up to 2 m tall, usually less, bearing stiff pubescence or glabrate. Leaves obovate to narrowly lanceolate, entire, 4–12 cm long, acute or acuminate at the tip, sessile or nearly so. Flowers conspicuous, solitary in the upper axils; sepals four, ovate, petals four, yellow, 1–2 cm long, hypanthium cylindrical, four-angled. Capsule 3–5 cm long, cylindrical, seeds brown. $2n = 32,48$. Low moist soils, glades, pineland along roadsides, Fla. to Tex., tropical America. All year. *Jussiaea angustifolia* Lam.; *J. scabra* Willd.; *J. suffruticosa* L.

The leaves are highly variable in size and width in this species.

2. L. leptocarpa (Nutt.) Hara—Herbs or shrubs with stems erect up to 1 m tall with short hirsute pubescence. Leaves lanceolate to narrowly elliptic, 4–20 cm long. Flowers in the upper leaf axils, short pedunculate or nearly sessile, hypanthium hirsute, sepals narrowly lanceolate, petals conspicuous, corolla about 1–2 cm wide. Capsule narrowly cylindrical, 4–6 cm long. Low ground, ditches, margins of swamps and lakes, Fla. to Tex., Ga. on the coastal plain.

3. L. peruviana (L.) Hara, PRIMROSE WILLOW—Stems herbaceous or often woody below, mostly erect and up to 1–2 m tall or branching, spreading and decumbent, densely pubescent to barely pubescent. Leaves usually 8–15 cm long, ovate to narrowly elliptic, acutely acuminate at both the apex and the base, puberulent, scabrous or pilose beneath, entire, sessile or nearly so. Flowers conspicuous, solitary in the upper axils; sepals four to five, lanceolate, petals four to five, rounded, 1–3 cm long, yellow, hypanthium four- to five-angled. Capsule obconic-cylindrical, 1–3 cm long, seeds tan. $2n = 96$. Moist soil, ditches, canal banks, south Fla., Mexico, tropical America, introduced into Old World tropics. All year. *Jussiaea peruviana* L.

Leaves are variable in width and may appear as narrowly lanceolate to lance-ovate.

4. L. erecta (L.) Hara—Erect annual herb up to 1 m tall, much branched, glabrous or barely pubescent, stem sharply four-angled. Leaves lance-ovate to narrowly lanceolate, 5–12 cm long, on definite petioles 3–10 mm long or more, acute or acuminate at the apex. Flowers in the upper leaf axils, sessile or nearly so; sepals four, lance-ovate, petals four, yellow, 4–5 mm long, hypanthium oblong-linear, four-angled. Capsule four-angled, puberulent, 1–1.5 cm long, seeds yellow brown, cylindric-ovoid. $2n = 16$. Moist soil, wet prairies, south Fla., Mex., pantropical. All year. *Jussiaea erecta* L.

5. L. repens Forst.—Smooth, prostrate, often mat-forming stems, usually aquatic and floating in fresh water. Leaves 2–4 cm long, elliptic to ovate, opposite and definitely petioled. Flowers inconspicuous, the hypanthium bearing two small narrow bracts; sepals four; petals 2–3 mm long. Capsule 4–6 mm long, four-angled. $2n = 48$. In fresh water or wet soil, inland swamps, glades, Fla. to N.C., Tenn., Tex., Calif., W.I., Mexico. Spring, fall. *L. repens* Ell.; *Isnardia natans* Kuntze; *Ludwigia natans* Ell.; *Isnardia repens* (Sw.) DC.

6. L. arcuata Walt.—Stems creeping, prostrate, somewhat fleshy, with numerous linear to oblanceolate leaves. Blades mostly 10–15 mm long, sessile or nearly so. Flowers conspicuous on elongate peduncles longer than the leaves, sepals 5–10 mm long, linear-lanceolate, petals spreading, obovate, corolla 2–3 cm wide. Capsule 8–10 mm long, somewhat curved. Margins of pools, ponds, and lakes, inland swamps and wet soil, Fla. to Va. on the coastal plain. Spring, fall.

Natural hybrids of **L. arcuata** × **L. repens** have been identified.

7. **L. linifolia** Poir.—Stems glabrous, erect, four-angled, up to 5 dm tall, branched. Leaves 3–6 cm long, alternate, linear to narrowly oblanceolate. Flowers inconspicuous, solitary in the upper leaf axils; sepals four, petals four, yellow, as long as the sepals. Capsule about 1 cm long, cylindrical, not angled, terete. $2n = 16$. Flooded grounds, often in brackish marshes near the coast, Fla. to Miss., N.C., on the coastal plain. Spring, fall. Reported only for Collier County in our area.

8. **L. virgata** Michx.—Erect or reclining stems, smooth to puberulent with numerous branches. Leaves 2–4 cm long, alternate, smooth, linear to narrowly lanceolate, sessile or nearly so. Flowers conspicuous, in terminal racemes; hypanthium short, smooth; sepals four, reflexed, about three times as long as hypanthium; petals four, yellow, about twice as long as the sepals. Capsule 4–5 mm long, cylindrical, slightly winged on the angle. $2n = 16$. Sandy pineland, Fla. to Miss., N.C. on the coastal plain. Spring, fall.

9. **L. maritima** F. Harper—Stems glabrous or nearly so, erect, branched. Leaves alternate, variable, narrowly linear to lanceolate, subsessile or short petioled, smooth, acutely acuminate at both the apex and base. Flowers in leafy terminal branches, axillary, solitary, conspicuous; sepals four, reflexed, longer than the hypanthium; petals four, yellow. Capsule four-angled, slightly winged. $2n = 16$. Wet pinelands, sandy soil, disturbed sites, ditches, Fla. to La., N.C. Spring, fall.

10. **L. hirtella** Raf.—Stems erect, up to 6 dm tall, with enlarged fleshy roots. Leaves 2–6 cm long, numerous or few, lanceolate to ovate-lanceolate or elliptic-lanceolate. Flowers conspicuous in upper leaf axils, sepals narrowly deltoid, petals longer than the sepals. Capsule with spreading pubescence, about 5–8 mm long. Moist soil, Fla. to Tex., N.J. Spring, fall.

11. **L. suffruticosa** Walt.—Erect or reclining stems up to 1 m tall. Leaves 2–8 cm long, alternate, smooth, linear to narrowly lanceolate or elliptic. Flowers in dense terminal spikes; hypanthium smooth, sepals four, broadly triangular, petals absent. Capsule obpyramidal, about 5 mm long. $2n = 32$. Low grounds, Fla. to N.C. on the coastal plain. Not found commonly in our area.

12. **L. microcarpa** Michx.—Stems angled, glabrous, erect, slender, up to 6 dm tall. Leaves mostly 1–2 cm long, alternate, narrowly ovate-spatulate to obovate, upper ones entire, glandular dotted around the margins. Flowers inconspicuous, solitary in upper leaf axils; hypanthium about 2 mm long; sepals four, broadly triangular, petals absent.

Capsule obpyramidal, 1.5 mm long. $2n = 16$. Low ground, inland swamps, glades, pineland, Fla. to Miss., N.C. on the coastal plain, W.I. All year.

13. L. curtissii Chapm.—Stems erect or ascending, smooth. Leaves mostly 1–3 cm long, alternate, narrowly ovate or ovate-spatulate, the upper leaves obscurely serrate. Flowers small, inconspicuous, solitary in the axils of the upper leaves; hypanthium small, sepals four, somewhat triangular, petals absent or minute. Capsule somewhat top shaped, 1–3 mm long. $2n = 48,64$. Ponds, ditches, open pineland, moist sand and low ground, endemic to south Fla.

14. L. alata Ell.—Stems winged or angled, erect or ascending, smooth, slender, up to 1 m tall, loosely branching. Leaves 3–10 cm long, alternate, linear-lanceolate to lanceolate stolons with leaves opposite or subopposite, narrowly to broadly ovate or spatulate. Flowers inconspicuous, in the axils of upper leaves; hypanthium with small bracts near the top; sepals four, triangular-ovate; petals minute, greenish. Capsule cupulate, 3–4 mm long, sharply four-angled or four-winged. $2n = 32,48$. Low ground, cypress swamps, ponds, Fla. to Va. on the coastal plain. *L. lanceolata* Nutt.

144. HALORAGÀCEAE WATER MILFOILS FAMILY

Herbs, chiefly aquatic with alternate, opposite, or verticillate leaves, the submersed leaves usually much divided. Flowers small and inconspicuous, bisexual or unisexual; sepals free, two to four, fused to the ovary, or absent; petals two to four or absent; stamens two to eight. Ovary inferior, one- to four-locular. Fruit a nut or drupe. About 7 genera and 160 species, widely distributed throughout the world. *Gunneraceae*

1. Perianth parts in fours, stigmas four; submerged leaves filamentous or featherlike 1. *Myriophyllum*
1. Perianth parts in threes, stigmas three; submerged leaves foliaceous or deeply pinnatifid . . . . 2. *Proserpinaca*

1. MYRIOPHÝLLUM L. WATER MILFOILS

Smooth aquatic herbs with elongate, simple or branched stems submersed or creeping on wet soil. Leaves alternate to whorled, the emersed blades entire, serrate, or toothed, the submersed blades somewhat larger and with finer divisions. Flowers small, bisexual or unisexual, usually solitary in the upper leaf axils; calyx four-lobed, deciduous, petals four, stamens eight or four. Ovary four-locular, styles four. Fruit dry, nutlike. About 35 species, widely distributed, chiefly in the temperate zones.

1. Leaves filamentous, irregularly scattered
    along the stem . . . . 1. *M. pinnatum*
1. Leaves deeply pinnately dissected, distinctly
    whorled . . . . 2. *M. brasiliense*

1. **M. pinnatum** (Walt.) BSP, PINNATE MILFOIL—Stems lax, bearing submersed leaves chiefly alternate or scattered, composed of capillary or filamentous segments mostly 1–3 cm long, the emersed leaves often developed into bractlike blades. Flowers in emersed spikes, bracts whorled, exceeding the flowers; petals elliptic, about 1 mm long. Fruit ovoid, 2 mm long, each carpel bearing a longitudinal ridge with sharply tuberculate edge. Ponds, ditches, wet soil, Fla. to Mass., Kans., Tex., W.I. Spring, summer.

2. **M. brasiliense** Camb. BRAZILIAN WATER FEATHER—Stems stout, very leafy, bearing distinctly whorled leaves 2–5 cm long, each with ten to eighteen segments on each side, pinnately dissected. Flowers axillary to submersed leaves, bracts exceeding the flowers; petals narrowly ovate, obtuse, about 1 mm long. Fruit ovoid, 2 mm long, minutely granular. Cultivated in aquariums, escaped and naturalized from South America in south U.S. Spring, fall.

2. PROSERPINÀCA L. MERMAID WEED

Smooth aquatic or marsh herbs with stems arising from creeping rhizomes. Leaves alternate, subsessile, lanceolate in outline, dentate or pinnatifid, often both kinds on the same plant. Flowers very small, bisexual, solitary in upper leaf axils; calyx fused, three-lobed, sepals ovate-acuminate; petals three; stamens six, three epipetalous and abortive, three episepalous and fertile. Ovary three-locular, styles three. Fruit a three-angled, three-locular nut. About three species, chiefly in temperate North America.

1. Emersed leaves merely serrate, submersed
    leaves usually pinnatifid toothed . . . 1. *P. palustris*
                                             var. *palustris*
1. Emersed leaves deeply pinnatifid, similar in
    appearance to submersed leaves . . . 2. *P. pectinata*

1. **P. palustris** L. var. **palustris**, SWAMP MERMAID—Stems simple, decumbent, succulent, often partially submersed with suberect tips, or growing on moist soil and rooting near the base. Submersed leaves pin-

natifid, lobes linear, emersed leaves 3–6 cm long, narrowly lanceolate, to elliptic, serrate. Flowers sessile in leaf axils, solitary or in clusters of two to three; calyx three-angled, lobes about 1 mm long, greenish petals very small. Nutlets three-angled, pyramidal or subglobose, 4–5 mm long. Ponds, moist soil, ditches and swamps, Fla. to La., Va. on the coastal plain. Spring, fall. *P. platycarpa* Small

    2. **P. pectinata** Lam. MERMAID WEED—Stems simple, decumbent, stout, often partially submersed or growing on moist soil. Submersed leaves pinnatifid, the lobes filamentous to linear, emersed leaves 2–3 cm long, oblong-ovate, deeply pinnatifid, with about six to twelve pairs of lateral segments. Flowers sessile in upper leaf axils, minute; calyx three-angled, greenish; nutlet 3–4 mm wide, irregularly ridged. Moist soil, swamps, Fla. to La., Nova Scotia on the coastal plain. Spring, fall.

145. ARALIÀCEAE   GINSENG FAMILY

    Trees, shrubs, woody vines, rarely herbs, with alternate simple or compound leaves. Flowers small in much branched inflorescence; flowers bisexual or unisexual, radially symmetrical; calyx minute, petals three to five, stamens as many as the petals and alternate with them. Ovary inferior, one- to several-locular. About 65 genera and over 800 species, chiefly in the tropics. *Hederaceae*

1. ARÀLIA L.   SARSAPARILLA

    Herbs, shrubs, or trees prickly armed or smooth with alternate, pinnately compound leaves. Flowers in umbels or panicles; calyx very small, petals five, white or greenish; stamens five. Ovary five-locular, styles free. Fruit a black berry tipped with persistent styles. About 30 species in North America and Asia.

    1. **A. spinosa** L. DEVIL'S WALKING STICK—Small tree or shrub up to 12 m tall, bearing stout prickles on the stems, branches, and petioles. Leaves twice or thrice compound, the leaflets 5–10 cm long, ovate, serrate, acute or acuminate at the tips. Flowers small, in umbels of a terminal compound panicle. Berry subglobose about 6–7 mm wide. $2n = 48$. Low ground, Fla. to Tex., Del., Mo. Sometimes cultivated in our area, apparently not well established in south Fla. Summer.

146. APIÀCEAE   CELERY FAMILY

    Herbs with stems having abundant pith, internodes often hollow. Leaves alternate, sheathing at the base, simple to much divided or dissected. Flower bisexual, rarely unisexual, in simple or compound umbels or in headlike inflorescence; calyx fused to the ovary, five-lobed, petals five, stamens five, alternate with the petals. Ovary inferior, twolocular, styles two. Fruit dry, dividing into two mericarps, these usually prominently ribbed. Cosmopolitan family of about 200 genera and 3000 species, chiefly in temperate regions. Nomen alt. Umbelliferae. *Ammiaceae*

1. Flowers in headlike inflorescence.
   2. Involucral bracts four or more; leaves
      ovate-lanceolate, often lobed . . . 1. *Eryngium*
   2. Involucral bracts two, conspicuous; leaves
      ovate-oblong . . . . 2. *Centella*
1. Flowers in umbels or whorled, not in heads.
   3. Leaves simple or reduced to phyllodia, not
      compound or dissected.
      4. Leaves cylindrical or linear, much reduced . . . . 3. *Oxypolis*
      4. Leaves orbicular, reniform-cordate, or
         elliptic.
         5. Leaves peltate; involucre absent or
            minute . . . . 4. *Hydrocotyle*
         5. Leaves oblong-cordate or elliptic, not
            peltate; involucre of two conspicuous bracts . . . . 2. *Centella*
   3. Leaves compound or finely dissected.
      6. Corolla yellow . . . . . . 5. *Foeniculum*
      6. Corolla white or greenish white.
         7. Leaves at least the lower trifoliolate,
            biternate, or pinnate.
            8. Basal leaves biternate or pinnate;
               umbels short pedunculate . . 6. *Apium*
            8. Basal leaves trifoliolate, leaflets
               serrate . . . . 7. *Cicuta*
         7. Leaves finely dissected, the segments
            filiform or linear.
            9. Involucres present; fruit ribbed but
               without tubercles . . . . . 8. *Ptilimnium*
            9. Involucres absent; fruit not ribbed
               but with tubercles . . . . 9. *Spermolepis*

1. ERÝNGIUM L. Eryngos

Perennial or biennial, usually smooth herbs with stout taproots, rootstocks, or fibrous roots. Leaves entire to lobed or divided. Inflorescence a solitary, involucrate head, or heads several; flowers bisexual, blue, purple, or white, each one subtended by a small bract; calyx often

Family 146. **APIACEAE.** *Eryngium aromaticum:* a, flowering plant, × ½; b, inflorescence, × 4¼; c, flower and subtending bract, × 15½; d, mature fruit, × 10½.

spinelike, petals five with incurved tips. Fruit scaly or tuberculate, globose to obovoid. About 250 species, cosmopolitan.

1. Leaves pinnately parted or divided, the
 lobes spine tipped; heads spiny-bristly . 1. *E. aromaticum*
1. Leaves entire, toothed, or lobed but not pinnatifid, spine tipped; heads not spiny-
 bristly.
 2. Leaves with parallel veins, margins with
 sharp bristles . . . . 2. *E. yuccifolium*
 var. *synchaetum*
 2. Leaves with net veins, not parallel veined.
 3. Floral bracts longer than the flowers;
 some of the basal leaves deeply lobed
 or divided . . . . 3. *E. baldwinii*
 3. Floral bracts shorter than the flowers;
 basal leaves entire, shallowly serrate,
 or palmately lobed . . . . . 4. *E. prostratum*

1. **E. aromaticum** Baldw. FRAGRANT ERYNGIUM—Herbs with spreading, decumbent or suberect stems to 6 dm long. Lowermost leaves oblong, 3–4 cm long, deeply pinnatifid or pinnately parted, the lobes spine tipped. Heads spiny-bristly, 8–10 mm long; sepals lanceolate, 2.0–2.6 mm long, acuminate; petals 1.5 mm long or longer. Fruit granular, 1–2 mm long. Low ground, pineland, glades, ditches, Fla. to Ga. on the coastal plain. Spring.

2. **E. yuccifolium** Michx. var. **synchaetum** Gray, BUTTON SNAKE-ROOT—Erect herbs to 9 dm tall, usually less. Lowermost leaves narrowly linear with distinct parallel veins, margins bearing sharp bristles. Heads not spiny or bristly; sepals broadly ovate to deltoid-ovate, 1–2 mm long; petals 1–1.5 mm long. Fruit 2–3 mm long, smooth or obscurely tubercular. Low ground, margins of inland swamps, glades, ditches, Fla. to Tex., Ga., on the coastal plain. Spring, fall.

3. **E. baldwinii** Spreng.—Prostrate herbs with creeping stems. Lowermost leaves mostly 3–9 cm long, often deeply lobed or divided, or elliptic to elliptic-ovate. Heads dense, cylindric to obovoid in outline; floral bracts longer than the flowers, awl shaped. $2n = 16$. Low ground, Fla. to La., Ga., on the coastal plain. Spring, fall.

The lower leaves are highly variable and may be obovate and merely toothed to deeply lobed, possibly resulting from hybridization with the following species.

4. **E. prostratum** Nutt.—Prostrate herbs, slender, rooting at the nodes, or with ascending branches. Lowermost leaves entire, or shallowly serrate to palmately lobed, lanceolate to ovate in outline. Heads 4–8 mm

long, cylindric, solitary, arising from leaf axils; bracts shorter than the flowers, minute, subulate. $2n = 16$. Fla. to Tex., Va., Mo. Reported from Keys in southwest Fla. Summer.

2. CENTÉLLA L.

Perennial herbs with simple leaves, usually of moist areas. Leaves clustered at the nodes. Umbels simple, arising from the base, with two to five short-pedicelled or subsessile flowers; corollas white; bracts ovate, two; sepals obsolete. Fruit orbicular or ellipsoidal to reniform, flattened laterally, with conspicuous ribs. About 20 species, widely distributed.

1. **C. asiatica** (L.) Urban, Coinwort—Stems creeping or prostrate, producing a cluster of petioled leaves at each node. Leaves 1–6 cm long, ascending, oblong to broadly ovate, entire, truncate or shallowly cordate at base, obtuse at the apex, petioles pubescent, highly variable in length. Umbels with peduncles 2–10 cm long, variable; fruit flat-ellipsoidal, 3–5 mm wide. Low ground, disturbed areas, wet soil, Fla., Fla. Keys to Tex., Del. on the coastal plain, tropical America. *C. erecta* (L.f.) Fern.; *C. repanda* (Pers.) Small

3. OXÝPOLIS Raf. Dropwort

Smooth, erect herbs arising from a cluster of tuber-bearing roots. Leaves pinnately divided or reduced to phyllodes. Flowers in open, compound umbels, bracts few, linear, or none; sepals minute or none, petals white. Fruit subglobose or elliptic, flattened dorsally and bearing five filiform ribs, with oil tubes solitary in the intervals. About five species in North America.

1. **O. filiformis** (Walt.) Britt. Water Dropwort—Slender herbs to 2 m tall, usually less. Leaves reduced to terete, filamentous phyllodia 3–6 dm long. Flowers few in compound umbels; involucral bracts many, threadlike. Fruit subglobose to broadly elliptic, 4–6 mm long, notched at the summit, broadly winged. $2n = 28$. Glades, margins of swamps, low ground, Fla. to La., S.C. All year.

4. HYDROCÓTYLE L. Water Pennywort

Perennial herbs of moist soil with slender stems prostrate or arching and rooting at the nodes. Leaves simple, with long petioles and ovate to orbicular, usually peltate blades. Flowers white, greenish, or yellowish green on axillary pedicel; umbels simple or compound; involucre minute or absent. Fruit orbicular to ellipsoidal, flattened laterally. About 100 species, generally cosmopolitan.

1. Flowers in umbels.
　　2. Umbels simple; leaf blades 2–5 cm wide .　1. *H. umbellata*
　　2. Umbels compound or proliferous; leaf
　　　　blades 3–10 cm wide . . . . . .　2. *H. bonariensis*
1. Flowers in spikelike racemes or whorled, not
　　in simple umbels . . . .　3. *H. verticillata*

1. H. umbellata L. MARSH PENNYWORT—Stems slender, succulent, floating or creeping, rooting at the nodes and branching freely. Leaves 5–7 cm wide, erect, orbicular, peltate, crenate or shallowly lobed, smooth and glossy green with long petioles. Umbels simple with usually ten or more flowers, peduncles mostly longer than the leaves. Fruit 2–3 mm broad with distinct notch or cordate at the base, margins acute. Low ground, glades, ditches, Fla. to Tex., Nova Scotia, Minn., Pacific coast and tropical America. All year.

2. H. bonariensis Lam. WATER PENNYWORT—Stems slender, succulent, creeping or floating, rooting at the nodes and branching freely. Leaves 3–10 cm wide, erect, blades orbicular to reniform-orbicular, shallowly crenate-lobed. Umbels simple or more often compound, proliferous, peduncles much longer than the leaves. Fruit about 3 mm wide, margins acute, conspicuously ribbed. Ponds, margins of swamps, ditches, moist sandy soil, Fla. to Tex., N.C. on the coastal plain, W.I., tropical America. Often found growing in large mats in shallow water. All year.

3. H. verticillata Thunb. WHORLED PENNYWORT—Stems filamentous, creeping, smooth. Leaves 3–6 cm wide, orbicular to elliptic, peltate, crenate, or lobed. Flowers in a simple or one to two spicate raceme to 15 cm or more long, bearing two to many whorls of flowers. Fruit 3–4 mm broad, rounded or truncate at the base, margins acute. Low ground, Fla. to Tex., Mass. near the coast, western U.S., tropical America. All year.

5. FOENÍCULUM Adans. FENNEL

Perennial herbs with leaves dissected into linear or filiform segments. Flowers in large, compound umbels on stout peduncles. Bracts absent, sepals none, petals yellow. Fruit oblong, somewhat flattened laterally, ribs prominent with oil tubes solitary in the intervals. About four species in the Old World.

1. F. vulgare Mill. FENNEL—Stout perennial herb up to 2 m tall, glabrous and glaucous. Leaf segments 1–3 cm long, less than 1 mm wide, petiolar sheaths of larger leaves 3–10 cm long. Compound umbels large, 4–10 cm wide, flowers yellow; fruit 3–4 mm long, ribs acute. $2n = 22$.

Sandy, disturbed areas, reported for Key West; native of the Mediterranean regions, introduced and naturalized throughout much of the U.S. and tropical America. Summer, fall.

6. ÀPIUM L.

Annual, biennial, or perennial glabrous herbs with taproots or creeping rootstocks. Stems usually much branched, leaves petioled, sheathing pinnately to ternate-pinnately twice compound. Flowers white or greenish in simple or compound umbels, bracts present or none. Fruits compressed laterally, ribs conspicuous, filiform. About 20 species, chiefly Northern Hemisphere. *Celeri* Adans.; *Cyclospermum* Lag.; not *Apium* sensu Small, *Petroselinum* Hoffm.

1. Lower leaves once-pinnate, leaflets three to
   nine, often serrate to deeply three-lobed
   or incised . . . .   1. *A. graveolens*
1. Lower leaves biternate, lowest leaves with
   wide segments, upper leaves with threadlike segments . . . .   2. *A. leptophyllum*

   1. A. graveolens L. GARDEN CELERY—Perennial herb to 1 m tall, with taproots branched above. Leaves once-pinnate, leaflets three to nine, suborbicular, lanceolate to triangular, often serrate to deeply three-lobed, or incised into cuneate-obovate segments. Umbels short-peduncled, bracts none; flowers usually greenish white. Fruit ovoid or compressed laterally 2–3 mm long. Disturbed soil, native of Eurasia, cultivated and locally established in south Fla. Spring, summer. *Celeri graveolens* (L.) Britt.

   2. A. leptophyllum (Pers.) F. Muell. MARSH PARSLEY—Annual with glabrous stems up to 6 dm tall, usually less, branched above. Lower leaves biternate, the lowest with wide segments, upper leaves with threadlike segments, filiform or narrowly linear, highly variable. Umbels short peduncled, bracts none, flowers minute, white. Fruit ovoid or oval, 1–2 mm long, prominently ribbed. Disturbed sites, lawns, ditches, Fla. Keys, Fla. to Tex., N.Y., Mo., W.I., Mexico, Central and South America. Spring, summer. *Cyclospermum ammi* (L.) Britt.

7. CICÙTA L.   WATER HEMLOCK

Perennial or biennial herbs with tuberous roots and ternately or pinnately compound leaves. Umbels compound, spreading, involucre of few, small bracts, or absent; involucels of several narrow bractlets; flowers white, sepals acute. Fruit ovoid to subglobose, smooth, with prominent corky ribs, oil tubes conspicuous. About ten species of the north temperate regions.

The roots, resembling small potatoes, are deadly poisonous.

1. **C. mexicana** C & R—Biennial herb bearing a cluster of somewhat elongated, tuberlike roots, stems to 2 m tall, stout. Leaves twice or thrice pinnately compound, leaflets oblanceolate to narrowly ovate, coarsely toothed. Larger umbels 8–20 cm broad; fruits subglobose, 2–3 mm long, often with a dark line between the ribs. Low ground, margins of inland swamps, wet soil, Fla. to Tex., N.J., Tenn., Mexico. Spring, summer. *C. maculata* L. var. *curtissii* (C & R) Fern.; *C. curtissii* C & R

8. PTILÍMNIUM Raf.   Mock Bishop Weed

Erect, branching, annual herbs with leaves much dissected into filiform segments. Umbels compound, pedunculate, terminal or lateral umbels; involucral bracts well developed and conspicuous, cleft or filiform; flowers small, white. Fruit ovoid to subglobose, somewhat flattened laterally, dorsal ribs threadlike, lateral ribs thick and corky. *Harperella* Rose; *Discopleura* DC.

1. **P. capillaceum** (Michx.) Raf.—Slender, erect herb to 8 dm tall, freely branching above. Leaves 5–10 cm long, pinnately dissected into filiform segments 5–30 cm long, with mostly three divisions at a node along the rachis. Umbels 2–5 cm broad, taller than the leaves; bracts 1–2 cm long, few, simple or divided into filiform segments; sepals minute, petals white, minute; fruit ovoid, 2–3 mm long. $2n = 14$. Disturbed sites, moist glades, margins of swamps, Fla. to Tex., Mass., on the coastal plain, to Mo., Kans. Spring, fall.

9. SPERMÓLEPIS Raf.

Annual herbs with glabrous stems and leaves. Leaves ternately twice compound, finely dissected with filiform or linear segments. Compound umbels terminal or from the upper leaf axils, peduncled, open; involucral bracts none, umbels two- to six-flowered, flowers with a few linear bractlets; sepals none or obsolete, petals white. Fruit ovoid, laterally flattened, ribbed. Four species in North America.

1. **S. divaricata** (Walt.) Raf.—Erect herb up to 7 dm tall, much branched above, very slender. Leaves much dissected, segments linear or filamentous. Pedicels variable, to 2 cm long, central flower sometimes sessile; petals white. Fruit 1–2 mm long, tuberculate or smooth. Glades, scrub vegetation, sandy or dry soil, disturbed sites, Fla. to Tex., Va. chiefly on the coastal plain, Kans., Mo. Spring.

Apparently this species is more abundant in north and central Fla. than in south Fla.

147. CORNÁCEAE Dogwood Family

Trees, shrubs, or rarely perennial herbs with simple leaves. Flowers small, in heads or branching panicles, bisexual or unisexual; heads may be surrounded by a conspicuous, leafy or petaloid involucre resembling a large flower; calyx tube fused to the ovary, four- to five-lobed, petals four to five, free; stamens as many as the petals and alternate with them, disk present, flattened. Ovary inferior, one- to four-locular. Fruit a drupe or berry. About 10 genera and 115 species, chiefly north temperate regions.

1. CÒRNUS L. Dogwood

Trees or shrubs with opposite or rarely alternate, simple, entire leaves. Inflorescence a corymbose or umbellate cyme or headlike and subtended by four involucrate bracts. Flowers small, bisexual; sepals four, petals four, stamens four. Ovary inferior, two-locular, style arising from a fleshy disk. Fruit a drupe with one seed. About 40 species, widely distributed. *Svida* Opiz; *Cynoxylon* Raf.

1. **C. foemina** Mill. Stiff Cornel—Shrub, stems grayish, up to 3 m tall or more. Leaves lanceolate to elliptic, 5–15 cm long. Flowers in corymbose cymes; sepals minute, triangular, petals white, oblong to oblong-lanceolate. Drupe subglobose, 4–5 mm wide. Wet soil, shallow water, Fla. to Me., Nebr. Spring.

Cornus florida L., the Flowering Dogwood, with four conspicuous, white, involucral bracts 3–6 cm long, occurs in north and central Fla., but it is not well established in south Fla. It is sometimes found in cultivation in our area. Spring.

148. ERICÁCEAE Heath Family

Shrubs, trees, or herbs, usually evergreen, bearing simple, alternate or sometimes opposite to whorled leaves. Flowers mostly bisexual, radially symmetrical or nearly so, solitary or in various kinds of inflorescences; calyx four- to seven-parted, corolla four- to seven-parted or sometimes petals free, often funnelform, companulate or urceolate, stamens eight to ten, anthers usually opening by terminal pores, or slits, sometimes with appendages or with tubular tips. Ovary inferior or occasionally superior, three- to seven-locular or twice the number by the formation of false septa, style one, stigma simple, often lobed. Fruit a capsule, drupe, or berry, the calyx usually persistent. About 75 genera and 2000 species of cosmopolitan distribution mostly in acid soils. *Vacciniaceae*

1. Leaves lanceolate to lance-ovate, sessile, less
   than 1 cm long; ovary inferior.
   2. Corolla urceolate; ovary four- to five-
      locular . . . .   1. *Vaccinium*
   2. Corolla campanulate; ovary ten-locular .   2. *Gaylussacia*
1. Leaves lance-ovate to elliptic, petiole over 1
   cm long; ovary superior.
   3. Flowers in terminal racemes, petals dis-
      tinct; upper stems often hirsute . . .   3. *Befaria*
   3. Flowers axillary, petals fused to form
      urnlike or globular corolla; upper stems
      glabrous or scurfy, not hirsute . . .   4. *Lyonia*

1. VACCÌNIUM L.   BLUEBERRY

Erect, evergreen or deciduous, often stoloniferous shrubs or rarely trees from rhizomes. Leaves alternate, short petioled, entire or serrulate. Flowers solitary in the upper axils or in terminal racemes; calyx fused to the ovary, four- to five-parted, corolla four- to five-lobed, tubular, ovoid, or campanulate, stamens eight or ten, opening by a terminal pore. Ovary inferior, fruit a many-seeded berry, crowned by the persistent calyx lobes. About 150 species, with many hybrids, chiefly in the Northern Hemisphere. *Cyanococcus* (Gray) Rydb.; *Batodendron* Nutt.; *Polycodium* Raf.; *Herpothamnus* Small; *Hugeria* Small; *Oxycoccus* Hill

1. **V. myrsinites** Lam. SHINY BLUEBERRY—Usually low, erect shrubs up to 6 dm tall, much branched. Leaves 0.5–2 cm long, obovate, ovate, elliptic or oblanceolate with bristly and glandular teeth, glossy green above, somewhat paler and sparingly pubescent beneath. Flowers in umbellate clusters of fascicles; sepals ovate to deltoid, spreading; corolla 6–7 mm long. Berry 5–8 mm wide, globular, blue black. Acid sandy soils, Fla. to N.C., La. Spring. *Cyanococcus myrsinites* (Lam.) Small

2. GAYLUSSÀCIA HBK   HUCKLEBERRY

Branching shrubs with deciduous or evergreen alternate leaves, the blades conspicuously resin dotted. Flowers in terminal racemes, corolla campanulate or tubular, five-parted, stamens ten. Ovary inferior, ovoid. Fruit a berrylike drupe with ten seedlike nutlets. About 50 species in Western Hemisphere, mostly in Brazil.

1. **G. dumosa** (Andrews) T & G, DWARF HUCKLEBERRY—Shrubs with slender stems up to 6 dm tall from creeping rhizomes. Leaves 2–6 cm long, obovate or spatulate, mucronate, short petioled or subsessile,

glandular beneath. Racemes elongate, leaf bracted, bracts persistent; corolla campanulate, pinkish white or white. Ovary glandular-pubescent, ten-locular. Fruit black, glandular. Low pinelands, scrub vegetation, Collier County and possibly elsewhere in south Fla., to N.J., Tenn., Miss. Spring. *Lasiococcus dumosus* (Andrews) Small

3. BEFÀRIA Mutis ex L.

Erect shrub with alternate, evergreen, thick leaves. Flowers showy, in 'elongate spikelike racemes; calyx six- to seven-parted, petals six to seven, spreading, stamens twelve to fourteen, anthers with terminal pores. Calyx campanulate, seven parted, petals usually seven, free, stamens twelve to twenty. Ovary superior, seven-locular, lobed. Fruit a lobed, septicidal, subglobose capsule, seeds numerous. About 15 species in tropical and subtropical America.

1. **B. racemosa** Vent. TARFLOWER—Shrubs up to 2.5 m tall, usually less, with stems bearing coarse pubescence. Leaves ovate to elliptic 2–7 cm long, usually with hirsute pubescence. Flowers white, conspicuous, fragrant in elongate racemes or panicles; calyx lobes 4–5 mm long, petals 2–3 cm long, white, sometimes pink tinged, spatulate to narrowly so, sticky; filaments pubescent. Capsule 6–9 mm wide, depressed-globose. Sandy soils, pine flatwoods, scrub vegetation, Fla. to Ga. on the coastal plain. Spring.

4. LYÒNIA Nutt.

Shrubs with evergreen or deciduous alternate leaves. Flowers in umbel-like lateral clusters on leafless branches, or axillary and in racemelike naked or leafy inflorescence; flowers white to pink, five-parted; calyx campanulate to saucer shaped, deeply lobed, usually persistent in fruit; corolla lobes short; stamens ten, included, filaments pubescent, anthers opening by terminal pores. Ovary superior, four- to eight-locular; fruit a globose five-ribbed capsule. About 40 species in America and Asia. *Xolisma* Raf.; *Desmothamnus* Small, *Neopieris* Britt.; *Arsenococcus* Small

1. Upper stems, pedicels, and leaves glabrous, never scurfy; corolla 7–10 mm long . . 1. *L. lucida*
1. Upper stems, pedicels, and leaf undersurfaces rusty scurfy; corolla less than 6 mm long.
   2. Corolla 3–5 mm long; leaves tending to be much reduced toward the end of branches . . . . 2. *L. fruticosa*
   2. Corolla 2–3 mm long; leaves only slightly reduced toward the end of branches . 3. *L. ferruginea*

1. **L. lucida** (Lam.) K. Koch, SHINY LYONIA, FETTERBUSH—Erect shrubs up to 2 m tall with evergreen leaves, branches smooth. Leaves leathery, glossy, elliptic to oblong-ovate, entire, with a distinct vein parallel to the slightly revolute margin. Flowers in umbel-like clusters arising from the leaf axils, lax; calyx lobes oblong, 3–5 mm long; corolla 7–9 mm, white or rose white. Capsule subglobose, 5 mm thick. $2n = 24$. Pinelands, Fla. to La., Va., Cuba. Spring. *Lyonia nitida* (Bartr.) Fern.; *Desmothamnus lucidus* (Lam.) Small

2. **L. fruticosa** (Michx.) G. S. Torr. STAGGERBUSH—Erect shrub or small tree with rusty pubescence, up to 3 m tall. Leaves 4–6 cm long, evergreen, margins not revolute, ovate to obovate or oblanceolate, much reduced in size toward the ends of branches, pubescent especially beneath. Flowers in umbel-like clusters; corolla 4–5 mm long, white or pinkish, urn shaped. Acid sandy soils, Fla. to S.C. on the coastal plain, W.I. Spring.

3. **L. ferruginea** (Walt.) Nutt. RUSTY LYONIA—Erect shrub or small tree up to 5 m tall, with rusty pubescence. Leaves 3–7 cm long, evergreen, margins rolled, elliptic to ovate or obovate, not much reduced toward the ends of branches. Flowers in umbel-like inflorescence; corolla 2–3 mm long, white or pinkish white, globular, angled at the base. Capsule 4–5 mm long. Moist or dry acid soil, Fla. to S.C. on the coastal plain, W.I., Mexico. Spring. *Xolisma ferruginea* (Walt.) Heller

MONOTROPA BRITTONII Small, a saprophytic herb with whitish or pale pinkish stems, leaves reduced to scales, and colorless or pinkish flowers, has been reported as occurring in our area in scrub vegetation. No recent collections have been seen. The plant is commonly known as INDIAN-PIPES.

149. THEOPHRASTÀCEAE JOEWOOD FAMILY

Shrubs or trees with opposite, leathery, evergreen leaves. Flowers bisexual, radially symmetrical, in racemes, corymbs, or panicles; sepals five, imbricate, corolla five-parted, campanulate or rotate-salverform; stamens five, partly fused to the corolla tube, staminodia five. Ovary inferior, of five fused carpels. Fruit a drupelike berry. About 5 genera and 50 species of tropical distribution.

1. JACQUÍNIA L.

Erect shrubs or trees with numerous entire, thick leaves. Flowers in corymbose erect racemes; calyx persistent, corolla deciduous, con-

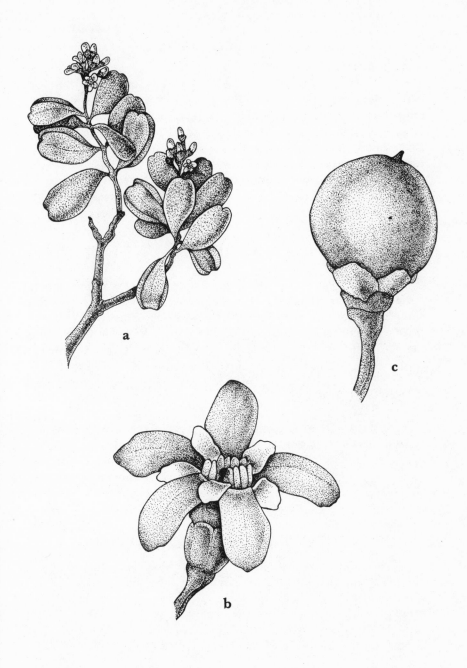

Family 149. **THEOPHRASTACEAE.** *Jacquinia keyensis:* a, flowering branch, × ½; b, flower, × 5; c, mature fruit, × 5.

spicuous; staminodia broad. Fruit an erect berry. About 25 species in tropical America.

1. J. keyensis Mez. Joewood—Shrub or tree up to 5 m tall with brittle stems and pale bark. Leaves 2–5 cm long, cuneate-spatulate or elliptic-ovate, glossy green. Flowers fragrant, showy; calyx lobes 2–3 mm long, orbicular-ovate; corolla yellow white, the lobes longer than the tube; staminodia elliptic. Berry 8–10 mm wide, subglobose. Coastal marl soils, south Fla., Fla. Keys, and W.I. All year.

This is one of our characteristic south Fla. species and is found widely in our area, especially in the Fla. Keys near the beaches. The flowers are very fragrant.

150. MYRSINÀCEAE Myrsine Family

Trees or shrubs with alternate, simple, punctate leaves. Flowers radially symmetrical, small, bisexual or unisexual by abortion and dioecious, in racemes or panicles; calyx four- to five-parted, sepals free or fused, corolla rotate or tubular, four- to five-lobed, stamens as many as the corolla lobes and opposite them, epipetalous. Ovary superior to semiinferior, one-locular, style simple. Fruit a drupe or berry, sometimes dry and dehiscent. About 40 genera and 1000 species, widely distributed in the tropics. *Ardisiaceae*

1. Flowers axillary on short spurlike peduncles,
    solitary or in small clusters; fruit less than
    5 mm in diameter . . . . 1. *Myrsine*
1. Flowers in terminal panicles or axillary
    long-pedunculate cymes; fruit over 5 mm
    in diameter . . . . 2. *Ardisia*

1. MYRSÍNE L.

Trees or shrubs with leaves often lepidote and usually entire. Flowers small, unisexual or polygamous in subglobose clusters or umbels in leaf axils or above leaf scars; sepals free or fused, dotted with many glandular points or lines; petals somewhat fused at base, dotted with glands; stamens without filaments. Ovary globose, style very short or absent. Fruit dry or berrylike. About 135 species in tropics and subtropics.

1. M. guianensis (Aubl.) Kuntze—Shrubs or low trees with rather thick branches. Leaves 6–10 cm long, obovate-oblong to elliptic, tending to cluster near the ends of stems, entire, usually recurved margins.

Family 150. **MYRSINACEAE.** *Myrsine guianensis:* a, fruiting branch, × ½;
b, flower, × 20; c, flowering branch, × 12½; d, mature fruit, × 20.

Flowers in umbels, clustered along the branches, sessile, about three to ten flowers per umbel, 2 mm long, glabrous. Stigma distinctly lobed in pistillate flowers. Berry about 4 mm wide. Hammocks, Fla. Keys, south Fla., American tropics. All year. *Rapanea guianensis* Aubl.

2. ARDÍSIA Sw.

Trees or shrubs with entire or serrate leaves. Flowers bisexual or polygamous in cymes, panicles, or clusters; calyx minute, campanulate, sepals deltoid, corolla rotate, tube very short, five-lobed; stamens included, filaments very short. Fruit a drupe, globose, one-seeded. About 200 species, pantropical. *Icacorea* Aubl.

1. Inflorescence a terminal panicle or raceme . 1. *A. escallonioides*
1. Inflorescence an axillary cyme . . . . 2. *A. solanacea*

1. A. escallonioides Schlecht. & Cham. MARLBERRY—Shrub or sometimes a small tree with light colored smooth stems. Leaves mostly 10–15 cm long on stout petioles, oblanceolate to narrowly ovate or elliptic. Flowers in terminal, dense panicles about as long as the leaf; calyx about 2 mm long, lobes ovate-deltoid, corolla lobes elliptic, white, somewhat punctate. Drupe 7–8 mm wide, black at maturity. Hammocks, pinelands, south Fla., Mexico, W.I. All year. *Icacorea paniculata* (Nutt.) Ludw.

2. A. solanacea Roxb.—Shrub or small tree with smooth stems. Leaves mostly 10–15 cm long, oblong-obovate or elliptical-oblong, entire, tapering to a short petiole or subsessile, punctate beneath. Flowers mostly in axillary corymbs, pedicels spreading; calyx lobes rounded to deltoid, corolla lobes ovate-acute, white, anthers with a cordate base. Hammocks, disturbed sites, introduced from East Indies and naturalized in south Fla., W.I. All year. *A. humilis* Vahl, *A. polycephala* Wight, not Wall.

151. PRIMULÀCEAE PRIMROSE FAMILY

Herbs with alternate, opposite, or basal leaves. Flowers bisexual; calyx of four to nine partially fused sepals, corolla of four to nine partially fused petals or sometimes absent; stamens as many as the petals and opposite them, epipetalous, staminodia sometimes present. Ovary superior or partially inferior, one-locular, with free central placentation, style one, stigma simple. Fruit a capsule, two- to four-valved. About 28 genera and 400 species, chiefly in the Northern Hemisphere.

1. Ovary partially inferior . . . . . .   1. *Samolus*
1. Ovary superior.
  2. Corolla longer than the calyx; style
    longer than the ovary . . . . .   2. *Anagallis*
  2. Corolla shorter than the calyx; style
    shorter than the ovary . . . . .   3. *Centunculus*

1. SÀMOLUS L.

Smooth perennial herbs with alternate, entire leaves, or basal leaves rosulate. Flowers small, white, in terminal racemes or panicles; calyx persistent, five-cleft, corolla subcampanulate, five-lobed, stamens five, sometimes alternating with as many staminodia, anthers cordate. Ovary partially inferior, capsule five-valved. About ten species of wide distribution. *Samodia* Baudoin

1. Larger basal leaves elongate-spatulate,
  mostly over 10 cm long; flowers 5–7 mm
  wide . . . .   1. *S. ebracteatus*
1. Larger basal leaves ovate or ovate-spatulate,
  mostly less than 10 cm long; flowers
  smaller, 2–3 mm wide . . . . . .   2. *S. parviflorus*

1. **S. ebracteatus** HBK, WATER PIMPERNEL—Erect herbs up to 4 dm tall, usually less. Leaves spatulate or obovate, 3–12 cm long, rather thick. Flowers in simple racemes or in panicles, pedicel ascending; calyx lobes ovate, corolla 6–7 mm wide, staminodia absent. Capsule 3–4 mm wide. Moist soil, south Fla. to Tex., N. Mex., W.I. All year. *Samodia ebracteata* (HBK) Baudoin

2. **S. parviflorus** Raf. PINELAND PIMPERNEL—Erect, slender, smooth herbs up to 6 dm tall. Leaves spatulate to ovate-elliptical, mostly 4–15 cm long, the basal ones rosultate. Flowers in simple, open racemes or sometimes panicles; sepals ovate, about 1 mm long, corolla about 3 mm wide, lobes elliptic. Capsule 2–3 mm long. $2n = 24$. Moist soil, pinelands, south Fla. to Tex., Canada, Calif., Central and South America. Spring, summer. *S. floribundus* HBK

2. ANAGÀLLIS L.   PIMPERNEL

Procumbent or spreading, usually annual herbs. Leaves opposite or whorled, entire. Flowers solitary, peduncles axillary; corolla rotate, longer than the calyx, tube very short; stamens five, filaments conspicuously pubescent. Capsule membranaceous, many-seeded. About 15 species, mostly in the Old World.

1. **A. pumila** Sw.—Stems mostly 1–2 dm tall. Leaves up to 1 cm long, apiculate. Flowers on slender pedicels, calyx lobes ovate, abruptly pointed, corolla greenish or white. Capsule 1–2 mm wide. Moist pinelands, low ground, south Fla., tropical America, Old World. Spring, fall.

3. CENTÚNCULUS L.   Chaffweed

Low annuals. Leaves mostly alternate, or lower ones opposite, somewhat fleshy. Flowers solitary, axillary, inconspicuous; corolla rotate, shorter than the calyx, tube very short, stamens four to five, filaments glabrous or nearly so. Capsule many-seeded. Two species widely distributed.

1. **C. minimus** L. False Pimpernel—Stems ascending up to 1 dm tall or more. Leaves less than 1 cm long, ovate, obovate to spatulate. Flowers sessile or nearly so, sepal lobes four, corolla lobes four, pink or whitish. Capsule about 2 mm wide. Low ground, sandy soil, south Fla. to Tex., Nova Scotia, Minn., B.C., W.I., Mexico. Summer.

152. PLUMBAGINÀCEAE   Leadwort Family

Herbs, shrubs, or climbing plants. Flowers bisexual, radially symmetrical, usually in bracteolate, umbel-like, or in one-sided inflorescences; sepals five, fused, often ribbed, petals five, mostly fused, stamens five, opposite the petals or lobes of the corolla. Ovary superior, one-locular, styles five, free or variously fused. Fruit a nut or eventually dehiscent. About 10 genera and 350 species, cosmopolitan. *Armeriaceae*

1. Scapose plants, larger leaves all basal, upper
     leaves reduced to mere scales; flowers
     small, less than 1 cm long . . . . .   1. *Limonium*
1. Leafy plants with alternate leaves, not
     scapose; flowers over 1 cm long . . .   2. *Plumbago*

1. LIMÒNIUM Mill.   Sea Lavender

Scapose perennial herbs from woody roots. Basal leaves narrow, upper leaves reduced to scales or absent. Flowers small, violet to lavender, in erect, branching, naked inflorescence, floral clusters subtended by two to three scarious bracts; calyx tubular, five-lobed, petals five, nearly free, long clawed; stamens five, epipetalous. Fruit exserted. About 150 to 300 species, cosmopolitan. *Statice* L.

1. **L. carolinianum** (Walt.) Britt. Sea Lavender—Erect stems up to 6 dm tall. Basal leaves 6–16 cm long, linear-lanceolate, elliptic-obovate

to spatulate. Flowers in large, branching, open panicles, clustered in sessile umbels along the inflorescence; bractlets 1–2 mm, calyx 5–6 mm long, glabrous or hirsute, acute or obtuse, corolla lavender. Fruit a capsule, mostly 5–6 mm long. Coastal areas, Fla. to Mexico, Nova Scotia on the coastal plain. Two varieties are found in our area:

1. Leaves lanceolate-elliptic to obovate-spatulate; flowers rather dense, internodes less than 5 mm . . . . 1a. var. *carolinianum*
1. Leaves linear-lanceolate to narrowly elliptic; flowers rather loose, internodes about 5 mm . . . . 1b. var. *angustatum*

1a. **L. carolinianum** var. carolinianum—Leaves chiefly lanceolate-spatulate, obtuse or retuse. Flowers usually dense. Fla. to Nova Scotia, Mexico. All year. *L. nashii* Small; *L. trichogonum* Blake

1b. **L. carolinianum** var. **angustatum** (Gray) Blake—Leaves chiefly linear-lanceolate, acute, often dwarfish plants. Flowers usually loose. South Fla. All year. *L. angustatum* (Gray) Small; *L. carolinianum* var. *obtusilobum* (Blake) Ahles, *L. nashii* var. *angustatum* (Gray) Ahles

2. PLUMBÀGO L. Leadwort

Perennial herbs or subshrub, erect, reclining, or climbing, bearing alternate leaves. Flowers in terminal, spikelike panicles; sepals fused, tubular, with stalked glands and unevenly five-lobed; petals fused, corolla trumpet shaped, tube elongate. Fruit a capsule, included. About 20 species, chiefly tropical and subtropical in distribution.

1. Corolla blue, the tube twice or more the length of the calyx; calyx with both glandular and eglandular hairs . . . 1. *P. capensis*
1. Corolla white, the tube less than twice the length of the calyx; calyx with only glandular hairs . . . . 2. *P. scandens*

1. **P. capensis** Thunb.—Stems erect, often much branched. Leaves mostly 3–9 cm long, variable, elliptic to elliptic-spatulate or ovate. Flowers blue, in short, dense panicles; bracts acuminate, calyx 1 cm long, corolla tube twice or more the length of the calyx, corolla mostly 4–5 cm long. $2n = 28$. Disturbed sites, dry soil, naturalized from south Africa.

2. **P. scandens** L.—Stems erect, reclining, or climbing. Leaves 2–10 cm long, elliptic-lanceolate to ovate. Flowers white in elongate, inter-

rupted panicles; bracts acuminate, calyx less than 1 cm long, corolla tube less than twice the length of the calyx; corolla mostly 3–4 cm long. Disturbed sites, dry soil, south Fla., tropical America.

153. SAPOTÀCEAE SAPODILLA FAMILY

Trees or shrubs armed or unarmed with milky sap and alternate, simple, usually entire leaves. Flowers small, bisexual, radially symmetrical, often in axillary, cauliflorous cymose or umbellate clusters or flowers solitary; calyx four- to nine-parted or sepals free, corolla four- to nine-lobed, fertile stamens as many as the corolla lobes or more numerous, epipetalous opposite the petals. Ovary superior, one to fourteen, usually four- or five-locular; fruit a one- to many-locular, rather hard indehiscent berry. About 40 genera and 600 species, widely distributed in the tropics.

1. Some or all the branches with thorns; leaves
   oblanceolate, mostly less than 7.0 cm long   1. *Bumelia*
1. Branches without thorns; leaves lance-ovate,
   mostly over 7.0 cm long.
   2. Undersurface of leaf densely copper
      brown or silvery brown tomentose . .   2. *Chrysophyllum*
   2. Undersurface of leaf glabrous or nearly so.
      3. Young stems and pedicels with rusty
         colored or brown pubescence; pedi-
         cels as long or longer than the pet-
         ioles; ovary ten- to twelve-locular .   3. *Manilkara*
      3. Young stems and pedicels without con-
         spicuous brown pubescence; pedicels
         shorter than petioles; ovary two- to
         five-locular.
         4. Flowers in dense clusters, ovary gla-
            brous; leaves lanceolate-acuminate,
            mostly less than 12 cm long . . 4. *Dipholis*
         4. Flowers solitary or in loose clusters,
            ovary pubescent; leaves ovate-
            lanceolate or elliptic, mostly over
            12 cm long.
            5. Young stems and pedicels pale pu-
               bescent or velutinous; fruit 5–7
               cm long; petioles mostly less than
               2.0 cm long . . . . . 5. *Pouteria*
            5. Young stems and pedicels glabrous;
               fruit 2–2.5 cm long; petioles
               mostly over 2.0 cm long . . . 6. *Mastichodendron*

1. BUMÈLIA Sw. MILK BUCKTHORNS

Shrubs or low trees usually but not always with stout thorns or spines. Leaves alternate or subopposite, and somewhat pubescent beneath, narrowly elliptic to oblanceolate. Flowers three to many in cymes or clusters; calyx five-lobed, corolla five-lobed, each lobe with a pair of lateral appendages, white; staminodes present, petaloid. Ovary usually pubescent and with five ovules. Fruit a small black berry with a large, ovoid seed. About 25 species in the tropics, subtropics, and warm temperate regions of America.

1. Leaf blades conspicuously reticulate veiny, veins apparently raised.
  2. Leaves tomentose pubescent beneath with often tawny or rusty hairs, very rarely becoming glabrate with age . . . . 1. *B. tenax*
  2. Leaves loosely pubescent beneath or soon glabrate . . . . 2. *B. reclinata*
1. Leaf blades not conspicuously reticulate . 3. *B. celastrina*

1. B. tenax (L.) Willd. TOUGH BUCKHORN—Thorny trees or shrubs up to 9 m tall with silky pubescent stems. Leaves evergreen, oblanceolate, obovate, or obovate-spatulate, 3–7 cm long, densely tomentose beneath. Flowers few to several in axillary cymes; calyx lobes 1–2 mm long, orbicular-ovate, corolla 4–5 mm wide, lobes ovate-orbicular, 1.5 mm long; staminodia deltoid-ovate, about 2 mm long, obtuse. Styles 1–2 mm long. Berry obovoid 12–14 mm long. Coastal hammocks, Fla. to S.C. on the coastal plain. All year. *B. megacocca* Small

A widely distributed form with leaf pubescence silvery is f. **anomala** (Sarg.) Cronquist

2. B. reclinata (Michx.) Vent. var. **reclinata**—Erect, diffuse or low spreading, spiny shrub with smooth stems. Leaves 2–5 cm long, evergreen, oblanceolate, spatulate to obovate or ovate, pubescent when young but soon glabrous. Flowers in axillary cymes, calyx lobes ovate to orbicular, about 2 mm long; corolla about 4 mm; staminodia ovate, somewhat lacerate at the margin, 1.5–2.0 mm long. Styles 1–2 mm long. Berry ovoid or globose, 4–7 mm long. Low pineland and sandy soil, glades, Fla. to Ga. on coastal plain. *B. microcarpa* Small; *B. macrocarpa* Nutt.

B. reclinata var. **rufotomentosa** (Small) Cronquist, an endemic variation occurring as far south as Hillsborough County, may reach our area, but no specimens have been seen. It has young stems and leaves rufous tomentose, becoming glabrate with maturity.

3. B. celastrina HBK, SAFFRON PLUM—Trees or shrubs up to 8 m tall with glabrous stems. Leaves evergreen, fascicled, oblanceolate to oblanceolate-spatulate, 1–4 cm long, glabrous. Flowers few to several in axillary cymes or umbels; calyx lobes 2 mm long, ovate, corolla 3–4 mm wide, lateral lobes of the corolla irregularly toothed; staminodia 2–3 mm long, lacerate. Styles 2.5–4 mm long. Berry cylindrical 7–13 mm long. Two varieties occur in our area:

1. Leaf blades oblanceolate to spatulate, 2–4
   cm long, slightly pubescent when young .   3a. var. *celastrina*
1. Leaf blades linear-oblong, obtuse, mostly
   1–2 cm long, glabrous when young . .   3b. var. *angustifolia*

3a. B. celastrina var. **celastrina**—Coastal hammocks, south Fla., Fla. Keys, Tex., W.I., Central and South America. All year. Not common in our area.

3b. B. celastrina var. **angustifolia** (Nutt.) R. W. Long—Dry pinelands, hammocks, south Fla., especially lower Fla. Keys, endemic. All year. *B. angustifolia* Nutt.

2. CHRYSOPHÝLLUM L.

Trees or shrubs with evergreen alternate leaves that often bear abundant lustrous pubescence beneath. Flowers in axillary fascicles or sometimes solitary, sepals five, corolla lobes five to seven, campanulate-cylindric, often pubescent; stamens epipetalous, as many as the corolla lobes, staminodia absent. Ovary five- to eleven-locular, with a short style and a five- to seven-lobed stigma. Fruit variable, berrylike, seeds one or few. About 90 species in tropical and subtropical areas.

1. C. oliviforme L. SATIN LEAF—Shrubs or trees up to 10 m with pubescent stems. Leaves evergreen, ovate to elliptic, 5–10 cm long, shiny green above, lustrous light brown or copper colored pubescence beneath. Flowers several to many in each cluster, sepals suborbicular, mostly 1.5 mm long; corolla about 5 mm wide, white, the lobes suborbicular. Fruit a dark purple, ovoid berry about 2 cm long, one-seeded. $2n = 52$. Hammocks, thickets, and pinelands, south Fla., north to Lee and Brevard Counties and Fla. Keys, W.I. All year.

The hard, light brown wood is sometimes used in cabinet work. **C. cainito**, the star-apple, is cultivated in south Fla. for its fruit.

3. MANÍLKARA Adans. SAPODILLA, WILD DILLY

Trees with alternate, leathery, parallel-veined leaves that are often clustered at the ends of stout branchlets. Flowers usually in axillary fascicles or solitary, bisexual; sepals six, biseriate, persistent and reflexed in age, corolla lobes six, smooth; stamens six, staminodes six, petaloid or fleshy or nearly obsolete, epipetalous. Ovary six- to fourteen-locular, conical, style linear. Fruit a subglobose or ellipsoid berry, fleshy pericarp firm, one- to several-seeded, topped by the persistent style. About 85 species, pantropical. *Sapota* Mill.; *Mimusops* sensu Sarg., Small; *Achras* L.

1. Young fruit tipped with conspicuous long
   spinelike style; leaves lance-elliptic, many
   conspicuously notched at the tip . . . 1. *M. bahamensis*
1. Young fruit very short style tipped; leaves
   lance-ovate, few or none notched at the
   tip . . . . 2. *M. zapota*

1. **M. bahamensis** (Baker) Lam. & Meeuse, WILD DILLY, WILD SAPODILLA—Small trees or shrublike, bearing evergreen leaves that tend to cluster toward the ends of branches; blades 4–10 cm long, elliptic or lanceolate-elliptic, many conspicuously notched at the top, tapering at the base. Flowers in small axillary fascicles; corolla lobes pale yellow, flower about 2 cm wide; staminodia somewhat deltoid. Berry almost spheroidal, brownish, about 3 cm wide. Hammocks, south Fla., especially the Fla. Keys, Bahama Islands, W.I. Winter, spring. *Mimusops emarginata* (L.) Britt.; *Achras emarginata* (L.) Little

2. **M. zapota** (L.) Royen, SAPODILLA—Trees often large with widespreading branches, younger stems and petioles usually with brown pubescence. Leaves evergreen, tending to cluster at the ends of branches, lance-ovate to lance-oblong, apex acute, seldom notched, 6–14 cm long, glabrous except that a brown puberulous line along the midrib sometimes occurs. Flowers usually solitary in leaf axils, corolla white, staminodes oblong. Berry ellipsoid, globose or depressed-globose, 5–8 cm wide, brownish mottled, short style persistent. $2n = 26$. Coastal areas, hammocks, old fields, south Fla., Fla. Keys, naturalized from Central America, cultivated throughout the tropics. All year. *Achras zapota* L.; *M. zapotilla* (Jacq.) Gilly

Harvested in tropical America, the latex is the source of chicle for chewing gum.

4. DÍPHOLIS A. DC.

Shrubs or small trees with alternate, pinnately veined, smooth leaves. Flowers in axillary fascicles or above leaf scars; calyx lobes five, pubescent, obtuse corolla lobes five, obtuse, each three-lobed; stamens five, filaments very short, epipetalous, staminodia five, petaloid, alternate with the lobes of the corolla. Ovary five-locular. Fruit a small berry, with a short, persistent style. About 14 species in tropical America.

1. **D. salicifolia** (L.) A. DC. Bustic—Shrubs or trees up to 25 m tall. Leaves evergreen, elliptic or elliptic-oblanceolate, acute or acuminate at both ends, 6–12 cm long; flowers white, numerous; calyx lobes ovate-elliptic, 1.5 mm long; corolla 4 mm wide, lobes lanceolate and finely toothed. Berry black, ellipsoid, ovoid or subglobose, about 6–10 mm long. $2n = 36$. Hammocks and pinelands, south Fla. and Fla. Keys, W.I., Mexico. All year.

The reddish brown, dense wood is sometimes used locally in cabinet work.

5. POUTÉRIA Aubl.

Trees or shrubs with alternate, somewhat obovate or oblanceolate leaves. Flowers solitary or in small axillary fascicles; sepals four to six, distinct, corolla lobes four to seven, white, yellow, or green, entire or nearly so, stamens epipetalous, as many as the lobes of the corolla, staminodia slender. Ovary five-locular, pubescent stigma somewhat broadened. Fruit a smooth berry, one- to several-seeded. About 150 species, chiefly tropical America. *Lucuma* sensu authors

1. **P. campechiana** (HBK) Baehni, Egg Fruit—Trees up to 25 m tall with elliptic to obovate smooth leaves, 9–20 cm long, obtuse or acute tipped. Flowers few in axillary fascicles or at defoliated nodes, sometimes solitary; sepals five, 3–5 mm long, sericeous, inner rounded at the base; corolla white, 7–9 mm long, lobes ovate; staminodia subulate, slender. Berry ovoid or subglobose, edible, 5–7 cm in diameter, smooth, yellow, green, or brownish. Hammocks south Fla. and Fla. Keys, naturalized from South America. *Lucuma nervosa* A. DC., misapplied

Common name is based on the resemblance of the fruit pulp to a hard-boiled egg.

P. domingensis (Gaertn. f.) Baehni also occurs rarely in southern Fla. It is a small tree up to 10 m tall with solitary or few-flowered clusters of yellow or white flowers and yellow fruits 3–6 cm in diameter. It is naturalized from the W.I.

Family 153. **SAPOTACEAE.** *Dipholis salicifolia:* a, flowering branch, × ½; b, flower, × 10; c, mature fruit, × 5.

6. MASTICHODÉNDRON Cronquist MASTIC

Shrubs or trees often large, with alternate, evergreen, smooth, usually ovate leaves. Flowers numerous in axillary fascicles, sepals five, corolla lobes five, smooth, stamens five, epipetalous, anthers shallowly notched, staminodia five, short. Ovary usually five-locular, stigma truncate. Fruit fleshy, usually one-seeded. About seven species in tropical or warm regions of America. *Sideroxylon* sensu Small, not L.

1. **M. foetidissimum** (Jacq.) Cronquist, WILD MASTIC—Trees up to 25 m tall. Leaves 4–15 cm long, thin, yellowish green, glossy, coriaceous, broadly elliptic to ovate, glabrous at maturity, the leaf margins are characteristically obscurely puckered. Flowers light yellow, calyx lobes suborbicular, about 2 mm long, corolla 6–7 mm wide, lobes elliptic to ovate-elliptic; staminodia lanceolate. Berry ovoid or globose, 1.5–3 cm long, juicy, yellow at maturity. Coastal hammocks, south Fla. to Brevard County, Fla. Keys, W.I., Bahama Islands.

The orange colored, hard heartwood is sometimes used locally in cabinet work.

154. EBENÀCEAE EBONY FAMILY

Trees or shrubs with alternate, entire leaves. Flowers radially symmetrical, mostly unisexual; pistillate flowers solitary, calyx three- to seven-lobed, persistent; corolla three- to seven-lobed; stamens two to four times as many as the lobes of the corolla and opposite the lobes. Ovary superior, two- to sixteen-locular, fruit a fleshy or leathery berry; staminate flowers with rudimentary ovary. About 5 genera and 450 species, chiefly in tropical and subtropical regions.

1. DIOSPÝROS L. PERSIMMON

Dioecious trees or shrubs with alternate, entire leaves. Flowers unisexual or rarely polygamous, axillary, cymose, or the pistillate solitary, calyx deeply four- to six-lobed, accrescent in fruit, corolla urn shaped or rotate, greenish white or yellow, of different sizes in staminate and pistillate flowers; stamens three to many, usually in two rows, unequal, filaments short and hairy. Ovary four- to eight-locular, styles four distinct above. Berry leathery, subglobose, succulent within. About 400 species, widely distributed in the tropics and warm regions.

1. **D. virginiana** L.—Deciduous trees mostly 10–15 m tall. Leaves 8–15 cm long, ovate to oblong, on distinct petioles 1–2 cm long, glabrous

or sometimes sparsely pubescent below. Flowers greenish yellow; staminate solitary, 1 cm long, usually with sixteen linear stamens; pistillate solitary, 1.2–2 cm long, with eight sterile stamens; berry yellowish brown, 2–4 cm thick, edible in later maturity. $2n = 60,90$. Drier soils, Fla. to Tex., Conn., Iowa. *D. mosieri* Small

The wood is dense and hard; the subglobose fruits are bitter and astringent when immature, becoming sweet, soft, and pulpy when ripe.

155. OLEÀCEAE Olive Family

Trees, shrubs, or woody vines with opposite, seldom alternate, simple or pinnately compound leaves. Flowers usually bisexual and radially symmetrical; calyx four- to five-lobed or toothed, petals four to five, free or fused, or corolla absent; stamens usually two, epipetalous. Ovary superior, two-locular, style simple or absent, stigma two-lobed or simple. Fruit various, mostly a berry, drupe, or capsule. About 22 genera and 450 species of wide distribution in both temperate and tropical regions.

1. Corolla 1.5–3.5 cm long; diffuse shrubs or woody vines; fruit a two-lobed berry, occurring in pairs . . . . 1. *Jasminum*
1. Corolla smaller or absent, less than 1.0 cm long; trees or shrubs.
  2. Leaves pinnately compound; flowers apetalous; fruit a samara . . . . . . 2. *Fraxinus*
  2. Leaves simple; flowers with a corolla or apetalous; fruit a drupe or berry.
    3. Larger leaves oblanceolate to lance-ovate, blades acuminate, over 10 cm long; fruit a drupe approximately 1 cm wide . . . . 3. *Osmanthus*
    3. Larger leaves less than 7 cm long; fruit approximately 0.6 cm wide or less.
      4. Flowers in terminal erect panicles; leaves broadly lanceolate to ovate, over 2 cm wide . . . . . 4. *Ligustrum*
      4. Flowers in axillary clusters, apetalous, minute; leaves narrowly lanceolate-acuminate, less than 1.5 cm wide . 5. *Forestiera*

1. JASMÍNUM L. Jasmine, Jessamine

Erect deciduous or evergreen shrubs or woody vines. Leaves opposite or sometimes alternate, pinnately compound with three to seven leaflets or reduced to one leaflet and appearing simple. Flowers in simple or branching terminal cymes or rarely solitary; calyx campanulate or funnelform with four to nine lobes or teeth; corolla salverform with cylindrical tube, four- to nine-lobed; stamens two included. Ovary two-locular, style single, stigma two-lobed. Fruit a two-lobed berry. About 200 species in Old World tropics and subtropics.

Many species are cultivated as garden ornamentals, and some have escaped and become naturalized in tropical and subtropical regions.

1. Leaves mostly simple or apparently so.
  2. Calyx pubescent or with scattered hairs.
    3. Calyx lobes mostly seven to twelve;
      stems puberulent . . . . . .   1. *J. sambac*
    3. Calyx lobes five to seven; stems glabrous   2. *J. nitidum*
  2. Calyx glabrous or nearly so, the lobes very
    short, less than 3 mm . . . . . .   3. *J. dichotomum*
1. Leaves trifoliolate . . . .   4. *J. fluminense*

1. J. sambac Solander, ARABIAN JASMINE—Stems climbing and bearing fine pubescence. Leaves apparently simple, blades 5–12 cm long, ovate to elliptic, mostly acute at the apex and rounded or obtuse at the base. Flowers in cymes, calyx lobes mostly seven to twelve, pubescent or with scattered hairs, narrowly linear; corolla white, tube 10–15 mm long, lobes ovate or round, turning pink with age, very fragrant. $2n = 26,39$. Hammocks in south Fla., naturalized from tropical Asia. All year.

2. J. nitidum Skan—Stems climbing and bearing fine pubescence or more commonly glabrous. Leaves apparently simple blades 5–7 cm long, pubescence beneath on the veins, acuminate, ovate-lanceolate or lanceolate. Flowers one to few in cymes, very fragrant; calyx lobes five to seven, pubescent, corolla white, lobes narrow and acute, becoming bright pink where exposed to the sun. Hammocks, disturbed sites, south Fla., naturalized from Admiralty Islands. All year. *J. amplexicaule* misapplied

3. J. dichotomum Vahl, GOLD COAST JASMINE—Stems climbing, glabrous. Leaves apparently simple, ovate to suborbicular, evergreen, 5–7 cm long, acuminate, glossy green. Flowers produced abundantly, very fragrant, opening at night; calyx lobes very short, less than 3 mm, glabrous; corolla white, pink in bud, about 3 cm long. Berry black. Disturbed sites, hammocks, south Fla., naturalized from Ghana and west coast of Africa. All year.

4. J. fluminense Vellozo—Stems climbing, evergreen, tomentose when young, becoming glabrous. Leaves trifoliolate, leaflets 5–7 cm long, broadly ovate, acute, terminal leaflet long stalked. Flowers in open cymes, very fragrant; calyx lobes very short, corolla lobes oblong, tube long and slightly curved; berries black. Disturbed sites, old fields, a somewhat troublesome weed, naturalized from Brazil. All year. *J. azoricum* misapplied

2. FRÁXINUS L. Ash

Dioecious, polygamodioecious, or monecious trees, rarely shrubs, with opposite, deciduous, pinnately compound leaves. Flowers unisexual or occasionally bisexual in dense clusters, axillary racemes or panicles from previous years' growth; calyx small, four-lobed, apetalous, stigma two-lobed, stamens two, ovary with single style. Fruit a samara. About 65 species, widely distributed, chiefly in the temperate zone.

1. **F. caroliniana** Mill. WATER ASH—Trees often shrubby, up to 15 m with terete stems, smooth or thinly pubescent. Leaves with five to nine leaflets, these 5–10 cm long, lanceolate to elliptic, entire or shallowly serrate. Samara flat, 4–5 cm long, lanceolate to elliptic, subtended by a minute calyx, wings pinnately netted. Lowlands, cypress heads, swamps, Fla. to Tex., Va. on the coastal plain. Spring. *F. pauciflora* Nutt.

This species is extremely variable, but variations scarcely deserve taxonomic recognition.

3. OSMÁNTHUS Lour.

Dioecious or polygamodioecious small trees or shrubs with evergreen, simple, opposite, entire leaves. Flowers small, unisexual or bisexual, in axillary or terminal panicles of white or greenish white flowers; calyx four-lobed, corolla funnel shaped, four lobed, stamens two, included. Ovary two-locular, ovules two per locule, style single. Fruit an ovoid or globose, one-seeded, bitter drupe. About 30 species, widely distributed. *Amarolea* Small

1. **O. americanus** (L.) Gray, WILD OLIVE—Low tree up to 15 m or shrub with pale, smooth stems. Leaves lance-oblong to obovate, 5–15 cm long, entire, shining, coriaceous, distinctly petioled, 1–2 cm. Flowers in axillary panicles; corolla 3–4 mm, dull white, ovate-spreading lobes nearly as long as the tube; drupe 1–1.5 cm, dark purple. Swamps and woods, Fla. to La., Va. Spring. *Amarolea americana* (L.) Small

Apparently considerable variation in fruit size and color occur in Fla., but it is problematical as to whether varieties should be recognized.

4. LIGÚSTRUM L. PRIVET

Shrubs or low trees deciduous or evergreen with simple, entire, opposite leaves. Flowers bisexual, in small panicles of white flowers at the end of main stems and on lateral branches; calyx small, campanulate truncate or shallowly four-lobed; corolla funnelform, the four lobes spreading or recurved; stamens two, epipetalous. Ovary two-locular, two ovules per locule. Fruit a small black berry, one- to four-seeded. About 30 species, chiefly in Asia. A number of species are cultivated as ornamentals, and those in our area have escaped.

1. **L. ovalifolium** Hassk. CALIFORNIA PRIVET—Shrub or low tree with glabrous stems. Leaves 4–7 cm long, evergreen or eventually deciduous, elliptic to ovate, shiny, coriaceous, acute tipped. Flowers in short panicles, corolla 7–8 mm long, lobes ovate, filaments exserted from the tube, anthers surpassing the corolla lobes. Drupe 6–8 mm wide, subglobose, black. Hammocks, roadsides, and disturbed sites, Fla. to Tex. on coastal plain, naturalized from Japan. Summer.

L. vulgare L. with corolla tube only 2–3 mm long and stems minutely pubescent is also reported for our area and may be naturalized. $2n = 46$.

5. FORESTIÈRA Poir.

Polygamodioecious or dioecious trees or shrubs with simple, opposite, deciduous or rarely evergreen leaves. Flowers small, in lateral clusters of nearly sessile flowers or in small panicles, greenish or whitish; calyx very minute or none, corolla absent, staminate flowers with one-pistillate flowers with stamens four. Ovary two-locular with two ovules per cell, absent or vestigial in the staminate flowers, stigma simple or two-lobed. Fruit a small, black, one-seeded drupe. About 20 species in tropics and subtropics of America. *Adelia* P. Browne

1. **F. segregata** (Jacq.) Krug & Urban, FLORIDA PRIVET—Small tree or shrub usually up to 3 m tall. Leaves 1–6 cm long, evergreen, elliptic to elliptic-spatulate, entire, thinly coriaceous. Flowers in short panicles, bracts ovate or suborbicular, 1–2 mm long, pubescent along the margins. Drupe ovoid to ellipsoidal, 5–8 mm long, black. Two varieties have been reported for our area:

1. Stems glabrous or nearly so; leaf blades 3–6
    cm long, dark green above . . . . .   1a. var. *segregata*
1. Stems puberulent; leaf blades 1–3 cm long,
    gray green above . . . .   1b. var. *pinetorum*

1a. **F. segregata** var. **segregata**—Small tree or shrub with glabrous or subglabrous stems. Leaves 3–6 cm long. Drupe 7–8 mm long. Coastal hammocks, Fla. to Ga., W.I. *F. porulosa* (Michx.) Poir.; *Adelia segregata* Jacq.

1b. F. segregata var. **pinetorum** (Small) M. C. Johnston—Shrubs with short, rigid, puberulent stems. Leaves 1–3 cm long, gray green above. Drupe 5–7 mm long. Pinelands, endemic to south Fla. *F. pinetorum* Small

156. LOGANIÀCEAE Logania Family

Trees, shrubs, vines, or herbs with opposite, entire leaves, stipules usually present. Flowers usually bisexual, radially symmetrical, in terminal cymes or solitary; calyx four- to five-lobed, corolla tubular, four- to five-lobed, stamens epipetalous, four to five. Ovary superior or sometimes half-inferior, two- to three-locular, style two-lobed. Fruit a capsule, berry, or drupe. About 33 genera and 600 species, chiefly tropical. *Spigeliaceae*

1. Woody twining vines . . . .   1. *Gelsemium*
1. Herbs or shrubs, not vinelike.
  2. Erect or spreading herbs.
    3. Leaves linear-subulate; stems spreading, decumbent . . . .   2. *Polypremum*
    3. Leaves narrowly to broadly lanceolate or ovate, not linear; stems erect.
      4. Capsule with a cupulate base; flowers in terminal spikes or spikelike racemes . . . .   3. *Spigelia*
      4. Capsule without a defined base; flowers in terminal cymes . . . .   4. *Cynoctonum*
  2. Shrubs, leaves pubescent below . . .   5. *Buddleja*

1. GELSÈMIUM Juss. Yellow Jessamine

Twining woody vines with opposite, narrow, evergreen leaves and conspicuous, fragrant yellow flowers crowded in short axillary cymes or solitary, calyx five-lobed, sepals elliptic, corolla funnelform, ovate lobes imbricate; stamens five, epipetalous. Ovary superior, two-locular, styles dimorphic, stigmas four, linear. Fruit a septicidal capsule, seeds narrowly winged, three species, North America and Asia.

1. **G. sempervirens** (L.) Ait. f. Yellow Jessamine—Climbing, slender vines, freely branching and tangling up to 5 m long. Leaves 4–7 cm long, lanceolate, entire. Flowers on very short pedicels bearing two or more scalelike bracts; corolla 2–3 cm, bright yellow; capsule oblong, 1–2 cm long, short beaked. $2n = 16$. Hammocks and low ground, Fla. to Mex., Va., Tenn., Ark. Winter, spring.

    Plant contains poisonous alkaloids.

2. POLYPRÈMUM L.

Glabrous, diffusely branching annual 1–3 dm long with narrow leaves connected at the base by stipular membranes. Flowers small, white, and sessile in leafy terminal cymes; sepals four, about 3 mm long, subulate, corolla funnelform, pubescent within the throat, stamens four, included. Ovary two-locular, style short, stigma capitate. Fruit an obovoid or subglobose capsule, about 2 mm long. A single species, east U.S., W.I., Central and South America.

1. **P. procumbens** L.—Characters of the genus. Disturbed sites, roadsides, dry sandy soil, Fla. to Mexico, Panama, Mo., tropical America. Spring, fall.

3. SPIGÈLIA L.

Erect herbs with flowers in terminal spicate inflorescence. Flowers five-parted; sepals narrow, glandular-punctate within, corolla with a long tube, lobes short; stamens five, epipetalous. Ovary three-locular, style one, slender, stigma pubescent. Fruit a septicidal capsule, few-seeded. About 80 species, chiefly tropical and subtropical America. *Coelostylis* T & G

1. **S. anthelmia** L. WEST INDIAN PINKROOT—Glabrous herb, stems nearly leafless except near the tip, weak, succulent, with lower leaves opposite, upper ones larger in whorls of four, subsessile. Upper leaf blades 3–9 cm long, stipules truncate, glabrous beneath. Inflorescence of one to six usually branched spikes from upper leaf axils; calyx deeply five-parted, sepals narrow-linear, subequal, corolla pale straw colored, lilac toward the tips, lobes ovate; stamens included, filaments very short. Stigma very pubescent; capsule scabrous to muricate. $2n = 32$. Hammocks, south Fla., Fla. Keys., tropical America.

4. CYNÓCTONUM J. F. Gmel. MITERWORT

Annual herbs with erect stems and bearing broad leaves connected at the base by a stipular line. Flowers small, sessile, in terminal, peduncled, branched, spikelike cymes, these eventually secund; sepals five, ovate, corolla funnelform, five-lobed, pubescent within the throat, stamens included. Ovary two-locular, style very short; fruit a capsule, deeply two-lobed. *Mitreola* R. Brown

1. Leaves narrowly lanceolate or ovate to ovate-
      orbicular, sessile . . . .   1. *C. sessilifolium*
1. Leaves narrowly to broadly lanceolate or

ovate, short petiolate.
2. Lower leaves ovate-lanceolate or ovate,
   fleshy; inflorescence mostly 2–4 cm long   2. *C. succulentum*
2. Lower leaves lanceolate to elliptic or
   ovate, thinly membranous; inflorescence
   mostly 4–6 cm long . . . . . .   3. *C. mitreola*

1. C. sessilifolium (Walt.) J. F. Gmel.—Annual herbs with stiffly erect stems, simple below and inflorescence branched above, up to 6 dm tall. Leaves narrowly lanceolate to elliptic or ovate, blades 3–6 cm long, sessile, at least the lower ones. Inflorescence becoming 5 cm long, bracteal leaves resembling sepals; corolla 3 mm long, the lobes ovate. Capsule 3–4 mm long. Low ground, Fla. to Va., Tex. Summer. Three varieties are present in our area:

1. Leaf blades elliptic to ovate or orbicular.
   2. Leaf blades mostly 2–4 cm long, apex ob-
      tuse . . . .   1a. var. *sessilifolium*
   2. Leaf blades mostly 1–1.5 cm long, apex
      acute . . . .   1b. var. *microphyllum*
1. Leaf blades narrowly lanceolate . . . .   1c. var. *angustifolium*

    1a. C. sessilifolium var. **sessilifolium**—Stems up to 6 dm tall. Leaf blades 2–5 cm long, elliptic, ovate to orbicular, apex obtuse. Ditches, low ground, wet places, coastal plain Fla. to La., Tex., Va.
    1b. C. sessilifolium var. **microphyllum** R. W. Long—Stems simple up to 4 dm tall, mostly less. Leaf blades 1–1.5 cm long, mostly less than 1 cm long, apex acute to broadly so, elliptic to broadly lanceolate. Inflorescence few branched. Pine flatwoods, moist sandy soil, Fla. to La., N.C.
    1c. C. sessilifolium var. **angustifolium** T & G—Stems mostly simple up to 6 dm tall, narrowly lanceolate to almost linear. Inflorescence few branched. Wet soil, Fla. to Ga. on the coastal plain. *C. angustifolium* (T & G) Small
    2. C. succulentum R. W. Long—Annual herbs up to 4 dm tall, simple. Leaves 3–5 cm long, narrowly elliptic to ovate, lower ones tending to cluster near the base, fleshy, short petioled. Inflorescence a branching cyme of short spikes, these mostly 2–4 cm long, flowers dense, corolla lobes ovate. Capsule 3–4 mm long. Pinelands, dry soil, rarely moist sandy places, endemic to peninsular Fla. All year.
    3. C. mitreola (L.) Britt. MITERWORT—Annual herbs with erect stems, simple below and branched above, 2–6 dm tall. Leaves mostly 3–6 cm long, membranous, lanceolate to elliptic or ovate, the lower ones distinctly petioled, the upper sessile or nearly so. Inflorescence becoming 5 cm long, crowded, bracteal leaves resembling sepals, corolla lobes

ovate or ovate-lanceolate. Capsule 3 mm long. $2n = 20$. Low ground, wet places, Fla. to Va., Ark., Mexico, W.I. All year. *Mitreola petiolata* T. & G.

This is a highly variable species particularly in leaf morphology. Apparent hybrids with C. sessilifolium occur in Fla.

5. BUDDLÈJA L. BUTTERFLY BUSH

Trees or shrubs, rarely herbs, with glandular, scaly, or stellate pubescence. Leaves opposite or rarely alternate, entire or dentate, bases usually connected by a stipular line. Flowers in racemes, panicles, or heads, calyx four-lobed, corolla funnelform or campanulate, four-lobed. Ovary two-locular, style simple, stigma two-lobed. Fruit a septicidal capsule. About 100 species in tropics and subtropics. *Buddleia*

1. **B. lindleyana** Fort.—Shrub up to 2 m or more tall, stems angular or somewhat winged. Leaves 5–10 cm long, ovate to ovate-lanceolate, tapering to the apex, dark green above, pale or slightly pubescent below. Flowers purplish violet, in dense, erect racemes, 6–20 cm long or more. Pinelands, disturbed sites, south Fla. to Tex., Ga., W.I., naturalized from China. *Adenoplea lindleyana* (Fort.) Small

157. GENTIANÀCEAE GENTIAN FAMILY

Perennial or annual herbs with entire, opposite or verticillate leaves that are often fused at the base. Flowers bisexual, radially symmetrical, usually cymose or solitary, conspicuous; calyx tubular or composed of free sepals; corolla cylindrical to rotate, fused with four to twelve lobes, stamens epipetalous, as many as the corolla lobes and alternate with the lobes. Ovary superior, usually one-locular. Fruit mostly a two-valved septicidal capsule. About 70 genera and 1100 species, widely distributed. *Menyanthaceae*

1. Aquatic herbs with orbicular or reniform
   leaves . . . . 1. *Nymphoides*
1. Terrestrial herbs, leaves not orbicular or
   reniform.
   2. Leaves reduced to scales or wanting.
      3. Stems yellowish green to purple tinged;
         stigma two-lobed . . . . . . 2. *Bartonia*
      3. Stems colorless or yellowish white, not
         green, saprophytes; stigma not lobed 3. *Leiphaimos*
   2. Leaves linear-subulate to lanceolate-ovate,
      not scalelike.

4. Style filiform; corolla rotate or campanulate, not plaited.
  5. Leaves clasping, fleshy, obovate to
    elliptic-lanceolate; style thick . . 4. *Eustoma*
  5. Leaves sessile, not conspicuously
    clasping or fleshy, linear-subulate,
    linear-lanceolate to lance-ovate;
    style filiform . . . . 5. *Sabatia*
4. Style stout, short or none; corolla funnelform, plaited . . . . . . 6. *Gentiana*

1. NYMPHOÌDES Hill. FLOATING HEARTS

Aquatic herbs, sterile stems with a single, floating, broad, cordate leaf, fertile stems bearing one to several leaves and an umbel of white or yellowish flowers. Calyx divided into oblong lobes, corolla campanulate, deeply lobed, tube short. Ovary tapered to a short style with a broad two-lobed stigma. Fruit a hard, ellipsoidal capsule. About 25 species, chiefly tropical. *Limnanthemum* Gmel.

1. **N. aquatica** (J. F. Gmel.) Kuntze, FLOATING HEARTS—Stems rather thick and coarse, 1–3 mm thick near the top. Leaves 5–12 cm long and wide, broadly ovate, deeply cordate, petioles purplish and glandular. Pedicels and calyx flecked with purple, corolla white, 10–14 mm long; stamens fused to the middle of the corolla tube. Capsule 10–14 mm, seeds conspicuously papillate. Ponds and quiet waters, Fla. to Tex., N.J. on the coastal plain. Spring, summer.

2. BARTÒNIA Muhl. ex Willd.

Annual or biennial herbs with often spiral or slender twining stems. Leaves reduced to minute scales. Inflorescence slender terminal panicles or racemes of small white, purplish, or yellowish flowers; calyx four-cleft, corolla campanulate, four-lobed, stamens inserted between the corolla lobes. Style short, stigma usually two-lobed. Fruit a capsule. Five species in North America.

1. **B. verna** (Michx.) Muhl.—Stems simple up to 3 dm tall. Leaves about 1–2 mm long, scalelike, opposite. Inflorescence racemose or paniculate; calyx lobes linear-lanceolate, corolla 6 mm long or more, the lobes spatulate to obovate-spatulate, white. Capsule 5–6 mm long. Wet soil, moist pineland, Fla. to La., Va. Spring.

3. LEIPHÀIMOS Cham. & Schlecht.

Erect, sometimes branched, perennial, delicate saprophytic herbs. Stems terete, without chlorophyll. Leaves very small, scalelike or absent,

Family 157. **GENTIANACEAE.** *Sabatia grandiflora:* a, upper stem and inflorescence, × ½; b, lower stem and roots, × ½; c, flower, × 1¼; d, mature fruit, × 3.

opposite and usually fused at the base. Flowers solitary or in loose cymose inflorescence; calyx four- to five-parted, corolla four- to five-lobed, contorted. Ovary subsessile, style filiform, stigma capitate. Fruit an ellipsoidal capsule. About 40 species, chiefly tropical America.

1. **L. parasitica** Schlecht. & Cham. GHOST PLANT—Stems pale 1–4 dm tall with opposite scalelike leaves 3–5 mm long. Calyx lobes lanceolate, acute tipped; corolla white or pinkish, 6–8 mm long, lobes deltoid to lanceolate. Capsule 5–6 mm long. Plants saprophytic on leaf mold in hammocks, south Fla., Fla. Keys, W.I. All year.

4. EÙSTOMA Salisb.

Rather tall annual herbs. Flowers solitary or in open racemes or panicles; calyx lobes five to six, narrow-lanceolate; corolla campanulate or funnelform, five- to six-lobed, lobes usually with an irregularly toothed margin. Ovary terminating in a filiform style, stigma rounded. Fruit a capsule. Five species in North America.

1. **E. exaltatum** (L.) Griseb.—Stems up to 1 m tall. Leaves 2–7 cm long, becoming elliptic-lanceolate to elliptic toward the top of the stem. Flowers conspicuous in open racemes or panicles; calyx lobes 10–12 mm long, corolla rose purplish, lobes elliptic or ovate, 18–20 mm long. Capsule 2–3 cm long. Dry soil, disturbed sites, pinelands, hammocks, south Fla., Fla. Keys, W.I. All year.

5. SABÀTIA Adans. MARSH PINK

Herbs with slender, smooth stems and linear to ovate, often sessile leaves. Flowers solitary or in loose terminal cymes, conspicuous and brightly colored; calyx lobes narrow, short, five to twelve-parted; corolla rotate, five- to twelve-parted, white, pink, or lilac, tube very short; stamens inserted at the top of the corolla tube. Style two-branched or lobed. Capsule eventually two-valved. About 15 species in North America. *Sabbatia* Adans

1. Corolla white or cream colored . . . . 1. *S. brevifolia*
1. Corolla rose or pink colored, never white throughout.
   2. Calyx and corolla lobes eight to twelve . 2. *S. bartramii*
   2. Calyx and corolla lobes mostly four to six.
      3. Calyx lobes foliaceous, over 2 mm wide; upper leaves lance-ovate, mostly over 1 cm wide . . . . 3. *S. calycina*
      3. Calyx lobes linear, less than 1 mm wide; upper leaves linear to linear-lanceolate, less than 1 cm wide.

4. Calyx lobes about as long as the co-
     rolla lobes, up to 1 cm long . . 4. *S. stellaris*
4. Calyx lobes shorter than corolla
     lobes, less than 1 cm long . . . 5. *S. grandiflora*

1. **S. brevifolia** Raf.—Stems 3–7 dm tall, erect. Leaves mostly 1–2 cm long, elliptic or lance-elliptic to linear. Flowers white or cream colored; calyx 7–8 mm long, lobes about twice as long as the tube; corolla lobes spatulate to oblanceolate, 10–12 mm long, longer than the calyx lobes. Capsule 4–5 mm long. $2n = 32$. Low ground, Fla. to Ala., Va. on the coastal plain. Spring. *S. elliottii* Steud.

2. **S. bartramii** Wilbur—Stems 6–9 dm tall, erect. Lower leaves spatulate, upper ones narrowly linear, 4–10 cm long. Flowers deep rose to whitish; calyx lobes awl shaped; corolla with a yellow eye, lobes ovate-spatulate, 3–3.5 cm long. Capsule about 10 mm long. Shallow pineland ponds, Fla. to Ala., S.C. on the coastal plain. Summer, fall. *S. decandra* (Walt.) Harper

3. **S. calycina** (Lam.) Heller—Stems up to 4 dm tall, perennial, solitary or clustered, widely branching. Leaves 3–6 cm long, especially the upper ones elliptic to narrowly elliptic, obtuse tipped and narrowed at the base. Flowers pink or with whitish lobes; calyx 15–25 mm long, lobes oblanceolate foliaceous; capsule 7–8 mm long. Low ground, Fla. to Tex., Va., W.I. Spring, summer.

4. **S. stellaris** Pursh —Stems slender, up to 6 dm tall, perennial, solitary, widely branching. Leaves 2–4 cm long, oblong to lanceolate-oblong below, rounded to a sessile base, three-veined, upper leaves somewhat narrower. Flowers pink to dark rose with a yellow eye; calyx ten-veined, lobes linear 7–15 mm long; corolla lobes oblanceolate, 10–15 mm; style deeply cleft. Capsule 6–8 mm long. Various sites, sandy soil, swamps, Fla. to La., Mass., Ind., Ark. Summer, fall.

5. **S. grandiflora** (Gray) Small—Stems up to 1 m tall or more, erect, slender. Upper leaves narrowly linear or filiform, mostly 4–10 cm long. Flowers dark rose or magenta with a yellow eye; calyx with very narrow or filiform lobes; corolla lobes elliptic-ovate to ovate or rhombic. Capsule 8–10 mm long. Low ground, peninsular Fla., Cuba. All year.

6. GENTIANA L. Gentian

Herbs often fleshy with opposite leaves. Flowers solitary or cymose, conspicuous, calyx small, lobes minute, corolla funnelform or salverform, four- to five-lobed, usually with intermediate plaited folds that have teeth or appendages at the sinuses, stamens epipetalous, as many as the lobes of the corolla. Style short or absent, stigmas two. Capsule ellipsoidal, two-valved. About 400 species of wide distribution, chiefly of temperate regions.

1. G. pennelliana Fern.—Stems mostly 1–2 dm tall or more, simple with fleshy roots. Leaves 1–3 mm wide, about 1–3 cm long, linear-obtuse, succulent. Flowers solitary, calyx lobes linear-subulate, corolla 6–7 cm long, white or greenish white, lobes ovate, the plaits laciniate. Capsule about 1.5 cm long. Hammocks, Lignum Vitae Key, pinelands, endemic to Fla. *Dasystephana tenuifolia* (Raf.) Pennell; *G. tenuifolia* (Raf.) Fern.

A single record from the Fla. Keys made a number of years ago has not been verified by recent collections, and the species may not be presently in our area.

158. APOCYNÀCEAE OLEANDER FAMILY

Trees, shrubs, vines, or herbs almost always with a milky sap and bearing opposite, whorled, or sometimes alternate entire leaves. Flowers bisexual, radially symmetrical, solitary, or in various kinds of inflorescences; calyx four- to five-lobed, often glandular within, corolla tubular, four- to five-lobed, convolute or usually twisted in bud, stamens four to five, epipetalous, filaments free, anthers appressed to stigma; hypogynous disk usually present. Ovary superior to partially inferior, one- to two-locular, style one, entire or split at the base. Fruit dry or fleshy, dehiscent or indehiscent. Seeds sometimes comose. About 200 genera and 2000 species, chiefly in the tropics.

1. Leaves alternate, blades narrowly to broadly lanceolate.
   2. Corolla large, 5–6 cm long; leaves linear to narrowly lanceolate, up to 14 cm long; fruit dry . . . . 1. *Thevetia*
   2. Corolla small, 1–1.2 cm long or less; leaves ovate to ovate-lanceolate, up to 7 cm long; fruit one- to two-seeded drupe . 2. *Vallesia*
1. Leaves opposite or whorled, blades lanceolate to elliptic, ovate.
   3. Erect or trailing herbs or shrubs.
      4. Corolla large, 7–10 cm long, yellow; larger leaves oblanceolate, mostly 10–12 cm long . . . . 3. *Allamanda*
      4. Corolla shorter, mostly less than 6 cm long; leaves various, narrowly lanceolate to elliptic, less than 8 cm long.
         5. Tall shrubs or trees up to 10 m tall;

leaves lanceolate up to 15 cm long;
corolla showy with toothed scales at
the mouth . . . .   4. *Nerium*
5. Low subshrubs, erect or trailing
herbs.
6. Pedicels mostly 1–2 cm long; corolla limb yellow . . . . .   5. *Angadenia*
6. Pedicels mostly less than 5 mm
long; corolla red, pink, white, or
variously colored . . . . .   6. *Catharanthus*
3. Twining vines, herbaceous or woody.
7. Corolla white or white with yellow
tube.
8. Leaves elliptic-ovate, cordate or truncate at base; fruit 15–20 cm long;
corolla white or greenish white   .   7. *Echites*
8. Leaves ovate-lanceolate, tapering to
base or cuneate; fruit 10–15 cm
long; corolla white with tube yellow within . . . .   8. *Rhabdadenia*
7. Corolla yellow . . . .   9. *Urechites*

1. THEVÈTIA L.

Small trees or shrubs with leathery, often shiny, alternate, spirally
arranged, narrow leaves with long intrapetiolar glands. Inflorescence
terminal, cymose; flowers conspicuous, long peduncled, large, funnel
shaped, yellow; calyx lobes five, long-acute with numerous glands
within; corolla with five obtuse lobes; stamens covered with five lanceolate, pubescent scales. Ovary surrounded by a somewhat five-lobed disk.
Fruit a drupe with a very hard endocarp said to be very poisonous.
About nine species in tropical America. *Cerbera* L.

1. **T. peruviana** (Pers.) Schum. LUCKY NUT, TRUMPET FLOWER—
Small tree or shrub with slender, terete, gray stems, glabrous throughout
with bitter latex. Leaves 6–12 cm long, linear to narrowly lanceolate.
Calyx lobes about 4–7 mm long, ovate to elliptic, corolla saffron, yellow, or sometimes white, tube mostly 2–4 cm long, lobes about 3–4 cm
long, very broad. Drupe 3–4 cm broad, bright red, depressed. Coastal
areas, south Fla. and Fla. Keys and in cultivation, naturalized from
tropical America. All year. *Cerbera thevetia* L.

2. VALLÈSIA Ruiz & Pav.

Small trees or shrubs with alternate rather broad leaves each having a stipular dentate gland. Inflorescence axillary, lateral, in cymose
umbels of few to many small flowers; calyx lobes short, equal, corolla

salverform, lobes shorter than tube; anthers cordate, not connivent, disk absent. Ovary two-locular, carpels free. Fruit a drupe. Two species in tropical America.

1. Pedicels over 5 mm long . . . . . . 1. *V. glabra*
1. Pedicels less than 5 mm long . . . . . 2. *V. antillana*

1. V. glabra (Cav.) Link, PEARL BERRY—Shrub or small tree 2–3 m tall, the branches sometimes elongate and vinelike. Leaves 2.5–7 cm long, distant, elliptic to elliptic-lanceolate, membranaceous to subcoriaceous. Calyx lobes deltoid to ovate-deltoid, 1–1.5 mm long; corolla white, tube 3–4 mm long, lobes lanceolate-ovate, 1.5–4 mm long. Drupe 8–10 mm long. Coastal hammocks, Fla. Keys, tropical America. Spring, fall. *Vallesia laciniata* Brand

2. V. antillana Woodson—Shrubs up to 4 m tall with membranaceous or subcoriaceous leaves. Blades elliptic to obovate-elliptic, 2–8 cm long. Calyx lobes ovate-deltoid, corolla tube 6–7 mm long, lobes oblong 4–5 mm long. Drupe oblong-ovoid, about 10 mm long. Coastal hammocks, south Fla., Fla. Keys, tropical America. Spring, fall.

In the past there has been some confusion over the identity of the above two species; **V. glabra** is distinguished from **V. antillana** essentially by having a smaller corolla, shorter lobes, and leaves which are usually elliptic-lanceolate.

3. ALLAMÁNDA L.

Shrubs with opposite or whorled, pinnately veined leaves and intrapetiolar glands. Flowers large, showy, funnelform; calyx lobes narrow, deeply five-parted, outer lobes often larger; corolla with a cylindrical tube, filaments very short. Ovary one-locular, simple, ovules many. Fruit a two-valved soft capsule, considered to be very poisonous, seeds winged. About 18 species in tropical America.

1. A. cathartica L. YELLOW ALLAMANDA—Shrub up to 3 m tall with puberulent or glabrous stems. Leaves 6–12 cm long, mostly in whorls of four, oblanceolate to elliptic, undulate margins mostly short acuminate. Flowers large, short pedicelled; calyx lobes narrow, elliptic, corolla 7–10 cm long, yellow. Capsule 4–6 cm in width, prickly, the spines about 1 cm long. $2n = 18$. Disturbed sites and in cultivation, naturalized from tropical America. All year. Plants may be almost herbaceous and clambering.

Plumeria rubra L., the FRANGIPANI, a small tree with alternate or spiraled leaves near the ends of branches, and large, salverform, rose, yellow, or white fragrant flowers in cultivation in south Fla., but it is apparently not found as an escape.

4. NÈRIUM L.

Shrubs or trees with whorled, usually ternate or opposite coriaceous leaves. Flowers showy, funnelform; calyx lobes narrow, glandular within; corolla with a campanulate throat bearing conspicuous five-cleft petaloid appendages near the orifice, corolla limb regularly five-lobed or parted; stamens inserted near the mouth of the tube. Ovary with carpels free, pubescent, disk absent, stigma head globose. Fruit composed of two elongate follicles, free, three-angular. About three species, Old World.

1. N. oleander L. OLEANDER—Tree or bushy shrub mostly 3 m tall or more, glabrous. Leaves 6–12 cm long, lanceolate or narrowly elliptic. Calyx lobes 5–6 mm long, lanceolate or narrowly ovate, pubescent and rather foliaceous; corolla conspicuous and showy, white, crimson, or rose purplish, glabrous. Corolla limb 3–5 cm wide, lobes obovate to ovate. Follicles 8–15 cm long. Cultivated and probably not established as an escape, persistent around former habitations and waste places, native of the Mediterranean regions. All year. The sap is very poisonous.

5. ANGADÈNIA Miers ex Woodson

Subshrubs with erect or decumbent stems containing milky sap. Leaves opposite, coriaceous, entire, petioled or almost sessile. Flowers mostly in a lateral scorpioid inflorescence, calyx five-parted, stamens five and agglutinated to the stigma. Ovary with two carpels fused at the apex, ovules many. Fruit of two terete follicles, with many comose seeds.

1. A. berterii (A. DC.) Miers, PINELAND ALLAMANDA—Usually erect or suberect shrubs with opposite, short-petioled or subsessile, ovate to oblong leaves. Leaf blades 1–3 cm long, somewhat revolute and rounded at the base. Calyx lobes deltoid-ovate, 2–3 mm long, corolla yellow or cream colored, 2–3 cm long, tube 5–6 mm long, throat campanulate, squamellae in calyx one to three. Follicles 5–10 cm long. Pinelands, south Fla., Fla. Keys, W.I. All year. *Rhabdadenia corallicola* Small

6. CATHARÁNTHUS G. Don

Trailing or erect annual or perennial herbs often with woody base and with opposite, deciduous leaves; interpetiolar glands present. Flowers rather large, two to three in cymes from the leaf axils; calyx small, the lobes acuminate; corolla salverform, with a callous annulus and pubescence in the throat, variously colored or white, tube narrowly cylindrical, five-lobed, salverform stamens inserted just below the throat. Ovary of two free carpels. Fruit of two cylindrical erect follicles, fifteen- to thirty-seeded or more. About six species, Old World. *Lochnera* Reichenb.

1. **C. roseus** (L.) G. Don, MADAGASCAR PERIWINKLE—Pubescent herb or subshrub with stems 2–7 dm tall. Leaves 4–8 cm long, oblong, rounded at the apex, narrowed at the base into short petioles. Pedicels short, mostly 1–3 mm, calyx lobes 3–4 mm long, linear-subulate, the corolla salverform, white, pink, or rose purple; tube 2–3 cm long; lobes abruptly pointed; follicles cylindric, 2–3 cm long. $2n = 16$. Disturbed sites, south Fla., pantropical, naturalized from Africa. Spring, summer. *Lochnera rosea* (L.) Reichenb., *Vinca rosea* L., *Ammocallis rosea* (L.) Small

Vinca minor L. COMMON PERIWINKLE with dark blue corollas and persistent, leathery, dark green leaves is widely cultivated and has been collected as an escape in Palm Beach County. It may also occur locally in our area.

7. ECHÌTES P. Browne

Twining, rather woody vines, with opposite, petioled, entire leaves and milky sap. Flowers large, usually in lateral cymes; calyx five-lobed, with a single squamella, glandular; corolla salverform, five-parted, the throat narrowed to the mouth. Stamens five, inserted near the base of the tube. Ovary of two carpels united at the apex. Fruit of two spreading follicles, many-seeded. About six species in tropical America.

1. **E. umbellata** Jacq. var. **umbellata**, DEVIL'S POTATO, RUBBER VINE —Smooth, often intricately twining vines. Leaves 5–9 cm long, ovate-mucronate. Cymes axillary, pedicels 1–1.5 cm, umbelliform, three- to seven-flowered; calyx segments glandular, squamellae deeply dissected; corolla white or greenish white, 2–6 cm long, tube cylindrical; disk present, five-lobed, stamens glabrous. Follicles 15–20 cm long, compressed-cylindrical seeds with tawny coma. Pinelands, south Fla., W.I., Mexico, South America. All year. *Echites echites* (L.) Britt.

8. RHABDADÈNIA Muell. Arg.

Climbing vines, or sometimes erect, with opposite, petioled leaves. Flowers in small racemes or solitary; calyx lobes rather broad and five-cleft; corolla tube broadened into a funnelform throat; stamens short, inserted near the top of the corolla tube. Ovary two-carpellate, nectaries five, style slender, stigma thick. Fruit composed of linear, parallel or somewhat divergent many-seeded follicles. About ten species in tropical America.

Family 158. **APOCYNACEAE.** *Urechites lutea* var. *sericea:* a, flowering branch, × ½; b, flower, × 1¼; c, mature fruit, × ½.

1. R. biflora (Jacq.) Muell. Arg. RUBBER VINE—Stems smooth, elongate, bearing obovate or oblong leaves 5–9 cm long, apiculate at the tip. Inflorescence long pedunculate; calyx lobes elliptic, 5–10 mm long, glabrous, corolla 5–6 cm long, white, the tube yellow within, lobes 2–3 cm wide; anthers oblong-lanceolate. Follicles 12–15 cm long. Mangrove areas, coastal hammocks, south Fla., Fla. Keys, W.I., tropical America. All year. *R. paludosa* (Vahl) Miers, *Echites biflora* Jacq.

9. URECHITÈS Muell. Arg.

Twining vines rarely suberect with opposite entire leaves and milky sap. Flowers large, showy in lateral, scorpioid inflorescence; calyx lobes five, narrow, with two to many squamellae, corolla with a short tube dilated into a campanulate throat, limb strongly five-lobed; stamens with sagittate anthers with narrow appendages, agglutinated to the stigma. Ovary with two carpels united at the apex, stigma capitate. Fruit of elongate, slender incurved follicles, free, seeds comose. About six species in tropical America.

1. U. lutea (L.) Britt. WILD ALLAMANDA—Vines or scrambling shrubs, bright green, glabrous or pubescent. Leaves usually 5–7 cm long, oblong to suborbicular, shiny above, glabrescent, puberulent to copiously pubescent below. Cymes few to many, pedicels 1–1.5 cm long, several to many-flowered, bracts foliaceous; calyx lobes 9–10 mm long, narrow, glabrous or pubescent, corolla 4–5 cm wide, bright yellow. Follicles 8–20 cm long, 4–5 mm wide. South Fla., W.I. An extremely variable species; two varieties occur in our area:

1. Leaf undersurfaces, sepals, and follicles glabrous or barely pubescent . . . . . 1a. var. *lutea*
1. Leaf undersurfaces, sepals, and follicles copiously pubescent . . . . 1b. var. *sericea*

1a. U. lutea var. **lutea**—Twining vines, generally glabrous. Mangrove belt, hammocks, range of the species. All year.

1b. U. lutea var. **sericea** R. W. Long—Often erect or scrambling shrubs, less commonly twining vines, generally pubescent. Pinelands, less commonly near seastrand than is var. *lutea*, range of the species. All year. *U. pinetorum* Small

159. ASCLEPIADÀCEAE MILKWEED FAMILY

Erect perennial herbs, shrubs, vines, rarely trees with opposite, whorled, or sometimes alternate, entire leaves and usually containing a milky sap. Flowers bisexual, radially symmetrical, mostly in cymes, umbels, or racemes; calyx short tubular, five-parted, corolla often rotate, five-lobed, corona sometimes present; stamens five, epipetalous near the base of the corolla tube, filaments very short or absent, anthers usually closely attached to the pistil forming a gynostegium. Ovary superior, composed of two free carpels; fruit two follicles, seeds generally with a tuft of long, silky hairs. About 280 genera and 1800 species, mainly in the tropics and subtropics.

1. Twining or sprawling vines, woody or herbaceous.
　2. Stout woody vines; corolla funnelform 4–6 cm long . . . . 1. *Cryptostegia*
　2. Slender herbaceous or only partly woody vines; corolla rotate, 2 cm long or less.
　　3. Larger leaves 6–7 cm long, lance-ovate to ovate-elliptic, 1–4 cm wide, rounded or subcordate at base . . 2. *Sarcostemma*
　　3. Larger leaves mostly less than 0.5 cm wide, linear to linear-lanceolate or, if ovate-elliptic, then less than 4 cm long and truncate at base . . . . 3. *Cynanchum*
1. Erect or decumbent perennial or annual herbs, not twining . . . . 4. *Asclepias*

1. CRYPTOSTÈGIA R. Brown

Woody vines with rather broad, coriaceous leaves. Flowers conspicuous in narrow cymes; calyx lobes ovate, erect; corolla funnelform, lobes elliptic or ovate, smooth within, corona present with five entire or two-lobed scales attached to the corolla; anthers fused around the stigma. Ovary smooth, follicles strongly divergent. Three species in Old World tropics.

1. C. grandiflora R. Brown, PINK ALLAMANDA—Stout vines bearing ovate, elliptic, or broadly lanceolate coriaceous leaves, 6–10 cm long with rounded apex. Calyx lobes lanceolate to ovate-acuminate; corolla lobes ovate, pink, rose purple, or white, corona appendages shorter than the corolla throat. Follicles 10–15 cm long, seeds 7–8 mm long. Established in shell mounds, disturbed sites, hammocks, south Fla., W.I., naturalized from Africa. All year.

2. SARCOSTÉMMA R. Brown

Slender herbaceous or only partly woody vines with lanceolate or ovate leaves. Flowers generally cymose, calyx five-lobed, corolla rotate, lobes rather broad, smooth within; anthers fused to the pistil. Follicles flask shaped. About 15 species, Old World tropics. *Funastrum* Fourn.; *Philibertia* HBK; *Philibertella* Vail

1. S. clausa (Jacq.) R & S, WHITE VINE—Herbaceous vines mainly with lance-ovate or ovate leaves 6–7 cm long, rounded or subcordate at the base. Flowers on long peduncles, twice or more as long as the leaves; calyx lobes elliptic-lanceolate, corolla white, lobes elliptic to ovate, 4–5 mm long. Follicles 5–8 cm long, long acuminate. Shell mounds, coastal hammocks, peninsular Fla. and Fla. Keys, tropical America, introduced from Old World. All year. *Funastrum clausum* (Jacq.) Schlecht.

3. CYNÁNCHUM L. VINE MILKWEEDS

Slender herbaceous vines with broad or very narrow leaves or stems leafless. Flowers in cymes or in umbel-like clusters, corolla obconic, lobes ovate-triangular, somewhat longer than the tube, corona present, forming a fleshy cup with shallow-toothed margin ascending around the gynostegium. Ovary smooth, follicles glabrous, lance-linear. About 150 species in warmer regions. *Ampelamus* Raf.; *Amphistelma* Griseb.; *Lyonia* Ell.; *Seutera* Reichenb.; *Metastelma* R. Brown; *Epicion* Small

1. Larger leaves ovate-elliptic, mostly 1.0–1.5
   cm wide . . . . 1. *C. northropiae*
1. Larger leaves linear-subulate to linear-
   lanceolate, mostly less than 0.6 cm wide,
   or branches leafless.
   2. Peduncles short, less than 0.5 cm or want-
      ing; cymes less than 1 cm in diameter.
      3. Corolla lobes glabrous within; branches
         usually leafless when mature . . . 2. *C. scoparium*
      3. Corolla lobes pubescent within; leaves
         usually persistent on older branches 3. *C. blodgettii*
   2. Peduncles longer, mostly 1–2 cm or
      longer; cymes over 1 cm in diameter . 4. *C. palustre*

1. C. northropiae (Schltr.) Alain, FRAGRANT CYNANCHUM—Twining vines with ovate-elliptic or elliptic leaves 2–4 cm long and mostly 1–1.5 cm wide. Flowers cymose, calyx lobes ovate, about 1.5 mm long; corolla white, lobes 2–3 mm long, obtuse, somewhat longer than the tube; corona lobes 1.5 mm long. Follicles narrowly lanceolate, 5–6 cm long. Hammocks, pinelands, south Fla., Fla. Keys, W.I. All year. *Epicon northropiae* (Schltr.) Alain; *Metastelma bahamense* Griseb.

2. C. scoparium Nutt. LEAFLESS CYNANCHUM—Twining, diffusely branching vine. Leaves narrowly linear 1–5 cm long, branches usually leafless at maturity. Flowers cymose, calyx lobes 1 mm long, deltoid,

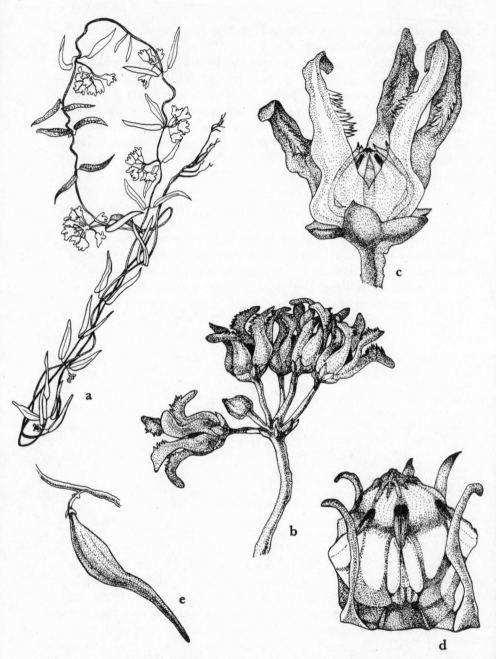

Family 159. **ASCLEPIADACEAE.** *Cynanchum blodgettii:* a, flowering branch, × 1; b, inflorescence, × 10; c, flower, × 30; d, gynostegium, × 40; e, mature fruit, × 1½.

corolla glabrous, greenish white, lobes 1-2 mm long, corona lobes less than 1 mm long. Follicles narrow 3-5 cm long. Hammocks and pinelands, south Fla., Fla. Keys, W.I. Spring, fall. *Amphistelma scoparis* (Nutt.) Small

3. **C. blodgettii** (Gray) Shinners—Twining vines with leaves usually persistent on stems. Leaves mostly 10-20 mm long, linear or linear-lanceolate. Flowers cymose; calyx lobes ovate, about 1 mm long, corolla whitish, pubescent within, lobes 2-3 mm long, corona lobes subulate. Follicles linear, 4-5 cm long. Hammocks, sandy soil, south Fla. and South Tex. Spring, fall. *Metastelma blodgettii* Gray

4. **C. palustre** (Pursh) Heller—Twining vines with linear, drooping leaves. Leaves mostly 2-7 cm long, acute tipped. Flowers in cymes over 1 cm wide, peduncles mostly 1-2 cm or more; calyx lobes lanceolate, about 2-3 mm long; corolla lobes purplish or greenish white, mostly 3-4 mm long, corona lobes 1-2 mm long, slightly notched at the apex. Follicles narrow, 4-6 cm long. Hammocks and marshes Fla. to Tex., N.C. on the coastal plain. Spring, fall. *Lyonia palustris* (Pursh) Small; *Seutera maritima* Decne.; and possibly including *C. angustifolium* Pers.

4. ASCLĒPIAS L. Milkweed

Perennial herbs from rhizomes or stout roots and with a milky sap at least for most species. Leaves opposite or alternate, or whorled in some. Flowers in peduncled terminal or axillary cymes; calyx lobes four to five, persistent, corolla rotate, deeply five-lobed, spreading or reflexed; corona segments formed into hoods attached at the bottom of the gynoecium, straight, curved, erect, or spreading at the base and upcurved at the tip, usually bearing a slender horn near the base; filaments fused. Follicles usually erect, acuminate, seeds with thick silky hairs. About 100 species in America. *Acerates* Ell.; *Biventraria* Small; *Oxypteryx* Greene; *Asclepiodora* Gray; *Asclepiodella* Small

1. Larger leaves linear or filiform, mostly less than 1 mm wide, opposite or whorled.
  2. Leaves whorled; corolla lobes 3-4 mm long, greenish white or white . . . .   1. *A. verticillata*
  2. Leaves opposite; corolla lobes about 10 mm long, white . . . . . .   2. *A. feayi*
1. Larger leaves narrowly lanceolate to broadly ovate, more than 1 mm wide, opposite or alternate.
  3. Stems hirsute, not containing milky sap; hoods bright orange; leaves mostly alternate . . . .   3. *A. tuberosa* ssp. *rolfsii*

3. Stems glabrous or appressed pubescent,
   often containing milky sap; leaves
   mostly opposite.
   4. Leaves sessile or sessile-clasping; stems
      erect, or decumbent.
      5. Leaves sessile-clasping, veins reddish
         or purple; stems decumbent . .   4. *A. humistrata*
      5. Leaves sessile, veins not reddish or
         purple; stems erect . . . . .   5. *A. pedicellata*
   4. Leaves petioled; stems erect.
      6. Corolla lobes scarlet, red, or orange
         red; hoods orange or yellow.
         7. Leaves linear to narrowly lanceo-
            late, less than 1 cm wide; hoods
            orange . . . .   6. *A. lanceolata*
         7. Leaves lanceolate, mostly over 1 cm
            wide; hoods yellow, pink, or
            rarely white.
            8. Corolla dark red or scarlet,
               showy; annual . . . . .   7. *A. curassavica*
            8. Corolla pink or white, small;
               perennial . . . . . .   8. *A. incarnata*
      6. Corolla lobes greenish, yellowish,
         purplish, or white.
         9. Corolla lobes mostly 0.9–1.0 cm
            long, erect or spreading, green-
            ish or greenish white; leaves
            lanceolate . . . .   9. *A. viridis*
         9. Corolla lobes mostly less than 0.8
            cm long, reflexed; leaves nar-
            rowly lanceolate to ovate-elliptic.
            10. Larger leaves linear-lanceolate,
                mostly 3–6 mm wide   . . .   10. *A. longifolia*
            10. Larger leaves ovate or obovate,
                over 10 mm wide.
                11. Stems erect or sometimes de-
                    cumbent; flowers greenish
                    white, corolla lobes nar-
                    rowly lanceolate 5–6 mm
                    long; leaves ovate-elliptic,
                    3–5 cm long . . . . .   11. *A. curtissii*
                11. Stems erect; flowers pale yel-
                    lowish green, corolla lobes
                    broadly lanceolate, truncate,
                    9–11 mm long; leaves ovate-
                    elliptic, 5–8 cm long   . . 12. *A. tomentosa*

   1. A. verticillata L.—Stems slender, erect, linear filiform up to 9
dm tall, glabrous or commonly pubescent on the angled stem in lines.
Leaves 2–5 cm long, numerous, sessile or subsessile, whorled, linear to

linear-filiform, revolute. Umbels few to several from the upper nodes, peduncles 1–3 cm long; flowers small, greenish white or white, corolla lobes 3–4 mm long; hoods divergent, 1–2 mm, horns subulate. Follicles erect, narrowly fusiform, slender, 7–10 cm long. Dry soil, Fla. to Tex., Mass., Saskatchewan, Ariz. All year.

2. **A. feayi** Chapm.—Stems very slender, simple, up to 3.5 dm tall. Leaves 7–9 cm long, opposite, sessile, filiform, 1–2 mm wide, glabrous. Umbels few-flowered, pedicels very slender, flowers rather large, calyx lobes 2–3 mm long, ovate, purplish, corolla white, the lobes about 1 cm long, hoods broadly oval, 3 mm long. Follicles erect on erect pedicels, narrowly fusiform, 9–12 cm long, glabrous. Pinelands, disturbed sites, endemic to peninsular Fla. Spring, fall. *Asclepiodella feayi* (Chapm.) Small

3. **A. tuberosa** L. ssp. **rolfsii** (Britt.) Woodson, BUTTERFLY WEED— Stems erect or ascending up to 7 dm tall, hirsute and branched above. Leaves 6–10 cm long, alternate, or opposite on the lateral branches, hastate, linear to lanceolate, sessile or barely petioled. Umbels terminal, solitary, or numerous, flowers bright orange red or yellow red; corolla 5–10 mm, hoods 5–7 mm nearly straight, erect. Follicles 8–12 cm erect. Sap of this species is not milky. Sandy soil, pinelands, Fla. to Miss., S.C. Summer, fall.

4. **A. humistrata** Walt.—Stems prostrate or decumbent, glaucous and glabrous, mostly 3–10 dm long. Leaves 3–10 cm long, opposite, sessile, somewhat fleshy, glaucous, ovate to broadly so, the veins usually reddish or purplish. Umbels greenish purple or grayish, peduncles 4–6 cm long, corolla lobes 6–7 mm long; hoods mostly 3–4 mm long, white, broadly rounded, horn flat, scarcely exserted. Follicles 9–10 cm long, erect, narrowly fusiform. Sandy soil, pinelands, scrub vegetation, Fla. to La., N.C. on the coastal plain. Spring, summer.

5. **A. pedicellata** Walt.—Low, erect, slender herbs with simple stems up to 5 dm tall, minutely puberulent. Leaves 2–5 cm long, opposite, sessile, linear to lance-ovate, minutely puberulent. Umbels few-flowered, peduncles slender 1–2.5 cm long, lobes 2 mm, corolla greenish cream, lobes about 1 cm long, hoods about 3 mm long. Pinewoods, Fla. to N.C. on the coastal plain. Spring, summer.

6. **A. lanceolata** Walt.—Stems slender up to 1 m tall or more, simple, glabrous, from tuberous rootstalks. Leaves opposite, distant, narrowly lanceolate or linear, long acuminate and separated by internodes 1–2 dm long. Umbels terminal, one to four, few-flowered; corolla red or reddish purple, 9–12 mm long; hoods ovate, 5–6 mm long, the lateral margin with a small tooth near the middle, horns subulate, somewhat incurved. Follicles erect on deflexed pedicels, narrowly fusiform, 8–10 cm

long. Glades, swamps, Fla. to Tex., N.J. on the coastal plain. Summer.

7. A. curassavica L. SCARLET MILKWEED—Annuals with slender stems up to 1 m tall or more. Leaves opposite, broadly lanceolate to elliptic, 6–12 cm long or more, glabrous or nearly so. Umbels terminal or subterminal, several- to many-flowered, peduncles 3–6 cm long, flowers large and showy, calyx lobes 2–3 mm long, corolla lobes dark red or scarlet, 5–10 mm long; hoods ovate, 3–5 mm long, yellow, horn stout, incurved, exceeding the hood. Follicles erect 7–10 cm long, smooth. $2n = 22$. Sandy disturbed soil, Fla. to Tex. on coastal plain, Calif., naturalized from tropical America, pantropical weed. All year.

8. A. incarnata L. ssp. **incarnata**—Stems stout, up to 15 dm tall, simple or much branched. Leaves 5–15 cm long, opposite, petioled, narrowly lanceolate to ovate-elliptic. Inflorescence usually paired in upper nodes, flowers small, calyx lobes 1–1.5 mm long, corolla bright pink or rarely white, reflexed-rotate, lobes 3–4 mm long, hoods about 1.5 mm long. Follicles erect on erect pedicels, 7–9 cm long, seeds oval, 7–10 mm long. Moist soil, Fla. to N. Mex., Canada, Utah. Summer.

9. A. viridis Walt.—Stems erect, ascending, or somewhat decumbent, 3–6 dm tall. Leaves 6–12 cm, oblong, lance-oblong, or elliptic, rounded or obtuse, and abruptly narrowed at the base to a short definite petiole. Umbels terminal, peduncles 3–6 cm long, flowers large and showy, calyx lobes 4–5 mm long, corolla greenish, 2–3.5 cm wide, hoods violet or purplish, horns rudimentary. Follicles erect in deflexed pedicels. Hammocks, pinelands, Fla. to Tex., Tenn., Nebr. Spring, summer. *Asclepiodora viridis* (Walt.) Gray

10. A. longifolia Michx.—Stems slender up to 6 dm tall, glabrous or nearly so, with a stout tuberous rootstalk. Leaves linear, numerous, 6–18 cm long, sessile or subsessile, glabrous or somewhat pilose. Umbels terminal or lateral, ten- to thirty-flowered on peduncles 2–6 cm long, rather lax; calyx lobes 2 mm long, corolla pale greenish white tinted with purple, lobes about 5 mm long, hoods 2–3 mm long, the base separated from the corolla by a distinct column, horns absent. Follicles erect on deflexed pedicles, 8–12 cm long. Low, moist soil, Fla. to La., Del. on the coastal plain. All year. *Acerates floridana* (Lam.) Hitchc. misapplied

11. A. curtissii Gray—Decumbent stems slender up to 7 dm tall, minutely pubescent and often divergently branching from fleshy rootstocks. Leaves 3–6 cm long, elliptic to ovate, opposite, glabrous or nearly so, petioled. Umbels many-flowered, flowers rather small, sparingly pubescent; corolla lobes pale greenish white, lanceolate, 5–6 mm long; hoods oblong-lanceolate, 4–5 mm long, over twice as long as the androecium. Sandy soil, endemic to peninsular Fla. Spring, fall. *Oxypteryx curtissii* (Gray) Small

12. A. tomentosa Ell. VELVET-LEAF MILKWEED—Stems erect, up to 1 m tall, softly pubescent, leaves 5–8 cm long, opposite, elliptic to ovate, petioled, ovate-elliptic. Umbels terminal or subterminal, several- to many-flowered, lax, softly pubescent; corolla lobes yellowish green, 9–11 mm long, hoods about 4 mm long, truncate, about equaling the androecium, horn somewhat incurved, broad. Follicles erect, narrowly fusiform 10–12 cm long, puberulent to glabrate. Sandy soil, pineland, Fla. to N.C., east Tex., on the coastal plain. Spring, summer.

160. CONVOLVULÀCEAE MORNING-GLORY FAMILY

Herbaceous or woody, usually twining or trailing plants without tendrils, often with milky sap and alternate, simple leaves. Flowers bisexual, solitary or in cymes, radially symmetrical, subtended by two bracts, sepals five, mostly free, corolla fused, often showy, generally funnel shaped, five lobed, stamens five, epipetalous near the base of the corolla tube. Ovary superior, one- to four-locular, often surrounded by a disk. Fruit a capsule or berry. About 50 genera and 2000 species, chiefly warm regions. *Dichondraceae, Cuscutaceae*

1. Nongreen parasitic twining herbs, not rooting in ground at maturity; leaves reduced to mere scales . . . . 1. *Cuscuta*
1. Green, erect, creeping, trailing or twining herbs or shrubs, rooting in ground; leaves normal.
  2. Creeping, often matted herbs; larger leaf blades orbicular to reniform, cordate, 1–2 cm wide; corolla deeply five-cleft . 2. *Dichondra*
  2. Erect, trailing or twining herbs or shrubs; larger leaf blades various, not small, orbicular or reniform; corolla not deeply cleft.
    3. Styles two, distinct.
      4. Styles two-cleft, stigmas four . . . 3. *Evolvulus*
      4. Styles not cleft, stigmas two . . . 4. *Bonamia*
    3. Styles fused, apparently one.
      5. Stem and leaves densely tomentose.
        6. Corolla showy, purple, up to 5 cm long . . . . 5. *Argyreia*
        6. Corolla small, white, less than 1 cm long . . . . 6. *Porana*

5. Stem and leaves glabrous or pubescent, not densely tomentose.
7. Sepals becoming coriaceous or cartilaginous, especially in fruit, lanceolate-elliptic to broadly ovate, or orbicular.
8. Corolla red; sepals broadly ovate to orbicular, becoming accrescent in fruit . . . . .   7. *Stictocardia*
8. Corolla white; sepals lanceolate, not accrescent in fruit.
9. Sepals becoming 2–4 cm long; fruit a four-valved dehiscent capsule, one- to four-seeded   8. *Merremia*
9. Sepals becoming 0.8–1.2 cm long; fruit a leathery or woody, indehiscent capsule, usually one-seeded . . .   9. *Turbina*
7. Sepals herbaceous or membranous, green, acute, acuminate-lanceolate.
10. Stigma one, globose, ovoid or two- to three-lobed.
11. Outer sepals about as long as inner ones . . . . .   10. *Ipomoea*
11. Outer sepals much longer than inner ones . . . . .   11. *Aniseia*
10. Stigmas two, flattened or linear .   12. *Jacquemontia*

1. CÚSCUTA L. Dodder

Parasitic twining vines with yellowish or brownish stems and alternate leaves reduced to mere scales. Flowers white or yellow in small cymose clusters; sepals five, fused, corolla campanulate to cylindric, four- to five-lobed, stamens inserted at corolla sinuses. Ovary two-locular, styles one to two. Fruit a capsule or indehiscent. About 167 species of wide distribution.

1. Each flower subtended by several small bracts, flowers sessile or subsessile . . .   1. *C. compacta*
1. Each flower without bracts, flowers with short pedicels.
2. Corolla lobes acute . . . .   2. *C. campestris*
2. Corolla lobes obtuse . . . . . .   3. *C. americana*

1. C. compacta Juss.—Twining, often matted yellowish brown stems. Flowers sessile or subsessile in dense masses mostly 4–5 mm long, bracts three to five, rounded and appressed, corolla lobes elliptic-ovate, shorter than the tube. Capsule globose, 3–5 mm wide. Pinelands, swamps, Fla. to Tex., Mass., Ill. All year. Found on shrubs, especially in moist areas, often growing in dense masses.

2. **C. campestris** Yuncker—Twining, slender yellowish brown stems growing in separate strands or more often matted. Flowers in loose clusters, each short pedicelled without subtending bracts, calyx lobes broadly ovate, corolla lobes acute, spreading, five-toothed. Capsule sub-globose, 3–4 mm wide. Generally throughout Fla. and the U.S., cosmopolitan, and found parasitic on a wide variety of host plants. All year. *C. pentagona* Engelm.

3. **C. americana** L.—Twining, slender yellowish brown stems growing in separate strands or occasionally matted. Flowers single or in clusters, short-pedicelled, calyx somewhat inflated on one side, corolla campanulate, lobes obtuse, five-toothed, erect. Capsule globose, 4–5 mm wide. Pinelands, hammocks, growing over low shrubs, south Fla., tropical America. All year.

2. DICHÓNDRA Forst. & Forst.

Prostrate, creeping pubescent herbs with orbicular or reniform leaves. Flowers small, axillary, solitary on short peduncles; sepals five, oblong, corolla campanulate or rotate, deeply five-lobed, stamens included, shorter than the corolla. Ovary deeply divided into almost distinct carpels, styles two, basal. Capsule irregularly dehiscent or indehiscent, two-seeded. Five species, mostly tropical and subtropical America.

1. **D. caroliniensis** Michx. FALSE PENNYWORT—Small creeping herbs, pubescent. Leaves 1–2 cm wide, reniform to orbicular, often notched at the apex and cordate at the base. Sepals becoming 3 mm long, spatulate-oblong; corolla greenish white, solitary. Capsule about 2–3 mm long, hirsute. Disturbed sites, low moist ground, lawn weed, Fla. to Tex., Va. on the coastal plain, Bermuda, Bahama Islands. All year.

3. EVÓLVULUS L. CREEPING MORNING GLORIES

Erect or prostrate herbs or subshrubs, never twining. Leaves small, entire. Flowers axillary, sessile or pedicelled, solitary or in cymes or spikes; sepals five, corolla rotate or funnel shaped, five-lobed or almost entire; stamens five, epipetalous. Ovary two-locular each with two ovules, styles two, free, disk small or absent. Fruit a two- to four-valved capsule one- to four-seeded. About 100 species in tropical and subtropical America.

1. Peduncles as long as the leaves or longer.
   2. Undersurfaces of mature leaves hirsute
      or villous . . . . 1. *E. alsinoides*
      var. *grisebachianus*
   2. Undersurfaces of mature leaves glabrous
      or nearly so . . . . 2. *E. glaber*
1. Peduncles shorter than the leaves, or flowers sessile.
   3. Larger leaves ovate to orbicular, less than
      1 cm long . . . . 3. *E. grisebachii*
   3. Larger leaves linear-subulate to linear-oblong, over 1 cm long . . . . . . 4. *E. sericeus*
      var. *sericeus*

1. E. alsinoides L. var. **grisebachianus** Meissner—Stems slender, spreading, prostrate or ascending up to 4 dm long, thinly pubescent. Leaves 1–2 cm long, somewhat distant, oblong-lanceolate to ovate, pubescent. Flowers solitary or few-clustered on the ends of filiform peduncles; calyx lobes lanceolate, 2–3 mm acute; corolla pale blue to white, rotate, 6 mm wide or more. Capsule 3–4 mm wide. Coastal areas, hammocks, disturbed areas, shell mounds, Fla. Keys, W.I. All year.

The species is a pantropical weed; it is apparently polymorphic, although few of the numerous described varieties appear to be well defined.

2. E. glaber Spreng.—Stems diffusely branching, prostrate and often creeping, silky pubescent. Leaves 1–2 cm long, oblong-ovate or obovate, becoming glabrous at maturity. Flowers mostly solitary, calyx lobes lanceolate-ovate, 3–4 mm long, acute; corolla blue, pink, or white, about 1 cm wide. Capsule ovoid, 3–4 mm wide. Coastal areas, on limestone or coral rock, south Fla., Fla. Keys, tropical America. All year.

3. E. grisebachii Peter—Stems and branches suberect, tufted, mostly 1 dm tall or less, silvery pubescent. Leaves about 1 cm long, ovate to suborbicular, acute tipped. Flowers mostly solitary or few-flowered; calyx-lobes linear, 4–5 mm long, corolla blue or white, 8–9 mm wide. Glades, pinelands, south Fla., W.I. All year. *E. wrightii* House

This species has been reported for the Fla. Keys but may be rare.

4. E. sericeus Sw. var. **sericeus**—Stems and branches filiform, spreading, ascending or decumbent about 1–3 dm long. Leaves 1–3 cm long, linear to linear-subulate, acute at both ends, silky pubescent to subglabrous. Corolla white or blue, about 8–10 mm wide, calyx lobes 4–6 mm long, acute, flowers mostly solitary. Capsule subglobose, about 4 mm in diameter. Pinelands, low ground, south Fla., Fla. Keys, to Ariz., Ga., Mexico, W.I., Central America, South America. All year.

A glabrous form has been named var. **glaberrimus** Robinson (*E.*

macilentus Small), and a form with leaves pubescent on both sides has been called var. **averyi** Ward; both from the Fla. Keys.

4. BONÁMIA Thouars

Perennial herbs rarely woody near the base, prostrate, ascending, or twining vines with alternate, entire leaves. Flowers axillary, solitary, or cymose, small, funnelform or rotate, sepals five, corolla five-lobed, stamens five, epipetalous, included or slightly exserted. Ovary two-locular, two ovules per locule, style deeply two-cleft or styles two, free, stigmas peltate. Fruit an ovoid or globose capsule, two- to eight-valved. About 40 species, pantropical. *Breweria* R. Brown; *Stylisma* Raf.

1. Larger leaves over 2 cm long; corolla over
   twice the length of the sepals . . . . 1. *B. villosa*
1. Larger leaves usually less than 2 cm long;
   corolla twice the length of the sepals or
   less . . . . 2. *B. abdita*

1. **B. villosa** (Nash) Wilson—Stems prostrate or twining, up to 2 m long, villous with brownish or gray hairs. Leaves 3–7 cm long, narrowly oblong-lanceolate, densely villous, corolla white, 1–2.5 cm long, stamens included. Ovary villous, capsule mostly one- to two-seeded. Dry sandy soil, hammocks, Fla. Spring, summer. *Stylisma villosa* (Nash) House; *Breweria villosa* Nash

2. **B. abdita** (Myint) R. W. Long—Prostrate vines up to 3 dm long, densely pubescent. Leaves about 1–2 cm long, narrowly linear-elliptic, densely villous with silvery or brownish hairs. Flowers solitary or with short peduncles, sepals 4–7 mm long, mostly half the corolla length or longer, corolla white, 8–13 mm long, stamens included. Ovary densely villous, styles free almost to the base. Capsule mostly one- to two-seeded. Dry sandy soil, peninsular Fla. Spring, summer. *Stylisma abdita* Myint

Bonamia patens (Desc.) Shinners var. **angustifolia** (Nash) Shinners with smooth linear leaves has been collected on the lower Fla. east coast and may occur in our area.

5. ARGYRÈIA Lour. WOOLY MORNING GLORY

Perennial twining vines pubescent with silky hairs and broad, cordate, entire leaves. Inflorescence an axillary, peduncled cyme, calyx often colored within, sepals five, coriaceous, corolla funnelform, campanulate or tubular, stamens five, included. Ovary two- to four-locular, stigma two-lobed. Fruit berrylike, leathery, indehiscent. About 90 species in tropical Asia and Australia.

1. **A. nervosa** (Burm.) Bojer—Stems and branches densely pubescent. Leaves 10–20 cm long, broadly ovate-cordate, glabrous above, pubescent beneath. Bracts ovate, unequal, glabrous within, sepals 4–5 mm long, corolla funnelform-cylindric, 5 cm long, corolla limb purplish. Hammocks, south Fla., naturalized from India. *A. speciosa* (L.f.) Sweet

Cultivated as an ornamental and apparently escapes locally and becomes established chiefly in hammocks.

6. PORÀNA Burm.

Climbing herbs with alternate, cordate-ovate entire leaves on definite petioles. Flowers small, in racemes, panicles, or cymes; sepals five, equal, free or fused, becoming much larger in fruit, corolla campanulate or funnelform, stamens five, included. Ovary two-locular, two to five ovules, style one, two-lobed or capitate. Fruit a small dehiscent or indehiscent capsule. About 15 species in the Old World tropics.

1. **P. paniculata** Roxb. CHRISTMAS VINE—Twining and trailing herbs with white pubescence, up to 10 m long. Leaves cordate-ovate, mostly 6–12 cm long, entire, acuminate, with prominent veins. Flowers abundant in numerous, axillary panicles, corolla about 1 cm wide, white. Capsule globose, pubescent, about 5 mm wide. $2n = 26$. Disturbed sites, local south Fla., naturalized from India. Winter.

7. STICTOCÁRDIA Hall. f.

Herbaceous vines with cordate to orbicular leaves, these often minutely glandular. Flowers in axillary cymes, pedunculate, one- to many-flowered, sepals elliptic to orbicular, obtuse to marginate, subcoriaceous, accrescent in fruit, corolla large, stamens and style included, epipetalous near the base of the corolla. Ovary four-locular, style one, stigma two-lobed. Capsule short, thick, globular, seeds four, pubescent. A few species in the tropics.

1. **S. tiliaefolia** Hall. f.—Climbing vines with glabrous or pubescent stems, leaves cordate to orbicular 6–20 cm long, apex obtuse. Flowers one to three on pubescent, axillary peduncles, pedicels 1–4 cm long, apex obtuse. Flowers one to three on pubescent, axillary peduncles, pedicels 1–4 cm long, sepals orbicular 12–18 mm, pubescent, corolla red, about 8–10 cm long. Capsule 2–3.5 cm long. Hammocks, disturbed sites, Key West, Fla., W.I., American tropics. *S. campanulata* (L.) Merrill

8. MERRÉMIA Dennst. ex Hall.

Twining herbs or sometimes prostrate or erect subshrubs. Leaves highly variable, entire to palmately lobed or compound. Flowers axillary, solitary or in few-flowered cymes, sepals five, mostly subequal, corolla campanulate or funnelform, white or yellow, usually with five distinct midpetal bands, stamens included. Ovary two- to four-locular. Capsule four-valved. About 80 species, tropical and subtropical regions. *Operculina* sensu Small

1. Leaves deeply parted or lobed; cymes few-
flowered or flowers solitary.
  2. Leaf blades five- to seven-parted, the segments coarsely toothed; corolla white
  with purple throat . . . . . .   1. *M. dissecta*
  2. Leaf blades five- to seven-lobed, the lobes
  lance-acuminate, not coarsely toothed;
  corolla yellow . . . .   2. *M. tuberosa*
1. Leaves simple, not deeply lobed or divided;
cymes umbellate, many-flowered . . .   3. *M. umbellata*

    1. M. dissecta (Jacq.) Hall. f.—Twining vines, glabrescent to pubescent, with leaves palmately parted nearly to the base with five to seven lanceolate, coarsely dentate segments. Peduncles 5–10 cm long, cymes five or few-flowered, sepals large, ovate-lanceolate, acute, 2–2.5 cm long, corolla white with purple or rose inside, 3–3.5 cm long, funnel shaped. Ovary glabrous, two-locular, capsule globose, four-seeded. $2n = 30$. Pinelands, Fla. to Tex., Ga. on coastal plain, tropical America. All year. *Ipomoea dissecta* Jacq.; *Operculina dissecta* (Jacq.) House
    2. M. tuberosa (L.) Rendle—Twining vines with leaves palmately five- to seven-lobed, lobes lance-acuminate, 4–6 cm long. Peduncles 5–7 cm long, cymes few-flowered or solitary, sepals large, ovate-lanceolate, 2 cm long, corolla yellow, 3–4 cm long. Capsule 3–4 cm wide, seeds black. $2n = 30$. Disturbed sites, pinelands, Fla. to Tex., naturalized from tropical America. All year. *Operculina tuberosa* (L.) Meissner
    3. M. umbellata (L.) Hall—Twining vines with simple, entire leaves, orbicular, oblong, cordate or cordate-sagittate, mostly 6–10 cm long. Flowers many in dense, umbel-like cymes, sepals about 1 cm long, becoming about 2 cm in fruit, corolla bright yellow about 3–4 cm long. Seeds black, pubescent. Disturbed sites, pinelands, Fla. Keys, W.I., tropical America. All year. *Ipomoea polyantha* R & S

9. TURBÌNA Raf.

    Twining or trailing herbaceous or somewhat woody vines with entire, ovate or cordate leaves. Flowers axillary in peduncled, many-flowered cymose or corymbose clusters; sepals unequal, coriaceous with thin margins, corolla campanulate, with distinct midpetaline bands, stamens unequal in length. Ovary two-locular with two ovules, stigma two-lobed.

Capsule turbinate, twice as long as broad, one-valved, indehiscent, one-seeded by abortion. About ten species, American and African tropics.

1. **T. corymbosa** (L.) Raf.—Trailing or twining herbs up to several m long. Sepals narrowly ovate or elliptic, 6–8 mm long, dark colored with whitish scarious margins; corolla white with greenish bands, 3–4 cm long. Capsule 8–10 mm long. Hammocks, shrubby areas, south Fla. and Fla. Keys, south Tex., W.I., tropical America. Fall, winter. *Rivea corymbosa* (L.) Hall.

This plant was used by Indians in religious and medicinal practices.

10. IPOMÒEA L.   Morning Glory

Herbs or shrubs, usually twining vines, or prostrate, creeping, or erect. Leaves highly variable, simple or divided. Flowers axillary, solitary or in few- to many-flowered cymes, often showy, sepals five, herbaceous, variable, corolla funnelform or campanulate, five-lobed or nearly entire, stamens five, epipetalous, usually included. Ovary two- to four-locular with mostly four ovules, stigma capitate, two- to three-lobed. Capsule globose or ellipsoid, four- to six-valved, seeds four to six or fewer. About 500 species, chiefly in the tropics and subtropics. *Calonyction* Choisy; *Exogonium* Choisy; *Pharbitis* Choisy; *Quamoclit* Moench

1. Style and stamens exserted; corolla with a
   long, slender tube.
   2. Corolla large, mostly over 10 cm long,
      white.
      3. Sepal tips, at least the outer, prolonged
         into   hooked   appendages;   seeds
         smooth or nearly so   .   .   .   .   .   1. *I. alba*
      3. Sepal tips obtuse, not prolonged into
         appendages; seeds pubescent .   .   .   2. *I. tuba*
   2. Corolla shorter, less than 8 cm long, red
      or scarlet.
      4. Corolla tube 4–5 cm long; stems becoming woody; seeds pubescent .   .   .   3. *I. microdactyla*
      4. Corolla tube 2–4 cm long; stems herbaceous; seeds smooth   .   .   .   .   .   4. *I. hederifolia*
1. Style and stamens included; corolla funnelform.
   5. Leaves entire or three-lobed or three-cleft.
      6. Corolla mostly 5–6 cm long or more
         7. Leaves sagittate, often deeply lobed
            and the lobes narrowly acuminate   5. *I. sagittata*
         7. Leaves entire to three-lobed, not sagittate.

8. Stems erect, plants bushy or subshrubs; leaves entire . . . . 6. *I. crassicaulis*
8. Stems trailing, twining, not erect
or bushy; leaves entire or often
three-lobed.
  9. Sepals hispid below, tips spread-
    ing . . . . 7. *I. hederacea*
  9. Sepals glabrous to merely puber-
    ulent below, tips not spread-
    ing.
    10. Corolla purplish; leaves fleshy,
      suborbicular, entire, notched
      at the apex . . . . . 8. *I. pes-caprae*
                    var. *emarginata*
    10. Corolla pink purple, crimson,
      or white; leaves entire
      to deeply three-lobed, not
      fleshy.
      11. Corolla white, leaves ovate
        to elliptic, some deeply
        lobed . . . . . 9. *I. stolonifera*
      11. Corolla pink purple or
        crimson; leaves broadly
        ovate, entire or three-
        lobed . . . . . 10. *I. acuminata*
6. Corolla mostly 3–4 cm long or less.
  12. Flowers solitary on peduncles up to
    2–3 cm long . . . . 11. *I. tenuissima*
  12. Flowers at least some in umbellate
    cymes, pedicels mostly 1 cm long
    or less.
    13. Corollas 2 cm long or less . . . 12. *I. triloba*
    13. Corollas at least some, 2–3 cm long 13. *I. trichocarpa*
5. Leaves palmately five- to seven-parted . 14. *I. cairica*

1. **I. alba** L. MOON FLOWERS—Twining vines often high in the trees. Leaves ovate to suborbicular, 5–15 cm long, entire to hastate, three- to five-lobed, slender acuminate, lobes acute or acuminate. Flowers white, showy, calyx lobes 2–3 cm, tips, at least the outer, prolonged into hooked appendages, corolla tube slender, 9–14 cm long, limb 10–14 cm wide. Capsule depressed, 3–3.5 cm wide, enveloped by the accrescent inner calyx lobes. $2n = 30$. Hammocks, south Fla., and pantropical. All year. *Calonyction aculeatum* (L.) House

The species is especially noticeable in an area that has been recently burned in south Fla.

2. **I. tuba** (Schlect.) G. Don, MOON VINE—Twining vines with broadly ovate, reniform, or reniform-ovate leaves 8–16 cm long, usually entire, broadly acuminate. Flowers white, showy, calyx lobes 1.5–2.5 cm

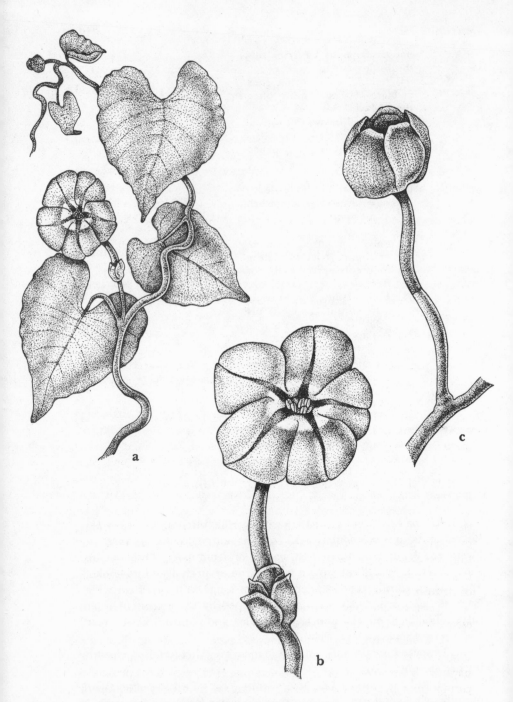

Family 160. **CONVOLVULACEAE.** *Ipomoea tuba:* a, flowering branch, × ½;
b, flower, × 1; c, mature fruit with enclosing bracts, × 1.

long, tips obtuse, not prolonged into appendages, corolla tube 6–12 cm long, limb up to 10 cm wide. Capsule ovoid 2.0–2.5 cm wide, much exceeding the calyx. Coastal hammocks, shrubby areas, south Fla., and Fla. Keys, pantropical. All year. *Calonyction tuba* (Schlect.) Colla; *C. grandiflorum* (Jacq.) Choisy

3. **I. microdactyla** Griseb. WILD POTATO—Trailing or often twining woody vines with elliptic, entire or lobed leaves 3–8 cm long, rather thick and obtuse at the base. Flowers crimson or scarlet, sepals ovate to elliptic, obtuse, corolla tube 4–5 cm long, limb 4–5 cm wide, slightly five-lobed. Capsule somewhat longer than the calyx, seeds with long hairs on the edges. Pinelands, south Fla., W.I. All year. *Exogonium microdactylum* (Griseb.) House

4. **I. hederifolia** L.—Twining vines with broadly ovate, acuminate leaves with cordate bases, entire, or coarsely toothed mostly 3–10 cm long. Flowers solitary, or in cymes, pedicels longer than the peduncles, calyx lobes oblong, obtuse, with elongate subulate appendages just below the apex, corolla scarlet or orange red, tube 2–4 cm long, stamens exserted. Disturbed sites, Fla. to Tex., Pa., Mo. All year. *Quamoclit coccinea* authors, *I. coccinea* authors, misapplied.

I. quamoclit L., a native of tropical America, characterized by pinnately dissected leaves, is cultivated in southeast U.S. It may occur in south Fla.

5. **I. sagittata** Lam. GLADES MORNING GLORY—Twining vines with smooth stems. Leaves 3–10 cm long, sagittate, the basal lobes divergent-spreading, segments linear to linear-lanceolate or lanceolate. Flowers solitary or in two- to three-flowered cymes, sepals 6–9 mm long, subequal, corolla 6–7 cm long, corolla limb purple. $2n = 30$. Wet land, glades, solution holes, Fla. to Tex., N.C. on the coastal plain. All year.

6. **I. crassicaulis** (Benth.) Robinson, BUSH MORNING GLORY—Erect stems up to 3 m high, subshrubby, branching. Leaves 10–15 cm long, ovate-acuminate, entire or nearly so. Flowers solitary, peduncles 2–5 cm long, corolla 6–7 cm long, pink purple. Coastal areas, Fla. Keys, naturalized from Brazil. All year. *I. fistulosa* Mart. ex Choisy; *I. leptophylla* of authors misapplied.

Plants are reported to produce large tubers. They may occur as an escape from cultivation in other areas of Fla.

7. **I. hederacea** (L.) Jacq.—Stems twining or climbing up to 3 m long. Leaves 4–12 cm long, ovate-cordate, deeply three-lobed, the lobes narrowly ovate to ovate-entire. Flowers one to three on short peduncles, corolla funnelform, the tube mostly white, the limb light blue, purple, or rose; sepals typically hairy pubescent, lanceolate, with long, usually recurved tips. $2n = 30$. Disturbed sites, land fills, Fla. to Panama, Nebr., Mexico, naturalized from tropical America. All year.

I. nil (L.) Roth, a native of Africa with leaves nearly entire, is also reported as an escape in our area.

8. I. pes-caprae (L.) R. Brown var. **emarginata** Hall, RAILROAD VINE—Trailing vines with very long, angular, glabrous stems. Leaves 2.5–10 cm long, thick, rather succulent, ovate or obovate, orbicular or kidney shaped, notched or two-lobed at apex. Flowers solitary or several flowers in cymes, sepals unequal, ovate, about 1 cm long, corolla funnelform, purple, 4–5 cm long. Ovary two-locular, capsule globose, 1–2 cm long. $2n = 30$. Sandy beaches, coastal areas, Fla. to Tex., Ga., tropical America. All year.

9. I. stolonifera (Cyr.) J. F. Gmel.—Trailing vines with glabrous stems rooting at the node. Leaves 4–10 cm long, rather succulent, highly variable in form, linear to ovate, obtuse or notched at the apex, sometimes deeply three-lobed. Flowers solitary or two- to three-flowered, cymose, sepals equal or nearly so, 1–2 cm long, corolla funnelform, white with purple center, 4–5 cm long. Ovary four-locular, capsule globose, 1.5 cm long. $2n = 30$. Sandy soil, Fla. to Tex., S.C., pantropical.

This species is probably not common in our area.

10. I. acuminata (Vahl) R & S, MORNING GLORY—Twining herbaceous vines with minutely pubescent or glabrescent stems. Leaves usually 6–9 cm long, broadly ovate, entire or three-lobed, cordate at the base. Flowers solitary or in few-flowered cymes, sepals ovate-lanceolate acuminate, 1–2 cm long, glabrous or nearly so, corolla limb purplish to white, 6–8 cm wide. Ovary three-locular, with six ovules. Capsule ovoid, seeds smooth. Various sites, hammocks, disturbed sites, peninsular Fla., Fla. Keys, W.I. All year. *Pharbitis cathartica* (Poir.) Choisy; *I. cathartica* Poir.

Cultivated forms may be densely pubescent.

11. I. tenuissima Choisy, ROCKLAND MORNING GLORY—Stems twining or trailing, leaves usually 3–5 cm long, lanceolate or linear-lanceolate, sagittate, acute or obtused tipped. Flowers solitary or few together, sepals ovate-lanceolate, equal, about 5–6 mm long, corolla limb purple, 3 cm long. Capsule globose. Pinelands, disturbed sites, south Fla., Fla. Keys, W.I. All year.

12. I. triloba L.—Twining or trailing herbaceous vines. Leaves 5–10 cm long, ovate and entire or becoming hastate three-lobed, glabrescent. Flowers solitary or few together, sepals pubescent, oblong to broadly so, 8–10 mm long, acute or acuminate, corolla purple, about 1.5 cm long, capsule subglobose, 7–8 mm wide, often pubescent, seeds glabrous. $2n = 30$. Coastal areas, disturbed sites, south Fla., pantropical. All year.

13. I. trichocarpa Ell.—Twining or trailing herbaceous vines. Leaves 3–12 cm long, ovate and entire or deeply three-lobed, cordate, highly variable. Flowers one to three, funnelform, sepals pubescent, corolla pink or purple, 2–4 cm long. Capsule pubescent, seeds smooth. $2n = 30$. Disturbed sites, sandy soil, Fla. to Tex., S.C. All year. *I. trifida* (HBK) G. Don

14. I. cairica (L.) Sweet—Twining or trailing herbaceous vines. Leaves 5–10 cm long, palmately five- to seven-parted, segments ellipticlanceolate, acute, or acuminate. Flowers solitary or few together, sepals 5–8 mm long, obtuse, corolla purplish, 5–6 cm long, capsule ovoid, about 1 cm long, seeds finely pubescent. $2n = 30$. Disturbed sites, roadsides, south Fla., tropical America, naturalized from Africa. All year.

I. batatas (L.) Lam., the Sweet Potato, may occur locally as an escape in our area. It is an allopolyploid with $2n = 90$.

11. ANISEÌA Choisy

Smooth or nearly smooth, turning or creeping herbs, becoming woody at the base. Leaf blades narrow. Flowers solitary or in reduced cymes, in leaf axils, sepals herbaceous, outer two much larger than inner three, corolla subentire or five-toothed. Ovary two-locular, ovules two per locule, style one, stigma globose, two-lobed. Capsule globose, four-valved, with four seeds. Five species chiefly in the American tropics.

1. A. martinicensis (Jacq.) Choisy—Smooth or glabrescent creeping stems often twining near the tip. Leaves 4–8 cm long, linear-oblong to narrowly lanceolate, bluntly mucronate, and tapering to the base, entire. Flowers axillary, peduncles exceeding the petioles, solitary or two-flowered, two outer sepals large, elliptical, about half as long as the corolla, three inner sepals short, ovate; corolla white. Capsule two-locular, seeds puberulous. Disturbed sites, Dade County, south Fla., naturalized from tropical America. All year. *Ipomoea martinicensis* Meyer

12. JACQUEMÓNTIA Choisy

Herbs or subshrubs usually twining or prostrate and often velutinous or tomentose. Leaves variable, often cordate at the base, entire, lobed, or dentate. Flowers in axillary, umbellate, or capitate cymes or solitary, sepals five, corolla funnelform or campanulate, five-toothed or lobed or almost entire, stamens five, styles simple, included, stigmas two. Ovary two-locular with four ovules. Fruit a globose capsule, four- to eight-valved, four-seeded. About 120 species, chiefly in the American tropics and subtropics.

1. Corolla blue or pinkish; annual . . . . 1. *J. pentantha*
1. Corolla white; perennial.
  2. Corolla over 2 cm wide; capsule obtuse.

3. Calyx becoming 0.2–0.3 cm long . . 2. *J. reclinata*
3. Calyx becoming 0.4–0.5 cm long . . 3. *J. curtissii*
2. Corolla smaller, less than 1.5 cm wide;
  capsule acute . . . . 4. *J. jamaicensis*

1. **J. pentantha** (Jacq.) G. Don—Stems twining and trailing, pubescent, becoming glabrescent, up to 1 m long. Leaves 3–5 cm long or more, ovate to ovate-lanceolate, acute at the apex, subcordate at the base. Calyx lobes equal, corolla light blue to blue, about 2 cm long, lobes slightly angular. Hammocks, Fla. Keys, south Fla., W.I.

2. **J. reclinata** House—Stems prostrate, reclining, or ascending, densely pubescent with matted hairs, becoming glabrescent, up to 2 m long or more. Leaves elliptic to ovate or suborbicular, blades 2–3 cm long or longer, apex rounded or slightly notched. Calyx lobes ovate, subequal, corolla white, 2–3 cm wide. Capsule ovoid, 4–5 mm long. Pinelands, hammocks, south Fla. and W.I. All year.

3. **J. curtissii** Peter ex Hall. f.—Stems up to 1 m, somewhat woody, prostrate, reclining, or erect, glabrous or pubescent with spreading hairs. Leaves 1–2 cm long or longer, elliptic or elliptic-spatulate, blades obtuse or acute pointed. Calyx lobes ovate or suborbicular, subequal, corolla white, 2–3 cm wide. Capsule 5–6 mm long. Pinelands, endemic to south Fla. All year.

The leaves are highly variable in this species.

4. **J. jamaicensis** (Jacq.) Hall. f.—Stems erect or ascending and covered with fine, soft pubescence. Leaves about 5–15 mm long, ovate to narrowly so, obtuse at the base, apex rounded or obtuse. Calyx lobes ovate, 1–2 mm long, corolla white, about 1 cm wide. Capsule acute-ovoid. Hammocks, Fla. Keys, W.I. All year.

161. POLEMONIÀCEAE Phlox Famliy

Annual or perennial herbs, few shrubs or trees, with opposite or alternate leaves. Flowers radially symmetrical or nearly so, bisexual, calyx five-parted, corolla five-lobed, stamens five, epipetalous and alternate with the corolla lobes. Ovary three-locular, superior, disk hypogynous, style one, stigmas three. Fruit a small capsule with numerous seeds. About 15 genera and 300 species, chiefly in North America.

1. PHLÓX L.

Annual or perennial herbs with entire, mostly opposite leaves. Flowers conspicuous in terminal or axillary cymes, calyx tubular, deeply parted into narrow lobes that are separated by a thin membrane, corolla salverform, tube very slender, lobes obovate to nearly orbicular, stamens five with short filaments inserted at different levels within the corolla tube. Ovary small, ovules one to four per locule. Capsule ovoid. About 50 species, mostly in west U.S.

1. **P. drummondii** Hook.—Annuals up to 5 dm tall, erect, simple or usually branching from near the base, stems villous glandular-pubescent. Lower leaves opposite, upper 3–8 cm long, alternate, clasping or narrowed to the base, broadly ovate or obovate to lanceolate. Flowers numerous, showy, in cymose clusters, calyx lobes long, narrow, corolla about 2 cm wide, in various colors and hues, purple, red, pink, white, and others. Disturbed sites, old fields, waste areas, widely distributed as a weed in Fla. Naturalized from Tex. Spring.

Plants often form large, brightly colored patches along the roadside or larger areas in fields and be locally common, especially along the lower Fla. west coast. Many forms have been described; our variety has been described as **P. drummondii** var. **peregrina** Shinners.

162. HYDROPHYLLÀCEAE Waterleaf Family

Perennial, biennial, or annual herbs usually pubescent or scabrous. Leaves basal, opposite or alternate, entire or pinnately or palmately lobed. Flowers bisexual, radially symmetrical, solitary in panicles or often in one-sided cymes; calyx lobes five, usually appendaged between the lobes, corolla fused, five-lobed, stamens five, epipetalous alternate with petals. Ovary superior, one- to two-locular, fruit a capsule. About 18 genera and 250 species, chiefly in North America.

1. Corolla funnelform or salverform; leaves
    spatulate; stems decumbent-spreading   .   1. *Nama*
1. Corolla rotate or campanulate; leaves
    lanceolate-ovate; stems erect .   .   .   .   2. *Hydrolea*

1. NÀMA L.

Prostrate or ascending perennial or annual herbs with alternate entire leaves. Flowers small, axillary, solitary, or paired in leaf axils; calyx lobes narrow, deeply five-parted, becoming somewhat accrescent; corolla five-lobed, usually blue, funnelform to slightly salverform, stamens epipetalous, filaments often unequal, filiform. Styles two, distinct. Ovary two-locular. Fruit a loculicidal, many-seeded capsule. About 35 species, chiefly North America. *Marilaunidium* Kuntze; not *Nama* sensu Small

1. **N. jamaicensis** L.—Perennial herb hirsute with branches 1–3 dm

long, prostrate or spreading. Leaves mostly 2–5 cm long, spatulate, entire. Flowers cymose, calyx lobes linear or nearly so, 6–8 mm long, ciliate along the margins, corolla salverform or funnelform, white or purplish, slightly longer than the calyx. Styles free or nearly so. Capsule about 1 cm long. $2n = 28$. Disturbed sites, roadsides and hammocks, Fla. to Tex. on coastal plain, tropical America. Spring, fall. *Marilaunidium jamaicense* (L.) Kuntze

2. HYDRÒLEA L.

Perennial or annual herbs, sometimes spiny with entire, alternate leaves. Flowers in terminal or axillary cymes; calyx lobes narrow, unequal, enlarging with age, corolla campanulate or rotate, deeply lobed, dilated at the base; corolla rotate to campanulate, five-parted; stamens epipetalous; styles two. Ovary two-locular, ovules many. Capsule globose, subtended by a persistent calyx. About 19 species in the tropical and subtropical regions in aquatic habitats.

1. **H. corymbosum** Macbride ex. Ell. SKY-FLOWER—Perennial herbs up to 6 dm tall with glabrous stems. Leaves elliptic to elliptic-lanceolate, mostly 3–6 cm long, acute tipped. Flowers corymblike, calyx lobes glandular-ciliate along the margins, shorter than the corolla, corolla azure blue, 10–15 mm long. Capsule 4–5 mm long. Wet land, Fla. to S.C. Summer. *Nama corymbosum* (Macbride) Kuntze

163. BORAGINÀCEAE Borage Family

Herbs with alternate or rarely opposite simple leaves, glabrous or often scabrous or hispid. Flowers bisexual, radially or rarely bilaterally symmetrical, often in scorpioid cymes; sepals five, free or partly fused, corolla tubular or funnelform, five-lobed or five-parted, stamens five, alternate with the petals, disk present. Ovary superior, two- to four-locular, entire or deeply four-lobed. Fruit a drupe or consisting of four nutlets. About 100 genera and 2000 species of wide distribution, chiefly in the Mediterranean region. *Ehretiaceae, Heliotropiaceae*

1. Flowers in racemes, cymes, panicles, or head-
    like inflorescence.
    2. Leaves and stems sparingly pubescent or
        glabrous; leaf blades obovate to broadly
        ovate, entire . . . . 1. *Bourreria*
    2. Leaves and stems strigose to scurfy pubes-
        cent; leaf blades lanceolate to broadly

ovate, serrate to undulate . . . . 2. *Cordia*
1. Flowers in one-sided spikes or spikelike racemes.
  3. Herbs or subshrubs; fruit dry . . . . 3. *Heliotropium*
  3. Shrubs or shrubby twiners; fruit fleshy . 4. *Tournefortia*

1. BOURRÈRIA Jacq.

Shrubs or trees with alternate leaves. Flowers in corymbose cymes; sepals partly fused, corolla deeply five-lobed, stamens epipetalous about midway the corolla tube. Ovary entire, fruit a drupe. About 25 species in tropical America.

1. Leaves broadly ovate up to 4–5 cm wide, some conspicuously notched at the tip . 1. *B. ovata*
1. Leaves narrowly ovate to obovate, mostly less than 4 cm wide, leaf tips rounded, obtuse, infrequently notched.
  2. Fruits small, 5–7 mm wide; leaves short petioled, less than 5 mm long . . . 2. *B. cassinifolia*
  2. Fruits larger, about 1 cm wide; leaves with petioles over 5 mm long . . . . . 3. *B. succulenta* var. *revoluta*

1. B. ovata Miers, STRONGBARK—Shrub or tree up to 10 m tall with glabrous stems. Leaves 5–12 cm long, ovate to elliptic, glabrous. Flowers in flat-tipped cymes, calyx 6–7 cm long, lobes triangular; corolla lobes somewhat shorter than the tube, white; stamens exserted, anthers about 1–2 mm long. Drupe 1 cm wide, orange. Hammocks, Fla. Keys, W.I. All year.

2. B. cassinifolia (A. Rich.) Griseb. SMOOTH STRONGBARK—Shrub or low tree up to 3 m tall with elliptic, ovate, or cuneate leaves mostly 1–3 cm long, glabrous, obtuse, becoming rather strongly veined. Flowers few in cymes, calyx 5–6 mm long, lobes ovate or triangular; corolla white, tube about as long or slightly longer than the calyx; anthers about 1–2 mm long. Drupe less than 1 cm wide, depressed. Pinelands, south Fla., W.I. All year.

3. B. succulenta Stahl. var. **revoluta** (HBK) O. E. Schulz, ROUGH STRONGBARK—Shrub or small tree with pubescent stems. Leaves 3–7 cm long, elliptic to ovate, acute, rounded or notched, scabrous and marginally pubescent or glabrous. Flowers several in cymes, calyx 6–7 mm long, lobes ovate; corolla white, tube exceeding the calyx; anthers over 2 mm long. Drupe about 1 cm, orange, barely depressed. Pinelands, hammocks, south Fla., Fla. Keys, W.I. All year.

2. CÓRDIA L.

Trees or shrubs with entire, persistent, petioled leaves and naked buds. Flowers terminal in scorpioid or branched cymes, calyx tubular or campanulate, corolla funnelform. Ovary four-locular, style two-branched above the middle, branches two-parted. Fruit a drupe, partly enclosed in the calyx. About 250 species, pantropical. *Sebesten* Adans.; *Varronia* Jacq.

1. Larger leaves mostly 2–4 cm long, lance-
   ovate; corolla white, 0.5–0.7 cm long . .   1. *C. globosa*
1. Larger leaves mostly 10–12 cm long; corolla
   with reddish lobes, 3–4 cm long . . .   2. *C. sebestena*

1. **C. globosa** (Jacq.) HBK—Shrub usually branched, up to 3 m tall. Leaves 2–4 cm long, elliptic to lance-ovate, serrate. Flowers in scorpioid cymes, calyx mostly 7–8 mm long; corolla 6–7 mm long, white. Drupe ovoid. Hammocks, south Fla., Fla. Keys, W.I. All year. *Varronia globosa* Jacq.

2. **C. sebestena** L. GEIGER TREE—Shrub or tree 10 m tall with rough, dark brown bark, stems with appressed hairs. Leaves mostly 10–12 cm long, ovate or elliptic-ovate, margins undulate or irregular. Flowers in cymes, calyx 1–2 cm long, corolla 3–4 cm long with reddish lobes. Drupe white, ovoid, 2–3 cm long. Hammocks, sandy soil, south Fla., Fla. Keys, W.I., tropical South America. All year. *Sebestena sebestena* Britt. & Small

3. HELIOTRÒPIUM L.   HELIOTROPE

Erect or prostrate herbs or low shrubs with entire leaves. Flowers small, usually in one-sided spikes or racemes or cymose; calyx lobes subequal, corolla funnelform or salverform, stamens epipetalous. Ovary unlobed, style short, stigma capitate. Fruit usually of four one-seeded nutlets or two-lobed and separating into two-seeded carpels. About 150 species, chiefly of tropical and subtropical regions. *Lithococca* Small; *Tiaridium* Lehmann; *Schobera* Scop.

1. Larger leaves lanceolate, 1–4 cm wide . .   1. *H. angiospermum*
1. Larger leaves narrowly lanceolate or spatulate-lanceolate, less than 0.6 cm wide.
   2. Undersurfaces of leaves glabrous or
      nearly so . . . .   2. *H. curassavicum*
   2. Undersurfaces of leaves densely pubescent   3. *H. polyphyllum*

1. **H. angiospermum** Murray—Annuals with stems usually up to

1 m tall or more. Leaves narrow to elliptic-ovate, 3–8 cm long. Flowers in scorpioid spikes; calyx lobes linear to lanceolate, 1–2 mm long; corolla white, tube 1–2 mm long. Fruit about 2 mm wide. $2n = 26$. Coastal hammocks, south Fla., Fla. Keys, tropical America. All year. *Schobera angiosperma* (Murray) Britt.; *H. parviflorum* L.

2. **H. curassavicum** L. SEASIDE HELIOTROPE—Annual or mostly perennial with stems up to 4 dm tall, ascending or forming mats. Leaves lance-linear, spatulate, or ovate, pale green, 3–6 dm long. Flowers in scorpioid spikes, the spikes paired; calyx lobes lanceolate to elliptic-lanceolate, corolla white or bluish, tube about 2 mm long. Fruit ovoid-depressed, about 2 mm wide. $2n = 28$. Sandy seashores, borders of marshes, coastal areas, Fla. to Del., N. Mex., Mexico, tropical America. All year.

3. **H. polyphyllum** Lehmann—Perennial with stems erect or spreading-decumbent. Leaves linear, linear-lanceolate to narrowly elliptic, densely pubescent underneath, chiefly 1–2 cm long. Inflorescence bracteate, calyx lobes ovate-lanceolate, 2–3 mm long, corolla white or yellow, tube 3–4 mm long. Fruit about 1 mm wide. Coastal areas, south Fla., Fla. Keys, South America. Two apparently ecotypic varieties occur in our area:

1. Stems erect, often strict . . . . 3a. var. *polyphyllum*
1. Stems spreading-decumbent or prostrate . 3b. var. *horizontale*

3a. **H. polyphyllum** var. **polyphyllum**, PINELAND HELIOTROPE—Corolla white or yellow; stems up to 1 m tall, often less. Low hammocks, pineland, moist soil. All year. Includes *H. leavenworthii* Torr.

3b. **H. polyphyllum** var. **horizontale** (Small) R. W. Long—Corolla bright yellow; stems 1–6 dm long, growing radially flat on the surface, usually matted. Pinelands, dry soil, endemic to south Fla. All year. *H. horizontale* Small

H. fruticosa L., a low annual with narrow leaves, was collected in Key West in the 19th century, but no recent collections have been seen. *H. phyllostachyum* Torr.

H. indicum L., an annual with hispid stems up to 1 m tall and pale blue flowers, may occur in our area. It is naturalized from the Old World tropics and may occur in Fla. in disturbed sites. *Tiaridium indicum* (L.) Lehmann

4. TOURNEFÒRTIA L.

Trees, shrubs, or vines with alternate entire leaves. Flowers in dense, one-sided cymes or cymes forked and branches elongate; calyx

Family 163. **BORAGINACEAE.** *Tournefortia gnaphalodes:* a, flowering branch, × ½; b, inflorescence, × 3½; c, flower, × 10; d, mature fruit, × 5.

five-lobed, corolla salverform, tube cylindric, lobes broad, spreading; stamens five, included, epipetalous. Ovary four-locular, style simple or two-lobed. Fruit a drupe. About 100 species in tropical and subtropical areas. *Mallotonia* Britt.; *Myriopus* Small

1. Stems and leaves densely pubescent; fruit entire, not lobed.
   2. Leaves silky tomentose, narrowly spatulate, crowded; maritime shrubs . . . 1. *T. gnaphalodes*
   2. Leaves hispid hirsute, ovate, distant; hammock vines or shrubs . . . . . . 2. *T. hirsutissima*
1. Stems and leaves sparingly pubescent to appressed-tomentose; fruit lobed.
   3. Leaves lanceolate-acuminate, undersurfaces appressed tomentose . . . . 3. *T. poliochros*
   3. Leaves ovate-obtuse, undersurfaces glabrous to puberulent . . . . . . 4. *T. volubilis*

1. **T. gnaphalodes** (L.) R. Brown, SEA LAVENDER—Shrub rather fleshy up to 2 m tall or usually less with pale silky tomentose pubescence, much branched and forming dense stands. Stems densely leafy, the leaves 5–10 cm long, linear to spatulate, obtuse. Flowers in dense one-sided cymes, cymes with two to four recurved branches, calyx campanulate, lobes 2–3 mm long, corolla white. Drupe ovoid, black, about 5 mm long. Beaches, south Fla., Fla. Keys, W.I. All year. *Mallotonia gnaphalodes* Britt.

2. **T. hirsutissima** L.—Shrubby or partly woody vine with stems pubescent. Leaves 10–15 cm long, broadly elliptic to ovate, hispid hirsute, distant, tapering gradually to the apex. Flowers in corymbose spikes, many-flowered; sepals ovate-acute, pubescent,~corolla white, lobes 2–3 mm long, ovate to deltoid. Drupe ovoid-globose or subglobose, 4–5 mm long. Hammocks, south Fla., tropical America. All year.

3. **T. poliochros** Spreng.—Slender woody vines up to 2 m long with canescent or rusty pubescence. Leaves 4–6 cm long, ovate, elliptic, or lanceolate, obtuse or acute at the apex, dark green and densely appressed tomentose below. Flowers in few to several secund spikes 4–7 cm long; calyx 1–2 mm long, pubescent, lobes lanceolate or ovate-lanceolate. Drupe 3–4 mm wide. Hammocks, south Fla., W.I. All year. *Myriopus poliochros* (Spreng.) Small

4. **T. volubilis** R & S—Slender woody vines often up to 3 m long or more, stems rusty canescent or glabrate. Leaves 3–7 cm long, ovate to oblong-lanceolate, acute or acuminate at the apex or obtuse, under-surfaces glabrous to puberulent. Flowers in several slender secund spikes, mostly 3–4 cm long; calyx 1 mm long, lobes ovate-lanceolate, corolla greenish white, about 2–3 mm long, lobes linear-subulate. Drupe

2–3 mm wide. Hammocks, south Fla., Fla. Keys, south Tex., W.I., tropical America. All year. *Myriopus volubilis* (L.) Small

164. AVICENNIÀCEAE   BLACK MANGROVE FAMILY

Shrubs or trees of maritime habitats producing numerous characteristic pneumatophores. Leaves opposite, decussate, persistent, entire, and thick textured. Flowers bisexual, radially symmetrical in axillary or terminal cymes or spicate clusters, opposite; sepals five, corolla four-parted, campanulate-rotate, stamens four, epipetalous. Ovary superior, one-locular with free central placentation, ovules four. Fruit a compressed oblique capsule, pubescent, somewhat fleshy, one-seeded. A single genus, tropics and subtropics in saline or brackish coastal areas.

1. AVICÉNNIA L.   BLACK MANGROVE

Characters of the family. About 16 species, pantropical, with a single species occurring in south Fla.

1. **A. germinans** (L.)—Trees up to 20 m tall with dark, scaly bark, young stems finely pubescent. Leaves 3–12 cm long, oblong or lanceolate-elliptic, dark green and often shiny above, tomentulose beneath, margins entire and often slightly revolute, lateral veins uniting at the edge. Flowers clustered in few-flowered spikes, calyx 3–4 mm long, corolla 1.5 cm wide, white. Capsule 3–5 cm long, ellipsoidal, acute. Coastal swamps, Fla., La., tropical America, west Africa. All year. *A. nitida* Jacq.

One of the most common trees in the coastal swamps of south Fla., characterized by the production of long horizontal roots that have short, vertical aerating branches (pneumatophores) that form a close network in the moist saline soil.

165. VERBENÀCEAE   VERBENA FAMILY

Trees, shrubs, or herbs with opposite, simple or compound leaves. Flowers bisexual, usually in terminal heads or spikes, or in axillary cymes; calyx mostly four- to five-toothed, campanulate or tubular, corolla four- to five-lobed, generally two-lipped or bilaterally symmetrical, disk present, stamens four, didynamous, rarely five or two. Ovary superior two- to five-locular, entire or two- to four-lobed. Fruit dry, a berry or drupelike, often splitting when mature into two to four one-seeded nutlets. About 100 genera and 2600 species of wide distribution, primarily in the tropics and warm temperate regions.

1. Trees or shrubs usually more than 1 m high, stems woody.
  2. Margins of some or all the leaves serrate or dentate.
    3. Flowers in heads or flat-topped spikes .   1. *Lantana*
    3. Flowers in axillary or terminal cymes.
      4. Inflorescence an axillary cyme, flowers less than 5 mm wide; leaf margins finely to coarsely serrate . .   2. *Callicarpa*
      4. Inflorescence a terminal cyme, flowers over 2.0 cm wide; leaf margins coarsely to broadly dentate . .   3. *Clerodendrum*
  2. Margins of leaves or leaflets mainly entire.
    5. Leaves simple; flowers in drooping axillary racemes . . . . . .   4. *Citharexylum*
    5. Leaves trifoliolate or palmately compound; flowers in erect, branching, or panicled cymes . . . . .   5. *Vitex*
1. Herbs or low shrubs, usually less than 1 m high, stems not woody or woody only below.
  6. Flowers in heads or flat-topped spikes .   1. *Lantana*
  6. Flowers in racemes, compacted or elongate spikes.
    7. Inflorescence a terminal, axillary, or spicate raceme.
      8. Corolla five-lobed or slightly two-lipped.
        9. Fruit fleshy; corolla salverform 1 cm or more long . . . .   6. *Duranta*
        9. Fruit dry; corolla tubular 0.5–0.7 cm long . . . . .   7. *Priva*
      8. Corolla two-lipped . . . .   1. *Lantana*
    7. Inflorescence an elongate or compacted spike.
      10. Flowers in long, quill-like spikes, mostly over 20 cm long; flowers embedded in stout peduncle . .   8. *Stachytarpheta*
      10. Flowers in spikes, less than 20 cm long; flowers not embedded in peduncle.
        11. Calyx five-lobed, erect, reclining, or procumbent herbs . . .   9. *Verbena*
        11. Calyx two-lobed, creeping or procumbent herbs . . . .   10. *Lippia*

1. LANTÀNA L. Lantanas

Shrubs or sometimes herbs, occasionally prostrate or trailing with stems often prickly. Leaves pubescent and mostly serrate. Flowers in

Family 165. **VERBENACEAE.** *Lantana ovatifolia* var. *ovatifolia:* a, flowering branch, × ½; b, flower, × 7; c, mature fruit, × 4½.

axillary spikes, dense terminal heads, or in flat-topped spikes; calyx membranous, corolla four- to five-lobed, tube slender, limb more or less two-lipped, stamens four, didynamous, epipetalous about the middle of the tube. Ovary two-locular. Fruit a small, fleshy drupe with two nutlets. About 70 species, chiefly in the tropics and warm temperate regions; several are escapes from cultivation in our area.

1. Flowers in dense axillary spikes . . . . 1. *L. microcephala*
1. Flowers in heads or flat-topped spikes.
  2. Stems with small prickles . . . . . 2. *L. camara*
  2. Stems without prickles or nearly so.
    3. Heads with involucral bracts.
      4. Leaf margins crenate-serrate; leaf tips obtuse.
        5. Corolla tube 2–5 mm long, white or pale blue . . . . . . 3. *L. involucrata*
        5. Corolla tube 8–20 mm long, dark red or lilac . . . . . . 4. *L. montevidensis*
      4. Leaf margins entire to obscurely serrulate; leaf tips acute . . . . 1. *L. microcephala*
    3. Heads with bracts but these not formed as an involucre . . . . 5. *L. ovatifolia*

1. **L. microcephala** A. Rich.—Stems erect, pale gray or whitish. Leaves 3–6 cm long, narrowly lanceolate to elliptic-lanceolate, pubescent, serrate. Flowers in depressed, axillary spikes, peduncles about 4–5 mm long, bracts ovate, calyx about 1 mm long; corolla limb about 3 mm wide. Nutlets about 1 mm wide. Hammocks, south Fla., W.I. All year. *Goniostachyum citrosum* Small; *L. citrosa* (Small) Moldenke

2. **L. camara** L. SHRUB VERBENA—Shrub up to 2 m tall, smooth or usually armed with small or stout, often recurved prickles. Leave 3–12 cm long, ovate, acute acuminate, crenate-dentate, rugose or scabrous above, strigose beneath. Flowers in dense heads or flat-topped spikes or stout axillary peduncles often longer than the leaves, bracts narrow about half as long as the corolla, corolla whitish, yellow, or pink, changing to orange or scarlet, limb about 6–7 mm long. Pinelands, coastal plain, Fla. to Tex., Ga., tropical America. All year. *L. aculeata* L.

This is a highly variable species and several varieties have been described: var. **aculeata** Moldenke has stout, recurved prickles; var. **mista** Bailey has flowers in globular heads that change from white to rose or blue. All may occur in our area.

3. **L. involucrata** L.—Shrub much branched up to 1 m tall or more, stems without prickles or nearly so. Leaves 2–4 cm long, narrowly elliptic to ovate, crenate or serrate, apex obtuse, scabrous-pubescent above,

pubescent beneath. Flowers in involucrate heads on peduncles 2–4 cm long, bracts ovate or lance-ovate, 3–6 mm long, corolla white or bluish white, tube 7–8 mm long. Drupes about 3–4 mm wide. $2n = 36$. Pinelands, hammocks, south Fla., Fla. Keys, south Tex., tropical America. All year. *L. odorata* L.

4. L. montevidensis (Spreng.) Briq. WEEPING OR TRAILING LANTANA—Shrub up to 2 m tall, with trailing or decumbent pubescent stems. Leaves mostly 2–3 cm long, ovate, base cuneate or truncate, dentate or crenate. Flowers in heads on elongate peduncles, red to lilac, corolla tubes 8–20 mm long. Drupes ovoid, about 5 mm long. Disturbed sites, pinelands, roadsides, south Fla., W.I., naturalized from South America. All year. *L. sellowiana* Link & Otto

5. L. ovatifolia Britt.—Shrubs erect or reclining, 2 m tall or less, often diffusely spreading, stems up to 7 dm long, somewhat four-angled, hirsute. Leaves 2–9 cm long, ovate, rather thick, mostly scabrous above, hirsute beneath, apex acute, shallowly crenate. Flowers in heads on slender axillary peduncles about 4–5 cm long, bracts not forming an involucre, corolla tube 5–6 mm long, yellow, orange to red, limb yellow. Drupe subglobose. Two well-marked varieties occur in our area:

1. Stems erect; leaves up to 9 cm long .   .   .   5a. var. *ovatifolia*
1. Stems decumbent or reclining; leaves up to
   4 cm long .   .   .   .   5b. var. *reclinata*

5a. L. ovatifolia var. **ovatifolia**—Shrubs erect, often diffusely spreading. Leaves rather thick, scabrous above, hirsute beneath, shallowly crenate. Hammocks, south Fla., W.I. All year.

5b. L. ovatifolia var. **reclinata** R. W. Long—Shrubs with reclining or decumbent stems. Leaves elliptic to ovate, serrate or crenate, scabrous, pubescent beneath. Pinelands, endemic to south Fla. All year. *L. depressa* Small

L. tiliaefolia Cham., with coarsely hirsute stems and leaves and orange or red flowers, may occur as an escape from cultivation in our area.

2. CALLICÁRPA L.

Trees or shrubs with scurfy pubescence and simple leaves. Flowers small, in axillary or terminal cymes; calyx four- to five-lobed or nearly truncate, corolla nearly radially symmetrical, funnelform or tubular, usually four-lobed, stamens four, anthers opening at the apex. Ovary four-locular, style thickened upward. Fruit a fleshy drupe with a hard endocarp, spreading into two to four nutlets. About 140 species in the tropics and subtropics.

1. C. americana L. BEAUTY-BERRY—Shrub up to 2 m tall, with stellate-pubescent stems. Leaves opposite, ovate-oblong or elliptic, finely to coarsely serrate, acuminate, whitish tomentose beneath, especially on younger blades. Flowers in axillary or terminal many-flowered cymes, calyx obscurely four-toothed, corolla bluish, lobes apiculate. Drupes 4–5 mm wide, densely clustered together. $2n = 36$. Pinelands, Fla. to Tex., Md., Okla., Mexico, W.I. Spring, fall.

3. CLERODÉNDRUM L. GLORY-BOWERS

Trees, shrubs, vines, or herbs with simple, opposite leaves. Flowers in terminal cymes; calyx distinctly five-lobed; corolla brightly colored or white, funnelform or salverform, five-lobed, tube slightly curved; stamens four, didynamous, exserted. Ovary imperfectly four-locular with four ovules. Fruit a globose or obovoid drupe. About 400 species in the tropics and subtropics. *Siphonanthus* L.

1. Corolla tube about 3 cm long or less; leaves
broadly ovate.
  2. Calyx short, about 5–8 mm long, flowers
    in erect panicles . . . . 1. *C. speciosissimum*
  2. Calyx long, spreading, about 1 cm long or
    more, flowers in dense corymbs . . . 2. *C. philippinum*
1. Corolla tube about 10 cm long or more;
leaves lanceolate . . . . 3. *C. indicum*

1. C. speciosissimum Geert—Shrub or small tree up to 4 m tall. Leaves often up to 30 cm long, ovate to cordate, entire or toothed, acute acuminate, densely pubescent. Flowers in erect compound cymes sometimes 40 cm long; calyx 4–6 mm long, five-toothed, corolla 2–3 cm long, scarlet. Drupe ovoid. Disturbed sites, vacant lots, south Fla., Fla. Keys, W.I. All year *C. fallax* Lindl.

2. C. philippinum Schauer—Shrub up to 3 m tall. Leaves 10–25 cm long, broadly ovate to deltoid-ovate, coarsely dentate, apex acute, truncate or somewhat cordate at the base, green above, densely pubescent beneath. Flowers in cymes, calyx elongate, 15–20 mm long, lobes subulate, corolla 22–25 mm long, white, rose, or bluish tinged, fragrant. $2n = 52$. Disturbed sites, vacant lots, Fla., Fla. Keys, W.I., tropical America, naturalized from Asia. All year.

Flowers may occur in normal form or "double"-flowered.

3. C. indicum Kuntze—Shrub up to 3 m tall. Leaves about 12–15 cm long, usually whorled, entire, smooth, lanceolate to lance-oblong. Flowers in axillary cymes or panicles, closing in the morning, calyx lobes mostly 8–10 cm long, ovate, corolla white, tube about 10 cm long or more. Fruit red or purple. Fence rows, disturbed sites, locally escaped in southeast U.S., naturalized from east India. All year.

C. bungei Steud. (*C. foetidum* Bunge), resembling **C. speciosissimum** except for having smaller flowers and leaves smooth above, and **C. glabrum** E. Mey., with small, whitish or pinkish flowers, have also been reported as escapes in our area and may be locally established.

4. CITHARÉXYLUM L.

Trees or shrubs with coriaceous, shiny, alternate leaves and somewhat angled stems. Flowers small in axillary or terminal spikes or slender racemes; calyx campanulate, obscurely five-lobed; corolla salverform, limb five-lobed, somewhat oblique; stamens four to five, epipetalous, one usually sterile or rudimentary, included. Ovary incompletely four-locular, stigma two-lobed. Fruit a berrylike drupe, the stone later separating into two nutlets. About 112 species with numerous varieties and hybrids, chiefly in tropical America.

1. **C. fruticosum** L. FIDDLEWOOD—Trees up to 10 m tall usually with smooth light brown bark, stems slender and angled. Leaves 5–14 cm long, oblong-obovate to broadly lanceolate, variable, with slightly revolute margins, shiny above, dull green beneath. Flowers fragrant, in drooping axillary pubescent racemes; corolla white 2–3 mm wide. Drupe reddish brown, subglobose. Hammocks, pineland, south Fla., Fla. Keys, W.I. All year.

5. VÌTEX L.

Trees or shrubs with trifoliolate or palmately compound leaves, rarely unifoliolate. Leaves deciduous or persistent. Flowers bisexual, few to many in cymes or panicles, calyx five-lobed, campanulate, corolla five-lobed, tubular or funnelform, slightly two-lipped, stamens four, didynamous. Ovary superior. Fruit a small drupe. About 100 species in the tropics and subtropics.

1. Leaves chiefly trifoliolate, leaflets sessile or
    subsessile . . . . 1. *V. trifolia*
1. Leaves palmately compound, mostly five-
    foliolate, leaflets with short stalks . . . 2. *V. agnus-castus*

1. **V. trifolia** L. f.—Shrub up to 3 m tall, erect or sometimes prostrate, upper stems appressed pubescent. Leaves trifoliolate or sometimes unifoliolate, sessile or subsessile, oblong-lanceolate or obovate, blades 5–7 cm long, entire, smooth and dark green above, gray white tomentulose beneath. Flowers in terminal panicles 8–12 cm long, corolla

bluish with white mark near the base of the lip. Disturbed sites, especially near the coast, south Fla., naturalized from Asia. Spring, summer.

Var. **variegata** Moldenke with leaves white along the margins may also be found in cultivation and is possibly a local escape.

2. V. agnus-castus L. CHASTE TREE—Shrub or low tree up to 4 m tall, stems grayish tomentulose with a strong, aromatic odor. Leaves long petioled, palmately compound with five to seven leaflets, blades mostly 4–8 cm long, lanceolate-acuminate, subentire, dark green above, grayish tomentulose beneath. Flowers in dense cymes arranged in terminal or axillary spikes 8–10 cm long, corolla gray blue without, lilac or pale blue within, about 8–10 mm long, style exserted. Hammocks, disturbed sites, south Fla., naturalized from Europe. Spring, summer.

A single specimen identified as **V. glabrata** R. Brown, with smooth trifoliolate leaves and axillary inflorescence, has been collected from the vicinity of Miami. It is also a native of Asia.

6. DURÁNTA L.

Shrubs or small trees, often spiny. Leaves whorled or opposite. Flowers small, in short or elongated terminal or axillary racemes; calyx tubular or campanulate, obscurely five-lobed, corolla funnelform or salverform, limb spreading, oblique or five-lobed, stamens four, didynamous, included. Ovary partially eight-locular, stigma unequally four-lobed. Fruit a drupe with four nutlets. About eight species of tropical America.

1. D. repens L. GOLDEN DEWDROP—Shrub or low tree up to 6 m tall, unarmed or spiny, stems slender and often drooping. Leaves 1.5–5 cm long, ovate-elliptic, numerous, coarsely serrate to occasionally entire, apex acute or obtuse, short petiolate. Flowers in recurving racemes 6–15 cm long; calyx 3–4 mm long, acute lobed, corolla lilac, limb 8–9 mm wide, ciliolate. Drupe yellow, globose, 7–12 mm in diameter, enclosed by the yellowish, beaked calyx. Hammocks, pinelands, south Fla., Fla. Keys, W.I., tropical America.

The white flowered form has been named **D. repens** var. **alba** (Mast.) Bailey and has been reported from the lower Fla. Keys. Both the lilac and the white flowered varieties are found in cultivation.

7. PRÌVA Adans.

Erect perennial herbs with opposite, serrate leaves. Flowers in slender spikes or racemes; calyx tubular, five-lobed, larger in fruit, corolla salverform or narrowly cylindric, limb spreading, somewhat two-lipped, five-lobed, stamens four, didynamous, included. Ovary two-locular, two ovules in each locule. Fruit dry, enclosed by calyx, separating into two nutlets. About ten species in the tropics.

1. **P. lappulacea** (L.) Pers. VELVET BURR—Stems up to 6 dm tall, pubescent, branching. Leaves 3–10 cm long, ovate, acute acuminate, serrate. Racemes 5–20 cm long, calyx 2–3 mm long, cylindric, obscurely lobed, pubescent, corolla salverform with rounded lobes. Fruit 5–6 mm long, ovoid-pyramidal, nutlets tuberculate 3–4 mm long, included in the calyx. Disturbed sites, south Fla., Fla. Keys, W.I., Mexico, tropical America.

8. STACHYTARPHÈTA Vahl

Herbs or shrubs with opposite or alternate, serrate or dentate leaves. Flowers sessile in axils of bracts or embedded in cavities of a thick, stout, spicate rachis; calyx membranous, five-lobed; corolla limb five-lobed, stamens two, staminodes two. Ovary two-locular, ovules solitary within each locule. Fruit separating into two nutlets, included in the calyx. About 40 species in tropical and subtropical America. *Valerianoides* Medicus

1. **S. jamaicensis** (L.) Vahl, BLUE PORTERWEED—Glabrous herb or shrub up to 2 m tall but usually less, with spreading or decumbent, branching, four-angled stems. Leaves mostly 2–8 cm long, lanceolate to ovate-lanceolate, punctate, serrate above the middle, acute at apex, acuminate to the base. Spikes quill-like, mostly 10–15 cm long, bracts lanceolate, corolla blue violet, 8–10 mm long, limb 7–8 mm wide. Fruit obpyriform, 5–6 mm long. Moist or dry soil, shell mounds, beaches, disturbed sites, roadsides, south Fla., Fla. Keys, W.I., pantropical. *Valerianoides jamaicensis* (L.) Kuntze

9. VERBÈNA L. VERVAIN

Shrubs or herbs with opposite leaves. Flowers in dense or loose usually slender spike terminating the stems; calyx tubular, unequally five-toothed, corolla salverform or funnelform, limb obscurely two-lipped, five-lobed, stamens four, included. Ovary four-locular, often shallowly four-lobed, style two-lobed. Fruit composed of four nutlets. About 230 species in tropical and subtropical regions. *Glandularia* J. F. Gmel.

1. Corolla tube 1–2 cm long; leaves dentate,
   incised, or deeply toothed or lobed, sparingly pubescent or smooth.

2. Calyx lobe tip subulate, up to 1 mm long;
   leaves toothed, lobed, or crenate-lobed   1. *V. maritima*
2. Calyx lobe tip bristlelike, up to 3 mm
   long; leaves deeply toothed or incised,
   sharply lobed . . . .   2. *V. tampensis*
1. Corolla tube less than 0.5 cm long; leaves
   serrate or pinnate-dissected, glabrous to
   hirsute or strigose.
3. Leaf blades pinnate-dissected, segments
   linear to subulate . . . . . .   3. *V. tenuisecta*
3. Leaf blades lanceolate to lance-ovate,
   serrate or toothed.
4. Leaves clasping or sessile; inflorescence
   a crowded spike . . . . . .   4. *V. bonariensis*
4. Leaves petioled; inflorescence an elongate, narrow spike . . . . . .   5. *V. scabra*

1. **V. maritima** Small—Stems and branches diffuse and creeping, mostly 10–20 dm long, somewhat short pubescent. Leaves 2–4 dm long, ovate, obovate-orbicular, or cuneate, incised or deeply toothed or lobed, sparingly pubescent to glabrate. Flowers in slender spikes, calyx about 1 cm long, lobes lanceolate or subulate, with tip up to 1 mm long, corolla tube about 2 cm long, rose purplish. Nutlets about 4 mm long. Pinelands, coastal dunes, endemic to south Fla. All year. *Glandularia maritima* Small

2. **V. tampensis** Nash—Stems and branches usually diffuse and erect, ascending or decumbent, 3–6 dm long, glabrate or sparingly pubescent. Leaves 2–8 cm long, lanceolate to ovate-lanceolate, elliptic or rhombic-ovate, toothed or incised or sharply lobed, serrate. Flowers in short spikes or corymbose spikes, calyx about 10–15 mm long, the longer lobes bristle tipped up to 3 mm long, corolla tube about 1.5 cm long, purple. Nutlets about 4 mm long, pitted. Hammocks, sandy soil, disturbed sites, endemic to peninsular Fla. Spring, fall. *Glandularia tampensis* (Nash) Small

3. **V. tenuisecta** Briq.—Stems and branches decumbent or ascending, 1–4 dm long, short pubescent. Leaves 1–4 cm long, pinnate-dissected, segments linear to subulate. Flowers in short spikes or corymbose spikes, calyx about 1 cm long, lobes short setaceous, corolla tube about 3–5 mm long, rose purple, white, or pink. Nutlets about 3 mm long, pitted. Roadsides, disturbed sites, sandy soil, Fla. to Ga., La., naturalized from South America. All year. *Glandularia tenuisecta* (Briq.) Small

4. **V. bonariensis** L.—Stems erect to 2 m tall, usually less,. villous

hirsute above. Leaves elliptic to elliptic-lanceolate, clasping or sessile, serrate at least above the middle of the blade, hirsute or strigose pubescent beneath. Spikes densely flowered, calyx becoming 3–4 mm long; corolla purplish or occasionally white, limb about 2 mm wide. Nutlets narrowly ellipsoidal, 2 mm long. Disturbed sites, roadsides, coastal and low pineland, coastal plain, Fla. to La., W.I., naturalized from South America. Spring, summer.

5. V. scabra Vahl—Erect herb or shrub with a main stem up to 1.5 m tall or more. Leaves 5–15 cm long, ovate to elongate-ovate, petiolate, serrate, very scabrous above, less so beneath. Spikes elongate, branched, calyx up to 2 mm long, corolla white, pink, or blue, limb about 2–3 mm wide. Nutlets 1–1.5 m, reticulate above, within the mature calyx 2–3 mm long, hispidulous. Low pineland, sandy soil, Fla. to Va., Tex., tropical America.

10. LÍPPIA L.   Capeweed

Herbs or shrubs often with creeping or procumbent stems and opposite or rarely alternate leaves. Flowers in dense heads or axillary spikes, bracts cuneate or suborbicular; calyx small, two- to four-lobed; corolla two-lipped, four-lobed or cleft, tube very slender, stamens four, didynamous. Ovary two-locular, ovules one in each chamber. Fruit separating into four nutlets. About 100 species, chiefly in tropical America. *Phyla* Lour.

1. Leaves narrowly lanceolate, acuminate to
   petiole, rugose; lower stems woody . .   1. *L. stoechadifolia*
1. Leaves ovate, cuneate to petiole, not rugose;
   stems herbaceous.
   2. Leaves ovate, whitish canescent when
      young, margins evenly serrate . . .   2. *L. geminata*
   2. Leaves obovate, pilose, serrate only above
      the middle . . . .   3. *L. nodiflora*

1. L. stoechadifolia HBK—Shrubs suberect or ascending up to 0.5 m tall with appressed pubescence. Leaves 3–6 cm long, narrowly lanceolate or ioblanceolate, evenly serrate, conspicuously pinnately veined, rugose, acuminate or tapered to the short petiole. Heads globose, or oblong-cylindric on axillary peduncles, bracts suborbicular, calyx two-lobed, corolla purplish, mostly 3 mm long. Fruit about 2 mm long. Glades, low pineland, south Fla., Fla. Keys, W.I., Mexico. All year. *Phyla stoechadifolia* (L.) Small

2. L. geminata HBK—Herbs or sometimes shrubs with numerous slender branches, densely pubescent and aromatic with a mintlike odor.

Leaves 2–7 cm long, ovate or obovate, cuneate to the petiole, evenly serrate, young leaves whitish canescent. Heads subglobose or short-oblong, on axillary peduncles, bracts about 3 mm long, ovate-puberulent, calyx two-lobed, corolla purplish, violet, or white. Fruit about 2–3 mm long. Disturbed sites, low ground, Key West, W.I. All year.

3. L. nodiflora Michx.—Herbs prostrate with creeping stems or branches ascending, up to 1 m long, rather densely puberulent. Leaves up to 6 cm long, obovate, spatulate or oblanceolate, pilose, serrate above the middle of the blade. Heads subglobose, becoming cylindric, very densely many-flowered, bracts closely imbricate, corolla purplish to white, 2–2.5 mm long. Fruit about 1 m long. Hammocks, beaches, disturbed sites, low areas, Fla. to Tex., N.C., W.I., tropical America. All year. *Phyla nodiflora* (L.) Greene

166. LAMIÀCEAE Mint Family

Herbs or shrubs mostly with four-angled stems and opposite, aromatic leaves. Flowers bisexual, bilaterally symmetrical, chiefly in axillary cymose clusters, often aggregated in terminal racemes or spikes, calyx equally five-lobed or cleft or oblique, corolla usually more or less two-lipped, stamens four, epipetalous, didynamous, or only two. Ovary four-lobed, style two-lobed. Fruit of four nutlets or achenes in the base of the persistent calyx. About 160 genera and 3200 species of wide distribution. Nomen alt. Labiatae

1. Ovary of four fused carpels, merely four-lobed, although sometimes deeply so; nutlets laterally attached.
  2. Stamens and style straight or nearly so; undersurfaces of leaves and bracts densely white pubescent . . . . .   1. *Teucrium*
  2. Stamens and style curved or coiled; undersurfaces of leaves and bracts glabrous to hirsute but not densely pubescent . . . .   2. *Trichostema*
1. Ovary of four distinct or nearly distinct carpels, deeply four-lobed; nutlets basally attached.
  3. Calyx with a distinct, transverse protuberance on the upper side . . .   3. *Scutellaria*
  3. Calyx without a transverse protuberance.
    4. Corolla strongly two-lipped with the upper lip sometimes more or less concave.
      5. Calyx strongly two-lipped.

6. Calyx five-lobed, the upper lip
three-lobed, broad, the lower lip
two-lobed, not bristle tipped .   4. *Prunella*
6. Calyx eight- to ten-lobed, the lobes
unequal, bristle tipped . . .   5. *Leonotis*
5. Calyx weakly two-lipped or nearly
radially symmetrical.
7. Anther-bearing stamens two . .   6. *Salvia*
7. Anther-bearing stamens four.
8. Stamens ascending along the upper lip of the corolla; flowers
mostly 1.5–3.0 cm long . .   7. *Physostegia*
8. Stamens descending along lower
lip of corolla; flowers usually
less than 1.0 cm long . . .   8. *Hyptis*
4. Corolla weakly two-lipped or nearly
radially symmetrical.
9. Stamens ascending under the upper
lip of the corolla; leaves linear-
lanceolate, mostly less than 1 cm
long, sessile.
10. Corolla tube straight or nearly so,
1.0–1.5 cm long . . . . .   9. *Satureja*
10. Corolla tube curved upwardly, 1.5–
2.0 cm long . . . . .   10. *Conradina*
9. Stamens descending along the lower
lip of the corolla; leaves broadly
lanceolate to ovate, petioled . .   11. *Ocimum*

1. TEÙCRIUM L.   Wood Sage

Perennial herbs with serrate or dentate leaves. Flowers in racemes
or panicles, calyx five-toothed, two-lipped, corolla with four upper
lobes nearly equal, lower lobe much larger, stamens four, exserted,
lying against the upper lobes. Ovary deeply four-lobed, nutlets smooth.
About 100 species of wide distribution.

1. **T. canadense** L. var. **hypoleucum** Griseb.—Stems up to 1 m tall
or more. Leaves 5–15 cm long, lanceolate to elliptic-lanceolate, whitish
pubescent beneath, serrate, acuminate. Flowers in branching racemes,
calyx becoming 7 mm long, oblique, corolla pinkish, about 1.5–2 cm
long. Low ground, hammocks, Fla. to Tex., S.C. on the coastal plain,
W.I. All year. *T. nashii* Kearney

T. canadense L. var. **canadense**, with spicate racemes and densely
pubescent calyx, has been reported for our area and may occur locally.
T. littorale Bicknell

2. TRICHOSTÈMA L. BLUE CURLS

Herbs usually glandular-pubescent with branched stems and entire leaves. Flowers axillary, terminating the stems and branches, calyx campanulate, deeply five-toothed, oblique, the upper three lobes elongate, the lower two very short; corolla lobes unequal, upper four ascending, stamens with long, curved, thin filaments. Ovary deeply lobed, nutlets reticulate. About ten species in North America.

1. Perennial; some or all the leaves obovate,
   usually less than 2 cm long . . . . . 1. *T. suffrutescens*
1. Annual; leaves lanceolate to lance-ovate,
   larger leaves usually over 2 cm long . . 2. *T. dichotomum*

1. T. suffrutescens Kearney—Perennial or biennial, with woody roots, stems up to 4 dm tall, bushy, pubescent. Leaves 1.5–2.0 cm long or more, lanceolate to lance-ovate or obovate. Flowers in racemes or bushy, paniclelike inflorescence, calyx becoming 5–6 mm long, corolla blue, mostly 6–8 mm long. Nutlets about 1 mm long. Pineland, endemic to Fla. Spring.

2. T. dichotomum L.—Annual; stems up to 8 dm tall, bushy and viscid pubescent with glandular trichomes. Leaves 2–7 cm long, oblong, obovate, to elliptic-lanceolate. Flowers solitary or few in upper axils to racemose or paniculate calyx becoming 5–6 mm long, lobes acuminate, corolla blue to bluish white or white, about 5 mm long. Nutlets 1–3 mm long. Dry soil, Fla. to Tex., Va., Mo.

3. SCUTELLÀRIA L. SKULLCAP

Perennial herbs or subshrubs usually with one-sided axillary racemes or solitary or terminal spikes or racemes. Calyx campanulate, becoming two-lipped, the upper with a crest or bump on its back, usually falling away in fruit; corolla recurved, elongate, expanded above the throat, upper lip arching, lower lip deflexed or spreading, stamens four, didynamous, anthers ciliate. Ovary deeply four-lobed, nutlets papillose or tuberculate. About 200 species, widely distributed.

1. S. havanensis Jacq.—Herbs with slender stems pubescent, usually branched from near the base, erect, up to 3 dm tall. Leaves ovate or orbicular-ovate, few-toothed to subentire. Flowers solitary, on short peduncles in upper axils, calyx about 1.5 mm long, corolla dark blue, about 1.5 cm long, lower lip 7–8 mm wide, three-lobed. Nutlets 1 mm long, fruiting calyx about 3 mm long. Pinelands, Fla., W.I. All year.

4. PRUNÉLLA L. SELF-HEAL

Perennial herbs with usually simple stems. Flowers in a dense spike or head, consisting of three-flowered clusters sessile in the axils of imbricate bractlike floral leaves, calyx campanulate-cylindric, upper lip broad, truncate, corolla ascending, two-lipped, the upper erect, arched, lower reflexed, spreading, three-cleft, filaments two-toothed at the apex, lower one bearing on another. Nutlets smooth. Five species native to the Old World.

1. **P. vulgaris** L. HEAL-ALL—Erect, tufted or diffusely ascending herbs up to 6 dm tall. Leaves mostly 2–8 cm long, few, lanceolate, elliptic to ovate, undulate. Flowering spikes mostly 2–6 cm long, floral bracts very ciliate, suborbicular or reniform, calyx 8–11 mm long, the upper lip somewhat mucronate, corolla blue or purple, sometimes whitish, usually 1.5–2.0 cm long. Nutlets about 2 mm long. Disturbed sites, roadsides, North America, naturalized from Europe. Spring, fall.

Several forms and varieties have been described for this species on the basis of pubescence and of pigmentation of corolla and calyx. These other forms are of doubtful taxonomic value.

5. LEONÒTIS R. Brown

Herbs or subshrubs with opposite, broad, dentate or serrate leaves. Flowers in rather dense globular whorls or cymes, calyx oblique, eight- to ten-lobed, bristle tipped, corolla two-lipped, upper erect, lower three-lobed, stamens four, filaments minutely puberulent. Nutlets three-angled, smooth. About 12 species in Africa. *Leonotis* L.

1. **L. nepetaefolia** (L.) R. Brown—Annual with simple or branched stems up to 2 m tall, softly pubescent. Leaves 5–12 cm long, ovate to ovate-deltoid, dentate or crenate, tapering or subcordate at the base. Flowers in dense clusters, 4–6 cm wide, calyx pubescent, lobes bristle tipped, corolla scarlet or orange yellow, 2–3 cm long, villous pubescent. Nutlets about 3 mm long, sharply angled. $2n = 28$. Disturbed sites, roadsides, Fla. to Tex., N.C., naturalized from South Africa. Summer, fall.

6. SÁLVIA L. SAGE

Herbs or shrubs with clustered flowers arranged in spikes, racemes, or panicles. Calyx two-lipped, upper lip three-toothed or entire, lower two-cleft or two-toothed, corolla strongly two-lipped, the upper lip entire or two-lobed, lower spreading, three-lobed or three-cleft, anther-bearing stamens two, ascending under the upper lip. Nutlets smooth. Over 500 species of wide distribution.

1. Leaves mostly basal, some deeply lobed or
      pinnatifid . . . .   1. *S. lyrata*
1. Leaves chiefly cauline, not lobed or pin-
      natifid.
     2. Corolla red or scarlet . . . . . .   2. *S. coccinea*
     2. Corolla blue or white.
       3. Calyx lobes over 5 mm long when
         plant in fruit.
         4. Calyx lobes awn tipped . . . .   3. *S. privoides*
         4. Calyx lobes merely acute, not awn
           tipped . . . .   4. *S. serotina*
       3. Calyx lobes less than 5 mm long when
         plant in fruit . . . .   5. *S. occidentalis*

    1. S. lyrata L.—Perennial herb scapose up to 6 dm tall, with basal leaves petioled, obovate, lyrate-pinnatifid or deeply lobed to subentire, often purplish shaded. Stem naked or with one to two pairs of leaves. Flowers in several whorls, bracts linear-oblong, upper lip of calyx truncate with three lobes; corolla dark blue, 2–3 cm long. Nutlets obovoid, about 2 mm long. Sandy soil, Fla. to Tex., Conn., Mo. Spring.

    2. S. coccinea Buchoz—Annual with pubescent erect stems up to 7 dm tall. Leaves 3–6 cm long, ovate or deltoid-ovate, crenate or crenate-serrate; acute or obtuse at the apex, truncate or subcordate at the base. Flowers in narrow racemes or panicles, 5–20 cm long, calyx minutely pubescent, about 1 cm long, lower lip 7–8 mm wide. Nutlets 2–3 mm long. Disturbed sites, Fla. to Tex., S.C., Mexico, W.I., tropical America. Spring, fall.

    3. S. privoides Benth.—Perennial up to 18 dm tall with erect pubescent stems. Leaves mostly 1–4 cm long, ovate, serrate, acute. Flowers in a narrow spicate raceme, becoming distant, calyx glandular-pubescent, 5–7 mm long, lower lobes awn tipped; corolla bluish, or whitish, mostly 5–6 mm long. Nutlets about 2 mm long. Dry ground, sandy soil, south Fla., Fla. Keys, W.I., Mexico, Central America. Spring, fall.

    4. S. serotina L.—Perennial herb with finely pubescent stems up to 7 dm tall, usually much branched. Leaves 1–4 cm long, ovate to deltoid-ovate or orbicular-ovate, serrate-crenate, apex obtuse, subcordate or rounded at the base. Flowers small in panicles 5–10 cm long, calyx becoming 5–7 mm long, glandular-hirsute, corolla 6–10 mm long, white or sometimes bluish. Nutlets about 2 mm long. Disturbed sites, roadsites, Fla., W.I., tropical America. Spring, fall. *S. micrantha* Vahl

    5. S. occidentalis Sw. West Indian Sage—Annual with erect, decumbent or prostrate densely pubescent stems. Leaves 2–6 cm long, elliptic to ovate, serrate, apex acuminate or at least acute, narrowed or subtruncate at the base. Flowers in narrow panicles up to 3 dm long, clusters becoming close together near the top, calyx about 3 mm long, glandular-pubescent, corolla blue, about 5 mm long. Nutlets about 2 mm long. Disturbed sites, Fla., W.I., tropical America. Spring, fall.

S. **blodgettii** Chapm., similar to S. **occidentalis** but with filiform petioles, was long ago reported for hammocks in Key West, but no recent collections have been seen.

7. PHYSOSTÈGIA Benth. FALSE DRAGONHEAD

Perennial herbs with erect, smooth stems and short-petioled or sessile leaves. Flowers conspicuous, opposite, scattered or numerous in simple or branching leafless spikes, calyx campanulate or short-cylindrical, corolla funnelform, throat inflated, two-lipped, upper erect, subentire, lower three-parted, stamens four, ascending under upper lip. Nutlets ovoid, smooth. Fifteen species in North America. *Dracocephalum* sensu Small, not L.

1. **P. denticulata** (Ait.) Britt.—Herbs with numerous stolons, stems smooth, up to 1 m tall. Lower leaves 2–14 cm long, oblong to oblanceolate, undulate, toothed, short petioled; upper leaves sessile, lanceolate or oblanceolate, acute tipped. Flowers in slender spikes, interrupted, becoming 1–3 dm long and flexuous, corolla purplish or magenta, about 2–3 cm long. Low ground, Fla. to Va. Summer. *Dracocephalum denticulatum* Ait.

8. HÝPTIS Jacq. BITTERMINTS

Erect herbs usually branched with mostly dentate leaves. Flowers often in dense, axillary cymes or in heads, calyx tubular or campanulate, subequally five-lobed, corolla two-lipped, the upper erect or spreading, the lower saccate and drooping, stamens four, declined, didynamous. Ovary four-carpellate, nutlets smooth or rough, oblong or ovoid. About 350 species, chiefly in tropical America.

1. Flowers in involucrate heads . . . . 1. *H. alata*
1. Flowers spicate or in axillary sessile or sub-
    sessile clusters.
    2. Stems strongly hirsute pubescent . . . 2. *H. mutabilis*
        var. *spicata*
    2. Stems glabrous or sparingly pubescent.
        3. Leaves lanceolate, leaf base tapering;
            floral bracts inconspicuous . . . 3. *H. verticillata*
        3. Leaves ovate, leaf base truncate-
            cuneate; floral bracts conspicuous . 4. *H. pectinata*

1. H. alata (Raf.) Shinners, Musky Mint—Stems up to 2 m tall, simple or occasionally branched, glabrous, puberulent to finely pubescent. Leaves 2–8 cm long, linear-lanceolate, lanceolate to ovate, subentire, serrate or dentate, cuneate or tapered to the base. Flowers in an involucrate head, calyx 6–8 mm long, lobes pubescent, corolla greenish or white, 8–10 mm long, stamens declined on the lower lip. Nutlets ovoid, about 1 mm long. Low ground, moist soil, Fla. to Tex., N.C., W.I. All year. *H. radiata* Willd. Two varieties are found in our area:

1. Leaves lanceolate to ovate, serrate or
   dentate . . . . 1a. var. *alata*
1. Leaves linear-lanceolate to lanceolate, subentire or few-toothed . . . . . . 1b. var. *stenophylla*

1a. H. alata var. **alata**—Stems puberulent to finely pubescent. Coastal plain, Fla. to Tex., N.C., W.I.

1b. H. alata var. **stenophylla** Shinners—Stems glabrous or glabrate. Endemic to lower peninsular Fla., in moist to wet sandy peat and grass–sedge meadows.

2. H. mutabilis (A. Rich.) Briq. var. **spicata** (Poir.) Briq.—Stems up to 2 m tall, scabrous, hirsute pubescent or appressed pubescent. Leaves mostly 3–7 cm long, ovate-reniform to orbicular-reniform, acuminate, unequally serrate or crenate. Flowers in spikes consisting of small axillary clusters subtended by foliaceous bracts, calyx tubular, becoming 6–7 mm long; corolla 3–4 mm long, bluish purple. Nutlets about 1 mm long. Moist sites, disturbed soil, Fla. to Va., naturalized from tropical America. All year.

3. H. verticillata Jacq.—Stems up to 2 m tall, glabrous or sparingly pubescent. Leaves mostly 6–12 cm long, lanceolate to narrowly so, serrate, acuminate to the apex, tapering to the base. Flowers in small axillary clusters, floral bracts inconspicuous, calyx 2–3 mm long, corolla mostly 3–6 mm long, white. Nutlets about 1.5 mm long. Hammocks, disturbed sites, south Fla., naturalized from tropical America. Summer, fall.

4. H. pectinata (L.) Poir.—Stems up to 1 m tall or more, sparingly pubescent. Leaves 2–8 cm long, ovate, irregularly serrate, acuminate to the apex, cuneate or truncate at the base. Flowers in axillary clusters forming a dense spike, bracts conspicuous, calyx 4–5 mm long, corolla about 4 mm long, bluish. Nutlets about 1 mm long. Moist sites, roadsides, naturalized from tropical America. All year.

9. SATURÈJA L. Savory

Shrubs or herbs usually with narrow leaves. Flowers often in dense racemes or panicles, calyx tubular or campanulate, oblique, corolla distinctly two-lipped, mostly purplish or whitish, stamens four. Ovary four-lobed, nutlets very small. About 130 species of wide distribution. *Pycnothymus* Small

1. **S. rigida** Bartr. ex Benth. PENNYROYAL—Stems erect or procumbent, diffusely branched, up to 7 dm long or more. Leaves about 1 cm long, narrowly lanceolate, entire, numerous. Flowers in dense racemelike panicles, calyx slightly oblique, upper three lobes narrow, the lower lip with two longer lobes, up to 3 mm long, corolla two-lipped, lower lip three-lobed, middle lobes somewhat larger than lateral ones, light purplish, about 8 mm long, stamens exserted. Ovary deeply four-lobed, nutlets about 1 mm long. Pinelands, endemic to peninsular Fla. All year.

10. CONRADÌNA Gray

Shrubs with slender branches and opposite, narrow, entire, revolute, often fascicled leaves. Flowers in small, axillary cymes, calyx strongly two-lipped, corolla bluish or purplish, two-lipped, upper lip erect or arched, lower lip three-lobed, stamens four, ascending under the upper lip. Nutlets very small. Four species in North America.

1. **C. grandiflora** Small—Shrub up to 1 m tall with few slender, usually curved branches. Leaves narrowly spatulate, margins revolute, mostly 1–3 mm long, punctate, white canescent beneath, glabrous above. Flowers bluish, calyx 6–7 mm long, finely pubescent, upper corolla lip 8–9 mm long. Nutlets about 1 mm long. Pineland, endemic to south Fla. All year.

This species is usually found in scrub vegetation in sandy soil and is north of our range but may enter the area on the East Coast.

11. ÒCIMUM L.

Low shrubs or herbs with erect or ascending stems. Flowers clustered, bracteate in rather dense racemelike panicles, calyx strongly two-lipped, five-lobed, corolla two-lipped, white or nearly so, the upper lip four-lobed, stamens four, didymous. Ovary deeply four-lobed, nutlets smooth or rugose. About 60 species in tropical and warm temperate zones.

1. **O. micranthum** Willd. WILD BASIL—Annual with pubescent, erect, branched stems up to 5 dm tall. Leaves 2–6 cm long, ovate to ob-

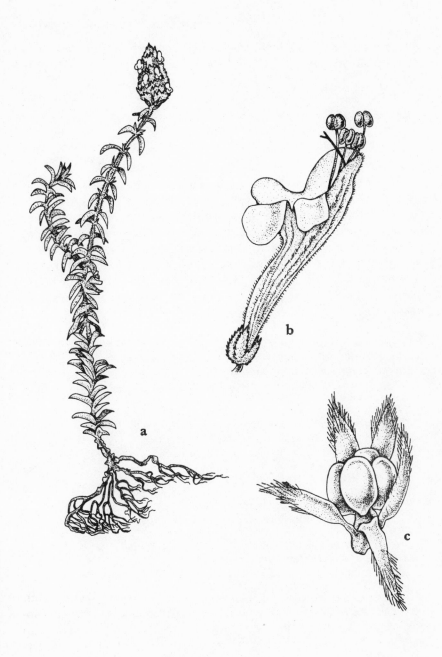

Family 166. **LAMIACEAE.** *Satureja rigida:* a, flowering plant, × ½; b, flower, × 6; c, mature fruit with calyx, × 15.

long, serrate, sometimes shallowly toothed with definite petioles. Flowers clustered, in panicles 3–10 cm long, calyx puberulent, eventually 6–7 mm long, oblique, corolla purplish, about 4 mm long. Nutlets about 1 mm long. Drier soil, disturbed sites, south Fla., W.I., Mexico, tropical America. All year.

O. basilicum L., BASIL, with white corollas much exceeding the calyx, may be found naturalized in waste places. It is a native of Asia.

167. SOLANÀCEAE NIGHTSHADE FAMILY

Small trees, shrubs, or herbs with alternate leaves. Flowers bisexual, radially or occasionally bilaterally symmetrical, terminal or axillary, solitary or cymose; calyx four- to six-lobed; corolla mostly five-lobed, the lobes folded, stamens five, epipetalous. Ovary two-locular, ovules very numerous. Fruit a two-chambered, many-seeded capsule or berry, often poisonous. About 85 genera and 2300 species, generally distributed but chiefly American tropics.

1. Leaves linear-lanceolate, less than 2 cm long; shrubs with stout thorns . . . . 1. *Lycium*
1. Leaves various, not linear-lanceolate, more than 2 cm long; herbs or shrubs unarmed or with spines.
  2. Corolla yellow or yellowish white; calyx inflated at maturity surrounding fruit . 2. *Physalis*
  2. Corolla white or colored, not yellow; calyx not inflated at maturity.
    3. Calyx distinctly five-lobed or parted.
      4. Corolla rotate, the lobes longer than the tube; fruit a berry . . . . 3. *Solanum*
      4. Corolla funnelform, tubular, or salverform, the lobes shorter than the tube; fruit a capsule.
        5. Flowers solitary, axillary . . . 4. *Petunia*
        5. Flowers in terminal racemes or panicles . . . . 5. *Nicotiana*
    3. Calyx truncate or obscurely lobed.
      6. Corolla rotate, the lobes longer than the tube . . . . 6. *Capsicum*
      6. Corolla funnelform or salverform, the lobes shorter than the tube . 7. *Cestrum*

1. LÝCIUM MATRIMONY VINE

Shrubs, vines, or small trees, usually spiny, with small leaves. Flowers small, in axillary clusters or solitary, calyx campanulate, three-to five-lobed or toothed, persistent, corolla usually five-lobed, funnel-form, tube slender, short, stamens five, filaments very slender. Ovary two-locular, stigma capitate or two-lobed. Fruit a small globose or oblong berry. About 110 species, widely distributed.

1. **L. carolinianum** Walt. CHRISTMAS BERRY—Shrub or low tree up to 2 m tall with curving, spiny branches. Leaves thickish, about 0.5–2 cm long, smooth, entire. Flowers in clusters or solitary, calyx lobes deltoid or ovate, acute, corolla lobes 3–6 mm long, bluish or lilac, rarely white. Berry bright red, about 1 cm long. Coastal dunes and hammocks, shell mounds, Fla. to Tex., S.C., W.I. All year.

2. PHÝSALIS L. GROUND CHERRIES

Herbs or subshrubs with entire or toothed leaves. Flowers mostly solitary, axillary, nodding, calyx five-lobed, campanulate, becoming enlarged and bladderlike, membranous, five-angled and ten-ribbed, completely enclosing the berry; corolla usually with a purplish or discolored center, campanulate or rotate, lobes plicate, stamens inserted at the base of the corolla. Stigma obscurely two-cleft. About 110 species from tropics to temperate regions.

1. Stems, leaves, or margins of sepals with branched or stellate hairs.
  2. Leaves linear to narrowly lanceolate, plant essentially glabrous . . . . 1. *P. angustifolia*
  2. Leaves linear-lanceolate to ovate; plant stellate pubescent . . . . . . 2. *P. viscosa*
1. Stems and leaves with simple hairs or glabrous.
  4. Plants annual; anthers mostly 1–2 mm long.
    5. Stems viscid pubescent . . . . . 3. *P. pubescens*
    5. Stems glabrous . . . . 4. *P. angulata* var. *angulata*
  4. Plants perennial; anthers mostly 3–5 mm long . . . . 5. *P. arenicola*

1. **P. angustifolia** Nutt.—Stems diffusely branching, scattered stellate pubescent when young, becoming glabrous at maturity. Leaves five to ten times longer than wide, linear to linear-lanceolate, entire, tapered to the petiole. Flowers solitary on nodding peduncles, calyx glabrous except stellate-ciliate margins of lobes; corolla about 2 cm wide, yellow with a purplish center. Fruiting calyx about 2 cm long, ovoid. Coastal areas, rich soil, south Fla. to La. All year.

2. **P. viscosa** L.—Stems creeping, slender, covered with a dense

grayish stellate pubescence or becoming glabrate at maturity. Leaves mostly 4–10 cm long, ovate or elliptic, white or grayish stellate pubescent, entire or undulate. Flowers mostly solitary, calyx stellate pubescent, lobes deltoid; corolla 1.5–2 cm wide, greenish yellow with a darker center. Fruiting calyx 2–3 cm long, ovoid. Coastal marl and sand, south Fla. to Tex., Va., tropical America. All year. Two varieties occur in our area:

1. Leaf blades ovate to spatulate . . . . 2a. var. *maritima*
1. Leaf blades lanceolate to linear-lanceolate . 2b. var. *elliottii*

2a. P. viscosa var. **maritima** (M. A. Curtiss) Waterfall—Leaves with petioles about one-third the blade length, blades 2–5 cm wide. Sandy soils, south Fla. to Va. on the coastal plain. *P. maritima* M. A. Curtiss

2b. P. viscosa var. **elliottii** (Kunze) Waterfall—Leaves lanceolate, tapering to the base, blades two to ten times longer than wide. Sandy soil, endemic to Fla. *P. elliottii* Kunze

3. P. pubescens L.—Annual with low, weak villous, viscid-pubescent annual up to 5 dm tall. Leaves mostly 3–6 cm long, ovate, rounded or cordate at the base, pale green, entire or with very few serrations. Flowers usually solitary, calyx lobes narrowly acuminate, corolla yellow with a darker center, about 1 cm wide, anthers usually purplish. Berry yellow, surrounded by the membranous calyx 2–3 cm long, subglobose and with prominent auricles near the base. Dry soil, sand dunes, Fla. to Tex., Va., Kans., south Calif., tropical America. All year. *P. floridana* Rydberg; *P. barbadensis* Jacq.

4. P. angulata L. var. **angulata**—Annual with stems erect up to 1 m tall, glabrous, angled. Leaves ovate, sinuate, and with acuminate teeth, blades 5–7 cm long. Flowers solitary, erect, calyx lobes triangular-lanceolate, glabrous; corolla yellow, 5–10 mm wide, anthers purplish. Berry yellow, surrounded by the fruiting calyx 3 cm long, not strongly angled. $2n = 48$. Coastal areas, Fla. to Tex., Va., Okla., tropical America. All year.

5. P. arenicola Kearney—Perennial with diffuse, puberulent stems. Leaves mostly 2–6 cm long, ovate to suborbicular, irregularly angulate-dentate, puberulent. Flowers solitary, calyx lobes deltoid-ovate, puberulent with simple hairs; corolla about 2 cm wide, pale yellow, anthers yellow. Berry yellow, surrounded by the fruiting calyx 3 cm long, strongly reticulate. Pinelands, scrub, Fla., Ga., Ala. Spring, summer.

P. turbinata Medicus, an annual up to 100 cm tall, hairy, and with fruiting calyx to 4 cm long, was reported by Small for the Fla. Keys. It is a native of Mexico, Central America, and the W.I.

3. SOLÀNUM Nightshade

Shrubs, herbs, or climbing vines usually stellate pubescent. Flowers variously arranged, cymose, umbellate, racemose or paniculate, calyx five-parted, campanulate or rotate, corolla rotate, five-lobed or five-angled, limb folded, tube very short, stamens epipetalous, filaments short, anthers linear or oblong, connivent around the style, opening at the tip by terminal pores or slits. Ovary mostly two-locular. Fruit a globose berry, calyx persistent. About 1500 species, chiefly in the tropics. Hybridization and polyploidy occur in a number of species.

1. Stems woody.
    2. Leaves all simple, undivided; stems erect.
        3. Inflorescence racemose, pedicels recurved . . . . 1. *S. bahamense*
        3. Inflorescence cymose, pedicels erect or straight.
            4. Leaves pubescent.
                5. Leaves and flowers wooly tomentose; fruits yellow . . . . . 2. *S. erianthum*
                5. Leaves and flowers short hirsute; fruits red . . . . . . 3. *S. blodgettii*
            4. Leaves glabrous or nearly so . . . 4. *S. diphyllum*
    2. Leaves some or all pinnately divided; stems trailing . . . . 5. *S. seaforthianum*
1. Stems herbaceous, sometimes woody near the base.
    6. Stems and leaves prickly.
        7. Leaves toothed or sinuately pinnatifid, sparingly pubescent . . . . . 6. *S. aculeatissimum*
        7. Leaves narrowly ovate or elliptic, silvery canescent . . . . . . 7. *S. elaeagnifolium*
    6. Stems and leaves not prickly.
        8. Stems glabrous or nearly so; corolla less than 8 mm wide . . . . . . 8. *S. americanum*
        8. Stems thinly pubescent; corolla more than 8 mm wide . . . . . 9. *S. ottonis*

   1. **S. bahamense** L.—Shrub up to 2 m, branched, prickly armed or smooth, stellate pubescent. Leaves mostly 6–12 cm long, lanceolate to oblong, margin undulate or entire. Inflorescence racemose, pedicels recurved, calyx lobes ovate, about 1.5 mm long; corolla blue, rarely white, about 1 cm wide. Berry red, globose about 7–8 mm wide. Various soils, hammocks, beaches, south Fla., Fla. Keys, W.I. All year.
   2. **S. erianthum** D. Don, POTATO TREE—Shrub or low tree up to 3 m tall or more, unarmed, stems stellate tomentose. Leaves 1–3 dm long, ovate or obovate, entire, woolly tomentose. Inflorescence cymose,

Family 167. **SOLANACEAE.** *Solanum blodgettii:* a, flowering branch, × ½; b, flower, × 4¼; c, mature fruit, × 3½.

pedicels erect or straight, calyx densely stellate pubescent, 5–6 mm long, the lobes deltoid-ovate; corolla white, about 1–1.5 cm wide, lobes ovate-oblong. Berry globose 1–2 cm wide, yellow. Hammocks, disturbed sites, south Fla., Fla. Keys, W.I., tropical America. All year. *S. verbascifolium* sensu Small

3. **S. blodgettii** Chapm.—Shrub up to 1.5 m tall with spreading branches, unarmed, stellate hirsute. Leaves 5–15 cm long, elliptic or oblong, entire or undulate, short hirsute. Inflorescence cymose or paniculate, sometimes densely flowered, calyx about 3 mm long, deltoid, hirsute; corolla white or bluish, 6–8 mm long. Berry globose, 4–6 mm wide, red. Pinelands and hammocks, south Fla., Fla. Keys, W.I. All year.

4. **S. diphyllum** L.—Low shrub up to 1 m tall or more, unarmed, stems smooth. Leaves 6–8 cm long, lanceolate to lance-elliptic, obtuse, entire, short petioled to subsessile. Flowers in cymes in upper leaf axils, corolla white, about 1 cm wide, lobes lance-acute, slightly reflexed. Style straight, stigma capitate. Fruit smooth, globose, yellow orange at maturity. Hammocks, south Fla., Mexico, introduced from tropical America. All year.

5. **S. seaforthianum** Andrews, BRAZILIAN NIGHTSHADE—Shrub or trailing vines with smooth or puberulent stems. Leaves broadly ovate, some or all pinnately divided, margins of leaves or leaf segments entire. Inflorescence cymose, pedicels drooping, widely divergent, calyx puberulent or subglabrous, minutely five-toothed; corolla blue or lavender, about 2 cm wide. Style incurved, berry subglobose, about 1 cm wide, red. Hammocks, disturbed sites, peninsular Fla., naturalized from South America. All year.

6. **S. aculeatissimum** Jacq. SODA APPLE—Perennial herbs or subshrubs up to 1 m tall or less, much branched, pubescent or subglabrous, armed with erect yellow prickles. Leaves mostly 8–16 cm long, ovate to suborbicular, pinnately lobed or angulately toothed. Flowers in few-flowered cymes, calyx covered with prickles, lobes ovate, about 2 mm long; corolla white about 1 cm wide or more, lobes lanceolate. Berry globose, about 1–3 cm wide, orange yellow or red. Beaches, sandy soil, Fla. to Tex., N.C. on the coastal plain, tropical America. Spring, fall.

7. **S. elaeagnifolium** Cav. WHITE HORSENETTLE—Perennial herbs or subshrubs to 1 m tall, stems woody below, herbaceous above, silvery canescent. Leaf blades 8–15 cm long, narrowly elliptic or ovate to broadly elliptic, margins undulate or repand. Flowers in the upper axils, umbellate or cymose, calyx segments linear-subulate, 5–10 mm long, corolla 2–3 cm wide, violet, purplish, or white. Berry globular, 10–15 mm in diameter, black or yellowish. $2n = 24$. Disturbed sites, various soils, Tex. to Mo., Ariz., naturalized in Fla. Spring, fall.

758   A FLORA OF TROPICAL FLORIDA

8. S. americanum Mill. COMMON NIGHTSHADE—Annual with smooth or slightly pubescent stems up to 1 m tall, usually less. Leaves 2–8 cm long, ovate, or ovate-deltoid, thin, acute acuminate to the apex. Inflorescence axillary, umbellate with mostly five to ten flowers per cluster, calyx lobes oblong, 1 mm long; corolla white, 7–9 mm wide, lobes lanceolate to oblong-lanceolate. Berry globose, shiny black, about 1 cm wide, in nodding clusters. $2n = 24,48,72$. Disturbed sites, various sites, Fla. to Nova Scotia, consisting of many ecotypic races.

9. S. ottonis Hylander—Annual with appressed pubescent stems up to 1 m tall, usually less. Leaves 3–10 cm long, lanceolate to ovate or elliptic, entire or undulate, appressed pubescent. Inflorescence axillary, umbellate, calyx lobes 1.5 mm long; corolla white or bluish white, 8–10 mm wide. Berry subglobose, black, about 1 cm wide in nodding clusters. Pinelands, sandy soil, disturbed sites, Fla. to La., N.C. on the coastal plain. All year.

This taxon possibly is an ecotypic race of **S. americanum.**

4. PETÙNIA Juss. PETUNIA

Herbs with entire leaves, the upper ones becoming opposite. Flowers large, conspicuous, axillary, calyx five-parted; corolla salverform or funnelform, five-lobed, the lobes folded in bud, stamens five, epipetalous low within the tube, included, disk present, fleshy. Ovary twolocular, stigma expanded or capitate. Fruit a two-valved capsule. About 14 species, principally in South America.

1. P. × hybrida Vilm. COMMON PETUNIA—Annual with viscid pubescence, weak, trailing or decumbent branching stems, mostly 6–9 dm long. Lower leaves mostly 4–10 cm long, narrow ovate or oblanceolate; upper leaves ovate or truncate, fleshy, soft. Flowers axillary, calyx 1–2 cm long; corolla 5–7 cm long, tube funnelform, limb broad, varying widely in size, color, and shape, white, pink, to dark purple, striped or otherwise marked. Cultivated plants apparently hybrids of **P. axillaris** BSP × **P. violacea** Lindl.; escaped into disturbed soil and locally established. All year.

5. NICOTIÀNA L. TOBACCO

Herbs, shrubs, or small trees with viscid pubescence and large leaves. Flowers in terminal racemes or panicles, calyx tubular or ovoidcampanulate, five-cleft; corolla funnelform or salverform, five-lobed, stamens five, filaments long. Ovary two-locular, style slender, stigma ovoid. Fruit a two-valved, many-seeded capsule. About 100 species, widely distributed, chiefly in the tropics and subtropics.

1. Capsule 2 cm long; corolla red or pink . .   1. *N. tabacum*
1. Capsule 1 cm long or less; corolla white or
     greenish white . . . .   2. *N. plumbaginifolia*

     1. N. tabacum L. Tobacco—Annual up to 2 m tall, mostly simple or few-branched. Leaves 1–3 dm long, oblong to lanceolate-oblong, lower leaves much larger than upper ones. Flowers in terminal racemes in panicles, calyx about 1 cm long, lobes ovate; corolla funnelform, mostly 5–6 cm long, red or pink, lobes deltoid-subulate. Capsule 2 cm long, ovoid. Disturbed sites, southeast U.S. to Fla. Keys, naturalized from South America. Summer.

     2. N. plumbaginifolia Viv.—Annual up to 1 m tall, simple or branched. Leaves 1–3 dm long, spatulate to elliptic, lower leaves longer and wider than upper ones. Flowers mostly in terminal racemes, calyx about 1 cm long; corolla funnelform, 3–5 cm long, white or greenish white, lobes ovate or narrowly ovate. Capsule ellipsoid, about 1–1.5 cm long. Disturbed sites, Fla. Keys, naturalized from tropical America.

6. CÁPSICUM L.   Cayenne Pepper

     Shrubs or herbs with branching stems and entire leaves. Flowers axillary, solitary, or in small cymes, calyx truncate or obscurely five-lobed; corolla five-lobed, nearly rotate, usually white, stamens five, fused to the base of the corolla. Ovary two- to three-locular, stigma ovoid, expanded, or minute. Fruit a berry. About 35 species, chiefly in tropical America.

     1. C. annuum L. var. minimum (Mill.) Heiser—Erect herbs or shrubs with forked stems up to 2 m tall with smooth stems. Leaves 1–3 cm long, ovate to lance-ovate, entire, acute or acuminate. Flowers mostly solitary in axils, calyx 1–2 mm long, toothed; corolla about 5 mm long, white, rotate. Berry red, globose or subglobose, 8–12 mm long. $2n = 24$. Hammocks, south Fla., Fla. Keys, W.I., tropical America.

     Lycopersicon esculentum Mill., the Common Tomato, is found outside of cultivation in various sites, especially in old fields, roadsides, and other disturbed areas and occasionally in hammocks. It is a native of South America. $2n = 24$.

7. CÉSTRUM L.

     Shrubs with branching stems and entire leaves. Flowers in axillary pedunculate cymes; calyx truncate or obscurely lobed or toothed; corolla funnelform or salverform, tube cylindrical, stamens five, anthers longitudinally dehiscent. Fruit a berry. About 150 species in the American tropics.

1. **C. diurnum** L. DAY JESSAMINE—Smooth shrub up to 2 m tall with numerous branches. Leaves 5–10 cm long, leathery, ovate-oblong, veins delicate. Flowers in contracted peduncled axillary cymes, calyx obscurely five-toothed or teeth deltoid or blunt; corolla white, tube clavate, tapering, lobes rounded, reflexed; stamens included. Berry subglobose, shiny blue black. Disturbed sites, dry soil, south Fla., introduced from tropical America.

This species is also found in cultivation throughout our area and has apparently escaped.

168. SCROPHULARIÀCEAE SNAPDRAGON FAMILY

Herbs or shrubs with alternate, opposite, or verticillate leaves. Inflorescence solitary, racemose, paniculate, or cymose; flowers bisexual, chiefly bilaterally symmetrical, calyx four- to five-parted; corolla fused, four- to five-lobed, often more or less two-lipped, stamens two or four, epipetalous, a fifth stamen may be represented by a staminode. Disk if present annular or one-sided. Ovary superior, two-locular, fruit a capsule or sometimes a berry. About 225 genera and 2600 species of wide distribution. *Rhinanthaceae*

1. Leaves alternate.
　2. Corolla nearly radially symmetrical; sta-
　　mens five . . . .　　1. *Capraria*
　2. Corolla two-lipped; stamens four . . .　2. *Linaria*
1. Leaves opposite or verticillate.
　3. Fertile stamens two.
　　4. Sepals five or calyx five-lobed.
　　　5. Corolla two-lipped or at least bilat-
　　　　erally symmetrical.
　　　　6. Staminodes two, represented by
　　　　　pair of slender filaments . . .　3. *Lindernia*
　　　　6. Staminodes absent or minute . .　4. *Gratiola*
　　　5. Corolla nearly radially symmetrical .　5. *Bacopa*
　　4. Sepals four or calyx four-lobed . . .　6. *Micranthemum*
　3. Fertile stamens four.
　　7. Corolla spurred . . . .　2. *Linaria*
　　7. Corolla not spurred.
　　　8. Leaves pinnatifid or twice-pinnatifid
　　　　with narrow lobes or segments.
　　　　9. Corolla yellow, rotate; stamens
　　　　　slightly exserted . . . . .　7. *Seymeria*
　　　　9. Corolla greenish white, tubular;
　　　　　stamens included . . . . .　8. *Conobea*
　　　8. Leaves entire or serrate, not pinna-
　　　　tifid.

10. Flowers in terminal inflorescence.
  11. Corolla tubular, two-lipped.
    12. Corolla white . . . . .   9. *Penstemon*
    12. Corolla red . . . . . .   10. *Russelia*
  11. Corolla salverform, nearly radially symmetrical . . . .   11. *Buchnera*
10. Flowers appearing axillary or in lateral racemes.
  13. Sepals fused forming a calyx tube   12. *Agalinis*
  13. Sepals nearly free, not fused forming a tube.
    14. Corolla two-lipped.
      15. Terrestrial plants.
        16. Pedicels without bracts .   3. *Lindernia*
        16. Pedicels with two bracts at base . . . . .   13. *Mecardonia*
      15. Aquatic or bog plants . .   14. *Limnophila*
    14. Corolla nearly radially symmetrical, not two-lipped.
      17. Stems creeping, often matted; leaves entire . .   5. *Bacopa*
      17. Stems erect; leaves serrate .   15. *Scoparia*

1. CAPRÀRIA L.

Perennial herbs or shrubs with alternate, serrate leaves. Flowers axillary, pedicelled, sepals five, free, equal; corolla campanulate, nearly radially symmetrical, white or bluish white, stamens usually five. Stigma two-lobed or expanded. Capsule ovoid, two-grooved, gland dotted, seeds reticulated. Four species in American tropics and subtropics.

1. C. biflora L.—Perennial herb with stems up to 1.5 m tall, pubescent to glabrate. Leaves 2–8 cm long, oblanceolate, cuneate, or elliptic. Pedicel 0.5–2 cm long, two per axil, sepals 5–7 mm long, linear or linear-lanceolate; corolla about 1 cm long, white, violet tinged or sometimes yellowish. Capsule 4–5 mm long, seeds brown. Many soils, disturbed sites, sandy soil, beaches, south Fla., tropical America. All year.

2. LINÀRIA Mill. TOADFLAX

Perennial or annual herbs with erect stems and many narrow leaves. Flowers in terminal racemes, calyx deeply five-lobed; corolla bilaterally symmetrical, strongly two-lipped, the upper lip two-lobed, the lower lip three-lobed, spurred at the base, stamens four, included. Stigma capitate, capsule globose or ovoid-cylindrical, seeds numerous, winged or wingless. About 510 species in the Northern Hemisphere.

1. L. canadensis (L.) Dum. BLUE TOADFLAX—Slender annual with a short taproot, stems erect up to 5 dm tall, smooth and also producing a small rosette of short, prostrate stems with ascending tips. Leaves 1–4 cm long, mostly opposite or three-whorled on prostrate-ascending stems, leaves on erect, main stems, chiefly alternate, linear. Racemes nearly bractless, sepals 2–3 mm; corolla pale to dark blue, 8–10 mm long, lower lip with two white ridges, spur curved, 2–6 mm long. Capsule 2–4 mm long, seeds wingless. Sandy soil, disturbed areas, south Fla. to Mass., Minn., Mexico, Pacific coast. Spring.

L. floridana Chapm., with corolla only 5–6 mm long and glandular-pubescent pedicels, may occur in our area. It is also found in dry sandy soil.

3. LINDÉRNIA All. FALSE PIMPERNEL

Low annual or biennial herbs with opposite, entire or denticulate leaves. Flowers small, solitary on axillary pedicels, calyx of five nearly free sepals; corolla two-lipped, upper lip narrower than the three-lobed lower lip, stamens two fertile and two sterile, the latter represented by a pair of slender filaments. Stigmas two, platelike, style elongate. Fruit an ovoid or ellipsoidal capsule, seeds numerous, yellow, winged or wingless. About 70 species in temperate and tropical regions. *Ilysanthes* Raf.

1. Stems prostrate or creeping; larger leaves broadly ovate to orbicular . . . . . 1. *L. grandiflora*
1. Stems erect or ascending; larger leaves narrowly elliptic to ovate, apex acute or obtuse.
  2. Leaves narrowly lanceolate to elliptic, 5–15 mm long . . . . 2. *L. anagallidea*
  2. Leaves lance-ovate to deltoid-ovate, 10–30 mm long . . . . 3. *L. crustacea*

1. **L. grandiflora** Nutt.—Stems prostrate, creeping, branched, and often forming mats. Leaves mostly 5–10 mm long, broadly ovate to orbicular, entire to crenate or serrate. Flowers on axillary pedicels 1–4 cm long, erect, sepals 2–3 mm long; corolla about 1 cm long, blue violet, appearing blue white mottled. Capsule 4–6 mm long, seeds brownish yellow, winged. Moist, usually sandy soil, roadsides and ditches, south Fla. to Ga. on the coastal plain. All year. *Ilysanthes grandiflora* (Nutt.) Benth.

2. **L. anagallidea** (Michx). Pennell—Stems erect or ascending, branching chiefly from near the base. Leaves mostly 5–15 mm long, entire to obscurely few-serrate, narrowly lanceolate to elliptic. Flowers on axillary pedicels 1–2 cm long, sepals 3–4 mm, narrow; corolla 6–8 mm long, white or faintly bluish white. Capsule 2–4 mm long, seeds

yellowish brown, wingless. Moist soil, Fla. to N.H., Minn., Tex., Colo. and on the Pacific coast. Summer, fall. *Ilysanthes inequalis* (Walt.) Pennell

3. L. crustacea (L.) F. Muell.—Stems erect or ascending, diffusely branching chiefly from near the base and sometimes rooting from the node. Leaves 1–3 cm long, lance-ovate to deltoid-ovate, subentire to serrate. Flowers on slender pedicels, 1–3 cm long, calyx 2–4 mm long, narrow, corolla purplish or white, 5–6 mm long. Capsule 4–5 mm long. Moist sand, pinelands, peninsular Fla., naturalized from Old World tropics where it is a common weed. All year. *Vandellia crustacea* Benth.

4. GRATÌOLA L. Hedge Hyssop

Perennial or annual herbs with opposite leaves. Flowers solitary on axillary pedicels, sepals six, unequal, corolla tubular or somewhat campanulate, two-lipped, yellow or white, stamens two. Style expanded or slightly two-branched at the tip, fruit a globose, many-seeded capsule. About 20 species of temperate and tropical distribution. *Sophronanthe* Benth.; *Tragiola* Small & Pennell

1. Leaves linear to narrowly lanceolate; corolla
   10–15 mm long.
   2. Leaves linear-lanceolate to lanceolate;
      corolla yellow with lobes white .   .   .   1. *G. ramosa*
   2. Leaves linear; corolla white .   .   .   .   2. *G. subulata*
1. Leaves lance-ovate to ovate; corolla 5–10
   mm long .   .   .   .   3. *G. pilosa*

1. G. ramosa Walt.—Perennial herb with smooth, mostly simple gland-dotted stems up to 3 dm tall. Leaves linear-lanceolate to lanceolate, with few serrations above the middle of the blade. Sepals linear, about 4 mm long, corolla 1–1.5 cm long, yellow with white lobes, throat with brownish veins. Capsule 1–2 mm long with brown finely net-veined seeds. $2n = 14$. Disturbed sites, sandy soil, Fla. to Md., La. Spring.

2. G. subulata Baldw.—Perennial herb with hispid hirsute, mostly simple stems up to 2 dm tall. Leaves about 1.5 cm long, linear, firm, glabrous, and whitened above, hispid on the midrib below. Sepals linear-lanceolate, subtended by bractlets, corolla 10–14 mm long, white. Capsule 4–5 mm long, seeds nearly black. Moist sand, pineland, Fla. to Miss., Ga. Spring, summer. *Sophronanthe hispida* Benth.

3. G. pilosa Michx.—Perennial herb with hirsute stems, mostly simple or few-branched up to 4 dm tall. Leaves mostly 1–2 cm long, lance-ovate to ovate, sessile, entire or obscurely serrate. Sepals 5–6 mm long, linear-lanceolate, subtended by bractlets, corolla 5–10 mm long, yellow, the throat with faint bluish purple lines. Capsule 4–5 mm long, seeds yellowish. Pinelands, Fla. to Va., Tex., Ky., Ark. Summer, fall. *Tragiola pilosa* (Michx.) Small & Pennell; *Sophronanthe pilosa* (Michx.) Small

5. BACÒPA Aubl. WATER HYSSOP

Low, often aromatic, succulent or creeping perennial herbs with opposite, entire, sessile leaves. Flowers axillary, sepals five, unequal, corolla radially symmetrical or nearly so, tubular or campanulate, stamens four, sometimes fewer. Stigmas two, fruit a capsule. About ten species in the American tropics. *Herpestis* Gaertn. f.; *Hydrotrida* Small; *Macuillamia* Raf.

1. Corolla white or pinkish white; plants not
     aromatic.
   2. Leaves with one main vein . . . . . . 1. *B. monnieri*
   2. Leaves with three to five palmate main
      veins . . . . 2. *B. cyclophylla*
1. Corolla blue or lavender; plants aromatic . 3. *B. caroliniana*

1. B. monnieri (L.) Pennell—Stems fleshy smooth, creeping, and forming mats. Leaves mostly 1–1.5 cm long, glandular dotted, cuneate-obovate to spatulate, with one main vein. Flowers solitary in leaf axils, pedicels 1–2 cm long, bracts linear, sepals ovate or narrowly so, 5–6 mm long, corolla white or pinkish white, campanulate, 8–10 mm long. Seeds grayish brown. Muddy shores, Fla. to Tex., Va., tropical America. All year. *Bramia monnieri* (L.) Pennell

2. B. cyclophylla Fern.—Stems finely pubescent, branched, creeping, about 1–2 dm long. Leaves 0.5–1.5 cm long, appearing to be whorled, oval or suborbicular with three to five palmate main veins. Flowers solitary in leaf axils, pedicels 2–8 mm long, puberulent, bracts narrow, sepals free, up to 5 mm long, corolla 3–4 mm long, pinkish white. Capsule about 2 mm long. Seeds dark brown. Low ground, Fla. to Va., American tropics. Apparently not common in our area. Summer. *Herpestis rotundifolia* Gaertn. f.

3. B. caroliniana (Walt.) Robinson—Stems aromatic, glabrous, becoming villous pubescent toward the tip, creeping or sometimes floating. Leaves 1–2.5 cm long, in subapproximate pairs, ovate, nearly cordate at base five- to nine-veined. Flowers solitary in upper axils, pedicels 5–15 mm long, bracts subulate, sepals glandular-punctate, ciliate, corolla blue or lavender, campanulate. Capsule 4–5 mm long. Seeds gray brown. Aquatic or wet shores, Fla. to Va., Tex. All year. *Hydrotrida caroliniana* (Walt.) Small

6. MICRÁNTHEMUM Michx.

Low creeping herbs much branched and forming mats, stems with opposite or whorled leaves. Flowers solitary, very small, in leaf axils, calyx four-parted, corolla salverform, upper three lobes subequal, lower lobe somewhat longer, stamens two, exserted. Style two-branched, divergent, fruit a globose smooth capsule, one-locular. Seeds numerous. About ten species, chiefly in W.I. *Globifera* J. F. Gmel.; *Hemianthus* Nutt.

1. **M. glomeratum** (Chapm.) Shinners—Stems slender, creeping, up to 5 cm long or somewhat longer. Leaves 2–3 mm long, or if submerged, up to 8–9 mm long, elliptic-cuneate, entire. Flowers minute, axillary, subsessile, calyx 1 mm long, lobes reflexed, acute, corolla 1–2 mm long, white. Capsule almost 1 mm long. Muddy and sandy shores of lakes and rivers, endemic to peninsular Fla. All year. *M. nuttallii* var. *glomeratum* Chapm.; *Hemianthus glomeratus* (Chapm.) Small

7. SEYMÈRIA Pursh.

Herbs with opposite leaves. Flowers solitary in upper leaf axils, calyx radially symmetrical, five-lobed; corolla slightly bilaterally symmetrical, lobes divergent; stamens four, equal, filaments villous at the base. Stigma one, minute, style about as long as the stamens. Fruit an ovoid, loculicidal capsule. About 22 species, in Mexico and south U.S. *Afzelia* J. F. Gmel.; *Dasystoma* Raf.

1. **S. pectinata** (Pursh) Kuntze—Stems 3–6 dm tall, divergently branched and pubescent. Leaf blades 2–3 cm long with three to four pairs of segments or lobes, mostly pinnatifid or twice-pinnatifid. Flowers solitary on distinct pedicels, calyx glandular-pubescent, lobes 3–4 mm long, linear-lanceolate; corolla about 1 cm long, stamens villous or hairy. Capsule 5–7 mm long, seeds winged. Pinelands, Fla. to Miss., S.C. on the coastal plain. Summer, fall. *Afzelia pectinata* (Pursh) Kuntze

8. CONÒBEA Aubl.

Erect annual herbs to 3 dm tall, much branched and diffusely spreading. Leaves opposite, blades 1–2 cm long, deeply pinnately parted or pinnatifid, petioled. Flowers axillary, solitary, sepals five, distinct; corolla about as long as the sepals, greenish white, stamens four, didynamous. Capsule ovoid, 3–4 mm long. One species, Fla. to Ontario, Kans., Tex. *Leucospora* (Michx.) Nutt.

1. **C. multifida** (Michx.) Benth.—Characters of the genus. Disturbed sites, Dade County, along streams and sandy shores north of our range. Summer, fall. *Leucospora multifida* (Michx.) Nutt.

9. PENSTÉMON Mitchell BEARDTONGUE

Perennial herbs with solitary or several erect stems rising from a rosette of petioled basal leaves, cauline leaves sessile and often clasping, progressively reduced in size. Flowers in terminal racemose or paniculate inflorescence, calyx five-parted, equal; corolla two-lipped, tubular, the tube being much longer than the lobes, fertile stamens four, staminode about as long as the fertile ones and often yellow pubescent toward the tip. Style long, stigma globose. Fruit a capsule, seeds many, angled. About 300 species, chiefly North America and northeast Asia.

1. **P. multiflorus** Chapm. ex Benth.—Stems up to 1.5 m tall, erect, smooth herbs. Cauline leaves glandular dotted, oblanceolate, entire or somewhat undulate-crenate, acute to the apex. Flowers conspicuous, nodding, in diffuse panicles; sepals covered with glandular hairs, 3–5 mm long, ovate-obtuse; corolla 2–2.5 cm long, white or slightly purplish white, throat inflated. Capsule 7–9 mm long. Dry sites, pinelands, scrub vegetation, Fla. to Ga. All year.

10. RUSSÉLIA Jacq.

Shrubs or subshrubs with opposite or whorled leaves; leaves sometimes scalelike. Flowers conspicuous in terminal cymes or panicles, calyx five-parted, sepals ovate; corolla two-lipped, funnelform or tubular, tube elongate, lobes short, stamens four, staminodes absent. Fruit a capsule, seeds many, cylindrical. About 15 species in tropical America.

1. **R. equisetiformis** Schlecht. & Cham. CORAL PLANT—Shrub up to 1 m tall or less, smooth, with numerous spreading branches. Cauline leaves 1–2 mm long, reduced to acute scales. Flowers in terminal cymes, sepals fused at the base, ovate, 4 mm long; corolla 2–2.5 cm long, bright red, lobes 4–5 mm long. Capsule ovoid, 5–6 mm long, mucronate. Disturbed sites, roadsides, peninsular Fla., W.I., naturalized from Mexico. All year. *R. juncea* Zucc.

11. BÚCHNERA L. BLUEHEARTS

Perennial herbs with sessile, opposite leaves progressively reduced toward the top. Flowers in terminal spike, calyx five-lobed, tubular; corolla salverform, almost radially symmetrical, pubescent within, the lobes spreading, stamens four. Stigma cylindric. Fruit an ovoid capsule. About 140 species, chiefly in the Old World tropics.

1. Corolla lobes over 5 mm long; larger leaves
linear to lanceolate . . . . 1. *B. elongata*
1. Corolla lobes less than 5 mm long; larger
leaves broadly lanceolate to ovate . . . 2. *B. floridana*

1. B. elongata Sw.—Stems pilose to rough pubescent up to 7 dm tall, slender, simple or branched above. Basal leaves oblong-obovate, 1-3 cm long, cauline leaves linear-lanceolate to lanceolate, 2-8 cm long, scabrous, entire or sparingly toothed. Spikes slender, calyx lobes deltoid-acute, about 1 mm long or less, pubescent; corolla violet blue or white, 9-12 mm long. Capsule ovoid, 5-6 mm long. Pineland, south Fla., W.I. All year.

2. B. floridana Gandoger—Stems glabrate or pilose pubescent up to 6 dm tall, slender, usually simple. Basal leaves 1-3 cm long, gland dotted, obovate, cauline leaves 3-6 cm long, broadly lanceolate to ovate. Spikes slender, calyx lobes deltoid-acute, about 1 mm long, glabrate, corolla violet or white, 5-8 mm long. Capsule ovoid, 4-5 mm long. Dry sites, roadsides, Fla. to Tex., N.C. on the coastal plain. Spring, fall.

12. AGALÌNIS Raf. FALSE FOXGLOVE

Erect herbs or subshrubs with linear, entire leaves. Flowers usually in racemes and axillary to reduced leaves, calyx five-lobed, radially symmetrical; corolla somewhat bilaterally symmetrical, tube campanulate and usually swollen beneath, stamens four, pubescent toward the base. Fruit a globose or subglobose capsule abruptly tipped, seeds many, reticulate. About 60 species, chiefly in the temperate regions of North and South America. *Gerardia* L. misapplied

1. Plants annual; corolla purplish.
  2. Fruiting pedicels over 6 mm long; plants
    succulent . . . . 1. *A. maritima*
  2. Fruiting pedicels up to 6 mm long; plants
    not succulent.
    3. Leaves and lower stems scabrous; leaves
      fascicled . . . . 2. *A. fasciculata*
    3. Leaves and lower stems glabrous or
      nearly so; leaves not fascicled.
      4. Flowers with pedicels, mostly less
        than 3 mm long . . . . . . 3. *A. harperi*
      4. Flowers with pedicels, 3-6 mm long.
        5. Leaves 1-2 cm long; corolla less
          than 2 cm long.
          6. Corolla with distinct yellow lines 4. *A. obtusifolia*
          6. Corolla purplish . . . . . 5. *A. filifolia*

5. Leaves 3–4 cm long; corolla more
    than 2 cm long . . . . .   6. *A. purpurea*
1. Plants perennial; corolla pink . . . .   7. *A. linifolia*

1. A. maritima Raf.—Annuals with succulent, glabrous, much branched stems up to 3.5 dm tall. Leaves up to 2–3 mm wide, mostly 2–3 cm long, linear to linear-lanceolate, scabrous above. Flowers in elongate racemes, calyx lobes deltoid, purplish, spreading; corolla pinkish purple, about 12–18 mm long, lobes fringed around the edges. Capsule globose, 5–6 mm long. Coastal areas, Fla. to Tex., Me., W.I. Fall.

Larger plants of southern and tropical distribution have been described as **A. maritima** var. **grandiflora** (Benth.) Pennell or **A. spiciflora** Engelm. These plants may be up to 6 dm tall and have leaves up to 4 cm long.

2. A. fasciculata (Ell.) Raf.—Annuals up to 1 m tall or more with angled, branched, scabrous stems. Leaves mostly 2–4 cm long, linear, scabrous above. Flowers in elongate racemes, several- to many-flowered, calyx lobes mucronate, lanceolate to subulate, 0.5–2 mm long, abruptly tipped; corolla 2.5–3.5 cm long, purplish, lobes spreading with posterior lobes pubescent near the base. Capsule 5–6 mm long, ovoid-globose. Pinelands, sandy soil, Fla. to Tex., Md., Mo. All year.

A. fasciculata var. **peninsularis** Pennell is a more slender plant with fascicles only moderately developed and stems scabrellous; it is found in the Everglades region. This plant may be conspecific with **A. filifolia.**

3. A. harperi Pennell—Annual up to 8 dm tall, smooth or nearly so and few-branched. Leaves 1–4 cm long, narrowly linear, scabrous above. Flowers in elongate racemes, calyx lobes deltoid-lanceolate to subulate with short, incurved hairs around the edges; corolla 15–18 mm long, pinkish purple with two yellow lines within, posterior lobes pubescent at the base. Capsule subglobose, 4–5 mm long. Low pinelands, Fla. to Ga., W.I. All year.

4. A. obtusifolia Raf.—Annual up to 8 dm tall, striate-angled, glabrous, and with stiffly ascending branches. Leaves 1–2 cm long, scabrous above. Flowers in elongate racemes, calyx lobes subulate, corolla pinkish purple, 1–1.5 cm long, posterior lobes pubescent at the base. Capsule 3–4 mm long, globose. Sandy soil, pinelands, reported from Lee County and presumably in south Fla.; Fla. to Miss., Pa., Ky. Fall. *Agalinis erecta* (Walt.) Pennell; *Gerardia parvifolia* (Benth.) Chapm.

5. A. filifolia (Nutt.) Raf.—Annual up to 8 dm tall, glabrous and branched. Leaves 1–2 cm long, filiform to narrowly linear, somewhat scabrous above. Flowers in racemes on pedicels 1–3 cm long, longer in fruit, calyx lobes linear, up to 1 mm long; corolla about 2 cm long, purple with spreading lobes. Capsule 4–5 mm long, ovoid-globose. Pinelands, Fla. to Ga. Fall. *A. keyensis* Pennell

6. A. purpurea (L.) Pennell—Annual with stems up to 1 m tall, much branched and slightly scabrous. Leaves 2–4 cm long, linear, scabrous above. Flowers in elongate racemes, calyx 2–4 mm long, lobes deltoid-acuminate; corolla 2–4 cm long, purplish, posterior lobe pubescent near the base. Capsule 5–7 mm long, globose. Glades, moist sandy soil, Fla. to Mexico, Nova Scotia, S. Dak., W.I. Summer, fall.

7. A. linifolia (Nutt.) Britt.—Perennial with elongated rootstocks, stems up to 16 dm tall, smooth, mostly simple or few-branched. Leaves 3–6 cm long, filiform to mostly linear, glabrous. Flowers in elongate racemes, calyx lobes subulate, corolla pink, 3–4 cm long. Capsule 6–7 mm long, globose to globose-ovoid. Low ground, Fla. to La., Del. on the coastal plain. Summer, fall.

13. MECARDÒNIA R & P

Perennial herbs with opposite, serrate leaves. Flowers small, yellow or white, solitary on long pedicels, axillary, each with two bractlets at the base, calyx five-parted, sepals elongate, unequal; corolla tubular-campanulate, bilaterally symmetrical, pubescent within, stamens four. Fruit an ovoid capsule, seeds many, reticulate. About four species in tropical and subtropical America.

1. Leaves lanceolate to oblanceolate; corolla
   white; plants erect . . . . 1. *M. acuminata*
1. Leaves broadly lanceolate to ovate; corolla
   yellow; plants procumbent-ascending.
   2. Sepals 6–10 mm long . . . . . . 2. *M. procumbens*
   2. Sepals 5–6 mm long . . . . . . 3. *M. tenuis*

1. M. acuminata (Walt.) Small—Stems erect or ascending up to 5 dm tall, glabrous, simple or few-branched. Leaves 2–4 cm long, lanceolate to oblanceolate, serrate above the middle. Flowers on filiform pedicels 1–3 cm long, mostly longer than the bractlets, sepals about 6–8 mm long; corolla mostly 1 cm long, white with longitudinal, purplish veins on the posterior side. Capsule 5–6 mm long, seeds dark gray. $2n = $ c. 42,44. Moist sites, sandy loam, Fla. to Tex., Del., Mo. Spring, fall. *Bacopa acuminata* (Walt.) Robinson

M. acuminata var. **peninsularis** Pennell, a smaller plant diffusely branched with leaves 1–2 cm long and corolla 7–8 mm long, is also reported for south Fla. but is apparently rare.

Family 168. **SCROPHULARIACEAE.** *Mecardonia acuminata:* a, flowering plant, × ½; b, flower, × 5; c, mature fruit and calyx, × 4.

2. M. procumbens (Mill.) Small—Stems procumbent and ascending, suberect, glabrous up to 4 dm tall, usually branched from the base. Leaves 1–2 cm long, ovate, serrate. Pedicels filiform, 1–2 cm long or longer, sepals 6–10 mm long, broadly lanceolate; corolla yellow or yellow white, about 1 cm long, sometimes shorter than the sepals. Capsule oblong, 4–6 mm long. Dry sites, sandy soil, south Fla., W.I., tropical America. All year.

3. M. tenuis Small—Stems prostrate or diffusely ascending up to 4 dm long. Leaves 1–2 cm long, ovate or lanceolate, serrate. Pedicels 1.5–3.5 cm long, sepals 5–6 mm long; corolla 7–8 mm long, lemon yellow with dark lines on the posterior side. Capsule 5–6 mm long. Dry sites, south Fla., Fla. Keys and W.I. All year. *Bacopa procumbens* var. *tenuis* Fern.; *B. tenuis* Standl.

14. LIMNÓPHILA R. Brown

Glabrous or somewhat pubescent herbs mostly in marshy or aquatic habitat. Leaves opposite, submersed blades divided into capillary segments. Flowers axillary, pedicelled or sessile, solitary, the upper ones forming a terminal raceme; bracts reduced, linear, calyx five-parted with narrow segments; corolla tube cylindric, two-lipped, stamens four, didynamous. Capsule oblong with four valves. About 30 species in warm regions of Asia, Africa, and Australia.

1. L. sessiliflora (Vahl) Blume—Annual herb with many glabrous or pubescent, partly submersed or floating stems. Leaves 1–3 cm long, emersed blades toothed, narrowly oblong, submersed leaves divided into numerous capillary segments. Flowers sessile, corolla 6–8 mm long, pale violet. Capsule somewhat compressed. Shallow water in ditches, wet places, often in dense colonies, peninsular Fla., collected in Glades County and presumably in our area, introduced and escaped from cultivation, from Old World tropics. All year.

The emersed tips of the plant produce flowers and fruits.

15. SCOPÁRIA L.

Herbs or subshrubs with opposite or verticillate gland-dotted leaves. Flowers pedicelled, solitary or in pairs in leaf axils, calyx four- to five-parted; corolla radially symmetrical, nearly rotate, four-cleft, densely pubescent within the throat, stamens four, about equal. Fruit a capsule, seeds many, angular. About six species in tropics and subtropics, America.

1. S. dulcis L. Sweet Broom—Annual usually branched up to 1 m tall with glabrous branches. Leaf blades 1–3 cm long, gland dotted, oblong-lanceolate to ovate, entire or serrate. Flowers solitary, pedicels 3–9 mm long, filiform, sepals obovate or oblong-acute, 1.5 mm long; corolla white, 5–10 mm long, lobes ovate with yellowish pubescence. Capsule ovoid-globose, bearing the persistent style about 2 mm long, seeds yellowish brown. $2n = 20,40$. Dry sites, Fla. to La., Ga., W.I., tropical America. All year.

169. BIGNONIÀCEAE   Bignonia Family

Trees and shrubs, often climbers with opposite or whorled mostly compound leaves. Flowers usually showy, bisexual, and bilaterally symmetrical; calyx fused, five-lobed or truncate; corolla funnelform or bell shaped, five-lobed, and sometimes two-lipped, stamens two or four fertile, alternate with the corolla lobes. Ovary superior, one- to two-locular. Fruit a capsule or fleshy. About 120 genera and 800 species, chiefly tropical.

1. Leaves all simple.
  2. Flowers solitary or clustered.
    3. Leaves oblanceolate to obovate, clustered two to four or more on short spurs . . . .   1. *Crescentia*
    3. Leaves oval to obovate, alternate, not clustered on short spurs . . . .   2. *Enallagma*
  2. Flowers racemose . . . .   3. *Tecoma*
1. Leaves trifoliolate, pinnately or palmately compound.
  4. Shrubby climbers, leaves with tendrils.
    5. Stamens somewhat exserted; corolla lobes with thin white margin . . .   4. *Pyrostegia*
    5. Stamens included; corolla lobes without white margin.
      6. Tendrils hooked or clawlike . . .   5. *Doxantha*
      6. Tendrils climbing with small disks, not clawlike . . . . . .   6. *Bignonia*
  4. Erect trees, shrubs, or vinelike without tendrils.
    7. Corollas yellow, red, or orange; vines or shrubs.
      8. Calyx lobes unequal . . . . .   7. *Tabebuia*
      8. Calyx lobes equal . . . . . .   3. *Tecoma*
    7. Corollas blue; trees . . . . . .   8. *Jacaranda*

1. CRESCÉNTIA L.

Trees with smooth, simple, often fascicled leaves. Flowers solitary or clustered, conspicuous, axillary or on lateral spurs, calyx membranous, two or five-parted; corolla bell shaped, five-lobed, unequal, stamens four, didynamous, disk annular. Ovary one-locular, fruit globose or ovoid with a hard pericarp, seeds many, wingless. About five species in tropical America.

1. **C. cujete** L. CALABASH TREE—Tree up to 10 m tall or more with spreading branches. Leaves 8–16 cm long, spatulate to oblanceolate, acute or short acuminate, fascicled. Flowers with stout peduncles, solitary, appearing spurred, calyx 2–3 cm long, lobes rounded or obtuse; corolla yellowish purple, 5–6 cm long, lobes undulate. Fruit subglobose to ellipsoid, 1–3 dm wide. Coastal hammocks, Fla. Keys, W.I., tropical America. Spring.

2. ENALLÁGMA BAILLON

Shrubs or trees with simple, alternate, entire leaves. Flowers terminal, on long peduncles, solitary or few-clustered; calyx membranous, two-lobed; corolla unequally five-lobed, stamens four, didynamous, included, disk annular. Ovary one-locular, ovules many. Fruit subcylindric or ellipsoidal, indehiscent. About three species in tropical America.

1. **E. latifolia** (Mill.) Small, BLACK CALABASH—Tree up to 10 m tall with smooth stems. Leaves 17–20 cm long, oval to obovate, abruptly acuminate. Flowers solitary on peduncles 2–3 cm long, calyx 2–3 cm long, lobed nearly to the middle or more; corolla purplish or yellowish, lobes toothed. Fruit subglobose, 6–8 cm long. Hammocks, Fla. Keys, W.I., tropical America. Spring.

3. TECÒMA Juss.

Shrubs, woody climbers, or trees with opposite, simple or pinnately compound leaves. Flowers conspicuous in terminal racemes or panicles, calyx five-toothed, cylindric-campanulate; corolla funnelform, five-lobed, somewhat two-lipped, stamens four, didynamous. Ovary two-locular, ovules numerous. Fruit an elongate capsule, seeds winged. About 16 species, warmer regions of Western Hemisphere. *Stenolobium* D. Don

1. Leaves simple . . . . 1. *T. gaudichaudii*
1. Leaves pinnately compound . . . . . 2. *T. stans*

1. **T. gaudichaudii** DC.—Smooth, erect shrubs up to 8 m tall, usually less. Leaves 10–15 cm long, opposite or alternate, simple, lanceolate to narrowly elliptic, serrate especially toward the apex. Flowers in terminal racemes, calyx five-toothed, 1–2 mm long, lobes deltoid; corolla yellow, 2–3 cm long. Capsule linear-acuminate, 8–10 cm long. Natural-

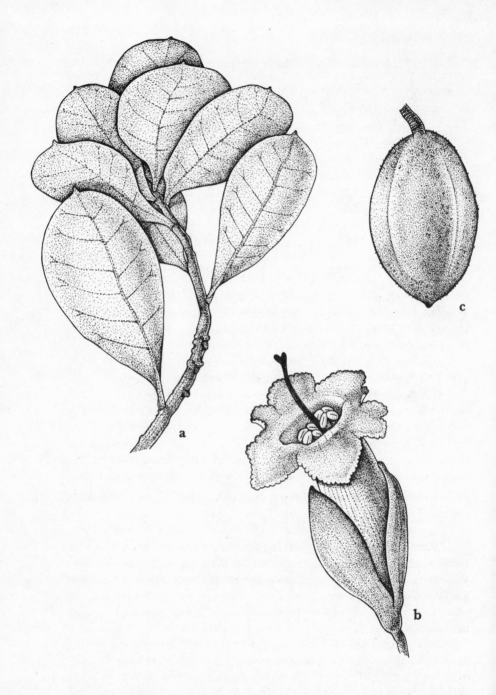

Family 169. **BIGNONIACEAE.** *Enallagma latifolia:* a, leafy branch, × ½; b, flower, × 1½; c, mature fruit, × ½.

ized in Dry Tortugas, possibly elsewhere in the Fla. Keys, from tropical America.

2. T. stans (L.) Juss. YELLOW ELDER—Smooth, erect shrubs up to 8 m tall. Leaves 10–30 cm long, pinnately compound, leaflets seven to thirteen, lanceolate to elliptic or oblong, serrate, acute acuminate at the apex. Flowers in terminal racemes, many-flowered, calyx 3–4 mm long, lobes deltoid or ovate, corolla 3–5 cm long, bright yellow, lobes broad. Capsule 1–2 dm long, linear, beaked. Hammocks, Fla. to Tex., W.I., tropical America, also in cultivation. All year.

Campsis radicans (L.) Seem. TRUMPET CREEPER, with leaves pinnately compound and nine to eleven leaflets, corollas orange or scarlet, occurs on the lower Fla. east coast and may be in our area.

4. PYROSTÈGIA Presl

Woody, climbing, evergreen vines with compound leaves of two to three leaflets and thin three-parted tendrils. Flowers in terminal panicles or branching inflorescence; calyx tubular or campanulate, often truncate; corolla tubular-funnelform, slightly curved, stamens exserted, disk annular. Ovary narrow, ovules many. Capsule linear, valves leathery. Four species in tropical America.

1. P. ignea Presl—Woody vines with striated branches. Leaflets 5–8 cm long, ovate-oblong, shallowly crenate-undulate, petioles slightly pubescent. Flowers crowded into pendulous panicles; calyx glandular, corolla reddish orange, 5–7 cm long, lobes obtuse, with conspicuous white, velutinous margins. Fruit to 30 cm long. Hammocks, south Fla., in Broward County and probably in our area, naturalized from Brazil. Spring, summer.

5. DOXÁNTHA Miers

Woody, climbing, evergreen vines with opposite, unifoliolate to trifoliolate leaves and a terminal clawlike, three-parted tendril. Flowers solitary or in short panicles; calyx truncate with regular or lobed margin; corolla funnelform, conspicuous, stamens included. Ovary somewhat four-angled with many ovules. Capsule linear, seeds with membranaceous wings. Two species in tropical America.

1. D. unguis-cati (L.) Rehder, CLAW VINE—Woody climbing vines to 10 m long or more with smooth stems. Leaflets 3–6 cm long, oblong-ovate or elliptic. Flowers clustered on slender peduncles 1–3 cm long; calyx 1–1.5 cm long, greenish; corolla 8–9 cm long, yellow with orange lines in the throat. Capsule 3–4 dm long, about 1 cm in diameter, seeds 2–3 cm long, winged. Persistent around former habitations, south Fla.,

W.I., probably not well established in our area, native to tropical America. Spring, summer.

6. BIGNÒNIA L.

Woody climbing vines with pinnately compound leaves bearing tendrils. Flowers solitary or cymose clustered, calyx five-toothed or truncate, somewhat tubular; corolla funnelform, five-lobed, two-lipped, stamens four, sometimes with an additional staminode. Ovary two-locular. Fruit a flattened capsule. About three species, chiefly tropical. *Anisostichus* Bureau; *Batocydia* Mart.

   1. **B. capreolata** L. CROSS VINE—Woody vines up to 10 m long or more with glabrous stems. Leaflets 5–10 cm long or longer, and a branched tendril lanceolate to elliptic. Flowers mostly in few-flowered clusters, peduncles one-flowered, calyx 5–8 mm long; corolla 4–5 cm long, red or orange without lines in the throat. Capsule 1–2 dm long, seeds winged, about 4 cm long. Climbing over walls, hammocks, Fla. to La., Md., Mo. Spring. *Anisostichus crucigera* (L.) Bureau

7. TABEBÙIA Gomez

Trees or shrubs with opposite, simple or palmately compound leaves. Flowers conspicuous, in terminal cymes or panicles, calyx tubular, lobed or toothed; corolla funnelform, five-lobed, nearly radially symmetrical or somewhat two-lipped, stamens four, didynamous. Ovary two-locular, ovules many. Fruit a linear, nearly terete capsule, seeds numerous, winged. About 100 species in tropical America.

   1. **T. pentaphylla** Hemsl.—Tree up to 20 m tall with grayish bark. Leaves 8–15 cm long, palmately three- to five-foliolate or sometimes apparently simple, leaflets oblong or elliptic, lepidote, shiny above, long petioled. Inflorescence many-flowered, pedicels about 2 cm long or less, calyx 1 cm long, lobes unequal; corolla 5–7 cm long, pink, rose, or white. Capsule 1–2 dm long. Hammocks and disturbed sites, south Fla., Broward County, W.I., tropical America. Spring, summer. *T. pallida* (Lindl.) Miers

   This species may not be well established in our flora.

8. JACARÁNDA Juss.

Trees with opposite, usually twice-pinnately compound leaves. Flowers in terminal or axillary panicles, calyx small, tubular, five-toothed or truncate; corolla campanulate, five-lobed, somewhat two-lipped, disk cushionlike, fertile stamens four, didynamous, staminode club shaped and pubescent at the top. Ovary two-locular. Fruit an oblong to orbicular woody capsule, seeds numerous and winged. About 40 species in tropical America.

    1. **J. acutifolia** H & B, JACARANDA—Trees up to 20 m tall or more. Leaves twice-pinnately compound, with sixteen or more pairs of pinnae and each of these further divided into fourteen to twenty-four pairs of oblong, cuspidate pinnules, the terminal one long acuminate, the foliage appearing fernlike. Flowers in pyramidal panicles up to about 2 dm long, corolla blue, mostly 5 cm long; corolla tube slightly curved, two-lipped. Capsule orbicular, about 5 cm wide. Persisting around former dwellings and locally escaped from cultivation in south Fla., naturalized from Brazil. Spring.

170. LENTIBULARIÀCEAE BLADDERWORT FAMILY

    Mostly insectivorus herbs, terrestrial, aquatic, or epiphytic. Leaves in rosettes or alternate or reduced. Flowers solitary or in racemes or spikes, scapose, bisexual, bilaterally symmetrical, calyx two- to five-parted, often two-lipped; corolla five-lobed, spurred, two-lipped, fertile stamens two, staminodes two. Ovary superior, one-locular. Fruit a capsule that opens irregularly. Five genera and about 300 species, widely distributed. *Pinguiculaceae*

1. Leaves ovate, rosulate, without small bladders; calyx five-lobed . . . . . . 1. *Pinguicula*
1. Leaves finely dissected, filamentous, or absent, bearing small bladders; calyx two-
   lobed . . . . 2. *Utricularia*

1. PINGUÍCULA L. BUTTERWORT

    Perennial, insectivorous herbs with basal rosettes of flat leaves and inflorescence scapose. Flowers solitary or few, calyx five-lobed, somewhat two-lipped; corolla spurred at the base, five-lobed, the lobes spreading unequal and two-lipped, with a pubescent or spotted palate. Style very short, capsule small. About 35 species in the Northern Hemisphere. Small insects are trapped by the viscid pubescence of the leaf surfaces and the inrolling of leaf margins, and later the insects are digested.

1. Corolla bluish, pink, or white, less than 2
   cm long . . . . 1. *P. pumila*
1. Corolla bright yellow, over 2 cm long . . 2. *P. lutea*

1. P. pumila Michx.—Small perennial herbs with moist or slimy pubescence. Leaves 1–3 cm long, ovate or elliptic, obtuse at the apex, gradually narrowed to the base, upper surfaces viscid pubescent. Scape 5–20 cm tall, slender, corolla pale blue, pinkish, violet, or white, 1–2 cm wide, spur subulate, about 3 mm long. Low pineland, Fla. to Tex., S.C., on the coastal plain, W.I. All year.

2. P. lutea Walt.—Perennial herbs up to 3 dm tall or more with moist or slimy pubescence. Leaves 2–6 cm long, ovate or narrowly so, obtuse at the apex, narrowed to the base, upper surfaces viscid pubescent. Scape slender, flowers solitary; corolla bright yellow, 2–4 cm wide, spur cylindric, 0.5–1 cm long. Low pineland, Fla. to N.C., La., on the coastal plain. Spring.

2. UTRICULÀRIA L. BLADDERWORT

Aquatic herbs or rooting in wet soil. Leaves linear or entire or variously dissected. Flowers each subtended by a small bract, in short racemes or appearing solitary and terminal, calyx two-parted into upper and lower segments; corolla yellow, white, or violet, two-lipped, the upper two-lobed or subentire, the lower three-lobed or entire, raised at the base into a conspicuous palate, tube prolonged into a spur or sac. Small, aquatic animals are trapped in little bladders on the submerged dissected leaves; the bladders have a valvular lid and often a few minute hairs at the opening. *Aranella* Barnhart; *Stomoisia* Raf.; *Calpidisca* Barnhart; *Lecticula* Barnhart; *Biovularia* Kam.; *Vesculina* Raf.

1. Corolla purple, blue, pinkish, or white.
  2. Branches conspicuously whorled and dissected; scape arising from axils of branches; aquatic . . . . 1. *U. purpurea*
  2. Branches alternate or opposite; scape arising from center of radiating branches; terrestrial . . . . 2. *U. resupinata*
1. Corolla yellow.
  3. Plants with whorled, leaflike structures modified as floats . . . . 3. *U. inflata*
  3. Plants with finely dissected leaves and stems, without leaflike floats.
    4. Plants aquatic.
      5. Scape one- to four-flowered; leaves rootlike . . . . 4. *U. biflora*
      5. Scape five- to fifteen-flowered; leaves matted-filamentous . . . . . 5. *U. foliosa*
    4. Plants terrestrial.
      6. Corolla 0.6–1.2 cm long; scape mostly 6–12 cm long.

Family 170. **LENTIBULARIACEAE.** *Utricularia inflata:* a, flowering plant, × ½; b, flower, × 4; c, mature fruit, × 4½.

7. Calyx fringed with marginal hairs 6. *U. fimbriata*
7. Calyx entire, without marginal
   hairs . . . . 7. *U. subulata*
  6. Corolla larger, 2.0–2.5 cm long; scape
    6–20 cm long . . . . . 8. *U. cornuta*

1. **U. purpurea** Walt.—Aquatic herbs with whorled long-petioled, compound leaves that are divided into many fine segments; leaves and stems slender, free-floating; bladders ovoid or subglobose, bearing short glandular hairs at the openings, hairs produced on the tips of some leaf segments. Flowering scape 5–15 cm tall with one to five flowers arising from axils of branches, corolla purple or pink, 1–1.3 cm wide, spur appressed to the lower lip. Ponds, ditches, or sluggish streams, Fla. to La., Nova Scotia, Wis., W.I., tropical America. Summer, fall. *Vesiculina purpurea* (Walt.) Raf.

2. **U. resupinata** Greene—Terrestrial herbs with filiform, underground stems and simple, or somewhat branching, linear-subulate leaves bearing minute bladders. Flowering scapes filiform, 2–15 cm tall, terminating in a single pedicel subtended by a cup-shaped bract; flower solitary; corolla purple, 6–12 mm long, lower lip much longer than the upper, spur curved. Wet pinelands, edges of ponds, Fla. to Nova Scotia, Wis. Fall. *Lecticula resupinata* (Greene) Barnhart

3. **U. inflata** Walt.—Aquatic herbs with whorled leaves having inflated petioles, forming floats; stems prolonged, submerged, bearing leaves dichotomously dissected into filiform segments that bear many small bladders. Scape arising from the floats, erect, mostly 10–15 cm wide, lower lip three-lobed. Ditches, ponds, canals, Fla. to Tex., N.J. Spring, fall.

U. inflata var. **minor** Chapm.—is a smaller form with scape only up to 5 cm tall, flowers one to five per raceme, and corolla about 1.5 cm wide. *U. radiata* Small

4. **U. biflora** Lam.—Aquatic herb with stems radiating from the base of scapes, floating or creeping. Leaves few, delicate larger ones with two to many segments up to 5 mm long, bladders none or scattered. Scape 5–10 cm tall, bearing one to four flowers; corolla yellow, 1–1.7 cm long, spur half as long to about as long as the lower lip. Shallow ponds, Fla. to Tex., Mass., Okla. Spring, fall. *U. pumila* Walt.; *U. macrorhyncha* Barnhart

5. **U. foliosa** L.—Aquatic herb with submerged stems. Leaves many, 3–6 cm long, the blades forked and dissected into very fine capillary segments bearing few to many bladders. Scapes erect up to 3 dm tall, racemes tightly five- to fifteen-flowered, pedicels recurving at maturity, calyx 3–5 mm long, corolla yellow, 1–2 cm wide, lower lip slightly

three-lobed, spur slender-conic, slightly curved. Ponds, brackish water, Fla. to La., W.I. All year.

6. U. fimbriata HBK—Terrestrial herbs with a basal rosette of linear leaves 5–6 mm long or leaves apparently absent. Scapes 5–15 cm tall, bearing one to seven flowers in a spike or subcapitate, calyx fringed with marginal hairs; corolla 6–8 mm wide, spur about as long as the lower lip. Low pineland, south Fla., W.I., tropical America. All year. *Aranella fimbriata* (HBK) Barnhart

7. U. subulata L.—Terrestrial herbs with very delicate, filiform underground parts that bear bladders on branchlets. Leaves if present up to 1 cm long, linear. Scape filiform, erect, up to 20 cm tall with one to ten flowers on elongate pedicels, bracts ovate, 1–2 mm long, peltate, calyx entire; corolla yellow, 0.5–1.0 cm long, lower lip 4–7 mm long with a prominent palate, spur about as long as the lip and appressed to its lower surface. Low wet ground, Fla. to Tex., Nova Scotia, Ark. All year. *Setiscapella subulata* (L.) Barnhart

8. U. cornuta Michx.—Terrestrial herb with finely branching roots. Leaves small, usually subterranean, linear to filiform, bearing minute bladders. Scape erect up to 20 cm tall, bearing one to six flowers in a spikelike raceme, bracts ovate, 1–2 mm long, pedicels 1–2 mm; corolla yellow, 2–2.5 cm long, lower lip much larger than upper with a raised palate, spur 7–14 mm, slightly curved downward. Margins of ponds, wet shores, Fla. to Tex., Newfoundland, Minn., W.I. Summer. *Stomoisia cornuta* (Michx.) Raf.

171. ACANTHÀCEAE Acanthus Family

Herbs, shrubs, and climbing vines with opposite leaves that often contain cystoliths. Flowers bisexual, bilaterally symmetrical or nearly radially symmetrical, axillary or terminal, solitary or more often in cymes, spike, racemes, or panicles, often with conspicuous bracts; calyx four- to five-lobed or toothed; corolla five-lobed or one- to two-lipped, stamens four, didynamous, or two, epipetalous, disk variously developed. Ovary superior, two-locular, ovules two to ten per locule. Fruit a capsule that dehisces explosively in most genera; seeds attached by characteristic hooked projections to capsule. About 250 genera and 2600 species, chiefly in the tropics and subtropics.

1. Climbing or creeping vines; leaves hastate
   or cordate . . . . 1. *Thunbergia*
1. Erect or reclining herbs; leaves various,
   linear, lanceolate, or ovate.

2. Plants scapose, larger leaves all basal.
  3. Basal leaves lance-ovate to ovate, more
    than 1 cm wide . . . .   2. *Stenandrium*
  3. Basal leaves linear-spatulate to nar-
    rowly oblanceolate, less than 1 cm
    wide . . . .   3. *Elytraria*
2. Plants not scapose, normal cauline leaves
  present.
  4. Corolla lobes mostly equal.
    5. Inflorescence spicate; conspicuous
      floral bracts broadly ovate-folia-
      ceous . . . .   4. *Blechum*
    5. Inflorescence racemose or axillary;
      bracts not ovate-foliaceous.
      6. Corolla tube yellow, lobes pink,
        lavender, or yellow . . . .   5. *Asystasia*
      6. Corolla lavender, purple, or white,
        tube not yellow.
        7. Floral bracts foliose, margins spi-
          nose; calyx four-parted, seg-
          ments unequal . . . . .   6. *Barleria*
        7. Floral bracts small, inconspic-
          uous, calyx five-parted, seg-
          ments equal or nearly so.
          8. Corolla over 2.5 cm long, tube
            longer than throat . . .   7. *Ruellia*
          8. Corolla less than 2.5 cm long,
            tube shorter than throat .   8. *Dyschoriste*
  4. Corolla strongly two-lipped.
    9. Leaves distinctly petioled; corolla red
      or crimson . . . .   9. *Dicliptera*
    9. Leaves sessile or subsessile; corolla
      lavender or purple . . . . .   10. *Justicia*

1. THUNBÉRGIA Retz.   CLOCK VINE

Shrubs or herbs mostly climbers with opposite, simple leaves. Flowers conspicuous, solitary on axillary peduncles or in racemes, subtended by two more- or less-foliaceous bracts; calyx truncate or ten- to fifteen-toothed, corolla funnelform or salverform, five-lobed, stamens four, didynamous. Ovary two-locular with two ovules in each locule. Capsule subtended by persistent bracts, seeds subglobose. About 200 species, chiefly in Old World tropics and subtropics.

1. Petioles conspicuously wing margined . .   1. *T. alata*
1. Petioles not wing margined . . . . .   2. *T. fragrans*

    1. T. alata Bojer ex Sims, BLACK-EYED SUSAN—Perennial, herbaceous. twining vine with pubescent stems up to 1 m long. Leaves 3–12 cm long, ovate to triangular-ovate, acute or obtuse at the apex, cordate or somewhat hastate at the base, petiole conspicuously winged. Flowers solitary on long axillary peduncles, calyx lobes subulate, corolla usually yellow, orange, or cream colored with a dark purple throat, five-lobed, bracts green. Capsule about 3 cm long, beak flattish. $2n = 18$. Disturbed sites, south Fla., especially near former habitations, widely distributed in the tropics, naturalized from Africa. All year.

    2. **T. fragrans** Roxb. WHITE THUNBERGIA—Perennial, herbaceous or somewhat woody twining vine with smooth stems up to 2 m long. Leaves 4–12 cm long, lanceolate to deltoid-ovate, apex acute, more or less hastate and coarsely one- to two-toothed at the base. Flowers solitary on axillary peduncles, calyx lobes lanceolate, unequal; corolla white, fragrant, five-lobed, the lobes toothed at the apex, bracts green, acute. Capsule about 2 cm long, beak flattened. Disturbed sites, south peninsular Fla., naturalized from India. All year. *T. volubilis* Pers.

    Thunbergia grandiflora, the LARGE-FLOWERED THUNBERGIA, a twining vine with broadly toothed, ovate leaves and blue or white, almost two-lipped corollas 7–9 cm long, is widely cultivated in south Fla. It is a native of India. Other Thunbergias are also popular ornamentals in our area. $2n = 28,56$.

2. STENÁNDRIUM Nees

    Mostly small, scapose herbs with basal rosettes. Flowers bracteate in terminal spikes, calyx five-lobed, sepals nearly equal, corolla funnelform, pink or purplish, four- to five-lobed, the lobes unequal, stamens four, disk inconspicuous. Ovary two-locular with two ovules in each locule, stigma two-lobed. Capsule oblong, seeds flattened. About 30 species in tropical and subtropical America. *Gerardia* sensu Small

    1. **S. dulce** (Cav.) Nees var. **floridana** Gray—Small scapose herb up to 6 cm tall. Leaves basal, ovate or elliptic, 2–3 cm long. Flowers in a terminal bracteate spike, calyx lobes subulate, about 1 cm long, corolla rose purplish, about 1.5–2 cm long. Capsule 9–12 mm long. Grassy roadsides, low pineland, endemic to south peninsular Fla., Fla. Keys. All year. *Gerardia floridana* (Gray) Small

3. ELYTRÀRIA Richard ex Michx.

    Stiff, erect, perennial herbs mostly with basal leaves and upper leaves reduced to rigid, sheathing scales. Flowers in terminal, bracteate spike; calyx scarious, four-lobed, slightly unequal; corolla blue or white, slightly two-lipped, stamens two, barely exserted. Ovary two-locular, ovules mostly six to ten in each locule. Capsule narrow with explosive dehiscence. About 15 species, chiefly in the American tropics. *Tubiflora* J. F. Gmel.

1. **E. caroliniensis** (J. F. Gmel.) Pers. var. **angustifolia** (Fern.) Blake —Stems up to 6 dm tall. Basal leaves 5–30 cm long, linear to linear-spatulate or narrowly oblanceolate. Flowers in terminal spikes, bracts apiculate, ciliate, calyx lobes 6–7 mm long, pubescent at their tips, corolla small, lobes shorter than the tube. Capsule 4–5 mm long. Low ground, endemic to south Fla. Spring, fall. *Tubiflora angustifolia* (Fern.) Small

4. BLÉCHUM P. Browne

Perennial herbs with opposite, petioled leaves. Flowers small, in dense terminal or axillary bracteate spikes; calyx deeply five-parted, the slightly unequal segments narrowly linear; corolla funnelform, five-lobed, the lobes rounded and almost equal, stamens four, didynamous. Ovary two-locular, each locule containing few to six ovules. Capsule ovoid to subspherical, seeds orbicular. About ten species in American tropics.

1. **B. brownei** Juss.—Herb with erect or ascending, puberulent stems up to 7 dm tall, branched above. Leaves 2–7 cm long, ovate, apex acute, glabrous or somewhat pubescent, cystoliths conspicuous. Flowers very reduced in dense, four-sided spikes 3–6 cm long; bracts ovate, 1–2 cm long, acute; corolla whitish, slightly longer than subtending bract. Capsule about 6 mm long, oblong, puberulent. $2n = 34$. Moist places, disturbed sites, margin of mangrove areas, naturalized from south Fla. and Fla. Keys, W.I., South America. All year. *B. pyramidatum* (Lam.) Urban

5. ASYSTÀSIA Blume

Perennial herbs or shrubs with entire or somewhat toothed leaves. Flowers in terminal racemes or panicles, bracts small, linear; calyx five-parted, the lobes linear or narrowly oblong, corolla funnelform, five-lobed, the lobes about equal, stamens four, didynamous, or sometimes two reduced to staminodes, anthers minutely spurred at the base, disk annular or cupulate. Ovary two-locular, two ovules in each locule. Capsule stipitate, oblong, two- to four-seeded. About 40 species in the Old World tropics.

1. **A. gangetica** (L.) T. Anders.—Perennial herbs with decumbent or reclining pubescent stems up to 1 m long. Leaves 3–10 cm long, ovate to broadly so, acute. Flowers in one-sided racemes 8–15 cm long, corolla 3–4 cm long, purplish, yellow, or white, funnelform. Cap-

sule 2–3 cm long, pubescent. $2n = 50,52$. Disturbed sites, lawns, margins of pinelands, south Fla., W.I., naturalized from India. Spring, summer. *A. coromandeliana* Nees

Odontonema strictum Kuntze, an erect shrub with bright red flowers, has escaped locally in Brickell Hammock, Miami, but apparently has not spread widely in south Fla. It is also cultivated.

6. BARLÈRIA L.

Shrubs or herbs with erect stems, armed or unarmed, and entire leaves. Flowers solitary or clustered in leaf axils or in terminal spikes; floral bracts large and often spinose, calyx four-parted, two outer segments large, interior ones small, narrow; corolla five-lobed, the lobes subequal, tube long, stamens four, didynamous. Ovary two-locular, ovules two in each locule, style entire. Capsule ovoid or oblong, four-seeded. About 250 species, chiefly in the Old World tropics.

1. **B. cristata L.**—Erect herb or subshrub up to 1 m tall, stems with scattered, yellowish pubescence. Leaves 5–10 cm long, broadly lanceolate, elliptic to narrowly ovate, entire, petioled, pubescent. Flowers axillary, sessile or nearly so, in bracteate clusters, bracts elliptic, whitish or greenish, strongly veined, soft pubescent, margins with conspicuous weak spines; corolla blue or white, lobes nearly equal, obtuse. Disturbed sites, persistent around former dwellings, locally escaped from cultivation in south Fla., naturalized from India. Spring, summer.

Possibly this species is not well established outside of cultivation in south Fla.; a second species also in cultivation, **B. nitida** Jacq., may be found locally as an escape. It is distinguished by its shiny bracts, glabrous except for pubescence along the veins, and leaves 3–5 cm long.

7. RUÉLLIA L.

Perennial herbs or shrubs with opposite mostly entire leaves. Flowers sessile or pedunculate, solitary or clustered, in leaf axils or in terminal racemes or panicles; calyx five-parted, segments equal, narrow; corolla funnelform, five-lobed, the lobes spreading and equal or nearly so, or the flowers small, tubular and cleistogamous; stamens four, didynamous, connivent or opposite the corolla lobes. Ovary two-locular, ovules mostly two to ten per locule, style long and slender, recurving at the apex and stigma simple with two equal branches or two unequal branches. Capsule narrowly cylindrical, opening sometimes suddenly and expelling the orbicular, compressed seeds. A large, complex genus with about 250 species, chiefly in the tropics and subtropics. *Dipteracanthus* Nees

786 A FLORA OF TROPICAL FLORIDA

1. Flowers borne in a terminal raceme or narrow panicle . . . . 1. *R. lorentziana*
1. Flowers borne in leaf axils, not in terminal racemes or spikes.
  2. Flowers distinctly peduncled, longer peduncles over 1 cm long; capsules over 2 cm long.
    3. Leaves lance-ovate to elliptic, larger blades over 2 cm wide . . . . . 2. *R. malacosperma*
    3. Leaves linear-lanceolate to narrowly lanceolate, blades mostly less than 2 cm wide . . . . 3. *R. brittoniana*
  2. Flowers subsessile or on very short peduncles; capsules less than 2 cm long . 4. *R. caroliniensis*

1. R. lorentziana Griseb.—Erect herb simple or divergently branching near the base up to 1 m tall, stems quadrangular and grooved, covered with pale cystoliths. Leaves up to 10 cm long, oblong-ovate to somewhat cordate, obscurely or irregularly serrate. Flowers in a narrow, terminal panicle, densely glandular-pubescent, corolla pale blue violet. Capsule 2–2.4 cm long, glandular-pubescent. Disturbed sites, south Fla., near Homestead, naturalized from Argentina. Spring, summer.

2. R. malacosperma Greenm.—Erect herbs with one to several stems to 1 m tall. Leaves to 12 cm long, larger blades over 2 cm wide, lance-ovate to elliptic, acute acuminate at the apex, cuneate to the base, margins undulate-dentate. Flowers solitary or in few-flowered cymes on axillary peduncles; calyx lobes 1–2 cm long, linear-lanceolate, corolla deep lavender or blue, 4–5 cm long. Capsule 2–2.5 cm long, cylindrical, brown, seeds suborbicular, about 2 mm wide. $2n = 34$. Disturbed sites, Key West, Tex., Mexico, naturalized from tropical America. Spring, summer.

3. R. brittoniana Leonard ex Fern.—Erect herb with stems one to several to 1 m tall, glabrous, usually branched near the top. Leaves to 30 cm long, linear-lanceolate to narrowly lanceolate, larger blades less than 2 cm wide, acuminate to the apex, tapering to the base, entire to undulate. Flowers solitary or in few-flowered cymes on axillary peduncles. Calyx lobes linear-lanceolate, 0.5–1 cm long; corolla lavender or blue, 2–4 cm long; cleistogamous flowers small, tubular, greenish brown. Capsule 2–2.5 cm long, cylindrical, seeds suborbicular, about 2 mm wide. $2n = 34$. Disturbed sites, peninsular Fla., scattered sites in southeast U.S., Tex., naturalized from Mexico. Spring, summer.

4. R. caroliniensis (J. F. Gmel.) Steud.—Erect or decumbent herbs with simple or divergently branching stems up to 9 dm tall, usually less, glabrous to villous. Leaves, at least the lower, elliptic to narrowly obovate, or spatulate, hispid to glabrate, mostly 2–10 cm long. Flowers in axillary clusters or solitary on very short peduncles or subsessile, calyx lobes linear, 1–2.5 cm long, glabrous to ciliate, corolla lavender, blue, or lilac, rarely white, 2–5 cm long. Cleistogamous flowers tubular, greenish brown. Capsule 1–1.8 cm long, glabrous to hirtellous. $2n = 34$. Fla. to Tex., N.J., Ind. Spring, summer, or all year. A highly variable species represented in Fla. by two distinctive subspecies:

1. Stems stiffly erect, simple, glabrous or nearly
   so; leaves glabrous or obscurely puberulent . . . .   4a. ssp. *caroliniensis*
1. Stems erect or decumbent, divergently
   branching, villous hirsute; leaves villous
   to strigose . . . .   4b. ssp. *ciliosa*

4a. **R. caroliniensis** ssp. **caroliniensis**—Pineland, dry soil, south Fla. to Va., Ind. Represented in our area by var. **caroliniensis** (*R. parviflora* sensu Small; *R. hybrida* Pursh) and by var. **succulenta** (Small) R. W. Long, an endemic with stiffly erect stems and purplish or reddish leaves. (*R. succulenta* Small).

4b. **R. caroliniensis** ssp. **ciliosa** (Pursh) R. W. Long—Pinelands, glades, south Fla., S.C., Miss. Represented in our area by endemic var. **heteromorpha** (Fern.) R. W. Long. (*R. heteromorpha* Fern.; *R. hybrida* sensu Small).

Hybrids occur commonly between these two varieties.

8. DYSCHORÍSTE Nees

Perennial low herbs sometimes woody at the base, with opposite, entire, sessile or subsessile leaves. Flowers axillary, solitary, or clustered; calyx lobes five, deeply cleft, linear, corolla five-lobed, somewhat two-lipped, usually blue or purplish, rarely white, tube short, stamens four, didynamous, anthers mucronate or appendaged at the base. Ovary with one to two ovules per locule. Capsule short-cylindrical with two to four seeds, not explosively dehiscent. About 100 species in the tropics and subtropics. *Calophanes* D. Don; *Apassalus* Kobuski

1. **D. oblongifolia** (Michx.) Kuntze, TWINFLOWER—Low erect herb up to 2 dm tall. Leaves 1–3 cm long, few, narrowly ovate to obovate. Flowers subsessile in leaf axils, calyx lobes subulate, about 1–1.5 cm long, corolla about 1 cm wide, blue or purplish. Capsule about 1 cm long. $2n = 30$. Low pinelands, south Fla. to N.C., La. on the coastal plain. Two varieties may occur in our area:

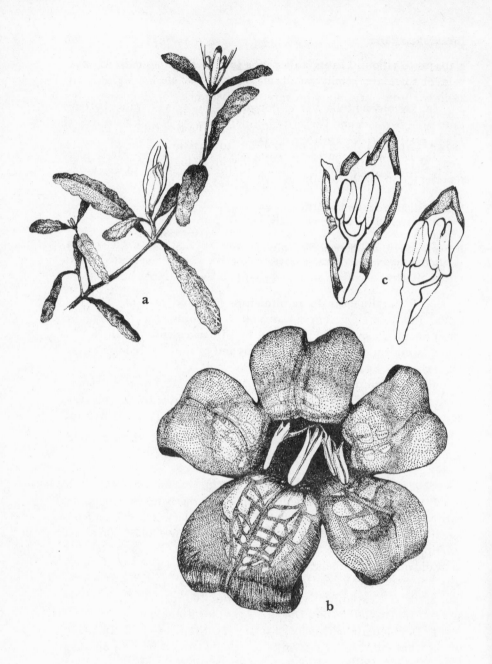

Family 171. **ACANTHACEAE.** *Dyschoriste oblongifolia:* a, flowering branch, × 1; b, corolla and stamens, × 2½; c, longitudinal section of floral bud showing didynamous stamens, × 2½.

1. Leaf blades elliptic; corolla about 2 cm long   1a. var. *oblongifolia*
1. Leaf blades linear; corolla about 1 cm long .   1b. var. *angusta*

1a. D. oblongifolia var. **oblongifolia**—Sandy soil, pine barrens, sand hills, Sarasota and Brevard Counties, north to N.C., La. on coastal plain, possibly in our area. Spring, fall.
1b. D. oblongifolia var. **angusta** (Gray) R. W. Long—Moist pinelands, disturbed sites, burned areas, roadsides, endemic to south Fla. and Fla. Keys. All year. *D. angusta* (Gray) Small

9. DICLÍPTERA Juss.

Perennial herbs with opposite, entire, petioled leaves. Flowers sessile in conspicuously bracteate spikes, calyx deeply five-parted, lobes equal, narrow; corolla usually strongly two-lipped, upper lip erect; stamens two, filaments elongate, but shorter than upper lip, disk usually cupular. Ovary with two ovules per locule. Capsule ovoid, short stipitate, seeds four or two. About 150 species in tropical and warm temperate regions. *Diapedium* Koenig; *Gatesia* Gray; *Yeatesia* Small
  1. D. assurgens (L.) Juss.—Perennial herbs up to 1 m tall, stiffly erect, often diffusely branching. Leaves 2–10 cm long, elliptic-lanceolate to ovate, margins entire. Flowers in terminal or axillary lax spikes, calyx 3–4 mm long, lobes narrowly lanceolate, corolla 2–3 cm long, red or crimson, two-lipped, the stamens somewhat exserted beneath the upper lip. Style elongate, stigma obscurely two-lobed, capsule ovoid, 7–8 mm long. $2n = 80$. The varieties occur in our area:

1. Outer bracts subulate-attenuate .   .   .   .   1a. var. *assurgens*
1. Outer bracts spatulate-mucronate   .   .   .   1b. var. *vahliana*

1a. D. assurgens var. **assurgens**—Disturbed sites, sandy soil, restricted to extreme south Fla. and Fla. Keys, W.I., not common in our area. All year.
1b. D. assurgens var. **vahliana** (Nees) Gomez—Disturbed sites, roadsides, shell mounds, coral soil, hammocks, peninsular Fla. to Fla. Keys, W.I., our common variety. All year. *D. vahliana* Nees
  Eranthemum pulchellum Andrews, a shrub to 1 m tall with long-petioled leaves and flowers in axillary and terminal spikes with blue corollas, is cultivated in south Fla. and may be locally escaped. It is native to India. *E. nervosum* (Vahl) R. Brown

10. JUSTÍCIA L. WATER WILLOW

Shrubs or herbs with opposite, entire leaves. Flowers on axillary peduncles, solitary, racemose, or spicate, sometimes with conspicuous bracts; calyx five-lobed, lobes subequal; corolla strongly two-lipped, straight or slightly curved, upper lip erect, lower three-lobed, stamens two. Stigma two-lobed or entire, capsule ovoid or oblong with two to four seeds. A large, complex genus with over 300 species, chiefly in the tropics. *Dianthera* L.

1. **J. ovata** (Walt.) Lindau—Perennial herb up to 5 dm tall. Leaves 3–10 cm long, linear to linear-elliptic, acuminate, sessile or nearly so. Flowers two to few or several in interrupted spikes on axillary peduncles, calyx lobes narrowly linear 5–7 mm long; corolla pale purplish, 1–2 cm long or less. Capsule 1–1.5 cm long. Two varieties occur in our area:

1. Corolla 1–2 cm long; leaf blades linear to
    linear-elliptic . . . . 1a. var. *lanceolata*
1. Corolla 1 cm long or less; leaf blades nar-
    rowly linear-lanceolate . . . . . . 1b. var. *angusta*

1a. **J. ovata var. lanceolata** (Chapm.) R. W. Long—Wet grounds, bottom lands, pineland ponds, south Fla. to Tex., Tenn. *J. lanceolata* (Chapm.) Small

1b. **J. ovata var. angusta** (Chapm.) R. W. Long—Low pinelands, marshy places, glades, endemic to peninsular Fla. *J. angusta* (Chapm.) Small

A large flowered species with corolla 2–3 cm long, **J. crassifolia** Chapm. occurs in Gulf County, Fla. The common SHRIMP PLANT, **Justicia brandegeana** Wasshausen & Smith (*Beloperone guttata* T. S. Brandg.) may persist for short periods around old dwellings, in citrus groves, and in waste places. It has conspicuous overlapping, reddish brown or greenish, ovate, pubescent bracts subtending the slender, two-lipped white corolla. It is a native of Mexico. All year.

172. PLANTAGINÀCEAE Plantain Family

Herbs, mostly scapose, with simple, basal or cauline leaves often sheathing at the base. Flowers radially symmetrical, usually bisexual, small and inconspicuous in bracteate spikes or heads; calyx four-parted; corolla three- or four-lobed, membranous or scarious, stamens four or fewer, epipetalous. Ovary superior, one- to four-locular, style simple, filiform. Fruit is circumscissile capsule or nutlike. Three genera and about 265 species, widely distributed.

1. PLANTÀGO L. Plantain

Scapose or leafy-stemmed herbs with strongly ribbed basal or cauline leaves, producing axillary or terminal bracteate spikes or sometimes heads of small, inconspicuous, unisexual or bisexual flowers. Calyx lobes equal or unequal, persistent; corolla four-lobed, membranous or scarious, salverform, stamens four or two. Ovary two-locular, ovules one to many in each locule. Fruit a membranous capsule. About 260 species, cosmopolitan.

1. Leaves lanceolate, oblanceolate, obovate, or ovate, more than 1 cm wide; floral bracts small, inconspicuous.
   2. Leaves ovate to suborbicular, leaf bases obtuse . . . . 1. *P. major*
   2. Leaves lanceolate, oblanceolate, or obovate, leaf bases narrowed to base, acute.
      3. Larger spikes over 10 cm long . . . 2. *P. virginica*
      3. Larger spikes less than 10 cm long . . 3. *P. lanceolata*
1. Leaves narrowly lanceolate-acuminate, usually less than 1 cm wide; floral bracts elongate, mostly 1 cm long or more . . . 4. *P. aristata*

1. P. major L. Common Plantain—Annual or perennial scapose herb with thick, basal, ovate or suborbicular, pubescent leaves to 4 dm long, entire, undulate, or coarsely toothed, obtuse, rounded or cordate at the base. Flowers in dense spikes up to 2 dm long or more, bracts ovate, about as long as the calyx; sepals 1–2 mm long, elliptic. Capsule purplish or brown, broadly conic to the rounded tip, seeds six to fifteen, reticulated, about 1 mm long. Disturbed sites, old fields, roadsides, throughout much of Western Hemisphere, naturalized from Eurasia.

2. P. virginica L. Southern Plantain—Annual or biennial, pubescent, scapose, dioecious or polygamodioecious herb with oblanceolate or obovate, spreading or ascending leaves 5–15 cm long, margins entire, undulate, or shallowly serrate, gradually narrowed to the base, subsessile. Flowers in dense spikes up to 2 dm long, bracts narrow, calyx lobes ovate, 2–3 mm long, longer than the bracts; corolla lobes spreading, of staminate flowers, lobes of pistillate flower unequally erect. Capsule ovoid or obovoid, seeds two to four, about 1 mm long, yellow. Old fields, disturbed sites, dry soil, Fla. to Tex., Me., Kans., Oreg., Calif.

3. P. lanceolata L. English Plantain, Rib Grass—Annual or perennial scapose herb with lanceolate or lance-oblong leaves, strongly ribbed, blades 10–15 cm long, gradually narrowed to the base. Flowers in dense spikes mostly 2–8 cm long, bracts ovate, membranous, calyx about 3 mm long, apparently two-lobed owing to fusion of two sepals; corolla about 5 mm wide, anthers exserted. Capsule one- to two-seeded. Disturbed sites, lawns, roadsides, old fields, widely distributed in U.S., naturalized from Europe.

4. **P. aristata** Michx. BRACTED PLANTAIN—Annual or short-lived scapose perennial with dark green, linear to lanceolate-acuminate, entire leaves, blades 5–20 cm long, entire, glabrous or thinly pubescent, gradually narrowed to a somewhat clasping base. Flowers cleistogamous in dense spikes up to 15 cm long or more, bearing conspicuous, green, linear-attenuate bracts that are several times longer than the flowers; sepals spatulate-oblong, 2–3 mm long; corolla glabrous, lobes ovate-rounded, about 2 mm long. Capsule somewhat narrowed toward the tip, 2–3 mm long, two-seeded. Dry soil, Fla. to Tex., Me., Ill.

173. RUBIÀCEAE MADDER FAMILY

Trees, shrubs, or herbs with opposite or whorled, simple leaves, interpetiolar or intrapetiolar stipules usually present, free or fused, sometimes leafy and similar to leaves. Flowers mostly bisexual, radially symmetrical; calyx four- to five-parted, fused to the ovary; corolla generally funnelform or tubular, four- to ten-lobed, stamens as many as the corolla lobes and alternate with them, epipetalous. Ovary inferior, mostly two- to four-locular, disk fleshy, ovules one to many, style filiform. Fruit a capsule, berry, or fruit. About 450 genera and 6000 species, widely distributed but chiefly in tropical regions.

1. Trees, shrubs, or woody vines.
     2. Flowers in dense, spherical heads . . .   1. *Cephalanthus*
     2. Flowers solitary or in a branching inflorescence, not in heads.
          3. Flowers or fruits sessile or subsessile.
               4. Branches with sharp thorns.
                    5. Larger leaves 2–3 cm long . . .   2. *Randia*
                    5. Larger leaves less than 1 cm long .   3. *Catesbaea*
               4. Branches without thorns, unarmed .   4. *Ernodea*
          3. Flowers or fruits peduncled or pedicelled.
               6. Flowers or fruits solitary, occasionally paired, borne on short, axillary peduncles.
                    7. Larger leaves 10–20 cm long, fruit pulpy.
                         8. Fruit obovoid, less than 3 cm long; shrubs . . . . .   5. *Morinda*

8. Fruit spherical, more than 5 cm
    long; trees . . . . . .   6. *Casasia*
7. Larger leaves mostly 2–8 cm long,
    fruit not pulpy.
    9. Undersurface of leaves with
        abundant appressed pubes-
        cence; fruit a capsule . . .   7. *Guettarda*
    9. Undersurface of leaves glabrous
        or nearly so; fruit a small
        drupe . . . .   8. *Exostema*
6. Flowers or fruits in racemes, panicles,
    or cymose clusters, axillary or ter-
    minal.
    10. Leaves with prominent lateral
        veins; larger leaves mostly over 8
        cm long.
        11. Undersurface of leaves often
            short appressed, pubescent,
            younger ones especially
            densely pubescent.
            12. Inflorescence axillary, few-
                flowered; leaf base rounded
                or truncate . . . . .   7. *Guettarda*
            12. Inflorescence terminal, many-
                flowered; leaf base tapering
                to petiole . . . . .   9. *Hamelia*
        11. Undersurface of leaves glabrous
            or sparsely pubescent.
            13. Fruits berrylike, mostly red or
                orange, in terminal or axil-
                lary cymes . . . . .   10. *Psychotria*
            13. Fruits coalescent in fleshy, yel-
                low syncarp 2–4 cm long .   5. *Morinda*
    10. Leaves with single prominent main
        vein, the lateral veins not con-
        spicuous; larger leaves mostly
        less than 8 cm long.
        14. Leaves linear-lanceolate with
            revolute margins, pubescent .   11. *Strumpfia*
        14. Leaves lanceolate-ovate to ovate,
            margins not revolute, glabrous.
            15. Flowers in axillary cymes or
                racemes; fruits white . .   12. *Chiococca*
            15. Flowers in terminal cymes;
                fruits dark purple . . .   13. *Erithalis*
1. Herbs.
    16. Some or all the larger leaves whorled,
        mostly three to six per node . . . .   14. *Galium*
    16. Larger leaves opposite, not whorled.
        17. Flowers sessile, borne singly, paired, or

in terminal or axillary clusters or in
headlike inflorescences; ovules one
per locule.
18. Flowers single or paired in axils, not
    in clusters or heads . . . . .   15. *Diodia*
18. Some or all the flowers in clusters or
    heads.
  19. Larger leaves or linear-lanceolate,
      less than 5 mm wide . . . .   16. *Borreria*
  19. Larger leaves lanceolate to ovate,
      more than 5 mm wide.
    20. Stems and leaves glabrous or
        nearly so.
      21. Sepals ovate-obtuse . . .   17. *Spermacoce*
      21. Sepals lanceolate-acuminate .   16. *Borreria*
    20. Stems and leaves hirsute or vil-
        lous . . . .   18. *Richardia*
17. Flowers mostly with distinct peduncles
    or pedicels, borne in axils or ter-
    minally; ovules many per locule.
22. Floral parts tetramerous . . . .   19. *Hedyotis*
22. Floral parts pentamerous . . . .   20. *Pentodon*

1. CEPHALÁNTHUS L. Buttonbush

Shrubs or small trees with opposite or whorled, entire, deciduous
leaves. Flowers densely crowded in spherical, peduncled heads; hypanthium obovoid, calyx four-parted, short; corolla four-parted, tubular.
Ovary two-locular, ovules one to three per locule, style filiform, stigma
capitate, fruit nutlike, slitting from the base upward into two to four
nutlets. About ten species, chiefly tropical and subtropical regions.

1. **C. occidentalis** L.—Shrub or small tree up to 3 m tall with opposite or whorled, short-petioled, entire leaves. Blades to 20 cm long,
ovate, lanceolate, or obovate, stipules deltoid, 2–3 mm long. Flowers in
dense, spherical heads about 2–3 cm wide on long axillary or terminal
peduncles; flowers arranged between filiform bractlets; corolla white,
5–8 mm long, lobes ovate or elliptic. Nutlets 5–6 mm long, obconical.
$2n = 44$. Moist sites, margins of swamps, ponds, or solution holes, Fla. to
Mexico, New Brunswick, Minn., W.I. All year.

Plants with young stems and lower surfaces of leaves soft pubescent
have been named **C. occidentalis** var. **pubescens** Raf.

2. RÁNDIA L.

Shrubs or trees, often spiny, with persistent opposite leaves. Flowers
axillary, solitary, calyx lobes four; corolla five-lobed, funnelform or
campanulate; stamens five, fused to near the top of the corolla, filaments

very short, disk annular or cushionlike. Ovary two-locular, style terminated by a club-shaped stigma. Fruit a berry. About 200 species, pantropical.

1. **R. aculeata** L. WHITE INDIGO BERRY—Branching spiny shrub or small tree to 3 m tall. Leaves 2–5 cm long, often fascicled or clustered, obovate, spatulate, or suborbicular, narrowed to a short petiole. Flowers axillary, sepals deltoid to ovate, about 1 mm long; corolla 5–7 mm long, white, tubular, lobes oblong to ovate. Berry about 1 cm long, ovoid to subglobose, white or greenish white. Moist or dry sites, scrub vegetation, south Fla., Fla. Keys, W.I., Mexico, Central America, South America. All year.

3. CATÉSBAEA L.

Spiny shrubs or small trees with small, fascicled, glabrous leaves. Flowers axillary, white, on very short peduncles, calyx four-lobed, corolla four-lobed, campanulate or funnelform, stamens four, epipetalous near the base of the corolla. Ovary two-locular, ovules few to several, stigma two-lobed. Fruit a white berry. About eight species, chiefly in the W.I.

1. **C. parviflora** Sw.—Shrub up to 2 m tall, much branched with long, slender, very spiny stems, the spines up to 2 cm long. Leaves mostly 2–10 mm long, numerous, coriaceous, glossy green, obovate to suborbicular, rounded at the apex and cuneate to the base, petiole short. Flowers sessile or subsessile, calyx lobes subulate to narrowly triangular-lanceolate, about 1 mm long, corolla about 5 mm long, tube campanulate, lobes obtuse. Berry globose or subglobose, white, about 2 mm wide. Pineland, margins of hammocks, Fla. Keys, W.I. All year.

4. ERNÒDEA Sw.

Smooth or sparingly pubescent, low shrubs with erect, reclining, or trailing stems. Leaves opposite, linear to lanceolate, subsessile, stipules fused to form a sheath. Flowers small, axillary, sessile, calyx four- to six-lobed, lobes subulate to deltoid; corolla four- to six-lobed, pinkish, red, or white, tube funnelform, stamens epipetalous near the top of the tube. Ovary two-locular, ovules one per locule, style slender, stigma capitate. Fruit a fleshy drupe. About six species in Fla. and the W.I.

1. **E. littoralis** Sw.—Shrub with prostrate or ascending, quadrangular glabrous stems to 1 m tall or more. Leaves 2–4 cm long, numerous, glossy green, linear to lance-ovate, lance-oblong to elliptic and somewhat fleshy, stipules 1–2 mm long. Calyx 5–8 mm long, lobes linear to

lanceolate; corolla tubular, funnelform, 5–10 mm long, usually reddish, pinkish white, or sometimes white. Drupe ovoid to globular about 5–6 mm long. Two varieties occur in our area:

1. Leaf blades linear-oblong to elliptic; calyx
   mostly 7–8 mm long . . . . . . 1a. var. *littoralis*
1. Leaf blades linear; calyx mostly 5–6 mm
   long . . . . 1b. var. *angusta*

   1a. E. littoralis var. **littoralis**—Fruit globular, corolla usually pinkish white or white, usually over 1 cm long. Coastal areas, sandy or rocky soil, south Fla., Fla. Keys, and W.I. All year.
   1b. E. littoralis var. **angusta** (Small) R. W. Long—Fruit ovoid, corolla usually pink or reddish, usually less than 1 cm long. Pinelands, south Fla., W.I. The two varieties appear to be ecotypically as well as generally morphologically distinguishable but are connected through occasional, intermediate forms. All year. *E. angusta* Small

5. MORÍNDA L. Indian Mulberry

Shrubs, trees, or vines with opposite or whorled leaves. Flowers in axillary or terminal capitate cymes, calyx truncate or minutely five-lobed; corolla funnelform or salverform, four- to seven-lobed; stamens as many as the lobes of the corolla, fused near the top of the tube. Ovary two-locular, ovules one per locule, stigma slender. Fruit consisting of drupes that become fused into a fleshy syncarp. About 60 species in the tropics.

1. Larger leaf blades 5–10 cm long . . . . 1. *M. royoc*
1. Larger leaf blades 25–30 cm long . . . 2. *M. citrifolia*

   1. M. royoc L.—Shrub or vine up to 3 m or more with slender, reclining stems, smooth or nearly so. Leaves 5–10 cm long, oblanceolate to elliptic-cuneate, acute acuminate to the apex, narrowed to the base, stipules broad, subulate tipped. Flowers in axillary, short-peduncled heads, corolla 6–8 mm long, white or reddish, lobes elliptic-lanceolate. Syncarp globose or subglobose, yellowish, fleshy, 2–3 cm in diameter. Many sites, common in hammocks, south Fla., W.I., tropical America. All year.
   2. M. citrifolia L.—Shrub or small tree with large, opposite leaves and smooth, tetragonal stems, blades 12–30 cm long, glabrous, glossy green, elliptic-acute, stipules broad, obtuse. Flowers in axillary, peduncled heads, corolla white. Syncarp about 3 cm wide, white or greenish, fleshy, globose. Lignum Vitae Key, W.I., naturalized from India.

6. CASÀSIA A. Rich.

Trees or shrubs with opposite, coriaceous leaves and deciduous stipules. Flowers in short-peduncled axillary cymes, calyx five- to six-lobed or truncate, campanulate; corolla salverform or almost rotate, five- to six-lobed, white or yellow; stamens five to six, epipetalous near the top of the corolla, disk present, cuplike. Ovary one- to two- locular, ovules many in each locule, fruit a pulpy berry, seeds flattened. About eight species, chiefly in the W.I.

1. C. clusiifolia (Jacq.) Urban, SEVEN-YEAR APPLE—Shrub or small tree up to 3 m with numerous branches. Leaves tending to cluster near the ends of branches, blades smooth, obovate to cuneate, 5–15 cm long, rounded or somewhat notched at the apex, tapering to the base, glossy green, entire, short petioled. Flowers in thick cymes, hypanthium 8–10 mm long, calyx lobes linear-subulate, corolla white, fleshy, tube 1–2 cm long, lobes lanceolate or oblong-lanceolate. Berry pulpy, large and conspicuous, ovoid to obovoid, 5–8 cm long. Coastal areas, south Fla., Fla. Keys, and Fla. Bay Keys, W.I. All year. *Genipa clusiaefolia* (Jacq.) Griseb.

7. GUETTÁRDA L. VELVETSEED

Shrubs or trees with opposite leaves and deciduous stipules. Flowers in axillary cymes, rarely solitary, bisexual or unisexual, hypanthium globose or ovoid, irregularly lobed or truncate; corolla salverform, tube elongate, four- to nine-lobed; stamens as many as the lobes, filaments none or very short. Ovary four- to nine-locular, ovules one per locule, style filiform, stigma two-lobed or capitate. Fruit a drupe. About 100 species, chiefly in tropical America.

1. Corolla 0.5–1.0 cm long; undersurfaces of
   leaves glabrous or sparingly pubescent .   1. *G. elliptica*
1. Corolla 2.0–2.5 cm long; undersurfaces of
   leaves with appressed pubescence, especially along veins . . . .   2. *G. scabra*

1. G. elliptica Sw.—Shrub or low tree, sometimes up to 6 m, usually less, with slender, pubescent stems. Leaves 2–8 cm long, chartaceous, elliptic to ovate or obovate, glabrate or puberulent above, appressed silky pubescent beneath, stipules 5–10 mm long, lanceolate. Cymes on slender pubescent peduncles, few- to several-flowered; calyx 2–3 mm long, truncate, corolla pink, pinkish white, or reddish, about 6 mm long, pubescent, usually four-lobed. Drupe globose, red changing to black, about

1 cm wide. Hammocks and pinelands, south Fla., W.I., Mexico, South America. All year.

2. G. scabra Vent.—Shrub or low tree up to 10 m tall, usually less, with young stems villous tomentose. Leaves mostly 5–15 cm long, coriaceous, elliptic to ovate or obovate, obtuse or mucronate at the apex, scabrous above, densely reticulate veined and finely pubescent beneath, stipules 2–3 mm, triangular-lanceolate. Cymes few-flowered, calyx finely pubescent, about 3 mm long, corolla white or reddish, 1–2 cm long, pubescent. Drupe globose, red, finely pubescent, about 5 mm wide. Hammocks and pinelands, south Fla., W.I., tropical America. All year.

8. EXOSTÈMA Richard ex HBK

Shrubs or trees with opposite leaves and deciduous stipules. Flowers solitary or in panicles, hypanthium cylindrical, calyx five-lobed, lobes narrow, corolla salverform, tube elongate, five-lobed, stamens five, epipetalous near the base of the tube, filaments long. Ovary two-locular, ovules many in each locule, style filiform. Fruit a two-valved capsule, seeds winged, many. About 25 species, chiefly in the W.I.

1. E. caribaeum (Jacq.) R & S, PRINCEWOOD—Shrub or small tree up to 8 m, usually less, with glabrous stems. Leaves 4–8 cm long, oblong-lanceolate to narrowly ovate, acuminate to the apex, narrowed at the base, midvein conspicuous, stipules ovate-acuminate, 1–2 mm long. Flowers borne solitary in leaf axils on slender peduncles, calyx cylindrical, 4–5 mm long, lobes short, corolla pinkish or white, tube 2–3 cm long, stamens exserted. Capsule oblong, woody, 1–1.5 cm long. Pineland and hammocks, Fla. Keys, W.I., tropical America. Spring, summer.

9. HAMÈLIA Jacq.

Shrubs or low trees with opposite or whorled leaves and narrow, deciduous stipules. Flowers secund in terminal,·compound cymes, hypanthium ovoid, five-lobed, the lobes short, corolla tubular, five-lobed, lobes short, stamens five, borne near the base of the tube, filaments short. Ovary five-locular, ovules many in each locule, style filiform. Fruit a small berry, five-lobed. About 25 species in tropical and subtropical America.

1. H. patens Jacq. FIREBUSH—Shrub or low tree up to 4 m tall with slender puberulent stems. Leaves 5–15 cm long, opposite or five- to seven-whorled, elliptic to ovate, acuminate to the apex, tapered to the base, stipules subulate, 2–3 mm long. Flowers in cymes, flowers numerous, subsessile, corolla bright red or scarlet, tubular, 1–2 cm long. Berry dark red or purplish, 5–6 mm long. Hammocks south Fla., W.I., tropical America. All year. *H. erecta* Jacq.

10. PSYCHÒTRIA L.   WILD COFFEE

Shrubs, trees, or perennial herbs with opposite leaves. Flowers small in terminal corymbs or panicles or in axillary clusters, calyx four- to five-lobed, corolla funnelform, four- to five-lobed, stamens as many as the corolla lobes, epipetalous, filaments short. Ovary two-locular, ovule one per locule, stigma two-cleft. Fruit a globose or oblong drupe. About 800 species in the tropics and subtropics.

1. Cymes or panicles sessile in upper leaf axils.
  2. Corolla green; petioles and main leaf
     veins pubescent . . . . 1. *P. sulzneri*
  2. Corolla white; petioles and veins glabrous 2. *P. nervosa*
1. Cymes or panicles with distinct peduncles
   in upper leaf axils.
  3. Leaves lanceolate to oblanceolate, acu-
     minate to the apex . . . . . . 3. *P. ligustrifolia*
  3. Leaves elliptic to ovate, obtuse or rounded
     at the apex, undersurface dotted with
     nodules . . . . 4. *P. punctata*

1. P. sulzneri Small—Branching shrub up to 2 m tall with pubescent stems. Leaves 8–15 cm long, narrowly oblong to elliptic-lanceolate, acuminate or acute at the apex, tapered to the base, entire, with prominent, lateral, pubescent veins. Flowers in cymes sessile in the upper leaf axils, corolla green, tube 2–3 mm long. Drupe red, orange, or yellow, about 5 mm long. Hammocks, peninsular Fla., Fla. Keys, W.I. Spring, summer.

2. P. nervosa Sw. —Branching shrub up to 3 m tall with glabrous or occasionally pubescent stems. Leaves 6–12 cm long, membranous, elliptic to oblong, acuminate at the apex, narrowed at the base, entire, strongly pinnately veined, glossy green above, paler beneath. Flowers in sessile panicles, in upper leaf axils, flowers sessile or subsessile, calyx about 1 mm long, truncate, corolla white, tube about 4 mm long. Drupe red, ellipsoid, 5–7 mm long. Hammocks, pinelands, south Fla., W.I., tropical America. Spring, summer. *P. undata* Jacq.

3. P. ligustrifolia (Northrop) Millsp.—Shrub up to 2 m tall with glabrous or nearly smooth stems. Leaves 5–12 cm long, lanceolate to oblanceolate, acuminate to the apex, gradually narrowed to the base, stipules orbicular, usually conspicuous. Flowers in peduncled panicles, calyx five-lobed, lobes deltoid, corolla white, tube 3–4 mm long. Drupe ellipsoidal, red, about 5 mm long. Hammocks, pineland, south Fla., W.I. Spring, summer. *P. bahamensis* Millsp.

4. P. punctata Vatke—Branching shrub up to 3 m tall with glabrous stems. Leaves narrowly elliptic-obovate to ovate, 4–8 cm long, obtuse or rounded at the apex, or sometimes short-acute, base attenuate, the blade spotted with nitrogenous nodules conspicuous on the undersurface, stipules bicuspidate. Flowers in rounded cymes or corymbs, corolla white, tube 4–5 mm long. Drupe ovoid, red or crimson, 5–6 mm wide. Disturbed sites locally and cultivated grounds, Key West, Fla. Keys, Miami, naturalized from Africa. All year. *P. bacteriophylla* Valet The leaf nodules contain nitrogen-fixing bacteria.

11. STRÚMPFIA Jacq.

Low shrub with numerous pubescent branches and bearing whorled, linear, coriaceous, revolute-margined leaves that are crowded near the ends of the stems, blades 1–2.5 cm long. Flowers in short, axillary racemes, calyx ovoid, five-lobed, corolla rotate, white, 3–4 mm long, deeply five-lobed, lobes lanceolate, stamens five, epipetalous near the base of the tube, filaments short. Ovary two-locular, ovules one per locule, style pubescent, stigma two-lobed. Fruit a small, fleshy drupe 4–6 mm long, red or white. One species in the W.I. and Fla. Keys.

1. S. maritima Jacq.—Characters of the genus. Low pinelands, maritime strand, lower Fla. Keys, Big Pine Key, Vaca Key. All year.

12. CHIOCÓCCA P. Browne SNOWBERRY

Shrubs or woody vines with opposite, leathery leaves and broad stipules. Flowers borne in axillary cymes, racemes, or panicles, calyx ovoid, five-lobed; corolla funnelform, five-lobed, stamens five, epipetalous near the base of the corolla tube, filaments fused below. Ovary two-locular, ovules one per locule, style filiform. Fruit a flattened, leathery, white drupe. About ten species, chiefly in Central America and South America.

1. Corolla yellow, 6–8 mm long; larger leaves
    mostly 5–8 cm long . . . . 1. *C. alba*
1. Corolla white, 4–5 mm long; larger leaves
    mostly 2–4 cm long . . . . 2. *C. pinetorum*

1. C. alba (L.) Hitchc.—Shrub with slender, usually spreading, diffusely branching or reclining stems, glabrous, up to 3 m tall. Leaves 2–8 cm long, elliptic, oblong, ovate-lanceolate, acute or obtuse at the apex,

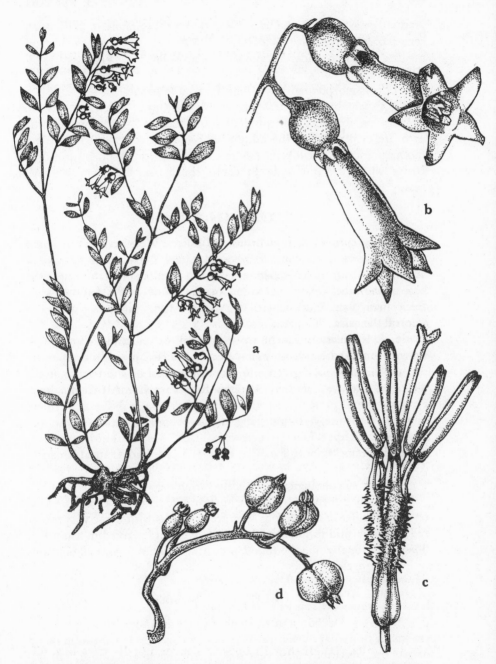

Family 173. **RUBIACEAE.** *Chiococca pinetorum:* a, flowering plant, × ½; b, flowers, × 4½; c, androecium and portion of gynoecium, × 12; d, mature fruits, × 2½.

gradually narrowed at the base. Flowers several to many in axillary racemes, corolla five-lobed, 6–9 mm long, yellow. Drupe bright white, orbicular, 5–7 mm wide. Hammocks, peninsular Fla., Fla. Keys, W.I. All year.

This is morphologically a highly variable species.

2. **C. pinetorum** Britt.—Shrub with trailing or vinelike stems, up to 1 m long, diffusely branching, glabrous. Leaves 1–4 cm long, elliptic or ovate, widest below the middle. Flowers few to several in axillary racemes, corolla five-lobed, 4–5 mm long, white or purplish white. Drupe white, 4–5 mm wide. Pineland, south Fla., Fla. Keys, W.I. All year.

13. ERITHÀLIS P. Browne

Shrubs or trees with glabrous, opposite, petioled leaves and fused stipules. Flowers small in corymbose panicles, hypanthium globose to ovoid, calyx four- to five-lobed or truncate; corolla epipetalous near the base of the tube. Ovary five- to ten-locular, ovules one per locule, style thick. Fruit a small drupe with five to ten nutlets. About six species in Central America, W.I., and Fla.

1. **E. fruticosa** L.—Shrub up to 3 m tall or sometimes a small tree up to 8 m tall with smooth stems. Leaves 3–6 cm long, elliptic to ovate or suborbicular, dark green and glossy, rounded or short-acute at the apex, narrowed at the base, stipules fused, persistent, 1–2 mm long. Flowers several to many in peduncled panicles, calyx 1–2 mm, denticulate, corolla white, 5–10 mm long, deeply lobed, the lobes elliptic, 3–5 mm long. Drupe globose, furrowed, black, 3–5 mm wide. Beaches, coastal hammocks, south Fla., Fla. Keys, W.I., Mexico, Central America. All year.

This is morphologically a highly variable species.

A number of species of Ixora including I. coccinea L. and I. grandiflora (Blume) Zoll. & Morr., evergreen shrubs with dense corymbs of red or yellow and tubular corollas, are commonly planted in south Fla. They occasionally may be found persisting around former habitations or in waste areas.

14. GÀLIUM L.   Bedstraw

Herbs with quadrangular, slender stems and whorled leaves. Flowers small, in axillary or terminal cymes or panicles, hypanthium ovoid or globose, obscurely lobed or truncate, corolla rotate, four-lobed; stamens four, filaments short. Ovary two-locular, ovules one per locule, styles two, stigma capitate. Fruit dry or sometimes fleshy, of two globose, indehiscent, one-seeded carpels. About 300 species, widely distributed.

1. Larger leaves mostly 4–5 mm wide, elliptic; stems and leaf undersurfaces hirsute; fruit fleshy . . . . 1. *G. hispidulum*
1. Larger leaves mostly 1–3 mm wide, narrowly lanceolate; stems and leaf undersurfaces glabrous; fruit dry . . . . 2. *G. obtusum* var. *floridanum*

1. G. hispidulum Michx.—Perennial herb with diffusely branching, hirsute stems up to 6 dm long, evergreen. Leaves 6–12 mm long, whorled in fours, elliptic or oblong, with an abrupt tip, hirsute underneath. Flowers in axillary cymes on pubescent peduncles, corolla 2 mm long, greenish white, lobes elliptic. Fruit a purple berry, about 3–4 mm wide. Pineland, sandy soil, Fla. to La., N.J. on the coastal plain. Spring, summer. *G. bermudense* misapplied

2. G. obtusum Bigel. var. **floridanum** (Wieg.) Fern.—Perennial herb with weak, often matted, diffusely branching stems up to 8 m long, pubescent at the nodes. Larger leaves 8–15 mm long, whorled in fours, narrowly lanceolate or oblanceolate, widest above the middle, glabrous beneath except the midvein hispid to scabrous. Flowers chiefly in terminal cymes with two to four flowers on short, ascending pedicels, corolla white, lobes 1 mm long, 2–3 mm wide. Fruit dry, smooth, about 3 mm wide. Low ground, Fla. to Tex., Va. All year. *G. tinctorium* L. misapplied

15. DIÓDIA L. Buttonweeds

Annual or perennial herbs or subshrubs with opposite leaves and stipules fused with the petioles to form a sheath. Flowers small, axillary, sessile or subsessile, hypanthium ovoid, calyx mostly two- to four-lobed, corolla funnelform or salverform, pink or white, tube slender, mostly four-lobed, stamens as many as the corolla lobes. Ovary two-locular or sometimes three- to four-locular, ovules one per locule, style two-branched or capitate. Fruit of two flattened nutlets. About 40 species, tropical America and tropical Africa. *Diodella* Small

1. Larger leaves less than 5 mm wide . . . 1. *D. rigida*
1. Larger leaves more than 5 mm wide.
  2. Sepals mostly four; filiform stipules mostly ten or more per node; fruit 2–3 mm long . . . . 2. *D. teres*

2. Sepals two; filiform stipules fewer than
   ten per node; fruit 5–6 mm long . . 3. *D. virginiana*

   1. D. rigida (Willd.) Cham. & Schlect.—Annual herb branching
with prostrate or procumbent finely pubescent stems. Leaves 1–4 cm
long, linear to linear-lanceolate, acute, closely hirsute, stipules elongate-
setose. Flowers axillary, in sessile, few-flowered clusters or solitary, sepals
1–2 mm long, ovate to ovate-lanceolate, corolla white or pinkish, 6–10
mm long, the lobes ovate-lanceolate or narrowly ovate. Nutlets obovoid,
3 mm long. Disturbed sites, pinelands, south Fla., W.I., South America.
All year. *Diodella rigida* (Willd.) Small
   2. D. teres Walt.—Annual herb branching with creeping or procumbent, loosely pubescent or puberulent stems up to 6 dm long.
Leaves mostly 4–8 cm long, linear-lanceolate to elliptic, rigid. Flowers
axillary in sessile, few-flowered clusters or solitary, sepals four, lanceolate about 2 mm long, corolla funnelform, 4–5 mm long, white, lobes
ovate to deltoid. Nutlets 2–3 mm long. $2n = 28$. Dry sites, Fla. to Tex.,
N.Y., Mo., W.I. A highly variable species. Spring, fall. *Diodella teres*
(Walt.) Small
   3. D. virginiana L.—Perennial herb with branching procumbent or
weak stems up to 1 m long, glabrous or somewhat pubescent especially
on the angles. Leaves mostly 4–8 cm long, lanceolate to oblanceolate or
narrowly spatulate. Flowers axillary, solitary or one- to three-flowered,
sessile, sepals two, linear, 4–6 mm long, corolla white, tube slender, 6–8
mm long. Nutlets 7–9 mm long. $2n = 28$. Moist sites, Fla. to Tex., N.J.,
Mo. All year.

16. BORRÈRIA Meyer

Annual or perennial herbs or subshrubs with opposite, entire leaves
and sheathing stipules. Flowers axillary or terminal, solitary or in compact cymes or clusters, hypanthium obovoid, calyx two- to four-lobed,
the lobes unequal, corolla salverform or funnelform, four-lobed, stamens four, epipetalous usually near the top of the tube. Ovary two-
locular, ovules one per locule. Fruit a small leathery or membranous
capsule. About 100 species in the tropics.

1. Inflorescence a terminal head and occasionally in penultimate axil; leaves linear-
   subulate; perennial . . . . 1. *B. terminalis*
1. Inflorescence headlike in most axils as well
   as terminal; leaves narrowly lanceolate to
   lance-ovate; annual.

2. Sepals ovate-acute; larger leaves mostly
over 5 mm wide . . . . 2. *B. laevis*
2. Sepals subulate; larger leaves mostly less
than 5 mm wide . . . . 3. *B. ocimoides*

1. **B. terminalis** Small—Perennial herb up to 3 dm tall, often diffusely branched and forming clumps or colonies. Leaves 1–3 cm long, linear-subulate to linear. Flowers in a compact terminal head and occasionally in the penultimate axil, sepals two to four, corolla white, usually 3 mm long, lobes ovate. Capsule about 2–3 mm long. Pinelands, coastal areas, endemic to south Fla., Fla. Keys. All year.

2. **B. laevis** (Lam.) Griseb.—Annual with spreading or ascending stems up to 4 dm tall, somewhat pubescent. Leaves 2–4 cm long, oblong, lanceolate to elliptic-lanceolate, acute acuminate at the apex, gradually narrowed at the base. Flowers in dense cymes or headlike clusters terminal and in many leaf axils, calyx four-lobed, ovate, corolla white, about 2–3 mm long. Capsule obovoid, about 2 mm long, seeds oblong. Pineland, Fla. to La., W.I., Mexico, tropical America.

3. **B. ocimoides** (Burm. f.) DC.—Annual with slender, erect, branching stems up to 6 dm tall. Leaves linear to narrowly lanceolate, 1–2.5 cm long, acute at the apex, tapered to the base, short petioled. Flowers in several- to many-flowered dense cymes or headlike clusters in leaf axils, calyx four-lobed, the lobes subulate, corolla white, lobes ovate. Capsule ellipsoidal, pubescent, about 1 mm long. Pinelands, coastal areas, south Fla., Fla. Keys, W.I., tropical America. All year.

17. SPERMACÒCE L.

Herbs with quadrangular stems and opposite, stipulate leaves. Flowers small, in dense axillary or terminal cymes, hypanthium obovoid, calyx four-lobed, corolla funnelform, four-lobed, stamens four, epipetalous. Ovary two-locular, ovules one per locule, style slender, stigma capitate or somewhat two-lobed. Fruit a two-lobed capsule. About five species in America.

1. Stems and leaves glabrous or nearly so . . 1. *S. tenuior*
var. *floridana*
1. Stems and leaves pubescent with whitish
hairs . . . . 2. *S. tetraquetra*

1. **S. tenuior** L. var. **floridana** (Urban) R. W. Long—Annual or sometimes perennial with glabrous or sparingly pubescent stems, prostrate or weakly ascending up to 4 dm long. Leaves 1–5 cm long, narrowly lanceolate to elliptic-ovate, acute acuminate at the apex, tapering

to the base. Flowers in dense axillary cymes or clusters, hypanthium glabrous or nearly so, corolla white, lobes ovate. Capsule 1-2 mm long, minutely hispidulous. Sandy soil, endemic to lower Fla. Keys, Dade County. Spring, summer. *S. keyensis* Small

The typical variety, **S. tenuior** var. **tenuior**, is also reported for Fla. but apparently is confined to the northern part. It has larger leaves on generally erect stems, and the hypanthium is hirsute.

2. S. tetraquetra A. Rich.—Annual with erect or decumbent densely pubescent stems up to 6 dm tall. Leaves 2-8 cm long, lanceolate to oblanceolate, acute at the apex, tapered or obtuse at the base. Flowers in dense axillary clusters, hypanthium hispid hirsute, calyx lobes lanceolate, acuminate, 1 mm long, corolla, white 2-3 mm long. Capsule ellipsoidal, 2 mm long, hispid pubescent. Disturbed sites, pinelands, south Fla., Fla. Keys, W.I. All year.

18. RICHÁRDIA L.

Annual or perennial herbs with opposite leaves. Flowers in involucrate cymes, axillary or terminal, sepals four to eight, fused at the base, foliaceous, corolla four- to eight-lobed, funnelform tube short, stamens as many as the corolla lobes, fused near the top of the tube. Fruit dry, mature carpels separating from one another. About ten species in tropical America. *Richardsonia* Kunth

1. Leaf blades with entire margins, or nearly
   so.
   2. Fruit bristly hirsute; perennial . . . 1. *R. brasiliensis*
   2. Fruit merely tuberculate, not hirsute; an-
      nual . . . . 2. *R. scabra*
1. Leaf blades serrate, at least the upper half . 3. *R. grandiflora*

1. R. brasiliensis (Moq.) Gomez—Perennial herb with diffusely branching stems from thick roots. Leaves 1-4 cm long, elliptic to narrowly ovate. Flowers in involucrate clusters, sepals ovate-alternate, about 1 mm long, corolla 3-4 mm long, white, lobes less than half as long as the spreading tube. Mature carpels obovoid, bristly hirsute. Waste places, pinelands, drier sites, peninsular Fla., naturalized from South America. All year.

2. R. scabra L.—Annual herb with branching stems, puberulent, up to 8 dm tall. Leaves 2-8 cm long, lance-ovate to elliptic, tending to be somewhat fleshy. Flowers in dense involucrate clusters, sepals narrowly ovate, 2-3 mm long, corolla 5-7 mm long, white, the lobes less than one-third as long as the tube. Mature carpels ellipsoidal, tuberculate. $2n = 28,56$. Disturbed sites, sandy soil, pinelands, Fla. to Tex., N.C. on the coastal plain, native of tropical America. All year.

3. R. grandiflora Steud.—Low perennial herb with branching stems, puberulent, up to 1 dm tall. Leaves 1–3 cm long, oblanceolate, serrate on the upper half of the blade. Flowers in involucrate clusters, terminal or axillary, calyx lobes subulate-linear, corolla white. Mature capsules ellipsoidal. Disturbed sites, lawn weed in Dade County, naturalized from tropical America. All year.

19. HEDYÒTIS L.

Small herbs or subshrubs with opposite leaves and entire, dentate or lacinate stipules connecting the petioles or leaf bases. Flowers solitary or cymose, peduncles often dimorphic, with some having exserted stamens and short included style, others with stamens included and style exserted, calyx four-lobed, corolla salverform or funnelform, stamens four, epipetalous above the middle of the tube. Ovary two-locular, style one or more, stigmas two. Fruit a globular or didymous capsule, seeds pitted. About 300 species in tropical and subtropical regions. *Houstonia* L.; *Oldenlandia* L.

1. Corolla 4 mm or more long; perennials.
   2. Leaves ovate or elliptical, mostly over 5
      mm wide; stems creeping, prostrate   .   1. *H. procumbens*
   2. Leaves filiform to linear, mostly less than
      3 mm wide; stems erect .   .   .   .   .   2. *H. nigricans*
1. Corolla to 2 mm long; annuals.
   3. Flowers on slender pedicels, usually one
      to three   .   .   .   .   3. *H. corymbosa*
   3. Flowers sessile or nearly so, numerous   .   4. *H. uniflora*
      var. *fasciculata*

1. H. procumbens (J. F. Gmel.) Fosberg, INNOCENCE—Perennial with prostrate or creeping stems up to 4 dm long, glabrate or sparingly pubescent. Leaves mostly 5–15 mm long, elliptic, ovate to suborbicular. Flowers solitary or in few-flowered cymes, sepals ovate or elliptic, about 1 mm long, corolla funnelform, four-lobed, white, smooth within. Capsules somewhat pubescent, 4 mm wide. Pinelands, sandy soil, roadsides, Fla. to La., S.C. on the coastal plain. Spring, fall.

2. H. nigricans (Lam.) Fosberg, DIAMOND FLOWERS—Perennial with tufted stems, erect, up to 7 dm tall or more. Leaves linear to linear-filiform, sessile, 1–5 cm long, somewhat rough to the touch, stipules bristle tipped. Flowers in branching cymes forming a panicle, sepals narrowly deltoid, 1–2 mm long, corolla white or purplish tinged, 3–6 mm long, deeply lobed. Capsule obovoid to turbinate, 1–3 mm long. Pinelands, Fla. to Tex., Ga., Nebr., Mexico. Spring, summer.

Plants become blackened on drying. Two well-marked varieties are recognized:

1. Leaves linear, 4–5 cm long; capsule 2–3
   mm long . . . . 2a. var. *nigricans*
1. Leaves filiform, 1–3 cm long; capsule 1 mm
   long . . . . 2b. var. *filifolia*

2a. H. nigricans var. **nigricans**—Leaves linear 4–5 cm; capsule 2–3 mm, sepals narrowly deltoid 1–2 mm, corolla tube 5–6 mm. Pinelands Fla., Tex., Ga., Nebr. *Houstonia angustifolia* Michx.; *H. nigricans* (Lam.) Fern.

2b. H. nigricans var. **filifolia** (Chapm.) Shinners—Leaves filiform to narrowly linear, 1–3 cm long. Sepals lanceolate to deltoid, about 1 mm long, corolla tube about 3 mm long. Capsule obovoid, about 1 mm long. Pinelands, sandy soil, south Fla. and Fla. Keys, Tex. *Oldenlandia angustifolia* var. *filifolia* Chapm.

3. H. corymbosa (L.) Lam.—Annual herb with erect or decumbent, diffusely branching stems up to 5 dm long. Leaves 1–4 cm long, linear to narrowly lanceolate. Flowers on slender pedicels, sepals 1 mm long, corolla 2 mm long, white, lobes ovate and puberulent within. Capsule ovoid, 2 mm long. Disturbed sites, roadsides, lawns, south Fla., W.I., tropical America, Old World. Spring, summer. *Oldenlandia corymbosa* L.

4. H. uniflora (L.) Lam. var. **fasciculata** (Bertoloni) W. H. Lewis— Annual with glabrous, erect, often diffusely branching stems up to 5 dm long. Leaves 0.5–2 cm long, elliptic to lanceolate or lance-ovate, sessile. Flowers sessile or nearly so, sepals 1 mm long, corolla lobes ovate-acute, less than 1 mm long, bluish white. Capsule ovoid, 1–2 mm long. Beaches, sandy soil, Fla. to Miss. on the coastal plain. Summer, fall. *Oldenlandia fasciculata* (Bertoloni) Small

Hedyotis callitrichoides (Griseb.) W. H. Lewis has been collected in Miami. It is a glabrous prostrate herb with leaves 1.5–3.5 mm long, suborbicular to elliptic, and solitary axillary flowers with white corollas about 2 mm long. It is found in the W.I., Mexico, and tropical Africa.

20. PÉNTODON Hochst.

Annual herbs with tender stems and opposite, broad leaves. Flowers in terminal or axillary cymes, calyx deeply five-lobed, corolla funnelform, the lobes slightly shorter than the tube, stamens five, epipetalous near the middle of the tube. Ovary two-locular, style thick.

Fruit a capsule, two-lobed, included within the hypanthium. Two species, one American and African, the other African.

1. P. pentandrus (Schum. & Thon.) Vatke—Annual with glabrous, diffusely branching tender stems, suberect or creeping. Leaves 1–5 cm long, elliptic to ovate or elliptic-lanceolate, entire. Flowers cymose, sepals deltoid to lanceolate, 2–3 mm long, corolla white, funnelform, tube about 3 mm long, lobes ovate-lanceolate. Capsule 3–4 mm long. Low ground, Fla. to La. on the coastal plain. Spring, fall. *P. halei* (T & G) Gray

174. CAPRIFOLIÀCEAE Honeysuckle Family

Shrubs or rarely herbs with opposite, simple, deeply divided or compound leaves. Flowers bisexual, radially or bilaterally symmetrical, often showy, usually in cymes; calyx five-lobed, fused to the ovary; corolla five-lobed, sometimes two-lipped; stamens five, epipetalous, alternate with the corolla lobes. Ovary inferior, one- to five-locular. Fruit a berry, drupe, or capsule. About 15 genera and 400 species, chiefly in the north temperate regions.

1. Leaves pinnately compound . . . . . . 1. *Sambucus*
1. Leaves simple . . . . 2. *Viburnum*

1. SAMBÙCUS L. Elderberry

Shrubs, small trees, or sometimes coarse herbs with pinnately compound leaves, leaflets serrate. Flowers many, small, in large, terminal compound cymes; calyx lobes minute or absent; corolla radially symmetrical, urceolate-rotate with spreading lobes, stamens five. Ovary with three to five locules, stigmas three, fruit a drupe containing three small nutlets. About 25 species in the north temperate regions.

1. S. simpsonii Rehder, Southern Elderberry—Shrub or small tree with soft stems and abundant white pith. Leaves opposite, pinnately compound, leaflets five to nine, 4–8 cm long, elliptic, serrate. Flowers small, densely clustered in large, terminal, flat-topped compound cymes; corolla white, 5–7 mm wide. Drupe blue black, juicy, about 5–6 mm wide. Low ground, disturbed sites, roadsides, margins of lakes and ponds, Fla. to La. on the coastal plain. All year.

2. VIBÚRNUM L. Arrowwood

Shrubs or trees with entire to lobed simple leaves. Flowers usually small in terminal or axillary cymes; sepals minute; corolla radially

Family 174. **CAPRIFOLIACEAE.** *Viburnum obovatum:* a, flowering branch, × ½; b, flower, × 8; c, mature fruit, × 5.

symmetrical, or sometimes marginal flowers bilateral, rotate to campanulate, five-lobed; stamens five, epipetalous, exserted. Ovary onelocular, stigmas three, sessile on a short stylopodium at the top of the ovary. Fruit a one-seeded drupe. About 120 species in the Northern Hemisphere and the Andes.

1. **V. obovatum** Walt. BLACK HAW—Shrub or low tree up to 4 m tall. Leaves 2–6 cm long, oblanceolate, cuneate to obovate, entire or barely serrate near the apex. Flowers small in rounded cymes; corolla white, 4–6 mm wide, stamens short. Drupe ovoid, black, about 6–7 mm long. Hammocks, margins of swamps, Fla. to Va. on the coastal plain. Spring.

Lonicera sempervirens L., the TRUMPET HONEYSUCKLE, is found in peninsular Fla. and may be expected to occur in our area. It is an evergreen, climbing shrub with opposite sessile or short-petioled, ovate leaves, the upper leaves connate at the base, and long, funnelform corolla, reddish orange outside, yellow within, about 3–5 cm long. Its range is from Fla. to Conn. and Tex. Spring.

175. VALERIANÀCEAE VALERIAN FAMILY

Herbs with opposite leaves. Flowers bisexual or unisexual in panicled or clustered cymes; calyx tubular; corolla tubular or funnelform, often bilaterally symmetrical; stamens free, one to three, epipetalous. Ovary inferior, style slender, stigmas one to three; locules three, one fertile with one ovule, the other two abortive or empty. Fruit indehiscent, one-locular or three-locular with two empty. About 10 genera and 370 species, chiefly in the north temperate regions.

1. VALERIÀNA L. VALERIAN

Perennial herbs with thick, strong-scented roots and simple or pinnately compound leaves. Flowers bisexual or unisexual or dimorphic; calyx composed of several plumose bristles, resembling a pappus which is rolled in flower, unrolled in fruit; corolla usually gibbous near the base, tube five-lobed, nearly radially symmetrical, stamens three. Fruits small, nutlike. About 210 species, widely distributed.

1. **V. scandens** L. var. **scandens**—Perennial herbaceous vines or clambering, stems leafy, much branched, glabrous or sparsely spreading, puberulent. Leaves 4.5–18 cm long, petiolate, three-parted, ovate-cordate, serrate to crenate. Inflorescence an aggregate dichasium, 12–40 cm long, bracts 1.7–2.5 cm long, flowers bisexual or unisexual, corolla

more or less campanulate 1–2.3 mm long, white, stamens exserted. Achenes oblong-linear to oval. Disturbed sites, sandy soil, pinelands, Fla., W.I., Mexico, Central America, South America. All year.

This species apparently is not common in our area.

176. CUCURBITÀCEAE   Gourd Family

Monoecious or dioecious herbs, or sometimes subshrubs, with prostrate or vinelike often rough-pubescent stems and with a watery sap. Leaves long petioled, palmately veined, often lobed, tendrils usually present. Flowers unisexual, radially symmetrical, staminate flowers with calyx tubular, lobes spreading; corolla free or fused, stamens three, usually fused by their anthers or sometimes by the filaments or free, pistillate flowers with calyx fused to the ovary. Ovary inferior with numerous ovules, stigmas two to three. Fruit usually soft, fleshy, indehiscent, berrylike or, if a hard rind forms, a form of berry known as a pepo; fruit sometimes membranous. About 100 genera and 850 species, chiefly in the tropics and subtropics.

1. Anthers linear or only slightly curved; leaves cordate in outline, three- to five-lobed; leaf blades and stems glabrous to puberulent . . . . 1. *Melothria*
1. Anthers sharply U-shaped; leaves and stems glabrous to harshly pubescent.
   2. Staminate flowers in elongate racemes; plants monoecious . . . . . . 2. *Luffa*
   2. Staminate flowers solitary or in small clusters or short racemes; plants monoecious or dioecious.
      3. Corolla whitish, cream colored, or greenish white; plants nearly glabrous or clammy pubescent.
         4. Corolla cream colored, or nearly white . . . . 3. *Cucurbita*
         4. Corolla greenish white . . . . 4. *Cayaponia*
      3. Corolla yellow or orange; plants glabrous or hirsute, not clammy pubescent.
         5. Larger leaves usually less than 5 cm wide; stems glabrous or sparingly pubescent . . . . 5. *Momordica*
         5. Larger leaves mostly over 6.0 cm wide; stems and petioles thickly pubescent with spreading hairs.

6. Leaves deeply pinnately lobed or
   pinnatifid, the lobes seven,
   lanceolate or pointed . . .   6. *Citrullus*
6. Leaves with prominent lobes often
   palmate, rounded, or ovate in
   outline . . . .   7. *Cucumis*

1. MELÒTHRIA L. MELONETTE

Perennial, monoecious or polygamous herbaceous vines, or sometimes annual, with slender stems and simple tendrils. Leaves rather
small, thin, lobed or toothed. Flowers small, staminate flowers clustered,
campanulate, stamens three, anthers more or less united, corolla five-
lobed; pistillate flowers solitary, calyx short-constricted above the ovary,
campanulate, five-lobed, corolla campanulate, deeply five-lobed. Ovary
ovoid, inferior, ovules many, style short, stigmas three. Fruit a small,
pulpy berry containing many flat seeds. About 60 species mainly in the
tropics and subtropics.

1. **M. pendula** L. CREEPING CUCUMBER—Herbaceous, slender, creeping or climbing vines. Leaves 2–8 cm wide, small, suborbicular, ovate,
cordate to reniform, angularly three to five shallowly to deeply lobed.
Flowers yellow or greenish, staminate flowers few in small racemes,
hypanthium glabrous to puberulent. Berry ovoid to ellipsoidal mostly
1–3 cm long, greenish becoming darker. $2n = 22$. Low ground, swamps,
hammocks, south Fla. to Mexico, Va., Mo. Spring, fall.

A highly variable species, especially in leaf and fruit morphology;
three weak varieties apparently occur in south Fla.:

1. Leaf blades thin, hirsutulous with short
   hairs, often 5–8 cm wide . . . . .   1a. var. *pendula*
1. Leaf blades thick, scabrous with stout hairs,
   often 2–4 cm wide.
   2. Leaf blades mostly 2–3 cm wide, deeply
      lobed; berries subglobose . . . .   1b. var. *aspera*
   2. Leaf blades mostly 3–4 cm wide, rough
      pubescent, shallowly lobed; berries el-
      lipsoidal to oblong . . . . . .   1c. var. *crassifolia*

1a. **M. pendula** var. **pendula**—Berry mostly 10–25 mm long,
swamps, bottom land, commoner farther north of our area.

1b. **M. pendula** var. **aspera** Cogn.—Berry mostly 10–15 mm in
diameter, sandy soil, disturbed sites, Fla., Ala. *M. nashii* Small; *M.
pendula* var. *microcarpa* Cogn.

1c. **M. pendula** var. **crassifolia** (Small) Cogn.—Berry mostly 15–20
mm long, moist soil, endemic to Fla. *M. crassifolia* Small

2. LÚFFA Adans.

Annual, monoecious, prostrate or climbing vines with lobed leaves. Flowers unisexual, calyx tubular, five-lobed, corolla campanulate, five-lobed, the lobes spreading, yellow, white, or pinkish, pistillate flowers with staminodia three. Styles stout, stigmas three, each three-lobed. Fruit an elongate ribbed berry. About eight species, all but one in the Old World tropics. *Luffa* L.

1. **L. cylindrica** (L.) Roem. VEGETABLE SPONGE—Vines with ribbed stems and branches. Leaves usually 1-3 dm wide, suborbicular, five- to seven-lobed, the lobes coarsely toothed. Calyx lobes lanceolate in pistillate flowers. Corolla yellow 10-12 dm wide, somewhat smaller in the staminate flowers. Berry cylindrical or club shaped. Disturbed sites, roadsides, naturalized from the Old World tropics, apparently not common in our area. All year.

3. CUCÚRBITA L. GOURD, PUMPKIN, SQUASH

Annual or perennial vines with large, cordate-angulate or lobed leaves. Flowers unisexual, solitary in leaf axils, corolla campanulate, deeply five-lobed, lobes recurved, anthers fused, one-locular, the other two-locular, pistillate flowers with staminodia, style short, with three-lobed stigmas. Fruit a fleshy berry with a tough rind. About 20 species in tropical America.

1. **C. okeechobeensis** Bailey, INDIAN PUMPKIN—Annual vine with herbaceous stems and unequally forking tendrils. Leaves 1-2 dm wide, orbicular to suborbicular or reniform, cordate, with five to seven shallow lobes, these irregularly toothed, covered with a clammy pubescence. Flowers peduncled, solitary, hypanthium pubescent, sepals lanceolate or subulate; corolla campanulate, cream colored, lobed crenulate or irregularly serrate. Berry globose, 7-9 cm wide, green streaked or flecked with white or dark green. Seeds ovoid, about 1 cm long. Hammocks, endemic to Fla., apparently rare. Spring, summer. *Pepo okeechobeensis* Small

C. pepo L., the FIELD PUMPKIN, with harsh, rough stems and bright yellow flowers, $2n = 40$, and **C. moschata** Duchesne, the CROOKNECK SQUASH, with soft pubescence and with fruits of various shapes and sizes, are occasionally found in disturbed areas.

4. CAYAPÒNIA S. Manso

Monoecious or dioecious herbaceous or somewhat woody vines with slender, climbing stems. Leaves lobed or palmately divided with

Family 176. **CUCURBITACEAE.** *Momordica charantia:* a, flowering branch, × ½; b, staminate flower, × 3½; c, mature fruit, × 1¼.

simple or divided tendrils. Flowers unisexual in racemes or panicles, calyx campanulate, five-lobed, corolla rotate or campanulate, five-parted, staminate flowers with three free stamens and a rudimentary three-lobed ovary. Pistillate flowers often with three staminodia, ovary three-locular, style three-branched, stigmas three. Fruit a small, fleshy berry. About 60 species in tropics and subtropics of America.

1. **C. racemosa** (Sw.) Cogn.—Smooth vine with herbaceous or woody stems. Leaves 6–12 cm long, orbicular or ovate, variously lobed, acute or tapering to the apex, cordate or somewhat reniform at the base, scabrous above, puberulent beneath. Flowers in racemes or panicles, calyx campanulate, 3–4 mm long, lobes deltoid; corolla greenish, about 1 cm wide. Berry oblong, 1.5–2 cm long, reddish. Hammocks, south Fla., W.I. Spring, summer.

No recent collections of this species from south Fla. have been seen, and it may no longer be in our area.

5. MOMÓRDICA L.

Monoecious or dioecious, annual or perennial, herbaceous, prostrate or climbing vines with simple or branched tendrils. Flowers unisexual, staminate solitary or clustered, calyx five-lobed; corolla five-lobed, stamens three, free. Pistillate flowers solitary, perianth similar to staminate one, staminodia glandlike or absent, ovary one-locular, ovules many, style slender, stigmas three. Fruit a cylindrical or ovoid berry with many elliptic or flattened seeds. About 60 species in the tropics and subtropics mainly in the Old World.

1. **M. charantia** L. WILD BALSAM APPLE—Creeping or climbing slender long stems often 2 m or more, tendrils filiform opposite the leaves. Leaves 4–10 cm wide, suborbicular or reniform, deeply palmately five- to seven-lobed, the lobes toothed, smooth or pubescent, acute to obtuse at the apex, petioles 3–6 cm long. Flowers mostly solitary on peduncles bearing an entire, cordate, or ovate bract near the middle or below, sepals ovate 3–4 mm long, corolla lobes 1–2 cm long, yellow. Berry 2–12 cm long, bright yellow, tubercled, ovoid or oblong. Hammocks, disturbed sites, Fla. to Tex. on the coastal plain, W.I., tropical America. All year.

6. CITRÙLLUS Schrad.

Annual or perennial, monoecious, prostrate, trailing vines with pubescent stems and bearing branched tendrils. Leaves deeply twice-pinnatifid. Flowers unisexual, solitary in leaf axils; corolla five-parted nearly to the base, staminate flowers with anthers free. Pistillate flowers with staminodia, ovary one-locular, ovules many, style short, stigmas three. Fruit a smooth berry. Four species in the Old World tropics.

1. C. vulgaris Schrad. WATERMELON—Annual with trailing hairy-pubescent vines. Leaves 10–18 cm long or more, ovate to oblong, bi-pinnatifid into lobed and toothed segments, broad at the tip. Flowers on axillary peduncles; corolla rotate, pale yellow, about 3 cm wide, five-lobed, the lobes ovate and obtuse. Berry up to 6 dm long, globose to oblong or cylindrical, smooth, soft fleshy within, red, green, or yellow, rind hard, striped or green, seeds flat, smooth, very numerous, white, black, or red. $2n = 22$. Disturbed sites, roadsides, escaped from cultivation, naturalized from tropical Africa. Spring, summer.

7. CÙCUMIS L.

Annual or perennial, monoecious or dioecious, trailing or climbing, herbaceous vines. Leaves entire or dissected, tendrils simple. Flowers simple, solitary or few-clustered in leaf axils, corolla campanulate to rotate, yellow, deeply five-lobed, staminate with anthers free. Pistillate with ovary one-locular, staminodia absent or rudimentary, stigma three to five. Fruit a fleshy, globular to cylindrical, smooth or prickly berry. About 40 species, all but one in warmer regions of the Old World. *Anguria* Jacq.

1. Corolla 1.0–1.5 cm wide; fruit ellipsoidal,
   prickly . . . . 1. *C. anguria*
1. Corolla wider, 3.0–5.0 cm; fruit elongate,
   not prickly . . . . 2. *C. melo*

1. C. anguria L. GOOSEBERRY GOURD—Trailing vine with slender, rough pubescent, angled stems and small tendrils. Leaves mostly 5–8 cm long, deeply three-lobed, the lobes again lobed, margins apiculate or serrate, very rough to the touch. Flowers 1–1.5 cm wide on slender peduncles, yellow. Berry ellipsoidal about 5 cm long, prickly on crooked peduncles, seeds many, smooth. Disturbed sites, south Fla. to Tex., tropical America.

2. C. melo L. MUSKMELON, CANTALOUPE—Trailing vine with striated or angled stems and pilose pubescence. Leaves about 7–15 cm long, orbicular to reniform, angled or shallowly lobed, margins sinuate-serrate, blades somewhat scabrous. Flowers 2.5–3 cm wide, corolla yellow, lobes obtuse. Berry highly variable in form, globular or oblong, somewhat furrowed, pubescent becoming glabrous, not spiny, flesh greenish or yellowish, seeds white. $2n = 24$. Disturbed sites, escaped from cultivation, naturalized from Asia.

177. CAMPANULÀCEAE   BLUEBELL FAMILY

Herbs, rarely shrubs or small trees, usually with alternate simple leaves. Flowers bisexual, radially or bilaterally symmetrical, often showy, calyx fused to the ovary or free, three- to ten-lobed; corolla fused, tubular, campanulate, often two-lipped, stamens as many as the corolla lobes and alternate with them. Ovary inferior, rarely superior. Fruit a capsule. About 70 genera and 2000 species of wide distribution. *Lobeliaceae*

1. Flowers radial, solitary or few, axillary or
    terminal . . . . 1. *Campanula*
1. Flowers bilateral, few to many flowers in
    terminal raceme . . . . 2. *Lobelia*

1. CAMPÁNULA L.

Annual or perennial herbs with simple, alternate leaves. Flowers showy, sepals five, corolla campanulate, funnelform, or rotate, five-lobed, stamens epipetalous near the base of the tube, included, anthers free. Ovary three- to five-locular with many ovules, style elongate. Capsule short, strongly ribbed, opening by three to five lateral pores. About 300 species in the Northern Hemisphere. *Rotantha* Small; *Campanulastrum* Small

1. C. floridana Wats. FLORIDA BLUEBELL—Small perennial herb diffusely branching and spreading, stems 2–4 dm long. Leaves 2–4 cm long, linear, linear-lanceolate to elliptic-lanceolate. Flowers pedunculate, hypanthium ovoid, sepals 6–10 mm long, linear, corolla blue violet to dark violet, lobes narrowly lanceolate. Capsule obovoid, 3–4 mm long. Low ground, endemic to peninsular Fla. All year. *Rotantha floridana* (Wats.) Small

2. LOBÈLIA L.

Annual or perennial herbs or shrubs with entire or serrate leaves. Flowers in terminal bracteate racemes or spikes, sometimes in the upper leaf axils, hypanthium usually ribbed, sepals narrow, entire, glandular-serrate or sometimes appendaged; corolla cleft to the base on the upper side, two-lipped, two lobes of upper lip erect, three lobes of the lower lip spreading, stamens exserted through the divided corolla, alternate with the corolla lobes, and fused to form a tube around the style. Ovary two-locular, ovules many. Fruit a capsule or berry. About 380 species mostly in tropics and subtropics, often containing a poisonous milky or colored sap.

1. Flowers small, mostly 5–8 mm long; leaves small, less than 2.0 cm long, orbicular or elliptical, distinctly petioled . . . . 1. *L. feayana*
1. Flowers larger, mostly over 1.0 cm long; leaves larger, 3 cm or longer, linear-lanceolate to lanceolate, sessile or nearly so.
   2. Plants scapose, the basal leaves oblanceolate or obovate; upper leaves much reduced, scalelike, or absent . . . . 2. *L. paludosa*
   2. Plants not distinctly scapose, leaves linear-lanceolate to lance-ovate; upper leaves smaller but not sharply reduced 3. *L. glandulosa*

1. **L. feayana** Gray, BAY LOBELIA—Small erect perennial herbs with glabrous stems up to 3 dm, usually less. Basal leaves 0.5–1.5 cm long, petioled, suborbicular, reniform or broadly ovate, upper leaves ovate or elliptic, entire or barely crenate. Flowers pedicelled, sepals 1–2 mm long, narrow, entire, corolla 7–9 mm long, blue. Capsule 3–4 mm long. Pinelands, endemic to Fla. Spring, fall.

2. **L. paludosa** Nutt. WHITE LOBELIA—Perennial erect herbs, scapose, up to 8 dm tall or less. Leaves 3–25 cm long, at least the lower narrowly spatulate to oblanceolate or elliptic, entire or crenate, sessile, the upper leaves absent or reduced. Flowers in spikelike racemes, sepals 3–5 mm long, narrow, entire or shallowly toothed, corolla pale blue or white, about 1 cm long. Capsule about 4 mm long. Low ground Fla. to La., Del. on the coastal plain. All year.

3. **L. glandulosa** Walt. GLADES LOBELIA—Perennial erect herbs, not scapose, up to 1 m tall, glabrous with slender stems. Leaves 3–15 cm long, linear-lanceolate to lance-ovate, entire to shallowly serrate, the teeth gland tipped, upper leaves not sharply reduced. Flowers appearing axillary, remote, hypanthium pubescent, sepals narrow, 6–10 mm long, glandular-serrate, corolla dark blue, lower lip pilose on the upper surface. Capsule 6–8 mm long. $2n = 28$. Glades, solution holes, Fla. to Miss., Va., on the coastal plain. All year.

L. floridana Chapm., with leaves 1–4 dm long, linear to somewhat linear-spatulate, and light blue flowers in a spikelike raceme, occurs in Charlotte County, Fla. and may extend into the range of this manual.

178. GOODENIÀCEAE Goodenia Family

Herbs or shrubs with alternate or sometimes opposite leaves. Flowers bisexual, radially or usually bilaterally symmetrical, calyx five-lobed, fused to the ovary, corolla five-lobed, split on one side, often one- to two-lipped, stamens five, free, alternate with the corolla lobes. Ovary one- to four-locular, inferior; the style produces a pollen cup or indusium under the stigma into which the pollen is shed before anthesis. Fruit a capsule or drupe or berrylike. About 14 genera and 320 species, chiefly Australia. *Brunoniaceae*

1. SCAÈVOLA L.

Stout succulent shrubs or herbs with mostly alternate, entire leaves. Flowers axillary in branching cymes or apparently solitary, bilaterally symmetrical, calyx five-lobed or represented by a mere border; corolla white or blue, tube split to the base on one side, pubescent within, lobes winged. Ovary inferior or nearly so, one- to two-locular, stigma subtended by a pubescent pollen cup. Fruit a drupe. About 100 species, pantropical but chiefly in Australia.

1. **S. plumieri** (L.) Vahl—Perennial herb or shrub up to 1.5 m tall, much branched. Leaves 3–6 cm long, alternate, obovate, entire, glossy green, gradually tapering to a short, winged petiole or subsessile, pubescent near the base. Flowers on short peduncles, calyx lobes obtuse or rounded; corolla about 2–3 cm long, white or pinkish white, glabrous on the outside, villous within the tube, lobes linear-lanceolate, stamens about as long as the corolla tube, pendent through the cleft in the tube. Drupe ovoid, black, about 1–1.5 cm long, two-seeded. Coastal areas, along beaches, peninsular Fla., Fla. Keys, W.I., tropical Africa. All year.

179. ASTERÀCEAE Aster Family

Annual or perennial herbs, shrubs, vines, rarely trees, with alternate or opposite, simple or compound leaves. Flowers small, bisexual or unisexual, crowded into heads on common receptacles that are surrounded by one to several series of bracts or phyllaries that compose the involucre; each flower sometimes provided with receptacular bracts that are termed paleae (chaff); calyx highly modified as a pappus of bristles, awns, or scales or pappus absent; corolla four- to five-parted, fused, tubular, filiform, two-lipped, or ligulate; tubular or disk flowers (florets)

Family 178. **GOODENIACEAE.** *Scaevola plumieri:* a, fruiting branch, × ½;
b, floral bud, × 2¼; c, flower, × 3½; d, mature fruits, × 1.

are usually radially symmetrical; strap-shaped or ligulate florets are bilaterally symmetrical. A head may be composed of only disk florets (discoid), or only of ligulate florets (ligulate), or of both together (radiate), wherein the disk florets always are found in the center of the head (disk) and the ligulate florets are marginal around the disk and are referred to as the rays; stamens five, united by their anthers forming a tube around the style, epipetalous. Ovary inferior, one-locular, style normally two-branched. Fruit one-seeded, dry, indehiscent, termed an achene often crowned by the pappus. About 1000 to 2000 genera and about 20,000 species of worldwide distribution. Nomen alt. Compositae. *Ambrosiaceae, Carduaceae, Cichoriaceae*

Usually considered the largest family of flowering plants, including many taxonomically difficult groups. Polyploidy, hybridization, and apomixis occur in numerous genera.

Key to Tribes

1. Heads radiate (composed of marginal ray florets and central disk florets) or discoid (composed only of disk florets); stems producing a clear, watery sap, not milky, or none.
    2. Heads radiate.
        3. Disk corollas radially symmetrical or nearly so, not two-lipped.
            4. Pappus of scales, awns, short bristles, or none, not soft capillary.
                5. Receptacle chaffy with pales or bracts subtending the florets . Tribe Heliantheae, Key 1, p. 824
                5. Receptacle naked or bristly, not chaffy . . . . Tribe Helenieae, Key 2, p. 841
            4. Pappus composed of capillary bristles (except short scales and bristles in *Boltonia*).
                6. Phyllaries equal in size, generally in one series or row . . . . Tribe Senecioneae, Key 3, p. 846
                6. Phyllaries unequal or equal, generally in several series or rows . . Tribe Astereae, Key 4, p. 849
        3. Disk corollas bilaterally symmetrical, two-lipped, the upper lip two-lobed, the lower lip three-lobed . . . . Tribe Mutisieae, Key 5, p. 863
    2. Heads discoid.
        7. Pappus of scales, awns, short bristles, or none.

8. Receptacle chaffy with pales or bracts subtending the florets . . . . Tribe Heliantheae, Key 1, p. 824

8. Receptacle naked or bristly, not chaffy . . . . Tribe Helenieae, Key 2, p. 841

7. Pappus composed of capillary bristles.

  9. Corollas of marginal florets filiform, pistillate; leaves and stems often white woolly pubescent . . . Tribe Inuleae, Key 6, p. 864

  9. Corollas of marginal florets not filiform; or, corollas filiform and stems and leaves not woolly pubescent.

    10. Some or all the corollas yellow, of bisexual florets, or, if greenish yellow or white, then the plants tall, shrubby . . . . . . Tribe Astereae, Key 4, p. 849

    10. Corollas of bisexual florets some other color than yellow, or whitish, or if yellow, then the leaves spiny or pinnately lobed.

      11. Phyllaries equal in size, generally in one series or row in the involucre . . . . . Tribe Senecioneae, Key 3, p. 846

      11. Phyllaries unequal, generally in several series or rows.

        12. Leaves and phyllaries not spiny; receptacle naked or nearly so; anthers with a short basal attachment, or none.

          13. Style branches thickened toward the ends, club shaped, covered with small papillae; anthers rounded at base; leaves opposite, whorled or alternate . . . . . Tribe Eupatorieae, Key 7, p. 867

          13. Style branches linear or filiform, not club shaped, bristly pubescent; anthers with a short basal appendage at base; leaves alternate or basal . . . Tribe Vernonieae, Key 8, p. 876

        12. Leaves and phyllaries spiny; receptacle bristly; anthers conspicuously sagittate at the base . . . . . . Tribe Cardueae, Key 9, p. 878

1. Heads ligulate (composed only of ray florets); stems usually containing a milky or colored sap . . . . Tribe Cichorieae, Key 10, p. 879

Key 1

TRIBE *Heliantheae*

1. Heads without ray florets; pistillate florets only one to five, corolla absent or much reduced.
  2. Leaves deeply lobed or divided; heads unisexual, staminate or pistillate . . 1. *Ambrosia*
  2. Leaves entire or obscurely to shallowly serrate; heads bisexual, with both pistillate and staminate florets present . . 2. *Iva*
1. Heads with ray florets (may be much reduced), or if discoid, the florets bisexual; disk floret corolla well developed, tubular or goblet shaped.
  3. Leaves all opposite.
    4. Middle and upper leaves much reduced, filiform to linear, less than 5 mm wide, entire . . . . . . 3. *Coreopsis*
    4. Middle and upper leaves narrowly lanceolate to broadly ovate, obscurely to deeply toothed or lobed.
      5. Some or all the leaves coarsely toothed, lobed, or divided (compound).
        6. Heads discoid, ray florets lacking.
          7. Disk florets white; leaf blades lobed or toothed . . . . 4. *Melanthera*
          7. Disk florets yellow; leaf blades pinnately dissected, toothed, or lobed . . . . . . 5. *Bidens*
        6. Heads with both disk and ray florets, often showy.
          8. Lower leaves divided, often trifoliolate compound, upper leaves simple or divided . . 5. *Bidens*
          8. Leaves simple, sometimes deeply lobed but not compound.
            9. Lower leaf blades three-lobed, sessile or nearly so; stems glabrous or with scattered hairs . . . . . . 6. *Wedelia*
            9. Leaf blades coarsely and irregularly toothed or incised, petioled; stems hirsute . . 7. *Tridax*
      5. Leaves barely to regularly shallowly serrate or toothed, not coarsely toothed or lobed.

10. Plants perennial; large involucres 6–15 mm wide, conical or hemispheric; leaf blades narrowly lanceolate to lance-ovate or obovate.

   11. Herbs; leaves petioled, blades narrowly to broadly lanceolate or lance-ovate; involucre conical, phyllaries foliaceous .   8. *Spilanthes*

   11. Maritime shrubs; leaves sessile or nearly so, blades oblanceolate or obovate; involucre hemispheric, phyllaries leathery or foliaceous . . .   9. *Borrichia*

10. Plants annual; involucres smaller, mostly less than 5 mm wide, not conical; leaf blades narrowly to broadly ovate.

   12. Phyllaries glabrous or with only scattered hairs; involucre broadly campanulate.

      13. Leaf blades ovate, serrate, with definite petioles; rays conspicuous . . . .   10. *Galinsoga*

      13. Leaf blades linear-lanceolate to lanceolate, sessile or subsessile; rays minute or absent . . . . . .   11. *Eclipta*

   12. Phyllaries hirsute to villous; involucre cylindric or narrowly campanulate.

      14. Phyllaries herbaceous or membranous, not prickly . .   12. *Calyptocarpus*

      14. Phyllaries prickly . . . .   13. *Acanthospermum*

3. Leaves alternate or lower leaves opposite and the upper alternate.

15. Leaves filiform to linear-lanceolate, not over 5 mm wide, entire.

   16. Receptacle with numerous pits, honeycombed . . . . .   14. *Balduina*

   16. Receptacle without pits.

      17. Stems and phyllaries hirsute; inflorescence a solitary head or short raceme . . . . .   15. *Phoebanthus*

      17. Stems and phyllaries glabrous or nearly so; inflorescence a spreading corymbose cyme . . .   3. *Coreopsis*

15. Leaves lanceolate to broadly ovate, over 5 mm wide, serrate or deeply divided.

18. Some of the leaf blades irregularly
lobed, deeply toothed, divided or
pinnatifid.
19. Ray corollas (ligules) yellow or
orange.
20. Tall, woody plants with large
showy heads; leaf blades
three- to five-lobed or entire .   16. *Tithonia*
20. Low, herbaceous plants with
heads terminal on long peduncles; leaf blades deeply
pinnatifid . . . . . .   17. *Berlandiera*
19. Ray corollas (ligules) cream colored or white.
21. Stems conspicuously winged;
disk florets fertile; leaves irregularly toothed or lobed .   18. *Verbesina*
21. Stems not conspicuously winged;
disk florets sterile; leaves
deeply lobed, incised or pinnatifid . . . .   19. *Parthenium*
18. Leaf blades lanceolate to broadly
ovate, obscurely serrate to serrate,
never lobed or divided.
22. Receptacle conic-elongate . . .   20. *Rudbeckia*
22. Receptacle flattish, convex.
23. Achenes flattened . . . .   18. *Verbesina*
23. Achenes turgid, not flat . . .   21. *Helianthus*

1. AMBRÒSIA L. Ragweed

Monoecious annual or perennial, rough or coarse herbs much
branched, usually with dissected or lobed leaves and many inconspicuous small heads in spicate or racemose inflorescences. Heads staminate
above with five- to twelve-lobed, flattened involucre containing five-
to twenty florets, corollas five-lobed. Pistillate heads one- to few-clustered, heads one-flowered, without corollas, below in axils of leaves,
each enclosed in an involucre, stigmas two, exserted. Pappus absent.
Achenes ovoid, lustrous, beaked. About 41 species, chiefly American.

Coarse herbs, including many obnoxious weeds in North America
which are hay fever plants.

1. Leaves finely pubescent; plants with leaves
predominately opposite . . . . .   1. *A. hispida*
1. Leaves glabrous to obscurely pubescent;
plants with leaves predominately alternate . . . .   2. *A. artemisiifolia*

1. A. hispida Pursh, Coastal Ragweed—Stems and numerous branches prostrate with ascending tips, up to two to many meters long, white, hirsute, bearing ovate tripinnatifid leaves. Staminate involucre about 1.5–2 mm high. Fruit 4 mm long, obovoid, with one to five conic spines. $2n = 144$. Sea beaches and coastal dunes, Fla., W.I., Central America, South America. Summer.

2. A. artemisiifolia L. Common Ragweed—Annual with erect stems up to 2 m tall, pubescent, bearing broadly ovate or lanceolate leaves deeply one to two pinnatifid, usually 5–10 cm long. Staminate involucres 2–4 mm wide, shallow. Pistillate involucres fascicled in upper leaf axils. Fruit beaked, 3–5 mm with short spines. $2n = 36$. Disturbed sites, beaches, Fla. and Tex., north to Canada, Calif., Wash. Summer, fall. *A. elatior* L. var. *elatior* (L.) Desc.; *A. artemisiifolia* var. *paniculata* (Michx.) Blankenship

This is a highly variable species, and numerous forms and varieties have been described of doubtful taxonomic value.

2. ÌVA L. Marsh Elder

Herbs or shrubs with opposite leaves, or the upper alternate, and small heads with greenish white florets. Heads borne in axils of leafy bracts, discoid and heterogamous, involucre imbricate, turbinate, or hemispheric, staminate flowers with undivided style, corollas distinctly five-lobed. Pistillate flowers few, corollas barely lobed, pappus none. Achenes slightly flattened. About 15 species, chiefly American.

1. Leaves linear, 0.1–0.2 cm wide . . . . . 1. *I. microcephala*
1. Leaves linear-lanceolate to lanceolate, more than 0.2 cm wide.
  2. Leaves sessile or subsessile . . . . . 2. *I. imbricata*
  2. Leaves with distinct petiole 0.3–1.0 cm long . . . . 3. *I. frutescens*

1. I. microcephala Nutt.—Erect annuals with smooth or pubescent erect branches up to 1 m tall. Leaves entire or shallowly toothed or serrate, gland dotted, lower opposite, upper alternate. Heads with threadlike bracts in racemes 5–20 cm long; involucres campanulate about 2 mm high. Achenes black, obovoid, about 1 mm long. $2n = 32$. Low pinelands, disturbed sites, wet fields, Fla. to S.C. on the coastal plain. Summer.

2. I. imbricata Walt.—Perennial, rather succulent, smooth shrub or suffruticose herb up to 1 m tall. Leaves 2–5 cm long, sessile, opposite below, alternate above, fleshy, lanceolate, entire or shallowly toothed.

Racemes 5–25 cm long, involucres 6–7 mm long with six to nine phyllaries; heads solitary, about 6–8 mm tall. Achenes about 2.5 mm, yellowish brown. $2n = $ c.68. Coastal dunes and beaches, south Fla. to La., north to Va.; W.I.

3. I. frutescens L.—Perennial shrub or herb up to 1 m or more tall, rather succulent, puberulent or hirsute. Leaves up to 10 cm long, 1–3 cm wide, opposite except uppermost, petiolate, lanceolate, sharply toothed. Involucres 4–5 mm wide, hemispheric, with four to six phyllaries in one series. Achenes 2–3 mm long, dark brown. Salt marshes and moist places along seashore, south Fla. and Tex., north to Va. Summer, fall.

3. COREÓPSIS L. Tickseed

Herbs or subshrubs with chiefly opposite entire to pinnatifid or ternate leaves. Heads radiate with conspicuous yellow, pink, or white neutral ligulate florets, mostly eight, and numerous disk florets. Involucral phyllaries dimorphic, in two rows, the outer foliaceous, the inner appressed membranous, receptacle almost flat. Achenes flattened, usually winged, pappus awnlike, persisting on the achene or a minute crown, or absent. About 120 species in America and tropical Africa.

1. Lower leaves oblong-lanceolate, over 1 cm wide; phyllaries broadly triangular, 3–4 mm long; plants mostly with one to three heads . . . . 1. *C. gladiata*
1. Lower leaves narrower, often lobed, oblanceolate to linear-lanceolate mostly less than 1.0 cm wide; phyllaries lanceolate-subulate, 1–2 mm long; plants with branching inflorescence one to many heads . . . . 2. *C. leavenworthii*

1. C. gladiata Walt.—Perennial herbs up to 1 m tall or more with entire, oblanceolate or spatulate, long-petioled leaves, up to 30 cm long. Upper and midcauline leaves much reduced. Heads several, disk 1.5–2 cm wide, dark purplish; phyllaries triangular. Pappus of two barbed awns, achenes 2–5 mm long. Low pinelands, swamps, Fla. and Miss., north to Va. on the coastal plain. Spring, summer.

2. C. leavenworthii T & G—Perennials up to 1.5 m tall with leaves usually entire, linear-lanceolate to oblanceolate, often lobed. Heads numerous, phyllaries subulate, about 2 mm long. Achene elliptical, about 3 mm long. Old fields, disturbed sites, roadsides, pinelands, moist areas, endemic to Fla. Spring, summer, or all year.

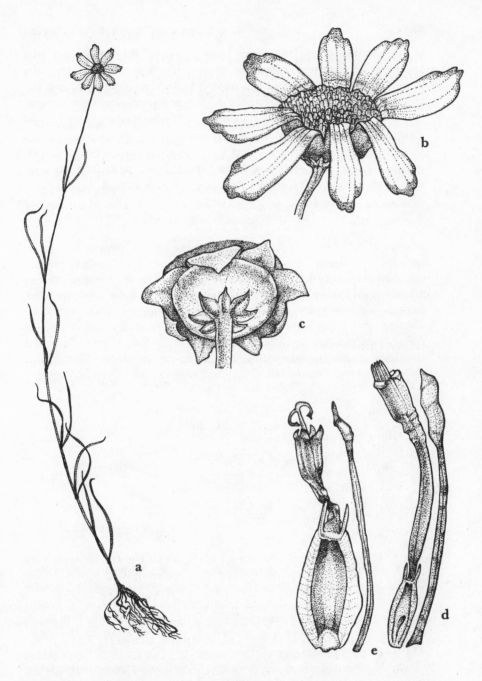

Family 179. **ASTERACEAE.** *Coreopsis lewtonii:* a, flowering plant, × ⅜; b, inflorescence, × 2½; c, involucre, × 3; d, disk floret and pale, × 10½; e, developing fruit and pale, × 10½.

A number of varieties have been described for this species that appear to be of minor taxonomic importance. A form with internodes longer than the leaves has been named **C. leavenworthii** var. **lewtonii** (Small) Sherff (*C. lewtonii* Small); **C. leavenworthii** var. **garberi** Gray has lower leaves with two or more broad lobes rather than narrow leaves characteristic of var. **leavenworthii**.

Cosmos bipinnatus Cav. and **C. sulphureus** Cav. with rays pink, purple, yellow, or white have been reported as naturalized in south Fla. They are natives of Mexico and tropical America.

4. MELANTHÈRA Rohrb.

Erect perennial herbs with opposite, entire, toothed or lobed leaves. Heads discoid, involucres hemispheric, with two to three rows of phyllaries. Corolla whitish, throat longer than tube; anthers dark or black tipped. Stigmas usually subulate-flattened. Achene compressed, four-angled. About 12 species in tropical America.

1. Leaves narrowly lanceolate to oblanceolate-
   ovate, usually less than 1.0 cm wide, serrate but never deeply lobed.
   2. Phyllaries linear-lanceolate acuminate, as
      long as the disk . . . . 1. *M. ligulata*
   2. Phyllaries ovate, shorter than the disk . 2. *M. angustifolia*
1. Leaves ovate-lanceolate, ovate, to broadly
   deltoid, sometimes deeply lobed.
   3. Leaves less than 5 cm long; low, sprawling
      herb . . . . 3. *M. parvifolia*
   3. Leaves over 5 cm long; erect, often bushy
      herbs.
      4. Phyllaries and chaff long acuminate . 4. *M. nivea*
      4. Phyllaries and chaff obtuse, blunt
         tipped, acute or short acuminate . . 5. *M. aspera*

1. **M. ligulata** Small—Stems decumbent, 5–10 dm long, pubescent. Leaves 10–16 cm long, linear to linear-lanceolate or elongate, irregularly serrate, highly variable. Outer phyllaries 10–15 mm long, linear-lanceolate, acuminate; corolla 5–6 mm long, cream colored. Pinelands, largely endemic to the area south of Lake Okeechobee to the Tamiami Trail. Spring, summer.

2. **M. angustifolia** A. Rich.—Stems hirtellous or puberulent, up to 1 m tall. Leaves about 5–8 cm long, linear-lanceolate or somewhat oblanceolate. Phyllaries rhombic-ovate, about 4–5 mm long; corolla 3.5–4.0 mm long. $2n = 30$. Wet pinelands, Everglades, south Fla., W.I.

3. M. parvifolia Small—Stems 3–10 dm long, slender, reclining, from a shallow, woody base; rough hispid. Leaves 3–5 cm long, occasionally somewhat longer, blades usually deeply lobed, the central lobe elongate, coarsely toothed to subentire; petioles very short. Heads long peduncled, few in number; phyllaries 5–7 mm long. Pinelands, lime rock, coastal strand, endemic to southern Fla. Hybrids with **M. angustifolia** have been reported. *M. radiata* Small

4. M. nivea (L.) Small—Stems stout, hirsute-scabrous up to 2 m tall. Leaves 5–20 cm long, ovate, deltoid or more often hastate-lobed, the margins serrate or crenate. Phyllaries narrowly lance-ovate, usually 2–3 mm wide, up to 1 cm long; corolla 3–4 mm long; chaff lance-acuminate. Moist soil, fresh-water shores, Fla. to La., north to S.C. on the coastal plain, tropical America. All year. *M. lobata* Small, *M. hastata* Michx.

5. M. aspera Jacq.—Stems stout, rough hirsute to short pubescent, up to 2.5 m tall, much branched, often bushy in appearance. Leaves about 6–12 cm long, variable, ovate to ovate-hastate or deltoid, serrate or dentate. Heads numerous or short peduncled; phyllaries ovate, mostly 0.5 mm wide. Pappus of two to three bristles with short fringed crown at the base; chaff acute. $2n = 30$. Two varieties occur in our area:

1. Blades rough hispid . . . . 5a. var. *aspera*
1. Blades short pubescent . . . . . . 5b. var. *glabriuscula*

5a. M. aspera var. **aspera**—Weedy, bushy plants with hispid, light green leaves. Dunes, beaches, and old reefs, southern Fla., W.I. All year.
5b. M. aspera var. **glabriuscula** (Kuntze) Parks—Strand plants with short-pubescent, dark green leaves. Fla. Keys, W.I. All year. *M. deltoidea* Michx.

5. BÌDENS L.   Beggar Ticks

Annual or perennial herbs with opposite, simple, dissected or compound leaves. Heads radiate or discoid, with many disk florets, ligulate florets three to eight when present, neutral; involucres dimorphic, with outer phyllaries foliaceous, inner appressed membranaceous, rays yellow or white or absent, disk florets usually yellow, bisexual receptacle flattened, style branches with hairy appendages; corolla tube shorter than throat; pappus of usually two to four awns. Achenes flattened or four-sided. About 240 species, cosmopolitan but chiefly in America.

1. Leaves divided, trifoliolate, or pinnately dissected.
   2. Leaf segments ovate, crenate-serrate . . 1. *B. pilosa*

2. Leaf segments linear or narrowly lanceo-
    late, serrate to entire . . . .   2. *B. mitis*
1. Leaves simple, sessile . . . .   3. *B. laevis*

1. B. pilosa L.—Stems up to 1 m or more, glabrous or nearly so. Leaves divided, often trifoliolate, the segments ovate and crenate-serrate. Involucre of short outer phyllaries, somewhat longer inner ones, heads discoid or radiate, rays white, disk corollas orange, pappus of two to four short awns, these barbed. Achene spindle shaped. Disturbed sites, roadsides, sandy soil, south Fla., Mexico, W.I., Central America, and Old World. All year. Two varieties occur in our area:

1. Heads discoid . . . .   1a. var. *pilosa*
1. Heads conspicuously radiate . . . . .   1b. var. *radiata*

1a. B. pilosa var. **pilosa**—Heads about 6–10 mm in diameter, with about thirty florets. *B. leucantha* L.

1b. B. pilosa var. **radiata** Sch.-Bip.—Ray florets five to six, white, sterile, ligules obovate, 8–10 mm long. Both varieties can be found commonly in south Fla., especially as weeds.

2. B. mitis (Michx.) Sherff—Stems erect up to 1 m tall, smooth or sparsely pubescent. Leaves pinnately dissected, segments linear or narrowly lanceolate 2–8 cm long, serrate to entire. Involucre 1–1.5 cm wide, ciliate; ray florets yellow or absent and the head discoid, chaff yellow. Pappus awns triangular. Wet pineland, borders of lakes and ponds, south Fla. to Miss., Va. All year.

3. B. laevis (L.) BSP—Stems up to 1 m or more, glabrous, often somewhat fleshy, branches few. Leaves simple, sessile, lanceolate-elliptic or oblanceolate, entire or finely serrate. Rays yellow, disk corollas 4–5 mm long, orange, pappus of two to four short awns retrorsely barbed. Achene 6–8 mm long, barely contracted at top. Moist soil, fresh or brackish water, swamps, south Fla. to Tex. on the coastal plain, north to N.H., Calif., South America. All year. *B. nashii* Small

6. WEDÈLIA Jacq.

Perennial, creeping or much branched herbs with opposite, usually cuneate, lobed or toothed leaves. Heads radiate; involucre of two to three unequal series of phyllaries, rays few, yellow, disk corollas with deltoid lobes; anthers with deltoid appendages at base, pappus a fimbriate crown. Achene three-angled, tuberculate. About 75 species, chiefly tropical regions. *Stemmodontia* Cassini; *Pascalia* Ortega

1. W. trilobata (L.) Hitchc.—Stems and numerous branches decumbent, creeping, somewhat fleshy. Leaves about 5–10 cm long, cuneate or elliptic-cuneate, three- to five-lobed or irregularly dentate. Outer phyllaries elliptic or ovate, ligules broadly ovate, about 10–12 mm long. Achenes of ray floret 4–5 mm, ovoid, coarsely tuberculate. Disturbed sites, pinelands, south Fla., W.I. All year.

This species is used in ornamental plantings in peninsular Fla. but apparently is naturalized only in south Fla.

7. TRÌDAX L.

Perennial decumbent herbs with opposite, toothed or pinnately dissected leaves. Heads radiate; involucre ovoid or campanulate, inner phyllaries somewhat broader than outer, rays few, pale yellow or yellow, disk corollas with funnelform throat longer than the tube; anthers narrow, pappus or plumose awns or scales. Achenes small, about 2–3 mm. Twelve species in tropical America.

1. T. procumbens L.—Stems branched from near the base, hirsute. Leaves usually 3–5 cm long, lance-ovate to ovate, broadly dentate or incised, lobed. Involucres about 7 mm long, densely pubescent, outer phyllaries lanceolate, inner broader; rays about as wide as long. Achene of disk floret 2 mm long. $2n = 36$. Disturbed sites, waste areas, margins of hammocks, south Fla., naturalized from W.I. All year.

8. SPILÁNTHES Jacq.

Annual or perennial herbs with opposite, usually toothed leaves. Heads small, radiate on long peduncles, many-flowered, involucres flat or campanulate, phyllaries in one to two series, subequal, receptacle columnar, ray florets yellow or white, pistillate, disk florets bisexual, lobes deltoid, pappus one to three slender awns. Ray achene three-angled; disk achene radially flattened. About 40 species, tropical and subtropical, mostly American.

1. S. americana (Mutis) Hieron.—Stems about 6 dm long, weak, often rooting at lower nodes, smooth or pubescent. Leaves 5–8 cm long, lanceolate to ovate, serrate or toothed. Heads few on naked peduncles, rays about 10 mm long, yellow, receptacle conic; disk about 9 mm wide or less. $2n = $ c.26,52–50. Low ground, moist sites, woods, south Fla. to S.C., Mo., tropical America. All year. *S. americana* var. *repens* (Walt.) A. H. Moore; *S. repens* (Walt.) Michx.

9. BORRÍCHIA Adans. SEA DAISIES

Maritime low shrubs with opposite, usually fleshy or coriaceous, pubescent leaves. Heads radiate, many-flowered, rays yellow, pistillate, phyllaries in two to three series, imbricate, receptacle flat or somewhat convex, chaff lanceolate, deciduous, disk florets bisexual; corolla throat much longer than tube; style with hairy appendages. Achenes quadrangular, smooth, pappus a short four-toothed crown. Seven species in warm temperate and tropical America and W.I.

1. Phyllaries spine tipped, reflexed . . . . 1. *B. frutescens*
1. Phyllaries not spine tipped, appressed . . 2. *B. arborescens*

1. B. frutescens (L.) DC.—Stems up to 1 m or more, with succulent, canescent, oblanceolate or spatulate, mucronate, aromatic leaves up to 7 cm long, sometimes finely toothed. Heads often solitary or few, disk about 1.5 cm wide or less; phyllaries canescent, appressed, becoming spine tipped, rays bright yellow, 1 cm or less. $2n = 56$. Sea beaches, dune hammocks, mangrove associations, salt marshes, low saline ground, south Fla. and Tex., north to Va., W.I., Mexico. Spring, summer.

2. B. arborescens (L.) DC. SEA OXEYE—Stems up to 1 m tall or more with short rhizomes. Leaves usually 3–6 cm long, gray green, spatulate-oblanceolate, fleshy, outer phyllaries acute, inner rounded, rays bright yellow, 1 cm long or less, chaff rigid pointed. $2n = 28$. Sea beaches, low hammocks, mangrove associations, shores, south Fla., W.I., Central America, South America. Spring, summer.

The two species sometimes are found growing intermixed.

SILPHIUM SIMPSONII Greene, ROSINWEED, has been reported as occurring in the lower Fla. Gulf coast, although no recent collections have been seen. It is a coarse perennial herb with oblong leaves, broad, flattened involucres or foliaceous phyllaries, yellow-flowered heads, and characteristic resinous juice in the stems and leaves.

10. GALINSÒGA Ruiz & Pav.

Annual herbs with opposite leaves and small heads. Heads several-flowered, radiate, involucre campanulate or hemispheric, rays few, mostly four to five, short, only slightly surpassing the disk, white or pink, pistillate, receptacle conic, disk florets yellow, bisexual; style with short, hairy appendage. Achenes four-angled. Pappus of few to many scales. Nine species, America, adventive elsewhere.

1. G. ciliata (Raf.) Blake, PERUVIAN DAISY—Diffusely branching annual up to 7 dm tall, often glandular-pubescent. Leaves about 3–7 cm long, petiolate, ovate, coarsely toothed or serrate. Involucre 2–3 mm tall, villous glandular-pubescent, phyllaries ovate; ray florets with well-developed pappus, pistillate, ligules white, 1–2 mm long, disk florets bisexual; pappus of disk florets tapered to awn tips. Achenes of disk pubescent. $2n = 32$. Disturbed sites, a cosmopolitan weed, naturalized from Central America and South America. All year.

11. ECLÍPTA L.

Annual herbs with opposite, simple leaves. Heads solitary, radiate, or discoid, small but with many florets, involucre hemispheric or campanulate, two rows of ten to twelve herbaceous phyllaries, ray florets pistillate, usually white, disk florets bisexual. Style branches with short, hairy appendages, pappus a very short crown. Achenes three- to four-angled. Four species, chiefly tropical regions and Australia. *Verbesina* sensu Small

1. **E. alba** (L.) Hassk.—Annual with pubescent, weak or decumbent stems, 2–9 dm tall, after rooting at the nodes. Leaves usually 5–10 cm long, lanceolate or lance-linear or lance-ovate, acute, sessile or subsessile. Heads one to three, phyllaries ten to twelve in one row, pubescent; disk about 5–6 mm wide, disk flowers many, bisexual, rays minute or absent. Achenes 2–3 mm long, warty. $2n = 22$. Disturbed sites, marshes, low ground, moist soil, south Fla. to Tex., north to Mass., Wis., pantropical regions. All year. *Verbesina alba* L.

12. CALYPTOCÁRPUS Lessing

Annual herbs with erect pubescent stems bearing opposite, dentate, or serrate leaves. Heads small, with many florets radiate; involucre with both herbaceous and membranous phyllaries; ray florets pistillate, ligule yellow, few; disk florets with funnelform corolla. Pappus of two to three awns. Achene turgid. Two species in U.S., Mexico and U.S.

1. **C. vialis** Lessing—Stems up to 5 dm tall with branches from near the base. Leaves 1–3 cm long, ovate or elliptic-ovate, stiffly hairy. Involucres 6–7 mm high, outer phyllaries ovate; disk corollas 4 mm long. Achenes muricate, about 5 mm long from ray and disk florets alike. $2n = 24,72$. Disturbed sites in dry soil, south Tex., W.I., Central America, adventive in south Fla. All year.

13. ACANTHOSPÉRMUM Schrank

Annual, coarse herbs with taproots. Leaves opposite, blades lance-ovate to cuneate-ovate, serrate or dentate. Heads radiate, ligules small; phyllaries few in one series, becoming prickly; outer whorl of florets pistillate, central florets discoid, sterile, corollas campanulate, tubes short. Pappus absent, achene somewhat compressed, angled, smooth. About eight species in tropical America, W.I., Madagascar, and Galapagos Islands.

1. Phyllaries more or less uniformly prickly
and bearing two large spines; leaf blades
shallowly serrate . . . . 1. *A. hispidum*
1. Phyllaries bearing prickles but not spines;
leaf blade deeply serrate or dentate . . 2. *A. australe*

1. **A. hispidum** DC.—Erect, branching, coarse herb up to 1 m tall, usually less. Leaves 3–6 cm long, ovate-elliptic, shallowly serrate, densely pubescent, sessile, base cuneate. Involucre narrowly campanulate, outer phyllaries elliptic, becoming spinescent and prickly at maturity; ligules short, yellow, disk corollas about 1 mm long. Disturbed sites, roadsides, waste places, Fla. to Ga., Ala. on coastal plain, tropical America, apparently not common in south Fla. Spring, fall.

2. **A. australe** (Loefl.) Kuntze—Decumbent or trailing coarse herb, usually rooting at the nodes. Leaves 1–3 cm long, rhombic-ovate, deeply serrate or dentate, or nearly so. Involucre narrowly campanulate, outer phyllaries lanceolate to elliptic, becoming prickly uniformly at maturity; ligules short, yellow, less than 2 mm long, disk corollas about 2 mm long. $2n = 20$. Disturbed sites, roadsides, sandy soil, Fla. to La., Va. on the coastal plain, tropical America, more abundant north of our area. Spring, fall.

14. BALDUÍNA Nutt.

Annual, biennial, or perennial herbs with numerous branches bearing alternate, linear, linear-lanceolate, or oblanceolate leaves. Heads radiate, florets many, involucre of three to five series of imbricate phyllaries, outer somewhat narrower than inner, ligules yellow, ray florets many, sterile; disk florets bisexual and fertile, with cylindric throat, tube absent, pappus of obovate scales, receptacle pitted. Achene puberulent. About three species, American. *Endorima* Raf.; *Baldwinia* Raf.; *Actinospermum* Ell.

1. **B. angustifolia** (Pursh) Robinson—Stems erect up to 1 m tall with numerous linear leaves. Involucre of ovate, acuminate phyllaries, ray florets ten to eighteen, bright yellow, conspicuous; disk corollas 5 mm; filaments shorter than anthers. Achene conical-turbinate, receptacle with characteristic pitted or honeycombed appearance. Pinelands, sandy soil, scrub areas, south Fla., to Miss., north to Ga.

15. PHOEBÁNTHUS Blake

Perennial erect herbs from horizontal tubers, leaves linear-lanceolate. Heads many-flowered, erect, with hemispheric involucre of numerous phyllaries having slightly reflexed tips, ray florets with long yellow ligules, disk corolla funnelform with short tube. Stigmas narrow, pappus of several scales between awns or teeth. Two species, North America.

1. **P. grandiflora** (T & G) Blake—Simple stems or with few branches above. Leaves about 3–6 cm long, alternate or mainly so, linear-lanceolate, entire. Heads conspicuous; phyllaries hirsute, ray florets usually sixteen to twenty, ligules mostly 3–4 cm long, disk about 2 cm wide. Achene ribbed and wing margined, pubescent along upper edge. $2n = 64$. Pinelands, oak scrub, sandy soil, endemic to central Fla. Summer. *Helianthella grandiflora* T & G

The occurrence of this species in south Fla. is infrequent, but it may occur in sandy scrub areas of north Collier County.

16. TITHÒNIA Desf. ex Juss.

Erect annual herbs with alternate, entire or three-lobed leaves. Heads erect, rather large, on long peduncles enlarged just below involucre, involucre companulate, phyllaries in two series, the outer shorter; ray florets neutral with bright yellow ligules; disk florets bisexual, corollas yellow, pappus of two awns or scales. Achene four-angled, flattened. About five species, tropical America.

1. **T. diversifolia** (Hemsl.) Gray—Stems up to 5 m tall, woody except near top, with numerous branches. Leaves ovate, entire or three- to five-lobed, serrate, pubescent beneath, tapering to petiole. Outer phyllaries ovate-appressed, inner oblong-foliaceous about 2 cm long; ligules about 5–6 cm long. Achene pubescent, about 6 mm long. Adventive in south Fla., pinelands, disturbed sites, peninsular Fla., naturalized from Mexico. Summer, fall.

17. BERLANDIÈRA DC.

Erect perennial herbs from enlarged roots. Leaves alternate, dentate-serrate or pinnatifid. Heads pedunculate, somewhat lax, phyllaries in two to three series, broad ovate; ray florets five to twelve, fertile, ligules yellow; or ray florets absent; disk florets with funnelform corolla, tube very short, filaments about as long as anthers. Style puberulent. Pappus absent or of two awns, achene ribbed. Eight species in North America.

1. **B. subacaulis** Nutt. GREEN EYES—Stems up to 5 dm tall, stiffly hairy. Leaves mostly 6–12 cm long, clustered at base of stem, elliptic-oblanceolate, pinnatifid. Involucre 1.5–2.0 cm broad with conspicuous elliptic or spatulate, foliaceous phyllaries, ligules 1–1.5 cm long; disk florets 3–3.5 mm long. Achene ovoid, about 5 mm long, pubescent. Dry pinelands, sandy soil, endemic to Fla. All year.

18. VERBESÌNA L. CROWNBEARD

Annual or perennial herbs with simple leaves and often winged stems. Heads radiate, many-flowered, ray florets neutral or pistillate with ligules yellow or white or sometimes absent, receptacle convex to conical; phyllaries of involucre herbaceous, in two rows, disk florets bisexual. Style branches with pubescent or papillate appendages, pappus usually of two awns. Achene flattened, receptacle chaffy. About 150 species, tropical America. *Ximenesia* Cav.; *Phaethusa* Gaertn.; *Pterophyton* Cassini

1. Annual, leaves gray canescent underneath;
   involucres mostly 2.0 cm or more wide;
   ray florets with yellow ligules . . . .   1. *V. encelioides*
1. Perennial, leaves pale green underneath,
   merely short hirsute; involucres smaller,
   mostly 5–8 mm wide. Ray florets with
   white ligules.
   2. Leaves pinnatifid or pinnately lobed . .   2. *V. laciniata*
   2. Leaves serrate, sinuate or undulate mar-
      gined but not lobed . . . . . .   3. *V. virginica*

1. **V. encelioides** (Cav.) Benth. & Hook. ex Gray, SKUNK DAISY —Erect pubescent annuals up to 7 dm tall with alternate, deltoid or ovate-deltoid leaves about 5–10 cm long, serrate or coarsely dentate. Heads on long peduncles; phyllaries linear-lanceolate, hirsute, ray florets pistillate, yellow, ligules 1–2 cm long, disk about 15–20 mm wide. Achenes winged. $2n = 34$. Pinelands, disturbed sites, western North America, adventive in Fla. Summer. *Ximenesia encelioides* Cav.

2. **V. laciniata** (Poir.) Nutt.—Erect pubescent perennials up to 1.5 m tall with pinnatifid or pinnately lobed leaves about 15–20 cm long, ovate-elliptic in shape. Phyllaries foliaceous about 6 mm long, ray florets white, four to five, pistillate, disk florets about 5 mm long. Achene 5–6 mm, obovate, or winged up to 7 mm long. Disturbed sites, old fields, sandy soil, Fla. to S.C. on the coastal plain. Summer. *Phaethusa laciniata* (Poir.) Small; *V. virginica* var. *laciniata* (Poir.) Gray

3. **V. virginica** L. FROSTWEED—Erect pubescent perennials up to 2 m tall with alternate ovate or lanceolate leaves up to 20 cm long, and serrate or undulate margins. Phyllaries up to 7 mm long, heads numerous, in a dense corymbose inflorescence, ray florets white, pistillate, one to five, disk small, 3–7 mm. Achenes about 5 mm, winged or without wings. Dry soil, Fla. and Tex. to Va. and Kans. Summer. *Phaethusa virginica* (L.) Small

19. PARTHÉNIUM L.

Herbs or shrubs with alternate, entire to pinnatifid leaves and small heads. Heads many-flowered, inconspicuously white or yellow, radiate, the ray florets pistillate and fertile, involucre hemispherical of two series of broad phyllaries; receptacle conical and chaffy; disk florets staminate, lobes ciliate. Pappus of scales, achenes enclosed by inner bracts, derived only from ray florets, ray corolla persistent. About 16 species, America and W.I.

1. **P. hysterophorus** L. SANTA MARIA—Annuals up to 1 m with pubescent stems, much branched. Leaves about 20 cm long, elliptic-ovate, pinnatifid or twice-pinnatifid. Involucre flat-hemispheric, heads numerous in usually leafy cymose inflorescence, disk small, about 3–5 mm wide, ligules minute, whitish. Achene black, ovate, less than 2 mm long. Pappus of two large rounded scales. $2n = 34,35$. Disturbed sites, roadsides, Fla. and Tex., Mass., Mo., naturalized from tropical America. Summer.

20. RUDBÉCKIA L. CONEFLOWER

Perennial, biennial, or annual herbs with alternate, entire or pinnatifid leaves. Heads with many florets, radiate, ray florets sterile, yellow, orange, or slightly purple, involucre of unequal green, foliaceous phyllaries in two to three rows, receptacle conic or columnar shaped, chaffy; disk florets bisexual, fertile, crowded. Pappus a short crown or absent. Achenes often four-angled, smooth, partially enclosed by bracts of receptacle. About 35 species, North America.

1. Pappus of short awns, chaff obtuse; lower
leaves glabrous to short pubescent . . 1. *R. fulgida*
1. Pappus absent, chaff acute; lower leaves and
stems scabrous . . . . 2. *R. hirta*
var. *floridana*

1. **R. fulgida** Ait.—Perennial with stems villous hirsute up to 1 m tall, usually stoloniferous. Leaves lanceolate to cordate; upper leaves reduced. Heads about 1–2 cm wide, long pedunculate; involucres reflexed, disk dark purple or brown, ligules 1–2 cm long, yellow to orange. Pappus a short crown barely toothed at angles. $2n = 38$, c. 76. Low woods or moist sites, Fla. and Tex., Va., Mo. Spring, summer. *R. spathulata* Michx.; *R. speciosa* Wenderoth

2. **R. hirta** L. var. **floridana** (Moore) Perdue, BLACK-EYED SUSAN—Biennial or perennial with hispid stems up to 1 m tall. Lower leaves 3–7

cm long, oblanceolate, elliptic-lanceolate, and often long petioled, scabrous; upper leaves reduced and sessile, sometimes clasping. Disk dark purple or brownish, about 2 cm wide; phyllaries hispid hirsute; ligules orange yellow, about 3–4 cm long. Pappus absent. Disturbed sites, roadsides, south Fla. and Mexico, Newfoundland, B.C. Spring, summer. Several varieties occur in this species, which is highly polymorphic, including **R. divergens** T. V. Moore.

21. HELIÁNTHUS L. Sunflower

Annual and perennial herbs or shrubs with simple leaves opposite, or the lower opposite and the upper alternate. Heads with many florets, radiate, involucre of several rows of green or colored phyllaries surrounding a flat or convex chaffy receptacle, ligules when present usually conspicuous, yellow, ray florets neutral, disk florets bisexual with tubular, five-lobed corolla. Achene thick, clasped by chaff, pappus of two awns. About 90 species, America. Hybridization and polyploidy occur widely in the genus.

1. Ray florets absent or very short, less than 3
   mm long; plant usually scapose, rough
   scabrous, upper leaves much reduced . . 1. *H. radula*
1. Ray florets conspicuous, yellow or orange;
   plant not scapose, glabrous to villous, but
   not scabrous nor with upper leaves much
   reduced.
   2. Stems low, spreading . . . . . . 2. *H. debilis*
   2. Stems erect
      3. Leaf blades lanceolate to elliptic-ovate,
         base gradually tapering, light green . 3. *H. agrestis*
      3. Leaf blades ovate to broadly ovate, base
         cuneate, dark green . . . . . 4. *H. annuus*

1. H. radula (Pursh) T & G—Perennial with erect stems up to 1 m tall, scapose, hirsute to scabrous. Leaves 5–20 cm long, opposite, the basal ones obovate or suborbicular forming a rosette, upper leaves much reduced, hispid. Peduncle elongate with a single large head, ligules minute, 1–2 mm long or apparently absent, disk about 1–3 cm wide, purplish. $2n = 34$. Sandy soil, pinelands, Fla. to Ala., Ga. Summer.

2. H. debilis Nutt.—Annual with numerous decumbent branches spreading from the stem base, up to 1 m tall, glabrous to scabrous. Leaves mostly 5–10 cm long, alternate, deltoid or deltoid-ovate, irregularly serrate or dentate, scabrous. Heads conspicuous at ends of branches, phyllaries narrowly lanceolate and extending beyond the disk, scabrous, ligules pale yellow, disk purplish red, about 1.5 cm wide. $2n = 34$. Coastal sandy places, sea beaches, Fla. to Tex., north to Ga. Spring, fall.

3. H. agrestis Pollard—Annual with fibrous roots, stems erect up to 2 m high, sometimes much branched, glabrous. Leaves alternate, or sometimes opposite below, lanceolate or lance-ovate, hispid, light green in color. Heads with phyllaries lanceolate, ligules bright yellow; disk purplish. Pinelands, roadsides, disturbed sites, endemic to south Fla. Summer.

4. H. annuus L. ANNUAL SUNFLOWER—Annual with taproots and stems erect up to 2 m tall or more, branched, more or less pubescent. Leaves 5–20 cm long, opposite below, alternate above, hispid pubescent, dark green in color, serrate, base cuneate or truncate. Involucre 2–10 cm wide or more, ligules 3–6 cm long; disk reddish purple or purplish. Achenes 5–15 mm long. $2n = 34$. Disturbed sites in south Fla., not common, adventive from western U.S. and also in cultivation. Summer, fall.

Key 2

TRIBE *Helenieae*

1. Corollas white, pink, purple, or red; stem
   not winged.
   2. Heads radiate, ligules purple or reddish
      brown with yellow tips . . . . .   22. *Gaillardia*
   2. Heads all discoid, corollas white or pink   23. *Palafoxia*
1. Corollas yellow or purplish; stem winged.
   3. Phyllaries and leaves with conspicuous
      gland dots . . . .   24. *Pectis*
   3. Phyllaries and leaves without glands.
      4. Heads all discoid, small, 3–5 mm high .   25. *Flaveria*
      4. Heads radiate, larger 5–15 mm high .   26. *Helenium*

22. GAILLÁRDIA Foug.   BLANKETFLOWER

Annual or perennial herbs with alternate, entire to pinnatifid leaves. Heads showy, fragrant, erect, many-flowered, radiate or discoid on long peduncles; phyllaries in two or three rows, the outer foliaceous, spreading and reflexed in fruit, ligules yellow, orange, purple, reddish, distinctly three-cleft or toothed; ray florets sterile, disk florets bisexual with funnelform corolla; receptacle convex to globose, with bristlelike or short chaff, pappus of six to twelve awn-tipped scales. Achene ribbed, villous. About 20 species, America.

1. G. pulchella Foug.—Annual or short-lived perennial up to 6 dm tall with stems freely branching at base and then ascending. Lower leaves 4–8 cm long, oblanceolate, coarsely toothed, hirsute below, upper leaves usually entire. Phyllaries lanceolate, pubescent, disk about 1.5–2.5 cm wide, purplish tipped, ligules usually dark purple or with the tip yellowish. Achenes strigose pubescent. $2n = 36$. Disturbed sites, roadsides, sandy soil, sea beaches, Fla. to north Mexico, Va., Minn., Colo. All year. *G. picta* Sweet

23. PALAFÓXIA Lag.

Tall, perennial or annual, erect herbs often with woody base and bearing alternate or opposite, entire leaves. Heads in corymbs, discoid, few-flowered, phyllaries in two series, involucre ellipsoidal, corollas white or pink, deeply five-cleft, pappus of several scales. Achene four-angled, pubescent. About seven species in North America. *Polypteris* Nutt.

1. Lower leaves ovate-lanceolate; pappus of
   short scales 1 mm long . . . . .   1. *P. feayi*
1. Lower leaves linear or linear-lanceolate;
   pappus of slender, pointed scales 4–5 mm
   long . . . .   2. *P. integrifolia*

1. P. feayi Gray—Narrow, tall, somewhat shrubby plant up to 2 m or more. Leaves about 5–8 cm long, elliptic or elliptic-lanceolate, puberulent. Involucre mostly 7–8 mm high, corollas white, anthers dark. Achene about 6 mm long, puberulent. Coastal pineland, scrub vegetation, endemic to Fla. Fall.

2. P. integrifolia (Nutt.) T & G—Slender, branching herbs up to 1 m or more tall. Leaves about 5–7 cm long, linear-lanceolate. Involucre about 1–1.5 cm high, phyllaries elliptic-spatulate, corollas white or pinkish, lobes linear. Achene about 5 mm long. Sandy soil, pinelands, Fla., Ga., on the coastal plain. Fall. *Polypteris integrifolia* Nutt.

24. PÉCTIS L.

Matted or slender gland-dotted herbs with pubescence formed in lines on the stems. Leaves opposite, linear or linear-lanceolate. Heads small, few-flowered, solitary or variously clustered, involucre narrow, phyllaries with glands, ligules yellow, few. Pappus of narrow scales with bristles. About 70 species, tropical America.

1. Heads with distinct peduncles borne singly
   in upper branches and terminally . .   1. *P. leptocephala*
1. Heads sessile or nearly so, borne in clusters
   in upper axils and terminally.

2. Leaves filiform to narrowly linear, mostly
    less than 2 mm wide, with two rows of
    glands beneath . . . .   2. *P. linearifolia*
2. Leaves linear-lanceolate to lanceolate, up
    to 5 mm wide, with glands scattered
    beneath . . . .   3. *P. prostrata*

    1. **P. leptocephala** (Cassini) Urban—Stems much branched, about
1–5 dm long, slender, the branches prostrate and forming mats. Leaves
1–4 cm long, less than 1 mm wide, linear, entire, acute tipped. Heads all
solitary on filiform peduncles 1–3 cm long. Involucre of about five to
six phyllaries, linear; ligules five to six, 2–3 mm long, disk florets few,
mostly four to eight, pappus bristles two to five. Achene thin, about 2–3
mm long. Disturbed sites, pinelands, sea beaches, south Fla., W.I. Summer, fall.

    2. **P. linearifolia** Urban—Stems erect or diffusely branching, 2–4
dm tall, pubescent. Leaves mostly less than 2 mm wide, filiform or
linear, with two rows of glands beneath, and with three to six pairs of
bristles near leaf base. Heads sessile or nearly so; involucre 5 mm high
or less, phyllaries four to six; ligules two to three, about 2 mm long, disk
florets six to seven, pappus scalelike, two to five, lance-attenuate. Achene
about 2–3 mm long. Disturbed sites, sandy pinelands, sea beaches, south
Fla., W.I. Summer, fall.

    3. **P. prostrata** Cav.—Stems procumbent or prostrate and much
branched from the base, about 1–2 dm long. Leaves usually 1–3 cm
long, up to 5 mm wide, linear lanceolate to linear-spatulate, with glands
scattered beneath, entire, and with five to nine pairs of bristles near the
base. Heads sessile or nearly so, involucre 5–7 mm high, phyllaries five,
linear-elliptic and keeled; ligules five, disk florets variable, usually five
to fifteen, corollas 2–3 mm long, pappus scales two to five, lance-attenuate. Achene 3–4 mm long. Disturbed sites, various soils, Fla. to Ariz.,
W.I., Central America, South America. Summer, fall.

25. FLAVÈRIA Juss.

    Annual or perennial herbs with slender, sessile leaves and many
small glomerate heads. Ray florets yellow, pistillate, inconspicuous,
often single or entirely absent, disk florets one to fifteen, minute, bisexual, yellow, involucre of two to nine phyllaries, pappus absent.
Achene about ten-ribbed, glabrous, narrow. About 20 species in U.S.,
Mexico.

1. Receptacle setose; heads with one to three
    florets; phyllaries one or two per head .   1. *F. trinervia*
1. Receptacle naked; heads with two to fifteen

florets; phyllaries three to nine per head.
2. Phyllaries three or four per head; inflorescence cymelike . . . . . .   2. *F. bidentis*
2. Phyllaries five to nine per head; inflorescence a dense or spreading corymb.
   3. Florets two to nine per head; bracts subtending the involucre less than the length of the phyllaries or lacking; leaf blades mostly 1–10 mm wide.
     4. Phyllaries keeled; leaf blades mostly not more than 2 mm wide . . .   3. *F. linearis*
     4. Phyllaries not keeled or obscurely so; leaf blades mostly 4–10 mm wide .   4. *F.* × *latifolia*
   3. Florets ten to fifteen per head; leaf blades mostly 5–15 mm wide; bracts subtending the involucre, conspicuous, equal to or exceeding the phyllaries in length . . . . . .   5. *F. floridana*

1. F. trinervia (Spreng.) Mohr—Stems widely branching, up to 1 m tall, smooth and bearing linear-elliptic or lanceolate, serrate, three-veined leaves mostly 3–10 cm long, 1–5 cm wide. Phyllaries one to two, ligule about 1 m long; corolla of single disk floret about 2 mm long; heads all sessile in axils of leafy bracts. Achene clavate, 1–2 mm long. $2n = 36$. Disturbed sites, roadsides, sandy soil, south Fla. and Fla. Keys, Ariz., Hawaii, Mexico, Central America, South America. All year.

2. F. bidentis (L.) Kuntze—Stems up to 10 dm, angled, smooth at nodes and along angles. Leaves about 5–8 cm long, narrowly elliptic, three-veined, glabrous or nearly so, margins serrate. Heads two- to nine-flowered on peduncles, arranged in cymelike clusters or corymbs. Phyllaries three to four, ovate, obtuse. Ligule narrow, about 1 mm long. Corolla of disk floret about 2–3 mm. Achene 1–2 mm. $2n = 36$. Disturbed sites, pinelands, scattered locales in Fla. to Ala., Ga. on the coastal plain, Mexico, and Central and South America. All year. Probably rare in our area, although it has been collected on Big Pine Key.

3. F. linearis Lag. YELLOWTOP—Stems mostly 3–8 dm tall, branching from the base, often shrubby, often decumbent, glabrous, and bearing many entire, linear or linear-lanceolate leaves 5–10 cm long, 1–2 mm wide. Heads many in spreading or dense corymbs; phyllaries five or six, keeled, about 3–4 mm long, elliptic or lanceolate; ray florets one, ligule ovate; disk florets about two to seven, the corollas 2–3 mm long. Achene about 2 mm long. $2n = 36$. Disturbed sites, roadsides, sandy soil, near marshes, hammocks, south Fla., Fla. Keys, W.I., Mexico. All year.

4. F. × latifolia (J. R. Johnston) Rydberg (pro var.)—Stems erect to 1 m tall, smooth and bearing linear-lanceolate to linear, entire or obscurely serrate leaves, about 2–10 cm long. Heads clustered in dense or open corymbs, phyllaries five, about 3–4 mm long, lance-elliptic. Ligule 2 mm long; disk florets usually two to nine per head, the corolla 3 mm long. Achene about 1.5 mm long. $2n = 36$. Low pinelands, marshes, especially on calcareous soils, swamps, endemic to south Fla. This plant apparently is a stabilized natural hybrid of **F. floridana** × **F. linearis**.

5. **F. floridana** J. R. Johnston—Stems erect up to 12 dm tall, smooth and bearing entire or obscurely serrate linear-lanceolate leaves about 4–8 cm long, 5–15 mm wide. Heads with ten to fifteen florets, in tight clusters; phyllaries five or six, mostly 4 mm long; ligule ovate, about 2 mm long; corolla a disk floret 4 mm long. Achene slender, 1–2 mm long. $2n = 36$. Sea beaches, dunes, endemic to south Fla., especially on the lower Gulf coast. Summer, fall.

26. HELÉNIUM L. SNEEZEWEED

Annual or perennial herbs with alternate, glandular-punctate leaves and solitary or numerous heads. Heads with many florets, usually radiate; involucres small with two to three rows of linear to spatulate phyllaries, receptacle mostly convex to conic, smooth; ray florets several to many, pistillate or rarely sterile, cuneate, three- to five-cleft, mostly yellow; disk florets many, bisexual, corolla lobes glandular-pubescent. Style branches with filamentous appendages. Pappus of awn-tipped scales. Achene four- to five-angled. About 40 species, chiefly in western U.S.

1. Larger leaves forming a basal rosette; stems with one to few heads.
  2. Achenes glabrous; leaves entire to shallowly dentate or sinuate . . . . .   1. *H. vernale*
  2. Achenes pubescent; leaves tending to be deeply lobed or toothed . . . . .   2. *H. pinnatifidum*
1. Larger leaves cauline, not forming a basal rosette; stems with numerous heads.
  3. Leaves filiform to narrowly linear, stems unwinged . . . .   3. *H. amarum*
  3. Leaves lanceolate, stems conspicuously winged . . . .   4. *H. flexuosum*

1. **H. vernale** Walt.—Stems erect up to 7 dm tall. Leaves to 15 cm long, linear to narrowly spatulate, pinnatifid-toothed, and forming a basal rosette, upper leaves reduced or absent. Phyllaries narrow-subulate, spreading, heads solitary or few; ligules bright orange yellow, ligules about 1 cm long, disk florets very numerous; corollas 4–5 mm. Achene 1–2 mm. Marshes, swamps, and low pinelands, Fla. to Miss., N.C. on the coastal plain. Spring.

2. H. pinnatifidum (Nutt.) Rydberg—Perennial with stems erect, up to 7 dm high, puberulent or hirsute. Leaves about 10–20 cm long, linear-oblanceolate, deeply lobed or pinnatifid, rather fleshy, and forming a basal rosette; upper leaves reduced, linear-lanceolate, toothed or entire. Heads solitary, phyllaries 5–7 mm long, spreading, ray florets neutral, ligules bright yellow, deeply cleft about 6–10 mm long; disk yellow, 1–2 cm wide, corollas 5 mm long. Achenes 2 mm long, hirsute along angles. Marshes, moist soil, Fla. and Ala., N.C. on the coastal plain. Spring.

3. H. amarum (Raf.) H. Rock, SPANISH DAISY—Smooth annuals up to 5 dm tall with numerous filiform or linear leaves 2–8 cm long. Heads on naked peduncles extending beyond the leafy stems; ligules about five to ten, up to 1 cm long; disk 6–12 mm wide, yellow. Pappus scales ovate with slender awn. $2n = 30$. Disturbed sites, roadsides in sandy soil, Fla. and Tex., Va., Kans. Spring, fall. *H. tenuifolium* Nutt.

4. H. flexuosum Raf.—Pubescent perennial up to 1 m with fibrous roots. Stems winged owing to decurrent leaf bases. Lower leaves oblanceolate, deciduous, upper lanceolate, mostly 5–12 cm long, to lance-linear, to lanceolate. Heads numerous in open corymbs, ray florets often purplish at base, neutral, disk florets purplish or brownish purple, disk ovoid-globose. Pappus scales ovate-lanceolate, awn tipped. $2n = 28$. Disturbed sites, low pinelands, Fla. and Tex., Mass., Mo. *H. nudiflorum* Nutt.; *H. polyphyllum* Small

Key 3

TRIBE *Senecioneae*

1. Heads radiate; florets yellow or orange . . 27. *Senecio*
1. Heads discoid; florets reddish, orange, purplish, or white.
  2. Florets white or whitish or sometimes pinkish.
    3. Leaves palmately veined or pinnately veined and hastate; all florets bisexual . . . . 28. *Cacalia*
    3. Leaves pinnately veined, oblanceolate, irregularly lobed; outer florets pistillate . . . . 29. *Erechtites*
  2. Florets red, lavender, or yellow orange . 30. *Emilia*

27. SENÈCIO L. GOLDEN RAGWORT, SQUAWWEED

Trees, shrubs, climbers, and herbs of diverse habit with alternate, often all basal leaves that are entire to variously lobed or divided.

Heads many-flowered, radiate or sometimes discoid; ray florets pistillate, ligules yellow, orange, or sometimes reddish, involucre campanulate to cylindrical with appressed more or less equal phyllaries in one or two rows, receptacle without chaff; disk florets bisexual, yellow to reddish. Pappus of numerous white, mostly smooth bristles. Achene five- to ten-ribbed. A very large genus with over 1500 species, cosmopolitan.

1. S. glabellus Poir.—Annual or biennial with fibrous roots and smooth, succulent, hollow stems up to 8 dm tall, mostly simple to the inflorescence. Larger leaves basal, up to 20 cm long, usually pinnatifid, deeply lobed, progressively much reduced upward on stem. Heads radiate, rather small, numerous, involucre about 6 mm long, phyllaries linear, ray florets orange yellow, about 12 mm long. Achene about 2 mm long, pubescent in ribs. $2n = 46$. Low ground, old fields, Fla. and Tex., N.C., S. Dak., Mexico. Spring.

2. S. confusus Britt.—Smooth herbaceous or woody vines with stems up to 4 m long or more, often high climbing. Leaves 3–10 cm long, lance-ovate to narrowly deltoid, coarsely dentate, petioled. Heads radiate, 2–3 cm wide or more, in terminal cymes or clusters; ligules ten to fifteen, orange or orange red. Cultivated and locally escaped in peninsular Fla. and Fla. Keys, popular in ornamental plantings, naturalized from Mexico. All year.

28. CACÀLIA L.   INDIAN PLANTAIN

Tall shrubs or herbs with alternate leaves. Heads discoid, cylindrical, whitish or pinkish in compound cymes; involucre cylindrical or campanulate, of a single row of herbaceous phyllaries often with scarious margins; receptacle flattish and without chaff; florets all bisexual, corolla deeply lobed. Achenes smooth, ribbed. About 40 species in America, Asia, Europe. *Mesadenia* Raf.; *Synosma* Raf.

1. C. lanceolata Nutt.—Stems glaucous up to 1.5 m tall bearing linear or narrowly lanceolate to ovate, entire to obscurely serrate leaves. Phyllaries narrow, about 1 cm long. Achene 5 mm long. *Mesadenia lanceolata* (Nutt.) Raf. Two rather well-marked varieties occur in our area:

1a. C. lanceolata var. **lanceolata**—Leaf blades mostly 3–4 cm wide or commonly less. $2n = 56$. Everglades, wet pinelands, Fla. to N.C., La. on the coastal plain. Summer, fall. *Mesadenia lanceolata* (Nutt.) Raf.

1b. C. lanceolata var. **elliottii** (Harper) Kral & R. K. Godfrey—Leaf blades mostly over 4 cm wide, ovate to broadly ovate. Low ground, pinelands, damp woods, Fla. to Ga., La., on the coastal plain. Summer, fall. *Mesadenia elliottii* Harper

29. ERECHTÌTES Raf.   Fireweed, Pilewort

Erect herbs, roots fibrous, with alternate leaves, entire to pinnately dissected, and usually cylindric heads in corymbs or panicles. Heads discoid, white or pale yellow, in panicles or cymes; involucre basally swollen, of a single series of narrow phyllaries, receptacle flat; marginal florets of head pistillate, inner florets bisexual or sometimes sterile, corolla four- to five-toothed. Achene five-angled or with many nerves. Pappus conspicuous, composed of numerous slender bristles. Five species, America.

1. E. hieracifolia (L.) Raf. var. **hieracifolia**—Annual, fibrous rooted weed up to 3 m tall with smooth or pubescent, grooved, somewhat fleshy stems. Leaves oblanceolate or lanceolate, sessile, coarsely toothed or lobed, often auriculate clasping. Heads corymbose or paniculate, rarely solitary; involucre cylindric about 1 cm long or more, phyllaries linear. Pappus white, abundant. Achenes with ten to twelve ribs, pubescent. $2n = 40$. Disturbed sites, old fields, south Fla. and Tex., Newfoundland to Nebr., W.I. All year.

This species is a very successful weed found abundantly in our area.

30. EMÍLIA Cassini

Perennial or annual herbs with alternate, mainly basal leaves. Heads discoid, solitary, or in lax corymbs; involucres somewhat swollen at base, phyllaries in a single row; disk florets bisexual, red, lavender, purple, or dark yellow; corolla tubular, with a cylindric throat, lobes narrow-lanceolate; pappus soft, white or purplish, setose. About five species, tropical Africa and Asia.

1. Corollas reddish or orange   .   .   .   .   .   1. *E. javanica*
1. Corollas purple or lilac .   .   .   .   .   .   2. *E. sonchifolia*

1. E. javanica (Burm.) C. B. Robinson, Tassel Flower—Stems up to 8 dm tall, slender with tender succulent leaves; lower ones spatulate-lanceolate, the upper lanceolate to linear with auricled bases. Heads small, on long peduncles; phyllaries linear, 10–12 mm long, involucre campanulate about 1 cm high; corolla reddish or orange, about 1 cm long. Achene about 4 mm long. $2n = 20$. Disturbed sites, pinelands, old fields, south Fla., W.I., Central America from Old World tropics. All year. *E. coccinea* (Sims) G. Don ex Sweet

2. **E. sonchifolia** (L.) DC. ex Wight—Stems up to 5 dm tall, slender. Leaves spatulate below with auricled bases, upper lanceolate to elliptic with auricled bases and somewhat pinnatifid. Heads small, on long peduncles; phyllaries linear 8–10 mm, involucre cylindric, less than 1 cm long, corolla purple or lilac, about 7–8 mm long. Achene 3 mm long. $2n = 10$. Disturbed sites, old fields, roadsides, south Fla., W.I., Central America, South America, from Old World tropics. All year.

GYNURA AURANTIACA (Blume) DC. VELVET PLANT, with stems having dense, long, purplish pubescence and discoid heads with bright yellow orange corollas, may escape locally from cultivation but apparently is not common in our area.

Key 4

TRIBE *Astereae*

1. Erect shrubs over 1 m tall; plants dioecious   31. *Baccharis*
1. Herbs or woody vines; plants not dioecious.
   2. Pappus composed of short awns . . .   32. *Boltonia*
   2. Pappus composed of capillary bristles.
     3. Ray corollas, if present, pale yellow, yellow, or orange.
       4. Larger leaves all basal, upper leaves much reduced, plants scapose.
         5. Heads in flat-topped corymbs, discoid . . . .   33. *Bigelowia*
         5. Heads in secund raceme or narrow panicle . . . .   34. *Solidago*
       4. Larger leaves cauline, not all basal, the upper leaves not conspicuously reduced.
         6. Involucres 2.0–2.5 cm broad; leaves pinnatifid-toothed, teeth bristle tipped . . . .   35. *Haplopappus*
         6. Involucres smaller, 0.2–1.5 cm broad; leaves merely serrate or entire.
           7. Involucres relatively small, 2–5 mm wide; inflorescence in dense, often secund racemes or panicles . . . . . .   34. *Solidago*

7. Involucres larger, 7–15 mm
    wide; inflorescence an open ra-
    ceme or panicle . . . .   36. *Heterotheca*
3. Ray corollas rose, pink, blue, or purple
    or white.
9. Involucres small, less than 5 mm
    wide; upper leaves linear-filiform .   37. *Conyza*
9. Involucres larger, more than 5 mm
    wide; upper leaves lanceolate to
    ovate.
10. Involucres cylindric, phyllaries car-
    tilaginous; disk corollas white .   38. *Sericocarpus*
10. Involucres saucer-shaped to cam-
    panulate, phyllaries not carti-
    laginous; disk corollas yellow,
    brown, or purplish.
11. Plants more or less scapose or
    with naked peduncles . . .   39. *Erigeron*
11. Plants with leafy peduncles . .   40. *Aster*

31. BÁCCHARIS L. Groundsel, Saltbush

Dioecious shrubs with alternate, fleshy, entire or toothed leaves.
Heads in panicles with many florets, unisexual; yellowish or greenish pistillate heads with tubular, filiform florets; staminate florets with funnelform corollas; receptacle without chaff, phyllaries imbricate, unequal,
involucre cylindric. Pappus of capillary bristles. Achenes smooth, ten-
ribbed. About 400 species, chiefly in tropical and subtropical regions.

1. Leaf blades narrowly linear to linear-lance-
    olate, mostly less than 5 mm wide . . .   1. *B. angustifolia*
1. Leaf blades lanceolate-ovate to spatulate,
    over 5 mm wide.
2. Leaf margin entire . . . .   2. *B. dioica*
2. Leaf margin lobed, toothed, or obscurely
    serrate.
3. Heads sessile or subsessile, in axillary
    or terminal clusters . . . . .   3. *B. glomeruliflora*
3. Some or all the heads distinctly pedun-
    cled in spreading panicles . . .   4. *B. halimifolia*
                                        var. *angustior*

1. **B. angustifolia** Michx. False Willow—Resinous shrubs up to 3
m tall or more. Leaves 3–8 cm long, linear to linear-lanceolate, entire or
nearly so. Pistillate involucres 4–5 mm with innermost phyllaries 4–5
mm, pedunculate. Achene 1–2 mm long. Salt marshes, coastal swamps,
and hammocks, Fla. to Tex., N.C., W.I. Fall.
2. **B. dioica** Vahl—Shrubs up to 3 m tall or more. Leaves 1–3 cm
long, entire, spatulate or broadly obovate-spatulate. Pistillate involucres

with distinct, ovate-obtuse outer phyllaries and linear acuminate inner ones. Achene 1–2 mm long. Coastal areas, hammocks, south Fla., central Fla., W.I. Fall.

3. **B. glomeruliflora** Pers. GROUNDSEL TREE—Shrubs up to 3 m tall or more. Leaves 3–5 cm long, cuneate-obovate to spatulate, usually toothed or serrate toward the apex. Pistillate involucres 5–6 mm long, usually in sessile glomerules, the inner phyllaries spatulate. Achene about 2 mm long. Well-drained soils, or salt marshes, hammocks, Fla. to N.C. on the coastal plain, W.I. Fall.

4. **B. halimifolia** L. var. **angustior** DC.—Shrubs up to 3 m or more with angled branches glabrous but somewhat resinous or scurfy. Leaves about 4–6 cm long, thick, elliptic to ovate, coarsely or distantly toothed or serrate. Panicles pyramidal, pistillate involucres 3–6 mm in pedunculate glomerules, phyllaries strongly imbricate, inner phyllaries linear. Achene 1–2 mm long, winged. $2n = 18$. Well-drained soils or marshes and sea beaches, hammocks, Fla. and Tex., Mass., Ark., W.I. Fall.

32. BOLTÓNIA L'Her. DOLL'S DAISY

Perennial, usually stoloniferous, smooth herbs with narrow, entire leaves. Heads radiate with many florets corymbose; ligules white, pink, or bluish, narrow, ray florets pistillate, phyllaries in two or more rows, strongly imbricate and subequal, receptacle hemispherical or conical, without chaff, disk florets bisexual, corollas yellow, style branches with hairy appendages, anthers as long as filaments or longer. Pappus of minute bristles and two to four awns. Achene flattened. Four species in North America.

1. **B. diffusa** Ell.—Stems up to 1 m or taller from narrow rhizomes and stolons, plants slender but with diffuse inflorescence. Leaves 3–15 cm long, about 5 mm wide or less, linear, leathery. Involucre of three to five unequal series, phyllaries imbricate, the outer subulate, about 1–2 mm long, the inner linear, ligules linear to spatulate, white or lilac. Achenes broadly winged. Pappus awns present. $2n = 18$. Low ground, Fla. to Tex., Ky., Mo. Fall.

33. BIGELÒWIA DC. RAYLESS GOLDENROD

Perennial with slender stems and alternate, narrow, entire leaves. Heads discoid in terminal corymbs; involucres narrow, phyllaries acuminate or acute; disk florets yellow with funnelform throat and lanceolate lobes. Achenes flattened, one to two ribs on a side. Pappus of bristles. About 40 species from North America to South America. *Chondrophora* Raf.

1. **B. nudata** (Michx.) DC. ssp. **australis** L. C. Anderson—Perennial

herb with stems smooth up to 5 dm tall. Leaves 9–14 cm long, spatulate or linear-spatule, basal leaves forming rosettes, cauline leaves few. Heads many in flat-topped corymbs, involucre cylindric, 6–8 mm long. Phyllaries 4–5 mm long, linear to linear-elliptic, acute. Pappus white, capillary. Low acidic soils, pinelands, peninsular Fla. All year.

34. SOLIDÀGO L. GOLDENROD

Perennial herbs with fibrous roots and simple, alternate, entire to toothed leaves and small heads. Heads with few to many florets, radiate; ray florets pistillate, yellow, phyllaries usually with a green tip, imbricate in several series; receptacle without chaff, small, disk florets bisexual, yellow. Pappus of white capillary bristles. Achene with several nerves. About 100 species, chiefly in North America. A complex genus with numerous hybrids. *Brintonia* Greene; *Euthamia* Nutt.; *Oligoneuron* Small; *Brachychaeta* T & G

1. Larger leaves elongate-spatulate forming a basal rosette; upper leaves reduced . .   1. *S. stricta*
1. Larger leaves cauline, not basal; upper leaves not greatly reduced.
  2. Inflorescence corymbose; leaf blades linear-filiform . . . .   2. *S. microcephala*
  2. Inflorescence racemose, thyrsoid, or paniculate, often with secund branches; leaf blades linear or lanceolate to ovate.
    3. Leaves glabrous or with merely scattered hairs.
      4. Leaves lanceolate-acuminate, margins obscurely to distinctly serrate . .   3. *S. leavenworthii*
      4. Leaves elongate-ovate, margins mostly entire, often revolute.
        5. Upper stem densely pubescent .   4. *S. fistulosa*
        5. Upper stem essentially glabrous.
          6. Margins of leaf blades strongly revolute . . . . . .   5. *S. chapmanii*
          6. Margins of leaf blades not revolute, occasionally obscurely serrate . . . . . .   6. *S. tortifolia*
    3. Leaves with pubescence on some or all the leaves.
      7. Leaves with three prominent veins, margins obscurely to distinctly serrate . . . .   7. *S. gigantea*
      7. Leaves with a single prominent vein, margins entire to obscurely serrate   8. *S. sempervirens* var. *mexicana*

1. S. stricta Ait.—Smooth perennial with solitary, slender stems up to 2 m, glabrous throughout. Lower leaves about 10–30 cm long, thick, firm, persistent, generally oblanceolate or elliptic-lanceolate, entire to shallowly or obscurely serrate, upper leaves reduced, entire, sessile. Phyllaries linear-acute, erose, ligules elliptic, five to seven, involucre 5–6 mm high. Achenes linear, pubescent. $2n = 54$. Low pinelands, glades, sandy soils, Fla. and Tex., N.J., W.I. Summer, fall. *S. petiolata* Mill.; *S. angustifolia* Ell.

A complex polyploid species composed of highly variable populations.

2. S. microcephala (Greene) Bush—Smooth or barely pubescent stems up to 1 m tall. Leaves 3–6 cm long, linear or filiform, glabrous, smaller ones subulate. Inflorescence corymbose, involucre slender, cylindric, about 4–5 mm long, glutinous, heads about ten to seventeen florets; ray florets seven to thirteen, disk florets three to four. Achenes pubescent. $2n = 18$. Sandy pineland, Fla. to Miss., N.J. *S. caroliniana* BSP, misapplied; *Euthamia minor* (Michx.) Greene, *S. minor* (Michx.) Fern. not Mill.; *Euthamia caroliniana* Greene

3. S. leavenworthii T & G—Stems smooth below, becoming short pubescent above, about 1 m tall or less, very leafy. Larger leaves 6–20 cm long, lanceolate or oblanceolate, obscurely to distinctly serrate, sessile, more or less glabrous. Inflorescence thyrsoid heads about 5–6 mm long, secund. Achenes minutely pubescent. $2n = 36$. Low ground on the coastal plain, Fla. to S.C. Summer, fall.

4. S. fistulosa Mill.—Stems up to 2 m tall from creeping rhizomes, smooth below, white pubescent or hirsute above, very leafy. Leaves 4–12 cm long, sessile and subclasping, lance-ovate to obovate, entire to obscurely serrate. Inflorescence thyrsoid, branches numerous, often recurved, dense; heads 5–7 mm long, secund, phyllaries thin, imbricate. Achenes pubescent. $2n = 18$. Low ground, pinelands, Fla. to La., north to N.J. Summer, fall.

5. S. chapmanii T & G—Stems smooth up to 1.5 m tall or more. Leaves 4–6 cm long, 1–2 cm wide, linear-lanceolate to obovate with margins strongly revolute and ciliolate. Heads 6–7 mm long, 2–3 mm wide; phyllaries pale yellow, glabrous, ciliate, much exceeded by pappus at anthesis. Achenes pubescent. Pinelands, sandy soil, hammocks, Fla. to south Ga. on the coastal plain. Summer, fall.

6. S. tortifolia Ell.—Stems up to 1 m tall or more with long, slender, creeping rhizomes, glabrous slender. Lower leaves reduced and early deciduous, upper leaves about 3–7 cm long, numerous, linear or narrowly elliptic, pubescent beneath, sessile. Inflorescence thyrsoid or paniculate, with recurved branches, involucre 3–4 mm long, phyllaries slender, glabrous, yellowish, heads five- to ten-flowered, ligules yellow, 2–4 mm long. Achenes about 1 mm long. $2n = 18$. Pinelands, sandy soil, Fla. and Tex., Va. on coastal plain. All year.

7. **S. gigantea** Ait.—Stems glabrous below, puberulent above, up to 2 m tall from creeping rhizomes. Lower leaves early deciduous, upper numerous, lance-linear, sessile, about 5–15 cm long, usually puberulent. Inflorescence thyrsoid with recurved spreading branches or paniculate. Phyllaries linear-acute, involucre 4–5 mm, yellowish. Ligules ten to seventeen. Achenes pubescent. $2n = 18,36$. Pinelands, old fields, damp woods, of wide distribution Fla. to Tex., New Brunswick, B.C. Summer, fall. *S. serotina* Retz.

A polyploid complex possibly related to another polyploid of wide distribution, S. altissima L.

8. **S. sempervirens** L. var. **mexicana** (L.) Fern.—Smooth or pubescent on upper stem, rather succulent perennials up to 2 m or more tall. Basal leaves about 10–40 cm long, oblanceolate and persistent, entire, upper leaves numerous, sessile, not much reduced in size. Inflorescence dense, paniculate or thrysoid, the lower branches recurved and secúnd; involucre hemispherical, about 3–5 mm long, phyllaries imbricate, acute; ligules seven to eleven. Achene pubescent. $2n = 18$, $18 + 1$. Low ground, coastal areas, Fla. to Canada; Central America, South America. All year. *S. mexicana* L.

35. HAPLOPÁPPUS Endl.

Herbs or shrubs from taproots with alternate simple to pinnatifid leaves and bearing rather large heads. Heads radiate with many florets in spreading panicles or racemes. Involucre hemispherical, phyllaries green at least in the tips; receptacle without chaff, flat; ray florets usually pistillate, yellow; disk florets bisexual, corolla lobes deltoid. Pappus of capillary bristles. Achene four- to five-angled. About 150 species in America. *Aplopappus* Cassini; *Sideranthus* Fraser; *Isopappus* T & G

1. **H. phyllocephalus** DC. var. **megacephalus** (Nash) Waterfall—Stems erect up to 1 m tall or more, much branched, suffrutescent, finely pubescent. Leaves 5–7 cm long, linear-elliptic to lanceolate-elliptic, broadly dentate or serrate with spreading spinelike teeth. Outer phyllaries 10–12 mm long, margins ciliolate. Sandy soil, disturbed sites, wet margins of hammocks, coastal areas, south Fla. and Fla. Keys to Tex. Summer, fall.

36. HETEROTHÈCA Cassini GOLDEN ASTER

Perennial or annual herbs with alternate entire or toothed leaves and often rather numerous heads. Heads usually radiate, many-flowered, in terminal racemes or corymbose panicles; ray florets pistillate, bright yellow, conspicuous; disk florets bisexual, corolla funnelform; receptacle without chaff, flat. Pappus of disk floret in two rows, the outer coarse short bristles or scales, inner of capillary bristles; pappus of ray floret present or absent. About 25 species in North America. *Pityopsis* Nutt.; *Chrysopsis* Ell.

1. Leaves linear to lance-ovate, lower ones not
    auricled at base.
    2. Leaves grasslike, linear to linear-lanceo-
        late, all or many of them with silvery
        white pubescence . . . . . .   1. *H. graminifolia*
    2. Leaves not grasslike, lanceolate to lance-
        ovate, not silvery white pubescent (may
        be woolly tomentose, hairs sometimes
        deciduous).
        3. Upper leaves conspicuously glandular-
            pubescent; lower leaves villous pubes-
            cent.
            4. Basal leaves forming a dense rosette .   2. *H. scabrella*
            4. Basal leaves not forming a rosette   .   3. *H. floridana*
        3. Upper stems glabrous or merely pubes-
            cent, eglandular; lower leaves not vil-
            lous pubescent . . . .   4. *H. hyssopifolia*
1. Leaves ovate or obovate, lower ones auri-
    cled-clasping at base . . . . . .   5. *H. subaxillaris*
                          var. *subaxillaris*

1. **H. graminifolia** (Michx.) Shinners, SILK GRASS—Stems up to 1 m tall, branched above from fibrous, stoloniferous rhizomes, silvery pubescent. Leaves up to 25 cm long, linear to lanceolate, silvery pubescent, grasslike. Panicles spreading, heads several to many; involucre 5–15 mm long, pilose, glands few, mostly covered by pubescence; ray florets 8–15 mm long, ligule 6–9 mm long. Pineland, sandy soil. Two varieties occur in our area:

1. Ray floret 8–10 mm long . . . . . .   1a. var. *graminifolia*
1. Ray floret 12–15 mm long   .   .   .   1b. var. *tracyi*

1a. H. graminifolia var. **graminifolia**—Leaves linear or linear-lanceolate. Sandy soil, pinelands, Fla. to Mexico, Del., Ohio. Fall. *Chrysopsis graminifolia* (Michx.) Ell.

1b. H. graminifolia var. **tracyi** (Small) R. W. Long—Leaves linear, the lower often elongate. Pinelands, endemic to south Fla., especially the lower Gulf coast. Fall. *Pityopsis tracyi* Small

2. H. scabrella (T & G) R. W. Long—Stems 1 m tall or more, upper portion glandular-pubescent. Leaves numerous, the basal ones in a dense rosette, spatulate or obovate, usually coarsely toothed, villous pubescent, the upper ones entire, linear to narrowly spatulate, glandular-scabrous. Involucre 6–9 mm long, phyllaries linear-subulate, acute. Achene 2–3 mm long. Pinelands, scrub vegetation, Fla. to Miss. on coastal plain. Fall. *Chrysopsis scabrella* T & G

3. H. floridana (Small) R. W. Long—Stems up to 4 dm tall, white cottony pubescent below, becoming glandular-pubescent above. Leaves densely white cottony pubescent, lower ones spatulate-lanceolate, about 8–10 cm long or less, upper leaves partially clasping, 1–3 cm long. Involucre 6–8 mm long, phyllaries glandular-pubescent. Sandy soil, pinelands, endemic to south Fla., lower Gulf coast. *Chrysopsis floridana* Small

Related to **H. scabrella** and possibly not specifically distinct from that species.

4. H. hyssopifolia (Nutt.) R. W. Long—Stems up to 1 m tall or less, smooth or somewhat cobwebby pubescent. Leaves rather numerous, basal ones forming a rosette, glabrous or somewhat white pubescent, linear-spatulate, upper ones crowded, filiform to linear, about 1–6 cm long. Involucre 6–8 mm long, phyllaries, lance-acuminate or acute. Achenes 1.5–2.5 mm long. Two varieties occur in our area:

1. Phyllaries lance-acute, not strongly reflexed; lower stems sparingly pubescent or glabrous . . . . 5a. var. *hyssopifolia*
1. Phyllaries elongate-subulate, often strongly reflexed; lower stems pubescent . . . 5b. var. *subulata*

4a. H. hyssopifolia var. **hyssopifolia**—Achenes mostly 2.0–2.5 mm long. Sandy soil, pinelands, Fla. to Ala. on the coastal plain. Fall. *Chrysopsis gigantea* Small

4b. H. hyssopifolia var. **subulata** (Small) R. W. Long—Achenes mostly 1.5–2.0 mm long. Sandy soil, endemic to south Fla. Fall. *Chrysopsis subulata* Small

5. H. subaxillaris (Lam.) Britt. & Rusby var. **subaxillaris**, CAMPHOR WEED—Stems up to 1 m tall, erect or ascending, glandular-hirsute, annual or short-lived perennial. Leaves 2–7 cm long, ovate or obovate, the lower ones auricled-clasping at the base, subentire to dentate. Heads many in diffuse corymbs or panicles, involucre 6–8 mm long, phyllaries stipitate glandular-pubescent; rays fifteen to thirty, ligules 1–2 cm long, bright yellow, those from ray florets glabrous or nearly so, those from disk florets densely pubescent with short bristles. Pinelands, disturbed sites near seashores, sandy soil, roadsides, Fla. to Del., Mexico, tropical America. Fall.

Plants with decumbent branches and lower leaves entire are var. **procumbens** Wagen.; it occurs in peninsular Fla. and may be expected in our area.

37. CONŸZA Lessing   Dwarf Horseweed

Herbs often weedy with numerous alternate leaves and many small heads. Heads discoid or inconspicuously radiate, phyllaries imbricate, the ligules white or purplish white, inconspicuous; marginal or ray florets pistillate, disk florets bisexual, few, less than twenty. Receptacle flat or nearly so, without chaff, pappus of capillary bristles. Achenes one- to two-nerved. About 60 species, temperate and subtropical regions. *Leptilon* Raf.

1. Stem simple or nearly so, central axis well
   defined.   1. *C. canadensis*
   2. Leaves 5–8 cm long . . . .   var. *pusilla*
   2. Leaves 3–4 cm long . . . .   2. *C. parva*
1. Stem diffusely branching, without a central
   axis . . . .   3. *C. ramosissima*

1. C. canadensis (L.) Cronquist var. **pusilla** (Nutt.) Cronquist— Annual to 1.5 m tall, simple, glabrous or nearly so. Leaves mostly 5–8 cm long, numerous, oblanceolate to linear-acute, serrate or entire, the lower early deciduous. Heads small and numerous in spreading panicle, involucre about 3–4 mm long, the phyllaries imbricate and minutely purple tipped; rays present, very short, white or pinkish. Disturbed sites, old fields, Fla. to Conn. on the coastal plain, tropical America. Summer, fall. *Leptilon pusillum* (Nutt.) Britt.; *Erigeron pusillus* Nutt.

2. C. parva Cronquist—Annual to 1 m tall, often less, simple, glabrous or nearly so. Leaves 3–4 cm long, linear-lanceolate to narrowly oblanceolate, entire or finely serrate, rameal leaves much reduced, linear. Heads small, numerous in narrow to spreading panicles or racemes; involucre 2–3 mm, phyllaries imbricate; ligules minute but evident, white; disk yellow. Disturbed sites, sandy soil, roadsides, wet pineland, Fla., tropical America. All year.

3. C. ramosissima Cronquist—Low annual, diffusely branching up to 3 dm tall without central axis. Leaves 3–4 cm long, linear to narrowly lanceolate, uppermost reduced to mere scales. Heads very numerous, involucre small, 3–4 mm high, phyllaries imbricate; ligules inconspicuous, purplish, about as long as the pappus. Disturbed sites, sandy soil near the coast, Fla. to Tex., Ala., inland to Minn., Ohio. All year. *Erigeron divaricatus* Michx.; *Leptilon divaricatum* (Michx.) Raf.

This species has been collected only in scattered locales in our area and is apparently not common.

38. SERICOCÁRPUS Nees White-Topped Aster

Herbs with alternate entire or serrate leaves. Heads corymbose, radiate, with many ten to twenty florets, ray florets usually five; involucre cylindrical or clavate, phyllaries at least the outer rather broad, imbricated in several rows, cartilaginous; ligules white or pinkish white, disk corolla with a long tube and funnelform throat; anthers somewhat longer than filaments, appendages elongate. Pappus of scabrous bristles. Achene one-nerved. Four species in North America, merged with Aster by some authors.

1. S. tortifolius (Michx.) Neu.—Stems slender to 1 m tall, densely puberulent, with leaves about 1–3 cm long, obovate to spatulate, thickly pubescent, sessile. Involucre about 6–9 mm long, cylindrical, outer phyllaries lance-ovate, densely pubescent, the inner ones linear or linear-lanceolate; pappus white; disk florets about 7 mm long. Achene 2–3 mm long. Dry, acidic pinelands or scrub vegetation, Fla. to La., north to N.C. Summer, fall. *Aster tortifolius* Michx.

39. ERÍGERON L. Fleabane

Annual or perennial herbs erect or ascending, with alternate leaves and one to many rather small heads. Heads in panicles or corymbs, radiate or discoid, many-flowered; ligules pistillate, numerous, usually white, pink, violet, or blue; involucre hemispheric or turbinate, phyllaries narrow, imbricate, receptacle flat, without chaff; disk florets yellow, minute, numerous, and bisexual. Pappus of capillary bristles. Achenes with one to five ribs. About 250 species, cosmopolitan but chiefly temperate America.

Taxonomically a difficult group in which hybridization, polyploidy, and apomixis occur.

1. Ray florets with pappus less than 1 mm
    long; plants annual . . . .   1. *E. strigosus* var. *strigosus*

1. Ray florets with long, bristlelike pappus
   over 1 mm long; plants perennial.
   2. Ray florets more than 100 per head; basal
      leaves often pinnately lobed or toothed 2. *E. quercifolius*
   2. Ray florets less than fifty per head; basal
      leaves entire or remotely serrate . . 3. *E. vernus*

1. E. strigosus Muhl. var. **strigosus,** DAISY FLEABANE—Annual or less commonly biennial up to 7 dm tall with stems pubescent. Lower leaves mostly 10–15 cm long, oblanceolate to elliptic, entire or toothed; upper leaves linear or linear-lanceolate, entire or nearly so. Heads several to many, involucre 2–5 mm long, pubescent; rays conspicuous, from 50 to 100, white, less commonly pink or bluish. Pappus of numerous bristles and several short scales; pappus of ray florets less than 1 mm long. $3n = 27$. Disturbed sites, waste places, hammocks, throughout much of North America. Spring, summer. *E. ramosus* (Walt.) BSP

2. E. quercifolius Lam. SOUTHERN FLEABANE—Perennial with stems up to 7 dm tall, simple, villous or hirsute. Basal leaves usually 9–15 cm long, oblanceolate or obovate, pinnately lobed to subentire. Upper leaves somewhat reduced, clasping. Heads one to few, involucre 3–4 mm long, viscid pubescent. Rays 100 to 150 per head, bluish, white, or pink. Pappus of ray florets over 1 mm long. $2n = 36$. Disturbed sites, various soils, Fla. to Tex., N.C. on the coastal plain. Spring, summer.

This is one of our most common roadside weeds in peninsular Fla.

3. E. vernus (L.) T & G—Perennial with stems up to 6 dm tall, simple, smooth or sparsely pubescent. Basal leaves 2–8 cm long, thick, elliptic-ovate to spatulate to suborbicular, entire to shallowly serrate, upper leaves reduced, few. Heads one to several, involucre 3–4 mm long, glutinous or sparsely pubescent. Rays about twenty-five to forty, often less than thirty, white. Pappus or disk florets simple, of ray florets over 1 mm long. Achene about 1 mm long, pubescent. $2n = 18$. Low pinelands, shady soil, ditches, Fla. to La., north to Va. on the coastal plain. Spring.

40. ÁSTER L.

Mostly perennial or annual herbs with simple, alternate, entire or variously toothed leaves and solitary or more frequently numerous hemispheric heads. Heads usually radiate, erect on ultimate branches or spreading, corymbose or paniculate; phyllaries imbricate in several series, the outer ones smaller than the inner; ray florets pistillate, ligules white, purplish, pink, or blue; receptacle flat, without chaff; disk florets bisexual, usually yellow or sometimes red, brownish, purple, or white.

Pappus of many capillary bristles. Achene with one to several ribs. About 500 species, worldwide, especially North America with many natural hybrids, taxonomically difficult. *Doellingeria* Nees; *Ionactis* Greene

1. Involucres mostly 8–12 mm or more high; middle to upper stem leaves more than 8 mm wide, lanceolate, lance-ovate, to cordate.
  2. Stems reclining or clambering . . . . . . 1. *A. caroliniensis*
  2. Stems erect.
    3. Leaves with short petioles or subsessile, but not clasping . . . . . . 2. *A. elliottii*
    3. Leaves sessile-clasping.
      4. Leaves rough pubescent . . . . 3. *A. patens*
      4. Leaves glabrous to minutely pubescent . . . . 4. *A. dumosus*
1. Involucres smaller, mostly 3–6 mm high; upper stem leaves and those in the inflorescence small, less than 8 mm wide, filiform, linear-lanceolate, scalelike, or apparently absent.
  5. Upper leaves and those of inflorescence appressed to the stems, linear-lanceolate, acuminate or scalelike, scabrous to villous pubescent.
    6. Leaves clasping, upper scalelike; phyllaries scabrous . . . . 5. *A. adnatus*
    6. Leaves sessile, not clasping, linear-lanceolate to acuminate; phyllaries white villous pubescent . . . . . . 6. *A. concolor*
  5. Upper leaves and those of inflorescence filiform, linear or linear-lanceolate, sometimes broader, not conspicuously appressed to stem, glabrous to minutely pubescent; or leaves much reduced and apparently absent.
    7. Plants perennial.
      8. Pappus bristles in one series; lower leaves linear to spatulate-lanceolate.
        9. Outer phyllaries with rounded or acute tips; lower leaves spatulate or linear-lanceolate to lanceolate 4. *A. dumosus*
        9. Outer phyllaries elongate-subulate; lower leaves filiform, wiry upper leaves much reduced or absent . 7. *A. tenuifolius* var. *aphyllus*
      8. Pappus bristles in two series; lower leaves elliptic to ovate . . . . 8. *A. reticulatus*

7. Plants annual, rarely perennial . . . 9. *A. subulatus*
var. *ligulatus*

1. **A. caroliniensis** Walt.—Stems woody, with soft pubescence, climbing or decumbent, diffusely branching and vinelike, up to 4 m long. Leaves 5–11 cm long, entire, acute or tapered at the apex, somewhat sagittate or clasping at the base. Heads one to several on leafy branches; outer phyllaries spatulate, inner ones linear, both with dark green tips, recurved or somewhat spreading; ligules pale purplish or pink, about 1.5–2.0 cm long. Achenes smooth. Low ground, swamps, hammocks, Fla. to N.C. on the coastal plain. Summer, fall. *A. carolinianus* Walt.

2. **A. elliottii** T & G—Stems 0.5–1.5 m tall from rhizomes, glabrous or with pubescence in lines. Leaves 25–30 cm long, numerous, the lower with long petioles, serrate, the upper ones reduced and becoming sessile. Heads several to many, corymbose or nearly so; involucre 8–11 mm long, glabrous or nearly so; linear-attenuate and lax or recurved. Ray florets numerous, twenty-five to fifty, pinkish or purplish. Achene smooth. Low ground, swamps, Fla. to Va. on the coastal plain. Fall.

3. **A. patens** Ait.—Stems one to several from creeping rhizomes, up to 1 m tall or more, hirsute. Leaves 5–15 cm long, more or less rigid, pubescent, cordate-clasping below, deciduous, oblong or ovate above. Heads numerous in a branching inflorescence; involucre glandular-puberulent, about 6–9 mm long, phyllaries linear-elliptic. Ray florets fifteen to thirty, blue or violet. Achenes pubescent. $2n = 20$. Sandy, acidic soils, pinelands, Fla. to Tex., Mass., Kans. Summer, fall.

4. **A. dumosus** L.—Stems slender up to 1 m tall, one to several from creeping rhizomes, smooth or puberulent toward the top. Leaves about 3–12 cm long, firm, linear–lance-linear or narrowly ovate, entire, scabrous on upper surface, rameal leaves scalelike. Heads in open, spreading, paniculate inflorescence, numerous, peduncles conspicuously elongate, up to 2 cm long or more. Involucre 4–5 mm high, glabrous, phyllaries linear-subulate, appressed-imbricate, whitish, with green tips. Ligules pale blue to occasionally white, disk corollas. Achene minutely puberulent. Gladeland, sandy places, moist or dry sites, Fla. and La., Mass., Wis. A highly variable species represented in our area by two well-marked varieties:

1. Lower cauline leaves 1–4 mm wide, linear to
   narrowly oblong . . . . 4a. var. *coridifolius*
1. Lower cauline leaves 3–10 mm wide, nar-
   rowly linear . . . . 4b. var. *subulaefolius*

4a. A. dumosus var. **coridifolius** (Michx.) T & G—Cauline leaves spreading or reflexed, mostly 3–9 cm long. Drier sites, Fla. to La., N.Y., Ill. Summer, fall.

4b. A. dumosus var. **subulaefolius** T & G—Cauline leaves ascending, 5–12 cm long; bracteal leaves on branches subulate tipped, linear. Various sites, moist soil, pinelands, Fla. to Tex., Mass. on the coastal plain. Summer, fall. *A. simmondsii* Small; *A. sulznerae* Small

5. A. adnatus Nutt.—Stems up to 8 dm tall, hispid pubescent, branched above. Lower leaves 2–3 cm long, usually obovate, upper leaves 3–10 mm long, scalelike, very numerous, more or less imbricate, scabrous and partly adnate to the stems. Heads terminating the scaly branches; involucre 4–6 mm long, phyllaries linear and scabrous; ligules dark blue or violet. Achene glabrous. Dry pinelands, Fla. to Miss., Ga. on the coastal plain. Fall.

This species is probably not as common in our area as it is farther north in peninsular Fla.

6. A. concolor L.—Stems one to several to 7 dm tall, leafy. Leaves about 5–6 cm long, linear-elliptic to elliptic-ovate, finely canescent on both sides, or the lower glabrate, sessile. Heads numerous, usually in a narrow raceme or panicle; involucre turbinate, phyllaries linear or linear-lanceolate, canescent; ligules blue to lilac, 5–10 mm long, disk corollas yellow; pappus tawny or white. Achene villous. Drier sites. Two varieties occur in our area:

1. Ligules 6–10 mm long, pappus tawny; up-
    per leaves linear . . . .   6a. var. *concolor*
1. Ligules 5–7 mm long, pappus white; upper
    leaves scalelike . . . .   6b. var. *simulatus*

6a. A. concolor var. **concolor**—Lower leaf blades narrowly elliptic to nearly ovate. Drier sites, Fla. to La., Mass., Tenn. Summer, fall.

6b. A. concolor var. **simulatus** (Small) R. W. Long—Lower leaf blades linear to linear-lanceolate or narrowly elliptic. Pinelands, endemic to south Fla. Summer, fall. *A. simulatus* Small

7. A. tenuifolius L. var. **aphyllus** R. W. Long—Stems glabrous, slender, somewhat wiry, about 1 m tall. Leaves linear-subulate to linear-lanceolate above, or absent, filiform below. Inflorescence corymbose, few to several heads; phyllaries elongate-subulate; ligules pale blue or whitish. Achene glabrous or nearly so. Gladelands, low ground, brackish marshes, south Fla., W.I. Summer, fall. *A. bracei* Britt.

8. A. reticulatus Pursh, WHITE-TOP ASTER—Smooth stems up to 1 m tall or more, stems one to many. Leaves 4–8 cm long, elliptic-obovate to ovate, somewhat conspicuously reticulate, serrate or short petioled. Heads infastigiate corymbs, involucre 6–7 mm long, pappus of two distinct series, the outer one of short bristles or scales, the inner of elongate bristles. Ligules white or cream colored, somewhat drooping, disk corollas yellow. Achene pubescent. Low pinelands, moist soil, Fla. to S.C. on the coastal plain. Spring, summer. *Doellingeria reticulata* (Pursh) Greene

9. A. subulatus Michx. var. **ligulatus** Shinners—Stems simple or many branches up to 1 m tall or less, smooth, from a short taproot, annual. Leaves 10–15 cm long, linear below, entire, sessile, rameal leaves much reduced, subulate. Heads few to numerous or solitary in open inflorescence; involucre of many imbricate, greenish or purplish phyllaries, linear-subulate or attenuate. Ligules reduced, 1–3 mm long, not conspicuous, bluish or whitish; disk corollas yellow; pappus very abundant, soft white. Achene pubescent. Pinelands, low ground, especially coastal areas, Fla. to Ala., Me., Mich. Summer, fall. *A. exilis* Ell.; *A. inconspicuus* Lessing misapplied

Key 5

TRIBE *Mutiseae*

41. CHAPTÀLIA Vent. Sunbonnets

Scapose perennial herbs with solitary, nodding head; stems with spreading woolly pubescence and bearing alternate basal leaves in rosettes. Leaves subentire to undulate or pinnatifid. Involucre narrow-cylindric with herbaceous phyllaries reflexed at the tips; ligules white, of marginal florets, disk corollas white, bilaterally symmetrical; anthers with lanceolate appendages, base tailed. Achene beaked or narrowed at the tip. About 25 species, chiefly Central and South America and W.I. *Thyrsanthema* Necker

1. Involucre becoming 2 cm long . . . . . 1. *C. dentata*
1. Involucre smaller, becoming 1–1.5 cm long   2. *C. tomentosa*

1. C. dentata (L.) Cassini, Pineland Daisy—Scapose herb to 3 dm tall, pubescent. Leaves about 5–12 cm long, oblanceolate or spatulate, undulate or denticulate, persistent, floccose pubescent above or eventually glabrate, lanate tomentose below. Involucre lanate pubescent, inner phyllaries usually 18–20 mm long, linear-filiform; disk corollas about 7–8 mm long. Achene 4–5 mm long, slender, beaked. $2n = 32$. Pinelands, south Fla., W.I. Spring. *Thyrsanthema dentata* (L.) Kuntze

2. C. tomentosa Vent.—Scapose herb to 3 dm tall with leaves 5–10 cm long, elliptic to oblanceolate, entire to obscurely denticulate, apiculate, densely white tomentose beneath. Phyllaries linear to subulate, inner to 1.5 cm long; ligules of pistillate marginal florets, purplish and white, 8–10 mm long. Achene 3 mm long, constricted into a short neck. Wet pineland, low ground, Fla. to Tex., N.C. on the coastal plain. Spring. *Thyrsanthema semiflosculare* (Walt.) Kuntze

Key 6

TRIBE *Inuleae*

1. Stems conspicuously winged; immature inflorescence an elongate, conical spike . .   42. *Pterocaulon*
1. Stems not winged.
   2. Stems and leaf undersurfaces densely white tomentose; phyllaries scarious .   43. *Gnaphalium*
   2. Stems and leaf undersurfaces glabrous or with merely scattered hairs; phyllaries not scarious.
     3. Larger leaves forming a basal rosette, small scapose plants . . . . .   44. *Sachsia*
     3. Larger leaves cauline, basal leaves not in rosettes . . . .   45. *Pluchea*

42. PTEROCÀULON Ell.   BLACKROOT

Perennial erect herbs with tomentose or woolly-pubescent stems and alternate leaves. Blades serrate or entire, decurrent on the petiole. Heads borne in an elongated spikelike inflorescence consisting of glomerate cymes; involucre cylindrical, phyllaries unequal; marginal florets pistillate, corollas filiform; inner florets bisexual with tubular, capillary bristles. About 25 species of wide distribution, America, Australia, Asia, Madagascar.

1. P. pycnostachyum (Michx.) Ell. RABBIT TOBACCO—Stems one to several up to 7 dm tall from large, tuberous black roots. Leaves about 6–12 cm long, narrowly elliptic or lanceolate, with undulate margins. Heads many in terminal spikes, 4–7 cm long; involucre 3.5–4 mm long, densely pubescent, corollas whitish, disk florets small, the lobes of the inner florets linear. Pinelands, disturbed sites, sandy soil, waste places, Fla. to Miss., N.C. on the coastal plain. Spring, fall. *P. undulatum* (Walt.) Mohr

43. GNAPHÀLIUM L.   EVERLASTING, CUDWEED

Woolly-pubescent herbs with alternate leaves, cauline or in basal rosettes or both. Heads discoidlike with many bisexual florets in glomerate cymes or corymbs or panicles, generally whitish in color. Involucre ovoid or campanulate, phyllaries scarious, imbricate in several series. Outer florets pistillate, corollas filiform, inner florets bisexual with corolla lobes deltoid or ovate. Pappus of capillary bristles. Achenes glabrous to resinous glandular-pubescent. About 200 species, cosmopolitan.

1. Inflorescence in terminal corymbs or corymbose panicles, involucre 6–7 mm long . 1. *G. obtusifolium*
1. Inflorescence spicate or in small axillary
   clusters, involucre 3–4 mm long . . . 2. *G. purpureum*

1. G. obtusifolium L. var. **obtusifolium,** RABBIT TOBACCO—Annual or biennial with stems white woolly especially above, to 8 dm tall. Leaves 3–8 cm long, lance-linear, sessile, white woolly beneath but upper surfaces green or barely pubescent. Heads in corymb or rounded panicle; involucre about 6–7 mm long, yellow white, campanulate, phyllaries oblanceolate, membranous pubescent at the base. Achene brown, smooth. Dry sites, disturbed areas, waste ground, sandy soil, Fla. to Tex., Nova Scotia, Manitoba. Spring, fall. *G. saxicola* Fassett

2. G. purpureum L. CUDWEED—Annual or biennial with thinly woolly stems up to 5 dm tall. Lower leaves 4.5–10 cm long, spatulate, oblong, or oblanceolate. Heads several to many in bracteate spikelike inflorescence or appearing in axillary clusters, involucres 3–4 mm long, campanulate, phyllaries acute, membranous, pale brown and woolly at base. Achenes distinctly resin dotted or papillate. $2n = 28$. Disturbed sites, especially sandy soil, tropical America north to Me., B.C.

1. Larger leaf blades narrowly spatulate, 1.5–
   2.0 cm wide, 4.5–5.0 cm long.
   2. Leaf blades densely white tomentose on
      lower surfaces, sparingly pubescent
      above . . . . 2a. var. *purpureum*
   2. Leaf blades equally pubescent on both
      surfaces, upper blades mostly linear-
      lanceolate . . . . 2b. var. *falcatum*
1. Larger leaf blades broadly spatulate, 1.5–2.0
   cm wide, 6.0–8.0 cm long . . . . . 2c. var. *spathulatum*

2a. G. purpureum var. **purpureum,** PURPLE CUDWEED—Flowering stem up to 5 dm tall, pubescence of tightly appressed hairs. $2n = 28$. Drier sites, sandy soil, Fla. to Tex., south Calif., Pa., tropical America. Spring, summer.

2b. G. purpureum var. **falcatum** (Lam.) T & G—Flowering stem up to 4 dm tall, pubescence densely white woolly. Pinelands, disturbed sites, Fla. to Tex. on coastal plain, inland to Tenn., South America. All year. G. *falcatum* Lam.

2c. G. purpureum var. **spathulatum** (Lam.) Ahles—Flowering stems 2–4 dm tall, pubescence thinner, greyish. Disturbed sites, road-sides, Fla. to Tex., coastal plain, Piedmont, W.I. G. *spathulatum* Lam.; G. *peregrinum* Fern.; G. *pensylvanicum* Willd.

44. SÁCHSIA Griseb.

Slender perennial herbs with mostly basal, alternate, serrate leaves. Heads in open corymbose inflorescence, involucres campanulate with inner phyllaries distinctly longer than outer ones. Marginal florets pistillate, corollas filiform, yellow; inner florets bisexual with tubular, yellow corollas. Pappus a single row of slender bristles. Four species in south Fla. and W.I.

1. S. bahamensis Urban—Stems up to 6 dm tall; basal leaves mostly 5–7 dm long, spatulate, upper leaves reduced and much smaller. Heads with distinct peduncles; inner phyllaries about 5–6 mm long, outer phyllaries 2–3 mm long, membranous or chaffy. Achene 2–3 mm long. Pinelands, south Fla., Fla. Keys, W.I. Spring, summer.

45. PLÙCHEA Cassini   MARSH FLEABANE

Shrubs or perennial or annual herbs with alternate, serrate leaves and pink, white, or yellowish heads. Inflorescence corymbose or glom-erate-cymose; involucre subovoid or campanulate, the phyllaries nar-row, attenuate, imbricate, receptacle flat, without chaff; marginal florets pistillate, corollas filiform; inner florets discoid, bisexual; anthers tailed, pappus a single row of bristles. Achene four- to five-angled. About 35 species in tropical and temperate regions.

1. Leaves sessile or sessile-clasping.
    2. Corollas rose or pink . . . .   1. *P. rosea*
    2. Corollas white . . . .   2. *P. foetida*
1. Leaves distinctly petioled.
    3. Shrubs; larger leaves entire to obscurely
       serrate ·. . . . .   3. *P. odorata*
    3. Herbs; leaves distinctly serrate.
       4. Corollas rose or pink . . . . .   4. *P. purpurascens*
       4. Corollas white . . . .   5. *P. camphorata*

1. P. rosea R. K. Godfrey—Perennial up to 9 dm tall, subglabrate below to often scurfy pubescent or puberulent above. Leaves 4–6 cm long, oblong to obovate, serrate, sessile and auriculate clasping, acute or obtuse at the apex. Heads sessile, few to many in cymes or terminal glomerules, pinkish; involucre 3–6 mm long, viscid pubescent or villous; corollas pale pink to rose. Achenes 1 mm long, black, thickly pubescent. Various sites, moist or dry pinelands, disturbed areas, Fla. to Miss., N.C. Spring, summer.

2. **P. foetida** (L.) DC.—Perennial with finely glandular-pubescent stems up to 1 m tall. Leaves usually 3–10 cm long, oblong or lanceolate to ovate, serrate or denticulate, sessile and partially clasping. Heads several to many, appearing whitish, involucre 5–7 mm long, glandular or viscid puberulent, corollas white or cream colored. Achene 1 mm long, pubescent only on the angles. $2n = 20$. Low ground and swamps, ditches, marshes, Fla. to Tex., N.J., Mo., W.I. Summer, fall.

3. **P. odorata** Cassini—Shrubs much branched up to 1 m tall, canescent pubescent above. Leaves mostly 10–15 cm long, lance-ovate to elliptic-ovate, entire, canescent. Heads in dense corymbs, pale pinkish, involucre 5–6 mm long, phyllaries obtuse, puberulent, corollas pinkish. $2n = 20$. Disturbed sites, margins of hammocks, south Fla., tropical America. Spring, summer.

4. **P. purpurascens** (Sw.) DC. CAMPHORWEED—Annual or perennial up to 6 dm tall with stems subglabrate below, glandular-puberulent above. Leaves 5–15 cm long, lanceolate to elliptic-ovate, serrate, blades decurrent to short petioles. Heads appearing pink purplish in corymbs; involucre 4–7 mm long, phyllaries puberulent, corollas pink. $2n = 20$. Salt marshes, pinelands, margins of hammocks, Fla. to Mass. on coastal plain, occasionally inland, W.I. Summer, fall.

5. **P. camphorata** (L.) DC.—Annual or short-lived perennial up to 1.5 m tall or more, glabrate or puberulent above. Leaves 10–25 cm long, elliptic-ovate or lanceolate, acuminate. Heads pedunculate, purplish, in a rounded, dense corymb or corymbose panicle; involucre 4–6 mm long, phyllaries acute, puberulent or glabrate, corollas pink to purplish. Achene 1 mm long, densely pubescent. $2n = 20$. Low ground, moist soil, Fla. to Tex., Del., Ill. Summer, fall. *P. petiolata* Cassini

Key 7

TRIBE *Eupatorieae*

1. Herbaceous vines; leaves cordate or sagittate . . . . 46. *Mikania*
1. Herbs erect, not vinelike; leaves linear to ovate.
    2. Larger leaf blades lanceolate-ovate, usually 15–30 mm wide or more; upper leaves sometimes much reduced.

3. Upper leaves sessile . . . . . .   47. *Carphephorus*
3. Upper leaves distinctly petioled.
   4. Pappus composed of short scales,
      cuplike in appearance . . . .   48. *Ageratum*
   4. Pappus composed of capillary bristles   49. *Eupatorium*
2. Larger leaf blades filiform, linear or lin
   ear-lanceolate, usually 1–10 mm wide;
   basal leaves sometimes elongate.
   5. Inflorescence an elongate spike or nar
      row raceme; heads sessile or on short
      peduncles up to 1.0 cm long . . .   50. *Liatris*
   5. Inflorescence branching, paniculate or
      corymbose, spreading.
      6. Pappus composed of plumose bris
         tles; phyllaries with green white
         striations . . . .   51. *Kuhnia*
      6. Pappus composed of capillary bris
         tles; phyllaries green or yellowish
         green . . . .   49. *Eupatorium*

46. MIKÀNIA Willd.   CLIMBING HEMPWEED

Perennial climbing vines, shrubs, or less commonly erect herbs, with opposite leaves and small heads. Heads discoid, of four tubular florets, bisexual; inflorescence corymbose, involucre narrow-cylindric of four larger phyllaries and sometimes several shorter ones, receptacle without chaff; pappus of capillary bristles; corolla whitish or pinkish, lobes lanceolate to deltoid. Achenes five-angled. About 250 species, chiefly tropical in distribution, especially in South America.

1. Phyllaries obtuse; stem and leaves often
   tomentose . . . .   1. *M. cordifolia*
1. Phyllaries acute; stem and leaves often
   glabrous or nearly so.
   2. Phyllaries mostly 3–4 mm long . . .   2. *M. batatifolia*
   2. Phyllaries mostly 4–6 mm long . . .   3. *M. scandens*

1. M. cordifolia (L.) Willd.—Herbaceous vines with stems thickly pubescent. Leaves mostly 5–15 cm long, ovate or hastate, tomentose, toothed or lobed, cordate at base. Corymbs often flat-topped, usually very large with many small heads; phyllaries about 6–8 mm long, linear-lanceolate, obtuse; corolla about 5–6 mm long, florets very fragrant. Low ground, hammocks, south Fla., W.I. All year.

2. M. batatifolia DC. HEMP VINE—Herbaceous vines with smooth stems. Leaves mostly 1–3 cm long, deltoid to ovate or hastate, glabrous, entire or repand. Corymbs often rounded, with heads in irregular clusters; phyllaries lanceolate-attenuate, about 3–4 mm long; corolla 3–4 mm long. Low ground, hammocks, south Fla., W.I. All year.

3. M. scandens (L.) Willd.—Herbaceous vines with finely pubescent stems or glabrate. Leaves mostly 3–8 cm long, deltoid or ovate-hastate, entire or with few teeth, cordate at base, acuminate at apex. Inflorescence corymbose, small, phyllaries acuminate, 4–5 mm long, corolla 4 mm long. $2n = 36$. Thickets, swamps, and moist woods, climbing on other vegetation, Fla. to Tex., Okla., Me., tropical America. Summer.

47. CARPHÉPHORUS Cassini

Perennial herbs with fibrous roots, alternate, entire or shallowly serrate leaves and heads in a corymbose inflorescence or narrow panicles. Heads with many florets, discoid, bisexual; involucre with appressed, imbricate, often glandular phyllaries, receptacle with marginal chaff; corolla purplish, funnelform, lobes lance-deltoid; pappus of unequal, capillary bristles. Achenes ten-ribbed. About seven species of North America. *Trilisa* Cassini.

1. Heads with twelve to forty florets; involucres 6–11 mm high . . . . 1. *C. corymbosus*
1. Heads with three to twelve florets; involucres 3–6 mm high.
  2. Stem glabrous; inflorescence a spreading
  corymbose cymes . . . . 2. *C. odoratissimus*
  2. Stem sparingly hirsute or viscid; inflorescence a thyrsoid panicle. . . . . 3. *C. paniculatus*

1. C. corymbosus (Nutt.) T & G—Stems up to 1 m tall, pubescent. Lower leaves spatulate or ovate-spatulate, upper leaves elliptic or cuneate. Phyllaries with conspicuous scarious margins, the outer ovate, the inner cuneate; lobes of the corolla narrowly deltoid. Achene about 3 mm long. Acidic soil, pinelands on coastal plain from Fla. to Ga.

2. C. odoratissimus (J. F. Gmel.) Herb.—Stems scabrous or hirsute to 1 m tall. Basal leaves usually 5–20 cm long, linear to oblanceolate or elliptic, upper leaves linear-lanceolate, sessile but not clasping. Inflorescence narrowly paniculate, lateral branches shorter than the terminal, phyllaries 3–5 mm long, corolla lobes ovate. Achene 3 mm long. Low pinelands, Fla. to La., N.C., on coastal plain. Fall.

3. C. paniculatus (J. F. Gmel.) Herb.—Stems to 1 m tall, viscid pubescent through internodes or glabrate in age. Basal leaves 5–20 cm long, linear-elliptic to oblanceolate, narrowed to broad petioles, cauline leaves scarcely smaller, floral bracts conspicuous. Inflorescence congested, narrowly paniculate, appearing cylindrical, densely hairy, inter-mixed with gland-tipped hairs; phyllaries 3–5 mm long, acute, glandular-puberulent; corolla lobes deltoid, purple. Achene 2 mm long, ten-ribbed, with slightly longer scabrous, purplish brown pappus. Low pinelands, Fla. to La., N.C., on coastal plain. Fall. *Trilisa paniculata* (Walt.) Cassini

48. AGERÀTUM L.

Annual or perennial herbs with erect stems and opposite dentate leaves. Inflorescence corymbose, heads appearing blue, white, or pink; corolla lobes deltoid. Pappus of obtuse or acute scales or cuplike. About 35 species, chiefly in the American tropics.

1. **A. littorale** Gray—Stems glabrous, usually branched near the base, up to 8 dm tall. Leaves 1–6 cm long, ovate or elliptic, crenate-serrate, acute at apex, cuneate at base. Inflorescence on long peduncles; involucre about 4 mm high, phyllaries barely pubescent; corolla blue, pappus reduced. Achene about 2 mm long. Shore hammocks, sea beaches, and sand dunes, endemic to south Fla.

A. conyzoides L. with leaves truncate or rounded at base may occur in our area; **A. houstonianum** Mill., a larger plant with cordate leaves, has been collected north of our area as an escape from cultivation.

49. EUPATÒRIUM L.

Perennial herbs or shrubs with erect or ascending stems and entire, serrate, or dissected leaves. Inflorescence usually corymbose, racemose, or paniculate; heads small, appearing white, pinkish, or purple; florets all discoid and bisexual; phyllaries imbricate in several series, unequal; receptacle flat without chaff; pappus a single row of capillary bristles. Achene black, usually five-angled. About 600 species, chiefly American. *Osmia* Sch.-Bip.; *Conoclinium* DC.

1. Upper leaves or leaf segments linear or linear-lanceolate, mostly 1–6 mm wide.
  2. Inflorescence distinctly corymbose.
    3. Larger leaves linear-lanceolate, 5–10 mm wide . . . . 1. *E. recurvans*
    3. Larger leaves filiform to narrowly linear, 1–3 mm wide . . . . . . 2. *E. hyssopifolium* var. *laciniatum*
  2. Inflorescence paniculate or racemose.
    4. Heads secund on branches, appearing sessile or subsessile . . . . . . 3. *E. leptophyllum*
    4. Heads not secund with short pedicels.
      5. Leaf segments filiform-subulate, mostly 1–2 mm wide or less . . 4. *E. capillifolium*

5. Leaf segments linear, occasionally linear-lanceolate, largest 3 mm or more wide . . . .   5. *E. compositifolium*
1. Upper leaves broad, mostly over 15 mm wide.
6. Shrubs; leaves ovate-cordate, entire to obscurely sinuate . . . . . .   6. *E. villosum*
6. Herbs; leaves lanceolate to ovate.
7. Leaf base cuneate, or gradually tapering to petiole.
8. Leaves lobed, deeply incised, or pinnatifid . . . .   7. *E. eugenei*
8. Leaves not lobed or incised.
9. Involucre cylindric, 5–10 mm long.
10. Leaf blades lance-ovate, 5–12 cm long . . . .   8. *E. odoratum*
10. Leaf blades deltoid-ovate, 2–5 cm long . . . . . .   9. *E. frustratum*
9. Involucre spreading, mostly 2–5 mm long.
11. Leaves lanceolate with tapering leaf bases; heads mostly with eight to ten florets . . . .   10. *E. serotinum*
11. Leaves ovate with cuneate leaf bases; heads mostly with five to eight florets . . . . .   11. *E. mikanioides*
7. Leaf base truncate or broadly rounded, not gradually tapered to petiole.
12. Corolla violet or purple; inflorescence cymose or umbellate . . .   12. *E. coelestinum*
12. Corolla white; inflorescence spreading, cymose or corymbose.
13. Leaves distinctly petioled, glabrous or nearly so . . . .   13. *E. aromaticum*
13. Leaves sessile; upper stems and leaves cinerous pubescent . .   14. *E. rotundifolium*

1. **E. recurvans** Small—Stems up to 1 m tall from tuberous rhizomes, branching from near the base, puberulent to hirsute. Leaves mostly 4–10 cm long, usually opposite below, alternate above, sessile, narrowly ovate or elliptic, often recurved or deflexed. Corymb broad, branches pubescent. Involucre 3–5 mm long, phyllaries closely imbricate; florets five per head, white, corolla 3–4 mm long. Achenes 3 mm long. $2n = 20,30$. Pinelands, sandy soil, Fla. to Miss., Va. Fall.

2. **E. hyssopifolium** L. var. **laciniatum** Gray—Stems in clumps up to 1 m tall, puberulent to scabrous; leaves mostly 2–10 cm long, opposite or verticillate in fours, or alternate above, glandular-punctate or glabrate, serrate or dentate, usually linear-oblong to narrowly lanceolate and with conspicuous axillary fascicles of reduced leaves. Involucre

cylindric, 4–7 mm long, canescent or puberulent, inner phyllaries scarious margined; florets five per head, corollas white, 3–4 mm long. Achene 2–3 mm long, glandular. Old fields, sandy soil, south Fla. to La., Mass., Ohio. Summer, fall. *E. torreyanum* Short

3. E. leptophyllum DC. FENNEL—Stems clustered with recurved-secund branches to 1 m tall or more, smooth. Leaves mostly 1–2 m wide, dissected with filiform segments. Phyllaries 3–4 mm long, with prolonged tips, heads secund on branches, white, corolla about 3 mm long. Achene 1–2 mm long. Moist soil, disturbed sites, margins of ponds or swamps, south Fla. to Miss., S.C. on the coastal plain. Fall.

4. E. capillifolium (Lam.) Small, DOG FENNEL—Stems clustered, to 2 m tall or more, branched, tending to be puberulent at least below. Lower leaves opposite, upper alternate, glabrous, glandular-punctate, pinnately divided or dissected with filiform divisions mostly 5–10 cm long, often in axillary fascicles. Heads many in paniculate inflorescence; involucre 2–4 mm long, phyllaries tipped; florets three to six per head, corollas white or greenish white. Achenes 1 mm long, smooth. Disturbed sites, old fields, south Fla. to La., N.J., Ark. Fall.

5. E. compositifolium Walt. DOG FENNEL—Stems clustered to 1 m tall or more, branched, puberulent to pilose below. Lower leaves mostly 6–12 cm long, opposite, upper alternate, pinnately divided or dissected with linear segments. Heads in panicle, fragrant, very numerous, three to six florets per head; involucre 4–5 mm long, phyllaries mucronate; corollas white, achene 1–2 mm long. Pinelands, old fields, dry soil, south Fla. to Tex., N.C. Fall.

6. E. villosum Sw.—Shrubs with stems up to 2 m tall, tomentose. Leaves usually 2–8 cm long, ovate or ovate-deltoid, entire with apex obtuse, entire or repand. Heads cymose or in small clusters, involucre campanulate, phyllaries obtuse or the inner acute, 3–4 mm long; corollas white or pinkish white about 3 mm long. Achene 1–2 mm long. Hammocks, pinelands, south Fla., Fla. Keys, W.I. All year.

7. E. eugenei Small—Stems up to 1.5 m tall, pubescent. Leaves one to two pinnately divided or parted, the segments linear-entire or toothed, upper leaves merely incised. Inflorescence paniculate; involucre campanulate, phyllaries mucronate, inner about 3–4 mm long. Corollas white; achene 1–2 mm long. Pinelands and woods, Fla. to Ala. on the coastal plain. Fall.

8. E. odoratum L.—Stems up to 3 m tall, pubescent, with numerous branches, shrublike. Leaves 4–12 cm long, lanceolate to ovate, tapered to apex, with prolonged petiolelike bases. Involucre 1 cm long, the phyllaries green tipped; corolla white or purplish white. Achene 4–5 mm long. $2n = 58$. Hammocks, thickets, south Fla., Fla. Keys, W.I., tropical America. All year. *Osmia odorata* (L.) Sch.-Bip.

9. **E. frustratum** Robinson—Stems up to 2 dm, puberulent, from thick fibrous roots. Leaves with definite slender petioles, blades 2–5 cm long, deltoid-ovate, ovate, or ovate-lanceolate, acute or narrowly obtuse, margin shallowly serrate, conspicuously palmately three-veined on the lower surface. Heads often few, involucre 6–7 mm high, phyllaries green tipped; corolla pale purplish or bluish, 4 mm long. Achene 3–4 mm long. Coastal hammocks, endemic to south Fla., Fla. Keys. All year.

10. **E. serotinum** Michx.—Stems up to 2 m tall, densely puberulent, especially above, glabrate below. Leaves mostly 6–20 cm long, lanceolate to ovate, serrate, acuminate, upper surface glabrous or barely puberulent, lower surface pubescent, petioles 1–3 cm long. Involucre 3–4 mm long, cylindric, phyllaries pubescent, imbricate, obtuse; florets white or pale purplish, about ten to fifteen per head. Achenes 2–3 mm long, smooth or somewhat atomiferous. Moist ground, disturbed sites, pineland, Fla. to Tex., N.Y., Ill. Fall.

This species apparently hybridizes with **E. mikanioides.**

11. **E. mikanioides** Chapm. SEMAPHORE EUPATORIUM—Stems up to 1 m tall or more, tomentose pubescent at least in upper stem. Leaves 4–6 cm long, ovate-deltoid, extended vertically, somewhat fleshy, margins repand, crenate, or dentate. Involucre 5–6 mm long, phyllaries 4–5 mm long, linear and conspicuously green margined, acute tipped; florets whitish or pinkish, about five per head. Achene 1 mm long or more. Glades, solution holes, salt marshes, low pinelands, endemic to Fla. Summer, fall.

12. **E. coelestinum** L. MISTFLOWERS, AGERATUM—Stems usually 6–9 dm tall, pubescent from narrow rhizomes, branched especially near the base. Leaves mostly 5–10 cm long, all opposite, ovate-deltoid, three-veined, margins crenate-serrate, puberulent or subglabrous. Involucre 3–5 mm long, phyllaries linear-acute, imbricate receptacle conical; florets blue or violet, rarely white, about forty to sixty per head. Achene 1–2 mm long. $2n = 20$. Moist soil, meadows, Fla. to Tex., N.J., Kans., W.I. Summer, fall. *Conoclinium coelestinum* (L.) DC.

13. **E. aromaticum** L.—Stems up to 1 m tall or more, smooth or barely puberulent. Leaves usually 3–6 cm long, ovate or ovate-deltoid, sharply serrate or crenate-serrate. Involucre 4–5 mm long, phyllaries acuminate or at least acute, linear-lanceolate. Florets thirteen to fourteen per head, white. Achene 2–3 mm long, glabrous. Pinelands, hammocks, south Fla. to Mass., Tenn., Miss. Summer, fall. *E. jucundum* Greene

Forms with sharply toothed leaves have been named var. **incisum** Gray.

14. **E. rotundifolium** L.—Stems up to 1.5 m tall, stiffly erect, densely pilose pubescent and resinous-glandular, especially above. Leaves usu-

ally 3–10 cm long, opposite, or the uppermost alternate, ovate-lanceo-late to suborbicular, serrate, sessile or subsessile. Corymbs broad, branches opposite, involucre 5–7 mm long, phyllaries sharply pointed, apex and margin white, florets white, about five to seven per head. Achene about 2 mm long. Pinelands, white sand, ditches, Fla. to Tex., Me., Okla. Summer, fall.

50. LIÀTRIS Schreb. Blazing Star

Perennial herbs mostly with cormlike underground parts, few with rhizome or rootstock with alternate leaves and small or moderate-sized heads in spicate or racemose inflorescence. Heads all discoid, the florets bisexual, flowering from the top to the bottom, pink, purple, reddish, or sometimes white. Pappus of one to two rows of barbellate or plumose capillary bristles. Achenes pubescent, tapered, ten-ribbed. About 32 species in North America. *Laciniaria* Hill; *Ammopursus* Small

1. Involucres usually 15–20 mm wide; larger
phyllaries 5–10 mm wide . . . . . 1. *L. ohlingerae*
1. Involucre much smaller, usually 3–5 mm
wide; phyllaries mostly less than 5 mm
wide.
　2. Larger phyllaries lanceolate-acuminate,
　10–15 mm long.
　　3. Pappus conspicuously plumose; achenes
　　small, 2–3 mm long . . . . . 2. *L. elegans*
　　3. Pappus merely barbellate; achenes 5–6
　　mm long . . . . 3. *L. chapmanii*
　2. Larger phyllaries ovate, apex acute or ob-
　tuse, mucronate, 5–8 mm long.
　　4. Heads sessile . . . . 4. *L. spicata.*
　　　　　　　　　　　　　　　var. *resinosa*
　　4. Heads with peduncles, often very short.
　　　5. Phyllaries glabrous or nearly so . . 5. *L. tenuifolia*
　　　5. Phyllaries ciliate, or margins ciliate,
　　　surface of phyllaries puberulent.
　　　　6. Larger involucres cylindric, 12–15
　　　　mm high . . . . . . 6. *L. garberi*
　　　　6. Larger involucres turbinate, 5–8
　　　　mm high . . . . . . 7. *L. gracilis*

1. L. ohlingerae (Blake) Robinson—Stems up to 1 m tall or more, usually one or few from top of root, minutely pubescent, usually branched above. Leaves linear-lanceolate, smooth, acute, somewhat thickish. Involucre about 2 cm long, phyllaries numerous, punctate, the outer orbicular, gradually narrowed to linear-spatulate inner ones. Corollas rose purplish, with slender tubes, lanceolate lobes acute. Pappus barbellate, white. Achene 8–9 mm long, densely pubescent.

White sand scrub, endemic in central Fla., possibly extending into our area to Collier County. Summer. *Laciniaria ohlingeri* Blake; *Ammopursus ohlingerae* Small

2. L. elegans (Walt.) Michx. var. **elegans**—Stems up to 1.5 m tall, solitary or few-clustered, finely pubescent. Lower leaves linear to linear-spatulate, upper somewhat reflexed. Phyllaries pubescent except for scarious or petallike tips, serrulate-acuminate. Pappus of two rows of plumose capillary bristles. Achene about 5 mm long. $2n = 18$. Sandy pinelands, Fla. to Tex., S.C., Ark. Fall. *Laciniaria elegans* (Walt.) Kuntze

3. L. chapmanii T & G—Stems 3–6 dm tall, stiff, usually single, very leafy tomentose pubescent especially above. Lower leaves linear, finely punctate, about 7–10 cm long, upper leaves reduced. Spike very dense, involucre smooth or nearly so, phyllaries mucronate or alternate, corolla purplish, rarely white. Pappus of one to two rows of barbellate capillary bristles. Achene 5–6 mm long. Sandy soil, pine barrens, Fla. to Ga. Fall. *Laciniaria chapmanii* (T & G) Kuntze

4. L. spicata (L.) Willd. var. **resinosa** (Nutt.) Gaiser—Stems up to 2 m tall, glabrous or only sparingly pubescent, from cormlike rhizomes. Leaves larger ones 10–40 cm long, linear or linear-lanceolate, very numerous, reduced upward. Inflorescence densely spicate; involucre cylindrical, 7–11 mm long; phyllaries suborbicular, often purplish colored, glabrous; florets rose purplish, about seven to nine per head. Pappus of capillary bristles barbellate. Achene 4–6 mm long. Pinelands, low ground, Fla. to La., Miss. Summer, fall.

5. L. tenuifolia Nutt.—Stems up to 1 m tall or more, smooth or nearly so, erect. Leaves larger ones 10–30 cm long, filiform or linear-filiform, very numerous, smaller near the base. Inflorescence a densely spicate raceme; involucre subcylindrical; phyllaries mucronate or acuminate; florets rose purplish. Pappus bristles barbellate. Achene 4–5 mm long. Pinelands, sandy soil, Fla. to Ga., S.C. on coastal plain.

Two varieties occur in our area:

1. Stems with scattered hairs along stem; leaves
   ciliate, filiform . . . . 5a. var. *tenuifolia*
1. Stems wholly glabrous; leaves eciliate, linear 5b. var. *laevigata*

5a. L. tenuifolia var. **tenuifolia**—Leaves dull, 10–25 cm long, less than 2 mm wide, Fla. to S.C. *Laciniaria tenuifolia* (Nutt.) Kuntze
5b. L. tenuifolia var. **laevigata** (Nutt.) Robinson—Leaves glossy green, 20–30 cm long, 2–8 mm wide, endemic to Fla. *Laciniaria laevigata* (Nutt.) Small
6. L. garberi Gray—Stems usually 3–5 dm tall, from fleshy, tuber-

ous roots, conspicuously hirsute. Leaves about 12–15 mm long, linear or linear-filiform, numerous. Inflorescence a spicate raceme; involucre cylindrical, phyllaries puberulent, viscid and glandular-punctate, margins ciliate, acuminate; florets rose purplish. Pappus bristles barbellate. Achene 3 mm long. Pinelands, endemic to south Fla. Summer, fall. *Laciniaria garberi* (Gray) Kuntze; *L. chlorolepis* Small; *L. nashii* Small

7. **L. gracilis** Pursh—Stems up to 1 m tall or less, glabrous. Leaves especially the lower linear to narrowly elliptic, numerous. Inflorescence spicate or racemous; involucre about 5–8 mm long, turbinate, phyllaries lanceolate, obtuse or blunt tipped, glabrous or finely ciliate; florets rose purplish; pappus bristles barbellate. Achene 2–3 mm long, coarsely ribbed. Pinelands, Fla. to Miss., Ga., on the coastal plain. Summer, fall. *Laciniaria laxa* Small

51. KÙHNIA L.

Perennial herbs with taproots and bearing opposite or alternate leaves and solitary or numerous moderate-sized heads. Heads discoid with bisexual florets, ten to twenty-five per head; involucre cylindrical with imbricate phyllaries obviously striate; receptacle without chaff; corollas white, yellow, or reddish. Pappus uniseriate of plumose bristles. Achenes ten-ribbed. About six species, North America and Mexico.

1. **K. eupatorioides** L. var. **floridana** R. W. Long—Stems up to 1 m tall or less, erect, slender, dark green. Leaves about 1–4 cm long, linear, entire to barely serrate, sometimes reflexed. Involucre pubescent, 8–9 mm long, the inner phyllaries ribbed. Florets dull white; corolla about 6–7 mm long. Pappus whitish. Endemic to pinelands, drier soil, south Fla. near Miami. Summer, fall. *K. mosieri* Small

Key 8

TRIBE *Vernonieae*

1. Leaves clustered at base of stem; upper
   stem and bracts villous pubescent . . .   52. *Elephantopus*
1. Leaves cauline; upper stems and bracts gla-
   brous or with only scattered short hairs .   53. *Vernonia*

52. ELEPHÁNTOPUS L. ELEPHANT'S FOOT

Perennial herbs erect, branching, with clusters of heads, each cluster subtended by one to three cordate leaflike bracts. Leaves basal and cauline, ovate to obovate, pubescent. Heads with florets all discoid, bisexual; involucre of four pairs of phyllaries, the two outer pairs shorter;

corolla blue purple, pink, rarely white, unequally five-cleft; pappus of five to eight flattened scales prolonged at tips into bristles. Achene ten-ribbed. About 32 species, chiefly tropical in distribution.

1. **E. elatus** Bertoloni—Stems up to 8 dm tall, erect, subscapose, stout, strongly pubescent to villous. Leaves mostly 10–20 cm long, 2–7 cm wide, rosulate, oblong-obovate to broadly ovate, sessile, densely pubescent on both sides, margins crenate. Inflorescence widely branched, bearing usually three to ten terminal clusters of heads; bracts subtending involucre ovate or triangular-ovate, about equaling the clusters; involucre 7–8 mm long, phyllaries pubescent. Achenes 4–5 mm long. Drier sites in pinelands, sandy soil, Fla. to Tex., S.C. Fall.

53. VERNÒNIA Schreb. IRONWEED

Perennial herbs with alternate, sessile or subsessile leaves and heads borne in cymose or corymbose inflorescence. Heads discoid, bisexual, with many purplish or white florets; receptacle without chaff, involucre of many imbricated phyllaries. Pappus in two rows, the outer of scalelike bristles, inner of many capillary bristles, both persistent. Achene ten-ribbed. A taxonomically difficult genus of about 600 species in tropical and subtropical America, Africa, and Asia. Natural interspecific hybrids are common.

1. Phyllaries linear, less than 1 mm wide;
   plants annual . . . .   1. *V. cinerea*
1. Phyllaries lanceolate, 2–3 mm wide; plants
   perennial.
   2. Leaf blades linear or linear-lanceolate,
      mostly less than 3 cm wide and entire .   2. *V. blodgettii*
   2. Leaf blades elliptic-ovate to obovate,
      mostly 3–6 cm wide, regularly sharply
      serrate . . . .   3. *V. gigantea*

1. **V. cinerea** (L.) Lessing—Stems up to 1 m tall, cinereous pubescent, from fibrous roots. Leaves usually 2–10 cm long, ovate or elliptic, shallowly serrate. Heads in corymbs, involucre 4–5 mm long, the phyllaries linear-subulate; corolla purplish, about 3–4 mm long. Pappus white. Achene about 1 mm long. Disturbed sites, margins of hammocks, south Fla., W.I., tropical America, naturalized from the Old World. All year. *Seneciodes cinerea* (L.) Kuntze

2. **V. blodgettii** Small—Stems up to 5 dm tall, usually branched near the base, smooth. Leaves 3–5 cm long, chiefly in lower stem, linear or linear-lanceolate, or basal leaves sometimes narrowly ovate, glabrous, entire, and slightly revolute. Involucre about 5 mm long, the phyllaries

narrowly elliptic, acute or acuminate; heads in corymb, corollas dark purplish, lobes narrowly lanceolate. Pappus stramineous. Achene 2–3 mm long. Pinelands, drier sites, endemic to south Fla. and the Fla. Keys. Summer, fall.

3. V. gigantea (Walt.) Trel.—Stems up to 2 m tall, glabrous or nearly so, branched. Leaves mostly 8–25 cm long, generally distributed on stem, elliptic to obovate blade gradually tapering to a short petiole or subsessile. Involucre about 5 mm long, the phyllaries lanceolate-elliptic, inner obtuse, outer acute; heads in spreading corymbs, corolla purplish, lobes narrowly lanceolate. Pappus purplish. Achene 3 mm long with scabrous or puberulent angles. Hammocks, low seasonally wet areas, margins of woods, swamps, south Fla. to Ala., Ga. on the coastal plain. Summer, fall.

Key 9

TRIBE *Cardueae*

54. CÍRSIUM Mill.　THISTLE

Herbs often spiny or bristly, with alternate, toothed or usually pinnatifid leaves, and moderate to large heads. Heads discoid, florets tubular, bisexual, or unisexual by abortion; phyllaries imbricate, usually some or all spine tipped and with a thick, glutinous dorsal ridge; receptacle densely bristly, florets numerous, purplish, yellow, or white with slender corollas, anthers tailed. Pappus of numerous plumose bristles. Achenes smooth. About 250 species of the Northern Hemisphere.

1. Heads surrounded by a whorl of spiny bracts　1. *C. horridulum*
1. Heads not surrounded by spiny bracts, not
　involucrate . . . .　2. *C. nuttallii*

1. C. horridulum Michx. PURPLE THISTLE—Stems up to 1 m tall or more. Leaves lance-ovate to pinnatifid, the lobes or margins with spines. Heads surrounded by very spiny bractlike leaves; involucre 3–5 cm long, sometimes shorter, phyllaries pubescent, the inner ones sometimes densely so; corollas yellow, cream, or purplish, less commonly white, 2–4 cm long, anthers as long or somewhat shorter than filaments. Two distinctive varieties are found in our area that intergrade locally:

1. Inner phyllaries eciliate, or with only few
　scattered hairs; blades pinnatifid . . .　1a. var. *horridulum*
1. Inner phyllaries densely ciliate with short
　marginal hairs . . . .　1b. var. *vittatum*

1a. C. horridulum var. **horridulum**—Leaves pubescent, becoming smooth, pinnatifid, the lobes with spines. $2n = 32$. Pineland, sandy fields, Fla. to Tex., Me. Spring. *Carduus spinosissimus* Walt., *Cirsium smallii* Britton

1b. C. horridulum var. **vittatum** (Small) R. W. Long—Leaves lance-ovate, the upper blades undulate, spiny. $2n = 32, 34$. Sandy soil, pine-lands, Fla. to S.C. on the coastal plain, W.I. Spring. *Carduus vittatum* Small, *Cirsium vittatum* Small

2. C. nuttallii DC.—Stems up to 4 dm tall, glabrous. Lower leaves variously ovate to elliptic-oblong, bipinnatifid. Outer phyllaries spine tipped, inner ones, short tipped, involucres 1.5–2.5 cm high, not sub-tended by bracts; corollas rose purple or lilac, mostly 2 cm or more long. $2n = 24$. Drier sites, Fla. to Miss., S.C. on the coastal plain. All year.

Key 10

TRIBE *Cichorieae*

1. Corollas rose, pinkish, or purplish; leaves linear, much reduced, or absent . . .	55. *Lygodesmia*
1. Corollas yellow, orange, or lavender, rarely white; leaves lanceolate to ovate, often deeply lobed.	
2. Involucres small, narrow, 4–5 mm wide .	56. *Youngia*
2. Involucres larger.	
3. Lower leaves and stems hirsute, corollas orange	57. *Hieracium*
3. Lower leaves and stems glabrous, or merely with scattered short hairs.	
4. Tips of inner phyllaries two-lobed or expanded	58. *Pyrrhopappus*
4. Tips of inner phyllaries acute or obtuse, not two-lobed.	
5. Phyllaries with conspicuous brown tips, sharply attenuate . . .	59. *Lactuca*
5. Phyllaries with green or yellow tips, gradually attenuate . .	60. *Sonchus*

55. LYGODÉSMIA D. Don RUSH PINK

Herbs with rushlike stems, milky juice, and with alternate, narrow leaves. Heads erect, pink or purplish, rarely white, all florets bisexual; involucre cylindric with four to eight larger phyllaries and with a few, much reduced outer ones. Pappus of numerous white or pale brown capillary bristles. About 12 species in North America.

1. L. aphylla (Nutt.) DC. ROSE RUSH—Stems erect up to 8 dm tall, rushlike, simple or branched at the base, leafless or nearly so. Leaves, if present, appearing as thin scales or one to few elongate, linear blades near base of stem. Involucre 2 cm long or more, cylindrical, outer phyllaries 1–3 mm long, inner phyllaries linear, 10–15 mm long. Florets rose, pinkish, or purplish, rarely white, the ligules 1.5–2 cm long. Pappus white, achene 10–12 mm long. Sandy soil, dry pinelands, Fla. to Ga. on coastal plain. Spring.

56. YOÚNGIA Cassini

Low herbs with milky juice and numerous basal leaves, upper leaves alternate and reduced. Heads several in corymbs or panicles, few- to many-flowered, bisexual, yellow; involucre cylindrical or campanulate, phyllaries in single series, not thickened on the back, with three to four very short bracts at the base. Pappus of abundant, white capillary bristles. Achene ten-ribbed, often beaked. About 35 to 40 species in temperate and tropical Asia.

1. **Y. japonica** (L.) DC.—Stems up to 6 dm tall, simple, or branched at the base, puberulent below, glabrate above. Basal leaves 5–15 cm long, pinnatifid, petioled, upper leaves absent or much reduced. Involucre 4–5 mm long, cylindrical, phyllaries acuminate-glabrous. Achene 1–2 mm long, ribbed. $2n = 16$. Lawn weed, disturbed sites, old fields, spreading, and locally established from Fla. to La., Pa., W.I., naturalized from Japan. Spring. *Crepis japonica* (L.) Benth.

57. HIERÀCIUM L. Hawkweed

Perennial herbs with milky juice, erect stems rhizomatous, bearing basal or alternate, toothed or entire leaves. Heads in corymbs, panicles, racemes, or solitary, erect; florets all bisexual, yellow to reddish orange; involucre hemispheric to narrowly campanulate, phyllaries imbricate in several series, often glandular or with stellate pubescence, pappus of one to two rows of pale brown or white capillary bristles. Achene narrow cylindrical, shiny black, usually strongly ten-ribbed. About 800 species, many species polymorphic because of natural hybridization and apomixis.

1. Lower stem densely pubescent, hairs long;
   lower leaf blades not purple veined . . 1. *H. gronovii*
1. Lower stem glabrous or nearly so, hairs scattered; lower leaf blades not purple veined 2. *H. venosum*

1. **H. gronovíi** L. Hawkweed—Stems pilose to densely pilose at base, less so above, to 1 m tall, usually less. Leaves 6–18 cm long, mostly

basal or near the lower part of the stem, reduced toward the top, oblanceolate to obovate, becoming narrowly lanceolate, base cuneate to attenuate. Heads few to many in narrow or spreading panicles, less commonly in corymbs, peduncles usually stipitate-glandular; ligules yellow; pappus tan. Achene fusiform, 3–4 mm long. Disturbed sites, old fields, roadsides, south Fla. to Tex., Mass., Ontario. Summer, fall. The species is known to hybridize with the next species. *H. megacephalon* Nash

A closely related endemic form with pappus white instead of tan colored has been named **H. argyraeum** Small. It may not be specifically distinct, however.

2. H. venosum L. RATTLESNAKE WEED—Stems to 6 dm tall, glabrous or nearly so. Leaves 5–15 cm long, mostly basal; elliptic-lanceolate to oblanceolate, veins usually purplish, obscurely denticulate or nearly entire, base attenuate, few reduced leaves above. Heads many in spreading, flat-topped corymbs, peduncles usually stipitate-glandular; ligules bright yellow, pappus yellowish. Achene 3 mm long, columnar. $2n = 18$. Oak scrub vegetation, prairies, dry sandy soil, disturbed sites, Fla. to Me., Nebr. Spring, summer.

Collected on lower Fla. Gulf coast and central Fla. and presumably in our area but probably infrequent; common north of our range.

58. PYRRHOPÁPPUS DC.

Biennial or perennial subscapose herbs. Leaves entire to irregularly toothed or pinnatifid. Herbs long pedunculate, involucre cylindrical to slightly ovoid; phyllaries in two unequal series, the inner longer and obliquely two-lobed or expanded at the tip; ligules yellow or less commonly white. Achenes cylindric-ellipsoid. About eight species in North America. *Sitilias* Raf.

1. P. carolinianus (Walt.) DC. var. **georgianus** (Shinners) Ahles, FALSE-DANDELION—Stems erect or ascending to 6 dm tall, glabrous below, puberulent above. Leaves mostly 10–25 cm long, mostly basal, usually dissected, cauline leaves few, sessile, reduced upward. Involucre 1.5–2.5 cm long; ligules 2–2.5 cm long. Achene 4–5 mm long, pubescent, pappus light tan in color. Roadsides, old fields, waste places, coastal plain Fla. to S.C., Ala. Spring.

Common along roadsides in central peninsular Fla. and extending into our area.

59. LACTÙCA L. LETTUCE

Annual, biennial, or perennial herbs with milky juice, erect stems, and alternate, entire to pinnatifid leaves. Heads many in spreading panicles, florets all bisexual, yellow, orange, blue, or white, the corolla tube usually over half as long as the ligule; involucre cylindrical, usually imbricate but sometimes only calyculate; pappus of capillary bristles. Achene ribbed, flattened, beaked or winged. About 100 species, chiefly Northern Hemisphere. *Brachyrhamphus* DC.; *Mulgedium* Cassini

1. Corolla lobes purple blue; leaves entire to
   pinnatifid; margins without spines.
   2. Achene with filiform neck attaching to
      pappus bristles . . . . 1. *L. graminifolia*
   2. Achene with short broad neck attaching
      to pappus bristles . . . . . 2. *L. floridana*
1. Corolla lobes yellow; leaves entire to pinnatifid; margins with delicate spines . . 3. *L. intybacea*

1. **L. graminifolia** Michx.—Biennial with stems up to 1.5 m tall or less, glabrous, greenish to reddish, subscapose. Leaves usually 10–40 cm long, mostly basal, entire or deeply pinnatifid lobed, elongate, pubescent at least along veins; upper leaves reduced, sinuate-pinnatifid. Heads in diffuse panicles, involucre 12–15 mm long, inner phyllaries linear; ligules purplish blue, rarely white or yellow, pappus white. Achene about 5 mm long. $2n = 34$. Disturbed sites, Fla. to Tex., S.C. on the coastal plain. Spring, summer.

2. **L. floridana** (L.) Gaertn. WILD LETTUCE—Biennial to 3 m, usually less, stems glabrous, often reddish. Leaves 1–2 dm long, dissected or unlobed, elliptic or deltoid-ovate. Heads in diffuse panicles, branches long and spreading; involucres 8–14 mm long, ligules blue or violet, pappus white. Achene elliptic, ribbed. Seasonally wet areas, margins of hammocks, Fla. to Tex., N.Y., Nebr., W.I. Summer, fall. *Mulgedium floridanum* (L.) DC.

3. **L. intybacea** Jacq. WILD LETTUCE—Annual with stems up to 1.5 m tall, smooth. Leaves 1–3 dm long, bright green, reduced above, margins or lobes with delicate spines. Heads solitary on narrow peduncles; involucre somewhat ovoid, phyllaries conspicuously scarious margined, outer series ovate, inner series linear-lanceolate; ligules yellow. Achene about 4 mm long, fusiform. Disturbed sites, south Fla., naturalized from tropical America. All year. *Brachyrhamphus intybaceus* (Jacq.) DC.

60. SÓNCHUS· L. Sow Thistle

Annual or perennial, stems erect, herbs with milky juice. Leaves basal or alternate, entire to dissected, usually auriculate. Heads cylindrical with many florets, all bisexual, yellow, in spreading corymbs;

involucre ovoid or campanulate, usually basally thicker with age; phyllaries lanceolate, generally imbricate; pappus of abundant, white, capillary bristles. Achene flattened, ribbed, beakless. About 50 species, almost cosmopolitan.

1. Leaves of upper stem with bases triangular
    or pointed; achenes transversely ridged .   1. *S. oleraceus*
1. Leaves of upper stem with bases rounded;
    achenes not transversely ridged .   .   2. *S. asper*

    1. S. oleraceus L. COMMON SOW THISTLE—Annual up to 2 m tall or more, smooth, with a short taproot. Leaves 8–30 cm long, runcinate-pinnatifid with weak-spined teeth or merely serrate, bases triangular or pointed, base rounded-auriculate, upper leaves reduced. Heads many in corymbose inflorescence, ligules pale yellow; involucre 10–13 mm long, phyllaries smooth. Achenes 2–3 mm long, transversely ridged. $2n = 16,32$. Disturbed sites, old fields, cosmopolitan weed, naturalized from Europe. Spring, summer.
    2. S. asper (L.) Hill, SPINY-LEAVED SOW THISTLE—Annual up to 2 m tall with taproot, generally smooth, or stipitate-glandular above. Leaves 6–30 cm long, lanceolate to oblanceolate, pinnatifid or sometimes without lobes, margins with rigid spines, prickly, bases of leaves rounded-auriculate, not acute. Corymbs with many heads; involucre 10–12 mm long, phyllaries obscurely stipitate-glandular; ligules light yellow. Achenes 2–3 mm long, not transversely ridged, merely several-nerved. $2n = 18$. Disturbed sites, old fields, cosmopolitan weed, naturalized from Europe. Spring, summer.

Appendix A

Flora Author List

Abbe	Ernst Cleveland Abbe, 1905–
P. Adams	William Preston Adams, 1930–
Adans.	Michel Adanson, 1727–1806
Ag.	Jacob George Agardh, 1813–1901
Ahles	Harry E. Ahles, 1924–
Ait.	William Aiton, 1731–1793
Ait. f.	William Townsend Aiton, 1766–1849
Alain	Herman Alain (Dr. E. E. Liogier), 1916–
Alexander	Edward Johnston Alexander, 1901–
All.	Carlo Allioni, 1725–1804
Alston	Arthur High Alston, 1902–
Ames	Oakes Ames, 1874–1950
Anders.	Thomas Anderson, 1832–1870
Anderson	Edgar Shannon Anderson, 1897–1969
L. C. Anderson	Loran Crittenden Anderson, 1936–
Anderss.	Niles Johan Andersson, 1821–1880
Andrews	Henry C. Andrews, 1796–1828
Arn.	George Arnold Walker Arnott, 1799–1868
Aschers.	Paul Fredrich August Ascherson, 1834–1913
Ashe	William Willard Ashe, 1872–1932
Aubl.	Jean Baptiste Christophe Fusée Aublet, 1720–1778
Baehni	Charles Baehni, 1906–1964
Bailey	Liberty Hyde Bailey, 1858–1955
Baillon	Henri Ernest Baillon, 1827–1895
Baker	John Gilbert Baker, 1834–1920
Balb.	Giovanni Battista Balbis, 1765–1831
Baldw.	William Baldwin, 1779–1819
Ball	Carleton Roy Ball, 1873–1935
Banks	Joseph Banks, 1743–1820
Barnhart	John Hendley Barnhart, 1871–1949

Bartr.	William Bartram, 1739–1823
Baudo.	A. Baudoin, *ca.* 1843
Beal	William James Beal, 1833–1924
Beauv.	Ambroise Marie Francois Joseph Palisot de Beauvois, 1752–1820
Becc.	Odoardo Beccari, 1843–1920
Beck	Lewis Caleb Beck, 1798–1853
Beer	Joseph Georg Beer, 1803–1873
Benedict	James Everard Benedict, 1854–1940
L. Benson	Lyman David Benson, 1909–
Benth.	George Bentham, 1800–1884
Berg	Otto Carl Berg, 1815–1866
Bergius	Peter Jonas Bergius, 1730–1790
Bernh.	Johann Jacob Bernhardi, 1774–1850
Bertero	Carlo Guiseppe Bertero, 1789–1831
Bertoloni	Antonio Bertoloni, 1775–1869
Beyr.	Heinrich Karl Beyrich, 1796–1834
Bicknell	Eugene Pintard Bicknell, 1859–1925
Bigel.	Jacob Bigelow, 1787–1879
Blake	Sidney Fay Blake, 1892–1959
Blanco	Francisco Manuel Blanco, 1778–1845
Blankenship	Joseph William Blankenship, 1862–1938
Blume	Carl Ludwig von Blume, 1796–1862
Boeckler	Johann Otto Boeckler, 1804–1899
Bogin	Clifford Bogin, 1920–
Boissier	Pierre Edmond Boissier, 1810–1885
Bojer	W. Bojer, 1800–1856
Bonpl.	Aimé Jacques Alexandre Bonpland, 1773–1858
Boomhour	Elizabeth Gregory Boomhour (Kerr), 1912–
W. Boott	William Boott, 1805–1887
Borbas	Vincenz von Borbas, 1844–1905
Bort	Katherine Stephens Bort, 1870–
Bory	B. M. Bory de Saint-Vincent, 1778–1846
Bosc	J. A. Bosc, 1764–1837
Bosch	Roelof Benjamin van den Bosch, 1810–1862
B & P	Friedrich Graf von Berchtold, 1781–1876, and Presl
Brackett	Amelia Ellen Brackett, 1896–1926
Brainerd	Ezra Brainerd, 1844–1924
Brand	August Brand, 1863–1930
Brand.	Townshend Stith Brandegee, 1843–1925
A. Braun	Alexander Carl Heinrich Braun, 1804–1877
Brenan	J. P. M. Brenan, 1917–
Briq.	John Isaac Briquet, 1870–1931
Britt.	Nathaniel Lord Britton, 1859–1934
Briz.	George K. Brizicky, 1901–1968
Brongn.	Adolphe Theodore Brongniart, 1801–1876
R. Brown	Robert Brown, 1773–1858
P. Browne	Patrick Browne, 1720–1790
Brueckner	Adolph Friederich Brueckner, 1781–1818
BSP	see Britt.; Stern; Poggenburg

Buch.	Franz Georg Philipp Buchenau, 1831–1906
Buchoz	Pierre Joseph Buchoz, 1731–1807
Buese	L. H. Buese, 1819–1888
Bunge	Alexander Andrejewicz von Bunge, 1803–1890
Burch	Derek G. Burch, 1933–
Bureau	Louis Edouard Bureau, 1830–1918
Burm.	Johannes Burman, 1706–1779
Burm. f.	Nicolas Laurens Burman, 1734–1779
Bush	Benjamin Franklin Bush, 1858–1927
Butt.	Frederick King Butters, 1878–1945
Camb.	Jacques Cambessedes, 1789–1863
A. Camus	Aimee Antoinette Camus, 1879–
Cassini	Alexandre Henri Cassini, 1781–1832
Cav.	Antonio Jose Cavanilles, 1745–1804
Cham.	Ludolf Adalbert von Chamisso, 1781–1838
Chapm.	Alvin Wentworth Chapman, 1809–1899
Chase	Mary Agnes (Merrill) Chase, 1869–1963
Ching	Ren-Chang Ching, 1899–
Chodat	Robert Hippolyte Chodat, 1865–1934
Choisy	Jacques Denis Choisy, 1799–1859
C. Chr.	Carl Fredrick Albert Christensen, 1872–1942
Christm.	Gottlieb Friedrich Christmann, 1752–?
Clarke	Charles Baron Clarke, 1832–1906
Clausen	Robert Theodore Clausen, 1911–
Cogn.	Celestin Alfred Cogniau, 1841–1916
Coker	William Chambers Coker, 1872–1953
Colla	Luigi Colla, 1766–1848
O. F. Cook	Orater Fuller Cook, 1867–1949
Copel.	Edwin Bingham Copeland, 1873–
Core	Earl Lemney Core, 1902–
Correa	Jose Francisco Correa da Serra, 1751–1823
Correll	Donovan Stuart Correll, 1908–
Cory	Victor Louis Cory, 1880–
Coulter	John Merle Coulter, 1851–1928
Coville	Frederich Vernon Coville, 1867–1937
C & R	see Coulter; Rose
Crantz	Heinrich Johann Nepomuk von Crantz, 1722–1797
Cronquist	Arthur John Cronquist, 1919–
Cunn.	Allan Cunningham, 1791–1839
M. A. Curtis	Moses Ashley Curtis, 1808–1872
Cyr.	Domenico Cyrillo, 1739–1799
Dandy	James Edgar Dandy, 1903–
DC.	Augustin Pyramus de Candolle, 1778–1841
A. DC.	Alphonse de Candolle, 1806–1893
C. DC.	Casimir de Candolle, 1836–1918
Decne.	Joseph Decaisne, 1807–1882
Degener	Otto Degener, 1899–
Dennst.	August Wilhelm Dennstedt, 17 ?–18 ?
Desc.	Michel Etienne Descourtilz, 1775–1836
Desf.	Rene Louiche Desfontaines, 1750–1833

Des Moulins	Charles Robert Alexandre Des Moulins, 1798–1875
Desv.	Nicaise Auguste Desvaux, 1784–1856
de Wild.	E. de Wildeman, 1866–1947
de Wit	H. C. D. de Wit, 1909–
Diels	Friedrich Ludwig Emil Diels, 1874–1945
Dietr.	Friedrich Gottlieb Dietrich, 1768–1850
A. Dietr.	Albert Dietrich, 1795–1856
Doell	Johann Christoph Doell, 1808–1885
D. Don	David Don, 1799–1841
G. Don	George Don, 1798–1856
Donn	James Donn, 1758–1813
Dorman	Keith William Dorman, 1910–
Dressler	Robert L. Dressler, 1927–
Druce	George Claridge Druce, 1851–1932
Dryand.	Jonas Carlsson Dryander, 1748–1810
Duch.	Duchassaing, ?–?
Duchesne	Antonine Nicolas Duchesne, 1747–1827
Dugand	Armando Dugand, 1906–
Dum.	Berthelemy Charles Joseph Dumortier, 1797–1878
Dunal	Michel Félix Dunal, 1789–1856
Dur.	Michel Charles Durieu de Maissonneurve, 1797–1878
Durand	Elie Magloire Durand, 1794–1873
Durazz.	Ippolito Durazzo, 1750–1818
A. A. Eaton	Alvah Augustus Eaton, 1865–1908
Eberm.	Carl Heinrich Ebermaier, 1802–1870
Ekman.	E. L. Ekman, 1883–1931
Ell.	Stephen Elliott, 1771–1830
Endl.	Stephen Ladislaus Endlicher, 1804–1849
Engelm.	George Engelmann, 1809–1884
A. M. Evans	Austin Murray Evans, 1932–
Fassett	Norman Carter Fassett, 1900–1954
Fawc.	William Fawcett, 1900–1954
Fee	Antoine Laurent Apollinaire Fee, 1789–1874
Fenzl	Eduard Fenzl, 1808–1879
Ferguson	Alexander McGowen Ferguson, 1874–
Fern.	Merritt Lyndon Fernald, 1873–1950
Fisch.	Friedrich Ernst Ludwig von Fischer, 1782–1854
Fluegge	Johann Fluegge, 1775–1816
Forsk.	Pehr Forskal, 1736–1768
Forst.	Johann Reinhold Forster, 1729–1798
Fort.	Robert Fortune, 1812–1880
Fosberg	Francis Raymond Fosberg, 1908–
Foster	Robert Crichton Foster, 1904–
Foug.	Auguste Denis Fougeroux de Bondaroy, 1732–1789
Fourn.	Eugène Pierre Nicolas Fournier, 1834–1884
Frapp.	Charles Frappier, 1853–1883
Fraser	P. Neil Fraser, 1830–1905
Gaertn.	Joseph Gaertner, 1732–1791
Gaertn. f.	Carl Friderich von Gaertner, 1772–1850
Gaiser	Lulu Odell Gaiser, 1896–1965

Gale	Shirley Gale (later Cross) 1915–
Gandoger	Michel Gandoger, 1850–1926
Garcke	Freiderich August Garcker, 1819–1904
Garden	Alexander Garden, 1730–1791
Gates	Frank Caleb Gates, 1887–
Gattinger	Augustin Gattinger, 1825–1903
Gaud.	Charles Gaudichaud-Beaupri, 1789–1854
Geert	August van Geert, ?
Gerrard	William Tyrer Gerrard, ?–1866
Gilly	Charles Louis Gilly, 1911–
Gl.	Henry Allan Gleason, 1882–
Gled.	Johann Gottlieb Gleditsch, 1714–1786
Gmel.	Johann Georg Gmelin, 1709–1775
J. F. Gmel.	Johann Friedrich Gmelin, 1748–1804
Glueck	C. M. H. Glueck, 1868–1940
R. K. Godfrey	Robert Kenneth Godfrey, 1911–
Gomez	Bernardino Antonio Gomez, 1769–1823
Gould	Frank W. Gould, 1913–
Graebn.	Karl Otto Peter Paul Graebner, 1871–1933
Graham	Robert Graham, 1786–1845
Gray	Asa Gray, 1810–1888
S. F. Gray	Samuel Frederick Gray, 1766–1828
B. D. Greene	Benjamin Daniel Greene, 1793–1862
Greene	Edward Lee Greene, 1842–1915
Greenm.	Jesse More Greenman, 1867–1951
Grev.	Robert Kaye Greville, 1794–1866
Griseb.	August Heinrich Rudolf Grisebach, 1814–1879
H & A	see Hook.; Arn.
Hack	Eduard Hackel, 1850–1926
Hall	Harvey Monroe Hall, 1874–1932
Hall. f.	Albrecht von Haller, 1758–1823
Ham.	Francis (Buchanon) Hamilton, 1762–1829
Hamet	M. Raymond Hamet, fl., 1910–1960
Hance	Henry Fletcher Hance, 1827–1886
Hara	Hiroshi Hara, 1911–
Harper	Roland McMillan Harper, 1878–1966
F. Harper	Francis Harper, 1886–
Harv.	William Henry Harvey, 1811–1866
Hassk.	Justus Carl Hasskarl, 1811–1894
Haw.	Adrian Hardy Haworth, 1772–1833
Hayek	August von Hayek, 1821–1928
H & B	see Humb.; Bonpl.
HBK	see Humb.; Bonpl.; Kunth
HBW	see Humb.; Bonpl.; Willd.
Hegelm.	Christoph Friedrich Hegelmaier, 1830–1906
Heimerl	Anton Heimerl, 1857–1942
Heiser	Charles Bixler Heiser, 1920–
Heist.	Lorenz Heister, 1683–1768
Heller	Amos Arthur Heller, 1867–1944
Hemsl.	William Botting Hemsley, 1843–1924

Henders.	Louis Forniquet Henderson, 1853–1942
Henr.	Johannes Theodoor Henrard, 1881–
Herb.	William Herbert, 1778–1847
Hermann	Frederick Joseph Hermann, 1906–
Hieron.	Georg Hans Emmo Wolfgang Hieronymus, 1846–1921
Hill	John Hill, 1716–1775
A. W. Hill	Arthur William Hill, 1875–1941
Hitchc.	Albert Spear Hitchcock, 1865–1935
Hochst.	Christian Fredrich Hochstetter, 1787–1860
Hodgdon	Albion Reed Hodgdon, 1909–
Hoffm.	Georg Franz Hoffman, 1761–1826
Holtt.	Richard Eric Holttum, 1895–
Hook.	William Jackson Hooker, 1785–1865
Hook. f.	Joseph Dalton Hooker, 1817–1911
Horton	P. M. Horton, ?
Host	Nicholas Thomas Host, 1761–1834
House	Homer Doliver House, 1878–1949
Hubb.	T. H. Hubbard, 1875–?
Hubbard	Fredrick Tracy Hubbard, 1872–1962
C. E. Hubbard	Charles Edward Hubbard, 1900–
Humb.	Friedrich Heinrich Alexander von Humboldt, 1769–1859
Hylander	Nils Hylander, 1904–
Jackson	Albert Bruce Jackson, 1876–1947
D. Jackson	Daniel Dana Jackson, 1870–1941
Jacq.	Nikolaus Joseph von Jacquin, 1727–1817
C. W. James	Charles William James, 1929–
Jenm.	George Samuel Jenman, 1845–1902
Jennings	Otto Emery Jennings, 1877–
Johnston	Ivan Murray Johnston, 1898–1960
J. R. Johnston	John Robert Johnston, 1880–
M. C. Johnston	Marshall Conring Johnston, 1930–
Jordan	Alexis Jordan, 1814–1897
Juss.	Antoine Laurent de Jussieu, 1748–1836
Kam.	Franz Kamienski, 1851–1912
Kearney	Thomas Henry Kearney, 1874–1956
Keng	Yi Li Keng, 1898–
Ker	John Bellenden Ker (before 1804, John Gawler), 1764–1842
Kiaersk.	Hjalmar Frederick Kiaerskon, 1835–1900
Kl. & Gke.	see Klotzsch; Garcke
Klotzsch	Johann Friedrich Klotzsch, 1805–1860
Kobuski	Clarence Emmeren Kobuski, 1900–1963
K. Koch	Karl Heinrich Emil Koch, 1809–1879
Koehne	Bernhard Adalbert Emil Koehne, 1848–1918
Koenig	Johan Gerhard Koenig, 1728–1757
Koern.	Friedrich Koernicke, 1828–1908
Kopp	Lucille E. Kopp (later Mrs. Robert F. Blum), 1926–
Kosterm.	Andre Joseph Guillaume Henri Kostermans, 1907–
Kral	Robert Kral, 1926–
Krug	Carl Wilhelm Leopold Krug, 1833–1898
Kuetz.	Friedrich Traugott Kuetzing, 1807–1893

Kuhn	Maximiliam Friedrich Adalbert Kuhn, 1842–1894
Kukenth.	Georg Kukenthal, 1864–1956
Kunth	Carl Sigismund Kunth, 1788–1850
Kuntze	Carl Ernst Otto Kuntze, 1843–1907
Kunze	Gustav Kunze, 1793–1851
Kurtz	Fritz (Federico) Kurtz, 1854–1920
Kurz	Sulpiz Kurz (J. Amann), 1834–1878
L.	Carolus Linnaeus, 1707–1778
L. f.	Carl von Linne, 1741–1783
Labille.	Jacques Julian Jouffou de Labillardière, 1755–1834
Lag.	Mariano Lagasca y Segura, 1776–1839
Lakela	Olga Lakela, 1890–
Lam.	Jean Baptiste Antoine Pierre Monnet de Lamarch, 1744–1829
Langsd.	Georg Heinrich von Langsdorff, 1774–1852
Laxm.	Erich Laxmann, 1737–1796
Leavenw.	Melines Conklin Leavenworth, 1796–1862
Le Conte	John Eatton Le Conte, 1784–1860
Leggett	William Henry Leggett, 1816–1882
Lehmann	Johann George Christian Lehmann, 1792–1860
Lemaire	Charles Antoine Lemaire, 1801–1871
Leonard	Emery Clarence Leonard, 1892–
Lessing	Christian Friedrich Lessing, 1809–1862
W. H. Lewis	Walter Hepworth Lewis, 1930–
L'Her.	Charles Louis L'Heritier de Brutelle, 1746–1800
Liebm.	F. M. Liebmann, 1813–1856
Lindau	G. Lindau, 1866–1923
Lindl.	John Lindley, 1799–1865
Link	Johann Heinrich Friedrich Link, 1767–1851
Little	Elbert Luther Little, 1907–
Lloyd	Francis Ernest Lloyd, 1868–1947
Lodd.	Conrad Loddiger, 1738–1826
Loefl.	Pehr Loefling, 1729–1756
Loesener	Ludwig Eduard Theodor Loesener, 1865–1941
R. W. Long	Robert William Long, 1927–
Lour.	Joao de Loureiro, 1715–1795
Ludw.	Christian Gottlieb Ludwig, 1709–1773
Lunan	John Lunan, fl., 1814–?
Macbride	James Francis Macbride, 1892–
Macf.	John Muirhead Macfarlane, 1855–1943
Macfadyen	James Macfadyen, 1798–1850
Mackenzie	Kenneth Kent Mackenzie, 1877–1934
McVaugh	Rogers McVaugh, 1909–
Maguire	Basset Maguire, 1904–
Makino	Tomitaro Makino, 1863–1957
Malme	Gustav Oskar Andersson Malme, 1864–1937
S. Manso	Antonio Luiz Particio da Silva Manso, 1788–1818
Marian	Joseph Eric Marian, 1901–
Marsh.	Humphrey Marshall, 1722–1801
Mart.	Carl Friedrich Philipp von Martius, 1794–1868
Martyn	John Martyn, 1699–1767

Mast.	Johannes Mattfeld, 1895–1951
Mattf.	Johannes Mattfeld, 1895–1951
Maxon	William Ralph Maxon, 1877–1948
Medicus	Friedrich Casimer Medicus, 1736–1808
Meeuse	Bastian Jacob Dirk Meeuse, 1916–
Meissner	Carl Friedrich Meissner, 1800–1874
Merrill	Elmer Drew Merrill, 1876–1956
Meyer	George Friedrich Wilhelm Meyer, 1782–1856
E. Mey.	Ernst Heinrich Friedrich Meyer, 1791–1858
Mez	Carl Christian Mez, 1866–1944
Michx.	Andre Michaux, 1746–1800
Michx. f.	Francois Andre Michaux, 1770–1855
Mieg.	Abbe Miegeville, 1814–1901
Miers	John Miers, 1789–1879
Mikan	Johan Christian Mikan, 1769–1844
Mill.	Philip Miller, 1691–1771
Millardet	Pierre Marie Alexis Millardet, 1838–1902
Millsp.	Charles Frederick Millspaugh, 1854–1923
Miq.	Friedrich Anton Wilhelm Miquel, 1811–1871
Mitchell	John Mitchell, 1680–1768
Moc.	Joseph Marian Mocino (Mozino), 1757–1820
Moench	Konrad Moench, 1744–1805
Mohr	Charles Theodore Mohr, 1824–1901
Moldenke	Harold Norman Moldenke, 1909–
Molina	Juan Ignacio Molina, 1737–1829
Moore	Thomas Moore, 1821–1887
A. H. Moore	Albert Hanford Moore, 1883–
T. V. Moore	Thomas Verner Moore, 1877–1926
Moq.	Christian Horace Benedict Alfred Moquin-Tandan, 1804–1863
Morelet	Arthur Morelet, 1809–1892
Morong	Thomas Morong, 1827–1894
Morr.	Charles Jacques Edouard Morren, 1833–1866
Morton	Conrad Vernon Morton, 1905–
F. Muell.	Ferdinand Jacob Heinrich von Mueller, 1825–1896
Muell. Arg.	Jean Mueller (of Aargau), 1828–1896
Muhl.	Gotthill Henry Ernest Muhlenberg, 1756–1817
Munro	William Munro, 1818–1880
Munson	Thomas Volney Munson, 1843–1913
Munz	Philip Alexander Munz, 1892–
Murray	Johan Andreas Murray, 1740–1791
Mutis	Jose Celestino Mutis, 1732–1808
Myint	Tin Myint, 1936–
Nash	George Valentine Nash, 1864–1919
Necker	Noel Joseph de Necker, 1730–1793
Nees	Christian Gottfried Daniel Nees von Esenbeck, 1776–1858
Nieuwl.	Julius Aloysius Arthur Nieuwland, 1878–1936
Northrop	Alice Belle Northrop, 1864–1922; John Isaiah Northrop, 1861–1891
Nutt.	Thomas Nuttall, 1786–1859

Oakes	William Oakes, 1799–1848
Oeder	George Christian von Oeder, 1728–1791
Opiz	Philip Maximilian Opiz, 1787–1858
Ortega	Casimiro Gomex [de] Ortega, 1740–1818
Osbeck	Pehr Osbeck, 1723–1805
Otto	Christoph Friedrich Otto, 1783–1886
G. B. Ownbey	Gerald Bruce Ownbey, 1916–
Parodi	Lorenzo Raimundo Parodi, 1895–
Pav.	Jose Antonio Pavon, 1750–1844
Pax	Ferdinand Albin Pax, 1858–1942
Pennell	Francis Whittier Pennell, 1886–1952
Perdue	Robert Edward Perdue, Jr., 1924–
Perrier	E. Perrier de la Bathie, 1825–1916
Perrine	Henry Perrine, 1797–1840
Pers.	Christiaan Hendrik Persoon, 1755–1837
Peter	Robert Peter, 1805–1894
Peterm.	Wilhelm Ludwig Petermann, 1806–1855
Pfeiffer	Norma Etta Pfeiffer, 1889–
Pfitz.	Ernst Hugo Heinrich Pfitzer, 1846–1906
Philippi	Rudolph Amandus Philippi, 1808–1904
Pilger	Robert Knud Friedrich Pilger, 1876–1953
Planchon	Julies Emile Planchon, 1823–1888
Plum.	Charles Plumier, 1646–1704
Poepp.	E. F. Poeppig, 1798–1868
Poggenburg	Justus Ferdinand Poggenburg, 1840–1893
Pohl	Johann Emanuel Pohl, 1782–1834
Poir.	Jean Louis Marie Poiret, 1755–1834
Poit.	Antoine Poiteau, 1766–1854
Pollard	Charles Louis Pollard, 1872–1945
Porter	Thomas Conrad Porter, 1822–1901
Prance	Ghillean T. Prance, 1937–
Prantl	Karl Anton Eugen Prantl, 1849–1893
Presl	Karel Boriwog Presl, 1794–1852
Proctor	George Richardson Proctor, 1920–
Pursh	Frederick Traugott Pursh, 1774–1820
Raddi	J. (G.) Raddi, 1770–1829
Radlk.	Ludwig Adolph Timotheus Radlkofer, 1829–1927
Raf.	Constantine Samuel Rafinesque-Schmaltz, 1784–1842
Raven	Peter Hamilton Raven, 1936–
Regel	Eduard August von Regel, 1815–1892
Rehder	Alfred Rehder, 1863–1949
Reichenb.	Heinrich Gottlieb Ludwig Reichenback, 1793–1879
Reichenb. f.	Heinrich Gustav Reichenback, 1824–1889
Rendle	Alfred Barton Rendle, 1865–1938
Retz.	Anders Johan Retzius, 1742–1821
A. Rich.	Achille Richard, 1794–1852
Richard	Louis Claude Marie Richard, 1754–1821
Richt.	Karl Richter, 1855–1891
Riddell	John Leonard Riddell, 1807–1865
Risso	J. Antoine Risso, 1777–1845

Robinson	Benjamin Lincoln Robinson, 1864–1935
C. B. Robinson	Charles Budd Robinson, 1871–1913
N. Robson	Norman K. B. Robson, 1928–
H. Rock	Howard Francis Leonard Rock, 1925–1965
Roem.	Johann Jacob Roemer, 1763–1819
C. M. Rogers	Claude Marvin Rogers, 1919–
Rogers	David J. Rogers, 1918–
Rohrb.	Paul Rohrbach, 1847–1871
Roland.	Daniel Rolander, 1725–1793
Rolfe	Robert Allen Rolfe, 1855–1921
Rose	Joseph Nelson Rose, 1862–1928
Rostk.	Friedrich Wilhelm Theophil Rostkovius, 1770–1848
Roth	Albrecht Wilhelm Roth, 1757–1834
Rottb.	Christen Friis Rottboell, 1727–1797
Roxb.	William Roxburgh, 1751–1815
Royen	Pieter van Royen, 1923–
Royle	John Forbes Royle, 1799–1858
R & S	see Roem.; Schultes
Rudge	Edward Rudge, 1763–1846
Ruhland	Eugen Otto Willy Ruhland, 1878–
Ruiz	Hipolito Ruiz, 1764–1815
Rumph.	Georg Eberhard Rumphias (Rumpf), 1627 (?)–1702
Rusby	Henry Hurd Rusby, 1855–1940
Russell	Norman H. Russell, 1921–
Rydberg	Per Axel Rydberg, 1860–1931
Sadebeck	R. Sadebeck, 1839–1905
St. Hil.	Auguste Francois Cesar Prouvencal de Saint-Hilaire, 1779–1853
E. St. John	E. P. St. John, 1866–?
R. St. John	R. P. St. John, 1869–1960
Salisb.	Richard Anthony Salisbury, 1761–1829
Sarg.	Charles Sprague Sargent, 1841–1927
Sauer	Jonathon Deininger Sauer, 1918–
Savi	Gaetano Savi, 1769–1844
Schauer	Johann Conrad Schauer, 1813–1848
Sch.-Bip.	Carl Heinrich Schultz-Bipontinus, 1805–1867
Scheele	Georg Heinrich Adolph Scheele, 1808–1864
Schelpe	E. A. C. L. E. Schelpe, 1924–
Schery	Robert Walter Schery, 1917–
Schiede	Christian Juilus Wilhelm Schiede, 1798–1836
Schindler	Anton Karl Schindler, 1879–
Schinz	Hans Schinz, 1859–1941
Schk.	Christian Schkuhr, 1741–1811
Schlecht.	Diederich Franz Leonhard von Schlechtendahl, 1794–1866
Schleiden	Matthias Jacob Schleiden, 1804–1881
Schltr.	Friedrich Richard Rudolf Schlechter, 1872–1925
Schmidel	Casimir Christoph Schmidel, 1718–1792
Schmidt	Wilhelm Ludwig Ewald Schmidt, 1804–1843
Schott	Heinrich Wilhelm Schott, 1794–1865
Schrad.	Heinrich Adolph Schrader, 1767–1836

Schrank	Franz von Paula von Schrank, 1747–1835
Schreb.	Johann Christian Daniel von Schreber, 1739–1810
Schubert	Bernice Giduz Schubert, 1913–
Schultes	Josef August Schultes, 1773–1831
O. E. Schulz	Otto Eugene Schulz, 1874–1936
Schum.	Henrich Christian Friederich Schumacher, 1757–1830
Schumann	Karl Mortiz Schumann, 1851–1904
Scop.	Johann Anton Scopoli, 1723–1788
Scribn.	Frank Lamson-Scribner, 1851–1938
Seem.	Berthold Carl Seeman, 1825–1871
Senn	Harold Archie Senn, 1912–
Sessé	Martino de Sessé y Lacasta, 1775–1809
Sherff	Earl Edward Sherff, 1886–1966
Shinners	Lloyd H. Shinners, 1918–1971
Short	Charles Wilkins Short, 1749–1863
Shuttlew.	Robert James Shuttleworth, 1810–1874
Sibth.	John Sibthorp, 1758–1796
Sieb.	F. W. Sieber, 1785–1844
Simpson	J. H. Simpson, 1841–?
Sims	John Sims, 1749–1831
Skan	Sidney Alfred Skan, 1870–1940
Skeels	Homer C. Skeels, 1873–1934
Sm.	James Edward Smith, 1759–1828
Small	John Kunkel Small, 1869–1938
Donnell Smith	John Donnell Smith, 1829–1928
J. G. Smith	Jared Gage Smith, 1866–1925
L. B. Smith	Lyman Bradford Smith, 1904–
Sod.	A. Sodiro, 1836–1909
Solander	Daniel Carl Solander, 1733–1782
Solms	Hermann Maximilian Carl Ludwig Friedrich zu Solms-Laubach, 1842–1915
Spach	Edouard Spach, 1801–1879
Spreng.	Curt Polycap Joachim Sprengel, 1766–1833
Spring	Anton Friedrich Spring, 1814–1872
Stahl	Christian Ernst Stahl, 1848–1919
Standl.	Paul Carpenter Standley, 1884–1963
Stapf	Otto Stapf, 1857–1933
Stearn	William Thomas Stearn, 1911–
Steenis	Cornelis Gijsbert Gerrit Jan von Steenis, 1901–
Stern	F. C. Stern, 1884–1967
Sterns	Emerson Ellick Sterns, 1846–1926
Steud.	Ernst Gottlieb Steudel, 1783–1856
Stokes	Jonathan Stokes, 1755–1831
Stuckey	Ronald L. Stuckey, 1938–
Stuntz	Stephen Conrad Stuntz, 1875–1918
Sudw.	George Bishop Sudworth, 1864–1927
Svenson	Henry Knute Svenson, 1897–
Sw.	Olof Peter Swartz, 1760–1818
Swallen	Jason Richard Swallen, 1903–
Sweet	Robert Sweet, 1783–1835

Swingle	Walter Tennyson Swingle, 1871–1952
Taubert	Paul Hermann Wilhelm Taubert, 1862–1897
Tenore	Michele Tenore, 1781–1861
T & G	see Torr.; Gray
Thellung	Albert Thellung, 1881–1928
Thompson	Charles Henry Thompson, 1870–1931
Thon.	Franz Thonner, 1863–?
Thouars	Louis Marie Aubert Du Petit Thouars, 1758–1831
Thunb.	Carl Peter Thunberg, 1743–1828
Tidestrom	Ivar Tidestrom, 1864–1956
Torr.	John Torrey, 1796–1873
G. S. Torr.	George Sanford Torrey, 1891–
Trel.	William Trelease, 1857–1945
Trin.	Carl Bernhard von Trinius, 1778–1844
Tryon	Rolla Milton Tryon, 1916–
Turcz.	Nicolaus Turczaninow, 1796–1864
Turner	Billy Lee Turner, 1925–
Turp.	Pierre Jean Francois Turpin, 1775–1840
Underw.	Lucien Marcus Underwood, 1853–1907
Urban	Ignatz Urban, 1848–1931
Vahl	Margin Vahl, 1749–1804
Vail	Anna Murray Vail, 1863–?
Valet	Friedrich F. Valet, 1847–1865
Van Ooststr.	S. J. Van Ooststroom, 1906–
Vasey	George Vasey, 1822–1893
Vatke	Georg Carl Wilhelm Vatke, 1849–1889
Vellozo	Jose Marianno Vellozo de Conceicao, 1742–1811
Vent.	Etienne Pierre Ventenat, 1757–1808
Vict.	Frere Marie-Victorin (Conrad Kirouac), 1885–1944
Vilm.	P. L. F. L. de Vilmorin, 1816–1860
Viv.	Domenico Viviani, 1772–1840
B. Vogel	B. C. Vogel, 1745–1825
Wagen.	Burdette Lewis Wagenknecht, 1925–
Wall.	Nathaniel Wallich, 1786–1854
Walp.	Wilhelm Gerhard Walpers, 1816–1853
Walt.	Thomas Walter, 1740–1789
Ward	Daniel Bertram Ward, 1928–
Wasshausen	Dieter Carl Wasshausen, 1938–
Waterfall	Umaldy Theodore Waterfall, 1910–
Wats.	Sereno Watson, 1826–1892
Watt	David Allan Poe Watt, 1830–1917
Weatherby	Charles Alfred Weatherby, 1875–1949
Webster	Grady Linder Webster, 1927–
Wedd.	Hugh Algernon Weddell, 1819–1877
Weigel	Christian Ehrenfried von Weigel, 1748–1831
Welw.	Friedrich Martin Joseph Welwitsch, 1806–1831
Wenderoth	Georg Wilhelm Franz Wenderoth, 1774–1861
Wendl.	Heinrich Ludolph Wendland, 1791–1869
L. Wheeler	Louis Cutter Wheeler, 1910–
Wheelock	William Efner Wheelock, 1852–1926

Wherry	Edgar Theodore Wherry, 1885–
Wieg.	Karl McKay Wiegand, 1873–1942
Wight	Robert Wight, 1796–1872
Wilbur	Robert Lynch Wilbur, 1925–
Willd.	Carl Ludwig Willdenow, 1765–1812
Willem.	P. R. F. de Paule Willemet, 1762–1790
Wilson	Kenneth Allen Wilson, 1928–
P. Wilson	Percy Wilson, 1879–1944
Winkl.	Hubert Winkler, 1875–1941
L. O. Wms.	Louis Otho Williams, 1908–
Wood	Alphonso Wood, 1810–1881
C. E. Wood	Carroll Emory Wood, Jr., 1921–
Woodson	Robert Everard Woodson, 1904–1963
Woodv.	W. Woodville, 1752–1805
C. Wright	Charles Wright, 1811–1885
Yuncker	Truman George Yuncker, 1891–1962
Zoll.	Heinrich Zollinger, 1818 (?)–1859
Zucc.	Joseph Gerhard Zuccarini, 1797–1848

Appendix B
Abbreviations

Ala.—Alabama
Ariz.—Arizona
Ark.—Arkansas
B.C.—British Columbia
Calif.—California
cm—centimeter
Colo.—Colorado
Conn.—Connecticut
Del.—Delaware
dm—decimeter
Fla.—Florida
Fla. Keys—Florida Keys
Ga.—Georgia
Ill.—Illinois
Ind.—Indiana
Kans.—Kansas
Ky.—Kentucky
La.—Louisiana
m—meter
Mass.—Massachusetts
Me.—Maine
Md.—Maryland
mm—millimeter
Minn.—Minnesota
Miss.—Mississippi

Mo.—Missouri
N.C.—North Carolina
Nebr.—Nebraska
N.J.—New Jersey
N. Mex.—New Mexico
N. Dak.—North Dakota
No.—Number
nos.—numbers
N.Y.—New York
Okla.—Oklahoma
Pa.—Pennsylvania
S.C.—South Carolina
S. Dak.—South Dakota
south Fla.—south Florida
sp.—species
ssp.—subspecies
Tenn.—Tennessee
Tex.—Texas
U.S.—United States
Va.—Virginia
var.—variety
Vt.—Vermont
W.I.—West Indies
Wis.—Wisconsin

abscíssus—cut off, steep, precipitous (L. adj.)
aculeàtus—prickly, spinelike (from L. *aculeus*)
adpréssus—appressed, lying flat against (from L. *ad* + *pressus*)
affìnis—neighboring, allied to, related to (from L. *adfinis*)
alàtus—winged (from L. *ala*)
álbicans—whitish, becoming white (from L. *albicare*)
albiflòrus—white flowered (from L. *albus* + *flos*)
álbus—white, particularly a dull rather than a glossy white (L. adj.)
alternifòlius—alternate leaves (from L. *alternus* + *folium*)
altìssimus—very tall, tallest (L. adj.)
amábilis—lovely (L. adj.)
ánceps—two-edged (L. adj.)
angustifòlius—narrow leaves (from L. *angustus* + *folium*)
ánnuus—annual, living but one year (L. adj.)
ápiculatus—apiculate, tipped with a point (from L. *apex*)
arboréscens—becoming treelike (from L. *arbor*)
arenicola—sand dweller (from L. *arena* + *-cola*)
argenteùs—silvery (from L. *argent*)
articulàtus—articulate, jointed (from L. *articulus*)
ascéndens—ascending (from L. *ascendere*)
atropùrpureus—dark purple (from L. *atrans* + *purpureus*)
aùreus—golden yellow (from L. *aurum*)
auriculàtus—auriculate, furnished with an earlike appendage (from L. *auris*)
austràlis—southern (L. adj.)
barbadénsis—of Barbados
barbàtus—bearded, provided with tufts of long weak hairs (from L. *barba*)
bicolor—two-colored (L. adj.)
biénnis—biennial, living only two years (from L. *biennium*)
biserràtus—doubly serrate, the teeth being themselves toothed (L. adj.)

brevicaùlis—short stemmed (from L. *brevis + caulis*)
brevifòlius—short leaved (from L. *brevis + folium*)
brévis—short (L. adj.)
bulbifera—bulb-bearing (from L. *bulbus + ferre*)
buxifòlius—with oval leaves, resembling those of boxwood (from L. *buxus + folium*)
callòsus—thick-skinned, with callosities (L. adj.)
campéstris—of the fields or plains (from L. *campester*)
canadénsis—Canadian, of Canada
cándidus—white, white hairy, shining (L. adj.)
carinàtus—keeled (from L. *carina*)
caroliniànus—Carolinian, pertaining to North or South Carolina
cathárticus—cathartic (from Gr. *katharos*)
caudàtus—caudate, having a tail-like appendage (from L. *cauda*)
cauléscens—having a stem or stems (from L. *caulis*)
cérnuus—slightly drooping (L. adj.)
chilénsis—Chilean, of Chile
chinénsis—Chinese, of China
chrysophyllus—golden leaved (from Gr. *chryso + phyllon*)
ciliàris—fringed with hairs on the margin (from L. *cilium*)
ciliatifòlius—bearing hairs on the leaf margin (from L. *ciliatus + folium*)
cinèreus—ash gray (L. adj.)
cirrhòsus—tendriled (from L. *cirratus*)
cláusum—closed (L. adj.)
clavàtus—clavate, club shaped (from L. *clava*)
coccifera—bearing berries (from L. *coccum + ferre*)
cochleàtus—spoonlike (from L. *cocleare*)
coloràtus—colored (other than green) (L. adj.)
condensàtus—condensed (used to describe inflorescences with numerous flowers on short pedicals, hence very close to the axis) (L. adj.)
confùsus—confused, uncertain as to characteristics (from L. *confundere*)
conjugàtus—joined, connected (L. adj.)
corallícola—living on coral or in soil derived from coral (from L. *coral + cola*)
corállinus—coral red (from L. *corallum*)
cordàtus—cordate, heart shaped (from L. *cor*)
cornùtus—horn-bearing (from L. *cornus*)
crassifòlius—with dense foliage (from L. *crassus + folium*)
crenulàtus—crenulate, having small rounded teeth (from L. *crena*)
cubénsis—Cuban, of Cuba
cymòsus—having a cyme or cymes (from L. *cymis*)
dactylòides—fingerlike (from Gr. *dactylos*)
débilis—weak (L. adj.)
dentàtus—toothed, usually with sharp teeth pointing outward (from L. *dens*)
dichótomus—forked, two-branched equally (from Gr. *dicho-*)
diffórmis—of different forms (L. adj.)
discolor—of two colors or of different colors (L. adj.)
dissèctus—dissected, deeply divided or cut into many segments (from L. *dissecare*)
distichus—distichous, arranged in two opposite rows, two-ranked (L. adj.)
divaricàtus—divaricate, spreading asunder at a wide angle (L. adj.)

diversifòlius—with leaves of different shapes of the same plant (from L. *diversus* + *folium*)

dùbius—doubtful (L. adj.)

dúlcis—sweet (L. adj.)

ebracteàtus—without bracts (from L. *bractea*)

echinàtus—armed with numerous rigid hairs or straight prickles or spines (from L. *echinus*)

elàtus—tall (L. adj.)

esculéntus—edible (from L. *esca*)

exaltàtus—raised high, lofty (L. adj.)

fàllax—deceptive (L. adj.)

fasciculàris—fascicled, clustered, brought together (from L. *fasciculus*)

fèrox—ferocious, very thorny (L. adj.)

ferrùgineus—rusty, light brown with a little mixture of red (from L. *ferrum*)

filifòrmis—threadlike (from L. *filum* + *forma*)

fimbriàtus—fringed (L. adj.)

flagellàtus—flagellate, whiplike (from L. *flagellum*)

flavéscens—yellowish, pale yellow (from L. *flavus*)

flàvidus—yellow, yellowish (from L. *flavus*)

flexuòsus—zigzag, bent alternately in opposite directions (L. adj.)

floribúndus—free-flowering, booming profusely (from L. *flors*)

flóridus—flowering, full of flowers (L. adj.)

flùitans—floating, swimming (from L. *fluere*)

fòetidissimus—extremely evil-smelling, stinking (from L. *foetens*)

foliòsus—leafy, full of leaves (from L. *folium*)

frutéscens—becoming shrubby (from L. *frutex*)

fruticòsus—shrubby, bushy (from L. *frutex*)

fúlgens—shining, glistening (from L. *fulgare*)

geniculàtus—geniculate, bent abruptly like a knee (from L. *geniculum*)

gigantèus—gigantic, very large (from Gr. *gigas*)

gìgas—of giants, immense (from Gr. *gigas*)

glabér—smooth, without hair (L. adj.)

glaùcus—with a bloom, grayish (L. adj.)

grácilis—slender, thin, slim, graceful (L. adj.)

gramíneus—grassy, grasslike (L. adj.)

hamàtus—hooked (L. adj.)

helvéticus—Swiss, of Switzerland (from L. *Helvetia*)

heteromórphus—various in form (from Gr. *heteros* + *morphe*)

heterophyllus—with leaves of more than one shape (from Gr. *heteros* + *phyllon*)

humifùsus—spread out over the ground, procumbent (from L. *humus* + *fundere*)

hùmilis—low-growing, dwarf (L. adj.)

hystrix—porcupinelike, bristly (from L. *hystrix*)

impléxus—implicated, interwoven (L. adj.)

incànus—hoary, gray (L. adj.)

indicus—Indian, of India or the East Indies

inflàtus—inflated, swollen up (L. adj.)

integér—entire (L. adj.)

involucràtus—with an involucre (L. adj.)

jamaicénsis—of Jamaica

japónicus—Japanese, of Japan

jasmíneus—jasminelike
laevigàtus—smooth; free from unevenness, hairs, or roughness (from L. *levis*)
lancifòlius—with lanceolate leaves (from L. *lancea*)
latifòlius—broad leaved (from L. *latus*)
làtus—wide, broad (L. adj.)
leiocárpus—smooth fruited (from Gr. *leios* + *carpos*)
lepidophýllus—scaly leaved (from Gr. *lepis* + *phyllon*)
léptopus—thin or slender stalked (from Gr. *leptos*)
leucánthus—white flowered (from Gr. *leucon* + *anthos*)
lineàris—linear (from L. *lineare*)
lineolàtus—marked by fine parallel lines (L. adj.)
littorális—of the seashore (from L. *litoralis*)
longiflòrus—long flowered (from L. *longus* + *flos*)
longifòlius—long leaved (from L. *longus* + *folium*)
lóngipes—long footed; long stalked (from L. *longus* + *pes*)
lùcidus—lucid, bright, shining, clear (L. adj.)
lùdovicianus—of Louisiana
lùridus—lurid, sallow, pale yellow (L. adj.)
lùtescens—yellowish, becoming yellow (from L. *luteus*)
lùteus—yellow (L. adj.)
lyràtus—lyrate, pinnatifid with large terminal lobe (from L. *lyra*)
macránthus—large flowered (from Gr. *macros* + *anthos*)
macrospérmus—large seeded (from Gr. *macros* + *sperma*)
maculàtus—spotted (from L. *macula*)
mágnus—large (L. adj.)
marínus—marine, of the sea (L. adj.)
marítimus—maritime, of the sea (L. adj.)
máximus—largest (L. adj.)
megacéphalus—large headed (from Gr. *mega* + *cephale*)
mellífera—honey-bearing (L. adj.)
mexicànus—Mexican, of Mexico
microcéphalus—small headed (from Gr. *micro* + *cephale*)
mìnor—smaller (L. adj.)
móllis—soft; soft hairy (L. adj.)
monostáchyus—single-spiked (from Gr. *monos* + *stachys*)
multiflòrus—many-flowered (from L. *multus* + *flos*)
nànus—dwarf (L. noun)
nìger—black (L. adj.)
nítidus—shining (L. adj.)
noctiflòrus—night-flowering (from L. *nox* + *flos*)
noctúrnus—nocturnal, of the night (L. adj.)
nodòsus—with nodes; jointed (L. adj.)
notàtus—marked (from L. *nota*)
nucíferus—nut-bearing (from L. *nux*)
nùdus—naked, nude (L. adj.)
nutáns—nodding (from L. *nutare*)
obscùrus—obscure, hidden (L. adj.)
occidéntalis—occidental, western (L. adj.)
odoràtus—odorous, fragrant (from L. *odor*)
officinàlis—officinal, medicinal (from L. *officium*)

oleràceous—vegetable-garden herb used in cooking (L. adj.)
ornithorhynchus—shaped like a bird's beak (from Gr. *ornis* + *rhynchos*)
ovàlis—oval (from L. *ovum*)
pacificus—of the Pacific, of regions bordering the Pacific Ocean
pàllens—pale (L. adj.)
pàllidus—pale (L. adj.)
palùstris—marsh-loving (from L. *palus*)
pàtens—spreading (from L. *patere*)
pauciflòrus—few-flowered (from L. *paucus* + *flos*)
pectinàtus—comblike, pinnatifid with many narrow divisions close together (from L. *pectinatim*)
pectoràlis—shaped like a breastbone (from L. *pectus*)
pedàtus—footed, of the foot (from L. *pes*)
perénnis—perennial, living three or more years (L. adj.)
pilòsus—pilose, shaggy with soft hairs (L. adj.)
pinetòrus—of pine forests (from L. *pinus*)
pinnàtus—pinnate, with leaves divided to the midrib into leaflets (from L. *pinna*)
plùmulus—feathery, plumed (from L. *pluma*)
polyánthes—many-flowered (from Gr. *poly* + *anthos*)
polydáctylus—many-fingered (from Gr. *poly* + *dactylos*)
polystáchyus—many-spiked (from Gr. *poly* + *stachyos*)
praècox—precocious, premature, very early (L. adj.)
propinquus—related, near to (L. adj.)
pudicus—bashful, retiring, shrinking (L. adj.)
pulchéllus—very pretty (L. adj.)
pùmilus—dwarf (L. noun)
punctàtus—punctate, dotted (from L. *pungere*)
purpuràscens—purplish, becoming purple (L. adj.)
pygmaeùs—pigmy (L. adj.)
quercifòlius—with leaves like those of oak (from L. *quercus*)
radìcans—rooting (from L. *radix*)
ramòsus—branched (L. adj.)
ràrus—rare, uncommon (L. adj.)
réctus—straight, upright (L. adj.)
recurvàtus—recurved (from L. *recurvare*)
religiòsus—used for religious purposes; venerated (L. adj.)
renifòrmis—reniform, kidney shaped (from L. *renes*)
rèpens—repent, creeping (from L. *rependere*)
rhizomàtous—with a well-developed rhizome (from Gr. *rhiza*)
rigidifòlius—with stiff leaves (from L. *rigidus* + *folium*)
rigidus—stiff, unbending (L. adj.)
ringens—gaping, open mouthed (from L. *ringor*)
ripàrius—of river banks (from L. *ripa*)
ròseus—rose, rosy (from L. *rosa*)
rotúndus—round, rotund (L. adj.)
rùber—red, ruddy (from L. *ruber*)
salicifòlius—willow leaved (from L. *salix* + *folium*)
sanguinàlis—blood red (from L. *sanguis*)
satìvus—cultivated (from L. *satio*)

scàber—scabrous, rough (L. adj.)

scándens—scandent, climbing (from L. *scandere*)

schizopétalus—with cut or incised petals (from Gr. *schistos* + *petalon*)

schizophýllus—leaves cut or incised (from Gr. *schistos* + *phyllon*)

scúlptus—carved (from L. *sculpere*)

secúndus—secund, side-flowering (L. adj.)

sempérvirens—evergreen (from L. *semper* + *virere*)

serìceus—silky (L. adj.)

serótinus—late; late-flowering or late-ripening (from L. *serus*)

serrulàtus—somewhat serrate (from L. *serrula*)

setàceus—bristlelike (from L. *saeta*)

siliceous—pertaining to or growing in sand (from L. *silex*)

simplex—simple, unbranched (L. adj.)

simulans—similar to, resembling (from L. *simulare*)

sinénsis—Chinese

speciòsus—showy, good-looking (L. adj.)

spectábilis—spectacular, showy (L. adj.)

sphaerocàrpus—spherical fruited (from Gr. *sphaera* + *carpos*)

spicifòrmis—shaped like a spike (from L. *spica*)

spinòsus—full of thorns, prickly (L. adj.)

squamòsus—full of scales (L. adj.)

stolónifer—bearing stolons or runners that take root (from L. *stolo* + *ferre*)

striàtus—striated, striped (from L. *stria*)

strictus—strict, upright, erect (L. adj.)

strìgosus—covered with sharp straight appressed hairs (L. adj.)

strobiliferus—cone-bearing (from Gr. *strobilus*)

succuléntus—succulent, fleshy (from L. *sucus*)

supìnus—bent backward, prostrate (L. adj.)

tènax—tenacious, strong (L. adj.)

tèner—tender, soft, delicate (L. adj.)

tenuíssimus—very slender, very thin (L. adj.)

ternifòlius—with leaves in three (from L. *terni*)

terréstris—of the earth, terrestrial (from L. *terra*)

tigrìnus—tiger striped (from L. *tigris*)

tinctòrius—of dyes (from L. *tingere*)

tinctús—dyed (from L. *tingere*)

tortuòsus—much twisted (L. adj.)

toxíferus—poison-producing (from L. *toxicum*)

trichophýllus—hairy leaved (from Gr. *trich* + *phyllon*)

trinérvis—three-nerved (from L. *tria* + *nervus*)

tripartìtus—three-parted (L. adj.)

triviàlis—common, ordinary, found everywhere (L. adj.)

tuberòsus—tuberous (from L. *tuber*)

tubiflòrus—tube flowered, trumpet flowered (from L. *tubus*)

tulipíferus—tulip-bearing (from L. *ferre*)

týpicus—typical, conforming to the standard or norm (from Gr. *typos*)

umbròsus—shaded, shade-loving (from L. *umbra*)

uncinàtus—hooked at the point (L. adj.)

undàtus—waved or wavy (from L. *unda*)

uniflòrus—one-flowered (from L. *unus* + *flos*)

ùrens—burning, stinging (from L. *urere*)

ùtilis—useful (L. adj.)

utriculàtus—with a utricle or small, bladdery, one-seeded fruit (from L. *utricularius*)

uvìferus—grape-bearing (from L. *uva*)

vaginàlis—vaginate, sheathed (from L. *vagina*)

vàlidus—strong (L. adj.)

variegàtus—variegated, in different colors (from L. *varius*)

venòsus—veiny, full of veins (L. adj.)

vernàlis—vernal, of the spring (L. adj.)

verrucòsus—warted, verrucose (from L. *verruca*)

vèrus—the true or genuine or standard (L. adj.)

villòsus—villous, soft hairy (L. adj.)

virgàtus—twiggy, long and slender (from L. *virga*)

viridis—green (L. adj.)

volùbilis—twining (L. adj.)

vulgàris—vulgar, common (L. adj.)

xanthocàrpus—yellow fruited (from Gr. *xanthos* + *carpos*)

zonàlis—zonal, zoned (from L. *zona*)

abaxial—the side of a lateral organ away from the axis; dorsal.

acaulescent—without a leafy stem; stemless or apparently so.

accrescent—increasing in size after flowering, as may the calyx in some plants; enlarging with age.

achene—a dry one-seeded indehiscent fruit; pericarp free from testa.

acicular—needle shaped.

acrostichoid—sporangia confluent, covering the fertile surface.

acuminate—tapering gradually to a pointed apex.

acute—tapering abruptly to a sharp point.

adaxial—the side of a lateral organ next to the axis; ventral.

adnate—grown together; the fusion of unlike parts.

adventitious—descriptive of buds, roots, which originate from an unusual location on the plant.

aerenchyma—a tissue of thin-walled cells and large intercellular spaces.

aestivation—the arrangement of flower parts in the bud.

aggregate fruit—one formed by the coherence of pistils that were distinct in the flower, as in the blackberry.

allopolyploid—a polyploid derived usually from interspecific hybridization; at least one set of chromosomes originated from an unrelated taxon.

alternate—borne one at a node, appearing on one side of the axis and then on the other.

alveolate—the surface marked as though honeycombed.

ament—a catkin; a close, bracteate spike of usually unisexual flowers, as in the willow.

anastomosing—connecting by cross veins and forming a network.

anatropous—the ovule inverted, with micropyle close to the side of the hilum and the chalaza at the opposite end.

androecium—a collective term for the stamens of a single flower.

androgynophore—a stalk bearing both stamens and pistil above the point of perianth attachment.

androphore—a stalk on which are borne all the stamens (androecium) of the flower, as in *Hibiscus*.

anemophilous—term applied to wind-pollinated flowers.

aneuploid—having a chromosome number which is not an exact multiple of the haploid number.

annual—completing a life cycle in one year.

annular—in the form of a ring; circular.

annulus—a group of thick-walled cells on the sporangium of ferns, the fleshy rim of the corolla in the milkweeds.

anther—the pollen-bearing part of the stamen, supported by the filament or sometimes sessile, and usually a bilocular sac.

antheridium—the organ which in many lower plants produces the male gametes or sperm.

anthesis—the expansion of the flower from the bud.

antrorse—directed upward, opposed to retrorse.

apetalous—lacking petals.

apiculate—terminating in a short, sharp flexible point.

apomict—in general, a plant produced without fertilization.

apomixis—a term for many types of reproduction in which there is no fusion of male and female gametes.

appressed—pressed flat against another organ.

approximate—drawn close together but not united.

arachnoid—cobwebby by soft and slender entangled hairs; also spiderlike.

archegonium—the multicellular, egg-containing female organ of many lower plants.

arcuate—bent like a bow, curved.

areola—a small depression or cavity in an exterior surface.

aril—a fleshy expansion of the funiculus, arising from the placenta and enveloping the seed.

aristate—awned; having bristlelike appendages.

articulate—jointed; having nodes or joints or places where two parts may easily separate.

attenuate—gradually becoming very narrow or slender.

auriculate—eared; auricled.

autopolyploid—a polyploid originating from the multiplication of the chromosome set of an individual.

awn—a bristlelike appendage.

axil—the upper angle formed between the axis and any organ which arises from it, e.g., of a leaf.

axillary—in or arising from an axil.

axile placentation—ovules borne on the central axis of a compound ovary, usually in vertical rows.

baccate—berrylike.

banner—the standard or the broad upper petal in a papilionaceous flower.

barbate—having long weak hairs in tufts; bearded.

barbed—bristles or awns provided with retrorse spinelike hooks.

barbellate—finely barbed.

basifixed—attached by the base.

berry—pulpy simple fruit developed from a single ovary and having one to many seeds.

biennial—completing life cycle in two years.

bifid—two-cleft.

bifurcate—twice-forked, as some Y-shaped hairs, stigmas, or styles.

bilabiate—two-lipped, often applied to a corolla or calyx.

bilateral symmetry—capable of bisection on one plane only to produce two mirror images.

binomial—scientific name of two terms, genus and species.

biotype—all the individuals having the same genotype; usually a single individual represents one biotype in cross-fertilizing populations.

bipartite—divided nearly to the base into two portions.

bipinnatifid—when the divisions of a pinnatifid leaf are themselves pinnatifid.

blade—the expanded part of a leaf or petal.

bract—modified leaf subtending a flower or belonging to an inflorescence or stem.

bractlet—a bract borne on a secondary axis, as on the pedicel or even on the petiole; bracteole.

bud—an undeveloped shoot or flower often surrounded by scales.

bulb—a short underground stem with enclosing fleshy leaves or scales, as an onion.

bulbil or *bulblet*—diminutive of bulb, especially those borne on a stem or in an inflorescence.

caducous—falling off of parts before the remaining ones mature.

callus—a hard swollen place, as the base of lemma in grass flowers or the base of the column in certain orchid flowers.

calyx—usually the lower and outermost series of floral parts; the sepals, collectively.

campanulate—bell shaped.

canaliculate—with one or more grooves running longitudinally.

cancellate—resembling latticework, as the pattern formed by the veins on the lower surfaces of the leaves in some water plants when the parenchyma is wholly absent.

canescent—gray pubescent or hoary, or becoming so.

capillary—hairlike, very slender, as capillary bristles.

capitate—compacted into a very dense cluster, or head, as the inflorescence of aster and clover.

capsule—a dry fruit resulting from the maturing of a compound ovary usually opening at maturity by one or more lines of dehiscence.

carinate—keeled; having a projecting longitudinal ridge on the undersurface.

carpel—a foliar unit of the angiosperm flower which bears the ovules; one or several comprises the ovary.

cartilaginous—tough and hard but not bony; gristly.

caruncle—a protuberance near the hilum of a seed.

caryopsis—a single-seeded, indehiscent fruit with fused testa and pericarp; term usually restricted to the fruit of grasses.

castaneous—chestnut brown.

catkin—an ament; a bracted spike of unisexual, apetalous flowers. (cf. *ament*)

caudate—bearing a tail-like appendage, as some leaf apices.

caulescent—having an evident stem above ground.

cauline—belonging to the stem, especially to an elongated stem or axis, as opposed to basal or rosulate.

cernuous—drooping or nodding.

cespitose—growing in clumps or tufts, especially of low plants, e.g., grasses.

channeled—hollowed out like a gutter, as in many leaf stalks.

chartaceous—papery.

chasmogamous—an open flower, opposite of cleistogamous.

chromosome—chromatin threads that bear genes (hereditary determiners) in the nucleus of a cell.

ciliate—bearing fine hairs uniformly, as on the margin of a leaf, petal, or flattened fruit.

ciliolate—minutely ciliate.

cinereous—ash colored; light gray.

circinate—coiled or rolled, with the apex at the center of the coil, as the immature fronds of many ferns.

circumscissile—dehiscing by a line around the entire circumference of a fruit or anther, the upper portion usually coming off as a lid.

cladode—a branch functionally replacing the green leaf.

clathrate—latticed, or pierced with apertures.

clavate—club shaped; a structure thickened at free end.

claw—the long narrow base of petals or sepals abruptly dilating to a limb.

cleft—incised, slashed sharply to or about the middle into divisions.

cleistogamous flower—a small, closed, self-fertilized flower.

clinandrium—in the orchids that part of the column in which the anther is embedded and concealed.

coalescent—union by growing together. See also *adnate*.

column—body formed by fusion of stamen, style, and stigmas in the orchids.

colliculate—surface covered with little rounded elevations.

coma—a tuft of soft hairs at the apices or edges of some seeds.

complanate—flattened, compressed.

compound leaf—a leaf of two or more leaflets or subdivisions on a common petiole.

compound pistil—two or more carpels coalescent into one body.

concolor—uniform in tint.

conduplicate—folded together lengthwise.

cone—a floral axis with imbricate scales (sporophylls) forming a strobilus; *staminate cone*—containing only those sporophylls associated with anther sacs; *ovulate cone*—containing only those sporophylls associated with ovules.

confluent—merging or blending together.

connivent—coming together or converging but not fused, the parts often arching.

conoidal—nearly conical.

contiguous—the condition in which parts, like or unlike, touch each other but do not fuse.

contorted—twisted; convolute (in aestivation).

cordate—heart shaped, with the notch and lobes basal, as in many leaves.

coriaceous—leathery.

corm—thickened base of a primary stem.

corolla—collective term for the petals of a flower; usually the inner whorl of perianth parts, sometimes absent.

corona—a crown within the corolla.

corrugated—wrinkled or minutely furrowed.

corymb—a flat-topped indeterminate inflorescence with the outer flowers opening first.

corymbose—arranged in corymbs.

costa—a rib, the midvein of a simple leaf.

crenate—scalloped or shallowly round toothed on the margin.

crested—with elevated and irregular or toothed ridge.

crustaceous—of brittle texture, as some lichens.

cucullate—hooded or hood shaped.

culm—the stems of grasses, sedges, and similar plants.

cultivar—a horticultural or garden variety.

cuneate—wedge shaped with the apex at the point of attachment.

cupulate—having a cupule or involucre of many imbricate, closely appressed
  whorls of bracts, as in the acorn.

cuspidate—tipped with a sharp, rigid point.

cymbiform—boat shaped.

cyme—a broad, more or less flat-topped determinate inflorescence with the central flowers opening first.

cymose—bearing cymes or relating to cymes.

cystolith—mineral concretions usually of calcium carbonate on a cellulose stalk,
  in the epidermis of some plants.

deciduous—falling after one season's growth; not persistent.

declinate—bent downward or forward, the tips often recurved.

decompound—more than once compound or divided.

decumbent—stems leaning or reclining on the ground with ascendent tips.

decurrent—the base extending below ordinary point of insertion and continu-
  ing down the stem as a wing or ridge or leaflike tissue.

decussate—opposite leaves alternating in pairs at right angles on stem.

deflexed—abruptly bent or turned downward; reflexed.

dehisce—to open spontaneously when ripe, as occurs in the anthers and in some
  seed capsules; *dehiscent*—opening or dehiscing when ripe.

deltoid—shaped like an equilateral triangle; deltalike.

dentate—with sharp, spreading, rather coarse indentations or teeth that are
  perpendicular to the margin.

denticulate—minutely or finely toothed, with teeth projecting outward.

depauperate—stunted; impoverished as if starved.

diadelphous—stamens fused by their filaments into two often unequal groups.

dichotomous—forking and reforking into two subequal or equal branches.

didymous—(1) divided into two lobes; (2) found in pairs.

didynamous—with four stamens, in two pairs, usually of different lengths.

digitate—with several distinct parts all arising from a common center and radiating out in one plane, as the fingers of a hand.

dimorphic—occurring in two forms as in some ferns which have sterile fronds
  and fertile fronds which are different in appearance.

dimidiate—halved, as when half an organ is so much smaller than the other
  half as to seem wanting.

diaphanous—translucent or transparent structure, as petals of *Callisia*.

dioecious—pertaining to a species having staminate and pistillate flowers borne
  on different individuals.

diploid—a species having the 2*n* complement of chromosomes in its somatic cells.

discoid—having only disk flowers in the head.

disk—a more or less fleshy or elevated part of the receptacle in the head of Asteraceae.

disk floret—the tubular central flowers on the receptacle of the head of Asteraceae.

dissected—divided rather deeply into narrow segments.

distichous—two-ranked; arranged vertically on opposite sides of a stem in the same plane.

divaricate—spreading very far apart; at a wide angle away from the axis.

dorsal—the back; the side away from the axis, as the lower side of a leaf; opposite of ventral.

dorsifixed—attached by the back.

downy—covered with very short, fine, and weak soft hairs.

drupe—fleshy, one-seeded, indehiscent fruit with the seed enclosed in a stony endocarp.

ebracteate—without bracts.

ecarinate—not keeled.

echinate—with stout, rather blunt prickles.

ecospecies—a group of plants comprised of one or more ecotypes and whose members are able to hybridize without detriment to the offspring.

ecotone—a transitional zone between two plant communities.

ecotype—the basic unit in biosystematic classification; a population distinguished by morphological and physiological characters that are frequently of a quantitative nature; interfertile with other ecotypes of the same ecospecies but prevented from fully exchanging genes by ecological factors.

edaphic—pertaining to the influence of the soil on the plants growing upon it.

ellipsoid—an elliptical solid.

elliptic—oblong with regularly rounded ends.

emarginate—with a shallow notch at the apex.

endemic—a plant having one restricted region or area of distribution.

endosperm—the reserve food which is stored around the embryo in the seed but is not part of embryo.

ensiform—sword shaped, as the leaves of *Iris*.

entire—descriptive of a margin without teeth, lobes, or divisions.

entomophilous—referring to insect-pollinated flowers.

ephemeral—lasting for a brief period, often one day or less.

epigynous—with flower parts borne on the ovary and fused to it.

epipetalous—adnate to or arising from the petals.

epiphyte—a plant growing upon another plant but deriving no nutrition from it.

epispore—spore provided with a roughened or smooth outer coat.

equitant—two-ranked and overlapping or sheathing, as the leaves of *Iris*.

erose—the margin appearing eroded or jagged or irregular.

evanescent—soon disappearing; lasting only a short time.

even-pinnate—referring to a leaf rachis terminating without a leaflet.

evergreen—remaining green in its dormant season; a plant which is green throughout the year; leaves persistent.

excurrent—extending beyond the margin or tip, as a midrib developing into a spine or awn; also, a plant with a continuous unbranched axis.

exindusiate—indusium lacking.

exocarp—the outer layer of the pericarp or fruit wall.

exserted—projecting beyond some surrounding structure, as stamens projecting beyond the corolla tube.

extrorse—facing outward from the axis, usually referring to anther dehiscence when away from the center of the flower.

falcate—sickle shaped.

farinose—covered with a white meal-like powder.

fascicle—a close cluster or bundle, e.g., a fascicle of pine needles.

fastigiate—with branches erect and more or less appressed.

ferruginous—rust colored.

fetid—having a disagreeable odor.

fibrillous—finely lined fibrous appearance to the surface.

filament—thread; referring particularly to the stalk of the stamen.

filiform—threadlike; long and very slender.

fimbriate—fringed along the margin.

fistulose—hollow; lacking pith.

flabellate—fan shaped, as a palmetto frond.

flaccid—lax, weak.

flavescent—yellowish; pale yellow.

flexuous—having a more or less zigzag or wavy form, usually referring to stems.

floccose—covered with tufts of soft woolly hairs.

floral tube—an elongated, slender floral cup bearing on its rim the sepals, petals, and stamens.

floret—a small flower, one of a cluster, as in Asteraceae and Poaceae.

floriferous—flower-bearing.

foliaceous—leaflike in texture and color, said particularly referring to sepals and calyx lobes or bracts that resemble small leaves in form.

foliolate—having leaflets, as in trifoliolate, of three leaflets.

follicle—a dry, dehiscent, unicarpellate fruit opening only on the dorsal suture.

free—distinct; not adnate or adherent to other organs, as petals free from the stamens or calyx.

frond—leaf of fern, also used for the foliage of palms.

fruit—the ripened ovary containing the seed(s).

frutescent—tending to be shrubby.

funnelform—pertaining to a gamopetalous corolla limb, gradually expanding outward.

fusiform—spindle shaped; long and narrowly elliptical with pointed ends.

gametophyte—the generation which produces the gametes; in ferns a minute body bearing antheridia and archegonia (sex organs) and in flowering plants reduced usually to the three-nucleate pollen tube and the eight-nucleate embryo sac and its contents.

gamopetalous—with a corolla consisting of petals that are fused at least at their base.

geniculate—bent abruptly, as a knee.

genus—a group of allied species under a single name, presumably of common origin, or an isolated species with marked differentiation (a monotypic genus).

gibbous—swollen on one side, usually basally.

glabrate—nearly glabrous, or becoming glabrous with maturity or age.

glabrescent—becoming glabrous or slightly so.

glabrous—without hairs or trichomes.

gland—a secreting organ or part; sometimes used in the sense of a glandlike body.

glandular-punctate—with translucent or colored dots or depressions or pits.

glaucous—covered with a waxy bloom.

globose—spheroidal; nearly spherical.

glochid—a barbed hair or bristle, often in tufts, as in many cacti.

glomerate—a compact cluster of flowers.

glomerule—a compacted cyme.

glume—one of the subtending bracts of the spikelet in grasses.

glutinous—sticky.

granular—covered with minute grains of hardened material; finely mealy.

gynandrous—when the stamens are adnate to the pistil, as in orchids.

gynoecium—a collective term for all the pistillate parts of a single flower.

gynophore—stipe or stalk of an ovary prolonged within the perianth.

gynostegium—the staminal crown in *Asclepias*.

habit—the general appearance of the plant, whether erect, prostrate, climbing, etc.

halophyte—a plant tolerant of various mineral salts, usually of sodium chloride in the soil solution.

haploid—an organism with a single set of chromosomes, the *n* generation.

hastate—having the shape of an arrowhead but with the basal lobes pointed or narrow and standing nearly or completely at right angles.

haustoria—the absorbing organs (often rootlike) of parasitic plants.

head—a right cluster of sessile or essentially sessile flowers or fruits at the apex of a peduncle; the inflorescence is indeterminate.

herb—a plant without persistent woody stems or without a definite woody structure.

herbaceous—not woody; stem dying back each year; also referring to soft branches before they become woody.

heterosporous—producing two kinds of spores, representative of two sexes.

hilum—in the seed the scar indicating the point of attachment to the funiculus.

hirsute—pubescent, the surface covered with long, somewhat coarse hairs.

hirtellous—hirsute, but with hairs shorter and softer.

hispid—rough pubescent, with stiff hairs standing more or less erect.

hispidulous—somewhat or minutely hispid.

hoary—covered with a close white or whitish pubescence.

homosporous—producing spores of one kind only, as in *Lycopodium*.

host—a plant which nourishes a parasite.

hyaline—colorless or translucent.

hybrid—a plant resulting from a cross between parents that are genetically unlike, as the offspring of hybridization of two different species.

hydrophyte—a water plant, partially or wholly immersed.

hypanthium—the enlarged cuplike receptacle in some plants in which sepals, petals, and stamens are inserted; in perigyny, the calyx tube.

hypogynium—the hypogynous disk subtending the ovary in certain rushes.

hypogynous—having the other parts of the flower free from and situated below the pistil on the receptacle.

imbricate—overlapping, as the shingles on a roof.

imparipinnate—odd-pinnate, tip of rachis bearing a leaflet.

immersed—(1) entirely underwater; (2) embedded in the substance of the leaf or thallus.

incised—jagged or sharply and rather deeply cut.

included—not protruded, as stamens not projecting from corolla; opposed to exserted.

indefinite—(1) too many for easy enumeration, as an abundance of stamens; (2) in a racemose inflorescence, the main axis being capable of constant extension, or seeming so.

indehiscent—not opening spontaneously at maturity, as in the fleshy fruits.

indigenous—native to the locality, not introduced.

indumentum—a rather dense pubescence or hairy covering.

induplicate—rolled or folded inward.

indurated—hardened, usually during the course of development.

indusium—(1) a ring of collecting hairs below the stigma that function in brushing the pollen off the legs of insects; (2) a flap of tissue or a modified scale which covers the sori of some ferns.

inequilateral—unequal-sided, as an oblique leaf base.

inferior—referring to the ovary, embedded into receptacle below the other floral parts.

inflorescence—any type of flower cluster or regular association of flowers.

insectivorous—insect-catching.

inserted—attached, as a stamen growing on the corolla.

interrupted—not continuous; with small leaflets or parts interposed or inserted between others.

internode—the part of an axis of stem between two nodes.

introrse—facing inward, usually referring to an anther whose dehiscence faces toward the center of the flower.

involucel—a secondary involucre; small involucre usually subtending the parts of a flower or flower cluster.

involucrate—having an involucre or series of phyllaries.

involute—rolled inward or toward the top side; as in some leaves.

keel—the two anterior united petals of a papilionaceous flower.

labellum—lip; an enlarged or otherwise modified petal.

labiate—(1) lipped; (2) a member of the Lamiaceae.

laciniate—slashed into narrow pointed lobes.

lacuna—a space, often used to refer to cavities or air spaces in immersed organs.

lamina—a blade or flat expanded portion.

lanate—woolly pubescent with long, intertwined, curly hairs.

lanceolate—lance shaped, much longer than broad, widest below the middle and tapering to the apex.

lateral—on the side or along the margins.

lax—loose; opposed to congested.

leaflet—one division of a compound leaf.

legume—simple fruit dehiscing on both sutures and the product of a simple unicarpellate ovary.

lemma—in grasses, the flowering glume, the lower of the two bracts immediately enclosing the flower.

lenticels—raised corky spots on young branches.

lenticular—lens shaped.

lepidote—surfaced with small scurfy scales.

liana—a woody climbing or twining plant.

ligneous—woody.

ligulate—strap shaped.

ligule—a strap-shaped organ or body; a strap-shaped corolla in the ray flowers of Asteraceae; an outgrowth between the blade and the sheath in grasses and sedges; hastula membrane in palms.

limb—(1) broad portion of a petal in particular, the expanding part of a gamopetalous corolla; (2) branch.

linear—long and narrow, the sides parallel or nearly so as blades of most grasses.

lip—the principal lobe of a bilabiate corolla or of a calyx.

littoral—belonging to or growing on the seashore.

lobe—a part of an organ such as petal, calyx, or leaf that represents a division to about the middle or halfway between the margin and the midrib or base.

lobulate—having small lobes.

locular—pertains to seed cavity of a carpel or of united carpels of ovary.

locule—the cavity of an anther or an ovary; pollen chamber or seed chamber.

loculicidal—dehiscence on the back, more or less midway between the partitions, into the fruit cavity.

lyrate—pinnatifid, but with an enlarged terminal lobe and smaller lower lobes.

mammillated—having nipple-shaped processes.

marcescent—withering without falling.

maritime—belonging to the sea or confined to the seacoast.

membranaceous—thin and membranelike or of parchmentlike texture.

mentum—an extension of the foot of the column in some orchids.

mesophyte—those plants which normally grow under moderate conditions of moisture, light, temperature, etc.; intermediate between *hydrophyte* and *xerophyte*.

microsporangium—the microspore-containing case; an anther sac in flowering plants.

microspore—the smaller of two kinds of spores in heterosporous plants, as *Selaginella*, in some ferns, and in seed plants.

midrib—the main vein of a leaf or leaflike part, a continuation of the petiole; costa.

monadelphous—having all the stamens of a single flower united by their filaments into a tube or single cluster, as in the mallow.

moniliform—cylindrical or terete, with beadlike constrictions.

monochasium—a cyme reduced to single flowers on each axis.

monoecious—with staminate and pistillate flowers on the same plant.

monopodial—the stem a single and unbranched axis.

mucronate—terminated abruptly by a short, sharp point.

multifid—cleft into many lobes or segments.

multiple fruit—the united product of several or many flowers, as in the pineapple and the mulberry.

muricate—roughened with sharp projection.

naked—without perianth or buds without scales.
navicular—boat shaped, as in glumes of most grasses.
nectariferous—producing nectar.
nectary—an organ or tissue that secrets nectar.
nerve—a slender vein or rib, especially if not branched.
nigrescent—blackish.
nodding—hanging; nutant.
node—a joint where a leaf is borne.
nut—an indehiscent, unilocular, one-seeded fruit, indurated pericarp; a large
    achene.
nutlet—diminutive of nut.
obcordate—heart shaped, with the point basal and the lobes apical.
oblanceolate—lance shaped, but with widest part above the middle.
oblique—slanting; unequal-sided.
oblong—longer than broad, with the sides and ends nearly or completely paral-
    lel.
obovate—inversely egg shaped in outline.
obovoid—inversely egg shaped.
obsolete—not evident or apparent; rudimentary; vestigial.
obsolescent—becoming rudimentary or almost vestigial.
obtuse—rounded or blunt at the ends.
ocrea (or *ochrea*)—a nodal sheath formed by fusion of stipules, referring to the
    sheathing stipules of Polygonaceae.
orbicular—of a flat body with a circular outline.
orifice—an opening.
ovate—shaped like a longitudinal section of a hen's egg, the broader end basal;
    egg shaped in outline.
ovoid—a solid that is oval in flat outline; egg-shaped solid.
ovary—the ovule-bearing part of a pistil that becomes a fruit.
ovule—undeveloped seed.
pale—the chaffy scales on the receptacle of many Asteraceae.
palea—in the grass flower, the upper of the two enclosing bracts.
paleaceous—chafflike texture of perianth segments.
paleate—scaly.
palmate—lobed or divided or ribbed in a palmlike or handlike fashion; digitate.
paludal—wet all through the year; swampy, of or growing in marshes.
palustrine—inhabiting boggy ground.
panduriform—fiddle shaped, drawn in at the middle.
panicle—an indeterminate branching raceme, an inflorescence in which the
    branches of the primary axis are racemose and the flowers pedicellate.
paniculate—having a panicle.
papillose—bearing minute pimplelike protuberances.
pappus—modified outer perianth series of Asteraceae borne on the ovary and
    often persisting in the fruit.
paracostal—rows of cells along the sides of midrib.
parasite—an organism which grows on and derives nourishment from another
    species called the host.
parietal—borne on or pertaining to the wall of the fruit.
pectinate—comblike or pinnatifid with very close, linear, equal divisions or
    parts.

pedate—said of a palmately lobed or divided leaf of which the two side lobes are again divided or cleft.

pedicel—stalk of one flower in a cluster.

peduncle—stalk supporting a flower cluster or a solitary flower.

pellucid—clear, almost transparent in transmitted light.

peltate—attached to its support by the lower surface umbrella fashion, instead of by a margin; shield shaped.

pendent—hanging from its support.

penninerved—with branch veins arising along the length of the midvein; pinnately nerved.

pepo—a fruit with leathery or indurated rind filled with soft placental tissue with numerous seeds, product of an inferior ovary, as in gourd, squash, and cucumber.

perennial—of three or more seasons' duration; persistent.

perfect flower—having both functional pistil and stamens; bisexual.

perfoliate—descriptive of a sessile leaf or bract whose base completely surrounds the stem, the latter seemingly passing through the leaf.

perianth—a collective term for the corolla and calyx, especially when they are not clearly differentiated.

pericarp—the wall of a ripened ovary, i.e., the wall of a fruit.

perigynium—a saccate scale enclosing the pistil in *Carex*.

perigynous—borne or arising from around the ovary and not beneath it, as calyx, corolla, and stamens borne on the rim of the hypanthium.

persistent—remaining attached; not falling off.

petal—a unit of the inner floral envelope (corolla) of a flower, often colored and showy.

petiole—leafstalk.

petiolule—stalk of a leaflet in a compound leaf.

phenotype—the form or appearance of an individual representing the result of interaction of external factors, as growing conditions, and the genotype.

phyllary—a bract of the involucre surrounding a common receptacle.

phyllode—a petiole taking on the form and functions of a leaf, as in some *Acacia*.

pilose—pubescent, with soft, straight, weak hairs.

pinna—a primary division or leaflet of a pinnate leaf.

pinnate—resembling a feather in general form with the divisions of the organ, usually a leaf, placed on either side of the central axis and arising along its length.

pinnatifid—cleft or parted in a pinnate (rather than palmate) way.

pinnule—a secondary pinna or leaflet in a pinnately compound leaf.

pistil—a unit of the gynoecium, typically comprised of the ovary, style (when present), and stigma.

pistillate—having pistils and no functional stamens.

pitted—having little depressions or cavities.

pith—the soft, spongy, central cylinder of most angiosperm stems, composed mostly of parenchyma tissue.

placenta—a place or part in the ovary where ovules are attached.

placentation—the arrangement of ovules within the ovary.

plicate—folded, as in a fan, or approaching this condition.

pod—a dry and dehiscent fruit; a rather general term used when no other more specific term is applicable; also commonly used for the fruit of the Fabaceae instead of the correct term *legume*.

pollen—spores or grains borne by the anther containing the male element.

pollinium—a coherent mass of pollen.

polygamous—bearing unisexual and bisexual flowers on the same plant.

polygamodioecious—pertaining to a species that is functionally dioecious but has a few flowers of the opposite sex or a few bisexual flowers on all plants.

polymorphous—with various or several forms; variable as to habit.

polyploid (also *polyploidy*)—a plant with a chromosome complement of more than two sets of the monoploid (haploid) or *n* number.

polystichous—when leaves are borne in many series.

pome—an accessory, as in fruit of apple, pear, quince, and related genera in the Rosaceae.

prickle—a small and weak spinelike body borne on bark or epidermis.

prismatic—angulate with flat sides.

procumbent—trailing or lying flat but not rooting.

proliferous—bearing offshoots or redundant parts; bearing other similar structures on itself.

prostrate—a general term for lying flat on the ground.

proximal—the part nearest the axis, as opposed to dital.

pruinose—having a bloom or powdery substance on the surface.

pseudobulb—the thickened or bulblike stems of certain orchids which are solid and fleshy and borne above ground.

puberulent—minutely pubescent, the hairs soft, straight, erect, and scarcely visible to the naked eye.

puberulous—slightly pubescent.

pubescent—covered with short soft hairs.

pulvinus—swollen base of a petiole.

punctate—with translucent or colored dots or depressions or pits.

pungent—(1) ending in a stiff sharp point or tip, as a holly leaf; (2) acrid to the taste.

pustulate—as though blistered.

pyrene—a seedlike nutlet of a small drupe.

pyriform—pear shaped.

quadrate—square.

raceme—a simple, indeterminate inflorescence of pedicelled flowers arranged along an elongated axis.

racemose—having racemes, or racemelike.

rachilla—a diminutive or secondary axis, or rachis; in particular, in the grasses and sedges the axis that bears the florets.

rachis—the primary axis of an inflorescence or of a compound leaf.

radial symmetry—capable of equal division in more than one direction through the center.

radical—arising from the root or its crown; referring to leaves that are basal or rosulate.

rameal—belonging to a branch.

ramose—branching, having many branches.

ray floret—the ligule or straplike corolla of many Asteraceae, borne marginally on the receptacle when differentiated from the disk florets.

receptacle—the more or less enlarged portion of an axis which bears the floral organs.

reclined—bent down or falling back from the perpendicular.

recurved—bent or curved downward or backward.

reflexed—abruptly bent or turned downward or backward.

reniform—kidney shaped.

repand—with a slightly uneven and somewhat sinuate margin.

repent—prostrate and rooting at the nodes.

resiniferous—containing or producing resin.

resupinate—turned upside down by a twisting of the pedicel, thus inverting the corolla.

reticulate—netted; in the form of a network; anastamosing irregularly in many places as a result of multiple branching.

retrorse—turned backward or downward.

retuse—with a shallow notch at a rounded apex.

revolute—rolled back from the margin or apex.

rhizome—underground stem; rootstock; distinguished from a root by presence of nodes, buds, or scalelike leaves.

rhombic—diamond shaped.

rib—the midvein or primary vein of a leaf or other organ; also, any prominent vein or nerve.

rosette—a whorl of leaves radiating from a crown or center and usually in a cluster just at or slightly above the ground, as in dandelion or pipewort.

rostellum—a small beak; a narrow extension of the upper edge of the stigma of certain orchids.

rostrate—beaked; narrowed into a slender tip or point.

rosulate—collected into a rosette, descriptive of a whorl of basal leaves.

rotate—wheel shaped; referring to a partially fused corolla with a flat circular limb at right angles to the very short or obsolete tube.

rotund—nearly circular; orbicular, inclining to be oblong.

ruderal—growing in waste places or among rubbish; a weed.

rudimentary—imperfectly developed or not developing fully and nonfunctional.

rufous—reddish brown.

rugose—with reticulate veins seemingly impressed into the surface, the tissue between veins bulging outward, giving a wrinkled appearance to the organ, usually a leaf.

runcinate—saw-toothed or sharply incised, the teeth turned backward.

runner—a slender trailing shoot taking root at the nodes.

saccate—bag shaped, pouchy, as lip of lady's slipper.

sagittate—arrowhead shaped with basal lobes pointing backward and sometimes curving toward the stalk.

salverform—descriptive of a corolla in which the petals are fused for most of their length into a slender tube, then the petals separate and expand abruptly into a flat limb.

samara—a dry, winged, indehiscent fruit, usually one-seeded, as in ash and elm, but sometimes two-seeded, as in maple.

saprophyte—a plant typically lacking chlorophyll, deriving nutritives from decaying organic matter.

scabrous—the surface rough or gritty to the touch.

scale—a variety of small, usually dry and appressed leaves or bracts, often only vestigial.

scape—flowering stalk arising from or below ground which bears no leaves or no normal leaves.

scarious—applied to dry, nongreen, leaflike, and membranaceous parts or bracts, as the wing of maple fruit.

scorpioid—descriptive of a coiling determinate inflorescence in which the flowers are borne alternately in two opposing rows, as in heliotrope.

scurfy—with scalelike or flaky particles on the surface, as in goosefoot.

secund—one-sided; referring to inflorescences when the flowers appear to be borne only on one side of the floral axis.

seed—the ripened ovule containing the embryo and usually a stored food supply within more or less hardened integuments.

sepal—one of the separate parts of a calyx, usually green and leaflike.

septicidal—dehiscence along or into the partitions (septa), not opening directly into the locule.

septum—a partition, as between locules of a seed capsule.

seriate—in series, usually in whorls or apparent whorls, as a perianth.

sericeous—silky.

serrate—saw-toothed along the margin, with the teeth pointed forward.

serrulate—minutely serrate.

seta—a bristle. (pl. *setae*)

setose—covered with bristles.

sheath—any long or more or less tubular structure surrounding an organ or part.

shrub—a woody plant with several stems branched from the base.

simple—pertaining to a leaf with a single undivided blade per petiole.

sinuate—with a deep wavy margin; also, *sinuous*.

sorus—a cluster of sporangia in ferns usually located on the undersurface of leaves.

spadix—a specialized, fleshy spike of many flowers found in certain plants subtended by a leaflike spathe.

spathe—the bract surrounding or subtending a flower cluster or a spadix; it is sometimes colored and flowerlike.

spatulate—spoon shaped.

species—a natural unit of classification composed of individuals and populations that exhibit characters that distinguish the members of this unit from other units within a genus.

spicate—spikelike.

spike—a usually unbranched, elongated, indeterminate inflorescence bearing sessile flowers.

spikelet—a secondary spike; a small spike in which each flower is subtended by a scale or bract, as in Poaceae.

spine—a strong and sharp-pointed woody body mostly arising from the wood of the stem.

spinulose—with small spines over the surface.

spore—an asexual, usually one-celled reproductive body.

sporangium—a spore case; a sac or body bearing spores.

sporocarp—a specialized organ containing sporangia.

sporophyll—a spore-bearing leaf; a leaflike organ bearing reproductive parts.

sporophyte—in ferns and seed plants the foliaceous vegetative plant with roots, stems, and leaves, normally the 2n diploid generation.

spur—a tubular or saclike projection from a blossom, as of a petal or a sepal; it usually contains a nectar-secreting gland.

stamen—the pollen-bearing unit of the flower typically composed of anther and filament.

staminate—having stamens and no pistils.

staminode—a sterile stamen, sometimes petal-like and showy.

standard—the broad, erect, upper petal of a papilionaceous flower.

stellate—starlike or with pointed, radiating branches; descriptive of a type of epidermal hair which is branched or merely forked.

sterile—lacking functional sex organs.

stigma—style apex or marginal lines receptive of pollen.

stipe—the stalk of a pistil or other small organ when axile in origin; also, the petiole of a fern leaf.

stipellate—the leaflets of a compound leaf when furnished with small stipule-like appendages on the petiolules.

stipitate—borne on a stipe or short stalk.

stipule—an appendage, usually green and leaflike, found at the base of a petiole, usually in pairs.

stolon—a horizontal stem which tends to touch the ground at intervals and root or to run underground for a distance and then send up a new plant at its tip.

stramineous—strawlike or straw colored.

striate—with fine longitudinal lines, channels, or ridges.

strict—upright, straight.

strigose—with sharp, appressed, straight hairs, stiff and often basally swollen; see *hispid*.

strobilus—a conelike structure containing the reproductive organs of one or both sexes, as in the horsetails, pines, magnolia.

style—the more or less elongated part of the pistil between the ovary and the stigma.

stylopodium—a disklike enlargement at the base of the style, as in some species in the Umbelliferae.

subshrub—an undershrub or small shrub which may have partially herbaceous stems.

subtend—to stand below and close to, as a bract underneath a flower.

subulate—awl shaped, tapering in a straight line from base to apex.

succulent—juicy, fleshy, and thickened; with a more or less soft texture.

sucker—a fast-growing, soft vegetative shoot arising from the basal part or root-stock of a woody plant.

suffrutescent—pertaining to a low shrubby plant, woody at the base and producing herbaceous shoots perennially.

sulcate—grooved or furrowed lengthwise.

superior—pertaining to an ovary free on a conelike receptacle (toras) and therefore positioned above the point of attachment of the floral envelopes.

suture—a line of splitting or dehiscence; a longitudinal seam.

sympatric—referring to a species growing in the same geographic locality as another closely related species.

sympetalous—with united petals.

synonym—a superseded or unused name of a taxon.

syncarp—a fleshy aggregate fruit.

synsepaly—with the sepals fused by their margins, at least basally.

taproot—the primary or main descending root in dicots, formed as a direct continuation of the embryonic radicle.

taxon—a general term applied to any taxonomic group, element, or population irrespective of its classification level, as in a species, division, genus, etc.

taxonomy—science of classification of organisms according to relation.

tendril—an elongated and twining, thin, modified leaf or branch by which a climbing plant clings to its support, as in the grape.

tepal—a sepal-like or petal-like unit of the floral envelope when calyx and corolla are not clearly differentiated from each other.

terete—a more or less perfectly cylindrical solid, not angular, as most plant stems which are not hollow.

ternate—in threes.

testa—the outer coat of a seed, developed from the integument of the ovule.

tetradynamous—an androecium of six stamens, four longer than the outer two, as in most Brassicaceae.

tetramerous—four-merous.

tetraploid—having four times the *n* or haploid complement of chromosomes in the somatic cells.

thallus—the whole plant body of a nonvascular plant, such as an alga or a liverwort; also a flat leaflike organ.

thorn—a pointed, stiff, woody body, simple or branched; morphologically an aborted branch.

throat—portion of the corolla where the limb joins the corolla tube.

thyrse—an inflorescence resembling a densely compacted panicle but with an indeterminate main axis and determinate lateral axes, as a lilac.

tomentose—covered densely with soft, matted, woolly hairs.

tortuous—bent or twisted many times in different directions.

torulose—with irregular swellings at close intervals; knobby.

torus—a receptacle.

trichome—a hair or bristle.

trigonous—three-angled, with plane faces.

truncate—not tapered, the base or apex more or less straight across.

tube—the united portion of a gamopetalous corolla or gamosepalous calyx.

tuber—thickened underground stem with eyes (buds) on the sides.

tubercle—a small tuber or rounded protuberance.

tumid—swollen, inflated.

turbinate—inversely conical; top shaped.

turgid—swollen tissue from internal water pressure.

umbel—an inflorescence whose pedicels or branches are of nearly equal length and arise from a common point.

umbonate—having embossed surface or umbo in the middle.

uncinate—shaped like a hook or hooked at the tip.

undulate—wavy margin or surface.

unguiculate—clawed; narrowed into a petiolelike base.

unifoliolate—pertaining to a compound leaf reduced to a single, terminal leaflet.

unisexual—of one sex; staminate only or pistillate only.

urceolate—hollow and shaped like an urn, widest below the open top.

utricle—a bladdery, one-seeded, usually indehiscent fruit.

valvate—meeting by the edges without overlapping; opening by valves or pertaining to valves.

valve—one of the carpels separating from a mature capsule by loculicidal or septicidal dehiscence.

variety—a subdivision of a species composed of individuals or populations differing from other members of the species in certain minor characters which are perpetuated in succeeding generations.

vein—a strand of vascular tissue which with others makes up the framework of leaves.

velutinous—velvety-textured covering composed of erect, straight, moderately firm hairs.

venation—veining; arrangement or pattern of veins in an organ.

ventral—front; relating to the inner face or part of an organ; opposed to dorsal.

ventricose—swelling unequally or inflated on one side.

vernation—the arrangement of leaves in the bud.

verrucose—having a warty or tubercled surface.

verticillate—arranged in whorls or seemingly so.

versatile—attached near the middle and turning freely on its support.

vescicle—a small bladdery sac or cavity filled with air or fluid.

vestigial—imperfectly developed; pertaining to an organ that is a much reduced and degenerate relic in the presently living plant.

vexillum—the standard or broad upper petal on a papilionaceous flower.

villous—covered densely with long and soft, but not matted hairs.

virgate—wand shaped; slender, straight, and erect.

viscid—sticky or viscous to the touch.

whorl—pertaining to a circular arrangement of leaves or flowers on an axis.

wing—a thin, dry, or membranaceous appendage, as of a fruit; also, the two lateral petals of a papilionaceous flower.

woolly—covered with long, soft, curly, more or less matted hairs; like wool; lanate.

xerophyte—a plant of a dry arid habitat.

Index to Scientific Names and Colloquial Names

Roman type is used for accepted specific and infraspecific names, colloquial names, genera, and families. *Italic type* is used to indicate synonyms.

NEW PORT RICHEY LIBRARY

3 2288 00041 2349